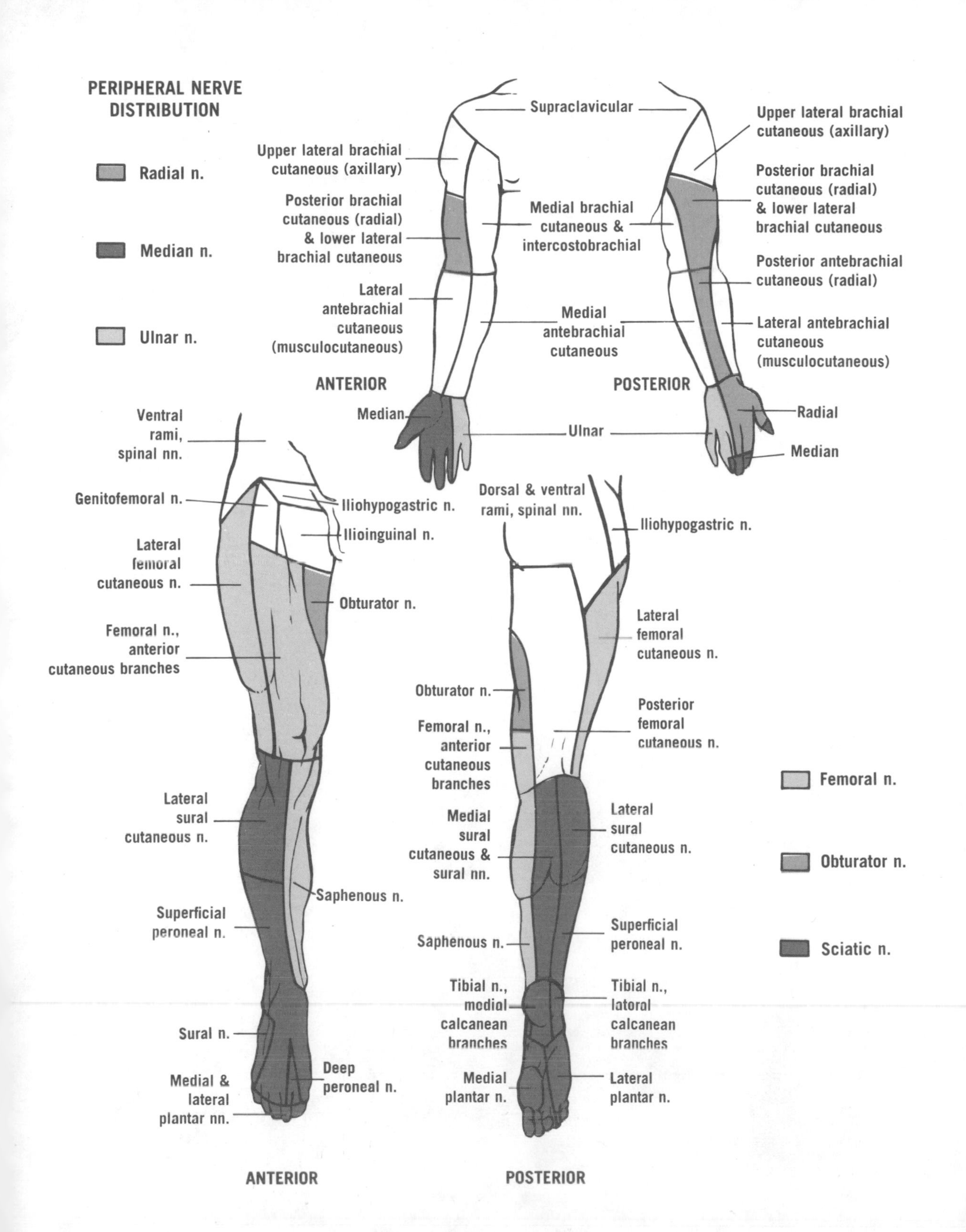
PERIPHERAL NERVE DISTRIBUTION
Radial n.
Median n.
Ulnar n.
Supraclavicular
Upper lateral brachial cutaneous (axillary)
Posterior brachial cutaneous (radial) & lower lateral brachial cutaneous
Medial brachial cutaneous & intercostobrachial
Lateral antebrachial cutaneous (musculocutaneous)
Medial antebrachial cutaneous
Posterior antebrachial cutaneous (radial)
ANTERIOR
POSTERIOR
Median
Ulnar
Radial
Median
Ventral rami, spinal nn.
Genitofemoral n.
Iliohypogastric n.
Ilioinguinal n.
Lateral femoral cutaneous n.
Obturator n.
Femoral n., anterior cutaneous branches
Dorsal & ventral rami, spinal nn.
Posterior femoral cutaneous n.
Lateral sural cutaneous n.
Medial sural cutaneous & sural nn.
Saphenous n.
Superficial peroneal n.
Sural n.
Deep peroneal n.
Medial & lateral plantar nn.
Tibial n., medial calcanean branches
Tibial n., lateral calcanean branches
Medial plantar n.
Lateral plantar n.
Femoral n.
Obturator n.
Sciatic n.
ANTERIOR
POSTERIOR

BASIC HUMAN ANATOMY

A Regional Study of Human Structure

RONAN O'RAHILLY, M.D.

with the collaboration of
FABIOLA MÜLLER, Dr. rer. nat.

University of California at Davis

1983

W. B. SAUNDERS COMPANY
Philadelphia, London, Toronto, Mexico City, Rio de Janeiro, Sydney, Tokyo

W. B. Saunders Company: West Washington Square
Philadelphia, PA 19105

1 St. Anne's Road
Eastbourne, East Sussex BN21 3UN, England

1 Goldthorne Avenue
Toronto, Ontario M8Z 5T9, Canada

Apartado 26370 - Cedro 512
Mexico 4, D.F., Mexico

Rua Coronel Cabrita, 8
Sao Cristovao Caixa Postal 21176
Rio de Janeiro, Brazil

9 Waltham Street
Artarmon, N.S.W. 2064, Australia

Ichibancho, Central Bldg., 22-1 Ichibancho
Chiyoda-Ku, Tokyo 102, Japan

Library of Congress Cataloging in Publication Data

O'Rahilly, Ronan.
Basic human anatomy.

1. Anatomy, Surgical and topographical. I. Müller, Fabiola.
II. Title. [DNLM: 1. Anatomy, Regional. QS 4 065b]
QM531.07 1983 611 82-47772
ISBN 0-7216-6990-5

Cover: Portrait of Andreas Vesalius of Brussels from his Fabrica (Basel, 1543). The drawing is generally attributed to Jan Stefan van Kalkar, and the original engraving of the woodcut to Francesco Marcolini da Forlí. This version has been modified for reproduction on cloth.

Basic Human Anatomy ISBN 0-7216-6990-5

Last digit is the print number: 9 8 7 6 5 4 3 2 1

in memory of

ERNEST GARDNER, M.D.,

colleague and friend

PREFACE

Vesalius concluded the preface of his *Fabrica* (1543) with the statement that anatomy "should rightly be regarded as the firm foundation of the whole art of medicine and its essential preliminary." Abraham Colles, after whom the most frequently seen fracture of the radius is named, stated in 1811 that it is a mistake to consider anatomy "in no other light than as a science in itself . . . instead of considering it as a science altogether subservient to the practice of medicine and surgery," that is, as far as students of the health sciences and related disciplines are concerned.

With those statements as guiding principles, it is the aim of *Basic Human Anatomy* to provide a solid morphological basis for a synthesis of anatomy, physiology, and clinical sciences. A deliberate effort has been made, using the words of Colles again, not to "describe the various parts of the human frame as of equal importance" but rather to give "to each part just that degree of attention it deserves." An obvious example is the difference in the importance of the small muscles of the hand and those of the foot. A deliberate effort has been made, too, to use the clinical sciences to improve understanding of basic anatomy. For instance, the main routes of surgical access to the prostate provide a better and more vivid concept of topography than does a traditional list of anatomical relationships. Clinical considerations are presented in this manner at appropriate points throughout the text. Moreover, historical perspective in such topics as cataract surgery, lithotomy, and the Hunterian significance of the adductor canal has been provided to foster interest in the appropriate morphology.

During the twentieth century, exciting developments such as endoscopy, electromyography, scintigraphy, and computerized scanning have contributed significantly to teaching and research in human gross anatomy. The improved understanding of anatomical structure made possible by the technologies of medicine is evident throughout *Basic Human Anatomy,* both in the text and in the illustrations. Also evident, however, is the belief that technology has in no way eroded the need for a foundation of knowledge of the gross structure of the body gained through the more traditional study of anatomy. After all, the present popularity of computerized scanning has sent anatomists and radiologists back to the study of cross-sectional anatomy, involving methods used at the end of the nineteenth century.

The goal throughout the work on *Basic Human Anatomy* has been to provide a well-illustrated text of intermediate length. It is shorter than, but is not meant to replace, *Anatomy: A Regional Study of Human Structure* by Gardner, Gray, and O'Rahilly (4th ed., W. B. Saunders Company, 1975), which is a more detailed book containing extensive references to the literature. Decidedly more than a primer but less than an advanced-level reference work, *Basic Human Anatomy* is particularly appropriate for use in current health sciences curricula. It is also especially suitable for review by preclinical students, clinical students, and postgraduate physicians.

A series of introductory chapters on systemic anatomy is followed by sections organized by body region. The terminology used is almost exclusively from the *Nomina anatomica,* 4th ed., 1977, translated into English where applicable. A glossary of eponyms appears at the end of the book, as does a table of measurements (end paper). The use of an anatomical atlas and a medical dictionary in conjunction with this work is strongly recommended.

A number of questions appear at the end of each chapter. They will serve as a guide for attentive reading if students scan them before reading the text or as a test for mastery of important points and applications if students attempt to answer them after reading the appropriate chapter. Nearly all of the questions can be answered after reading the text; however, many of the answers given at the end of the book include additional information. The answers are complete statements and hence can be used for review independently of the questions.

Additional reading recommended at the end of many chapters consists mostly of books and monographs. The few references given in footnotes and in the answers to questions concern points of special interest or of controversy and may serve to emphasize that, contrary to a common fallacy, publication of articles on gross anatomy continues and is "alive and well."

We have been most fortunate in the cordial relationship that has existed between us and the W. B. Saunders Company. In particular, Mrs. Roberta Kangilaski has spared no effort in bringing this work to successful fruition.

RONAN O'RAHILLY AND FABIOLA MÜLLER

CONTENTS

Part 1 GENERAL ANATOMY

Part 2 THE UPPER LIMB

Part 3 THE LOWER LIMB

Part 4 THE THORAX

Part 5 THE ABDOMEN

Part 6 THE PELVIS

Part 7 THE BACK

Part 8 THE HEAD AND NECK

APPENDIX

Part 1

GENERAL ANATOMY

1
INTRODUCTION

ANATOMY AND ITS SUBDIVISIONS

Anatomy is the science of the structure of the body. When used without qualification, the term is applied usually to human anatomy. The word is derived indirectly from the Greek *anatome*, a term built from *ana*, meaning "up," and *tome*, meaning "a cutting" (compare the words tome, microtome, and epitome). From an etymological point of view, the term "dissection" (*dis-*, meaning "asunder," and *secare*, meaning "to cut") is the Latin equivalent of the Greek *anatome*.

Anatomy, wrote Vesalius in the preface to his *De Fabrica* (1543), "should rightly be regarded as the firm foundation of the whole art of medicine and its essential preliminary." Moreover, the study of anatomy introduces the student to the greater part of medical terminology.

Anatomy "is to physiology as geography is to history" (Fernel); that is, it provides the setting for the events. Although the primary concern of anatomy is with structure, structure and function should be considered together. Moreover, by means of *surface* and *radiological anatomy*, emphasis should be placed on the anatomy of the living body. As Whitnall expressed it, "I cannot put before you too strongly the value and interest of this rather neglected [surface] aspect of anatomy. Many a student first realizes its importance only when brought to the bedside or the operating table of his patient, when the first thing he is faced with is the last and least he has considered." The classical methods of physical examination of the body and the use of some of the various "-scopes," e.g., the stethoscope and the ophthalmoscope, should be included. Radiological studies facilitate achievement of "an understanding of the fluid character of anatomy and physiology of the living" (A.E. Barclay), and the importance of variation should be kept in mind.

In relation to the size of the parts studied, anatomy is usually divided into (1) *macroscopic* or *gross anatomy*, and (2) *microscopic anatomy* or *histology* (now used synonymously in English). In addition, *embryology* is the study of the embryo and the fetus, that is, the study of prenatal development, whereas the study of congenital malformations is known as *teratology*.

In general, works dealing with human anatomy are arranged either (1) *systemically*, that is, according to the various systems of the body (skeletal, muscular, digestive, etc.) or (2) *regionally*, that is, according to the natural, main subdivisions of the body (head and neck, upper limb, thorax, etc.). In this book, after the general features of certain systems have been discussed in introductory chapters, the remainder of the work, in general, follows a regional approach. The regional plan has been adopted chiefly because the vast majority of laboratory courses in human anatomy are based on regional dissection. Anatomy considered on a regional basis is frequently termed *topographical anatomy*.

ANATOMICAL TERMINOLOGY

The following etymological works are recommended:

Field, E. J., and Harrison, R. J., *Anatomical Terms: Their Origin and Derivation*, 3rd ed., Heffer, Cambridge, 1968.

Skinner, H. A., *The Origin of Medical*

Terms, 2nd ed., Williams & Wilkins, Baltimore, 1961.

International agreement has been reached on a Latin nomenclature, the *Nomina anatomica.* A revision of this terminology, translated into English where applicable, is used in this book. Eponyms are avoided.

TERMS OF POSITION AND DIRECTION (fig. 1–1)

All descriptions in human anatomy are expressed in relation to the *anatomical position,* a convention whereby the body is erect, with the head, eyes, and toes directed forward and the upper limbs by the side and held so that the palms of the hands face forward. There is no implication that the anatomical position is one of rest. It is often necessary, however, to describe the position of the viscera also in the recumbent posture, because this is a posture in which patients are frequently examined clinically.

The *median plane* is an imaginary vertical plane of section that passes longitudinally through the body and divides it into right and left halves. The median plane intersects the surface of the front and back of the body at what are called the anterior and posterior me-

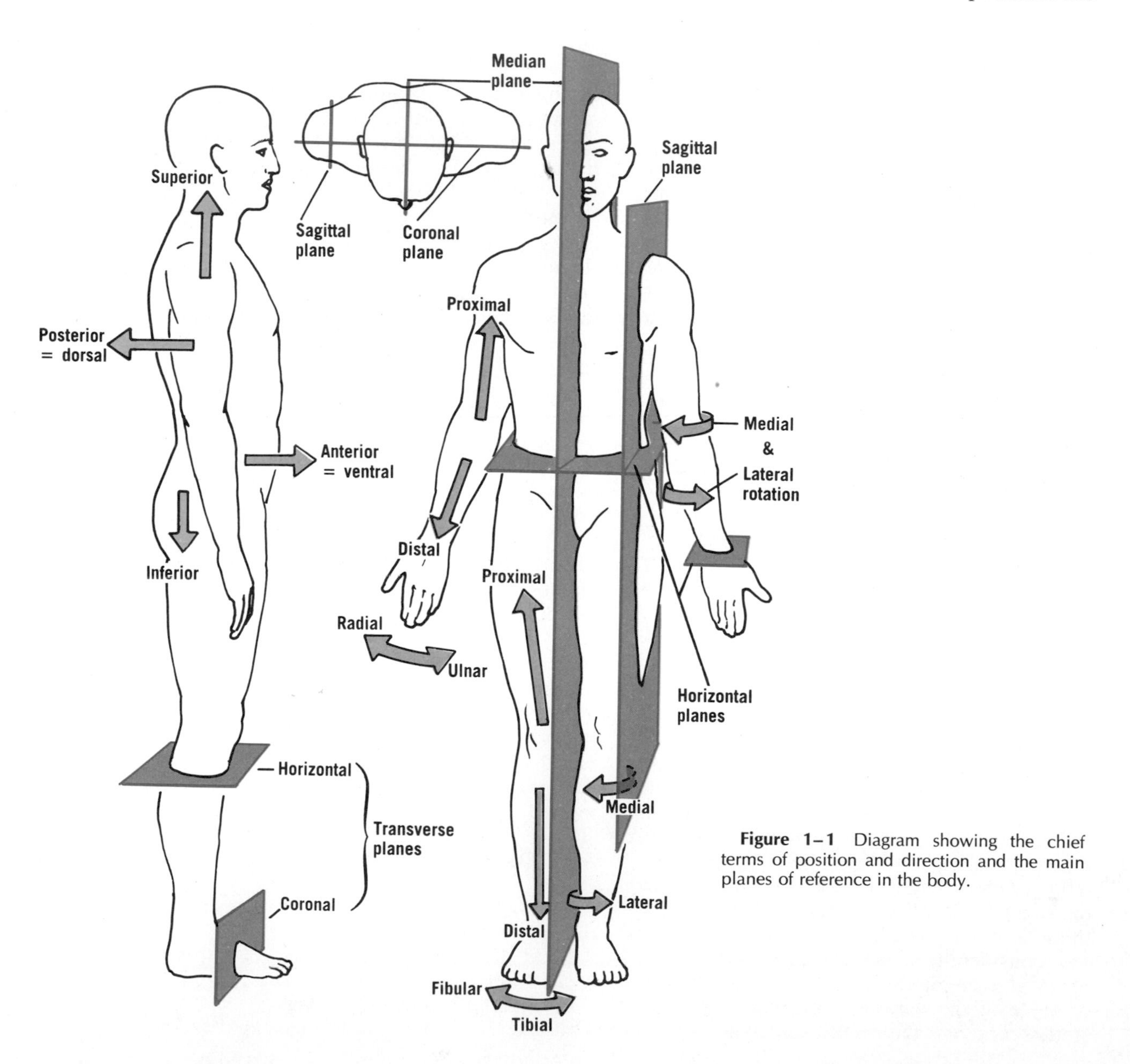

Figure 1–1 Diagram showing the chief terms of position and direction and the main planes of reference in the body.

dian lines. It is a common error, however, to refer to the "midline" when the median plane is meant.

Any vertical plane through the body that is parallel with the median plane is called a *sagittal plane*. The sagittal planes are named after the sagittal suture of the skull, to which they are parallel. The term "parasagittal" is redundant: anything parallel with a sagittal plane is still sagittal.

Any vertical plane that intersects the median plane at a right angle and separates the body into front and back parts is termed a *coronal*, or *frontal, plane*.

The term *horizontal plane* refers to a plane at a right angle to both the median and coronal planes: it separates the body into upper and lower parts.

The term *transverse* means at a right angle to the longitudinal axis of a structure. Thus, a transverse section through an artery is not necessarily horizontal. A transverse section through the hand is horizontal, whereas a transverse section through the foot is coronal (fig. 1–1).

The term *medial* means nearer to the median plane, and *lateral* means farther from it. Thus, in the anatomical position, the thumb is lateral to the little finger, whereas the big toe is medial to the little toe. *Intermediate* means lying between two structures, one of which is medial and the other lateral. In the upper limb *radial* means lateral and *ulnar* means medial: in the lower limb *fibular* or *peroneal* means lateral and *tibial* means medial. The border of a limb on which either the thumb or the big toe is situated is sometimes called *preaxial*, and the opposite border, *postaxial*. These two terms are based on the arrangement of the limbs in the embryo during the sixth postovulatory week, when the thumbs and the big toes are both on the rostral border of the limbs (see figs. 8–10 and 15–11).

Medial and lateral rotation (which should never be referred to as internal and external) means rotation (e.g., of the hip) around a vertical axis so that the front of the part moves medially or laterally, respectively.

Anterior or *ventral* means nearer the front of the body. *Posterior* or *dorsal* means nearer the back. In the upper limb the term *palmar* (formerly *volar*) means anterior. In the foot, *plantar* means inferior, and the term *dorsal* is commonly used for superior.

Superior means nearer the top or upper end of the body. *Inferior* means nearer the lower end. *Cranial* or *cephalic* is sometimes used instead of superior, and *caudal* instead of inferior. *Rostral* means nearer the "front end," that is, the region of the nose and mouth.

The suffix "-ad" is sometimes added to a positional term to indicate the idea of motion. Thus, *cephalad* means proceeding toward the head. Such terms are useful occasionally in describing growth processes, but their application is best limited.

In the limbs, *proximal* and *distal* are used to indicate, respectively, nearer to and farther from the root or attached end of a limb. (Proximal and distal have a special meaning in the case of the teeth.)

Internal and *external* mean, respectively, nearer to and farther from the center of an organ or a cavity. *Superficial* and *deep* mean, respectively, nearer to and farther from the surface of the body.

The term *middle* is used for a structure lying between two others that are anterior and posterior, or superior and inferior, or internal and external.

In addition to the technical terms of position and direction, certain common expressions are also used in anatomical descriptions: front, back, in front of, behind, forward, backward, upper, lower, above, below, upward, downward, ascending, descending. These terms are free of ambiguity provided that they are used only in reference to the anatomical position. A number of other common terms, such as "under," however, are generally best avoided.

HISTORY OF ANATOMY

Anatomy can be traced from the Greek period, B.C., and the Roman Empire, A.D., to Andreas Vesalius, who reformed the subject in his *De humani corporis fabrica* ("On the Workings of the Human Body") in 1543. Subsequent highlights include the discovery of the compound microscope (1590), the founding of microscopic anatomy by Malpighi (seventeenth century), the discovery of the circulation of the blood by Harvey (1628), the establishment of modern embryology by Wolff (eighteenth century), the gross classification of tissues by Bichat (1801), and many notable advances during the nineteenth and twentieth centuries.

The best general introduction to the history of anatomy is Singer, C., *A Short History of Anatomy and Physiology from the Greeks to Harvey*, Dover, New York, 1957. Two other interesting works are Saunders, J. B. de C. M., and O'Malley, C. D., *The Illustrations from the Works of Andreas Vesalius of Brussels,* World Publishing Co., Cleveland, 1950; and O'Malley, C. D., and Saunders, J. B. de C. M., *Leonardo da Vinci on the Human Body,* Schuman, New York, 1952.

ANATOMICAL LITERATURE

In addition to journals (such as *Acta Anatomica, American Journal of Anatomy, Anatomy and Embryology,* and the *Journal of Anatomy*), many detailed books are available. Some useful works are cited below.

Systemic Anatomy

Quain's Elements of Anatomy, 11th ed., Longmans, Green, London, 1908–1929, several volumes. The most detailed account in English.

Regional Anatomy

Gardner, E., Gray, D. J., and O'Rahilly, R., *Anatomy: A Regional Study of Human Structure,* 4th ed., W. B. Saunders Company, Philadelphia, 1975. Provides more detail than this book and includes extensive references to the literature.

Von Lanz, T., and Wachsmuth, W., *Praktische Anatomie,* Springer, Berlin, 1935–1979, several volumes. Contains superb illustrations.

Applied Anatomy

Abrahams, P., and Webb, P., *Clinical Anatomy of Practical Procedures,* Pitman, Tunbridge Wells, 1975.

Lachman, E., and Faulkner, K. K., *Case Studies in Anatomy,* 3rd ed., Oxford University Press, New York, 1981.

Schneider, L. K., *Anatomical Case Histories,* Year Book, Chicago, 1976.

Surface and Radiological Anatomy

Hamilton, W. J., Simon, G., and Hamilton, S. G. I., *Surface and Radiological Anatomy,* 5th ed., Heffer, Cambridge, 1971.

Systemic Atlases

Sobotta, J., *Atlas of Human Anatomy,* and Spalteholz, W., *Atlas of Human Anatomy,* various editions and publishers. Two well-known examples.

Regional Atlases

Bassett, D. L., *A Stereoscopic Atlas of Human Anatomy,* Sawyer's, Portland, Oregon, 1952–1962. Superb color transparencies.

Bo, W. J., Meschan, I., and Krueger, W. A., *Basic Atlas of Cross-sectional Anatomy,* W. B. Saunders Company, Philadelphia, 1980. Photographs and radiograms of anatomical sections. Other important cross-sectional atlases are those by Eycleshymer and Schoemaker (1911) and by Symington (1917).

Grant's Atlas of Anatomy, 7th ed., ed. by J. E. Anderson, Williams & Wilkins, Baltimore, 1978. An excellent, annotated atlas.

Jamieson's Illustrations of Regional Anatomy, 9th ed., rev. by R. Walmsley and T. R. Murphy, Churchill Livingstone, Edinburgh, 1971–1972, several volumes. Simple blackboard drawings.

McMinn, R. M. H., and Hutchings, R. T., *A Colour Atlas of Human Anatomy,* Wolfe, London, 1976 (distributed in the United States by Year Book, Chicago). Excellent color photographs.

Pernkopf, E., *Atlas of Topographical and Applied Human Anatomy,* ed. by H. Ferner and trans. by H. Monsen, W. B. Saunders Company, Philadelphia, 1980. Beautiful illustrations.

QUESTIONS

1–1 Which type of plane would include the entire length of the vertebral column?

1–2 Which types of planes would pass through both shoulder joints?

1–3 Which type of plane is transverse to (a) the little finger, (b) the big toe, and (c) the neck?

1–4 How is the thigh moved into a position of flexion, abduction, and lateral rotation?

1–5 Which is the most important book ever written on anatomy and when was it published?

1–6 Was the circulation of the blood appreciated at the time of Vesalius?

2 THE LOCOMOTOR SYSTEM

SKELETON

The skeleton consists of bones and cartilages. A *bone* is composed of several tissues, predominantly a specialized connective tissue known as *bone*. Bones provide a framework of levers, they protect organs such as the brain and heart, their marrow forms certain blood cells, and they store and exchange calcium and phosphate ions.

The term *osteology,* meaning the study of bones, is derived from the Greek word *osteon,* meaning "bone." The Latin term *os* is used in names of specific bones, e.g., *os coxae,* or hip bone; the adjective is osseous.

Cartilage is a tough, resilient connective tissue composed of cells and fibers embedded in a firm, gel-like, intercellular matrix. Cartilage is an integral part of many bones, and some skeletal elements are entirely cartilaginous.

BONES

The skeleton includes the *axial skeleton* (bones of the head, neck, and trunk) and the *appendicular skeleton* (bones of the limbs). Bone may be present in locations other than in the bony skeleton. It often replaces the hyaline cartilage in parts of the laryngeal cartilages. Furthermore, it is sometimes formed in soft tissues, such as scars. Bone that forms where it is not normally present is called *heterotopic bone*.

Types

Bones may be classified according to shape: long, short, flat, and irregular.

Long Bones (fig. 2–1). Long bones are those in which the length exceeds the breadth and thickness. They include the clavicle, humerus, radius, ulna, femur, tibia, and fibula, and also the metacarpals, metatarsals, and phalanges.

Each long bone has a shaft and two ends or extremities, which are usually articular. The shaft is also known as the *diaphysis*. The ends of a long bone are usually wider than the shaft, and are known as *epiphyses*. The epiphyses of a growing bone are either entirely cartilaginous or, if epiphysial ossification has begun, are separated from the shaft by cartilaginous *epiphysial discs*. **Clinically, the term epiphysis usually means bony epiphysis. The part of the shaft adjacent to an epiphysial disc contains the growth zone and newly formed bone and is called the metaphysis. The bony tissue of the metaphysis and of the epiphysis is continuous in the adult.**

The shaft of a long bone is a tube of *compact bone* ("compacta"), the cavity of which is known as a *medullary* (*marrow*) *cavity*. The cavity contains either red or yellow marrow, or combinations of both. The epiphysis and metaphysis consist of irregular, anastomosing bars or trabeculae, which form what is known as *spongy* or *cancellous bone*. The spaces between the trabeculae are filled with marrow.

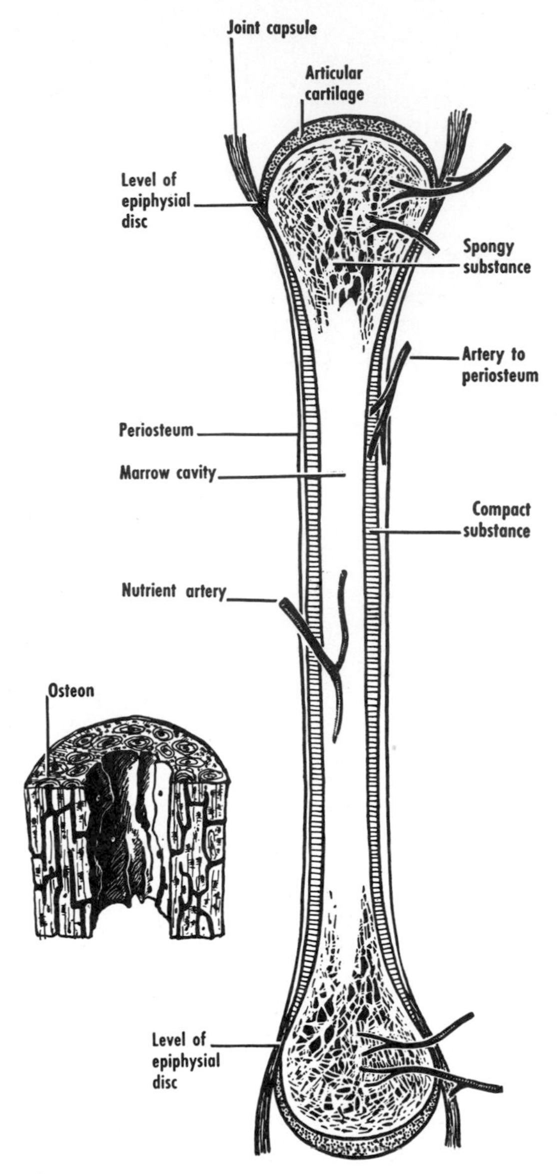

Figure 2–1 Diagram of a long bone and its blood supply. The inset shows the lamellae of the compacta arranged in osteons, i.e., vascular canals surrounded by concentric layers of bone.

The bone on the articular surfaces of the ends is covered by cartilage, which is usually hyaline.

The shaft of a long bone is surrounded by a connective tissue sheath, the *periosteum.* Periosteum is composed of a tough, outer fibrous layer, which acts as a limiting membrane, and an inner, more cellular osteogenic layer. The inner surface of compact bone is lined by a thin, cellular layer, the *endosteum.* At the ends of the bone the periosteum is continuous with the joint capsule, but it does not cover the articular cartilage. Periosteum serves for the attachment of muscles and tendons to bone.

Short Bones. Short bones occur in the hands and feet and consist of spongy bone and marrow enclosed by a thin layer of compact bone. They are surrounded by periosteum, except on their articular surfaces.

Sesamoid bones are a type of short bone and occur mainly in the hands and feet, embedded within tendons or joint capsules. They vary in size and number. Some clearly serve to alter the angle of pull of a tendon. Others, however, are so small that they are of scant functional importance.

Accessory, or supernumerary, bones are bones that are not regularly present. They occur chiefly in the hands and feet. They include some sesamoid bones and certain ununited epiphyses in the adult. They are of medicolegal importance in that, when seen in radiograms, they may be mistaken for fractures. Callus, however, is absent, the bones are smooth, and they are often present bilaterally.

Flat Bones. Flat bones include the ribs, sternum, scapulae, and many bones of the skull. They consist of two layers of compact bone with intervening spongy bone and marrow. The intervening spongy layer in the bones of the vault of the skull is termed *diploë:* it contains many venous channels. Some bones, such as the lacrimal and parts of the scapula, are so thin that they consist of only a thin layer of compact bone.

Irregular Bones. Irregular bones are those that do not readily fit into other groups. They include many of the skull bones, the vertebrae, and the hip bones.

Contours and Markings

The shafts of long bones usually have three surfaces, separated from one another by three borders. The articular surfaces are smooth, even after articular cartilage is removed, as in a dried bone. A projecting articular process is often referred to as a head, its narrowed attachment to the rest of the bone as the neck. The remainder is the body or, in a long bone, the shaft. A condyle (knuckle) is a protruding mass that carries an articular surface. A ramus is a broad arm or process that projects from the main part or body of the bone.

Other prominences are called processes, trochanters, tuberosities, protuberances, tubercles, and spines. Linear prominences are ridges, crests, or lines, and linear depressions are grooves. Other depressions are fossae or foveae (pits). A large cavity in a bone is termed a sinus, a cell, or an antrum. A hole or opening in a bone is a foramen. If it has length, it is a canal, a hiatus, or an aqueduct. Many of these terms (e.g., canal, fossa, foramen, and aqueduct) are not, however, limited to bones.

The ends of bones, except for the articular surfaces, contain many foramina for blood vessels. Vascular foramina on the shaft of a long bone are much smaller, except for one or sometimes two large nutrient foramina that lead into oblique canals, which contain vessels that supply the bone marrow. **The nutrient canals usually point away from the growing end of the bone and toward the epiphysis that unites first with the shaft. The directions of the vessels are indicated by the following mnemonic: To the elbow I go; from the knee I flee.**

The surfaces of bones are commonly roughened and elevated where there are powerful fibrous attachments but smooth where muscle fibers are attached directly.

Blood and Nerve Supply

Bones are richly supplied with blood vessels, and the pattern of supply of a long bone is illustrated in figure 2–1. The nutrient artery, as well as periosteal, metaphysial, and epiphysial vessels, is also shown. In a growing bone, the metaphysial and epiphysial vessels are separated by the cartilaginous epiphysial plate. Both groups of vessels are important for the nutrition of the growth zone, and disturbances of blood supply may result in disturbances in growth. When growth stops and the epiphysial plate disappears, the metaphysial and epiphysial vessels anastomose within the bone.

Many nerve fibers accompany the blood vessels of bone. Most such fibers are vasomotor, but some are sensory and end in periosteum and in the adventitia of blood vessels. Some of the sensory fibers are pain fibers. Periosteum is especially sensitive to tearing or tension. Fractures are painful, and an anesthetic injected between the broken ends of the bone may give relief. A tumor or infection that enlarges within a bone may be quite painful. Pain arising in a bone may be felt locally, or it may spread or be referred. For example, pain arising in the shaft of the femur may be felt diffusely in the thigh or knee.

Bone Marrow

Before birth, the medullary cavities of bones, and the spaces between trabeculae, are filled with red marrow, which gives rise to red blood corpuscles and to certain white blood cells (granulocytes). From infancy onward there is both a progressive diminution in the amount of blood cell–forming marrow and a progressive increase in the amount of fat (yellow marrow).

In the adult, red marrow is usually present in the ribs, vertebrae, sternum, and hip bones. The radius, ulna, tibia, and fibula contain fatty marrow in their shafts and epiphyses. The femur and humerus usually contain a small amount of red marrow in the upper parts of the shafts, and small patches may be present in their proximal epiphyses. The tarsal and carpal bones generally contain only fatty marrow. Loss of blood may be followed by an increase in the amount of red marrow as more blood cells are formed.

Development and Growth

All bones begin as mesenchymal proliferations that appear early in the embryonic period. In *membrane* bones (comprising the clavicle, mandible, and certain skull bones), the cells differentiate into osteoblasts that lay down an organic matrix called osteoid. Bone salts are then deposited in this matrix. Some osteoblasts are trapped in the matrix and become osteocytes. Others continue to divide and form more osteoblasts on the surface of the bone. Bone grows only by apposition, that is, by the laying down of new bone on free surfaces.

Most bones, however, develop as *cartilage* bones. The mesenchymal proliferations become chondrified as the cells lay down cartilage matrix and form hyaline cartilages that have the shapes of the future bones. These cartilages are then replaced by bone, as illustrated in figure 2–2.

Skeletal Maturation

Skeletal development involves three interrelated but dissociable components: increase in size (growth), increase in maturity, and aging. Skeletal maturation is "the metamorphosis of the cartilaginous and membranous skeleton of the foetus to the fully ossified bones of the adult" (Acheson). Skeletal status, however, does not necessarily correspond with height, weight, or age. In fact, the maturative changes in the skeleton are intimately related to those of the reproductive system. These in turn are directly responsible for most of the externally discernible changes on which

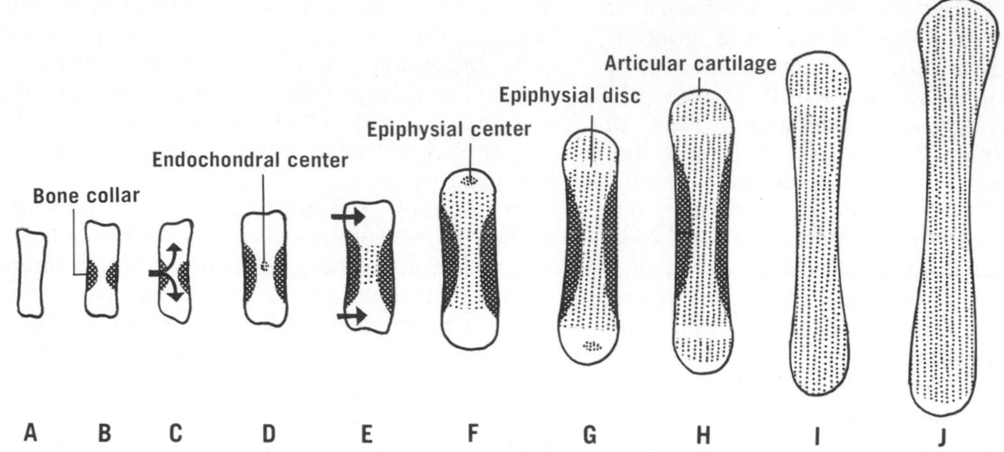

Figure 2–2 Diagrams of the development of a long bone. **A,** Cartilaginous model. **B,** Bone collar. **C,** Vascular invasion of bone collar and cartilage. **D,** Endochondral ossification begins. **E,** Cartilaginous epiphyses begin to be vascularized (*arrows*). **F,** An epiphysial center of ossification appears. **G,** An epiphysial center begins in the other end. **H,** Two epiphysial discs are evident. **I,** The last (second) epiphysial center to appear fuses first with the shaft. **J,** The first epiphysial center to appear (where most growth in length occurs) fuses last with the shaft.

the estimation of general bodily maturity is usually based. The skeleton of a healthy child develops as a unit, and the various bones tend to keep pace with one another. Hence, radiographic examination of a limited portion of the body is believed by some workers to suffice for an estimation of the entire skeleton. The hand is the portion most frequently examined: "As the hand grows, so grows the entire skeleton," it is sometimes stated.

The assessment of skeletal maturity is important in determining whether an individual child is advanced or retarded skeletally and, therefore, in diagnosing endocrine and nutritional disorders. Skeletal status is frequently expressed in terms of skeletal age. This involves the comparison of radiograms of certain areas with standards for those areas; the skeletal age assigned is that of the standard that corresponds most closely. Detailed standards have been published for the normal postnatal development of the hand, knee, and foot. Tables showing the times of appearance of the postnatal ossific centers in the limbs are provided inside the back cover.

Skeletal Maturation Periods. The following arbitrary periods are convenient.

EMBRYONIC PERIOD PROPER. This comprises the first eight postovulatory weeks of development. The clavicle, mandible, maxilla, humerus, radius, ulna, femur, and tibia commence to ossify during the last two weeks of this period.

FETAL PERIOD. This begins at eight postovulatory weeks, when the crown-rump length has reached about 30 mm. The following elements commence to ossify early in the fetal period or sometimes late in the embryonic period: scapula, ilium, fibula, distal phalanges of the hand, and certain cranial bones (e.g., the frontal).

The following begin to ossify during the first half of intra-uterine life: most cranial bones and most diaphyses (ribs, metacarpals, metatarsals, phalanges), calcaneus sometimes, ischium, pubis, some segments of the sternum, neural arches, and vertebral centra (C.V.1 to S.V.5).

The following commence to ossify shortly before birth: calcaneus, talus, and cuboid; usually the distal end of the femur and the proximal end of the tibia; sometimes the coracoid process, the head of the humerus, and the capitate and hamate; rarely the head of the femur and the lateral cuneiform.

CHILDHOOD. The period from birth to puberty includes infancy (i.e., the first one or two postnatal years). Most epiphyses in the limbs, together with the carpals, tarsals, and sesamoids, begin to ossify during childhood. Ossific centers generally appear one or two years earlier in girls than in boys. Furthermore, those **epiphyses that appear first in a skeletal element usually are the last to unite with the diaphysis. They are located at the so-called growing ends (e.g., shoulder, wrist, knee).**

ADOLESCENCE. This includes puberty and

the period from puberty to adulthood. Puberty usually occurs at 13 ± 2 years of age in girls, and two years later in boys. Most of the secondary centers for the vertebrae, ribs, clavicle, scapula, and hip bone begin to ossify during adolescence. The fusions between epiphysial centers and diaphyses occur usually during the second and third decades. These fusions usually take place one or two years earlier in girls than in boys. The closure of epiphysial lines is under hormonal control.

ADULTHOOD. The humerus serves as a skeletal criterion for the transitions into adolescence and into adulthood, in that its distal epiphysis is the first of those of the long bones to unite, and its proximal epiphysis is the last (at age 19 or later). The center for the iliac crest fuses in early adulthood (age 21 to 23). The sutures of the vault of the skull commence to close at about the same time (from age 22 onward).

CARTILAGE

Cartilage is a tough, resilient connective tissue composed of cells and fibers embedded in a firm, gel-like intercellular matrix.

A skeletal element that is mainly or entirely cartilaginous is surrounded by a connective tissue membrane, the perichondrium, the structure of which is similar to periosteum. Cartilage grows by apposition, that is, by the laying down of new cartilage on the surface of the old. The new cartilage is formed by chondroblasts derived from the deeper cells of the perichondrium. Cartilage also grows interstitially, that is, by an increase in the size and number of its existing cells and by an increase in the amount of intercellular matrix. Adult cartilage grows slowly, and repair or regeneration after a severe injury is inadequate. Adult cartilage lacks nerves, and it usually lacks blood vessels.

Types

Cartilage is classified into three types: hyaline, fibrous, and elastic.

Hyaline Cartilage. This is so named because it has a glassy, translucent appearance resulting from the character of its matrix. The cartilaginous models of bones in the embryo consist of hyaline cartilage, as do the epiphysial discs. Most articular cartilages, the costal cartilages, the cartilages of the trachea and bronchi, and most of the cartilages of the nose and larynx are formed of hyaline cartilage. Nonarticular hyaline cartilage has a tendency to calcify and to be replaced by bone.

Fibrocartilage. Bundles of collagenous fibers are the prominent constituent of the matrix of fibrocartilage. Fibrocartilage is present in certain cartilaginous joints, and it forms articular cartilage in a few joints, for example, the temporomandibular.

Elastic Cartilage. The fibers in the matrix are elastic, and such cartilage rarely if ever calcifies with advancing age. Elastic cartilage is present in the auricle and the auditory tube, and it forms some of the cartilages of the larynx.

ADDITIONAL READING

Enlow, D. H., *Principles of Bone Remodeling,* Thomas, Springfield, Illinois, 1963. An excellent review with original observations.

Frazer's Anatomy of the Human Skeleton, 6th ed., rev. by A. S. Breathnach, Churchill, London, 1965. A detailed synthesis of skeletal and muscular anatomy arranged regionally.

Vaughan, J. M., *The Physiology of Bone,* Clarendon Press, Oxford, 1970. An excellent account of bone as a tissue and of its role in mineral homeostasis.

JOINTS

A *joint* or *articulation* is "the connexion subsisting in the skeleton between any of its rigid component parts, whether bones or cartilages" (Bryce). *Arthrology* means the study of joints, and *arthritis* refers to their inflammation.

Joints may be classified into three main types: fibrous, cartilaginous, and synovial.

FIBROUS JOINTS

The bones of a fibrous joint (*synarthrosis*) are united by fibrous tissue. There are two types: *sutures* and *syndesmoses*. With few exceptions, little if any movement occurs at either type. The joint between a tooth and the bone of its socket is termed a *gomphosis* and is sometimes classed as a third type of fibrous joint.

Sutures. In the sutures of the skull, the bones are connected by several fibrous layers. The mechanisms of growth at these joints (still in dispute) are important in accommodating the growth of the brain.

Syndesmoses. A syndesmosis is a fibrous joint in which the intervening connective tissue is considerably greater in amount than in a suture. Examples are the tibiofibular syndesmosis and the tympanostapedial syndesmosis.

CARTILAGINOUS JOINTS

The bones of cartilaginous joints are united either by hyaline cartilage or by fibrocartilage.

Hyaline Cartilage Joints. This type (*synchondrosis*) is a temporary union. The hyaline cartilage that joins the bones is a persistent part of the embryonic cartilaginous skeleton and as such serves as a growth zone for one or both of the bones that it joins. Most hyaline cartilage joints are obliterated, that is, replaced by bone, when growth ceases. Examples include epiphysial plates and the spheno-occipital and neurocentral synchondroses.

Fibrocartilaginous Joints. In this type (*amphiarthrosis*), the skeletal elements are united by fibrocartilage during some phase of their existence. The fibrocartilage is usually separated from the bones by thin plates of hyaline cartilage. Fibrocartilaginous joints include the pubic symphysis and the joints between the bodies of the vertebrae.

SYNOVIAL JOINTS

Synovia is the fluid present in certain joints, which are consequently termed synovial. Similar fluid is present in bursae and in synovial tendon sheaths.

General Characteristics

Synovial (*diarthrodial*) joints possess a cavity and are specialized to permit more or less free movement. Their chief characteristics (fig. 2–3) are as follows:

The articular surfaces of the bones are covered with cartilage, which is usually hyaline in type. The bones are united by a *joint capsule* and ligaments. The joint capsule consists of a *fibrous layer,* the inner surface of which is lined by a vascular, connective tissue, the *synovial membrane,* which produces the synovial fluid (synovia) that fills the *joint cavity* and lubricates the joint. The joint cavity is sometimes partially or completely subdivided by fibrous or fibrocartilaginous *discs* or *menisci.*

Types

Synovial joints may be classified according to axes of movement, assuming the existence of three mutually perpendicular axes. A joint

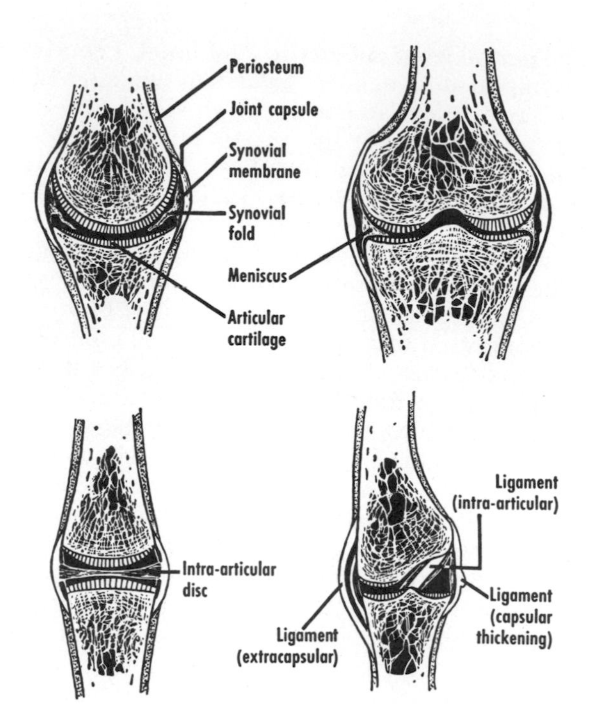

Figure 2–3 Synovial joints. The joint cavity is exaggerated. Articular cartilage, menisci, and intra-articular discs are not covered by synovial membrane, but intra-articular ligaments are.

that has but one axis of rotation, such as a hinge joint or pivot joint, is said to have one degree of freedom. Ellipsoidal and saddle joints have two degrees of freedom. Each can be flexed or extended, abducted or adducted, but not rotated, at least not independently. A ball-and-socket joint has three degrees of freedom.

Synovial joints may also be classified according to the shapes of the articular surfaces of the constituent bones. The types of synovial joints are plane, hinge and pivot (uniaxial), ellipsoidal and saddle (biaxial), condylar (modified biaxial), and ball-and-socket (triaxial). These shapes determine the type of movement and are partly responsible for determining the range of movement.

Plane Joint. The articular surfaces of a plane joint permit gliding or slipping in any direction, or the twisting of one bone on the other.

Hinge Joint, or Ginglymus. A hinge joint is uniaxial and permits movement in but one plane, e.g., flexion and extension at an interphalangeal joint.

Pivot, or Trochoid, Joint. This type, of which the proximal radio-ulnar joint is an example, is uniaxial, but the axis is vertical, and one bone pivots within a bony or an osseoligamentous ring.

Ellipsoidal Joint. In this type, which resembles a ball-and-socket joint, the articulating surfaces are much longer in one direction than in the direction at right angles. The circumference of the joint thus resembles an ellipse. It is biaxial, and the radiocarpal joint is an example.

Saddle, or Sellar, Joint. This type is shaped like a saddle; an example is the carpometacarpal joint of the thumb. It is biaxial.

Condylar Joint. Each of the two articular surfaces is called a condyle. Although resembling a hinge joint, a condylar joint (e.g., the knee) permits several kinds of movements.

Ball-and-Socket, or Spheroidal, Joint. A spheroidal surface of one bone moves within a "socket" of the other bone about three axes, e.g., as in the shoulder and hip joints. Flexion, extension, adduction, abduction, and rotation can occur, as well as a combination of these movements termed *circumduction*. In circumduction, the limb is swung so that it describes the side of a cone, the apex of which is the center of the "ball."

Movements

Active Movements. Usually one speaks of movement *of* a part or of movement *at* a joint; thus, flexion of the forearm or flexion at the elbow. Three types of active movements occur at synovial joints: (1) gliding or slipping movements, (2) angular movements about a horizontal or side-to-side axis (flexion and extension) or about an anteroposterior axis (abduction and adduction), and (3) rotary movements about a longitudinal axis (medial and lateral rotation). Whether one, several, or all types of movement occur at a particular joint depends upon the shape and ligamentous arrangement of that joint.

The range of movement at joints is limited by (1) the muscles, (2) the ligaments and capsule, (3) the shapes of the bones, and (4) the opposition of soft parts, such as the meeting of the front of the forearm and arm during full flexion at the elbow. The range of motion varies greatly in different individuals. In trained acrobats, the range of joint movement may be extraordinary.

Passive and Accessory Movements. Passive movements are produced by an external force, such as gravity or an examiner. For example, the examiner holds the subject's wrist so as to immobilize it and can then flex, extend, adduct, and abduct the subject's hand at the wrist, movements that the subject can normally carry out actively.

By careful manipulation, the examiner can also produce a slight degree of gliding and rotation at the wrist, movements that the subject cannot actively perform himself. These are called accessory movements (often classified with passive movements), and are defined as movements for which the muscular arrangements are not suitable, but which can be brought about by manipulation.

The production of passive and accessory movements is of value in testing and in diagnosing muscle and joint disorders.

Structure and Function

The mechanical analysis of joints is very complicated, and articular movements involve spherical as well as plane geometry.

The lubricating mechanisms of synovial joints are such that the effects of friction on articular cartilage are minimized. This is brought about by the nature of the lubricating fluid (viscous synovial fluid), by the nature of the cartilaginous bearing surfaces that adsorb and absorb synovial fluid, and by a variety of mechanisms that permit a replaceable fluid rather than an irreplaceable bearing to reduce friction.

Synovial Membrane and Synovial Fluid. Synovial membrane is a vascular connective tissue that lines the inner surface of the capsule but does not cover articular cartilage. Synovial membrane differs from other connective tissues in that it produces a ground substance that is a fluid rather than a gel. The most characteristic structural feature of synovial membrane is a capillary network adjacent to the joint cavity. A variable number of villi, folds, and fat pads project into the joint cavity from the synovial membrane.

Synovial membrane is responsible for the formation of synovial fluid, which is a sticky, viscous fluid like egg-white in consistency. The main function of synovial fluid is lubrication, but it also nourishes articular cartilage.

Articular Cartilage. Adult articular cartilage is an avascular, nerveless, and relatively acellular tissue. The part immediately adjacent to bone is usually calcified.

Cartilage is elastic in the sense that, when it is compressed, it becomes thinner but, on release of the pressure, slowly regains its original thickness.

Articular cartilage is not visible in ordinary radiograms. Hence the so-called radiological joint space is wider than the true joint space.

Joint Capsule and Ligaments. The capsule is

composed of bundles of collagenous fibers, which are arranged somewhat irregularly.

Ligaments are classified as *capsular, extracapsular,* and *intra-articular*. Most ligaments serve as sense organs in that nerve endings in them are important in reflex mechanisms and in the detection of movement and position. Ligaments also have mechanical functions.

The relationship of the epiphysial plate to the line of capsular attachment is important (see fig. 2–1). For example, the epiphysial plate is a barrier to the spread of infection between the metaphysis and the epiphysis. If the epiphysial line is intra-articular, then part of the metaphysis is also intra-articular, and a metaphysial infection may involve the joint. In such instances, a metaphysial fracture becomes intra-articular, always serious because of possible damage to articular surfaces. If the capsule is attached directly to the periphery of the epiphysial plate, damage to the joint may involve the plate and thereby interfere with growth.

INTRA-ARTICULAR STRUCTURES. Menisci, intra-articular discs, fat pads, and synovial folds (fig. 2–3) aid in spreading synovial fluid throughout the joint, and thereby assist in lubrication.

Intra-articular discs and menisci, which are composed mostly of fibrous tissue but may contain some fibrocartilage, are attached at their periphery to the joint capsule. They are usually present in joints at which flexion and extension are associated with gliding (e.g., in thc kncc), and that require a rounded combined with a relatively flattened surface.

PERIARTICULAR TISSUES. The fascial investments around the joint blend with capsule and ligaments, with musculotendinous expansions, and with the looser connective tissue that invests the vessels and nerves approaching the joint.

Joints are often injured, and they are subject to many disorders, some of which involve the periarticular tissues as well as the joints themselves. Increased fibrosis (adhesions) of the periarticular tissues may limit movement almost as much as does fibrosis within a joint.

Absorption from Joint Cavity. A capillary network and a lymphatic plexus lie in the synovial membrane, adjacent to the joint cavity. Diffusion takes place readily between these vessels and the cavity. Hence, traumatic infection of a joint may be followed by septicemia. Most substances in the blood stream, normal or pathological, easily enter the joint cavity.

Blood and Nerve Supply. The pattern is illustrated in figure 2–4. Articular and epiphysial vessels arise more or less in common and form networks around the joint and in the synovial membrane, respectively.

The principles of distribution of nerves to joints were best expressed by Hilton in 1863: "The same trunks of nerves, whose branches supply the groups of muscles moving a joint, furnish also a distribution of nerves to the skin over the insertions of the same muscles; and —what at this moment more especially merits our attention—the interior of the joint receives its nerves from the same source." Articular nerves contain sensory and autonomic fibers, the distribution of which is summarized in figure 2–4.

Some of the sensory fibers form proprioceptive endings in the capsule and ligaments. These endings are very sensitive to position and movement. Their central connections are such that they are concerned with the reflex control of posture and locomotion and the detection of position and movement.

Other sensory fibers form pain endings, which are most numerous in joint capsules and

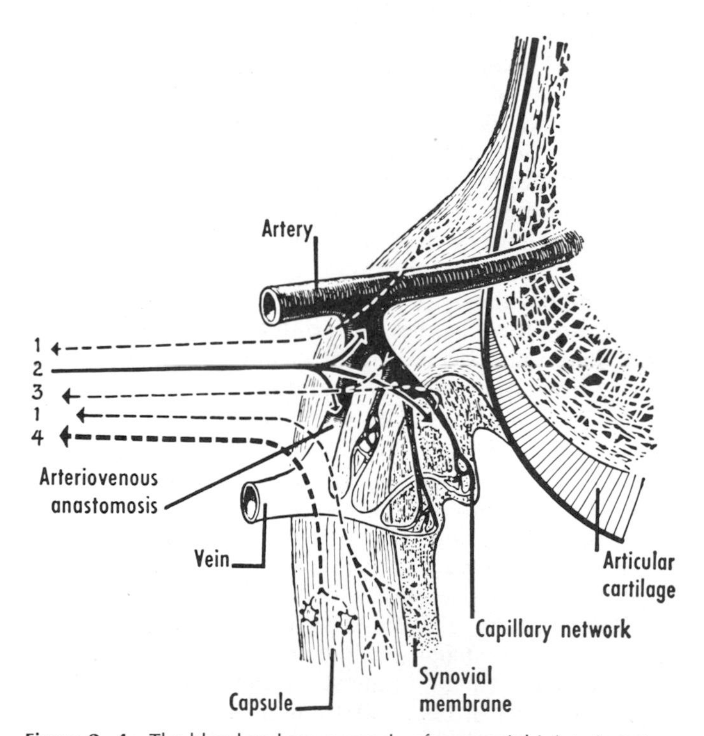

Figure 2–4 The blood and nerve supply of a synovial joint. An artery is shown supplying the epiphysis, joint capsule, and synovial membrane. The nerve contains (*1*) sensory (mostly pain) fibers from the capsule and synovial membrane, (*2*) autonomic (postganglionic sympathetic) fibers to blood vessels, (*3*) sensory (pain) fibers from the adventitia of blood vessels, and (*4*) proprioceptive fibers. Arrowheads indicate direction of conduction.

ligaments. Twisting or stretching of these structures is very painful. The fibrous capsule is highly sensitive; synovial membrane is relatively insensitive.

Use-Destruction (Wear and Tear, Attrition)

With time, articular cartilage wears away, sometimes to the extent of exposing, eroding, and polishing or eburnating the underlying bone. Use-destruction may be hastened or exaggerated by trauma, disease, and biochemical changes in articular cartilage.

ADDITIONAL READING

Barnett, C. H., Davies, D. V., and MacConaill, M. A., *Synovial Joints*. Longmans, London, 1961. A good account of the biology, mechanics, and functions of joints.

Freeman, M. A. R. (ed.), *Adult Articular Cartilage*, Pitman, London, 1973. A good account of lubrication and synovial fluid.

Gardner, E., The Structure and Function of Joints, in *Arthritis*, 8th ed., ed. by J. L. Hollander and D. J. McCarty, Lea & Febiger, Philadelphia, 1972.

MUSCLES

Movement is carried out by specialized cells called muscle fibers, the latent energy of which is, or can be, controlled by the nervous system. Muscle fibers are classified as skeletal (or striated), cardiac, and smooth.

Skeletal muscle fibers are long, multinucleated cells having a characteristic cross-striated appearance under the microscope. These cells are supplied by motor fibers from cells in the central nervous system. The muscle of the heart is also composed of cross-striated fibers, but its activity is regulated by the autonomic nervous system. The walls of most organs and many blood vessels contain fusiform (spindle-shaped) muscle fibers that are arranged in sheets, layers, or bundles. These cells lack cross-striations and are therefore called smooth muscle fibers. Their activity is regulated by the autonomic nervous system and certain circulating hormones, and they supply the motive power for various aspects of digestion, circulation, secretion, and excretion.

Skeletal muscles are sometimes called *voluntary* muscles, because they can usually be controlled voluntarily. However, many of the actions of skeletal muscles are automatic, and the actions of some of them are reflex and only to a limited extent under voluntary control. Smooth muscle and cardiac muscle are sometimes spoken of as *involuntary* muscle.

SKELETAL MUSCLES

General Characteristics

Most muscles are discrete structures that cross one or more joints and, by contracting, can cause movements at these joints. Exceptions are certain subcutaneous muscles (e.g., facial muscles) that move or wrinkle the skin or close orifices, the muscles that move the eyes, and other muscles associated with the respiratory and digestive systems.

Each muscle fiber is surrounded by a delicate connective tissue sheath, the *endomysium*. Muscle fibers are grouped into fasciculi, each of which is enclosed by a connective tissue sheath termed *perimysium*. A muscle as a whole is composed of many fasciculi and is

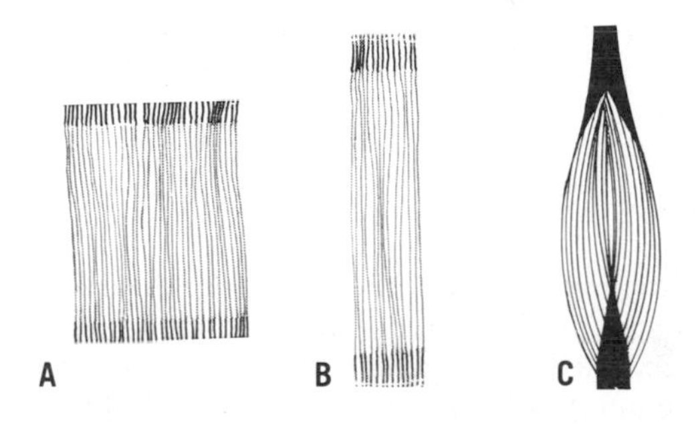

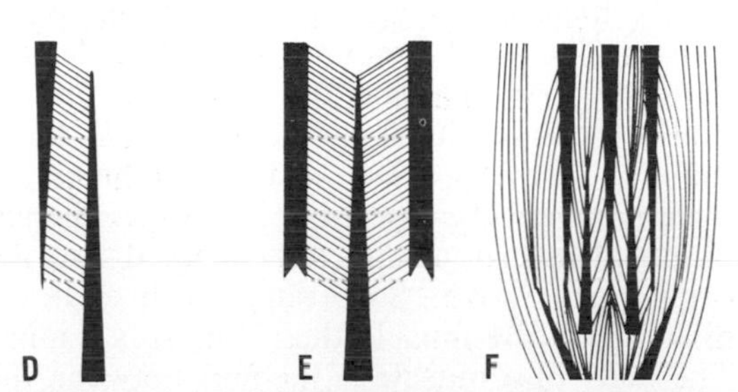

Figure 2–5 The arrangement of fibers in muscles. The fibers are basically parallel (*upper row*) or pennate (or penniform), i.e., arranged as in a feather (*lower row*). **A,** Quadrilateral, e.g., pronator quadratus. **B,** Straplike, e.g., sartorius. **C,** Fusiform, e.g., flexor carpi radialis. **D,** Unipennate, e.g., flexor pollicis longus. **E,** Bipennate, e.g., rectus femoris. **F,** Multipennate, e.g., deltoid. Pennate muscles usually contain a larger number of fibers and hence provide greater power.

surrounded by *epimysium,* which is closely associated with fascia and is sometimes fused with it.

The fibers of a muscle of rectangular or quadrate shape run parallel to the long axis of the muscle (fig. 2–5). The fibers of a muscle of pennate shape are parallel to one another, but lie at an angle with respect to the tendon. The fibers of a triangular or fusiform muscle converge upon a tendon.

The names of muscles usually indicate some structural or functional feature. A name may indicate shape, e.g., trapezius, rhomboid, or gracilis. A name may refer to location, e.g., tibialis posterior. The number of heads of origin is indicated by the terms biceps, triceps, and quadriceps. Action is reflected in terms such as levator scapulae and extensor digitorum.

Muscles are variable in their attachments: they may be absent, and many supernumerary muscles have been described. Variations of muscles are so numerous that detailed accounts of them are available only in special works.

Individual muscles are described according to their origin, insertion, nerve supply, and action. Certain features of blood supply are also important.

Origin and Insertion

Most muscles are attached either directly or by means of their tendons or aponeuroses to bones, cartilages, ligaments, or fasciae, or to some combination of these. Other muscles are attached to organs, such as the eyeball, and still others are attached to skin. When a muscle contracts and shortens, one of its attachments usually remains fixed and the other moves. The fixed attachment is called the *origin,* the movable one the *insertion.* In the limbs, the more distal parts are generally more mobile. Therefore the distal attachment is usually called the insertion. However, the anatomical insertion may remain fixed and the origin may move. Sometimes both ends remain fixed: the muscle then stabilizes a joint. The belly of a muscle is the part between the origin and the insertion.

Blood and Nerve Supply

Muscles are supplied by adjacent vessels, but the pattern varies. Some muscles receive vessels that arise from a single stem, which enters either the belly or one of the ends, whereas others are supplied by a succession of anastomosing vessels.

Each muscle is supplied by one or more nerves, containing motor and sensory fibers that are usually derived from several spinal nerves. Some groups of muscles, however, are supplied mainly if not entirely by one segment of the spinal cord. For example, the motor fibers that supply the intrinsic muscles of the hand arise from the first thoracic segment of the spinal cord. Not infrequently, muscles having similar functions are supplied by the same peripheral nerve.

Nerves usually enter the deep surface of a muscle. The point of entrance is known as the "motor point" of a muscle, because electrical stimulation here is more effective in producing muscular contraction than it is elsewhere on the muscle, nerve fibers being more sensitive to electrical stimulation than are muscle fibers.

Each motor nerve fiber that enters a muscle supplies many muscle fibers. The parent nerve cell and its motor fiber, together with the muscle fibers supplied, make up a *motor unit.*

Denervation of Muscle. Skeletal muscle cannot function without a nerve supply. A denervated muscle becomes flabby and atrophic. The process of atrophy consists of a decrease in size of individual muscle fibers. Each fiber shows occasional spontaneous contractions termed fibrillations. In spite of the atrophy, the muscle fibers retain their histological characteristics for a year or more, eventually being replaced by fat and connective tissue. Provided that nerve regeneration occurs, human muscles may regain fairly normal function up to a year after denervation.

Actions and Functions

In a muscle as a whole, gradation of activity is made possible by the number of motor units. If all the motor units are activated simultaneously, the muscle contracts once. But if motor units are activated out of phase or asynchronously (nerve impulses reaching motor units at different times), tension is maintained in the muscle.

Long and rectangular muscles produce a greater range of movement, whereas pennate muscles exert more force. Power is greatest when the insertion is far removed from the axis of movement, whereas speed is usually greatest when the insertion is near the axis.

The actions of muscles that cross two or more joints are particularly complicated. For instance, the hamstring muscles that cross the hip and knee joints cannot shorten enough to extend the hip and flex the knee completely at the same time. If the hips are flexed fully, as in bending forward to touch the floor, the hamstrings may not be able to lengthen enough to allow one to touch the floor without bending the knees. This is also known as the ligamentous action: it restricts movement at a joint. It is due in part to relative inextensibility of connective tissue and tendons, and can be modified greatly by training. The term *contracture* means a more or less permanent shortening of a muscle.

The pattern of muscular activity is controlled by the central nervous system. Most movements, even so-called simple ones, are complex and in many respects automatic. The overall pattern of movement may be voluntary, but the functions of individual muscles are complex, variable, and often not under voluntary control. For example, if one reaches out and picks something off a table, the use of the fingers is the chief movement. But in order to get the fingers to the object, the forearm is extended (the flexors relaxing), other muscles stabilize the shoulder, and still others stabilize the trunk and lower limbs so as to ensure maintenance of posture.

Muscles may be classified according to the functions they serve in such patterns, namely as prime movers, antagonists, fixation muscles, and synergists. A special category includes those that have a paradoxical or eccentric action, in which muscles lengthen while contracting (fig. 2–6). In so doing, they perform negative work. A muscle may be a prime mover in one pattern, an antagonist in another, or a synergist in a third.

Prime Movers. A prime mover (fig. 2–6) is a muscle or a group of muscles that directly brings about a desired movement (e.g., flexion of the fingers). Gravity may also act as a prime mover. For example, if one holds an object and lowers it to the table, gravity brings about the lowering (fig. 2–6). The only muscular action involved is in controlling the rate of descent, an example of paradoxical action.

Antagonists. Antagonists are muscles that directly oppose the movement under consideration. Thus, the triceps brachii, which is the extensor of the forearm when acting as a prime mover, is the antagonist to the flexors of the forearm. Depending on the rate and force of movement, antagonists may be relaxed, or, by lengthening while contracting, they may control movement and make it smooth, free from jerkiness, and precise. The term *antagonist* is a poor one, because such muscles cooperate rather than oppose. Gravity may also act as an antagonist, as when the forearm is flexed from the anatomical position.

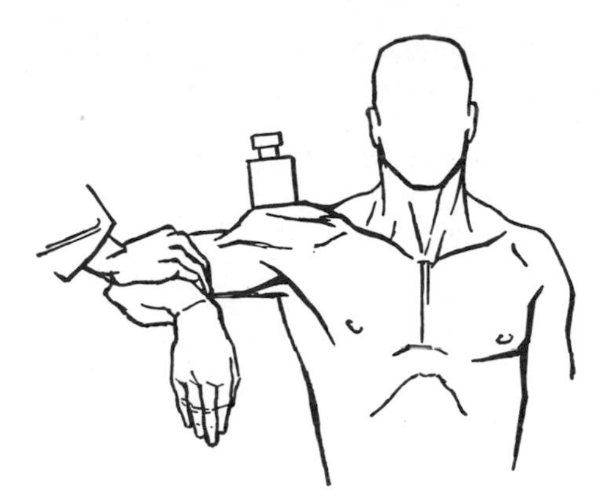

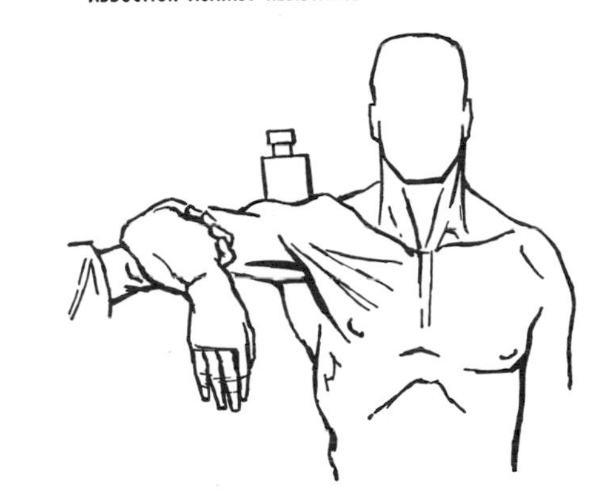

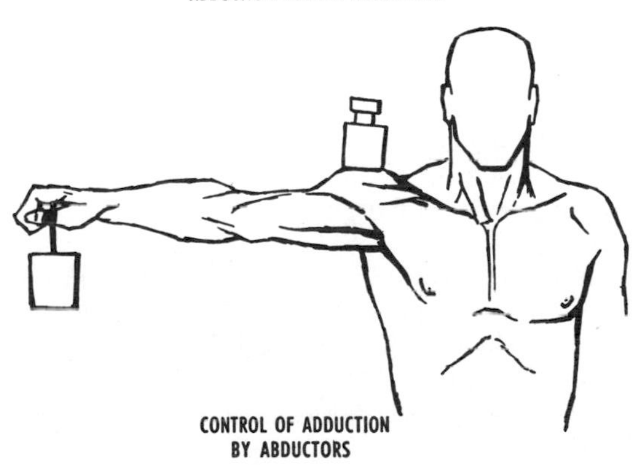

Figure 2–6 Muscular actions. When the arm is abducted against the examiner's resistance, the deltoid becomes tense. On adduction against resistance, the deltoid relaxes and the weight sinks into it. On adduction produced by lowering a pail from a horizontal position, the pectoralis major is relaxed. The contracted deltoid controls the descent by lengthening. The deltoid is now an antagonist to gravity, which is the prime mover, and is doing negative work (paradoxical action).

Fixation Muscles. Fixation muscles generally stabilize joints or parts and thereby maintain posture or position while the prime movers act.

Synergists. Synergists are a special class of fixation muscles. When a prime mover crosses two or more joints, synergists prevent undesired actions at intermediate joints. Thus, the long muscles that flex the fingers would at the same time flex the wrist if the wrist were not stabilized by the extensors of the wrist, these being synergists in this particular movement.

Testing of Muscles

Five chief methods are available to determine the action of a muscle. These are the anatomical method, palpation, electrical stimulation, electromyography, and the clinical method. No one of these methods alone is sufficient to provide full and accurate information.

Anatomical Method. Actions are deduced from the origin and insertion, as determined by dissection, and are verified by pulling upon the muscle, for example, during an operation. The anatomical method is usually the only way of determining the actions of muscles too deep to be examined during life. This method shows what a muscle can do, but not necessarily what it actually does.

Palpation. The subject is asked to perform a certain movement, and the examiner inspects and palpates the muscles taking part.

The movement may be carried out without loading or extra weight and with gravity minimized so far as possible by support or by the recumbent position. Alternatively, the movement may be carried out against gravity, as when flexing the forearm from the anatomical position, with or without extra load. Finally it may be tested with a heavy load, most simply by fixing the limb by an opposing force. For example, the examiner requests the subject to flex the forearm and at the same time holds the forearm so as to prevent flexion. Palpation of muscles that are contracting against resistance provides the best and simplest way of learning the locations and actions of muscles in the living body. Palpation is also the simplest and most direct method of testing weak or paralyzed muscles, and it is widely used clinically. When several muscles take part, it may not be possible to determine the functions of each muscle by palpation alone.

Electrical Stimulation. The electrical stimulation of a muscle over its motor point causes the muscle to contract and to remain contracted if repetitive stimulation is used. Like the anatomical method, electrical stimulation shows what a muscle can do, but not necessarily what its functions are.

Electromyography. The mechanical twitch of a muscle fiber is preceded by a conducted impulse that can be detected and recorded with appropriate instruments. When an entire muscle is active, the electrical activity of its fibers can be detected by electrodes placed within the muscle or on the overlying skin. The recorded response constitutes an electromyogram (E.M.G.). Records can be obtained from several muscles simultaneously. This makes electromyography valuable for studying patterns of activity. Electromyography, like palpation, may be classified as a natural or physiological method. Finally, the electromyographic pattern may be altered by nervous or muscular disease. Electromyography, therefore, can be used in diagnosis.

The disadvantage, as in palpation, is the difficulty in assessing the precise function of a muscle that is taking part in a movement pattern.

Clinical Method. A study of patients who have paralyzed muscles or muscle groups provides valuable information about muscle function, primarily by determining which functions are lost. But great caution must be exercised. In some central nervous system disorders, a muscle may be paralyzed in one movement yet take part in another. Even in the presence of peripheral nerve injuries or direct muscle involvement, patients may learn trick movements with other muscles that compensate for or mask the weakness or paralysis.

Reflexes and Muscle Tone

Many muscular actions are reflex in nature, that is, they are brought about by sensory impulses that reach the spinal cord and activate motor cells. The quick withdrawal of a burned finger and the blinking of the eyelids when something touches the cornea are examples of reflexes. It is generally held that muscles that support the body against gravity possess tone, owing to the operation of stretch reflexes initiated by the action of gravity in stretching the muscles. Whether this is strictly or always true in the human is open to question. There is evidence that when a subject is in an easy standing position, little if any muscular contraction or tone can be detected in human antigravity muscles.

The available evidence indicates that the only "tone" possessed by a completely relaxed muscle is that provided by its passive elastic tension. No impulses reach a completely relaxed muscle, and no conducted electrical activity can be detected.

Structure and Function

Each skeletal muscle fiber is a long, multinucleated cell that consists of a mass of myofibrils. Most muscle fibers are less than 10 to 15 cm long, but some may be more than 30 cm long.

Resting muscle is soft, freely extensible, and elastic. Active muscle is hard, develops tension, resists stretching, and lifts loads. Muscles may thus be compared with machines for converting chemically stored energy into mechanical work. Muscles are also important in the maintenance of body temperature. Resting muscle under constant conditions liberates heat, which forms a considerable fraction of the basic metabolic rate.

One of the most characteristic changes after death is the stiffening of muscles, known as rigor mortis. Its time of onset and its duration are variable. It is due chiefly to the loss of adenosine triphosphate from the muscles.

TENDONS AND APONEUROSES

The attachment of muscle to bone (or other tissue) is usually by a long, cord-like tendon or sinew or by a broad, relatively thin aponeurosis. Tendons and aponeuroses are both composed of more or less parallel bundles of collagenous fibers. Tendons and aponeuroses are surrounded by a thin sheath of looser connective tissue. Where tendons are attached to bone, the bundles of collagenous fibers fan out in the periosteum.

Tendons are supplied by sensory fibers that reach them from nerves to muscles. They also receive sensory fibers from nearby superficial or deep nerves.

Synovial Tendon Sheaths

Where tendons run in osseofibrous tunnels, for example, in the hand and foot, they are covered by double-layered synovial sheaths (fig. 2–7). The *mesotendineum,* which is the tissue that forms the continuity between the synovial layers, carries blood vessels to the tendon. The fluid in the cavity of the sheath is similar to synovial fluid and facilitates movement by minimizing friction.

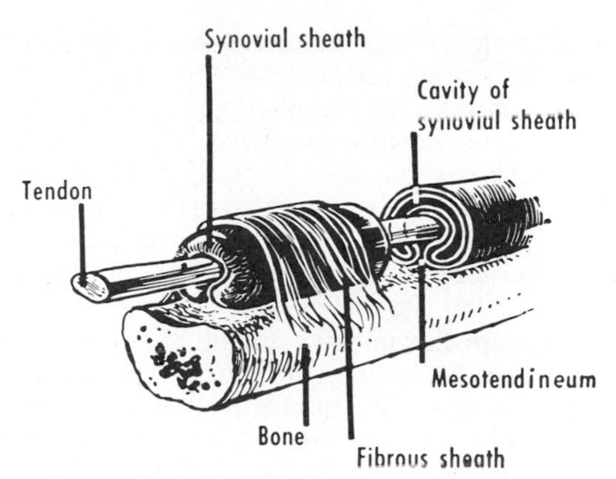

Figure 2–7 Synovial and fibrous sheaths of a tendon, and a section of the synovial sheath.

The lining of the sheath, like synovial membrane, is extremely cellular and vascular. It reacts to infection or to trauma by forming more fluid and by cellular proliferation. Such reactions may result in adhesions between the two layers and a consequent restriction of movement of the tendon.

BURSAE

Bursae (from L. *bursa,* a purse), like synovial tendon sheaths, are connective tissue sacs with a slippery inner surface and are filled with synovial fluid. Bursae are present where tendons rub against bone, ligaments, or other tendons, or where skin moves over a bony prominence. They may develop in response to friction. Bursae facilitate movement by minimizing friction.

Bursae are of clinical importance. Some communicate with joint cavities, and to open such a bursa is to enter the joint cavity, always a potentially dangerous procedure from the standpoint of infection. Some bursae are prone to fill with fluid when injured, for example, the bursae in front of or below the patella (housemaid's knee).

FASCIA

Fascia is a packing material, a connective tissue that remains between areas of more specialized tissue, such as muscle. Fascia forms fibrous membranes that separate muscles from one another and invest them, and as such it is often called deep fascia. Its functions include providing origins and insertions for muscles, serving as an elastic sheath for muscles, and forming specialized retaining bands (retinacula) and fibrous sheaths for tendons. It provides pathways for the passage of vessels and nerves and surrounds these structures as neurovascular sheaths. It permits the gliding of one structure on another. The mobility, elasticity, and slipperiness of living fascia can never be appreciated by dissecting embalmed material.

The main fascial investment of some muscles is indistinguishable from epimysium. Other muscles are more clearly separated from fascia, and are freer to move against adjacent muscles. In either instance, muscles or groups of muscles are generally separated by intermuscular septa, which are deep prolongations of fascia.

In the lower limb, the return of blood to the heart is impeded by gravity and aided by muscular action. However, muscles would swell with blood were it not for the tough fascial investment of these muscles, which serves as an elastic stocking. The investment also prevents bulging during contraction and thus makes muscular contraction more efficient in pumping blood upward.

Fascia is more or less continuous over the entire body, but it is commonly named according to region, for example, pectoral fascia. It is attached to the superficial bony prominences that it covers, blending with periosteum, and, by way of intermuscular septa, is more deeply attached to bone.

Fascia may limit or control the spread of pus. When shortened because of injury or disease, fascia may limit movement. Strips of fascia are sometimes used for the repair of tendinous or aponeurotic defects.

Proprioceptive endings in aponeuroses and retinacula probably have a kinesthetic as well as a mechanical function.

ADDITIONAL READING

Basmajian, J. V., *Muscles Alive,* 4th ed., Williams & Wilkins, Baltimore, 1978. An excellent study of muscle functions as revealed by electromyography.

Lockhart, R. D., *Living Anatomy,* 6th ed., Faber & Faber, London, 1963. Photographs showing muscles in action and methods of testing.

Rosse, C., and Clawson, D. K., *The Musculoskeletal System in Health and Disease,* Harper & Row, Hagerstown, Maryland, 1980. An attractive account of functional anatomy, clinical applications, and diseases.

Royce, J., *Surface Anatomy,* Davis, Philadelphia, 1965. Photographs and key drawings of the living body.

QUESTIONS

2–1 Is there a difference between membrane bones and cartilage bones in the adult?

2–2 Where is red marrow found in the adult?

2–3 Which portion of the body is examined most frequently in the assessment of skeletal maturation?

2–4 Are epiphysial centers visible radiographically in the knee at birth?

2–5 Which parts of the limb bones are cartilaginous in the adult?

2–6 What result would be expected from premature closure of epiphysial plates?

2–7 Provide examples of (a) plane, (b) hinge, (c) pivot, (d) ellipsoidal, (e) saddle, (f) condylar, and (g) ball-and-socket joints.

2–8 What are (a) the origin and (b) the functions of synovial fluid?

2–9 What is the importance of the relationship between the epiphysial plate and the line of capsular attachment?

2–10 What advantage have pennate muscles?

2–11 What is the total number of (a) bones and (b) muscles in the body?

3
THE NERVOUS SYSTEM

The nervous system comprises the central nervous system, consisting of the brain and spinal cord, and the peripheral nervous system, consisting of the cranial, spinal, and peripheral nerves, together with their motor and sensory endings.

CENTRAL NERVOUS SYSTEM

The central nervous system is composed of millions of nerve and glial cells, together with blood vessels and a little connective tissue. The nerve cells, or *neurons,* are characterized by many processes and are specialized in that they exhibit great irritability and conductivity. The glial cells, termed *neuroglia,* are characterized by short processes that have special relationships to neurons, blood vessels, and connective tissue.

BRAIN

The brain is the enlarged, head end of the central nervous system; it occupies the cranium, or brain case. The term *cerebrum* (L., *brain;* adjective cerebral) generally means brain, but sometimes is used for the forebrain and midbrain only. *Encephalon,* of Greek origin, is found in such terms as *encephalitis,* which means inflammation of the brain.

The brain presents three main divisions: forebrain (prosencephalon), midbrain (mesencephalon), and hindbrain (rhombencephalon) (fig. 3–1). The forebrain in turn has two subdi-

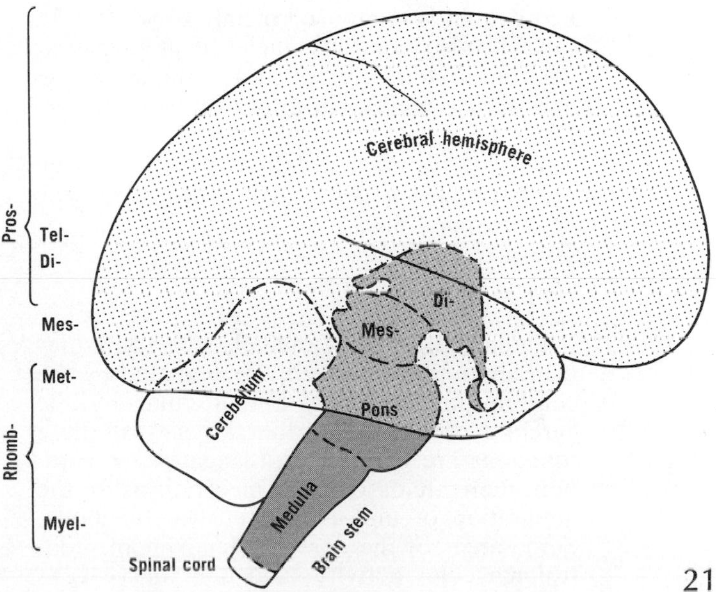

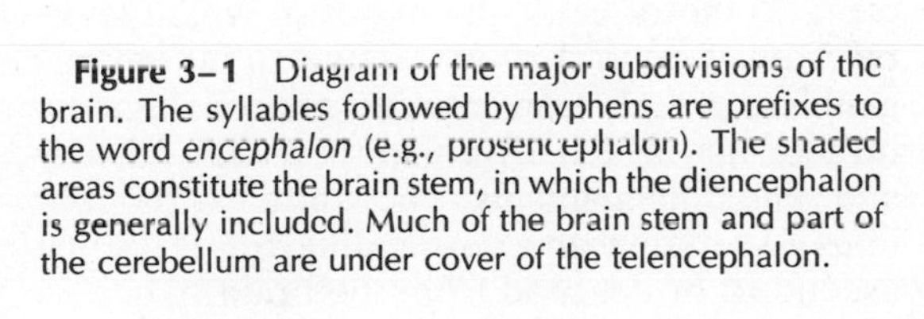

Figure 3–1 Diagram of the major subdivisions of the brain. The syllables followed by hyphens are prefixes to the word *encephalon* (e.g., prosencephalon). The shaded areas constitute the brain stem, in which the diencephalon is generally included. Much of the brain stem and part of the cerebellum are under cover of the telencephalon.

visions, telencephalon (endbrain) and diencephalon (interbrain). The hindbrain likewise has two subdivisions, the metencephalon (afterbrain) and the myelencephalon (marrowbrain). The bulk of the brain is formed by two cerebral hemispheres, which are derived from the telencephalon. The hemispheres are distinguished by convolutions, or gyri, which are separated by sulci. The unpaired diencephalon lies between the hemispheres. It forms the upper part of the *brain stem,* an unpaired stalk that descends from the base of the brain. The brain stem is formed by the diencephalon, midbrain, pons, and myelencephalon, or medulla oblongata. The last is continuous with the spinal cord at the foramen magnum. The *cerebellum* is a fissured mass of gray matter that occupies the posterior cranial fossa and is attached to the brain stem by three pairs of peduncles. Twelve pairs of cranial nerves issue from the base of the brain and from the brain stem.

The cerebral cortex, which is the outer part of the hemispheres and is only a few millimeters in thickness, is composed of gray matter, in contrast to the interior of the brain, which is composed partly of white matter. Gray matter consists largely of the bodies of nerve and glial cells, whereas white matter consists largely of the processes or fibers of nerve and glial cells.

The interior of the cerebral hemispheres, including the diencephalon, contains not only white matter but also masses of gray matter known collectively as basal nuclei.

The cerebellar cortex, like the cerebral, is composed of gray matter. The interior of the cerebellum is composed mainly of white matter, but also contains nuclei of gray matter. The brain stem, by contrast, contains nuclei and diffuse masses of gray matter in its interior.

The interior of the brain also contains cavities termed *ventricles,* which are filled with cerebrospinal fluid.

Functions

The highest mental and behavioral activities characteristic of humans are mediated by the cerebral hemispheres, in particular by the cerebral cortex. Important aspects of these functions are learning and language. In addition, there are association mechanisms for the integration of motor and sensory functions. Some areas of the cerebral hemispheres control muscular activity, and their nerve cells send processes to the brain stem and spinal cord, where they are connected with motor cells, the processes of which leave by way of cranial nerves or ventral roots in the spinal cord. Other areas are sensory and receive impulses that have reached the spinal cord by way of peripheral nerves and dorsal roots, and have ascended in the spinal cord and brain stem by pathways that consist of a succession of nerve cells and their processes. Fibers that ascend and descend in the brain and spinal cord are generally grouped into tracts. The tracts are usually named according to their origin and destination, e.g., corticospinal.

The brain stem contains, in addition to tracts that descend and ascend through it, collections of cells that (1) comprise major integrating centers for motor and sensory functions, (2) form the nuclei of most cranial nerves (all of the cranial nerves except the first are attached to the brain stem), (3) form centers concerned with the regulation of a variety of visceral, endocrinological, behavioral, and other activities, (4) are functionally associated with most of the special senses, (5) control muscular activity in the head and part of the neck, (6) supply pharyngeal arch structures, and (7) are connected with the cerebellum.

The cerebellum is concerned with the automatic regulation of movement and posture, and it functions closely with the cerebral cortex and the brain stem.

Spinal Cord

The spinal cord is a long, cylindrical mass of nervous tissue, oval or rounded in transverse section. It occupies the upper two-thirds of the vertebral canal. In contrast to the cerebral hemispheres, gray matter is found in the interior, surrounded by white matter (fig. 3–2).

The neurons of the spinal cord include (1) motor cells, the axons of which leave by way of ventral roots and supply skeletal muscles; (2) motor cells, the axons of which leave by way of ventral roots and go to autonomic ganglia; and (3) transmission neurons and interneurons, which are concerned with sensory and reflex mechanisms. The white matter contains ascending and descending tracts. Some ascend to or descend from the brain, whereas others connect cells at various levels of the cord.

Attached to the spinal cord on each side is a series of *spinal roots,* termed *dorsal* and *ventral* according to their position. Generally there are 31 pairs, which comprise 8 cervical, 12 thoracic, 5 lumbar, 5 sacral, and 1 coccygeal. Corresponding dorsal and ventral roots join to form a spinal nerve. Each spinal nerve divides into a dorsal and a ventral ramus, and these are distributed to various parts of the body.

The spinal cord carries out sensory, integrative, and motor functions, which can be categorized as reflex, reciprocal activity (as one activity starts, another stops), monitoring and modulation of sensory and motor mechanisms, and transmission of impulses to the brain.

Meninges and Cerebrospinal Fluid

The brain and spinal cord are surrounded and protected by layers of non-nervous tissue, collectively termed *meninges*. These layers, from without inward, are the *dura mater*, *arachnoid*, and *pia mater*, and are described in more detail elsewhere. The space between the arachnoid and the pia mater, the *subarachnoid space*, contains cerebrospinal fluid (C.S.F.).

The ventricles of the brain contain vascular choroid plexuses, from which C.S.F., an almost protein-free liquid, is formed. This fluid circulates through the ventricles, enters the subarachnoid space, and eventually filters into the venous system. C.S.F. protects the brain and serves to minimize damage from blows to the head and neck.

Blood Supply

The brain is supplied by the cerebral branches of the vertebral and internal carotid arteries, the meninges mainly by the middle meningeal branch of the maxillary artery. The spinal cord and spinal roots are supplied by the vertebral arteries and by segmental arteries. Peripheral nerves are supplied by a number of small branches along the course of the nerves.

PERIPHERAL NERVOUS SYSTEM

A *nerve* is a collection of nerve fibers that is visible to the naked eye. The constituent fibers are bound together by connective tissue. Each fiber is microscopic in size and is surrounded by a sheath formed by a neurilemmal cell (comparable to the glial cells of the central nervous system). Hundreds or thousands of fibers are present in each nerve. Thus, according to the number of constituent fibers, a nerve may be barely visible, or it may be quite thick. A nerve as a whole is surrounded by a connective tissue sheath, the *epineurium*. Connective tissue fibers run inward from the sheath and enclose bundles of nerve fibers. Such bundles are termed *funiculi* (fasciculi); the connective tissue that encloses them is called *perineurium*. Very small nerves may consist of only one funiculus derived from the parent nerve. Finally, each nerve fiber and its neurilemmal sheath are enclosed by a connective tissue sheath termed *endoneurium*.

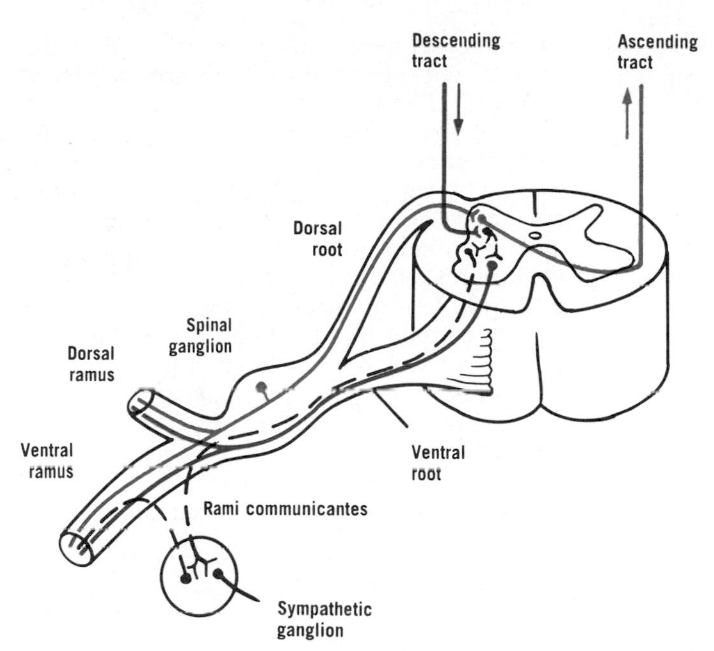

Figure 3–2 A horizontal section of the spinal cord, and dorsal and ventral roots and a spinal nerve. The arrangement of the rami communicantes is usually much more complicated.

Nerve fibers may be classified according to the structures they supply, that is, according to function. A fiber that stimulates or activates skeletal muscle is termed a *motor* (efferent) fiber. A fiber that carries impulses from a sensory ending is termed a *sensory* (afferent) fiber. Fibers that activate glands and smooth muscle are also motor fibers, and various kinds of sensory fibers arise from endings in viscera. Consequently, a more detailed classification of functional components is sometimes required.

Spinal Nerves

The spinal roots, which are anchored to the spinal cord, consist of a dorsal root, attached to the dorsal aspect of the spinal cord, and a ventral root, attached to the ventral aspect of the cord. Each dorsal root (which contains sensory fibers from skin, subcutaneous and deep tissues, and often from viscera also) is formed by neuronal processes that carry afferent impulses into the spinal cord and which arise from neurons that are collected together to form an enlargement termed a *spinal ganglion* (fig. 3–2). Each of the ventral roots (which contain motor fibers to skeletal muscle, and of which many contain preganglionic autonomic fibers) is formed by processes of neurons in the gray matter of the spinal cord. **Basically, dorsal roots are afferent, ventral roots efferent.** The corresponding dorsal and ventral roots join to form a spinal nerve. Each spinal nerve then divides into a dorsal and a ventral ramus.

Distribution of Spinal and Peripheral Nerves

The dorsal rami of spinal nerves supply the skin and muscles of the back. The ventral rami supply the limbs and the rest of the trunk. The ventral rami that supply the thoracic and abdominal wall remain relatively separate throughout their course. In the cervical and lumbosacral regions, however, the ventral rami intermingle to form plexuses, from which the major peripheral nerves emerge.

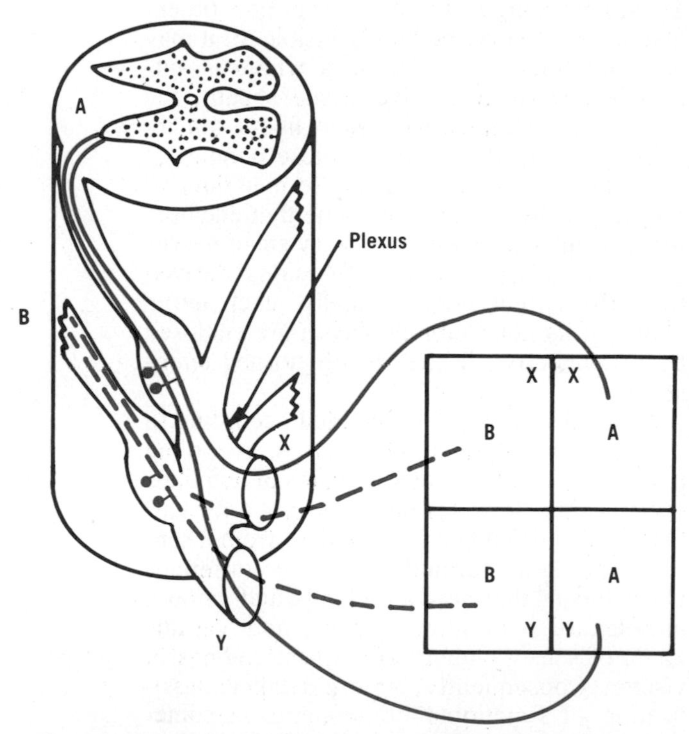

Figure 3–3 Spinal and peripheral nerve distributions. Of the two sensory fibers of spinal nerve *A*, one joins peripheral nerve *X* and the other joins *Y*. A plexus allows both fibers to enter spinal nerve *A*. Two fibers of spinal nerve *B* also join the two peripheral nerves. Thus, the areas (dermatomes) supplied by the two spinal nerves differ from those supplied by the two peripheral nerves, as is shown in the subdivided rectangle. Overlap is omitted.

When the ventral ramus of a spinal nerve enters a plexus and joins other such rami, its component funiculi or bundles ultimately enter several of the nerves emerging from the plexus. Thus, as a general principle, each spinal nerve entering a plexus contributes to several peripheral nerves, and each peripheral nerve contains fibers derived from several spinal nerves. This arrangement leads to two fundamental and important types of distribution (fig. 3–3). Each spinal nerve has a *segmental,* or *dermatomal,* distribution. **A dermatome is the area of skin supplied by the sensory fibers of a single dorsal root through the dorsal and ventral rami of its spinal nerve.** Dermatomes based largely on Foerster are shown inside the front cover.

The mixture of nerve fibers in plexuses is such that it is difficult if not impossible to trace their course by dissection; hence, dermatomal distribution has been determined by physiological experimentation and by studies of disorders of spinal nerves. Methods have included stimulation of spinal roots, study of residual sensation when a root is left intact after section of the roots above and below it, study of the diminution of sensation after section of a single root, and study of the distribution of the blisters that follow inflammation of roots and spinal ganglia in herpes zoster (shingles). Such studies have yielded complex maps, chiefly because of variation, overlap, and differences in method. Variation results from intersegmental rootlet anastomoses adjacent to the cervical and lumbosacral spinal cord and from individual differences in plexus formation and peripheral nerve distribution. Overlap is such that section of a single root does not produce complete anesthesia in the area supplied by that root: at most, some degree of hypalgesia may result. By contrast, when a peripheral nerve is cut, the result is a central area of total loss of sensation surrounded by an area of diminished sensation.

There is little specific correspondence between dermatomes and underlying muscles. The general arrangement is that the more rostral segments of the cervical and lumbosacral enlargements of the spinal cord supply the more proximal muscles of the limbs, and that the more caudal segments supply the more distal muscles. A muscle usually receives fibers from each of the spinal nerves that contribute to the peripheral nerve supplying it (although one spinal nerve may be its chief supply). Section of a single spinal nerve weakens several muscles but usually does not paralyze them. Section of a peripheral nerve results in severe weakness or total paralysis of the muscles it supplies. Moreover, autonomic dysfunction occurs in the area of its distribution.

CRANIAL NERVES

The 12 pairs of cranial nerves are special nerves associated with the brain. The fibers in cranial nerves are of diverse functional types. Some cranial nerves are composed of only one type, others of several.

Cranial nerves differ significantly from spinal nerves, especially in their development and their relation to the special senses and because some cranial nerves supply pharyngeal arch structures. They are attached to the brain at irregular rather than regular intervals; they are not formed of dorsal and ventral roots; some have more than one ganglion, whereas others have none; and the optic nerve is a tract of the central nervous system rather than a peripheral nerve.

CHARACTERISTIC FEATURES OF PERIPHERAL NERVES

The branches of major peripheral nerves are usually muscular, cutaneous (or mucosal), articular, vascular (to adjacent blood vessels), and terminal (one, several, or all of the foregoing types). Muscular branches are the most important: section of even a small muscular branch results in complete paralysis of all muscle fibers supplied by that branch and may be seriously disabling. The importance of sensory loss varies, but such loss is most disabling in the hand, head, and face.

Peripheral nerves vary in their course and distribution, but not as much as blood vessels do. Adjacent nerves may communicate with each other. Such communications sometimes account for residual sensation or movement after section of a nerve above the level of a communication.

AUTONOMIC NERVOUS SYSTEM

The autonomic nervous system regulates the activity of cardiac muscle, smooth muscle, and glands.

The autonomic system can be considered as a series of levels that differ in function in that the higher the level, the more widespread and general its functions. The highest level is the cerebral cortex, certain areas of which control or regulate visceral functions. These areas send fibers to the next lower level, the hypothalamus, located at the base of the brain. The hypothalamus is a coordinating center for the motor control of visceral activity. One of its many functions, for example, is the regulation of body temperature. The hypothalamus has nervous and vascular connections with the hypophysis, by virtue of which it influences the hypophysis and, through the hypophysis, the other endocrine glands. The hypothalamus also sends nerve fibers to lower centers in the brain stem that are concerned with still more specific functions, for example, the reflex regulation of respiration, heart rate, and circulation. These centers function through connections with still lower centers, which are collections of nerve cells in the brain stem and spinal cord that send their axons into certain cranial and spinal nerves. It is characteristic of these axons that, unlike motor fibers to skeletal muscle, they synapse with multipolar cells outside the central nervous system before they reach the structure to be supplied. These multipolar cells are collected into ganglia; the ganglionic is the lowest level. The axons that pass from the central nervous system to these ganglion cells are termed *preganglionic fibers*. The axons of ganglion cells are called *postganglionic fibers;* all such fibers from a particular ganglion supply a specific organ or region of the body.

SYMPATHETIC SYSTEM

The sympathetic, or thoracolumbar, part of the autonomic system comprises the preganglionic fibers that issue from the thoracic and upper lumbar levels of the spinal cord. These fibers reach spinal nerves by way of ventral roots and then leave the spinal nerves and reach adjacent ganglia by way of rami communicantes (see fig. 3–2). These ganglia are contained in long nerve strands, the sympathetic trunks, one on each side of the vertebral column, extending from the base of the skull to the coccyx. Some preganglionic fibers synapse in ganglia of the trunk, others continue to ganglia of the prevertebral plexuses, and still others synapse with cells in the medulla of the suprarenal glands. The postganglionic fibers either go directly to adjacent viscera and blood vessels or return to spinal nerves by way of rami communicantes and, in the area of distribution of these nerves, supply the skin with (1) secretory fibers to sweat glands, (2) motor fibers to smooth muscle (*arrectores pilorum*), and (3) vasomotor fibers to the blood vessels of the limbs.

Parasympathetic System

The parasympathetic, or craniosacral, part of the autonomic system comprises the preganglionic fibers that issue from the brain stem (cranial nerves 3, 7, 9, 10, 11) and sacral part of the spinal cord (segments S2,3 or S3,4). The ganglion cells with which these fibers synapse are in or near the organs innervated. The postganglionic fibers are very short: apparently none go to blood vessels, smooth muscle, or glands of the limbs and body wall. Most viscera, however, have a double motor supply, sympathetic and parasympathetic, sometimes with opposing roles.

Functions of the Autonomic Nervous System

By its role in central integrating mechanisms, the autonomic system is involved in behavioral and neuroendocrinological mechanisms, and in the processes whereby the body keeps its internal environment constant, that is, maintains temperature, fluid balance, and ionic composition of the blood. The parasympathetic system is concerned with many specific functions, such as digestion, intermediary metabolism, and excretion. The sympathetic system is an important part of the mechanism of reaction to stress.

ADDITIONAL READING

Bossy, J., *Atlas du système nerveux,* Éditions Offiduc, Paris, 1971. Beautiful photographs.

De Armond, S. J., Fusco, M. M., and Dewey, M. M., *Structure of the Human Brain,* 2nd ed., Oxford University Press, New York, 1976. A good example of several atlases that are now available.

Gardner, E., *Fundamentals of Neurology,* 6th ed., W. B. Saunders Company, Philadelphia, 1975. A concise account of the nervous system. Excellent for orientation and review.

QUESTIONS

3–1 Look up the origin of the word *encephalon.*

3–2 Is the diencephalon a part of the brain stem?

3–3 How are the 31 pairs of spinal nerves subdivided?

3–4 What is the basic functional distinction between ventral and dorsal roots of spinal nerves?

3–5 What are the two fundamental types of cutaneous distribution of sensory fibers in spinal nerves?

3–6 What is a dermatome?

3–7 In which disease may a dermatome be outlined?

3–8 Why are the cranial nerves so called?

3–9 What are the main differences between cranial and spinal nerves?

3–10 Is the autonomic system purely motor?

4
THE SKIN, HAIR, AND NAILS

LAYERS OF SKIN

Common integument refers to skin and subcutaneous tissue, hair, nails, and breast. The last-named is described with the upper limb. The *skin* (*cutis*) provides a waterproof and protective covering for the body, contains sensory nerve endings, and aids in the regulation of temperature. The skin is important, not only in general medical diagnosis and surgery, but also as the seat of many diseases of its own. The study of these is called *dermatology* (Gk *derma,* skin).

The area of the body surface is about 2 sq m. The temperature of the skin in general is normally about 32 to 36° C. (90 to 96° F.).

The skin (fig. 4–1) varies in thickness from about 0.5 to 3 mm. It is thicker on the dorsal and extensor than on the ventral and flexor aspects of the body. It is thinner in infancy and in old age. The stretching of the abdominal skin during pregnancy may result in red streaks (*striae gravidarum*) that remain as permanent white lines (*lineae albicantes*).

The skin consists of two quite different layers: (1) the epidermis, a superficial layer of stratified epithelium that develops from ectoderm, and (2) the dermis, or corium, an underlying layer of connective tissue that is largely mesodermal in origin. The dermis makes up the bulk of the skin.

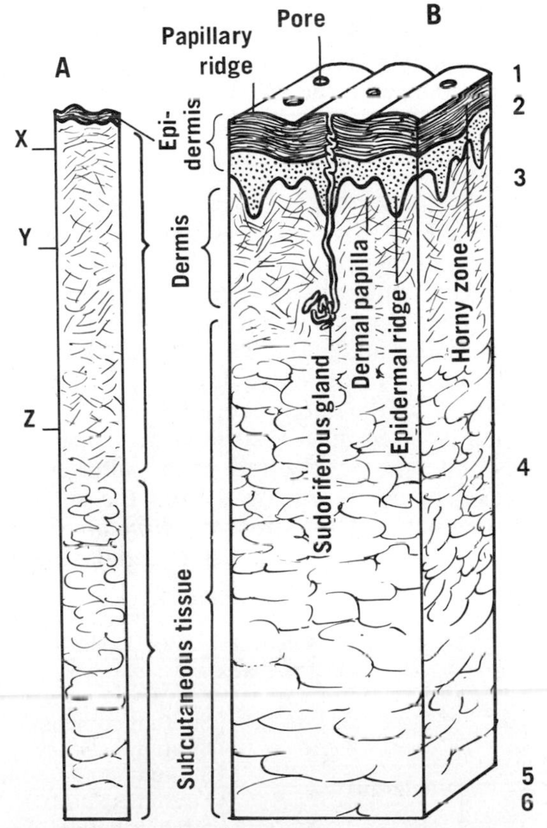

Figure 4–1 General view of skin and subcutaneous tissue. **A,** "Thin" skin from abdomen. **B,** "Thick" skin from palm of hand. *X, Y,* and *Z* represent the levels of a superficial (Thiersch) graft, a split thickness graft (including one third to one quarter of the dermis), and a full thickness (Wolfe) graft, respectively. The numerals 1 to 6 represent the levels of degrees of burns, according to Dupuytren's classification. Other classifications of burns are also used.

CORIUM (DERMIS)

The corium, or dermis, contains downgrowths from the epidermis, such as hair follicles and glands. It presents a superficial papillary layer of loose collagenous and elastic fibers, together with fibroblasts, mast cells, and macrophages. Elevations (papillae) project toward the epidermis. The thicker, deep reticular layer of the dermis consists of dense, coarse bundles of collagenous fibers. Some of the fibers enter the subcutaneous tissue, where they form bundles between lobules of fat. Smooth muscle is found in some regions (areola and nipple, scrotum and penis, and perineum). In some areas, muscle fibers of skeletal type (e.g., platysma) may be inserted into the skin. In tattooing, foreign particles, such as carbon, are introduced into the dermis.

The skin lies on the *subcutaneous tissue* (''superficial fascia''), a layer of fatty areolar tissue that overlies the more densely fibrous fascia. It should be remembered that fat is liquid, or nearly so, at body temperature. The subcutaneous tissue serves as a depot for fat storage and aids in preventing loss of heat. When a pinch of skin is picked up, subcutaneous tissue is included. A hypodermic injection is one given into the subcutaneous tissue.

Epidermis

The skin is covered by a film of emulsified material produced by glands and by cornification. The epidermis is an avascular layer of stratified squamous epithelium that is thickest on the palms and soles. The epidermis, where it is thick, presents five layers, as listed in table 4–1. In the outer layers, which may conveniently be grouped as the horny zone, the cells become converted into soft-keratin flakes that are worn away from the surface continuously. The *stratum corneum* is a tough, resilient, semitransparent cellular membrane that acts as a barrier to water transfer. Under normal conditions, mitotic figures are practically confined to the deepest layer, the *stratum basale,* which is, therefore, the normal germinative layer of the epidermis. The various layers show the stages through which the basal cells pass before their keratinization and shedding. The cells of the epidermis are replaced approximately once per month. Keratin is a protein that is present throughout the epidermis, perhaps in a modified form. It is readily hydrated—hence the swelling of skin on immersion in water—and dryness of the skin is due chiefly to a lack of water.

TABLE 4–1 Arrangement of Layers of Body Surface

Skin	Epidermis	Stratum corneum
		Stratum lucidum
		Stratum granulosum
		Stratum spinosum
		Stratum basale (or cylindricum)
	Corium, or dermis	Papillary layer
		Reticular layer
Subcutaneous tissue		
Fascia		

Human epidermis displays a rhythmic mitotic cycle. Mitosis is more active at night, and it is stimulated by a loss of the superficial, or horny, zone. A part, or the whole thickness, of the epidermis may be raised up in the form of blisters by plasma when the skin is damaged (e.g., by a second-degree burn), and prolonged pressure and friction result in callosities and corns.

Several pigments, including melanin, melanoid, carotene, reduced hemoglobin, and oxyhemoglobin, are found in the skin. Melanin, which is situated chiefly in the stratum basale of the epidermis, protects the organism from ultraviolet light.

When an area of epidermis, together with the superficial part of the underlying dermis, is destroyed, new epidermis is formed from hair follicles, and also from sudoriferous and sebaceous glands, where these are present. If the injury involves the whole thickness of the dermis (e.g., in a deep burn), however, epithelization can take place only by a growing over of the surrounding edge of the epidermis or alternatively by the use of an autograft. Free skin grafts of the epidermis and a part or all of the thickness of the dermis can be applied, and vascularization takes place through connections between the subcutaneous vessels and those in the graft. A defect of the skin that extends into the dermis is termed an ulcer.

Lines of thickened epidermis known as *papillary ridges* form a characteristic pattern on the palmar aspect of the hand and the plantar aspect of the foot. They are concerned with tactile sensation. They contain the openings of the sweat glands and overlie grooves in the dermis; these grooves are situated typically between rows of double ridges known as *dermal ridges* (fig. 4–1). The papillary ridges appear in fetal life in a pattern that remains permanently. They are especially well developed in the pads of the digits, and finger prints in adults and foot prints in infants are used as a means of identification of an individual.

SPECIALIZED STRUCTURES OF SKIN

SWEAT (SUDORIFEROUS) GLANDS

The sweat, or sudoriferous, glands regulate body temperature, because perspiration withdraws heat from the body by the vaporization of water. The sweat glands develop in the fetus as epidermal downgrowths that become canalized. They are simple tubular glands, each having a coiled secretory unit in the dermis or in the subcutaneous tissue and a long, winding duct that extends through the epidermis and opens by a pore on the surface of the skin (fig. 4–1). Sweat glands are particularly numerous in the palms and the soles, where they open on the summits of the papillary ridges. The chief stimuli to sweating are heat and emotion. Emotional perspiration occurs characteristically on the forehead, axillae, palms, and soles.

Large sweat glands in certain locations, such as the axilla, areola, external acoustic meatus, and eyelid, develop from hair follicles and differ from the more common (eccrine) glands in being apocrine; that is to say, portions of the secreting cells disintegrate in the process of secretion. The perspiration from the apocrine glands is rich in organic material that is susceptible to bacterial action, resulting in an odor.

Water passes through the epidermis also by diffusion. This is termed insensible perspiration because it cannot be seen or felt.

HAIRS

Hairs (or *pili; pilus* in the singular) are characteristic of mammals. The functions of hair include protection, regulation of body temperature, and facilitation of evaporation of perspiration; hairs also act as sense organs. Hairs develop in the fetus as epidermal down growths that invade the underlying dermis. Each downgrowth terminates in an expanded end that becomes invaginated by a mesodermal *papilla*. The central cells of the downgrowth become keratinized to form a hair, which then grows outward to reach the surface. The hairs first developed constitute the *lanugo*, or down, which is shed shortly before birth. The fine hairs that develop later constitute the *vellus*. Although hairs on many portions of the human body are inconspicuous, their actual number per unit area is large. In a few places (such as the palms and the soles and the dorsal aspect of the distal phalanges) the skin is *glabrous*, that is, devoid of hair.

The shaft of a hair consists of a *cuticle* and a *cortex* of hard-keratin surrounding, in many hairs, a soft-keratin *medulla* (fig. 4–2). Pigmented hairs contain melanin in the cortex and medulla, but pigment is absent from the surrounding sheaths. The color of hair depends mainly on the shade and the amount of pigment in the cortex and, to a lesser extent, on air spaces in the hair. In white hairs pig-

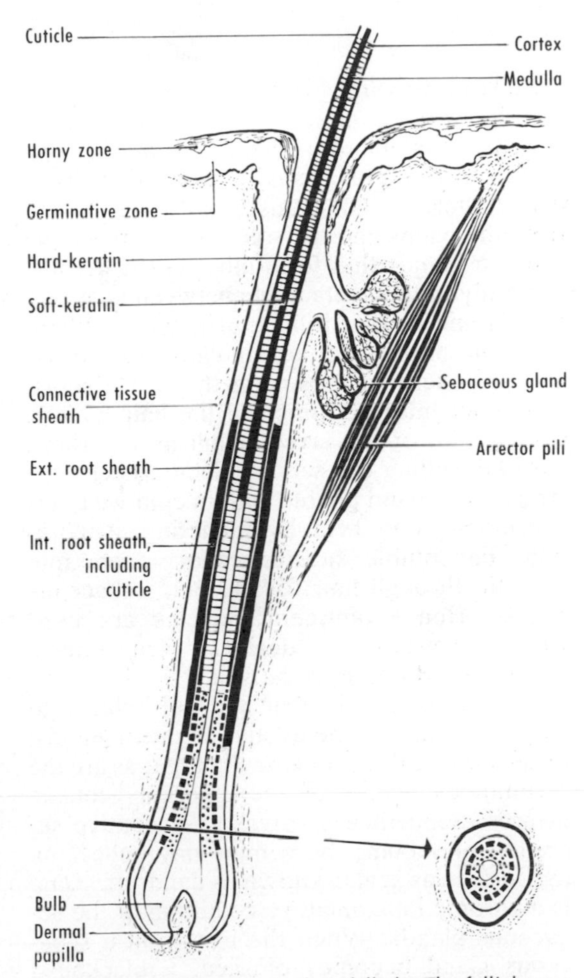

Figure 4–2 Diagram of a hair follicle. The follicle consists of an *external root sheath,* mainly the basal-cell layer of the epidermis, and an *internal root sheath* of soft-keratin, which includes a *cuticle* firmly anchored to that of the shaft of the hair.

ment is absent from the cortex, and the contained air is responsible for the whiteness; "gray hair" is generally a mixture of white and colored hairs.

The *root* of a hair is situated in an epidermal tube known as the hair *follicle,* sunken into either the dermis or the subcutaneous tissue. The follicle is dilated at its base to form the *bulb* (*matrix*).

In the obtuse angle between the root of a hair and the surface of the skin, a bundle of smooth muscle fibers, known as an *arrector pili muscle,* is usually found. It extends from the deep part of the hair follicle to the papillary layer of the dermis. On contraction it makes the hair erect. The arrectores pilorum are innervated by sympathetic fibers and contract in response to emotion or cold. This results in an unevenness of the surface called "goose pimples" or "goose skin."

Sebaceous Glands

Sebaceous glands develop from the epidermis in the fetus, usually from the walls of hair follicles. Sebaceous glands are absent from the palms and the soles. They are simple alveolar glands that form lobes in the dermis, generally in the acute angle between an arrector pili and its hair follicle. The basal cells of the gland proliferate, accumulate fat droplets, and are excreted as sebum through a short, wide duct into the lumen of the hair follicle. Contraction of the arrector pili may perhaps aid in expelling the sebum. Sebum keeps the stratum corneum pliable and, in cold weather, conserves body heat by hindering evaporation. Fat-soluble substances may penetrate the skin through hair follicles and sebaceous glands. Hence ointment vehicles are used when penetration is desired. Medicaments should be rubbed into the skin.

Sebaceous glands that are not related to hairs are found in the eyelids as tarsal glands; these are said to be apocrine in type, as are the ceruminous glands of the external acoustic meatus. Seborrhea involves an excessive secretion of sebum; the sebum may collect on the surface as scales known as dandruff. Acne is a chronic inflammatory condition of the sebaceous glands. When the exit from a sebaceous gland becomes plugged, a blackhead (comedo) forms; complete blockage may result in a wen (sebaceous cyst). At birth an infant is covered with *vernix caseosa,* a mixture of sebum and desquamated epithelial cells.

Nails

The nails (or *ungues; unguis* in the singular) are hardenings of the horny zone of the epidermis. They overlie the dorsal aspect of the distal phalanges (fig. 4–3). They protect the sensitive tips of the digits and, in the fingers, serve in scratching. Nails develop in the fetus as epidermal thickenings that undercut the skin to form folds from which the horny substance of the nail grows distally.

The horny zone of the nail is composed of hard-keratin and has a distal, exposed part, or *body,* and a proximal, hidden portion, or *root.* The root is covered by a distalward prolongation of the stratum corneum of the skin. This narrow fold is composed of soft-keratin and is termed the *eponychium.* Distal to the eponychium is the "half-moon," or *lunula,* a part of the horny zone that is opaque to the underlying capillaries.

Deep to the distal or free border of the nail, the horny zone of the fingertip is thickened and is frequently termed the *hyponychium.* The horny zone of the nail is attached to the underlying *nail bed.* The *matrix,* or proximal part of the bed, produces hard-keratin. Further distally, however, the bed may also generate nail substance. Moreover, the most superficial layer of the nail may be produced by the epithelium immediately dorsal to the root and proximal to the eponychium. The growth of the nail is affected by nutrition, hormones, and disease. Nail growth involves considerable protein synthesis, as a result of which nonspecific changes occur in the nails in response to various local and systemic disturbances. White spots indicate incomplete keratinization.

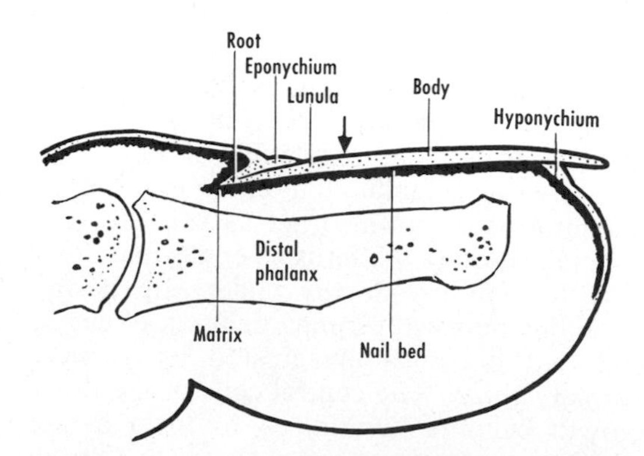

Figure 4–3 Diagram of a sagittal section of a fingernail. The arrow indicates the junction between the root and body of the nail.

BLOOD SUPPLY AND INNERVATION OF SKIN

The skin has a profuse blood supply, which is important in temperature regulation. The subcutaneous arteries form a network in the subcutaneous tissue, and from this is derived a subpapillary plexus in the dermis. Capillary loops in the dermal papillae arise from the subpapillary plexus, and from these loops the avascular epidermis is bathed in tissue fluid. A subpapillary plexus of venules gives the skin its pink color: the vessels become dilated when the skin is heated, and thereby make it look red. Most birthmarks consist of dilated capillaries (*hemangioma*). The dermis contains a lymphatic plexus that drains into the collecting vessels in the subcutaneous tissue. The cutaneous lymphatics can be shown *in vivo* by injecting vital dyes, and every intradermal injection is an intralymphatic one.

The skin has a rich sensory innervation (fig. 4–4). The cutaneous nerves pierce the fascia and ramify in the subcutaneous tissue to form plexuses both there and in the dermis. Finer axonal ramifications may run between the deeper cells of the epidermis. The cutaneous nerves supply both the skin and the subcutaneous tissue. The area of distribution of a given nerve, however, varies, and considerable overlapping of adjacent nerve territories takes place.

Nerves supplying the skin may form several different types of nerve endings, and these endings have been related in a general way to the basic types of sensations that can be appreciated in the skin and the subcutaneous tissue, namely, pain, touch, temperature changes, and pressure, or deep touch.

Hairy skin contains simple, free endings and plexuses around the hair follicles. Skin without hair, that of the palm, for example, presents the three types of sensory endings that are characteristic of the somatic nervous system: (1) free nerve endings arising from small myelinated fibers, (2) expanded tips, and (3) encapsulated endings. The basic types of sensation, however, can be elicited from both hairy and glabrous skin. Hence correlations between the type of sensation and a specific type of nerve ending are not justified. Lamellated corpuscles are particularly large, encapsulated endings that are found chiefly in the subcutaneous and deeper tissues.

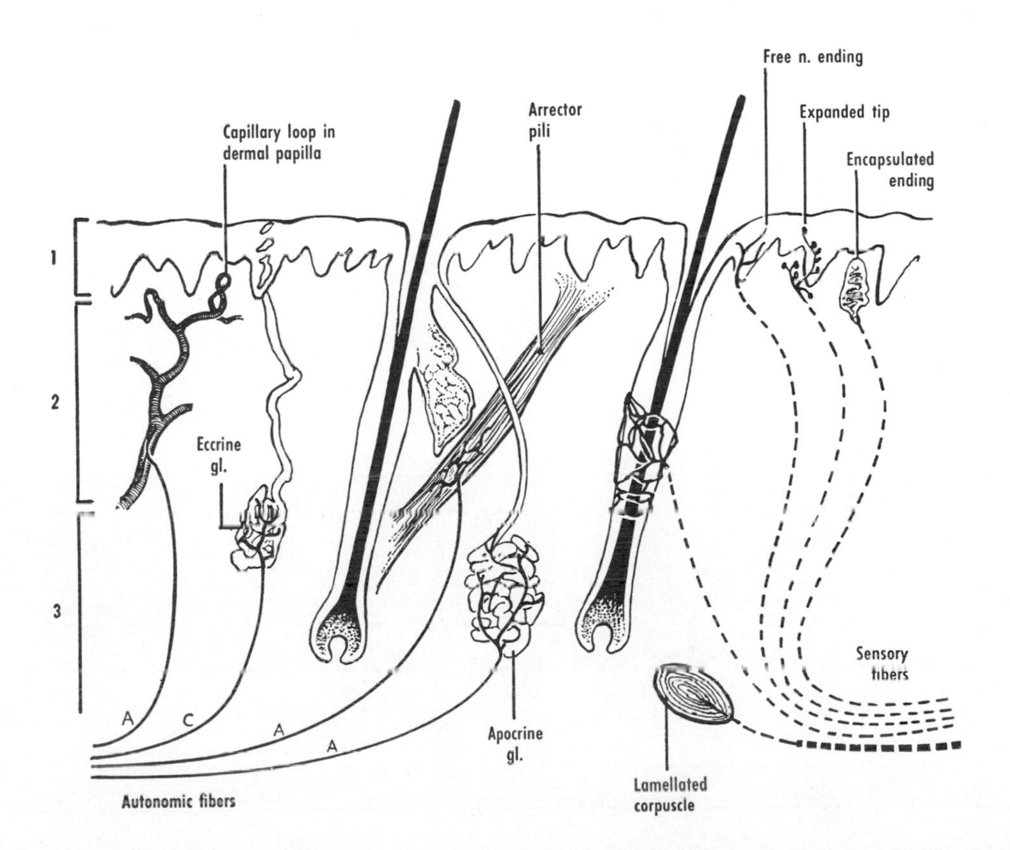

Figure 4–4 Diagram of the innervation of the skin. The numerals 1, 2, and 3 indicate the epidermis, dermis, and subcutaneous tissue, respectively. The letters A and C stand for adrenergic and cholinergic nerve fibers, respectively.

ADDITIONAL READING

Jarrett, A. (ed.), *The Physiology and Pathophysiology of the Skin,* Academic Press, New York, 1973, 1974. Volume 1 is on the epidermis, volume 2 on nerves and blood vessels, and volume 3 on the dermis.

Montagna, W., and Parakkal, P. F., *The Structure and Function of Skin,* 3rd ed. Academic Press, New York, 1974. A good introduction.

Pinkus, H., Die makroskopische Anatomie der Haut, in *Normale und pathologische Anatomie der Haut,* ed. by O. Gans and G. K. Steigleder, Springer, Berlin, vol. 2, 1964. An excellent account.

QUESTIONS

4–1 How do the two layers of the skin differ?

4–2 Into which layer is a hypodermic injection given?

4–3 What is normally the germinative layer of the epidermis?

4–4 What is a second-degree burn?

4–5 How does water leave the skin?

4–6 What is an arrector pili and how does it function?

4–7 What is the function of sebum?

4–8 Can the different types of nerve endings in the skin be related to the basic types of sensation (pain, touch, temperature changes, and pressure, or deep touch)?

5
RADIOLOGICAL ANATOMY

GENERAL ASPECTS

The details of radiographic processes are the province of radiographers, whereas the radiologist is concerned chiefly with the interpretation of radiograms and fluoroscopic images. This presupposes a knowledge of anatomy. Radiography has proved particularly valuable in the detection of the early stages of deep-seated disease, when the possibility of cure is greatest. There is little departure from the normal, however, during these early stages; hence knowledge of the earliest detectable variations, that is, of "the borderlands of the normal and early pathological in the skiagram" (Köhler), is of great medical importance. Radiodiagnosis is the most important method of non-destructive testing of the living body.

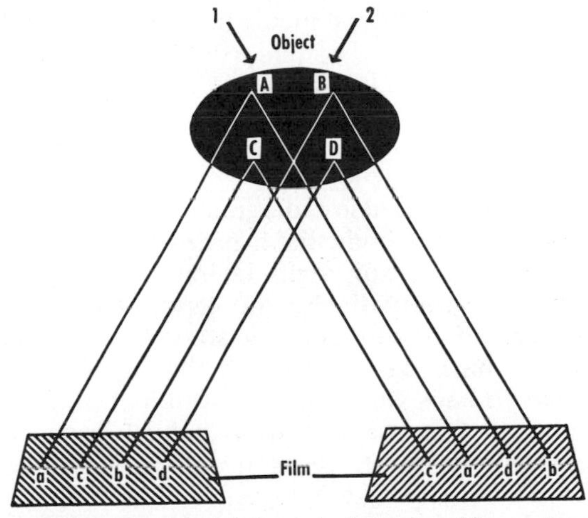

Figure 5–2 Dissociation of planes in oblique views (after Tillier). When the incident radiation is in the direction of arrow 1, the images of parts *A*, *B*, *C*, and *D* of the object are projected in the order *c, a, d, b* on the film. With incident radiation in the direction of arrow 2, however, the order is *a, c, b, d*. Thus when the tube is displaced to the right, the upper plane (*AB*) becomes displaced to the left in relation to the lower plane (*CD*).

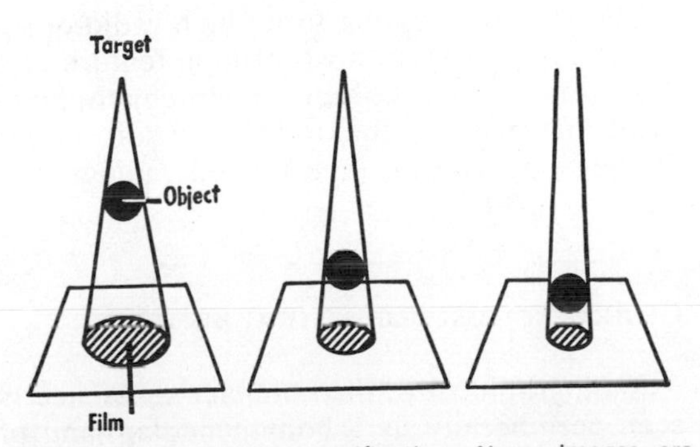

Figure 5–1 Image magnification. X-ray images are shadows, and the geometry of their formation is similar to that relating to shadows formed by ordinary light. Thus the image becomes more enlarged the nearer the object is placed to the target, or the source of radiation. In the second diagram, the image of the object is smaller than in the first diagram because the object is farther from the target. In the third diagram, the target is more than 2 meters from the film, and the magnification is negligible (teleradiography).

Some fundamental principles of radiography are illustrated in figures 5–1 and 5–2.

RADIO-OPACITY

The following structures produce the usual radiographic image, and they are arranged in order of increasing radio-opacity (i.e., whiteness on negative film, blackness on fluoroscopic screen or on positive print) for a constant thickness:

1. Air, as found, for example, in the trachea and lungs, the stomach and intestine, and the paranasal sinuses. Also oxygen when injected into the ventricles of the brain.
2. Fat.
3. Soft tissues, e.g., heart, kidney, muscles.
4. Lime (calcium and phosphorus), for example, in the skeleton.
5. Enamel of the teeth.
6. Dense foreign bodies, for example, metallic fillings in the teeth. Also radio-opaque contrast media, such as a barium meal in the stomach.

When the density of a structure is too similar to that of adjacent structures, it is possible to use contrast media in some sites. Contrast media are classified as radiolucent (e.g., oxygen) and radio-opaque (e.g., barium).

Positioning

The views used in radiography are named for the part of the body that is nearest the film, for example, anterior, right lateral, left anterior oblique. Alternatively, the terms anteroposterior and postero-anterior are used when the x-rays have passed through the object from front to back (tube in front of object, film behind) or from back to front (tube behind object, film in front), respectively. Details of the views commonly employed are given in special books on radiographic positioning.

Fluoroscopy

A fluorescent screen consists of cardboard coated with a thin layer of fluorescent material (phosphor), such as zinc cadmium sulphide. When the screen is activated by x-rays, light is emitted. Increase in brightness may be obtained by intensifying the image electronically. Moreover, the fluoroscopic image can be photographed. The great advantages of fluoroscopy are the ability to observe the motion of parts of the subject and the ability to change the position of the subject during the examination.

Special Procedures

The ordinary radiogram gives no impression of depth other than that dependent on anatomical expectation. In the case of many areas, it is necessary, therefore, to take at least two views, one at a right angle to the other (biplane radiography).

Stereoradiography. Stereoradiography is a procedure whereby dual radiograms are made, corresponding to the points of view of the two eyes. Two views of the same object are taken, each at a slightly different angle. This is accomplished by shifting the x-ray tube about 6 cm (the interpupillary distance) between the two exposures. The resultant films are placed in a special viewer (stereoscope) in order to examine them stereoscopically (in "3D").

Tomography. In tomography (Gk *tomos,* a section, as in microtome), a radiogram is made of a selected layer of the body. Both tube and film are rotated during the exposure, but in opposite directions, resulting in blurring of the tissue planes other than that in which the amount of motion is virtually nil.

Computerized Axial Tomography (C.A.T.). A mathematical reconstruction (computerized tomogram, or C.T. scan) of computer-assisted measurements of absorption density may be made from thousands of tomographic data obtained by means of a rapidly moving x-ray tube. The tube travels back and forth on a rail while hundreds of density measurements are made by detectors on the far side of the body. After each traverse, the unit rotates a degree until 180 degrees have been covered. The result appears on a cathode-ray tube, or as a photograph, or on a print-out sheet.

SKELETAL RADIOLOGY

The skeleton, owing to its high radio-opacity, is generally the most striking feature of a radiogram. It is important to appreciate, however, that many of the organs and soft tissues of the body can be investigated radiographically.

General Features of a Long Bone

Radiographically, the compact substance is seen peripherally as a homogeneous band of lime density. A nutrient canal may be visible as a radiolucent line traversing the compacta obliquely. In some areas the compacta is thinned to form a cortex. The cancellous, or spongy, substance is seen particularly toward the ends of the shaft as a network of lime density presenting interstices of soft-tissue density. Islands of compacta are visible occasionally in the spongiosa. The bone marrow and the periosteum present a soft-tissue density and are not distinguishable as such.

In many young bones the uncalcified portion of an epiphysial disc or plate can be seen radiographically as an irregular, radiolucent band termed an epiphysial line. When an epiphysial line is no longer seen, it is said to be closed, and the epiphysis and diaphysis are said to be united or fused. The radiographic appearance of fusion, however, precedes the disappearance of the visible epiphysial disc as seen on the dried bone.

The term *metaphysis* is used radiologically for the calcified cartilage of an epiphysial disc and the newly formed bone beneath it.

General Features of a Joint

The articular cartilage presents a soft-tissue density and is not distinguishable as such. The so-called radiological joint space, that is, the interval between the radio-opaque epiphysial regions of two bones, is occupied almost entirely by the two layers of articular cartilage, one on each of the adjacent ends of the two bones. On a radiogram the "space" is usually 2 to 5 mm in width in the adult. The joint cavity is rarely visible. The "radiological joint line," that is, the junction between the radio-opaque end of a bone and the radiolucent articular cartilage, is actually the junction between a zone of calcified cartilage over the end of the bone and the uncalcified articular cartilage.

Skeletal Maturation

The development of bones and skeletal maturation are discussed in Chapter 2. Tables showing the times of appearance of the postnatal ossific centers in the limbs are provided inside the back cover. The time of appearance that is given is the age at which 50 per cent of normal children show a certain center radiographically, while the remaining 50 per cent do not yet show it. It is important to appreciate that a considerable range of variation occurs on each side of the median figure.

ADDITIONAL READING

Hamilton, W. J., Simon, G., and Hamilton, S. G. I., *Surface and Radiological Anatomy,* 5th ed., Heffer, Cambridge, 1971.

Köhler, A., and Zimmer, E. A., *Borderlands of the Normal and Early Pathologic in Skeletal Roentgenology,* 3rd ed., Grune and Stratton, New York, 1968. Based on the 11th edition in German.

Tillier, H., *Normal Radiological Anatomy,* trans. by R. O'Rahilly, Thomas, Springfield, Illinois, 1968.

QUESTIONS

5–1 Look up the discovery of x-rays. When were they discovered and by whom?

5–2 Where would the film be placed for a left anterior oblique view of the thorax?

5–3 What are (a) the epiphysial line and (b) the joint space in terms of radiology?

Part 2

THE UPPER LIMB

The upper limb consists of four major parts: a girdle formed by the clavicles and scapulae, the arm, the forearm, and the hand. Although very mobile, the limb is supported and stabilized by muscles connected to the ribs and vertebrae. The following Latin words (given with their English equivalents) are the basis of many anatomical terms (e.g., axillary artery, extensor pollicis): *membrum* (limb; extremity in this sense is obsolescent), *humerus* (shoulder), *axilla* (armpit), *brachium* (arm), *cubitus* (elbow), *antebrachium* (forearm), *carpus* (wrist), *manus* (hand), *palma* (palm), *digiti manus* (fingers), *pollex, pollicis* (thumb).

ADDITIONAL READING

Castaing, J., and Soutoul, J. H., *Atlas de coupes anatomiques. I. Membre supérieur. II. Membre inférieur,* Maloine, Paris, 1967. Interesting didactic drawings of cross sections.

Frazer's Anatomy of the Human Skeleton, 6th ed., rev. by A. S. Breathnach, Churchill, London, 1965. A detailed synthesis of skeletal and muscular anatomy arranged regionally. A classic.

Haymaker, W., and Woodhall, B., *Peripheral Nerve Injuries: Principles of Diagnosis,* 2nd ed., W. B. Saunders Company, Philadelphia, 1953. Detailed, illustrated account, including tests of muscular actions. Important reference.

Henry, A. K., *Extensile Exposure,* 2nd ed., Livingstone, Edinburgh, 1957. Excellent account of applied anatomy.

Lockhart, R. D., *Living Anatomy,* 6th ed., Faber & Faber, London, 1963. Photographs showing muscles in action.

Medical Research Council, *Aids to the Examination of the Peripheral Nervous System,* H.M.S.O., London, 1976. Brief and valuable, including tests of muscular actions.

Royce, J., *Surface Anatomy,* Davis, Philadelphia, 1965. Photographs and key drawings of the living body.

6 THE BONES OF THE UPPER LIMB

CLAVICLE

The clavicle, or collar bone (figs. 6–1 to 6–4), connects the trunk to the upper limb by extending from the manubrium sterni to the acromion of the scapula. It is technically a long bone with a shaft and two ends, it can be readily palpated *in vivo*, and it is one of the most commonly fractured bones in the body (usually at the junction of its medial two thirds and lateral third).

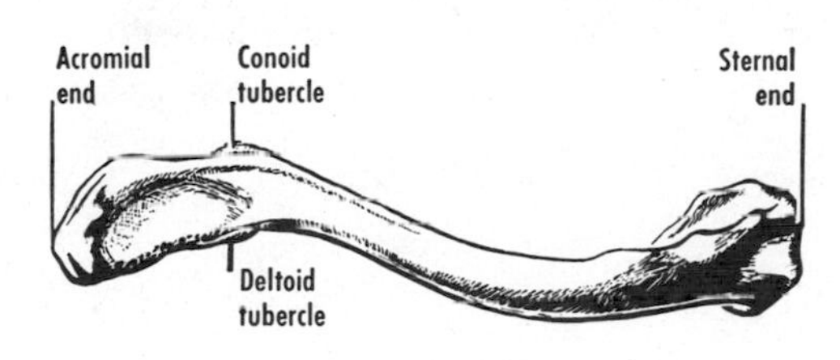

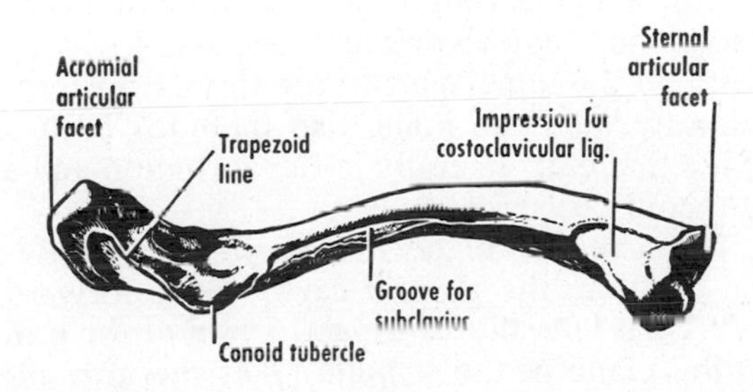

Figure 6–1 The right clavicle, viewed from in front and above and from below.

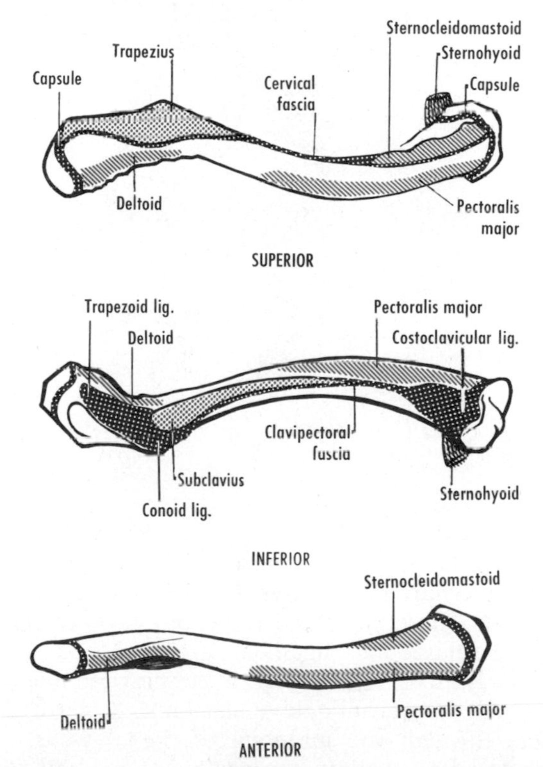

Figure 6–2 Muscular, ligamentous, and fascial attachments to the right clavicle.

The medial end is rounded and takes part in the sternoclavicular joint. The medial two thirds of the shaft is convex forward and arches in front of the brachial plexus and subclavian vessels. The costoclavicular ligament is attached inferiorly, and a shallow groove

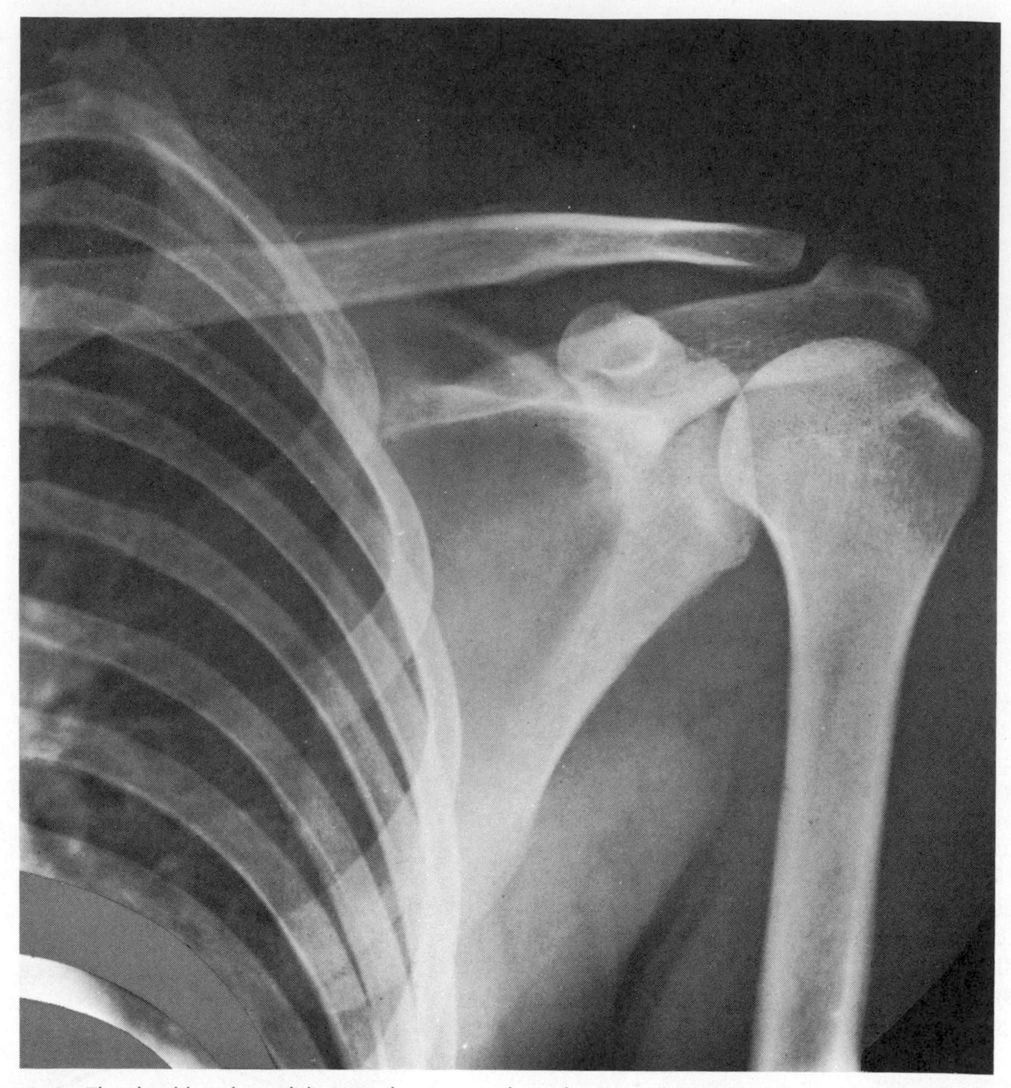

Figure 6–3 The shoulder of an adult. Note the acromioclavicular joint, glenoid cavity, coracoid process, and inferior angle of the scapula.

lodges the subclavius muscle. The lateral third of the shaft is concave forward and is flattened. The conoid and trapezoid parts of the coracoclavicular ligament are attached inferiorly. The lateral end of the clavicle takes part in the acromioclavicular joint. A vertical line through the midpoint of the clavicle is sometimes used in surface anatomy and is known as the *midclavicular line*.

The clavicle is the first bone to begin ossification, which it does in connective tissue ("membrane") during the seventh postovulatory week. The clavicle may be defective or absent in cleidocranial dysostosis. An epiphysial center usually develops at the medial end only.

SCAPULA

The scapula, or shoulder blade (figs. 6–3 to 6–10), is a large, flat, triangular bone that connects the clavicle to the humerus. Its body rests on the upper part of the thorax posterolaterally, and the bone also includes both a spine that ends laterally in the acromion and a coracoid process.

The scapula is highly mobile. In the anatomical position, the glenoid cavity looks forward as well as laterally. Thus, abduction of the arm in the plane of the scapula takes the arm anterolaterally.

The body of the scapula is triangular and presents a concave *costal surface* (*subscapu-*

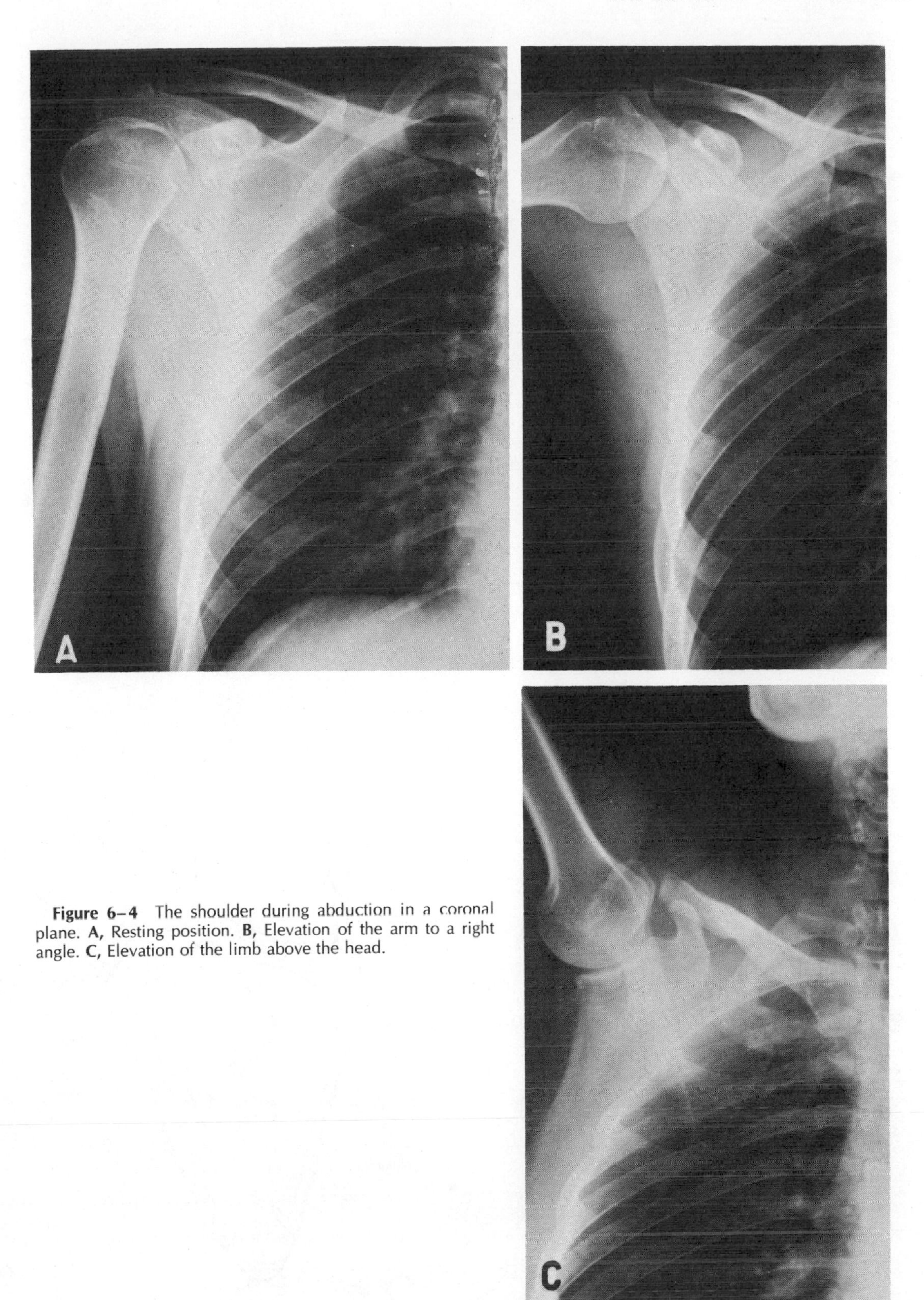

Figure 6–4 The shoulder during abduction in a coronal plane. **A,** Resting position. **B,** Elevation of the arm to a right angle. **C,** Elevation of the limb above the head.

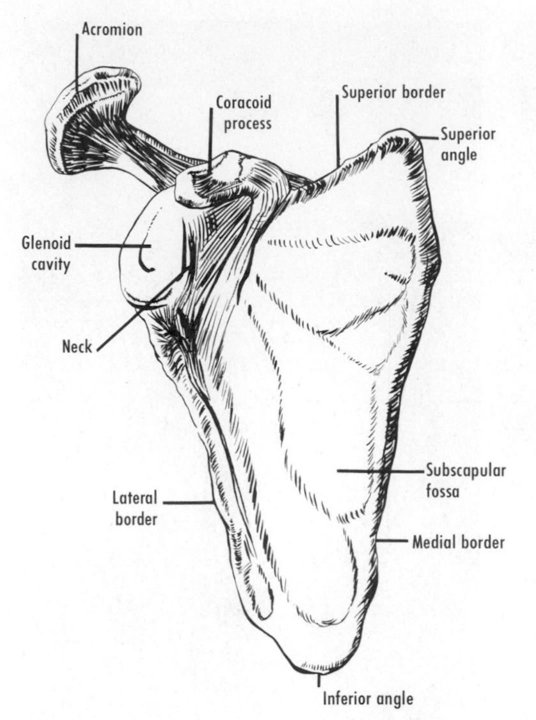

Figure 6–5 The right scapula, costal aspect, anatomical position.

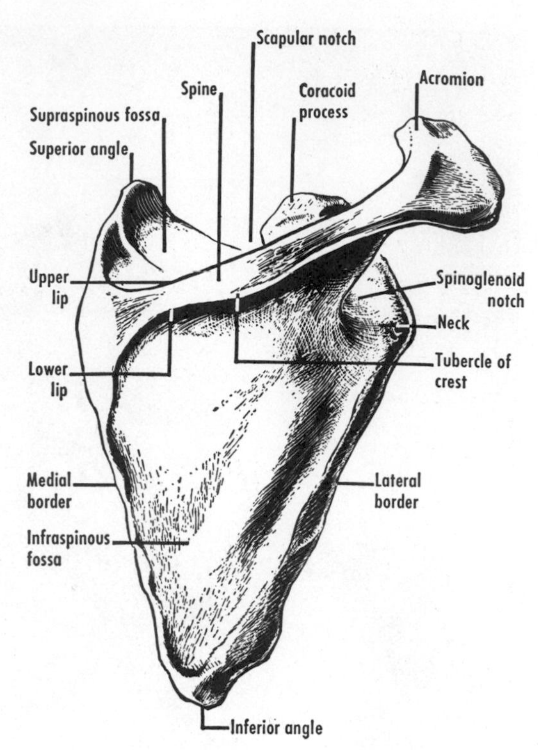

Figure 6–7 The right scapula, dorsal aspect, anatomical position.

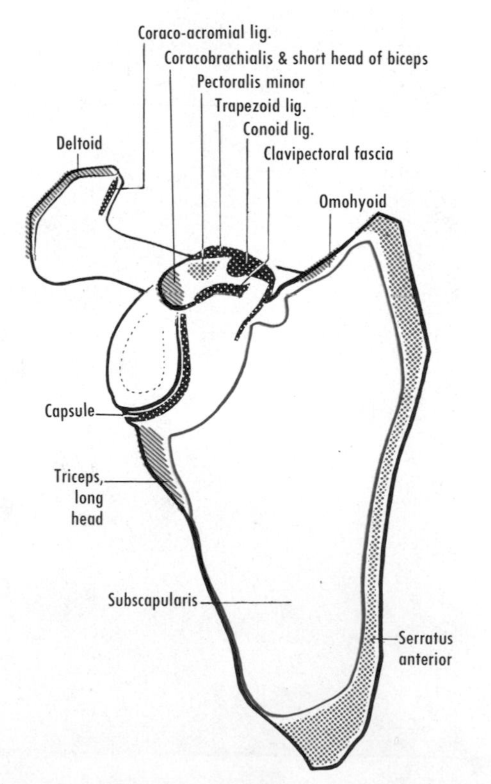

Figure 6–6 The right scapula, muscular and ligamentous attachments, costal aspect.

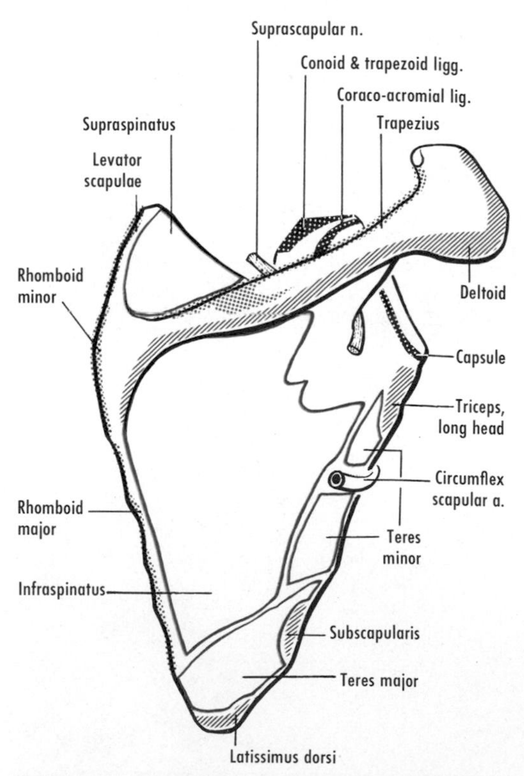

Figure 6–8 The right scapula, muscular and ligamentous attachments, dorsal aspect. The extension of the subscapularis origin to the dorsal aspect is inconstant.

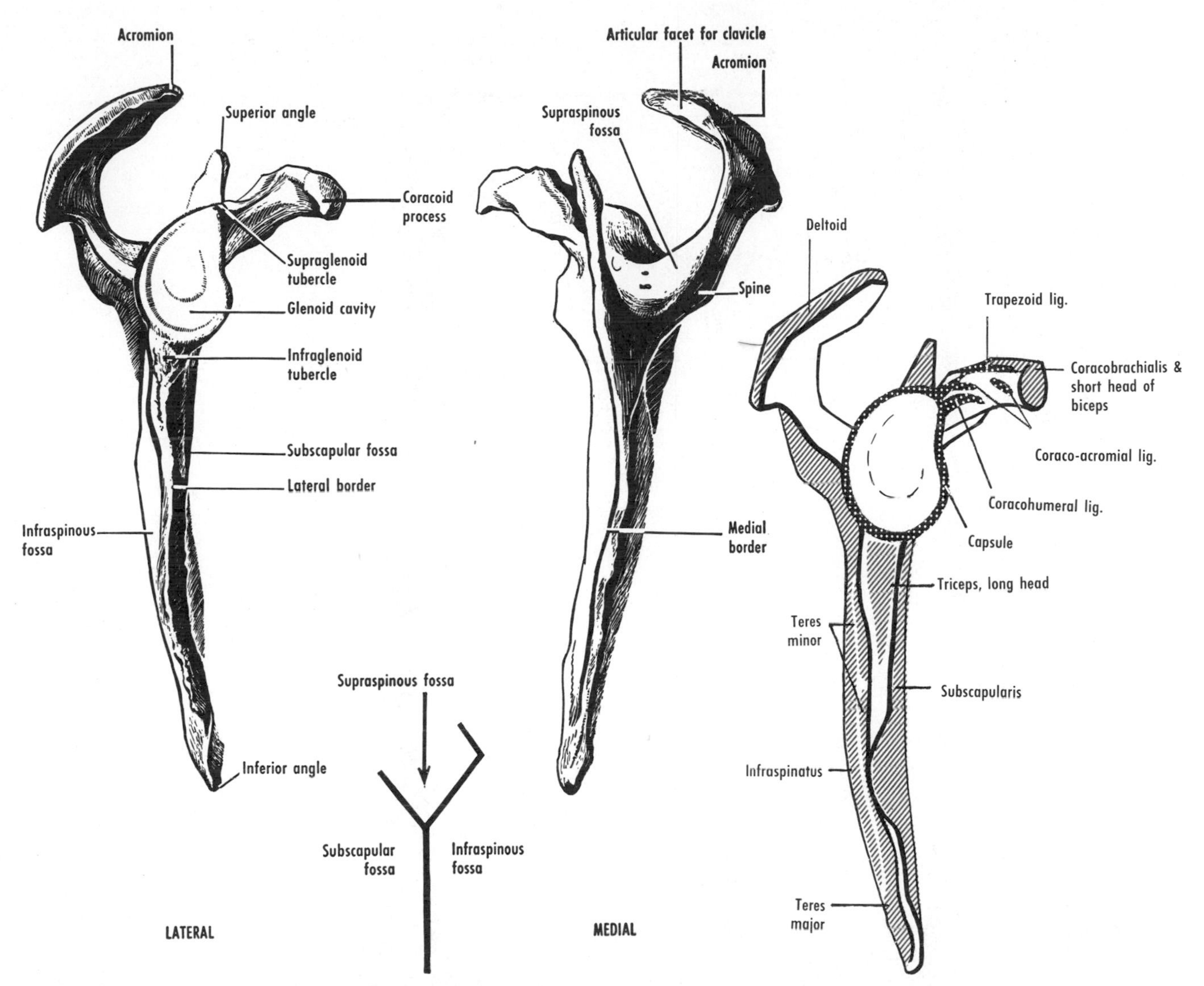

Figure 6–9 The right scapula from lateral and medial aspects. The inset illustrates that the upper and lower parts of the body form an angle, at the level of the spine, that contributes to the depth of the subscapular fossa.

Figure 6–10 The right scapula, muscular and ligamentous attachments, lateral aspect. The origin of the tendon of the long head of the biceps from the supraglenoid tubercle and the glenoid lip is not shown.

lar fossa) applied to the thorax and a *dorsal surface*, which is divided by the spine of the bone. The smaller upper part is the *supraspinous fossa*, and the lower portion is the *infraspinous fossa*. The superior border of the scapula presents the *scapular notch*. The *medial border*, usually convex, can be seen and felt. The *inferior angle* and the medial border usually ossify from separate epiphysial centers. The *lateral border* ends above in the *infraglenoid tubercle*. Superolaterally, the scapula presents the piriform *glenoid cavity* for articulation with the head of the humerus. The *supraglenoid tubercle* lies above the cavity.

The *spine* of the scapula projects horizontally backward from the body of the bone, and its crest can be felt subcutaneously. The trapezius and deltoid are attached to the crest. The spine continues laterally into the *acromion*, which articulates with the clavicle. The acromion is a subcutaneous process of the scapula, and it ossifies independently. **Clinically the arm is measured from the tip of the acromion to the lateral epicondyle of the humerus.**

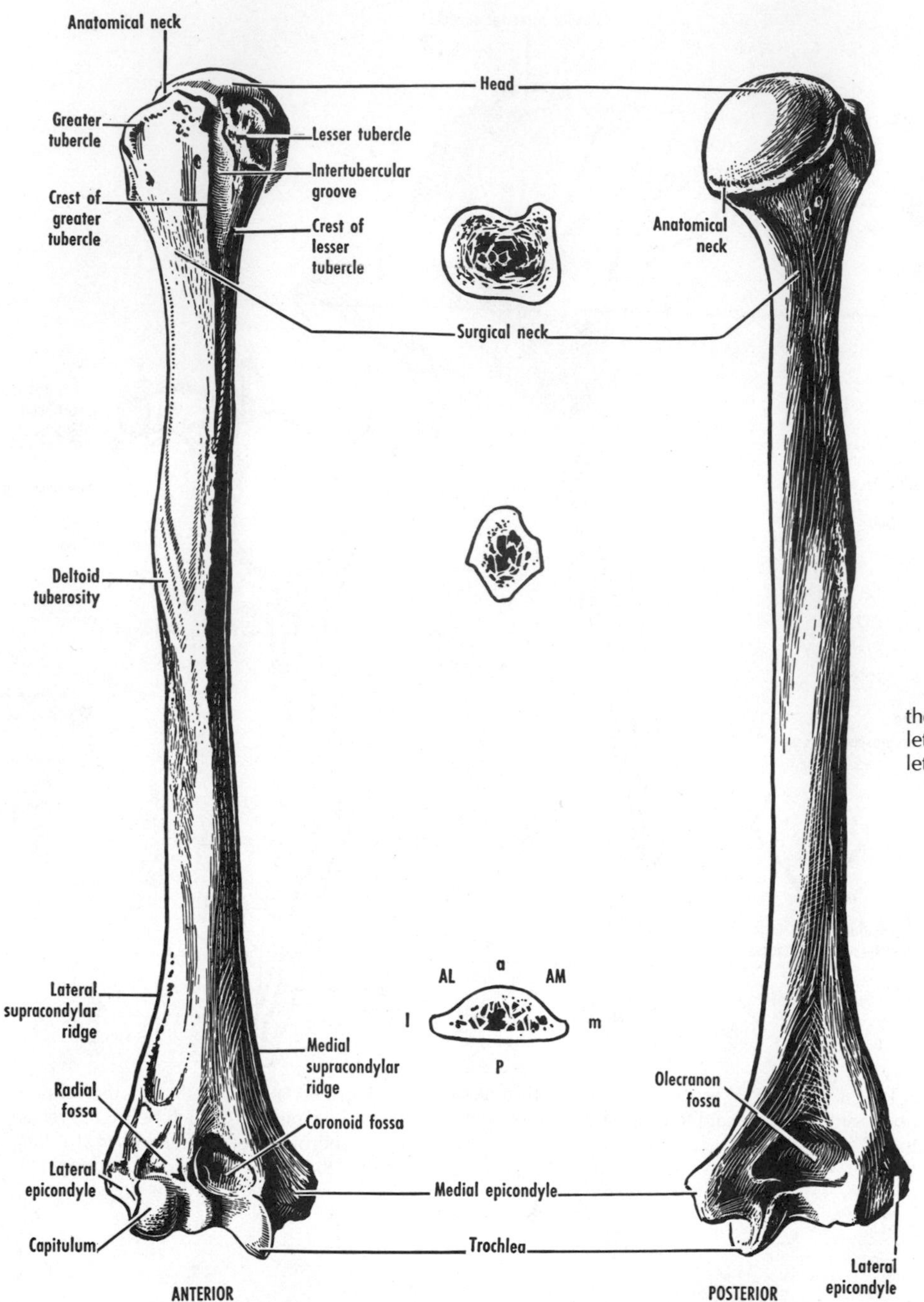

Figure 6–11 The right humerus. In the lowermost cross section, capital letters indicate surfaces and small letters indicate borders.

The *coracoid process* projects forward and can be felt indistinctly below the junction of the lateral and intermediate thirds of the clavicle. It is usually ossified from two epiphysial centers.

HUMERUS

The humerus (figs. 6–3, 6–4, and 6–11 to 6–17) is the bone of the shoulder and arm. It articulates with the scapula at the shoulder and with the radius and ulna at the elbow.

The upper end consists of the head, anatomical neck, and greater and lesser tubercles separated from each other by an intertubercular groove. The *head,* almost hemispherical, faces medially, upward, and backward. The *anatomical neck* is at the periphery of the head. The *greater tubercle* projects laterally, beyond the acromion. Unless the shoulder is dislocated, **a ruler will not make contact simultaneously with the acromion and the lateral epicondyle.** The greater tubercle is covered by the deltoid muscle, which is responsible for

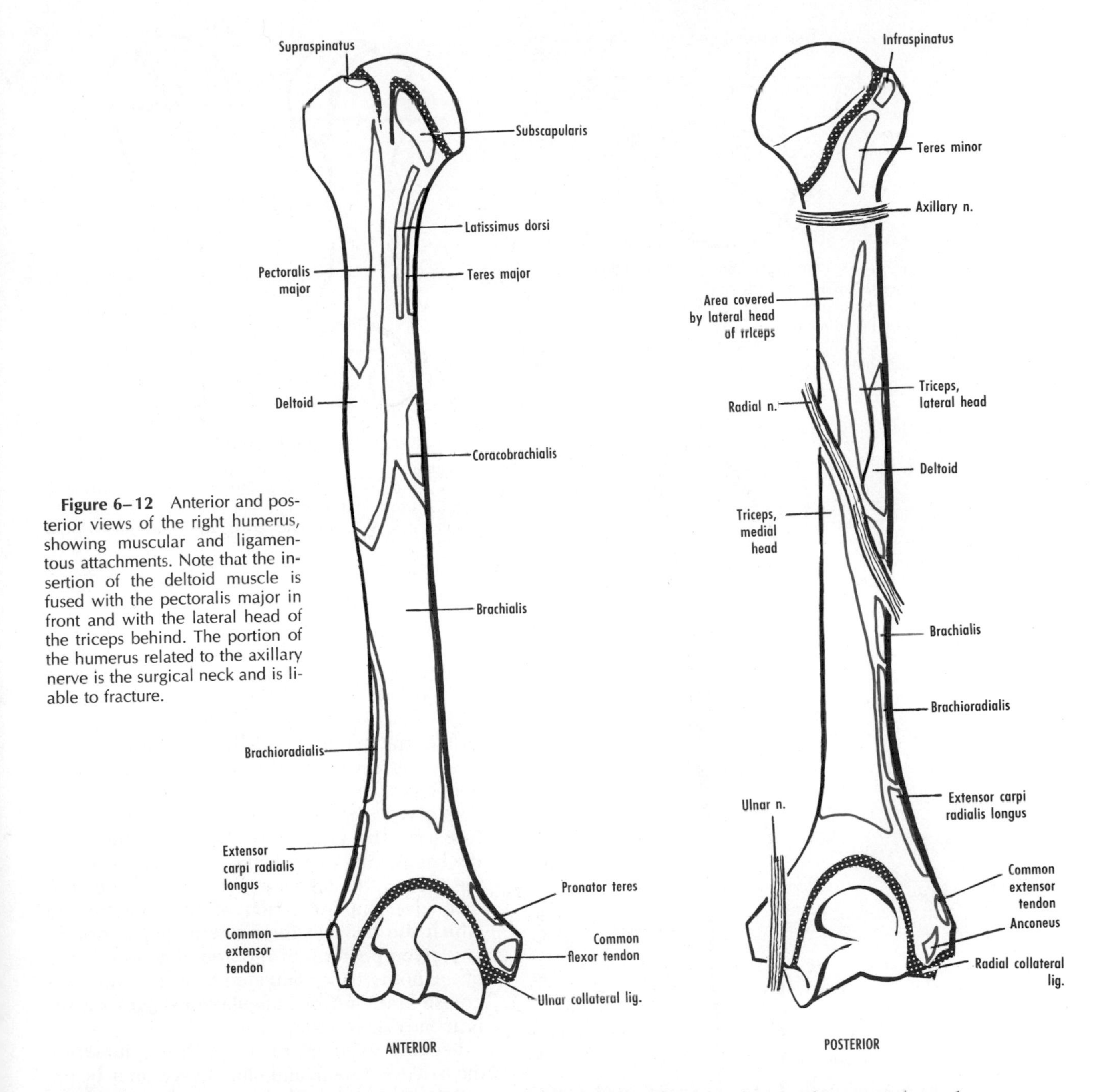

Figure 6–12 Anterior and posterior views of the right humerus, showing muscular and ligamentous attachments. Note that the insertion of the deltoid muscle is fused with the pectoralis major in front and with the lateral head of the triceps behind. The portion of the humerus related to the axillary nerve is the surgical neck and is liable to fracture.

the normal, rounded contour of the shoulder. The *lesser tubercle* projects forward (see fig. 6–13). The *intertubercular groove* lodges the tendon of the long head of the biceps. The *surgical neck,* a common site of fracture of the humerus, is the point at which the upper end of the bone meets the shaft. **The axillary nerve lies in contact with the surgical neck** (see fig. 6–12).

The *shaft* presents anterolateral, anteromedial, and posterior surfaces and lateral, anterior, and medial borders. The deltoid muscle is inserted into a tuberosity on the anterolateral surface at about the middle of the shaft. **The radial nerve runs downward and laterally on the posterior surface** (see fig. 6–12).

The lower end of the humerus includes the lateral and medial epicondyles and a condyle consisting of the capitulum and trochlea. The *lateral epicondyle* gives origin to the supinator and to the extensor muscles of the forearm. The *capitulum* articulates with the head of the radius. The *trochlea* is a pulley-shaped projection that articulates with the trochlear

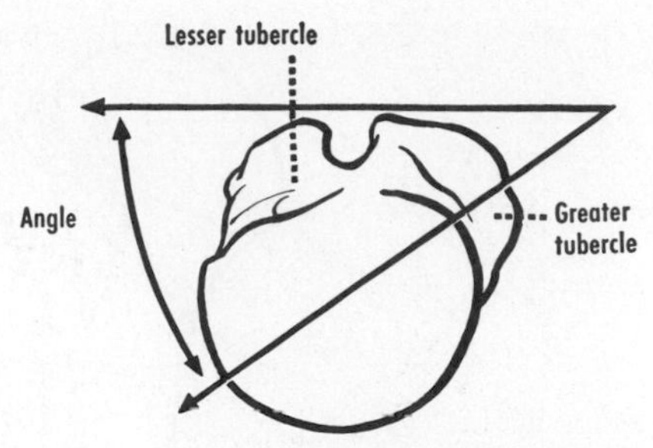

Figure 6–13 The right humerus from above, showing the lesser tubercle anteriorly, the greater tubercle laterally, and the intertubercular groove between. The upper arrow indicates the direction in which the medial epicondyle points. The lower arrow indicates the long axis of the head. The angle between the arrows shows the amount of torsion.

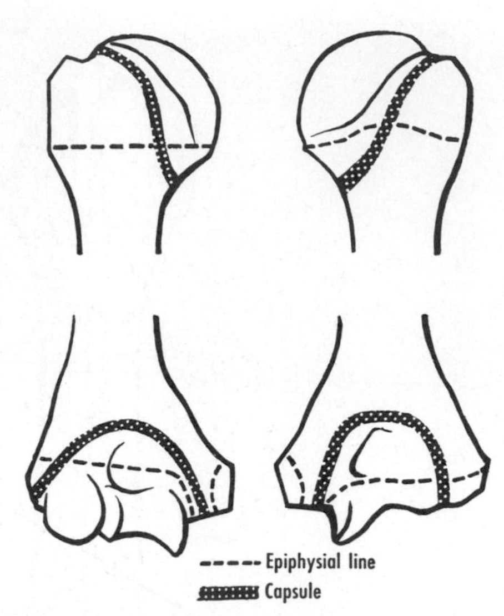

Figure 6–15 The upper and lower ends of the right humerus, showing the usual position of the epiphysial lines and the usual line of attachment of the joint capsule. The epiphysial lines at both ends are partly extracapsular. (Modified from Mainland.)

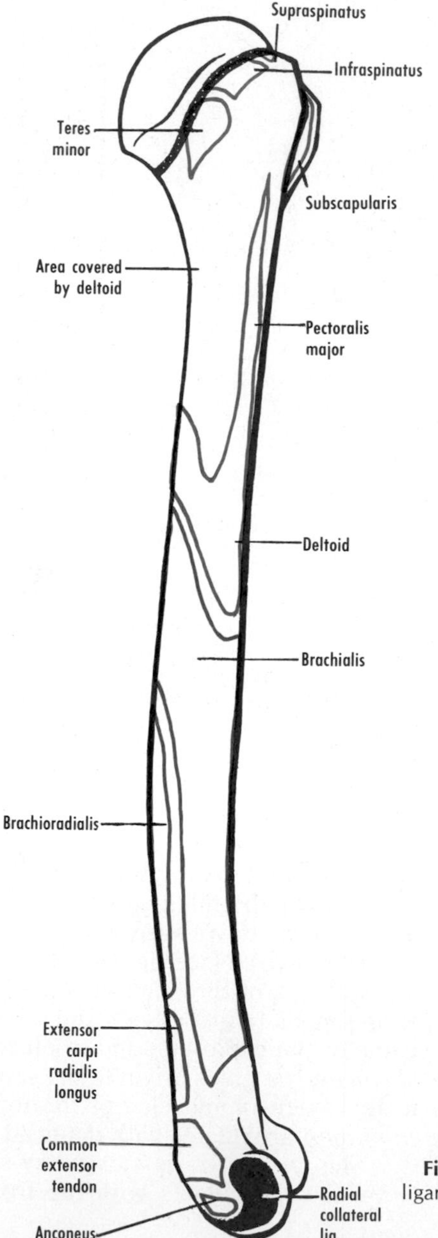

Figure 6–14 Lateral view of the right humerus, showing muscular and ligamentous attachments.

notch of the ulna. It is set obliquely, so that a "carrying angle" exists between the arm and the extended and supinated forearm. *Radial* and *coronoid fossae* are situated in front, above the capitulum and trochlea, respectively. A deeper *olecranon fossa* is located above the trochlea behind. The *medial epicondyle* gives origin to the flexor muscles of the forearm. **The ulnar nerve lies in a groove on the back of the medial epicondyle and is palpable there ("funny bone"). The medial epicondyle gives an indication of the direction in which the head of the humerus is pointing** in any given position of the arm. The lower end of the humerus is angulated forward, and a decrease in the normal angulation suggests a supracondylar fracture.

Because of their contact with the humerus, the axillary, radial, and ulnar nerves may be injured in fractures of the surgical neck, shaft, and medial epicondyle, respectively.

The shaft begins to ossify during the eighth postovulatory week, and a center is usually present in the head at birth. Centers for the greater and lesser tubercles appear postnatally, as do four centers for the lower end.

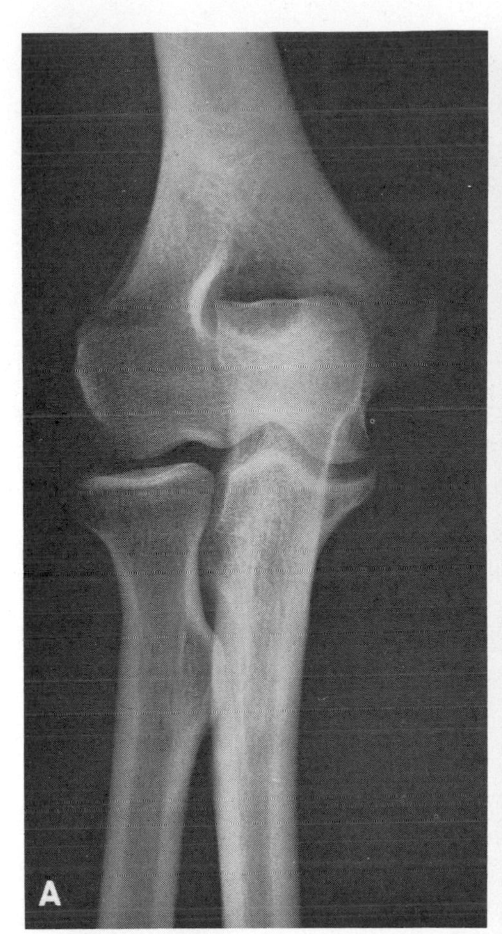

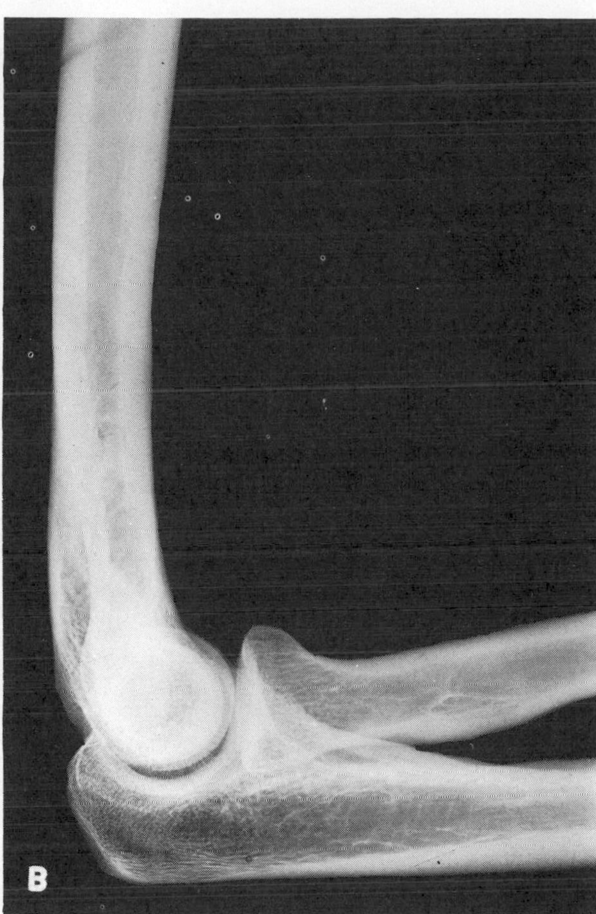

Figure 6–16 Elbows of adults. **A,** Anteroposterior view. Note the olecranon fossa, trochlea, and medial epicondyle of the humerus; the head and tuberosity of the radius; and the olecranon and coronoid process of the ulna. **B,** Lateral view. Note the olecranon and coronoid process of the ulna. (Courtesy of Sir Thomas Lodge.)

RADIUS

The radius (figs. 6–16 to 6–24) is shorter than and lateral to the ulna. It articulates with the humerus above, the ulna medially, and the carpus below.

The upper end consists of a head, neck, and tuberosity. The upper, concave surface of the *head* articulates with the capitulum of the humerus. The circumference of the head articulates with the ulna medially but is elsewhere covered by the annular ligament (see fig. 9–6). The head of the radius can be felt *in vivo* immediately below the lateral epicondyle (in the "valley" behind the brachioradialis), particularly during rotation. The *tuberosity* of the radius is situated anteromedially below the neck.

The *shaft* presents anterior, posterior, and lateral surfaces and anterior, posterior, and interosseous borders. The interosseous border is attached by the interosseous membrane to a corresponding border on the ulna (see fig. 6–23).

The lower end of the radius terminates in the *styloid process* laterally. The process is palpable between the extensor tendons of the thumb. It gives attachment to the radial collateral ligament. **The styloid process of the radius is about 1 cm distal to that of the ulna.** This relationship is important in the diagnosis of fractures and in the verification of their correct reduction. Medially, the lower end of the radius presents the *ulnar notch,* which articulates with the head of the ulna. At about the middle of the convex dorsal aspect of the lower end of the radius, a small prominence, the *dorsal tubercle,* may be felt (see fig. 6–22). The lower surface of the lower end articulates with the lunate medially and the scaphoid laterally.

A fall on the outstretched hand may result in a (Colles') fracture of the lower end of the radius, in which the distal fragment is displaced posteriorly and generally becomes impacted,

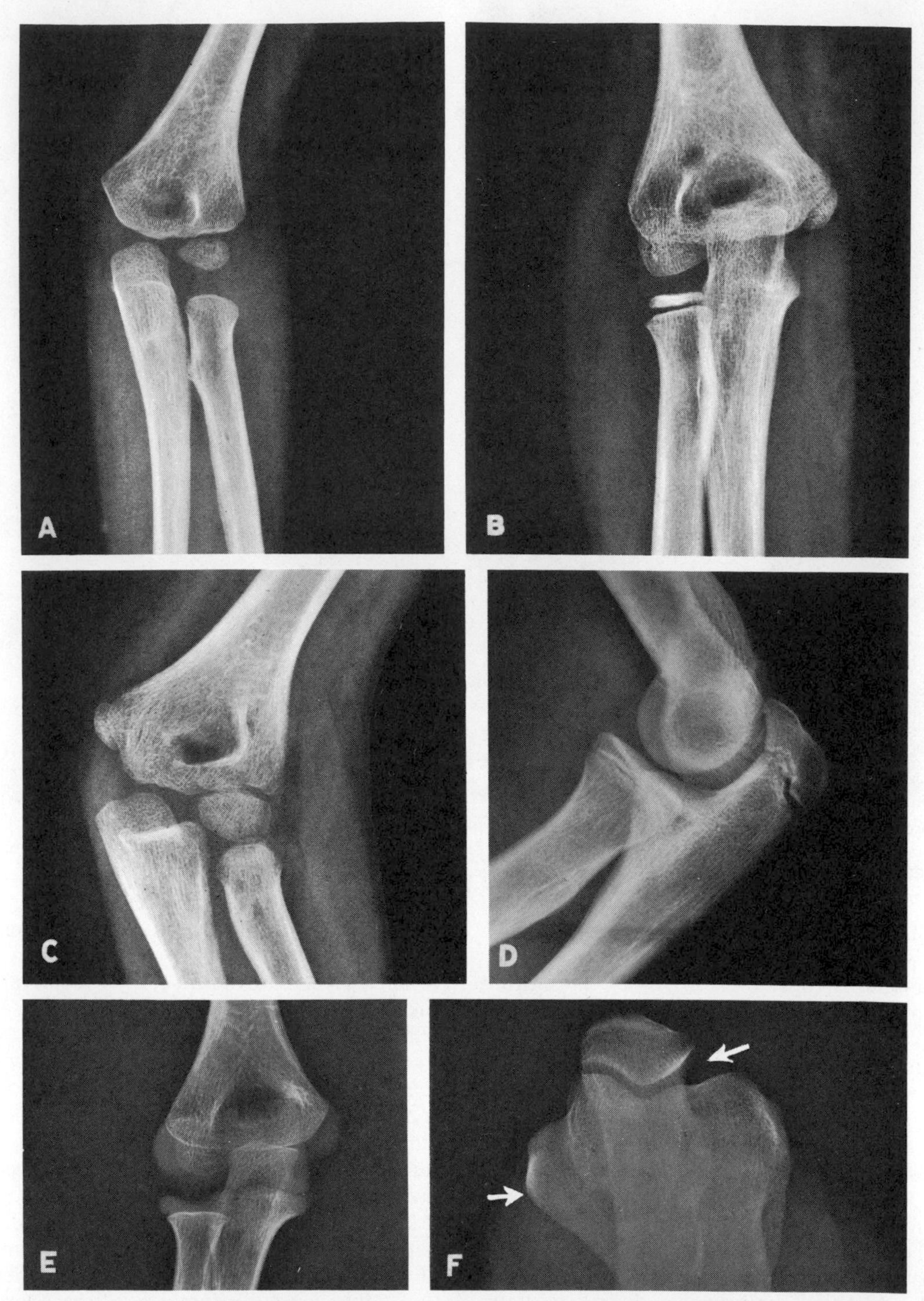

Figure 6–17 The elbow. **A,** The elbow of a child. Note the epiphysis for the capitulum and the lateral part of the trochlea of the humerus. The ulna is at left. **B,** The elbow of a child. Note the additional epiphyses for the medial epicondyle of the humerus and the head of the radius. **C,** The elbow of a child, oblique view, showing epiphyses for the capitulum, lateral part of the trochlea, and medial epicondyle. **D,** The epiphysis for the upper end of the ulna. Note also the epiphysis for the head of the radius. **E,** A radiograph of the dried bones of a 5-year-old boy. Note the outline of the cartilage. **F,** The flexed elbow of an adult. Note the medial epicondyle (*arrow on left*) and the joint line between the olecranon and the trochlea (*arrow on right*). (**A, B,** and **C,** Courtesy of S. F. Thomas, M.D., Palo Alto Medical Clinic, Palo Alto, California. **D,** Courtesy of G. L. Sackett, M.D., Painesville, Ohio. **F,** Courtesy of V. C. Johnson, M.D., Detroit, Michigan.)

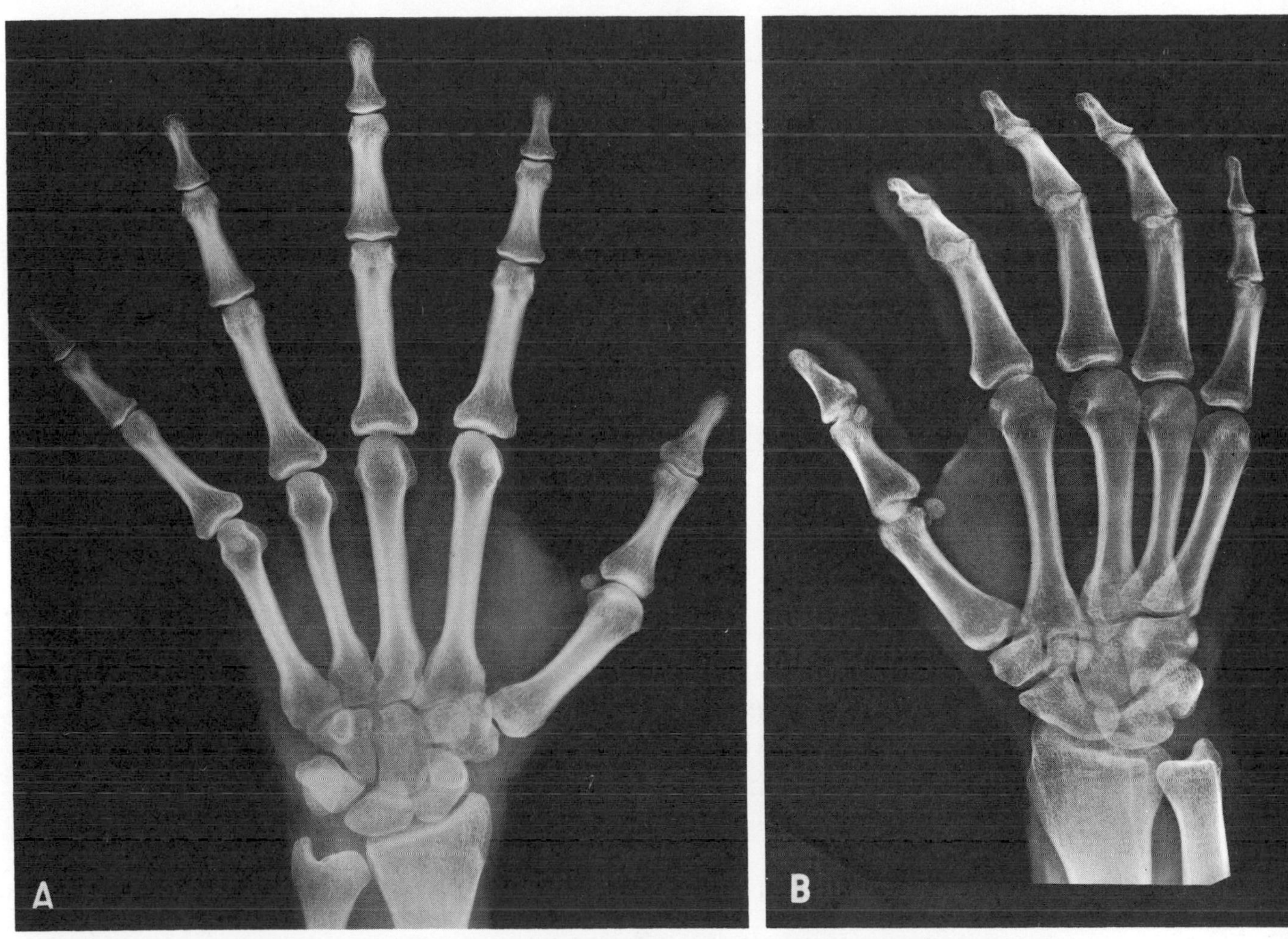

Figure 6–18 Hands of adults. **A,** Postero-anterior view. Note the hook of the hamate and the sesamoid bones of the first, second, and fifth fingers. **B,** Oblique view. (**A** and **B,** Courtesy of S. F. Thomas, M.D., Palo Alto Medical Clinic, Palo Alto, California.)

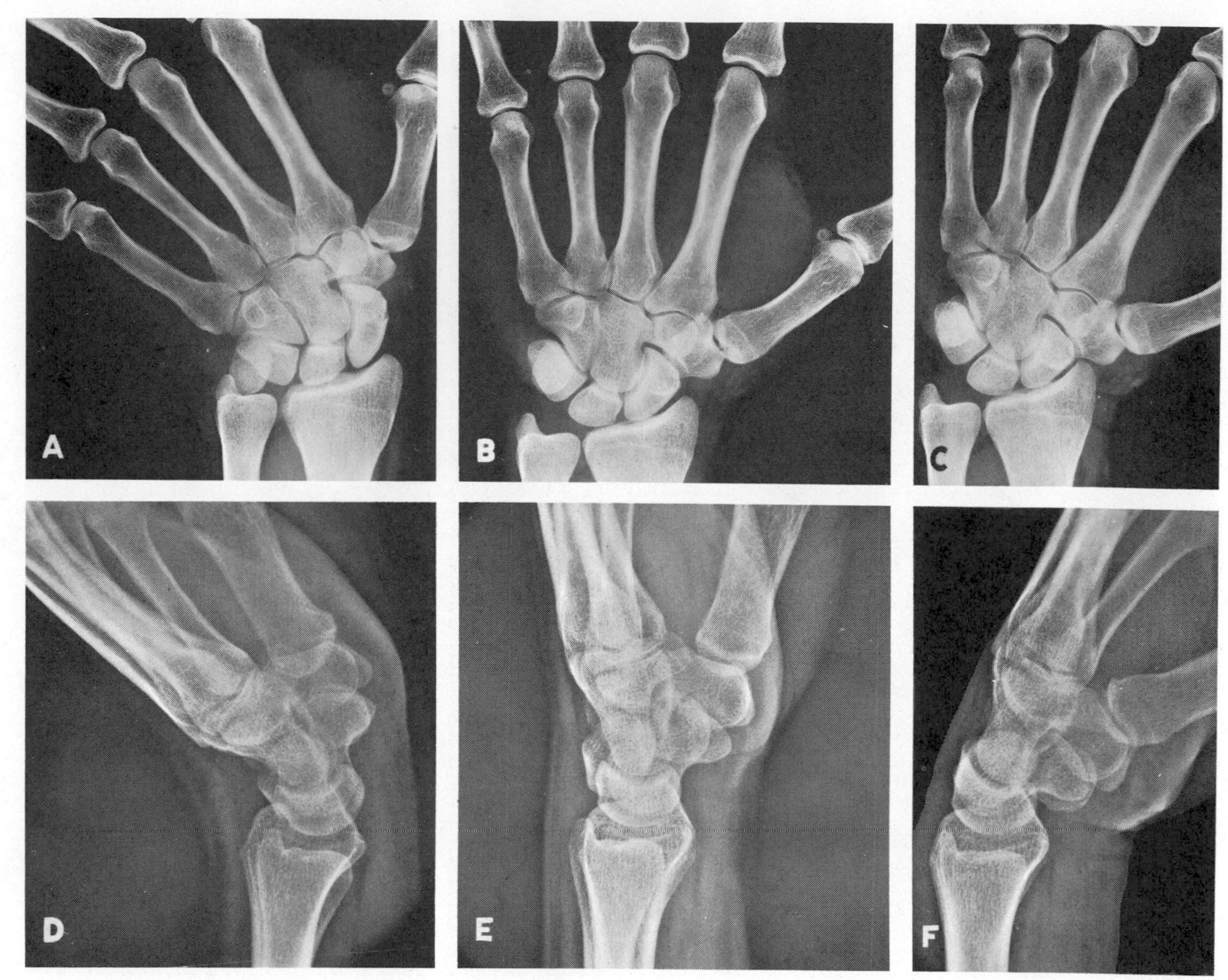

Figure 6–19 The hand in various positions. **A, B,** and **C** are postero-anterior views. (Note the relation to the radius of the joint line between the lunate and the triquetrum.) Cf. fig. 6–27. **D, E,** and **F** are lateral views. **A,** Adduction. **B,** Straight position. **C,** Abduction. **D,** Extension. **E,** Straight position. Note the lunate, capitate, scaphoid, and trapezium. **F,** Flexion.

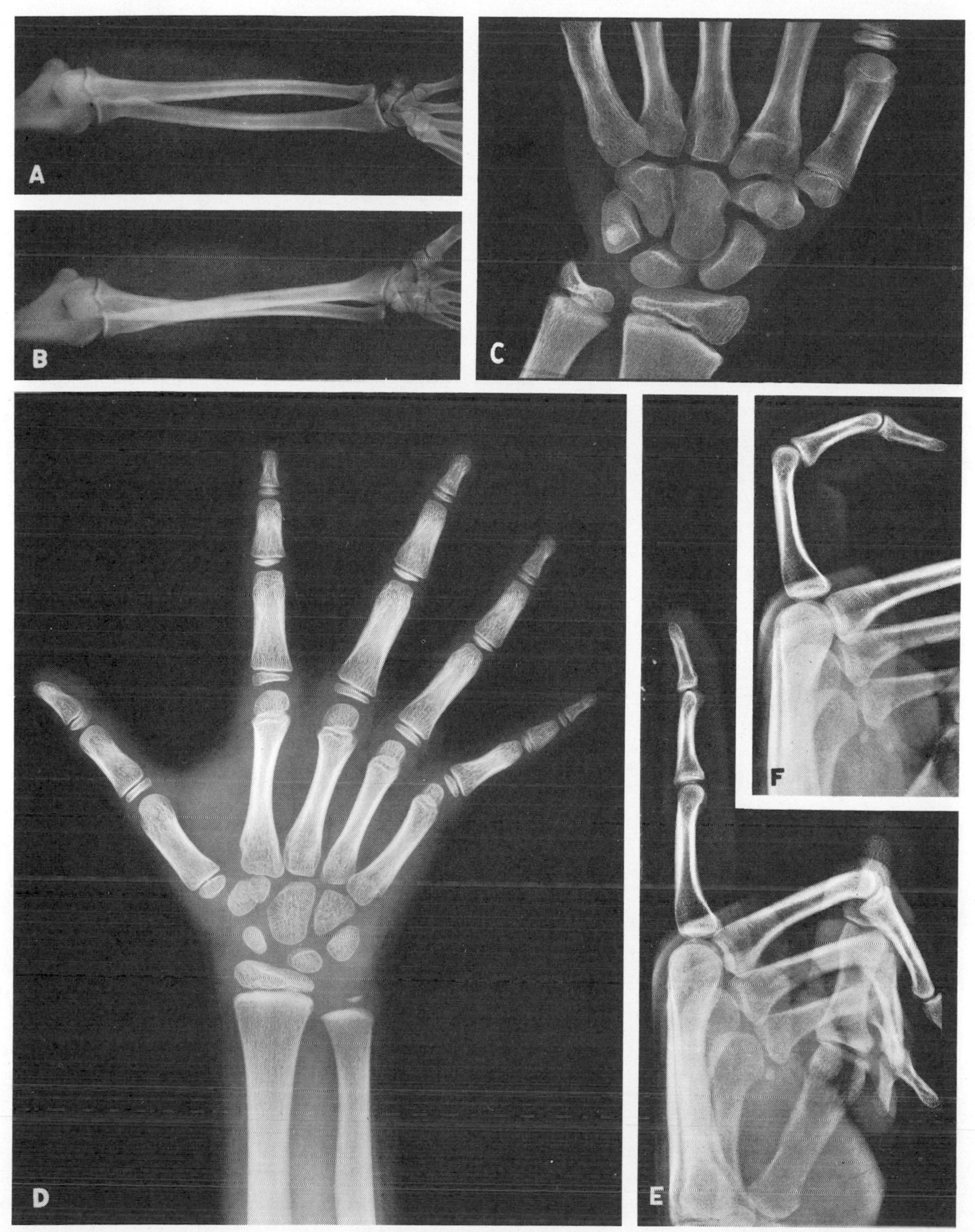

Figure 6–20 Various views of the hand. **A** and **B,** The forearm and hand in supination (**A**) and pronation (**B**). **C,** The hand of a child. Note the epiphyses for the lower ends of the radius and ulna and for the base of the first metacarpal and an accessory epiphysis for the base of the second metacarpal. **D,** The hand of a child. The pisiform does not yet show. Note the epiphyses for the metacarpals and phalanges. **E** and **F,** The index finger in extension (**E**) and flexion (**F**). Note the shift in position (relative to the heads of the proximal and middle phalanges) of the bases of the middle and distal phalanges. (**A, B, E,** and **F,** Courtesy of S. F. Thomas, M.D., Palo Alto Medical Clinic, Palo Alto, California. **C,** Courtesy of J. Lofstrom, M.D., Detroit Memorial Hospital, Detroit, Michigan.)

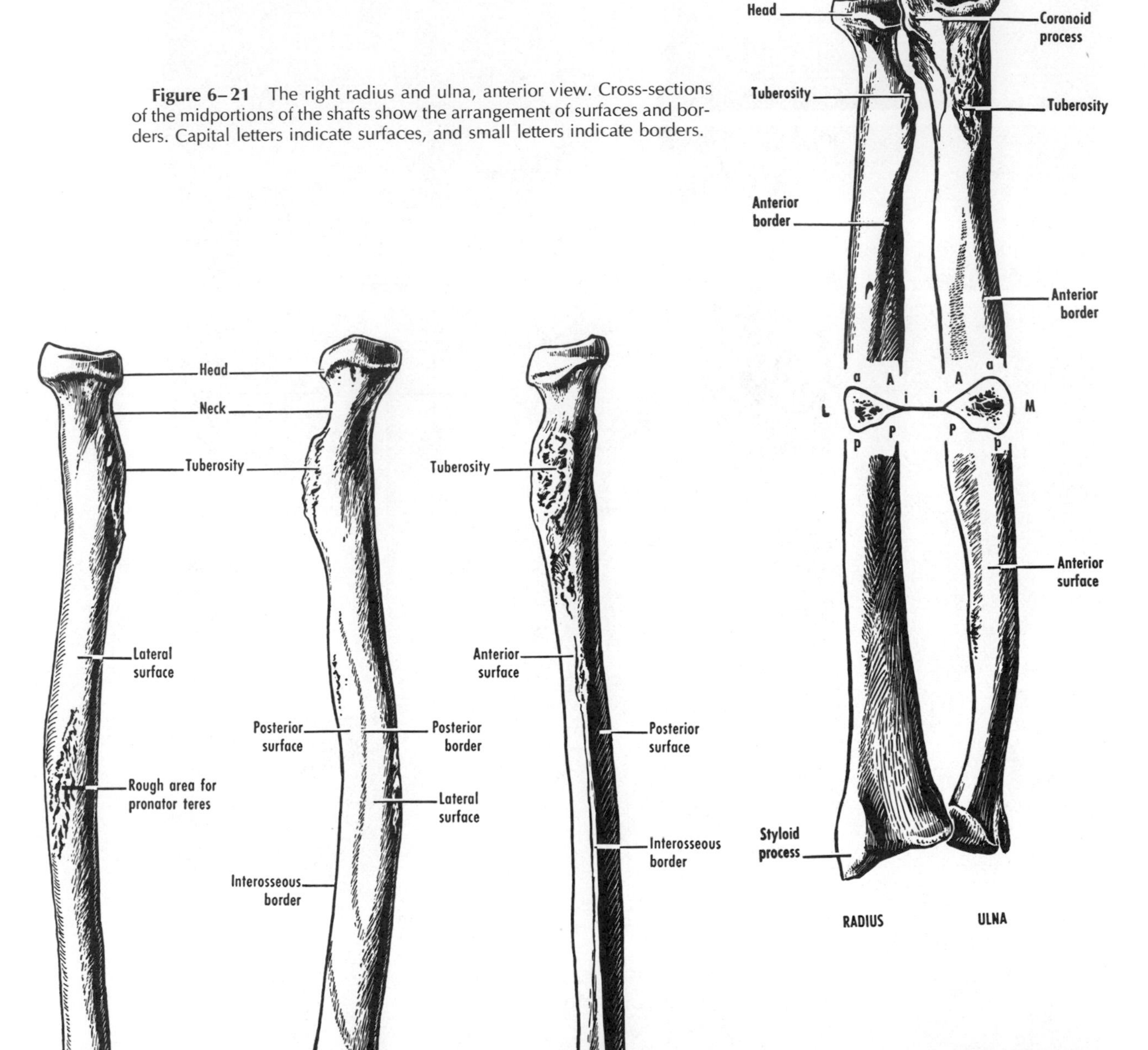

Figure 6–21 The right radius and ulna, anterior view. Cross-sections of the midportions of the shafts show the arrangement of surfaces and borders. Capital letters indicate surfaces, and small letters indicate borders.

Figure 6–22 The right radius. In the lateral view, note the shallow groove immediately to the right of the styloid process; this is occupied by the tendons of the abductor pollicis longus and extensor pollicis brevis. In the posterior view, note that the dorsal tubercle is grooved; the groove is occupied by the tendon of the extensor pollicis longus. The tendons of the extensor carpi radialis longus and brevis lie to the radial side of the tubercle; the tendons of the extensor indicis and extensor digitorum lie to the ulnar side.

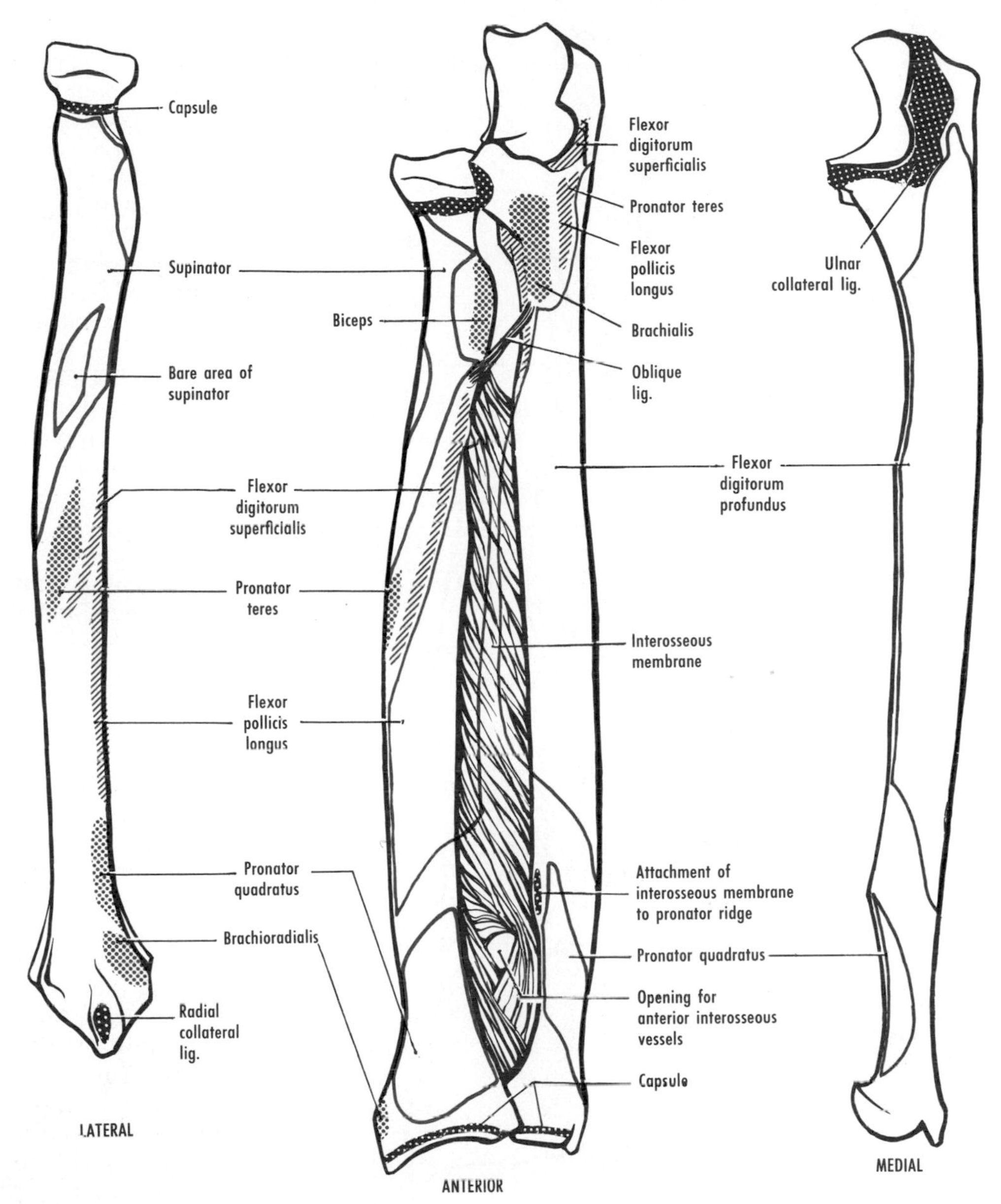

Figure 6–23 Muscular and ligamentous attachments to the right radius and ulna. About midway on the shaft of the radius is a rough area for the insertion of the pronator teres, below which the shaft is covered by the tendons of the brachioradialis and the extensor carpi radialis longus and brevis. The interosseous membrane gives origin in part to the flexor pollicis longus and flexor digitorum profundus.

bringing the styloid processes of the radius and ulna to approximately the same horizontal level.

The shaft begins to ossify during the eighth postovulatory week, and centers appear postnatally for the lower end and the head (see fig. 6–20).

ULNA

The ulna (figs. 6–16 to 6–21 and 6–23 to 6–26) is longer than and medial to the radius. It articulates with the humerus above, the radius laterally, and the articular disc below.

The upper end includes the olecranon and the coronoid process. The *olecranon* is the prominence on the back of the elbow, which rests on a table when a subject leans on his elbow. The lateral epicondyle, the top of the olecranon, and the medial epicondyle are in a straight line when the forearm is extended, but form an equilateral triangle when the forearm is flexed. Above, the olecranon receives the insertion of the triceps. The posterior aspect, covered by a bursa, is subcutaneous. In front, the olecranon forms a part of the *trochlear notch,* which articulates with the trochlea of the humerus. The *coronoid process,* which completes the trochlear notch, projects forward and engages the coronoid fossa of the humerus during flexion. It is prolonged downward as a rough area termed the *tuberosity of*

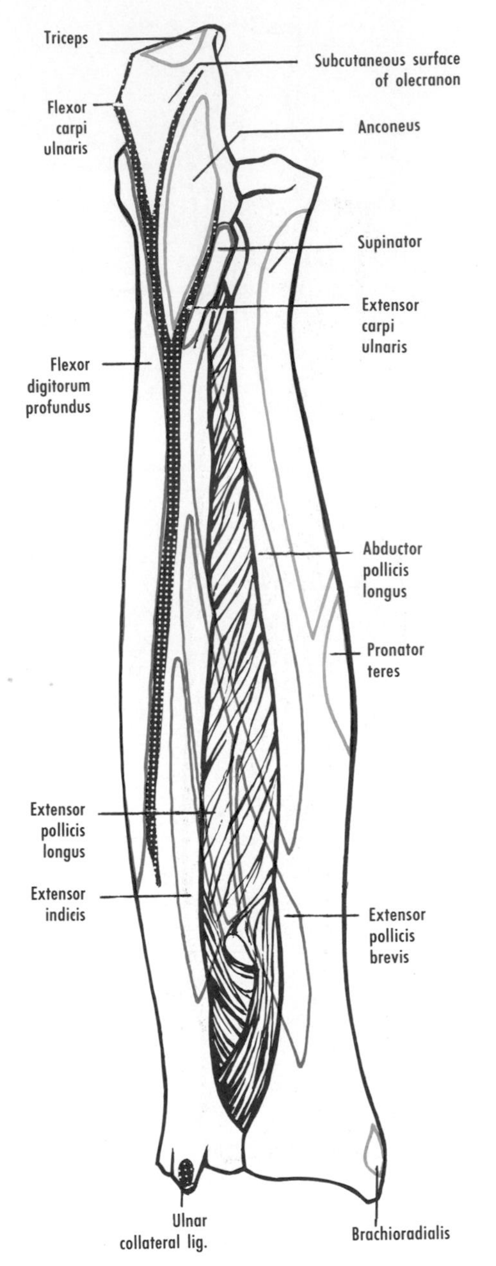

Figure 6–24 The right radius and ulna, showing muscular and ligamentous attachments, posterior aspect.

the ulna. The *radial notch* is on the lateral aspect of the coronoid process and articulates with the head of the radius.

The *shaft* presents anterior, posterior, and medial surfaces and anterior, posterior, and interosseous borders. The posterior border is completely subcutaneous and readily palpable *in vivo*. It separates the flexor from the extensor muscles of the forearm.

The lower end includes the *styloid process* and the head. The styloid process, small and conical, is situated posteromedially and is readily palpable *in vivo*. The *head* of the ulna articulates with the ulnar notch of the radius. The lower aspect of the head is separated from the carpus by the articular disc.

The shaft begins to ossify during the eighth postovulatory week, and centers appear postnatally for the lower and upper ends of the bone (see figs. 6–17 and 6–20).

The relationships of joint capsules to epiphysial lines (see figs. 6–15 and 6–26) are important, because epiphysial discs tend to limit the extent of infection, but it is possible for infection to spread from the shaft to the joint when part of the diaphysis is intracapsular.

CARPUS

The carpal bones, usually eight in number, are arranged in two rows of four (figs. 6–18 to 6–20 and 6–27 to 6–29). Their names are scaphoid, lunate, triquetrum (or triquetral), pisiform, trapezium, trapezoid, capitate, and hamate. (Other names, such as navicular, are now obsolete in the carpus.) The pisiform lies in front of the triquetrum, whereas each of the other carpals has several facets for articulation with adjacent bones.

The intact carpus is convex behind and concave in front, where it is bridged by the flexor retinaculum to form the *carpal canal* or *tunnel* for the flexor tendons and the median nerve. Hence, the posterior surfaces of the carpals are generally larger than the anterior, with the exception of the lunate, where the converse holds. The flexor retinaculum extends between the scaphoid and trapezium laterally and the triquetrum and hamate medially (see fig. 11–2). These four bones can be distinguished by deep palpation *in vivo*.

The *scaphoid* presents anteriorly a tubercle that can be felt under cover of and lateral to the tendon of the flexor carpi radialis. **A fall on the outstretched hand may result in fracture of the scaphoid,** generally across its "waist." In some fractures the blood supply of the proximal fragment may be cut off, resulting in aseptic necrosis. The *lunate* is broader in front than behind. **Anterior dislocation of the lunate is a fairly common injury of the wrist.** In adduction of the hand, the lunate articulates with the radius only, whereas in the straight position or in abduction, it articulates with the articular disc also (see fig. 6–19). The *pisiform*, the smallest of the carpals and the last to ossify, lies in front of the *triquetrum* and can be

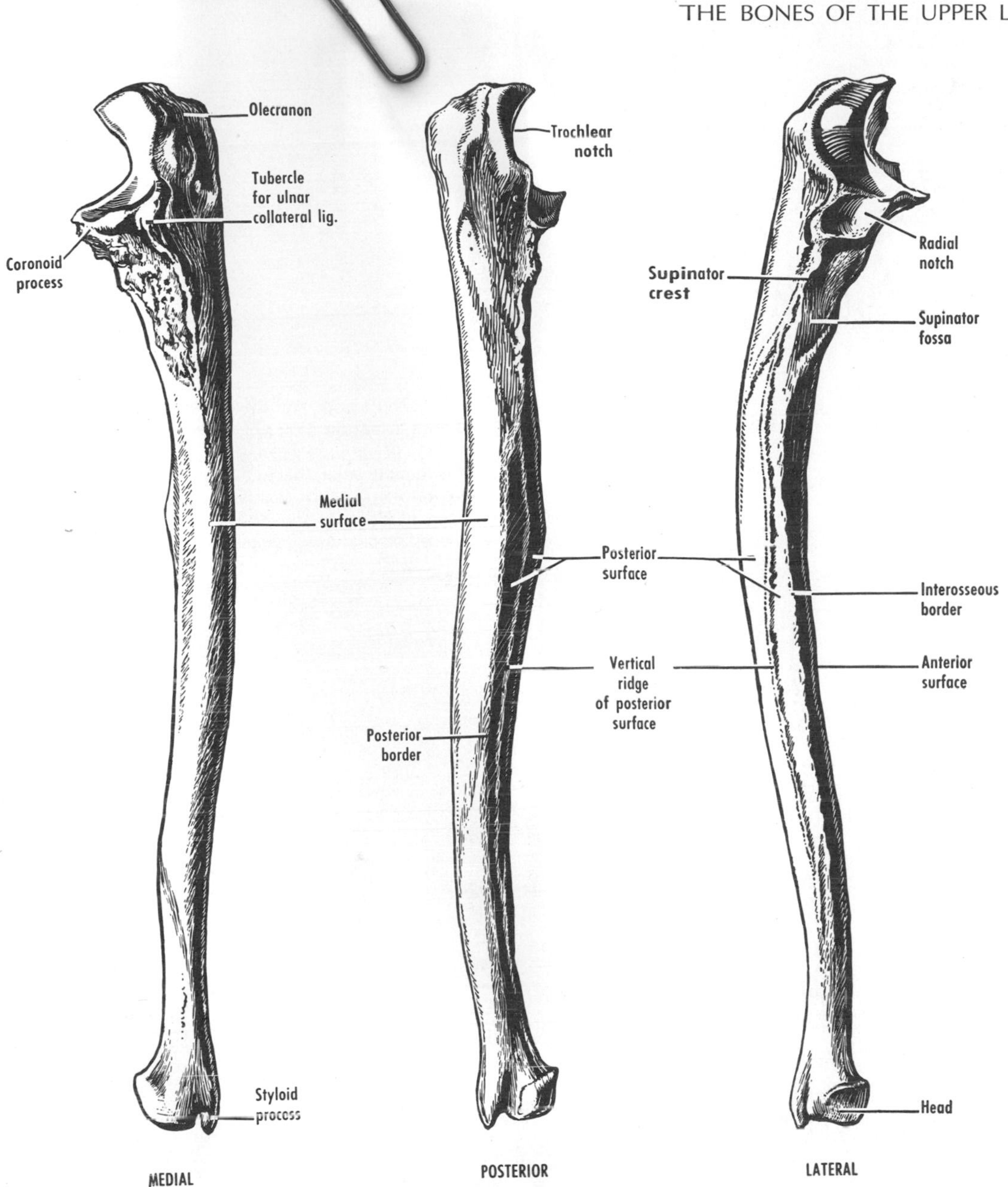

Figure 6–25 The right ulna.

moved passively from side to side when the flexor carpi ulnaris is relaxed.

The *trapezium* supports the thumb by means of a saddle-shaped facet for the first metacarpal. Like the adjacent scaphoid, it has a tubercle anteriorly. The *trapezoid* is associated with the index finger. The *capitate,* the largest of the carpals and the first to ossify, is placed centrally and is in line with the third metacarpal. It has a prominent head superiorly. The *hamate* sends a marked hook anteriorly, which gives attachment to the flexor retinaculum.

Accessory ossicles are sometimes found between the usual carpal bones, and their possible occurrence should be kept in mind in interpreting radiograms. Carpal fusions (e.g., between the lunate and triquetrum) may also occur.

Each carpal bone usually ossifies from one center postnatally. Those for the capitate and hamate develop first and may appear before birth. Radiography of the carpus is frequently used for the assessment of skeletal maturation: the carpus under consideration is compared with a series of standards.

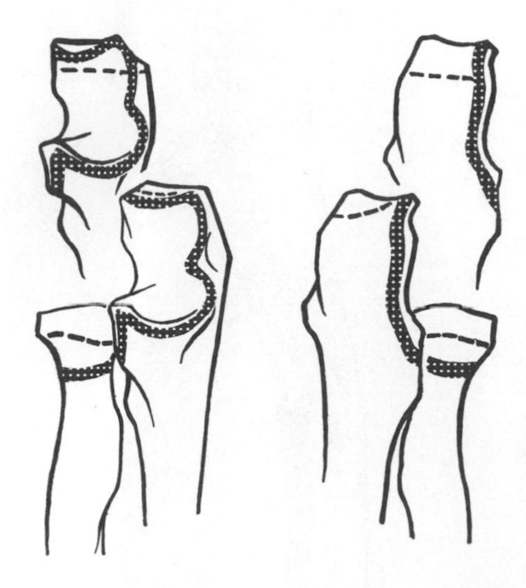

Figure 6–26 The upper and lower ends of the right radius and ulna, showing the usual position of the epiphysial lines and the usual line of attachment of the joint capsule. The epiphysial line of the head of the radius is intracapsular, that of the upper end of the ulna partly or entirely extracapsular, and those of the lower end extracapsular. The additional views of the ulna (*upper two figures*) show a variation in the position of the epiphysial line. (Modified from Mainland.)

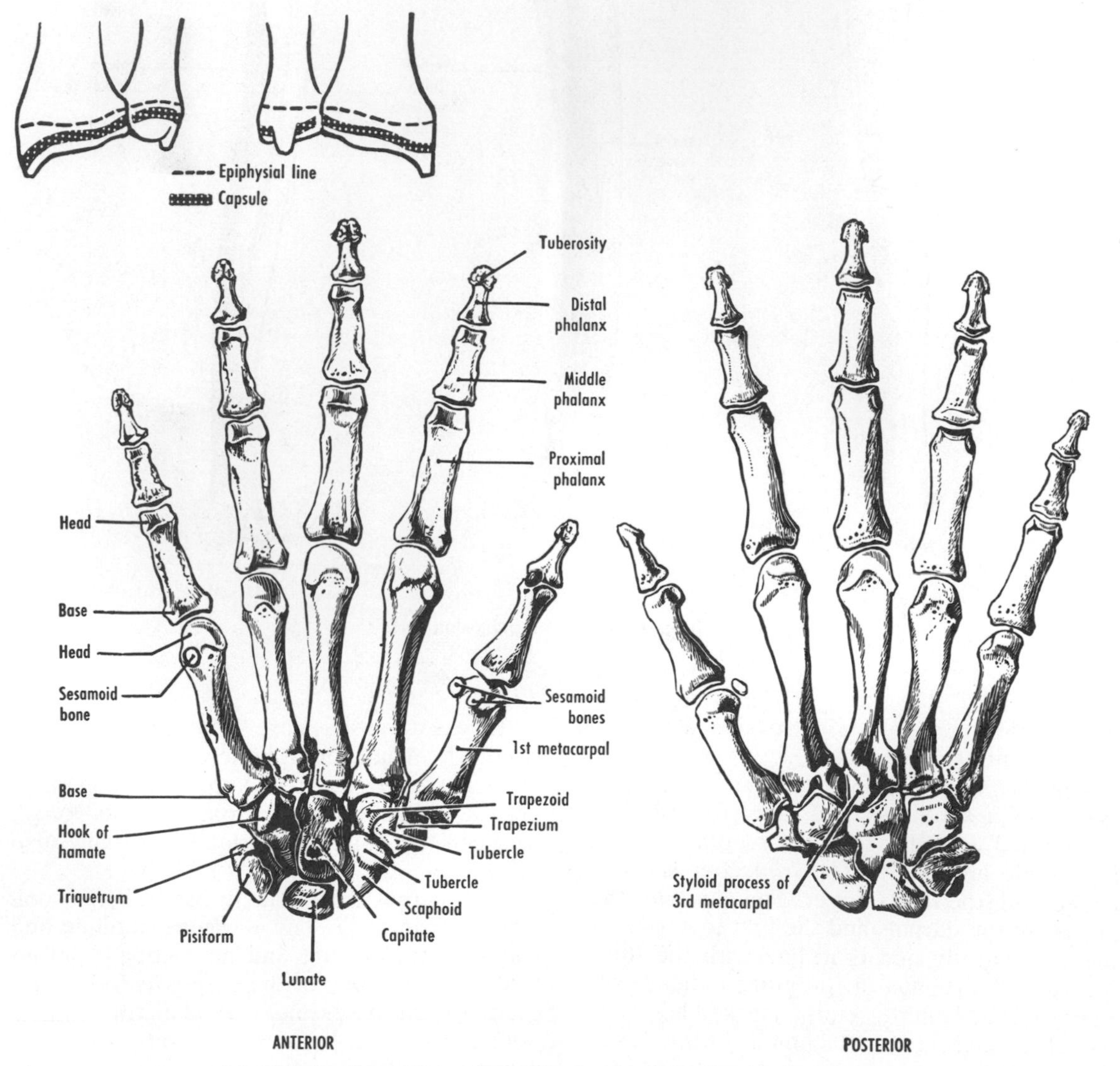

Figure 6–27 Bones of the right hand, anterior and posterior aspects. The sesamoids shown are those commonly present.

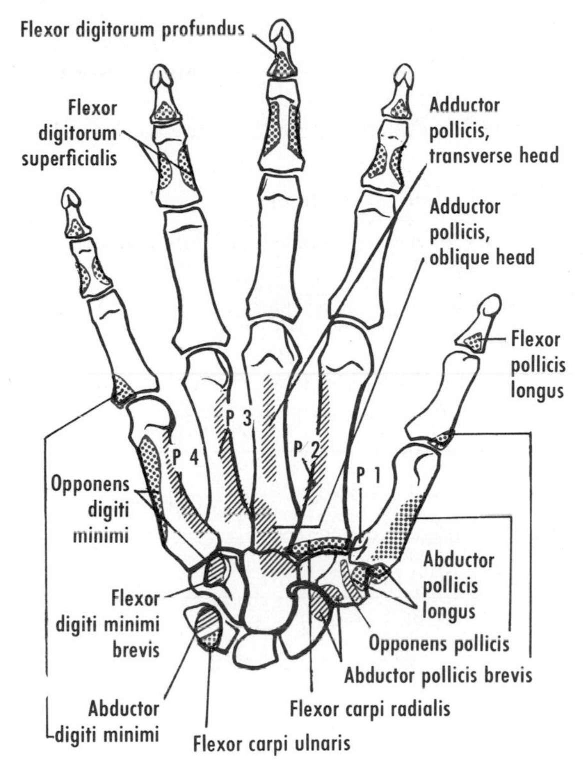

Figure 6–28 Bones of the right hand, showing muscular attachments, anterior view. The flexor pollicis brevis is not shown. Of the interossei, only the palmar (*P*) ones are shown.

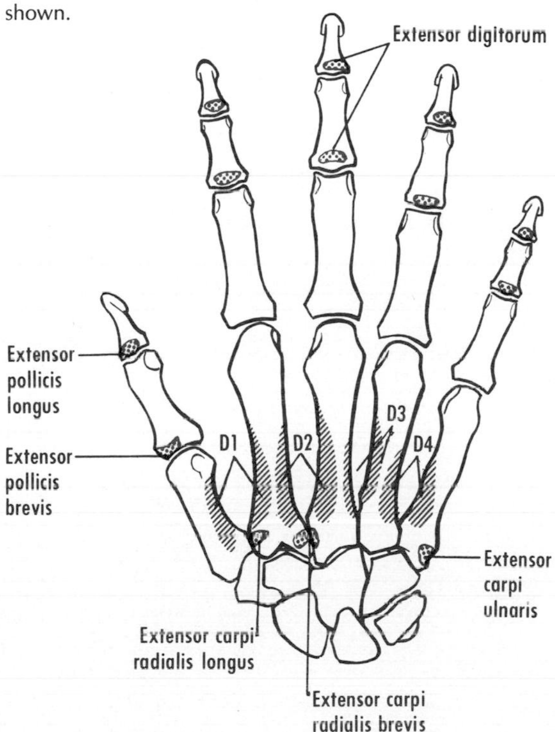

Figure 6–29 Bones of the right hand, showing muscular and tendinous attachments, posterior view. Each dorsal interosseous muscle (*D*) arises from the shafts of adjacent metacarpals.

METACARPUS

The carpus is connected to the phalanges by five metacarpal bones, referred to collectively as the metacarpus. They are numbered from 1 to 5, from the thumb to the little finger. The first is the shortest and the second the longest. They contribute to the palm, and they can be felt posteriorly under cover of the extensor tendons.

Each metacarpal is technically a long bone, consisting of a base proximally, a shaft, and a head distally. The *base* articulates with the carpus and, except for that of the first, with the adjacent metacarpal(s) also. The base of the first metacarpal has a saddle-shaped facet for the trapezium. The *head* of each metacarpal articulates with a proximal phalanx and forms a knuckle of the fist.

The *shaft* of each metacarpal begins to ossify during fetal life, and centers appear postnatally in the heads of the four medial bones and in the base of the first metacarpal. Accessory centers termed "pseudoepiphyses" are sometimes seen in the head of the first and in the base of the second metacarpal.

PHALANGES

The thumb has two phalanges, whereas each of the other fingers has three. They are designated proximal, middle, and distal. Each phalanx is technically a long bone, consisting of a base proximally, a shaft, and a head distally. The *base* of a proximal phalanx articulates with the head of a metacarpal, and the *head* of the phalanx presents two condyles for the base of a middle phalanx. Similarly, the head of a middle phalanx presents two condyles for the base of a distal phalanx. Each distal phalanx ends in a rough expansion termed its *tuberosity*.

Each phalanx begins to ossify during fetal life, and centers appear postnatally in their bases.

Sesamoid bones are found related to the front of some of the metacarpophalangeal and interphalangeal joints. Two in front of the head of the first metacarpal are almost constant.

ADDITIONAL READING

Frazer's Anatomy of the Human Skeleton, 6th ed., rev. by A. S. Breathnach, Churchill, London, 1965. A detailed, regional synthesis of skeletal and muscular anatomy.

Pyle, S. I., Waterhouse, A. M., and Greulich, W. W., (eds.), *A Radiographic Standard of Reference for the Growing Hand and Wrist,* Year Book Medical Publishers, Chicago, 1971.

QUESTIONS

6–1 Which is the first bone to ossify?

6–2 Where is the clavicle most likely to fracture from indirect violence to the hand or shoulder?

6–3 What is the most lateral bony point of the shoulder?

6–4 Which nerves are particularly prone to injury in fractures of the humerus?

6–5 What is the relationship of the epicondyles of the humerus to the top of the olecranon?

6–6 Why is it important to know that the styloid process of the radius ends more distally than that of the ulna?

6–7 What is the most famous fracture of the radius?

6–8 Which carpal bone is most frequently fractured?

6–9 Which carpal bone is most frequently dislocated?

6–10 Do any carpals show ossification at birth?

7
VESSELS, LYMPHATIC DRAINAGE, AND THE BREAST

VEINS

The blood from the upper limb is returned to the heart by two sets of veins, superficial and deep. Both sets have valves, and both drain ultimately into the axillary vein.

Superficial Veins (fig. 7–1). The superficial veins are highly variable, lie mostly in the subcutaneous tissue, and return almost all of the blood. Digital veins drain into a *dorsal venous network* in the hand, which leads to two prominent vessels, the cephalic and the basilic.

The *cephalic vein* begins immediately posterior to the styloid process of the radius, where it can be cut down on in an emergency. It then winds anteriorly around the lateral border of the forearm, reaches the front of the elbow, ascends lateral to the biceps, lies in the groove between the deltoid and pectoralis major muscles (where it can be exposed surgically), and ends in the axillary vein.

The *basilic vein* winds anteriorly around the medial border of the forearm, reaches the front of the medial epicondyle, ascends medial to the biceps, pierces the fascia, and accompanies the brachial artery to the axilla, where it joins the brachial veins and becomes the axillary vein.

In front of the elbow, the cephalic and ba-

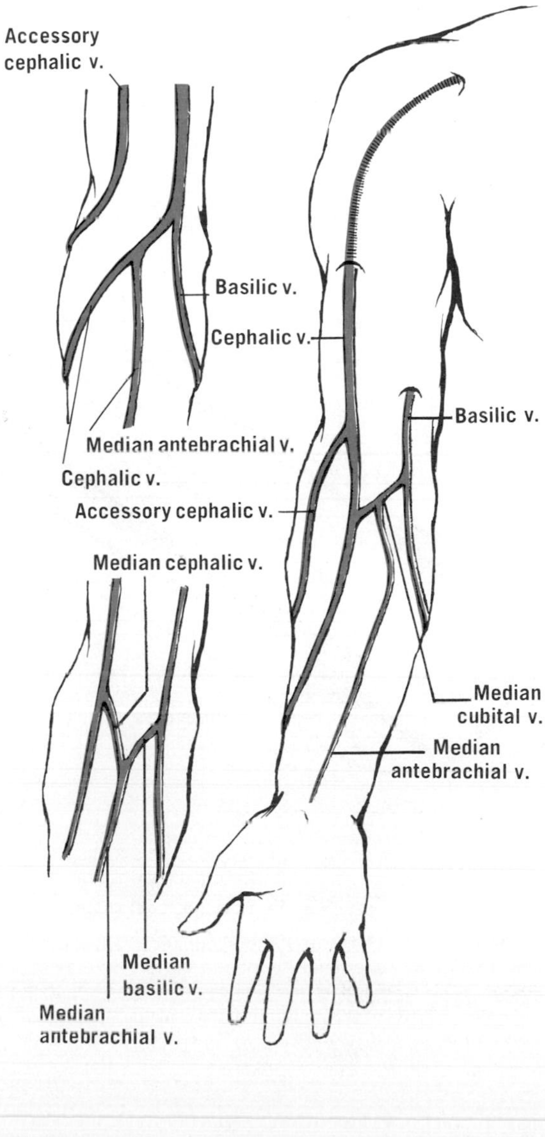

Figure 7–1 Diagram of some common patterns of the superficial veins of the upper limb. Only the larger channels at the elbow are shown; these are the ones most likely to be visible through the skin.

silic veins are frequently connected by the *median cubital vein,* which runs upward and medially from the cephalic to the basilic. It was formerly used for bleeding (*phlebotomy*). It may receive a vessel (*median antebrachial vein*) from the front of the forearm. The median cubital crosses the bicipital aponeurosis, which (*grâce à Dieu*) separates it from the underlying brachial artery and median nerve. **The median cubital vein or an associated vessel is used for taking blood samples, for intravenous injections, for blood transfusions, and for the introduction of catheters in cardiac catheterization.**

The valves in the superficial veins of the forearm can be demonstrated *in vivo,* as was recounted by Harvey in 1628.

Deep veins. The deep veins accompany the arteries, usually in cross-connected pairs (*venae comitantes*), contain valves, and ultimately reach the axillary vein, which continues as the subclavian.

LYMPHATIC DRAINAGE

The lymphatic vessels of the upper limb, most of those from the breast, and the cutaneous vessels from the trunk above the level of the umbilicus drain into the axillary nodes.

Lymphatics from the fingers reach vessels that accompany the cephalic and basilic veins and gain the *lateral axillary* and *deltopectoral* (or *infraclavicular*) nodes (fig. 7–2).

Axillary Nodes. These important nodes are arbitrarily divided into five groups (fig. 7–2).

1. *The lateral nodes* lie behind the axillary vein and drain the upper limb.
2. The *pectoral nodes,* at the lower border of the pectoralis minor, drain most of the breast.
3. The *posterior,* or *subscapular, nodes,* in the posterior axillary fold, drain the back of the shoulder.
4. The *central nodes,* near the base of the

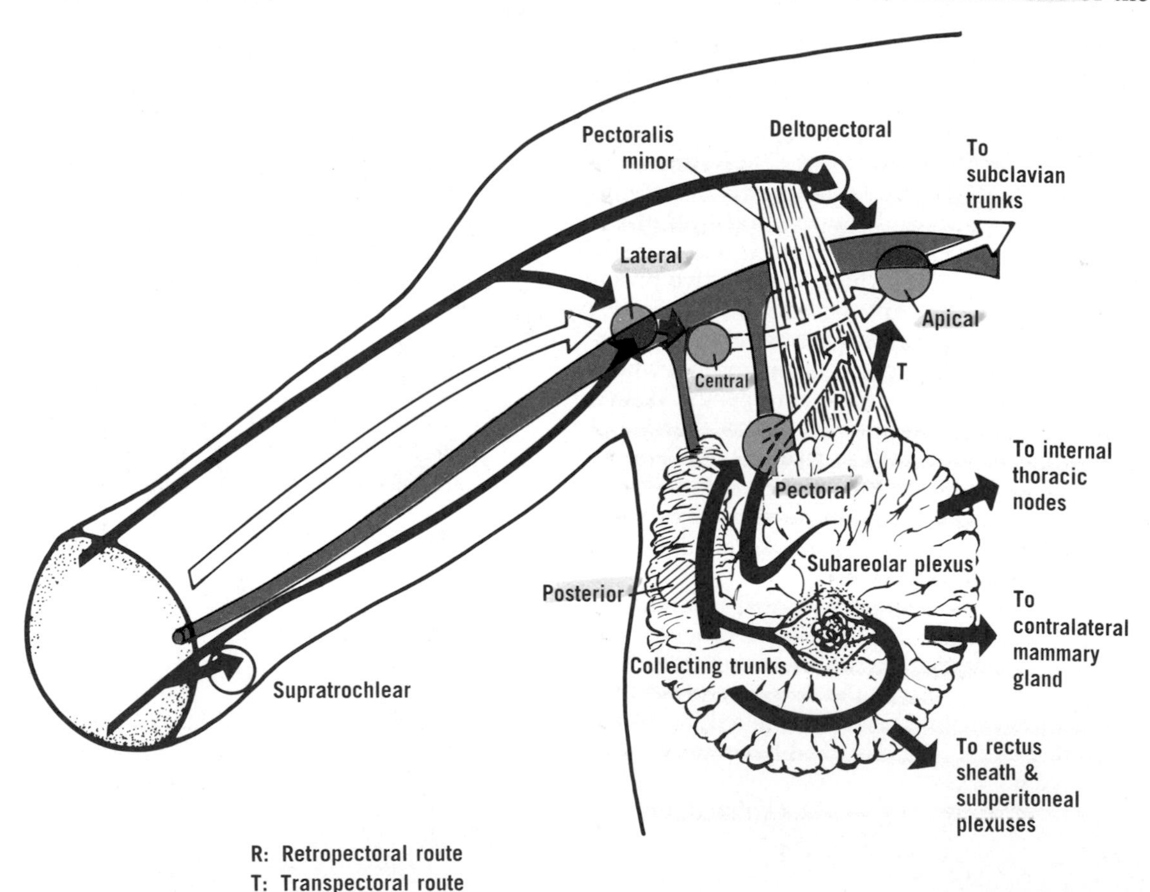

Figure 7–2 Diagram of the lymphatic drainage of the upper limb and breast. The supratrochlear and deltopectoral nodes receive many superficial lymphatic vessels. The axillary nodes are indicated by capital letters. The lateral nodes drain the upper limb. The subareolar plexus drains by collecting trunks into the axillary nodes. The pectoral nodes drain most of the breast. The apical nodes receive the lymph from the other axillary groups. Retropectoral (*R*) and transpectoral (*T*) routes are also shown.

axilla, receive the lymph from the preceding three groups. They form the group most likely to be palpable (against the lateral thoracic wall).

5. The *apical nodes* lie medial to the axillary vein and above the upper border of the pectoralis minor. **The apical nodes receive the lymph from all the other groups and sometimes directly from the breast,** and they drain into two or three *subclavian trunks,* which enter the jugular-subclavian venous confluence, or join a common lymphatic duct, or empty into lower, deep cervical nodes.

THE BREAST

The breast overlies the pectoralis major, serratus anterior, and external oblique muscles. It usually extends from the second to the sixth ribs but **the mammary gland is more extensive than the breast and generally extends into the axilla as an "axillary tail." The superolateral quadrant of the breast contains a large amount of glandular tissue and is the site of 60 per cent of carcinomata of the breast.**

The *mammary gland* is situated within the subcutaneous tissue, deep to which is the fascia covering the pectoralis major and serratus anterior. The gland is normally mobile on the fascia. The parenchyma is arranged in about 15 to 20 lobes, each of which is drained by a *lactiferous duct* opening on the nipple. The ducts, each of which may show a sinus near its termination, can be injected with a radio-opaque medium and then visualized radiographically (*mammography*). The stroma, which consists of adipose and fibrous tissue, is inseparably intermingled with the epithelial parenchyma. Anteriorly, the subcutaneous tissue sends, in the words of Sir Astley Cooper, "large, strong, and numerous fibrous . . . processes, to the posterior surface of the skin which covers the breast." These *suspensory ligaments* account for the dimpling of the skin seen in certain pathological conditions, such as carcinoma.

The *nipple,* which is often at the level of the fourth intercostal space, contains the minute openings of the lactiferous ducts. It contains smooth muscle, which compresses the ducts and renders the nipple erect. The nipple is surrounded by an *areola* of pigmented skin, which becomes brown during pregnancy and then remains so. The areola contains accessory mammary glands, sweat glands, and sebaceous glands that form tubercles during pregnancy and lubricate the nipple during lactation.

Development and Growth. Two vertical ectodermal thickenings, the *mammary ridges,* appear on the trunk during embryonic life, and these extend from the axillary to the inguinal region. The rostral part of each forms the nipple. The mammary glands develop from the nipples during fetal life. Accessory nipples (*polythelia*) or glands (*polymastia*) usually, but not invariably, develop on the line of the mammary ridges. At puberty, in the female, the breasts grow and the ducts bud and form lobules, but true secretory alveoli do not develop until pregnancy. The glandular tissue involutes after the menopause.

Blood Supply. The mammary gland is highly vascular and is supplied by branches of the internal thoracic, axillary, and intercostal arteries. Deep veins drain into the correspondingly named veins. Connections between the intercostal veins and the vertebral plexus allow metastasis to bones and to the nervous system.

Lymphatic Drainage. **The lymphatic and venous drainages of the breast are of great importance in the spread of carcinoma** (fig. 7–2). About three quarters of the lymphatic drainage is to the axillary nodes: (1) Lymphatics pass around the edge of the pectoralis major and reach the pectoral group of axillary nodes; (2) routes through or between the pectoral muscles may lead directly to the apical nodes of the axilla; (3) lymphatics follow the blood vessels through the pectoralis major and gain the parasternal (internal thoracic) nodes; (4) connections may lead across the median plane and hence to the contralateral breast; (5) lymphatics may reach the sheath of the rectus abdominis and the subperitoneal and subhepatic plexuses.

It should be noted that free communication exists between nodes below and above the clavicle and between the axillary and cervical nodes.

ARTERIES

The main artery carrying blood to the upper limb is concerned primarily with the supply of the vital centers of the medulla oblongata. The vessel is named successively subclavian, axillary, and brachial. The arrangement of this vessel and its branches is summarized in fig-

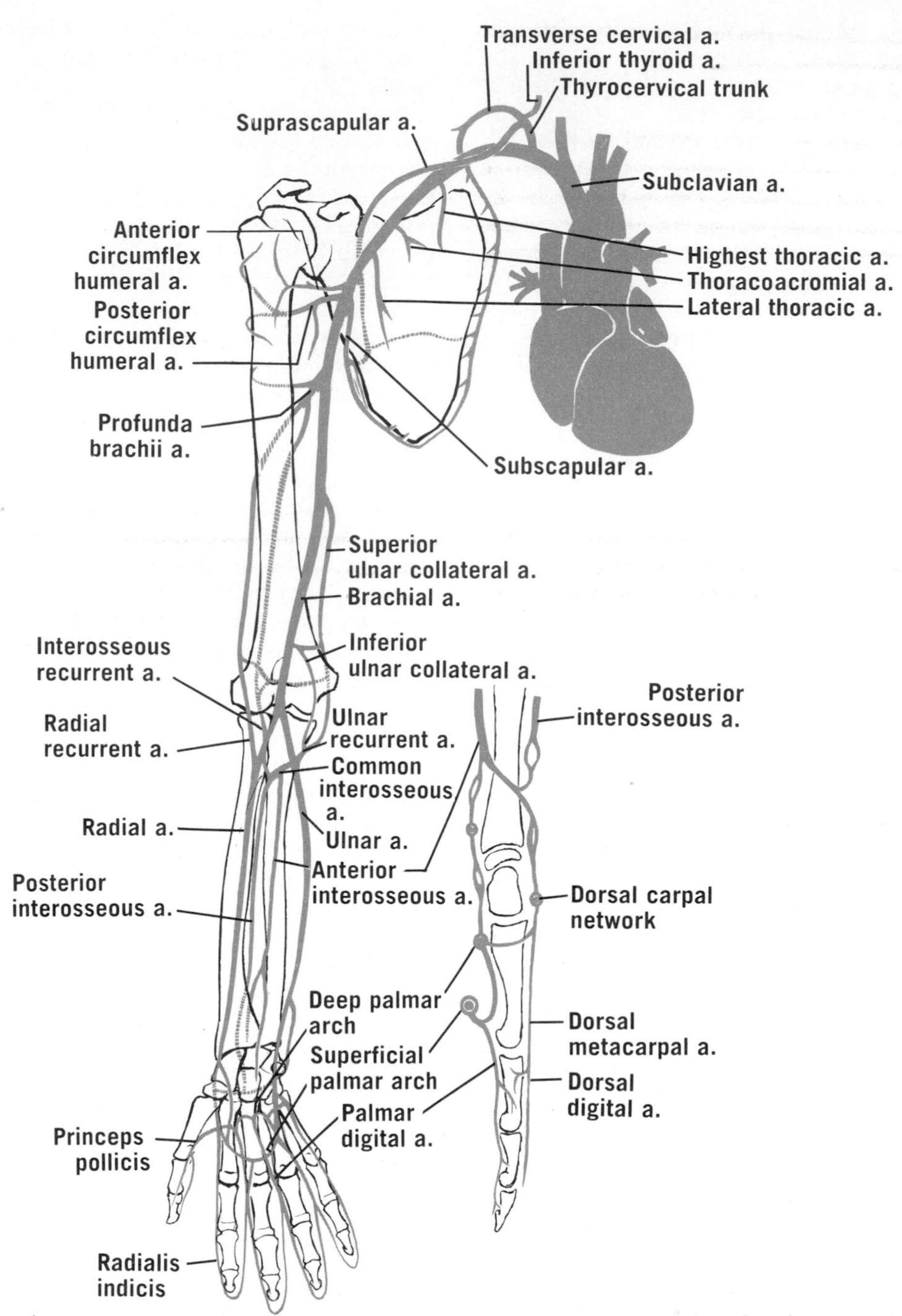

Figure 7–3 The arteries of the upper limb. The sagittal section (*lower right*) is based on Grant.

ure 7–3. The *subclavian artery* arches across the first rib and, at the outer border of that bone, its name is changed to *axillary*. At the base of the axilla, where the axillary artery leaves the lower border of the teres major, its name is changed to *brachial*. Immediately below the elbow joint, the brachial divides into the *radial* and *ulnar arteries*, which, in the hand, form anastomotic connections known as the superficial and deep palmar arches. A number of metacarpal and digital branches supply the fingers.

ADDITIONAL READING

Harvey, W., *Movement of the Heart and Blood in Animals*, trans. by K. J. Franklin, Blackwell, Oxford, 1957 (distributed in the United States by Charles C Thomas, Springfield, Illinois). Harvey's classic demonstration of the venous valves in 1628 is in Chapter 13.

QUESTIONS

7–1 Where was the usual site for bloodletting?

7–2 When a cardiac catheter is inserted at the right elbow, how does it reach the heart?

7–3 What is the effect on the superficial veins of opening and closing the fist?

7–4 In which layer is the mammary gland situated?

7–5 What are the suspensory ligaments of the breast?

7–6 What is the "axillary tail?"

7–7 Into which nodes do most of the lymphatics of the breast drain?

7–8 Does the lymphatic drainage of the breast involve the pectoral muscles?

8
THE SHOULDER AND AXILLA

MUSCLES OF PECTORAL REGION

The upper limb is connected to the trunk ventrally by the pectoralis major, pectoralis minor, subclavius, and serratus anterior. The pectoralis major is inserted into the humerus, the others into the shoulder girdle. They are all supplied by branches of the brachial plexus. The origin, insertion, innervation, and action of each muscle are listed in table 8–1. In this and subsequent tables of muscles, only the chief attachments and principal actions are given.

The fascia extending between the pectoralis major and the latissimus dorsi forms the floor of the axilla. This *axillary fascia* is suspended from the fascia around the pectoralis minor, and traction on it produces the hollow of the armpit. It is attached above to the clavicle as the *clavipectoral fascia,* which is also anchored to the first rib and to the coracoid process (fig. 8–1).

TABLE 8–1 Muscles of Pectoral Region

Muscle	Origin	Insertion	Innervation	Action
Pectoralis major	Medial half of clavicle (clavicular head), front of sternum, costal cartilages, & aponeurosis of external oblique (sternocostal head)	Lateral lip of intertubercular groove of humerus	Lateral & medial pectoral	Adducts arm
Pectoralis minor	Ribs 2-5	Coracoid process	Medial & lateral pectoral	Probably depresses point of shoulder
Subclavius	Junction of rib 1 & costal cartilage 1	Lower surface of clavicle	Nerve from upper trunk of brachial plexus	Protects vessels & may assist in depressing clavicle
Serratus anterior	Ribs 1-8	Costal aspect of superior angle, medial border, and inferior angle of scapula	Long thoracic nerve	Rotates scapula

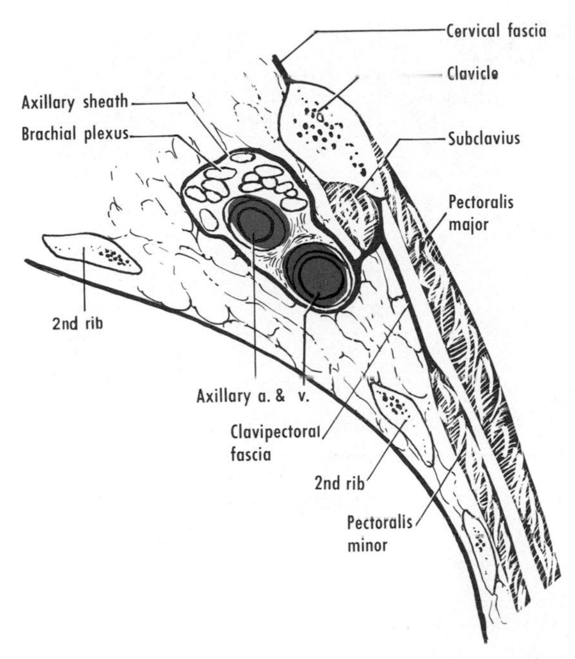

Figure 8–1 Diagram of the clavipectoral fascia and its relation to the axillary sheath. Sagittal section.

The *supraclavicular nerves* (lateral, intermediate, and medial) arise from the cervical plexus and cross the clavicle (where they may be rolled against the bone). They supply the skin over the shoulder and, because they arise from the same roots (C3, 4) as the phrenic nerve, diaphragmatic inflammation is one cause of pain referred to the shoulder.

The *pectoralis major* is a large, fan-shaped muscle, the rounded, lower border of which forms the anterior axillary fold (fig. 8–2). It functions mainly as an adductor of the arm and aids in throwing, pushing, and shoveling. When the arms are fixed in climbing, it draws the body upward. The pectoralis major covers the *pectoralis minor,* which, in turn, covers the second part of the axillary artery.

The *serratus anterior* (see fig. 8–5) is a large, fan-shaped muscle that forms the medial wall of the axilla. Its digitations of origin (which interdigitate with those of the external oblique muscle of the abdomen) can be seen in a muscular person *in vivo*. The serratus rotates the scapula so that the inferior angle moves laterally; it is thereby important in abduction of the arm above the horizontal. It pulls the scapula forward in throwing and pushing. Paralysis of the serratus anterior (e.g., from injury or inflammation of its nerve) is characterized by "winging" of the scapula, i.e., the medial border of the bone stands away from the chest wall.

SUPERFICIAL MUSCLES OF BACK (see fig. 8–4)

The upper limb is connected to the vertebral column by the latissimus dorsi and trapezius and, deep to these, by the levator scapulae, rhomboid minor, and rhomboid major (table 8–2). The latissimus is inserted into the humerus, the others into the shoulder girdle. Although the muscles are on the back, they are supplied from the ventral rami of cervical nerves. The trapezius also receives an important supply from the accessory nerve.

The *trapezius* (fig. 8–3) is a large, superficial, triangular muscle that is responsible for the sloping ridge of the neck. The muscles of the two sides together form a trapezoid or cowl. The region around C.V.7 is more aponeurotic than muscular, resulting in a slightly depressed area frequently visible *in vivo*. The uppermost part of the trapezius, together with the levator scapulae, elevates the shoulder. The muscles of the two sides together brace the shoulders by pulling the scapulae backward, and their weakness results in drooping shoulders.

The *latissimus dorsi* is a large, mostly superficial, triangular muscle. Its uppermost part is covered by the trapezius posteriorly, and its lateral part, together with the teres major, forms the posterior axillary fold. The trapezius and latissimus dorsi intersect near the medial border of the scapula, at a small area termed the *triangle of auscultation* (figs. 8–3 and 8–4), where the thoracic wall is minimally covered by muscles. The latissimus is a powerful adductor and extensor of the arm and is important in the downstroke in swimming as well as in rowing, climbing, and hammering, and in supporting the weight of the body on the hands.

MUSCLES OF SHOULDER

The deltoid, supraspinatus, infraspinatus, teres minor, teres major, and subscapularis arise from the shoulder girdle and are inserted into the humerus. They are supplied by C.N.5 and 6 through branches of the brachial plexus.

The *deltoid* is responsible for the roundness of the shoulder (see fig. 8–2). Its origin embraces the insertion of the trapezius. It is a powerful abductor of the arm in the plane of the scapula. The anterior and posterior parts of the muscle are involved, respectively, in flexion and medial rotation and in extension

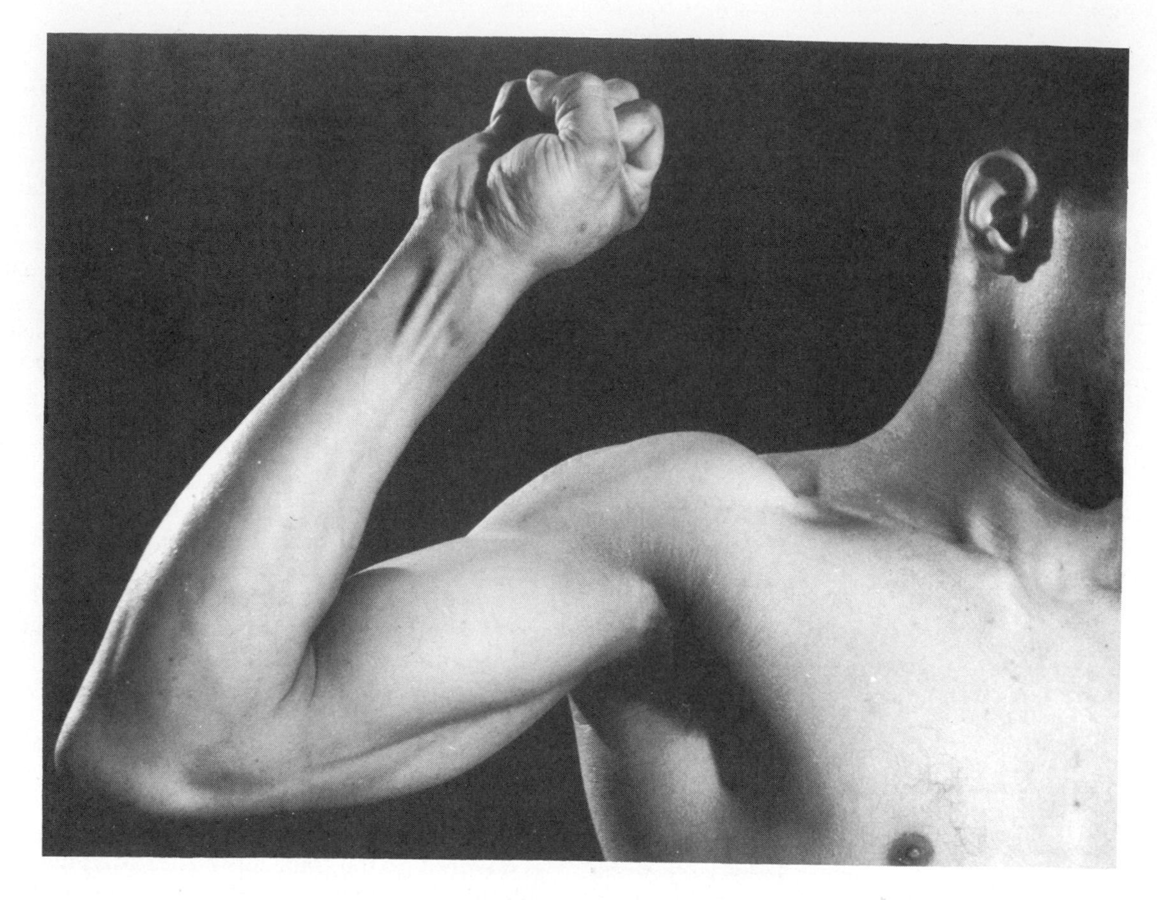

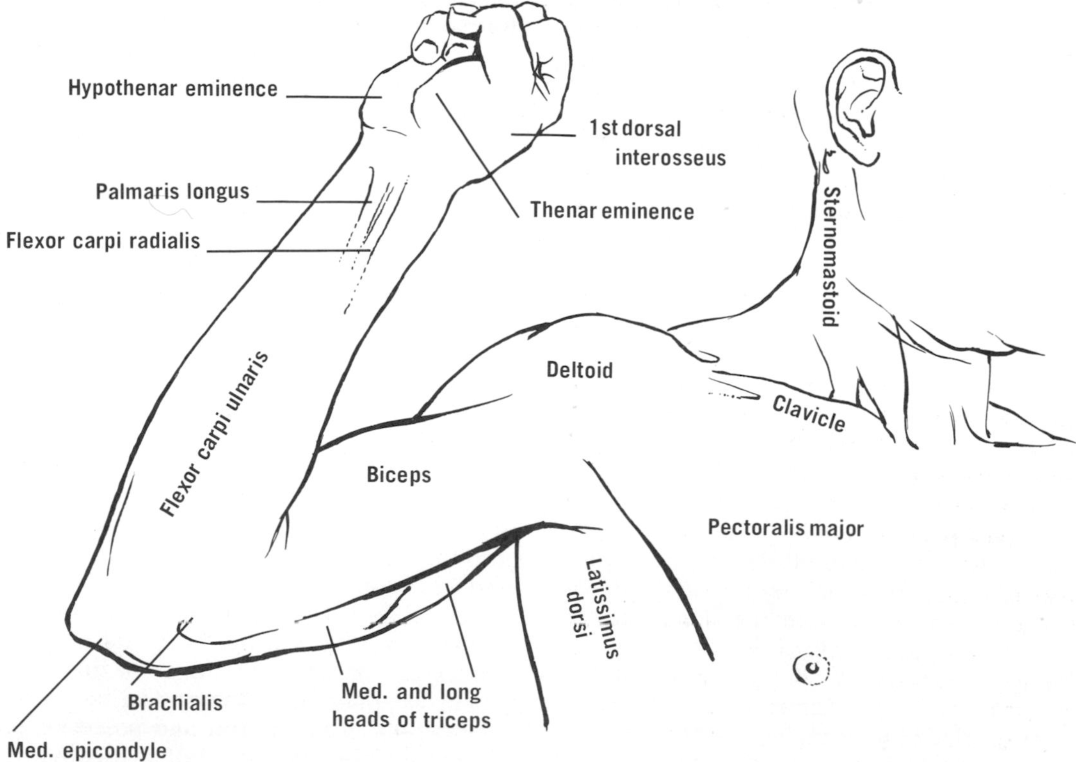

Figure 8–2 Surface landmarks of the upper limb. (From Royce, J., *Surface Anatomy,* Davis, Philadelphia, 1965.)

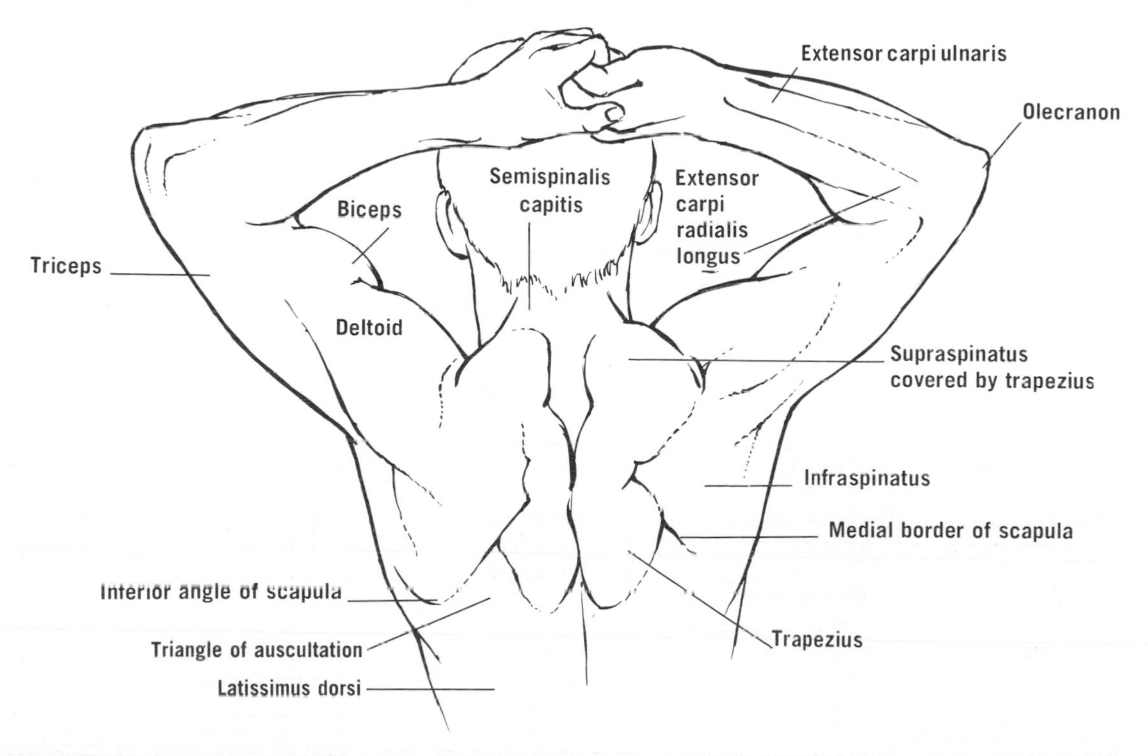

Figure 8–3 Surface landmarks of the back and upper limbs. (From Royce, J., *Surface Anatomy*, Davis, Philadelphia, 1965.)

TABLE 8–2 Superficial Muscles of Back

Muscle	Origin	Insertion	Innervation	Action
Trapezius	Spines of C.V.7 & T.V.1-12, ligamentum nuchae, & often occipital bone	Lateral third of clavicle, acromion, & spine of scapula	Accessory & C.N.3,4	Elevates shoulder, & retracts & rotates scapula
Latissimus dorsi	Spines of T.V.7-12, thoracolumbar fascia, iliac crest, lower ribs, & inferior angle of scapula	Intertubercular groove of humerus	Thoracodorsal	Adducts & extends arm
Levator scapulae	Transverse tubercles of C.V.1-4	Upper part of medial border of scapula	C.N.3,4	Elevates scapula
Rhomboid minor & major (often fused)	Spines of C.V.7 & T.V.1-5	Medial border of scapula	Dorsal scapular	Retract & fix scapula

and lateral rotation. The posterior and middle parts together abduct the arm in a coronal plane. The deltoid acts also as a stabilizer in horizontal movements, e.g., drawing a line across a blackboard. The *supraspinatus* aids the deltoid in abduction, which is usually incomplete in paralysis of either of these muscles (fig. 8–5). The interval between the *teres minor, teres major,* and the surgical neck of the humerus is divided longitudinally by the long head of the triceps into a *triangular space* medially and a *quadrangular space* (containing the axillary nerve) laterally (see fig. 9–2).

The tendons of the supraspinatus, infraspinatus, teres minor, and subscapularis blend with the capsule of the shoulder joint and form a musculotendinous (or rotator) cuff, which is incomplete below (fig. 8–6). The tendons of the cuff are prone to degenerative changes. Moreover, nipping of a tender structure (e.g., from a calcified deposit in the supraspinatus tendon) between the acromion and the greater

TABLE 8–3 Muscles of Shoulder

Muscle	Origin	Insertion	Innervation	Action
Deltoid	Lateral third of clavicle, acromion, & spine of scapula	Deltoid tuberosity on anterolateral surface of humerus	Axillary	Abducts arm
Supraspinatus	Supraspinous fossa of scapula	Greater tubercle of humerus	Suprascapular	Abducts arm
Infraspinatus	Infraspinous fossa of scapula	Greater tubercle of humerus	Suprascapular	Rotates arm laterally
Teres minor	Lateral part of dorsal surface of scapula	Greater tubercle of humerus	Axillary	Rotates arm laterally
Teres major	Dorsal surface of inferior angle of scapula	Medial lip of intertubercular groove of humerus	Lower subscapular	Adducts arm
Subscapularis	Subscapular fossa of scapula	Lesser tubercle of humerus	Subscapular	Rotates arm medially

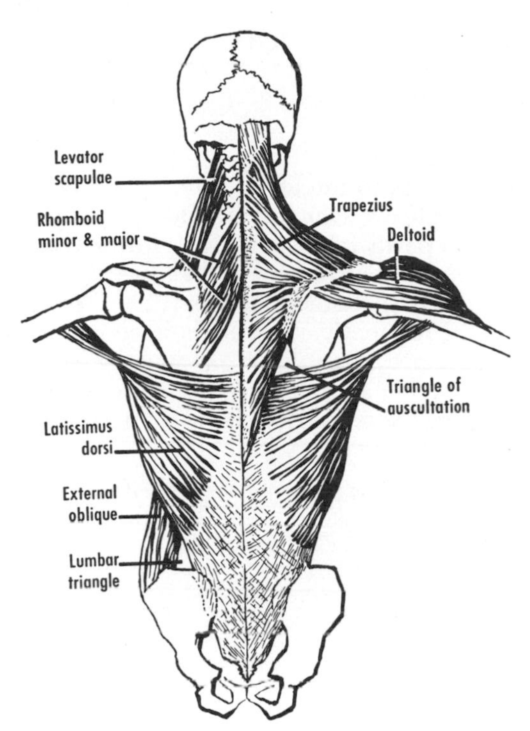

Figure 8–4 The superficial muscles of the back. Note how the latissimus dorsi covers the inferior angle of the scapula. The left trapezius is removed to show the rhomboids and levator scapulae. (Based on Mollier.)

tubercle of the humerus results in pain during the mid-range of abduction (painful arc syndrome).

AXILLA

The pyramidal interval between the arm and the chest wall is termed the axilla. The pectoralis major and latissimus dorsi form prominent anterior and posterior axillary folds, respectively, *in vivo*. A vertical line midway between the anterior and posterior axillary folds is referred to as the *midaxillary line*. The fascial base extends between these folds. **The apex of the axilla is the interval between the posterior border of the clavicle, the superior border of the scapula, and the external border of the first rib. Through the apex, the axillary vessels and their accompanying nerves pass from the neck to the arm.** The axilla is bounded medially by the upper ribs and their intercostal muscles and by the serratus anterior; it is limited laterally by the intertubercular groove of the humerus.

The chief contents of the axilla are the axillary artery and vein, a part of the brachial plexus and its branches, and the axillary lymph nodes.

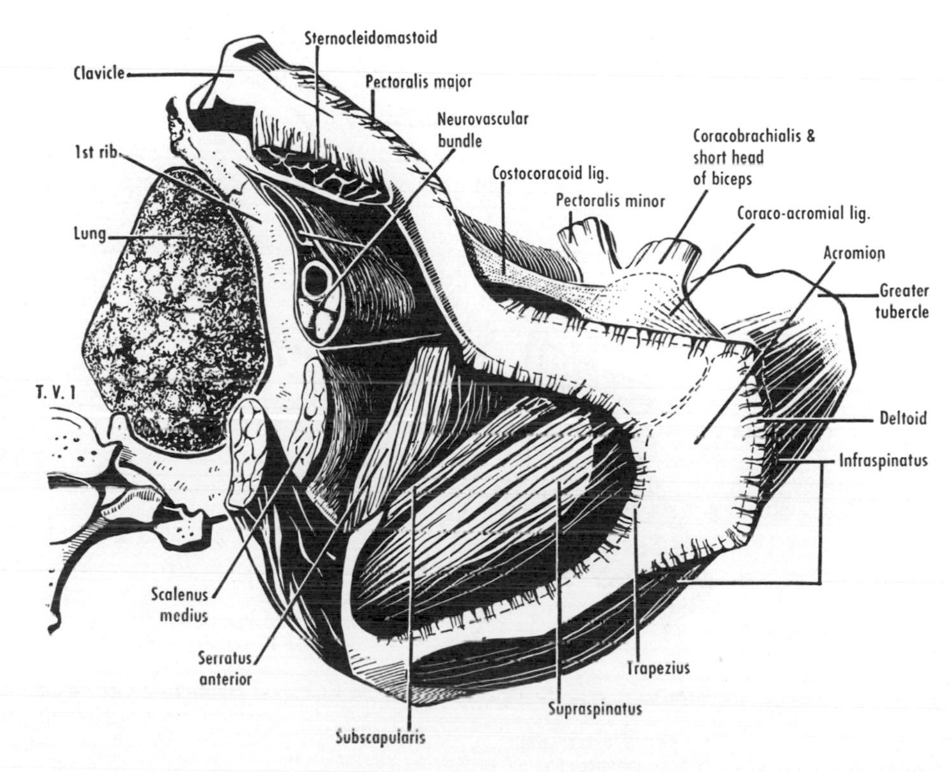

Figure 8–5 The shoulder from above. Note the relation of the neurovascular bundle (subclavian vessels and brachial plexus) to the clavicle. Note also that the insertion of the trapezius in the concavity formed by the spine, acromion, and clavicle is embraced by the origin of the deltoid muscle from the corresponding convexity.

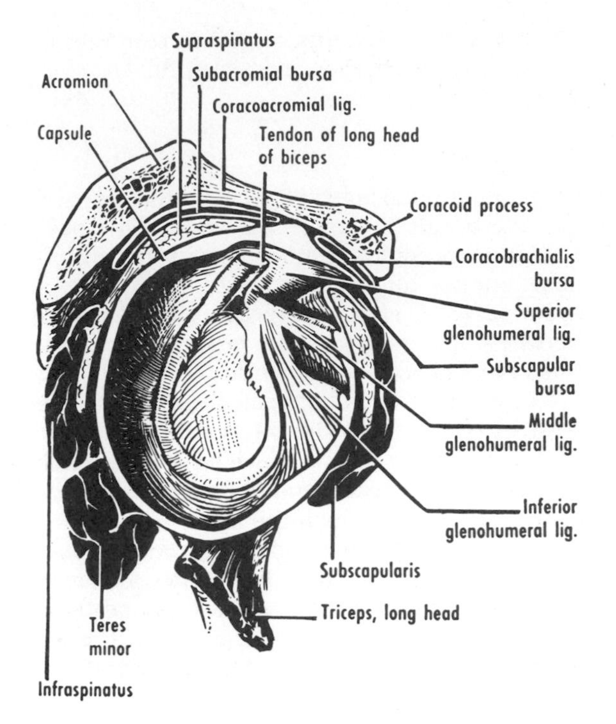

Figure 8–6 The musculotendinous (or rotator) cuff and the capsule of the shoulder joint, shown after cutting the cuff and removing the humerus. The tendons of the supraspinatus, infraspinatus, teres minor, and subscapularis blend with the capsule and form the cuff, which is incomplete below. The subscapular bursa communicates with the joint cavity (between the superior and middle glenohumeral ligaments).

NERVES OF UPPER LIMB

BRACHIAL PLEXUS (figs. 8–1 and 8–7 to 8–9)

The nerves to the upper limb arise from the brachial plexus, which is situated partly in the neck and partly in the axilla. It is formed by the union of the ventral rami of the lower four cervical and first thoracic nerves (C.N.5, 6, 7, 8; T.N.1). Frequently it receives a contribution from one nerve higher or one nerve lower.

The brachial plexus descends in the posterior triangle of the neck. Here it lies above the clav-

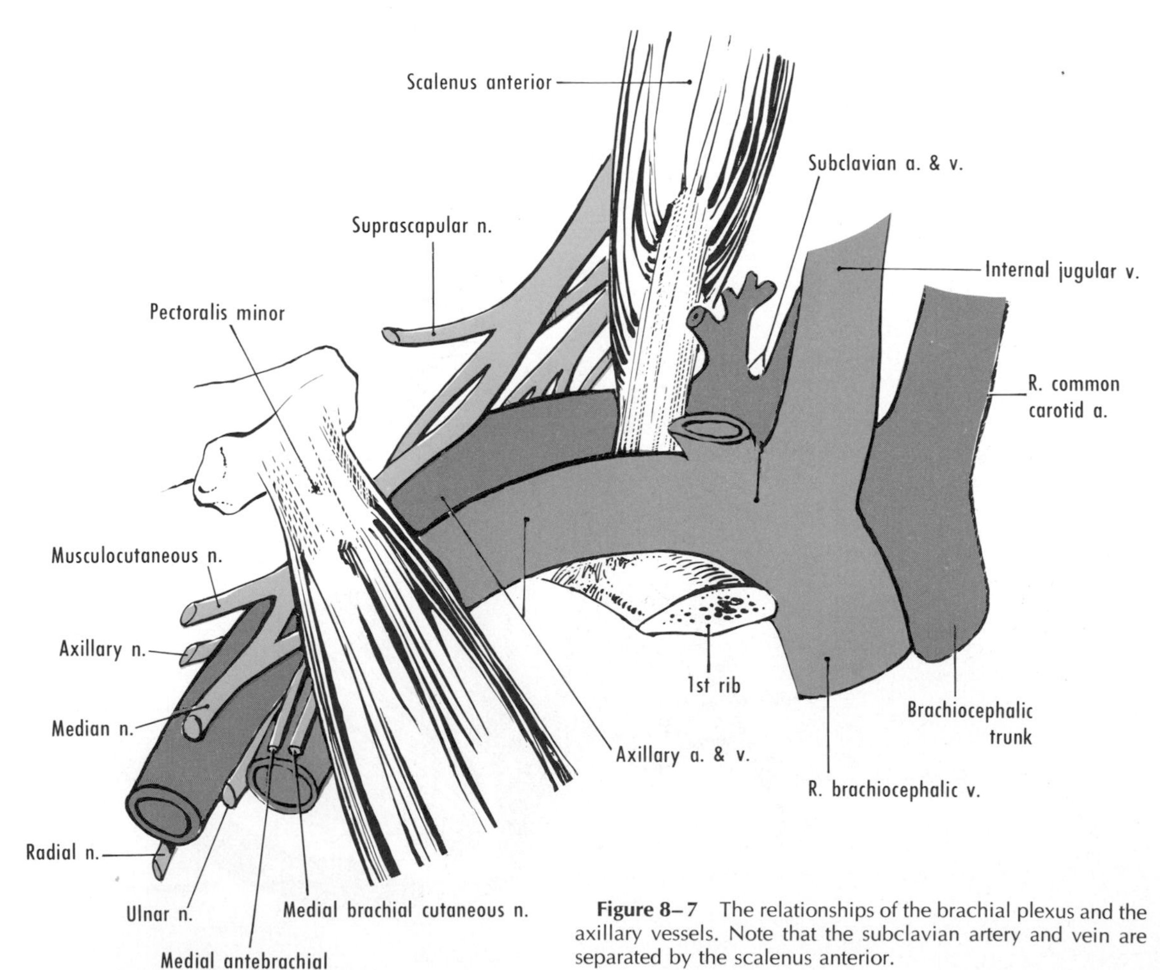

Figure 8–7 The relationships of the brachial plexus and the axillary vessels. Note that the subclavian artery and vein are separated by the scalenus anterior.

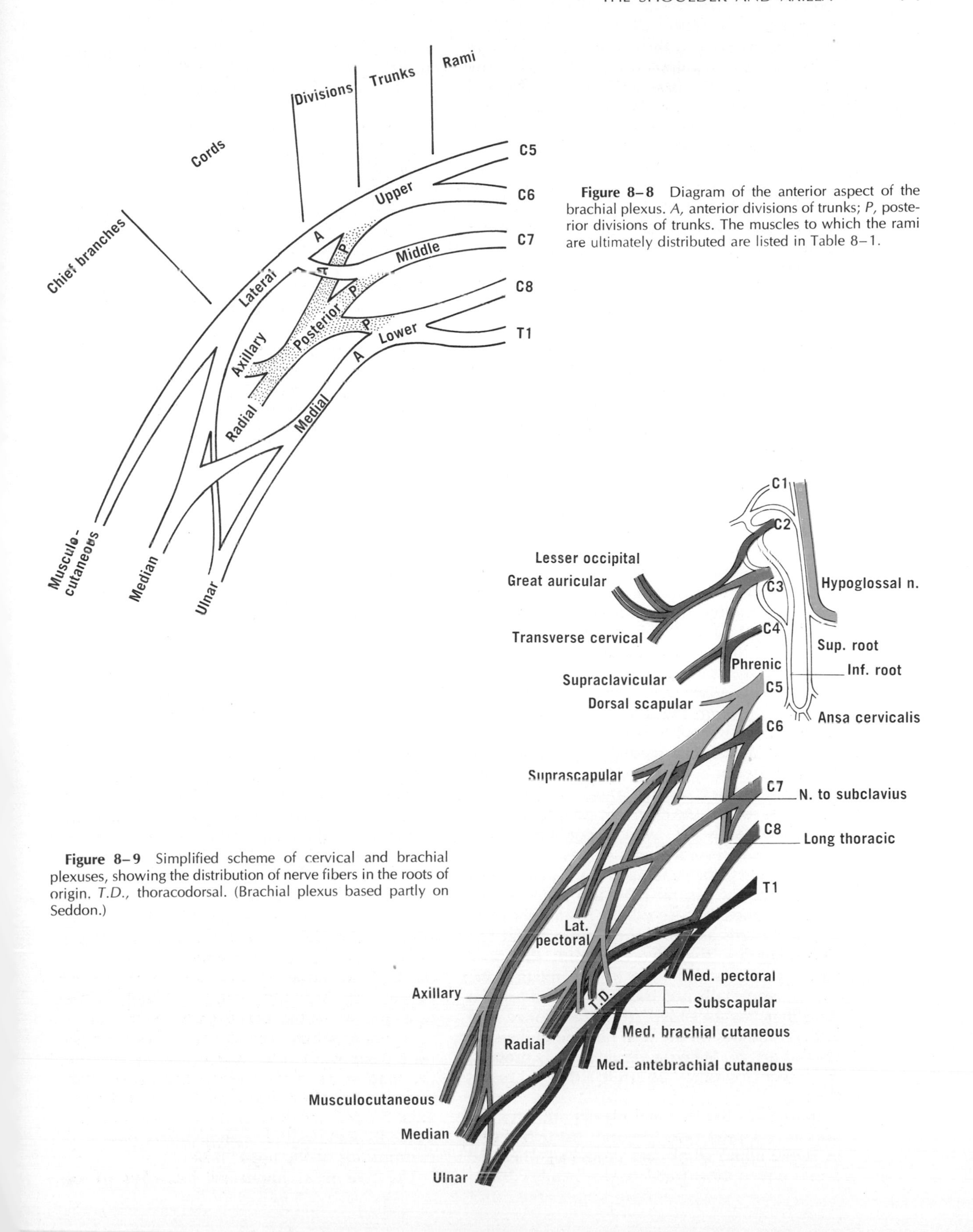

Figure 8–8 Diagram of the anterior aspect of the brachial plexus. *A*, anterior divisions of trunks; *P*, posterior divisions of trunks. The muscles to which the rami are ultimately distributed are listed in Table 8–1.

Figure 8–9 Simplified scheme of cervical and brachial plexuses, showing the distribution of nerve fibers in the roots of origin. *T.D.*, thoracodorsal. (Brachial plexus based partly on Seddon.)

icle and posterior to the sternomastoid muscle, where it can be palpated in the living. In surface anatomy, the brachial plexus in the neck lies below a line from the posterior margin of the sternomastoid at the level of the cricoid cartilage to the midpoint of the clavicle. Here the plexus can be injected with a local anesthetic (brachial block), the pulsations of the third part of the subclavian artery (situated below and in front of the plexus) being used as a guide (fig. 8–7).

The brachial plexus descends in the concavity of the medial two thirds of the clavicle (see fig. 8–5) and accompanies the axillary artery under cover of the pectoralis major. The plexus is enclosed with the axillary vessels in the *axillary sheath,* which is a prolongation of the cervical fascia downward behind the clavicle and into the axilla. The brachial plexus may be blocked in the axilla by injecting a local anesthetic into the axillary sheath.

A common arrangement of the brachial plexus is shown in figures 8–8 and 8–9. The first two ventral rami (C.N.5, 6) unite to form the *upper trunk,* the next (C.N.7) constitutes the *middle trunk,* and the last two (C.N.8; T.N.1) join to form the *lower trunk.* Each trunk divides into an *anterior* and *posterior division.* The anterior divisions of the upper and middle trunks unite to form the *lateral cord,* that of the lower trunk constitutes the *medial cord,* and the three posterior divisions join to form the *posterior cord.* **The cords are named from the positions that they occupy in relation to the second part of the axillary artery. In general, the lateral and medial cords supply the front of the limb, whereas the posterior cord supplies the back. At the lateral border of the pectoralis minor, the cords divide into terminal branches, each of which contains fibers derived from several spinal nerves.**

In summary, the brachial plexus is composed successively of ventral rami and trunks in the neck, divisions that are usually behind the clavicle, and cords and branches in the axilla. The nerve bundles that descend from the neck come to meet, and then to accompany, the more superficially and medially placed arterial tube that has ascended from the thorax. The lower trunk lies on the first rib behind the subclavian artery. When a cervical rib is present, the lower trunk may be stretched as it crosses the rib.

Injuries to the brachial plexus are very important. "Upper type" injuries (to C.N.5 or 6 or to the upper trunk) are produced when the arm is pulled downward and the head is drawn away from the shoulder. The upper limb tends to lie in medial rotation ("waiter's-tip hand"). Such injuries may occur during birth. "Lower type" injuries (to C.N.8, T.N.1, or to the lower trunk) are produced when the arm is pulled upward and may occur during birth also. The short muscles of the hand are affected and "claw hand" results (see fig. 8–14). Involvement of the brachial plexus may also be a part of the neurovascular compression syndrome.

Branches of the Brachial Plexus

Several branches arise *above the clavicle* (fig. 8–9). The ventral rami give rise to (1) the dorsal scapular nerve (chiefly C.N.5) to the rhomboids and (2) the long thoracic nerve (C.N.5-7) to the serratus anterior. The upper trunk gives off (1) the nerve to the subclavius (chiefly C.N.5), which frequently contributes to the phrenic nerve, and (2) the suprascapular nerve (C.N.5, 6) to the supraspinatus and infraspinatus.

The terminal branches of the cords arise *below the clavicle.* Several of them send articular twigs (e.g., to the shoulder joint), although these are not listed separately here.

The *lateral cord* gives origin to (1) lateral pectoral nerves (C.N.5-7) to the pectoralis major and minor, (2) the musculocutaneous nerve, (3) the lateral head of the median, and (4) usually a lateral head for the ulnar (fig. 8–9).

The *medial cord* (fig. 8–9) provides (1) medial pectoral nerves (C.N.8; T.N.1) to the pectoralis major and minor, (2) the medial brachial cutaneous nerve (T.N.1) to the medial side of the arm, (3) the medial antebrachial cutaneous nerve (C.N.8; T.N.1) to the medial side of the forearm, (4) the ulnar, and (5) the medial head of the median.

The *posterior cord* gives off (1) the upper subscapular nerve(s) (C.N.5) to the subscapularis, (2) the thoracodorsal nerve (C.N.7, 8) to the latissimus dorsi, (3) the lower subscapular nerve(s) (C.N.5, 6) to the subscapularis and teres major, (4) the axillary, and (5) the radial.

A simple scheme (fig. 8–10) shows the basic arrangement of the segmental innervation (*dermatomes*) of the skin of the upper limb, which is supplied successively down the preaxial border and then up the postaxial border. Table 8–4 shows the segmental innervation of the muscles of the upper limb.

The five most important branches of the

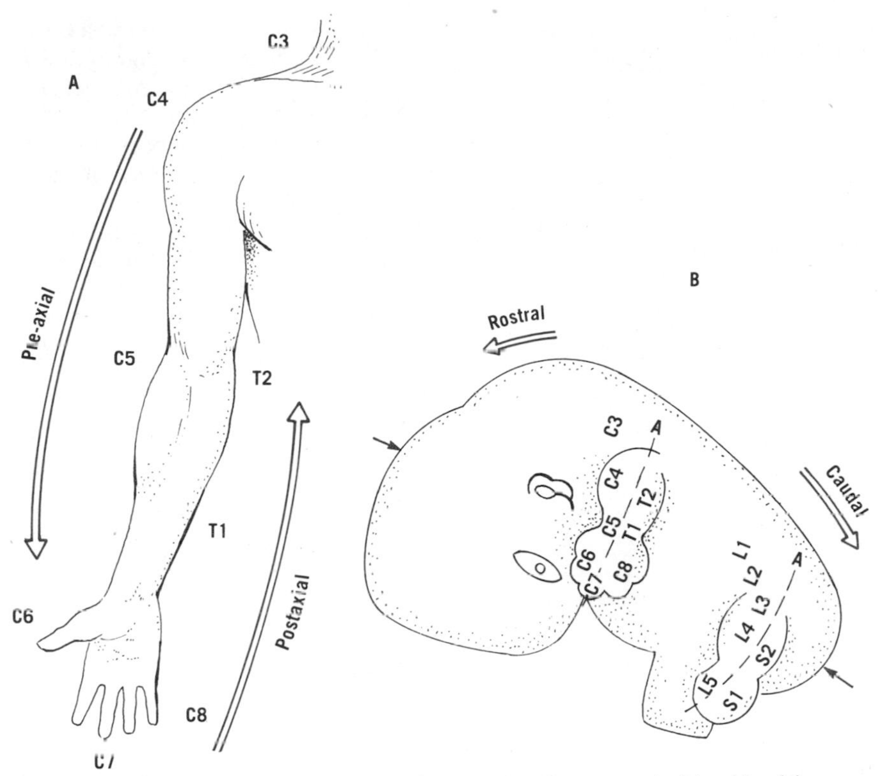

Figure 8–10 **A,** The basic arrangement of the segmental innervation (dermatomes) of the skin of the upper limb, which is supplied successively down the pre-axial border and then up the postaxial border. **B,** An embryo of 7 weeks (measuring 17 mm between small arrows), showing the longitudinal axes (*A*) of the upper and lower limbs. The parts of the limbs rostral to the axes are called pre-axial, those caudal to the axes are termed postaxial. The sequence of dermatomes from pre-axial to postaxial is shown on each limb. Cf. fig. 15–11.

TABLE 8–4 SEGMENTAL INNERVATION OF MUSCLES OF UPPER LIMB*

Muscles	C.3	4	5	6	7	8	T.1
Levator scapulae, trapezius	■	■					
Rhomboids			■				
Deltoid, supraspinatus, infraspinatus, teres major, subscapularis, biceps, brachialis, brachioradialis, supinator			■	■			
Serratus anterior, pectoralis major (lat.)			■	■	■		
Pronators				■			
Triceps				■	■	■	
Most extensors of hand and fingers					■	?	
Latissimus dorsi					■	■	
Flexores digitorum superficiales et profundi					■	■	■
Pectoralis major (med.), palmaris longus, flexor pollicis longus						■	■
Muscles of hand						?	■

* Adapted from various sources. It should be emphasized that, for many muscles, some of these figures are uncertain.

brachial plexus are the musculocutaneous, median, ulnar, axillary, and radial nerves.

The *musculocutaneous nerve* (C.N.5-7) (fig. 8–11), very variable, arises from the lateral cord and usually pierces the coracobrachialis. It may carry a part or all of the lateral head of the median nerve and send these fibers to the medial head in the arm. The musculocutaneous nerve supplies the flexor muscles on the

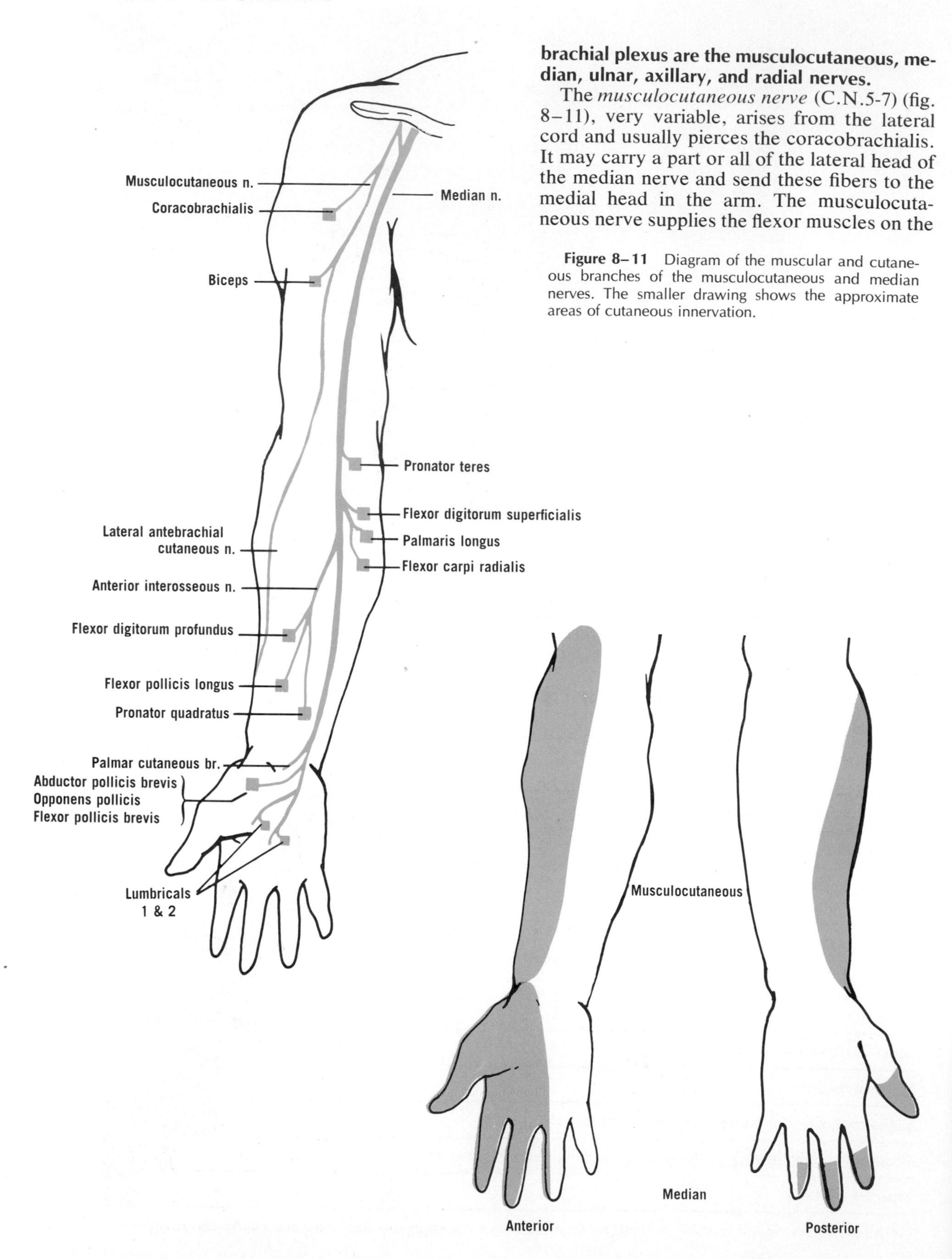

Figure 8–11 Diagram of the muscular and cutaneous branches of the musculocutaneous and median nerves. The smaller drawing shows the approximate areas of cutaneous innervation.

front of the arm and the skin on the lateral side of the forearm.

The *median nerve* (C.N.[5,] 6-8; T.N.1) (fig. 8–11) arises by lateral and medial heads from the lateral and medial cords, respectively. The median nerve supplies most of the flexor muscles on the front of the forearm, most of the short muscles of the thumb, and the skin on the lateral part of the front of the hand.

In injury to the median nerve (see fig. 8–14), anesthesia and proprioceptive loss in the digits impose a severe handicap on the proper use of the hand. In section above the elbow, pronation is lost, flexion and abduction of the hand are impaired, interphalangeal flexion is lost in the lateral two fingers, and thumb movements, especially opposition, are severely impaired. In section at the wrist, a serious condition, anesthesia and impaired thumb movements result. Median nerve lesions are commonly followed by painful disorders (e.g., *causalgia*).

The *ulnar nerve* (C.N.7, 8; T.N.1) (fig. 8–12) arises from the medial, and usually also from the lateral, cord. The ulnar nerve supplies some of the flexor muscles on the front of the forearm, many of the short muscles of the hand, and the skin on the medial part of the front and back of the hand.

In injury to the ulnar nerve, anesthesia and proprioceptive loss occur in the ulnar portion of the hand, including the little and ring fingers. In section above the elbow, adduction of the hand is impaired, the fingers cannot be adducted or abducted, the proximal phalanges cannot be flexed, and the middle and distal phalanges cannot be extended. The result is a "claw hand" (*main en griffe*) (see fig. 8–14), which is the opposite of the Z-position (see fig.

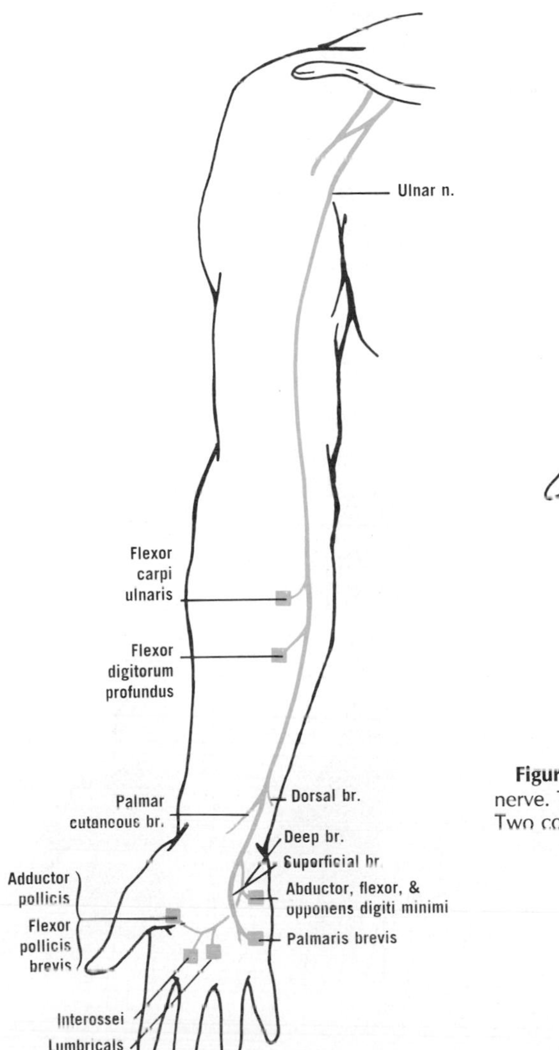

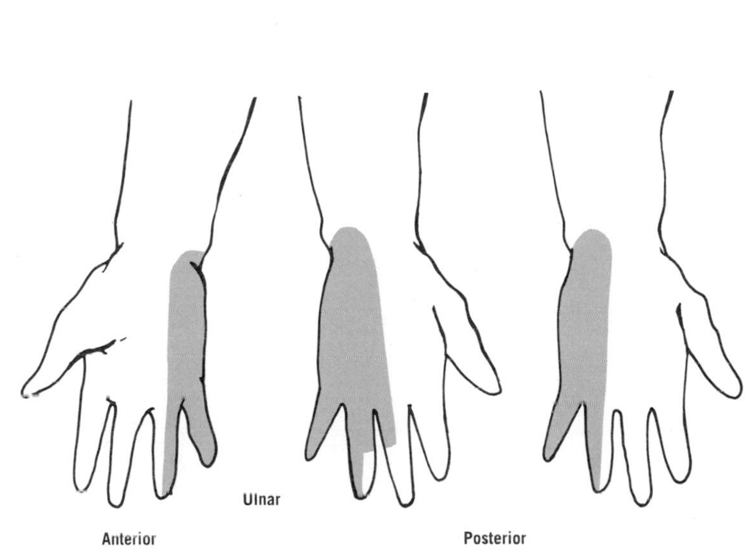

Figure 8–12 Diagram of the muscular and cutaneous branches of the ulnar nerve. The smaller drawing shows the approximate areas of cutaneous innervation. Two common varieties of distribution are presented for the back of the hand.

11–8). In section at the wrist, clawing is more marked. Recovery after ulnar nerve lesions is seldom complete.

The *axillary nerve* (C.N.5, 6) (fig. 8–13) is a branch of the posterior cord. The site of the axillary nerve may be represented by a horizontal line through the middle of the deltoid. At the lower border of the subscapularis, it turns back through the quadrangular space (with the posterior circumflex humeral artery), between the long and lateral heads of the triceps (see fig. 9–2*A*). It divides into anterior and posterior branches. The former winds around the surgical neck of the humerus (the axillary was formerly known as the circumflex nerve) and supplies the deltoid and gives some cutaneous twigs. The posterior branch supplies the teres minor and deltoid and becomes the upper lateral brachial cutaneous nerve.

Section of the axillary nerve results in paralysis of the deltoid (and incomplete abduction of the arm by the supraspinatus) and loss of sensation in a small patch of skin over the deltoid.

The *radial nerve* (C.N.[5,] 6-8; [T.N.1]) (fig. 8–13) may be regarded as the continuation of the posterior cord. It winds across the back of the humerus under cover of the lateral head of

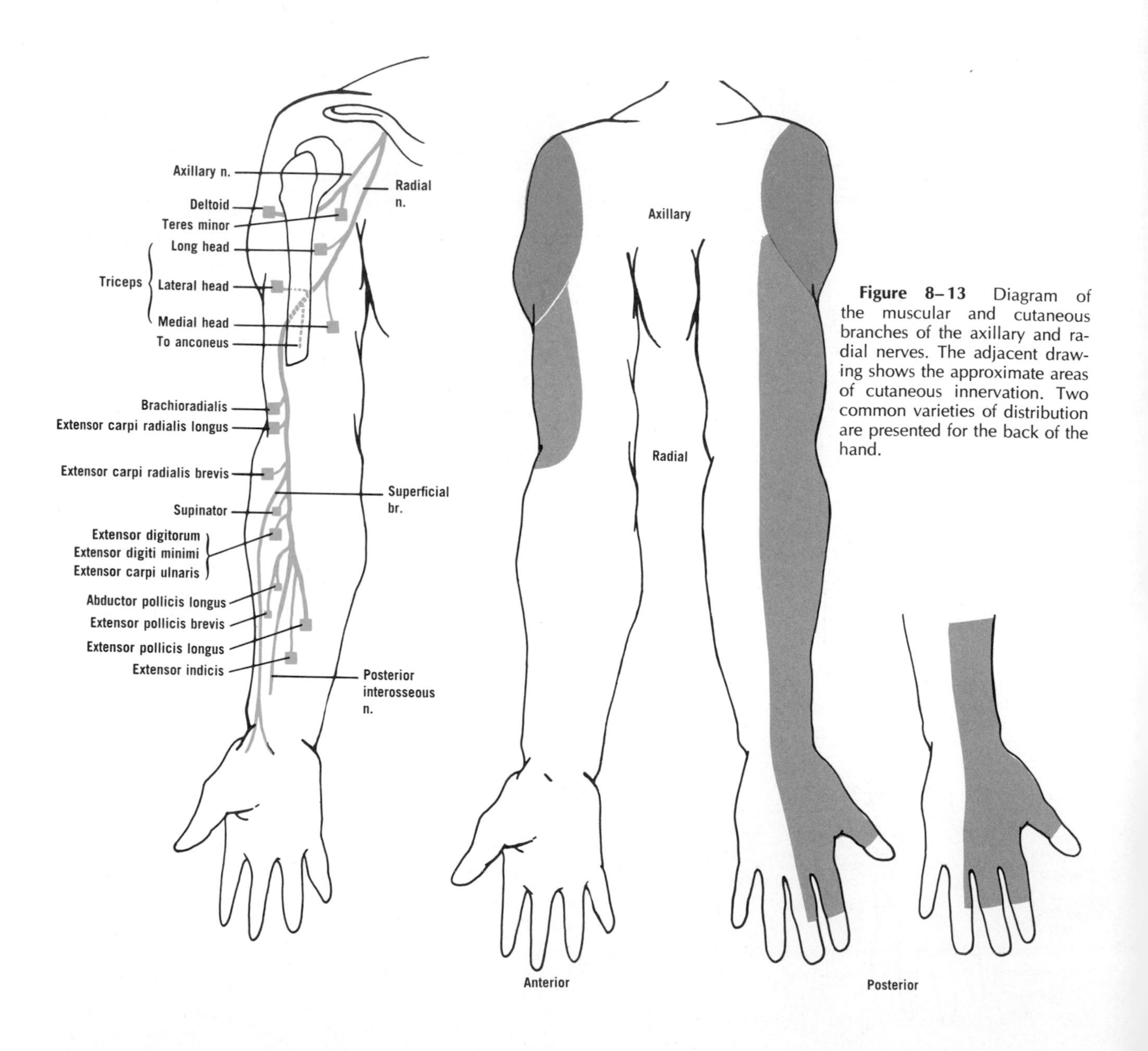

Figure 8–13 Diagram of the muscular and cutaneous branches of the axillary and radial nerves. The adjacent drawing shows the approximate areas of cutaneous innervation. Two common varieties of distribution are presented for the back of the hand.

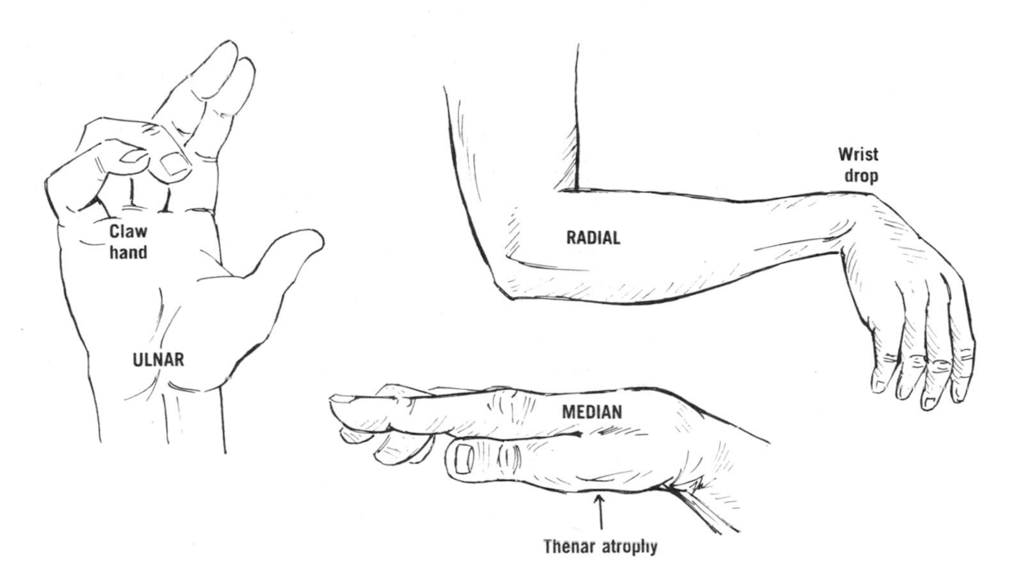

Figure 8–14 Drawings showing the main motor defects after section or other damage to the ulnar, radial, and median nerves.

the triceps (see figs. 6–12 and 9–2) (it was formerly known as the musculospiral nerve) and thereby gains the lateral aspect of the limb (hence the name *radial*). The radial nerve supplies the extensor muscles on the back of the arm and forearm and the skin on the back of the arm, forearm, and hand.

The radial nerve may be injured in the axilla by pressure of a crutch or by hanging the arm over the back of a chair ("Saturday night palsy"). Section of the radial nerve in the axilla causes extensive muscular paralysis, including loss of extension of the forearm, wrist, and proximal phalanges (fig. 8–14). In a lesion at the elbow, extension of the forearm and wrist are preserved; supination may or may not be weakened. Thumb movements are impaired. Anesthesia is slight, owing to overlapping of adjacent nerves. Radial nerve fibers regenerate well.

BLOOD VESSELS

Axillary Artery

The most important function of the main artery carrying blood to the upper limb is the supply of vital centers in the medulla oblongata. The vessel is named successively subclavian, axillary, and brachial (see figs. 7–3 and 8–7). The subclavian artery ascends into the neck to form an arch, which lies on the first rib and extends above the clavicle. **In severe arterial bleeding from the upper limb, the main vessel can be compressed downward against the first rib in the angle between the clavicle and the posterior margin of the sternomastoid.**

At the apex of the axilla, where the subclavian artery reaches the outer border of the first rib (see fig. 8–7), its name is changed to axillary. **The pectoralis minor crosses the axillary artery, and it is customary to consider the vessel in three parts: (1) above, (2) behind, and (3) below that muscle.** The first part is enclosed in fascia (*axillary sheath*) in common with the axillary vein and the brachial plexus (see fig. 8–1). The second part, being covered by the pectoralis minor, is below and medial to the coracoid process. The cords of the plexus are named from the positions that they occupy in relation to the second part. The third part is fairly superficial and is the one used for compression and ligation.

Branches of the Axillary Artery (see fig. 7–3)

The axillary artery provides half a dozen named but quite variable branches. The branches vary in their level of origin, but it is frequently stated that the first part gives one branch, the second part two, and the third part three. The significant point is the existence of an extensive arterial anastomosis around the scapula and on its costal and dorsal surfaces. This anastomosis usually enables a collateral circulation to become established after ligature of the third part of the subclavian or the first part of the axillary artery.

(1) The *highest*, or *superior thoracic, artery* supplies adjacent muscles. (2) The *thoracoacromial artery* divides into branches that ramify on the thoracic wall (supplying the pectoral muscles) and the acromion (hence the

name). (3) The *lateral thoracic artery* descends along the lateral border of the pectoralis minor and gives off mammary branches. (4) The *subscapular artery* descends along the lateral border of the subscapularis, gives off the *circumflex scapular artery* (which passes backward through the triangular space), and then accompanies the thoracodorsal nerve as the *thoracodorsal artery*. (5) The *anterior circumflex humeral artery* is inconstant. (6) The *posterior circumflex humeral artery* passes backward through the quadrangular space with the axillary nerve.

Axillary Vein (see fig. 8–7)

The basilic and brachial veins unite to form the axillary, which accompanies the artery on its medial side. It receives the cephalic vein and commonly the thoraco-epigastric veins (which provide a collateral route in obstruction of the inferior vena cava). At the outer border of the first rib, the axillary continues as the subclavian vein, which does not rise above the clavicle. It unites with the internal jugular to form the brachiocephalic vein.

JOINTS OF SHOULDER

Shoulder Joint (see figs. 6–3, 6–4, and 8–6)

The shoulder, or glenohumeral, joint is a large, freely movable, ball-and-socket articulation between the glenoid cavity of the scapula and the head of the humerus. The joint is strengthened and stabilized by adjacent muscles and tendons, especially by the musculotendinous cuff. The coracoid process, the coraco-acromial ligament, and the acromion form a protective arch above the supraspinatus tendon and the head of the humerus. The arch and the cuff are separated by a large, common subacromial and subdeltoid bursa. The painful arc syndrome has been mentioned already. The shallow glenoid cavity is slightly deepened by a fibrous or fibrocartilaginous lip, the *glenoidal labrum,* attached to its margin. The joint capsule, which is fused to the tendons of the cuff, is attached to the margin of the glenoid cavity and to the anatomical neck of the humerus, except medially, where it extends downward onto the shaft, thereby allowing abduction although permitting downward dislocation. The upper epiphysial line of the humerus is extracapsular, except medially. The capsule is thickened by various ligaments between the two bones. The joint is supplied by adjacent nerves, e.g., from the posterior cord.

The synovial membrane that lines the capsule provides a sheath for the biceps tendon as the latter traverses the joint and descends in the intertubercular groove of the humerus. **The joint cavity usually communicates with the subcoracoid bursa and the subscapular bursa (beneath the subscapularis tendon).** Several other bursae (e.g., the subacromial bursa and its continuation, the subdeltoid bursa) occur around the shoulder but usually do not communicate with the joint.

Sternoclavicular Joint

The sternum and first costal cartilage articulate with the medial end of the clavicle to form, on the basis of the movements that occur, a ball-and-socket joint. The capsule is strengthened by ligaments such as the *costoclavicular*. The articular surfaces are largely fibrocartilaginous and are separated by a fibrous or fibrocartilaginous *articular disc*.

Acromioclavicular Joint

The medial border of the acromion forms a plane joint with the lateral end of the clavicle (see fig. 6–3). The articular surfaces are largely fibrocartilaginous, and the joint may be divided. The *coracoclavicular ligament* (which consists of *conoid* and *trapezoid parts*) extends from the coracoid process to the conoid tubercle and trapezoid line on the inferior surface of the clavicle. This ligament reinforces the acromioclavicular joint.

Scapular Ligaments

The coraco-acromial ligament has already been mentioned. The (superior) *transverse scapular ligament,* which may be ossified, bridges the scapular notch. The suprascapular artery and nerve pass over and under the ligament, respectively. ("Army over and navy under the bridge.")

MOVEMENTS OF SHOULDER

Movements at Shoulder Joint (figs. 6–4 and 8–15)

The movements at a ball-and-socket joint are abduction and adduction, flexion and ex-

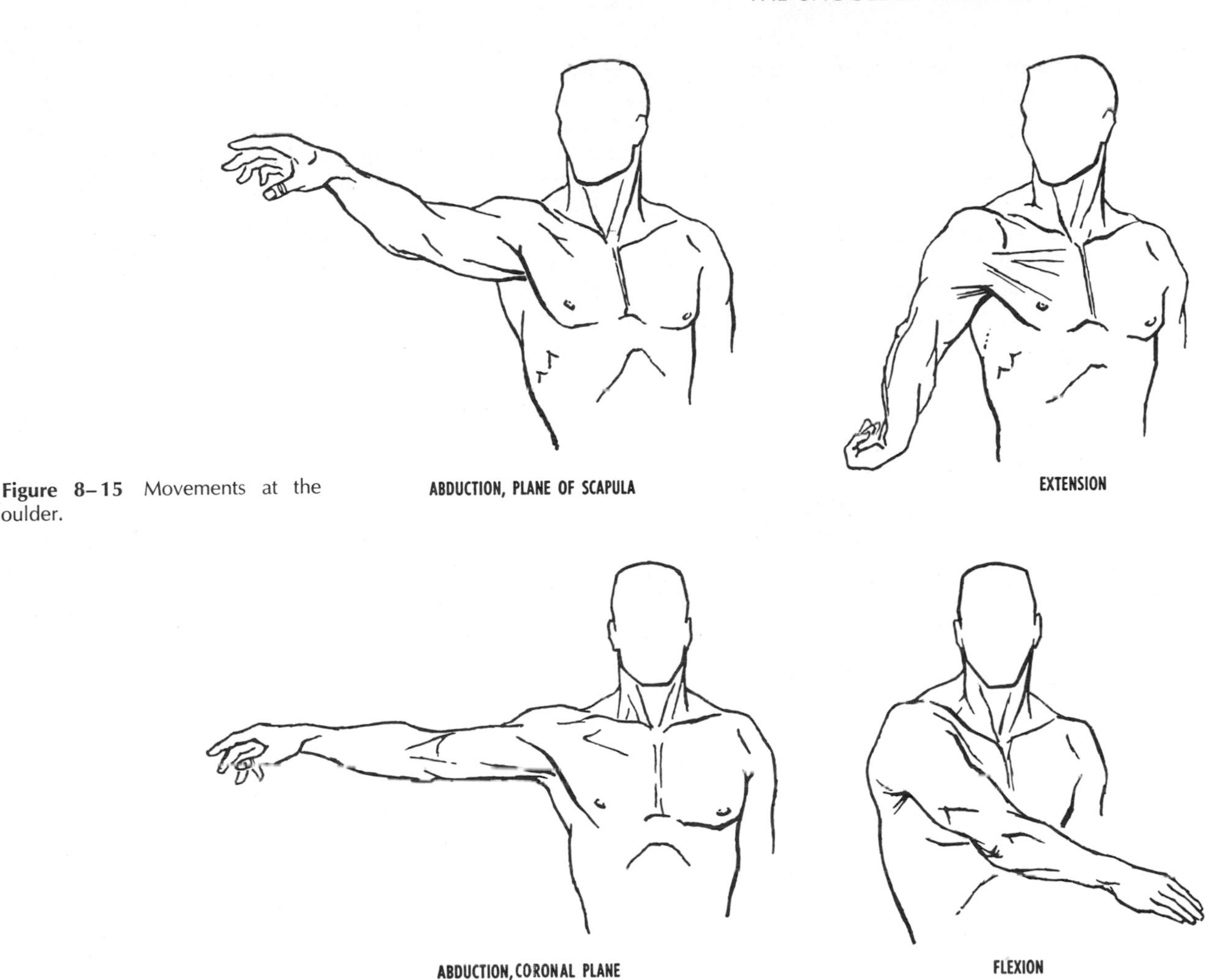

Figure 8–15 Movements at the shoulder.

tension, and rotation and circumduction. **The shoulder has the greatest freedom and range of movement of any joint, owing in large part to the addition of scapular movement. The glenoid cavity and the head of the humerus do not fit snugly, but the ever-present tendency for dislocation during movement is resisted by the muscles of the musculotendinous cuff, which hold the head of the humerus in place.**

The movements described here are those related to the plane of the body of the scapula. Hence abduction is movement laterally and forward, away from the trunk; adduction is the converse. Similarly, in flexion the arm is carried forward and medially across the front of the chest, and in extension it moves backward and laterally away from the chest. Lateral movement in a strictly coronal plane includes extension and lateral rotation as well as abduction. If the arm is fixed, as in climbing, movements of the scapula on the humerus occur.

Abduction is performed by the deltoid and supraspinatus, and, at least in many patients, this movement is impaired if either muscle is paralyzed. The chief adductors against resistance are the pectoralis major, latissimus dorsi, and teres major. The main flexors are the pectoralis major, anterior part of the deltoid, and biceps. The extensors are the latissimus dorsi and posterior part of deltoid. The subscapularis rotates the humerus medially, and the infraspinatus and teres minor rotate it laterally. Abduction and lateral rotation are mainly under the control of the C5 segment of the spinal cord, and adduction and medial rotation are controlled by C6-8 segments.

Movements of Shoulder Girdle

Little if any motion occurs at the shoulder joint without accompanying movement or displacement of the rest of the shoulder girdle.

Displacements of the scapula include (1) elevation and depression, (2) rotation, (3) lateral or forward movement, and (4) medial or backward movement. During movement, the acromion is kept away from the chest wall by the clavicle. The lateral end of the clavicle travels in an arc around the sternoclavicular joint, which acts as a pivot. The medial border of the scapula is held against the chest and travels in the arc of the chest wall. Movement at the acromioclavicular joint is continuous, and the clavicle must be free to rotate around its longitudinal axis.

Elevation of the scapula, as in shrugging the shoulders, is produced by the upper part of the trapezius and the levator scapulae. Depression is usually left to gravity. Forward movement of the scapula along the chest wall, as in pushing, is produced by the serratus anterior. Backward movement, as in bracing the shoulders, is caused by the trapezius muscles and the rhomboids. The serratus anterior rotates the scapula so that its inferior angle moves laterally, and this movement generally accompanies elevation of the arm. The opposite rotation can be carried out against resistance by the levator scapulae and rhomboids.

A movement such as elevation of the arm is highly complicated and requires precise integration. The deltoid and supraspinatus begin abduction, then the scapula commences to rotate. The inferior angle of the scapula travels laterally (serratus anterior), and the lateral angle proceeds upward and medially (trapezius). The clavicle moves also. The humerus can be elevated on the scapula to about 120 degrees, and the scapula on the chest about 60 degrees, so that the combined movements allow elevation of the arm to a fully vertical position.

Injury to the accessory nerve interferes with shrugging of the shoulder. Damage to the long thoracic nerve (which supplies the serratus anterior) results in "winging" (displacement of the medial border) of the scapula during movements of the shoulder girdle.

QUESTIONS

8–1 Name important muscles involved in (a) throwing and pushing, (b) climbing and hammering, (c) drawing a line across a blackboard.

8–2 What is "winging of the scapula" and how is it produced?

8–3 What is the chief effect of paralysis of the trapezius?

8–4 What gives roundness to the shoulder?

8–5 Which structures form the musculotendinous cuff and what is its clinical importance?

8–6 On what does the lower trunk of the brachial plexus lie?

8–7 Why is the upper limb in medial rotation ("waiter's-tip hand") after an "upper type" (Erb-Duchenne) injury to the brachial plexus?

8–8 Why may "claw hand" follow a "lower type" (Klumpke-Déjerine) injury to the brachial plexus?

8–9 What are the results of deltoid paralysis?

8–10 Which bursae usually communicate with the cavity of the shoulder joint?

8–11 What normally resists the tendency of the shoulder joint to become dislocated?

9 THE ARM AND ELBOW

MUSCLES OF ARM (table 9–1)

The muscles of the front of the arm are the biceps, coracobrachialis, and brachialis. They are supplied by the musculocutaneous nerve. The triceps is the muscle of the back of the arm, and it is supplied by the radial nerve. The muscles of the front and back are separated from each other by lateral and medial *intermuscular septa* (fig. 9–1).

The *biceps brachii* arises from the scapula by two heads. The *long,* or *lateral, head,* from the supraglenoid tubercle, descends within the capsule of the shoulder joint and lies in the intertubercular groove. The *short,* or *medial, head* arises from the coracoid process in common with the coracobrachialis. The bicipital tendons give way to muscular bellies that unite and continue as a readily palpable tendon. The insertion is into the tuberosity of the radius (its posterior part; hence the biceps can act as a supinator) and the fascia of the forearm (and ultimately the ulna) by means of the *bicipital aponeurosis* (see fig. 9–4). The biceps and brachialis are the chief flexors of the forearm. The origin of the brachialis embraces the insertion of the deltoid. The coracobrachialis is generally pierced by the musculocutaneous nerve.

The flexion of the forearm (or the twitch of the muscle without movement) that follows tapping of the tendon of insertion of the biceps is known as the biceps jerk. The reflex center is in segments C5 and 6 of the spinal cord.

The *triceps brachii* forms the bulk of the back of the arm. Its three heads are arranged

TABLE 9–1 MUSCLES OF ARM

Muscle	Origin	Insertion	Innervation	Action
Biceps brachii	Supraglenoid tubercle of scapula (long, or lateral, head) & tip of coracoid process (short, or medial, head)	Tuberosity of radius & fascia of forearm & ulna (by bicipital aponeurosis)	Musculocutaneous	Flexes & supinates forearm
Coracobrachialis	Tip of coracoid process	Medial border of humerus	Musculocutaneous	Assists in flexing arm
Brachialis	Anterolateral & anteromedial surfaces of humerus	Front of coronoid process & tuberosity of ulna	Musculocutaneous	Flexes forearm
Triceps brachii	Infraglenoid tubercle of scapula (long head) & posterior surface of humerus (lateral & medial heads)	Upper surface of olecranon & fascia of forearm	Radial	Extends forearm

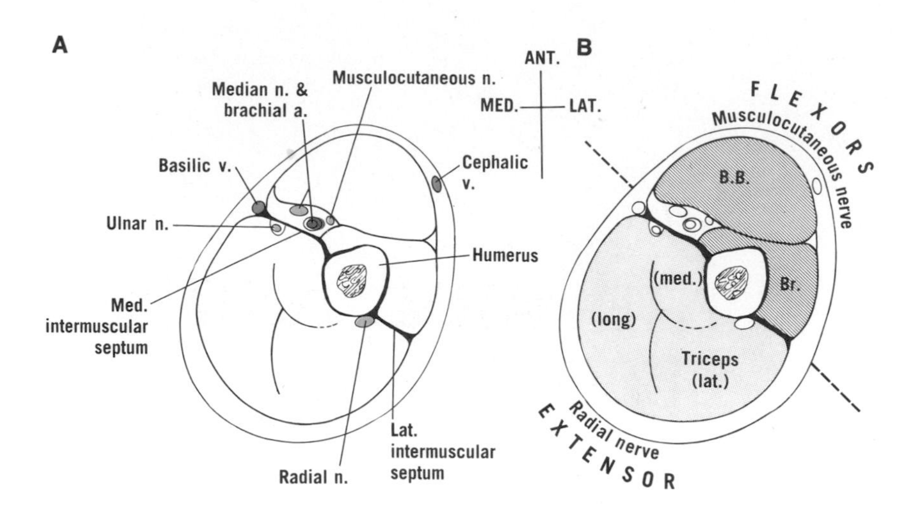

Figure 9–1 Horizontal section through the middle of the arm. In **A**, the nerves and vessels are identified. **B** shows the flexor muscles (supplied by the musculocutaneous nerve) anteriorly and the extensors (supplied by the radial nerve) posteriorly. *B.B.*, biceps brachii; *Br.*, brachialis. The three heads of the triceps are indicated.

in two planes (fig. 9–2): the *long* and *lateral heads* occupy a superficial plane, whereas the *medial head* is deeper. The radial nerve passes between the long and medial heads and then lies on the humerus under cover of the lateral head. The long head separates the triangular from the quadrangular space and the teres major from the teres minor. The triceps is the extensor of the forearm and takes part in pushing, throwing, hammering, and shoveling. A subcutaneous bursa over the olecranon and tendon of the triceps may become thickened ("miner's elbow").

The extension of the forearm (or the twitch without movement) that follows tapping of the tendon of insertion of the triceps is known as the triceps jerk. The reflex center is in segments C6 and 7 of the spinal cord.

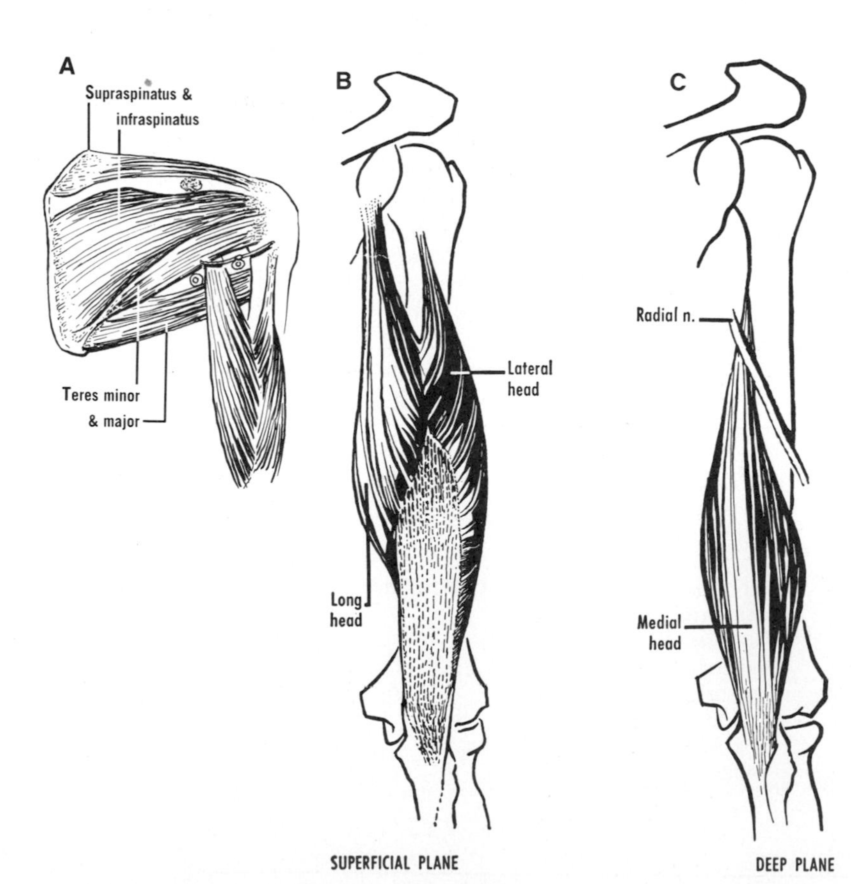

Figure 9–2 A, The triangular and quadrangular spaces, separated by the long head of the triceps. The subscapular and posterior circumflex humeral arteries, respectively, pass through these spaces, and the axillary nerve traverses the quadrangular space. The spine of the scapula has been cut. **B** and **C** show the superficial and deep planes of the triceps.

NERVES OF ARM

The muscles on the front of the arm are supplied by the musculocutaneous nerve; the triceps is supplied by the radial nerve (see fig. 9–1).

The *musculocutaneous nerve* (from the lateral cord) usually pierces the coracobrachialis and descends between the biceps and brachialis (see fig. 9–1) to gain the lateral side of the arm. **The musculocutaneous nerve supplies the biceps, coracobrachialis, brachialis, and elbow joint and becomes the lateral antebrachial cutaneous nerve.**

The *radial nerve* (the continuation of the posterior cord) dips backward with the profunda brachii artery, winds around the humerus (fig. 9–2) under cover of the long and lateral heads of the triceps, and comes forward into the cubital fossa, where it lies in a deep groove between the brachioradialis and the brachialis (figs. 9–3 and 9–4). **At or below the level of the lateral epicondyle, the radial nerve divides into superficial and deep branches. The nerve gives a number of cutaneous branches (posterior brachial, lower lateral brachial, posterior antebrachial) and muscular twigs to the triceps, anconeus, brachialis, bra-**

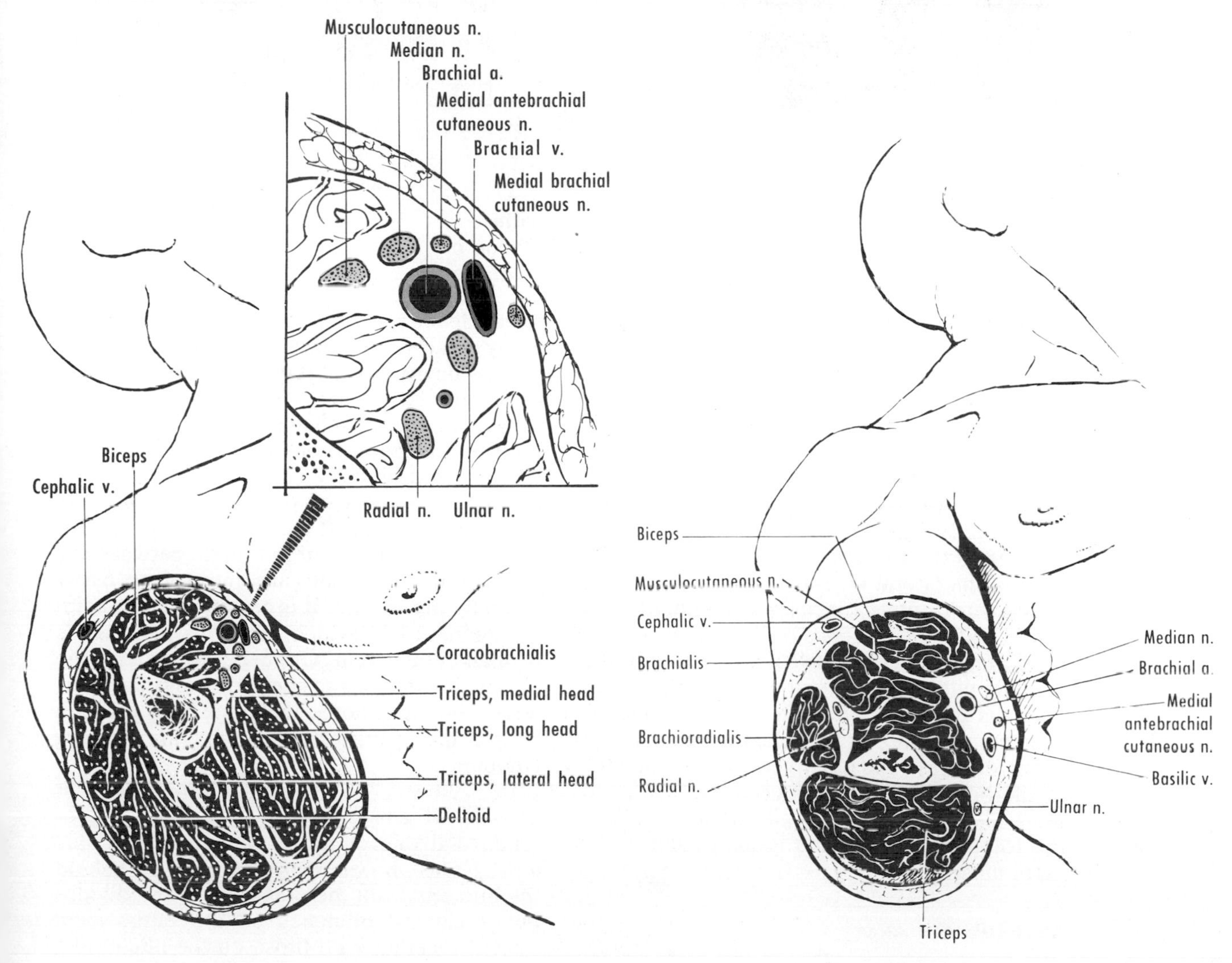

Figure 9–3 Horizontal sections through the upper and lower parts of the arm.

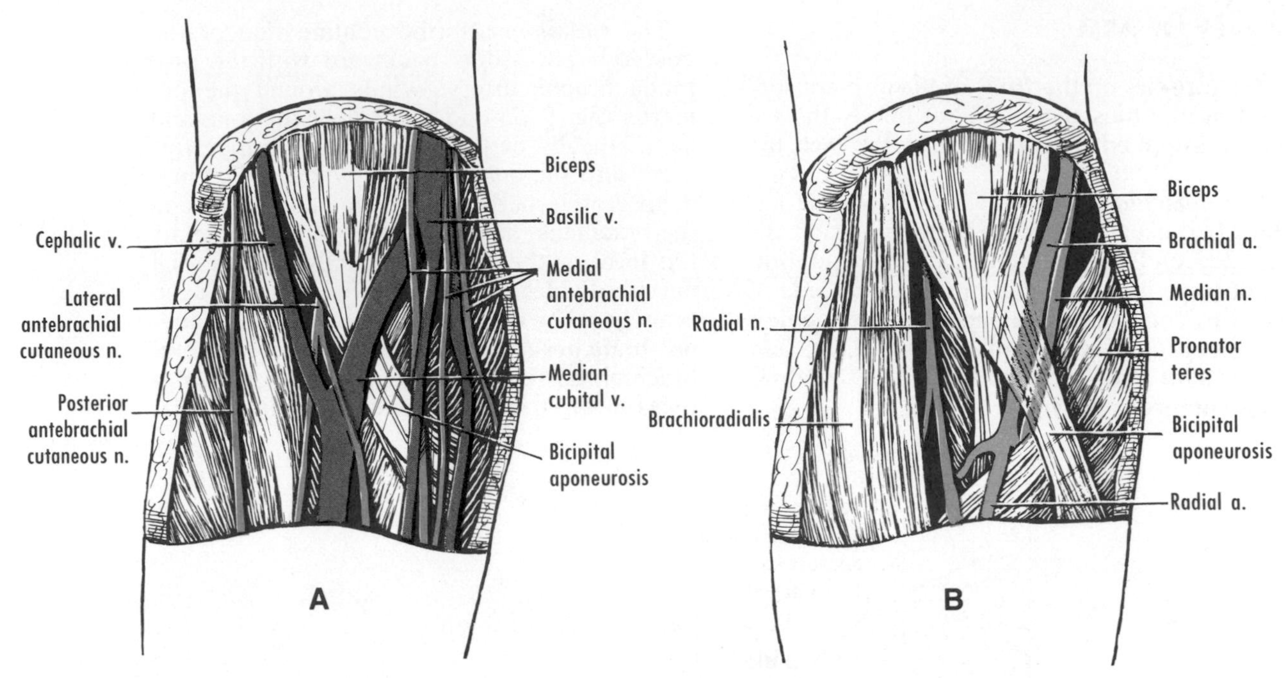

Figure 9–4 The cubital fossa. **A**, The superficial nerves and veins. **B**, The contents of the cubital fossa.

chioradialis, extensor carpi radialis longus and usually brevis, and branches to the elbow joint. The superficial branch is the direct continuation of the radial nerve into the forearm. The deep branch winds laterally around the radius between the layers of the supinator and continues as the posterior interosseous nerve to the muscles on the back of the forearm.

The *median* and *ulnar nerves* are merely in passage to the forearm. The median nerve, at first on the lateral side of the brachial artery, crosses in front of (occasionally behind) that vessel and then descends on its medial side. In the cubital fossa (fig. 9–4*B*), it lies deep to the median cubital vein and the bicipital aponeurosis and gives a branch to the elbow joint. It enters the forearm between the heads of the pronator teres. The ulnar nerve, at first medial to the brachial artery, proceeds backward to the back of the medial epicondyle. It enters the forearm between the heads of the flexor carpi ulnaris.

ARTERIES OF ARM

At the lower border of the teres major, i.e., the distal limit of the posterior axillary fold, the axillary becomes the brachial artery (figs. 7–3 and 9–4). **The brachial artery lies superficially on the medial side of the arm, at first medial to the humerus and then in front of the bone. Hence the artery can be compressed laterally against the humerus above and posteriorly below. The pulsations of the artery can be felt where it is partly overlapped by the biceps and coracobrachialis above. This vessel is used in sphygmomanometry.** The brachial artery lies successively on the triceps and brachialis and is crossed (usually anteriorly) by the median nerve. At the elbow the artery lies in the middle of the cubital fossa, between the biceps tendon laterally and the median nerve medially (fig. 9–4*B*). It is crossed by the bicipital aponeurosis, which separates it from the median cubital vein. Opposite the neck of the radius, it divides into the radial and ulnar arteries, but variations, including high division (in the upper third of the arm), are not uncommon.

In addition to muscular and nutrient branches, the brachial artery gives off the profunda and one or more *ulnar collateral arteries*. The *profunda brachii* crosses behind the humerus with the radial nerve and divides into collateral branches on the radial side. These, together with those on the ulnar side, form an extensive anastomosis around the elbow joint, which is completed below by recurrent branches derived from the radial and ulnar arteries. The existence of the anastomosis is important; the details of the individual branches are not.

CUBITAL FOSSA (fig. 9–4)

The brachioradialis laterally and the pronator teres medially form a V-shaped interval known as the cubital fossa. These muscles, which belong to the forearm, arise from the lateral and medial supracondylar ridges of the humerus, respectively, and descend to the radius. The floor of the fossa is formed by the brachialis and by the supinator laterally. The contents include the biceps tendon, brachial artery, and median nerve, from lateral to medial. The artery usually divides at the apex of the fossa. The radial nerve is in a deep groove between the brachioradialis and brachialis. The fossa is roofed by fascia, strengthened by the bicipital aponeurosis, and crossed by the superficial veins. **The median cubital vein is frequently used for intravenous injections and blood transfusions. Its close relationship to the underlying brachial artery and median nerve should be kept in mind.**

Elbow Joint. **The humerus, radius, and ulna form a hinge joint,** situated 2 or 3 cm below the epicondyles (figs. 6–16, 9–5, and 9–6). The capitulum of the humerus articulates with the upper aspect of the head of the radius (humeroradial joint), and the trochlea of the humerus articulates with the trochlear notch of the ulna (humero-ulnar joint). These two parts of the elbow joint are continuous with each other and share a common cavity with the proximal radio-ulnar joint. Effusions of the elbow joint generally occur posteriorly, as do dislocations, and it is from this aspect that the joint is most easily approached surgically.

The capsule is weak in front and behind but is strengthened on each side by ligaments. The *radial collateral ligament* extends fanwise from the lateral epicondyle to the annular ligament. The *ulnar collateral ligament* runs from the medial epicondyle to the coronoid process and the olecranon. The elbow joint is supplied by adjacent nerves.

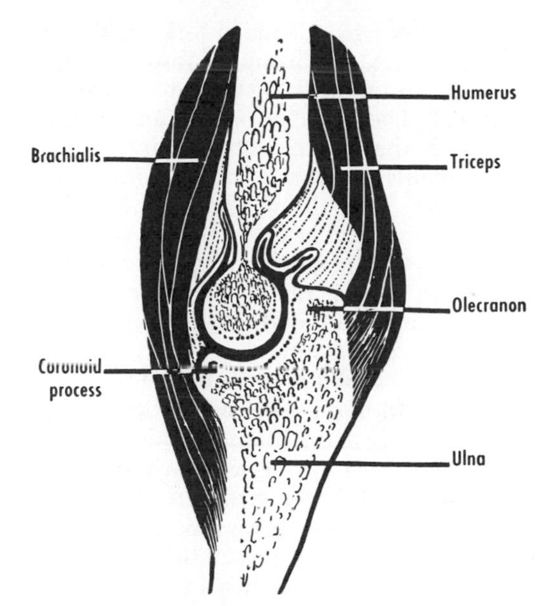

Figure 9–5 Schematic sagittal section through the humero-ulnar part of the elbow joint. The width of the joint cavity is exaggerated.

Proximal Radio-ulnar Joint. **The circumference of the head of the radius fits into the radial notch of the ulna to form a pivot joint.** It is surrounded by the strong *annular ligament,* which is attached to the anterior and posterior margins of the notch. **In children, the head of the radius may be subluxated through the an-**

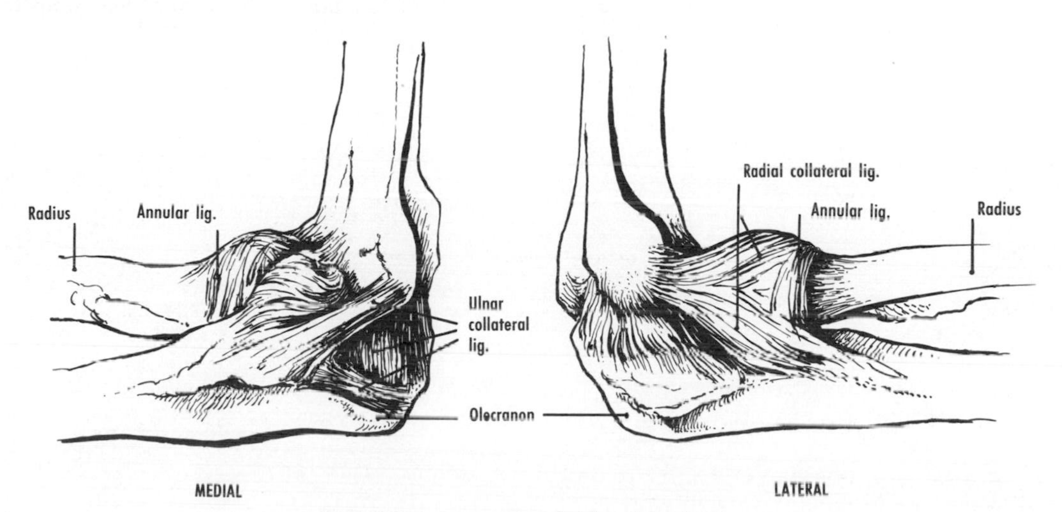

Figure 9–6 Ligaments of right elbow joint.

nular ligament by a sudden jerk on the limb. Some fibers extend from the lower margin of the notch to the neck of the radius (*quadrate ligament*). The synovial membrane is continuous with that of the elbow joint.

Movements at Elbow and Proximal Radioulnar Joints. The elbow is a hinge joint; hence voluntary movement is limited to flexion and extension. Owing to the shape of the medial part of the trochlea, however, the supinated forearm makes a "carrying angle" with the arm during extension.

The term "supination" is used for the position of the forearm and hand when the palm faces forward, as in the anatomical position. The term "pronation" is used when the palm faces backward. These terms are used also for the movements that bring about these positions. Supination and pronation are considerably more complicated than a mere lateral and medial rotation. The axis of movement extends from the middle of the head of the radius to the lower end of the ulna. Although the head of the radius merely rotates within the annular ligament, its lower end describes an arc around the lower end of the ulna and carries the hand with it. In pronation, the shafts of the radius and ulna cross each other (the main reason for selecting supination to be the anatomical position). The movements are usually accompanied by rotation of the humerus (minimal when the elbow is flexed), and the ulna does not remain fixed. Supination has been thought to be generally stronger than pronation, and the threads of screws are arranged to take advantage of this (for right-handed people).

Flexion is controlled by segments C5 and 6 of the spinal cord, pronation and supination by C6, and extension by C7 and 8. The flexors of the forearm are the brachialis, biceps, and brachioradialis. The extensor is the triceps, particularly the medial head. The pronators are the pronator quadratus and the pronator teres. The supinators are the supinator and the biceps.

QUESTIONS

9–1 What are the two main compartments of the arm?

9–2 What are the chief actions of the biceps brachii?

9–3 Which spinal segments are associated with (a) flexion and (b) extension at the elbow?

9–4 Where is the ulnar nerve in contact with the humerus?

9–5 Between which muscles, or heads of muscles, do the radial, median, and ulnar nerves enter the forearm?

9–6 Which important structures lie medial to the biceps tendon in the cubital fossa?

9–7 Under which conditions is subluxation of the head of the radius found?

9–8 Look up the origin of the word *cubital*.

10
THE FOREARM

The muscles of the forearm consist of (1) an anterior group, the flexors of the wrist and fingers and the pronators, and (2) a posterior group, the extensors of the wrist and fingers and the supinator (fig. 10–1).

MUSCLES OF FRONT OF FOREARM
(tables 10–1 and 10–2)

Five superficial and three deep muscles occupy the front of the forearm. The superficial group arises mostly from the front of the medial epicondyle of the humerus by a common tendon, and from adjacent fascia, and is supplied chiefly by the median nerve (fig. 10–2). The deep group is supplied mostly by the anterior interosseous nerve, a branch of the median. Those muscles in both groups that are not innervated by the median are supplied by the ulnar nerve. The tendons of the flexor carpi radialis, palmaris longus, and flexor carpi ulnaris are readily palpable. The *palmaris longus,* however, is often absent. The *flexor digitorum superficialis* is more deeply placed than the other superficial muscles. In a finger, as in the palm, the superficialis and profundus tendons are enclosed in a common synovial sheath. Opposite the proximal phalanx, the superficialis splits into two slips that embrace the profundus tendon and then reunite behind it and are inserted into the middle phalanx. The profundus continues to the distal phalanx. Both tendons are anchored to the phalanges and interphalangeal joints by fibrous bands termed *vincula,* which carry the blood supply to the tendons.

Injury to the brachial artery near the elbow (e.g., from a supracondylar fracture of the humerus) may result in damage to the deep flexors (pollicis longus and digitorum profundus) and flexion deformity of the wrist and fingers (Volkmann's ischemic contracture).

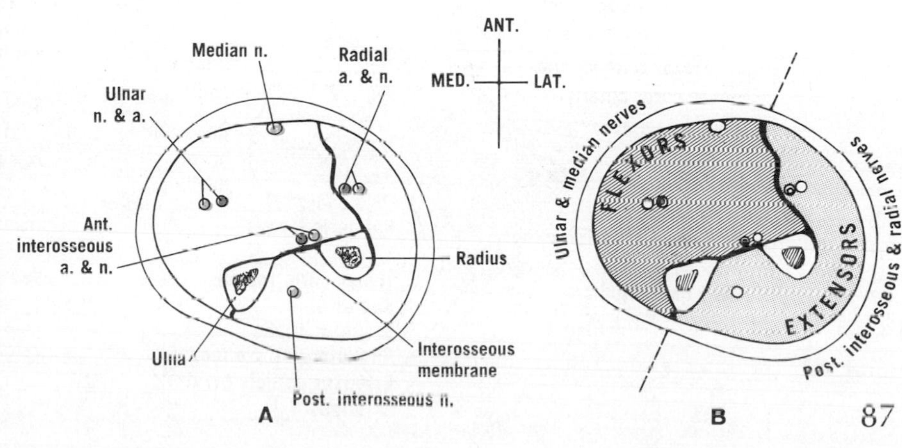

Figure 10–1 Horizontal section through the middle of the forearm. In **A**, the nerves and vessels are identified. **B** shows the flexor muscles (supplied by the ulnar but mainly by the median nerve) anteriorly and the extensors (supplied by the radial and posterior interosseous nerves) posteriorly. The individual muscles are not shown. Note that the flexors and extensors are separated by the subcutaneous posterior border of the ulna.

TABLE 10–1 SUPERFICIAL MUSCLES OF FRONT OF FOREARM

Muscle	Origin	Insertion	Innervation	Action
Pronator teres	Medial supracondylar ridge & coronoid process of ulna (deep head)	Lateral surface of radius	Median	Pronates & flexes forearm
Flexor carpi radialis	Medial epicondyle	Bases of 2nd & 3rd metacarpals	Median	Flexes & abducts hand
Palmaris longus	Medial epicondyle	Flexor retinaculum & palmar aponeurosis	Median	Probably tenses palmar aponeurosis
Flexor carpi ulnaris	Medial epicondyle & olecranon (2nd head)	Pisiform (& hamate & 5th metacarpal)	Ulnar	Flexes & adducts hand
Flexor digitorum superficialis	Medial epicondyle & anterior border of radius (2nd head)	Middle phalanges of fingers 2-5	Median	Flexes middle phalanges

TABLE 10–2 DEEP MUSCLES OF FRONT OF FOREARM

Muscle	Origin	Insertion	Innervation	Action
Flexor digitorum profundus*	Front of ulna & interosseous membrane	Distal phalanges of fingers 2-5	Anterior interosseous	Flexes distal phalanges
Flexor pollicis longus	Front of radius & interosseous membrane	Distal phalanx of thumb	Anterior interosseous	Flexes distal phalanx
Pronator quadratus	Front of distal part of ulna	Front of distal part of radius	Anterior interosseous	Pronates forearm

* Innervated by ulnar nerve also.

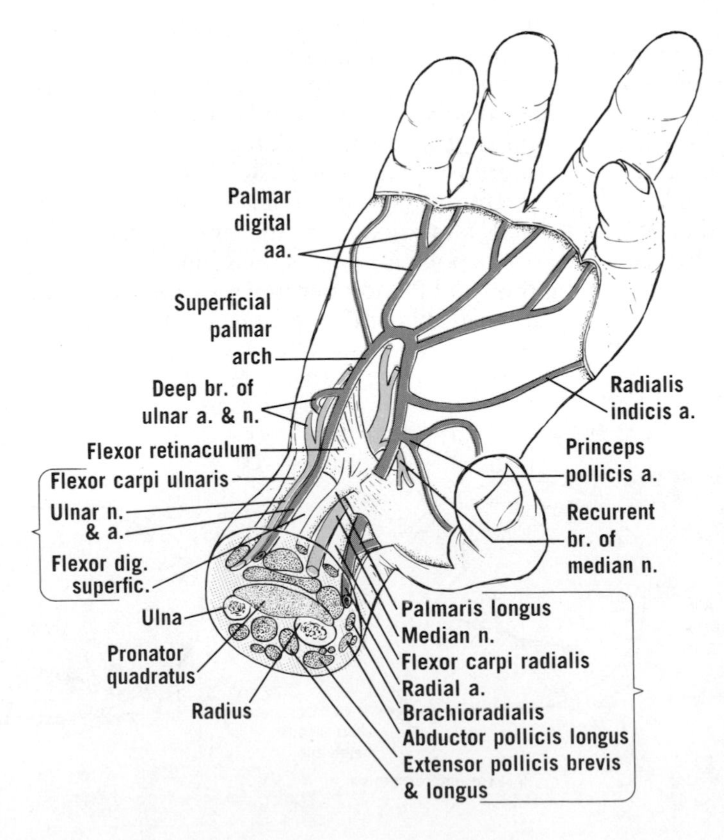

Figure 10–2 The superficial palmar arch (see also fig. 11–10), the flexor retinaculum, and the structures on the front of the wrist (see also fig. 10–5). The palmar aponeurosis has been removed. Note the median nerve passing deep to the flexor retinaculum (within the carpal tunnel) and, immediately on emerging, giving off its very important recurrent branch to the thenar muscles. (Based partly on Spalteholz.)

MUSCLES OF BACK OF FOREARM
(tables 10–3 and 10–4)

Seven superficial and five deep muscles occupy the back of the forearm. The superficial group arises mostly from the back of the lateral epicondyle of the humerus by a common tendon. The muscles are supplied by (1) the radial nerve or (2) its deep branch, which continues as (3) the posterior interosseous nerve.

The *brachioradialis* can be seen and felt on the lateral side of the elbow and forearm when these are flexed.

The insertion of the *extensor carpi radialis brevis* into the base of the third metacarpal is a common site for the development of a "ganglion." The term *ganglion* is used here for a cystic swelling adjacent to, and often communicating with, a synovial sheath or joint. A common site is the dorsum of the wrist.

On the back of the hand, the extensor tendons diverge but are connected by a variable arrangement of bands (see fig. 11–7). A fibrous sheet on the back of each finger is known as the *extensor expansion,* or *dorsal aponeurosis,* and it contains a hood of transverse fibers. The expansion is penetrated by the extensor tendon, which then divides into three slips: a central slip to the base of the middle phalanx and two collateral bands, which, fused with expansions from the interossei and lumbricals, unite and proceed to the base of the distal phalanx.

The *supinator* is arranged in two layers, separated by the deep branch of the radial nerve, which sometimes comes into direct contact with the radius. The nerve meets the posterior interosseous vessels at the lower border of the supinator.

The tendon of the *extensor pollicis longus* lies in a groove on the medial aspect of the dorsal tubercle of the radius. Below this, digi-

TABLE 10–3 SUPERFICIAL MUSCLES OF BACK OF FOREARM

Muscle	Origin	Insertion	Innervation	Action
Brachioradialis	Lateral supracondylar ridge	Lateral surface of distal part of radius	Radial	Flexes forearm
Extensor carpi radialis longus	Lateral supracondylar ridge	Base of 2nd metacarpal	Radial	Extend & abduct hand
Extensor carpi radialis brevis	Lateral epicondyle	Bases of 2nd & 3rd metacarpals	Radial / Deep branch of radial	Extend & abduct hand
Extensor digitorum	Lateral epicondyle	Middle & distal phalanges of fingers 2-5	Deep branch of radial	Extends proximal phalanges
Extensor digiti minimi	Lateral epicondyle	Extensor expansion of 5th finger	Deep branch of radial	Extends proximal phalanx
Extensor carpi ulnaris	Lateral epicondyle & posterior border of ulna	Base of 5th metacarpal	Deep branch of radial	Extends & adducts hand
Anconeus	Lateral epicondyle	Olecranon	Radial	Stabilizes elbow joint

TABLE 10–4 DEEP MUSCLES OF BACK OF FOREARM

Muscle	Origin	Insertion	Innervation	Action
Supinator	Lateral epicondyle chiefly	Upper third of radius	Deep branch of radial	Supinates forearm
Abductor pollicis longus	Interosseous membrane, radius, & ulna	Base of 1st metacarpal laterally	Posterior interosseous	Abducts 1st metacarpal
Extensor pollicis brevis	Interosseous membrane & radius	Proximal phalanx of thumb	Posterior interosseous	Extends thumb
Extensor pollicis longus	Interosseous membrane & ulna	Distal phalanx of thumb	Posterior interosseous	Extends distal phalanx
Extensor indicis	Interosseous membrane & ulna	Extensor expansion of index finger	Posterior interosseous	Assists in extension

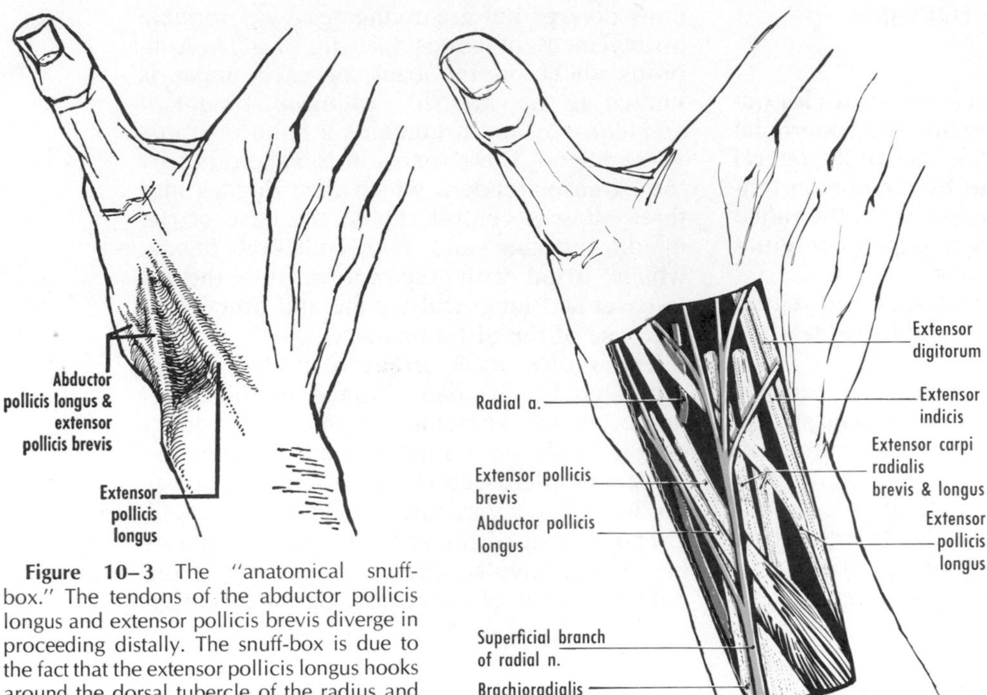

Figure 10–3 The "anatomical snuff-box." The tendons of the abductor pollicis longus and extensor pollicis brevis diverge in proceeding distally. The snuff-box is due to the fact that the extensor pollicis longus hooks around the dorsal tubercle of the radius and hence is separated from the extensor pollicis brevis.

Figure 10–4 The "anatomical snuff-box." The radial artery lies on the floor of the snuff-box, which is formed by the scaphoid and trapezium.

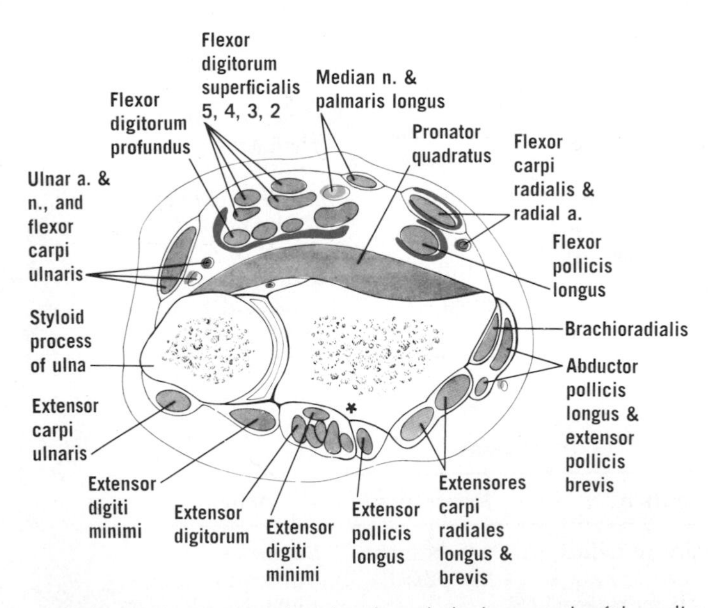

Figure 10–5 Horizontal section through the lower ends of the radius and ulna. Three synovial sheaths are shown anteriorly, but the six sheaths on the posterior aspect (see fig. 11–3) are not included. The radial artery and the styloid process of the ulna indicate the boundaries between the flexor and extensor muscles. The median nerve is overlapped by the palmaris longus (when present) and is at the lateral edge of the flexor superficialis. The radial artery lies lateral to the flexor carpi radialis. Note the relationship of the extensor pollicis longus to the dorsal tubercle of the radius (*asterisk*). (Modified from Castaing and Soutoul.)

tal twigs of the superficial branch of the radial nerve can be felt crossing the tendon. When the thumb is extended, a hollow known as the *anatomical snuff-box* appears between the tendon of the extensor pollicis longus medially and those of the extensor pollicis brevis and abductor pollicis longus laterally (figs. 10–3 and 10–4). Its floor is formed by the scaphoid and trapezium and is crossed by the radial artery. The hollow is limited above by the styloid process of the radius. The narrow tendon sheath of the abductor pollicis longus and extensor pollicis brevis is liable to inflammation, or stenosing tenosynovitis (de Quervain's disease).

The fascia on the back of the forearm is thickened distally to form the *extensor retinaculum,* which is anchored to the radius and ulna. From the deep aspect of the extensor retinaculum, septa produce half a dozen compartments for the extensor tendons (fig. 10–5). Each compartment contains a synovial sheath.

NERVES OF FOREARM

Radial Nerve (see fig. 8–13). **The radial nerve enters the forearm between the bra-**

chioradialis and brachialis and, at or below the level of the lateral epicondyle, divides into superficial and deep branches.

The superficial branch is the continuation of the radial nerve, and its distribution is cutaneous and articular. It accompanies the radial artery, lying lateral to it, and then winds dorsally deep to the brachioradialis, to become subcutaneous and supply the lateral part of the dorsum of the hand. Its terminal branches usually supply two and a half fingers but generally reach no further than the proximal phalanges of the index and middle fingers. (The dorsal innervation here is completed by digital branches of the median nerve.) Several cutaneous branches of the radial nerve can be felt by stroking a fingernail down the taut tendon of the extensor pollicis longus.

The deep branch of the radial nerve has a muscular and articular distribution. It winds laterally between the superficial and deep layers of the supinator and often makes direct contact with the radius, being vulnerable in fractures. At the lower border of the supinator, it meets the posterior interosseous vessels. For the rest of its course, the deep branch of the radial nerve is then termed the posterior interosseous nerve, a name that is sometimes applied to the entire course of the deep branch. The posterior interosseous nerve reaches the interosseous membrane distally and ends in an enlargement from which twigs are distributed to adjacent joints.

The brachioradialis and extensor carpi radialis longus (and often brevis) are supplied by the radial nerve. Its deep branch supplies the supinator (and often the extensor carpi radialis brevis), extensor digitorum, extensor digiti minimi, and extensor carpi ulnaris. The *posterior interosseous nerve* supplies the abductor pollicis longus, extensor pollicis brevis, extensor pollicis longus, and extensor indicis.

Median Nerve (see fig. 8–11). **The median nerve usually enters the forearm between the two heads of the pronator teres. It extends down the middle of the forearm to the midpoint between the styloid processes.** It is accompanied by the median artery, a branch of the anterior interosseous. The median nerve passes behind the tendinous arch connecting the two heads of the flexor digitorum superficialis and remains under cover of that muscle, adherent to its deep aspect and lying on the flexor digitorum profundus, until it approaches the wrist, where it becomes more superficial. It may give a cutaneous branch to the palm. **The median nerve is found between the flexor digitorum superficialis medially and the flexor carpi radialis laterally, and it may be partly covered by the palmaris longus. It enters the hand by passing through the carpal canal, behind the flexor retinaculum and in front of the flexor tendons.**

In the cubital fossa, branches are given medially to the pronator teres, flexor carpi radialis, palmaris longus, and flexor digitorum superficialis. The *anterior interosseous nerve* arises from the back of the median in the cubital fossa, and, in company with the anterior interosseous artery, descends along the interosseous membrane between the adjacent margins of the flexor pollicis longus and flexor digitorum profundus, both of which it supplies. It then passes behind, and supplies, the pronator quadratus and ends in branches to adjacent joints.

Communications between the median and ulnar nerves in the forearm may be important in that fibers to the small muscles of the hand may, in some people, be transferred from one nerve to the other, resulting in unusual signs in the event of injury.

Ulnar Nerve (see fig. 8–12). The ulnar nerve lies in a groove on the back of the medial epicondyle, where it can be felt against the "funny bone." **The ulnar nerve enters the forearm between the two heads of the flexor carpi ulnaris. Shortly thereafter it meets the ulnar artery, which lies on its lateral side. It then becomes superficial. The ulnar nerve lies between the flexor digitorum superficialis laterally and the flexor carpi ulnaris medially. The ulnar nerve and artery enter the hand by passing in front of the flexor retinaculum, lateral to the pisiform, and between that bone and the hook of the hamate.** The course of the ulnar nerve may be represented by a line from the medial epicondyle to the lateral edge of the pisiform.

Branches are given to the two muscles between which the ulnar nerve lies; the flexor digitorum profundus and the flexor carpi ulnaris. In the middle of the forearm, a large, cutaneous *dorsal branch,* which descends between the ulna and the flexor carpi ulnaris, turns dorsally and is distributed to the medial part of the dorsum of the hand. There may also be a variable *palmar branch.*

ARTERIES OF FOREARM

Radial Artery (see fig. 7–3). The smaller terminal division of the brachial artery begins in the cubital fossa, opposite the neck of the radius. Distally in the forearm, **the radial artery lies lateral to the flexor carpi radialis tendon, which serves as a guide to it. Its pulsations can readily be felt by three fingers here, thus supplying information of clinical importance, e.g., pulse rate and rhythm and compressibility and condition of the arterial wall.** The vessel leaves the forearm by winding dorsally across the carpus.

A *recurrent branch* is given to the anastomosis around the elbow joint, and there are also palmar and carpal branches.

Ulnar Artery (see fig. 7–3). The larger terminal division of the brachial artery begins in the cubital fossa, opposite the neck of the radius. It first runs downward and medially and then directly downward. Distally the ulnar artery and nerve emerge laterally from under cover of the flexor carpi ulnaris, and the pulsations can be felt at the wrist. The ulnar artery leaves the forearm by passing in front of the flexor retinaculum on the lateral side of the pisiform. After it gives off its *deep palmar branch,* it continues as the superficial palmar arch.

A *recurrent branch* is given to the anastomosis around the elbow joint. The *common interosseous artery* divides almost immediately into *anterior* and *posterior interosseous arteries,* which descend on the front and back of the interosseous membrane, respectively, and end dorsally on the carpus. The anterior vessel gives off the *median artery*.

QUESTIONS

10–1 Which type of contracture may result from injury to the brachial artery?

10–2 What are the two main compartments of the forearm?

10–3 What is the chief origin of the superficial extensors?

10–4 What is a "ganglion" in the clinical sense?

10–5 Which important bones form the floor of the anatomical snuff-box?

10–6 What is the characteristic feature of radial nerve palsy?

10–7 Where is the median nerve situated at the wrist?

10–8 Which muscles are supplied by the anterior interosseous nerve?

10–9 What is the clinical significance of communications between the median and ulnar nerves in the forearm?

10–10 What is the guide to the radial artery?

11
THE HAND

The hand is distal to the forearm, and its skeletal framework includes the carpus, or wrist. In lay usage, however, the word "wrist" is used for the distal end of the forearm, a wrist watch being worn over the lower ends of the radius and ulna. The position of the hand at rest is shown in figure 11–1. The hand is of enormous importance, particularly the thumb. Many hand injuries result in permanent disabilities.

The fingers (or digits of the hand) are numbered from one to five, beginning with the thumb. **In clinical notes the fingers should be identified by name rather than by number: thumb (pollex),** and **index, middle, ring,** and **little fingers.** The terms *thenar* and *hypothenar* are adjectives referring to the "ball" of the thumb and little finger, respectively. General features of the hand that should be noted include the lengths of the fingers, the thin and mobile skin of the dorsum (allowing accumulation of fluid subcutaneously), the thick and anchored skin of the palm, the flexure lines associated with cutaneous movement (but which do not necessarily indicate the sites of joints), the fingerprints, and the nails. The subcutaneous fat that constitutes the tips of the fingers is loculated by fibrous septa and occupies a closed *pulp space,* which is liable to infection (*felon,* or *whitlow*) from a penetrating wound followed by the possibility of damage to the distal phalanx owing to involvement of its blood supply.

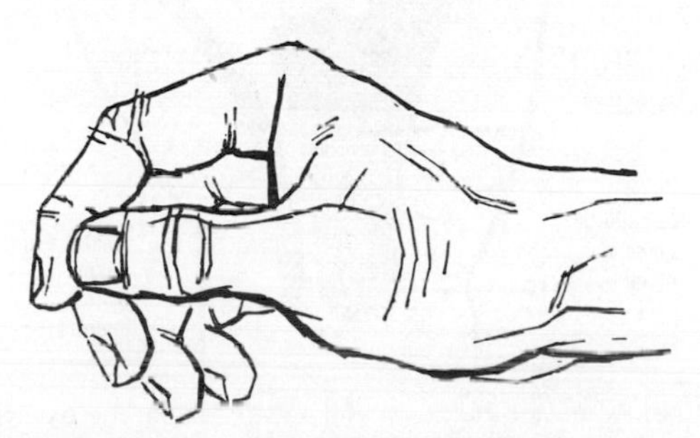

Figure 11–1 The position of the hand at rest.

The major structures that can be identified *in vivo* on the front of the wrist are, from lateral to medial: radial artery (pulse), flexor carpi radialis (medial to which is the median nerve), palmaris longus (when present), flexor digitorum superficialis, ulnar artery (medial to which is the ulnar nerve), and flexor carpi ulnaris (see fig. 10–2).

The flexor retinaculum is a strong, transverse, fibrous band that confines the flexor tendons of the five fingers, together with their synovial sheaths and the median nerve, to the arch of the carpus, which it converts into the carpal canal, or tunnel (figs. 11–2 to 11–4). The retinaculum is situated distal to the flexion creases in the skin of the wrist. It is attached laterally to the tubercles of the scaphoid and trapezium and medially to the triquetrum, pisiform, and the hook of the hamate. At least the proximal points of attachment can be felt *in vivo,* and, because the retinaculum measures about 3 cm proximodistally, the quadrilateral outline of the retinaculum can be mapped on the living hand. Owing to a variety of known and unknown causes, the median nerve may become constricted beneath the flexor retinaculum, resulting in tingling, numbness, and muscular signs (*carpal tunnel syndrome*). The condition is treated by dividing the flexor retinaculum.

The palmar aponeurosis is a strong, triangular membrane overlying the tendons in the palm (fig. 11–2). Its apex is continuous with the palmaris longus (when present) and is an-

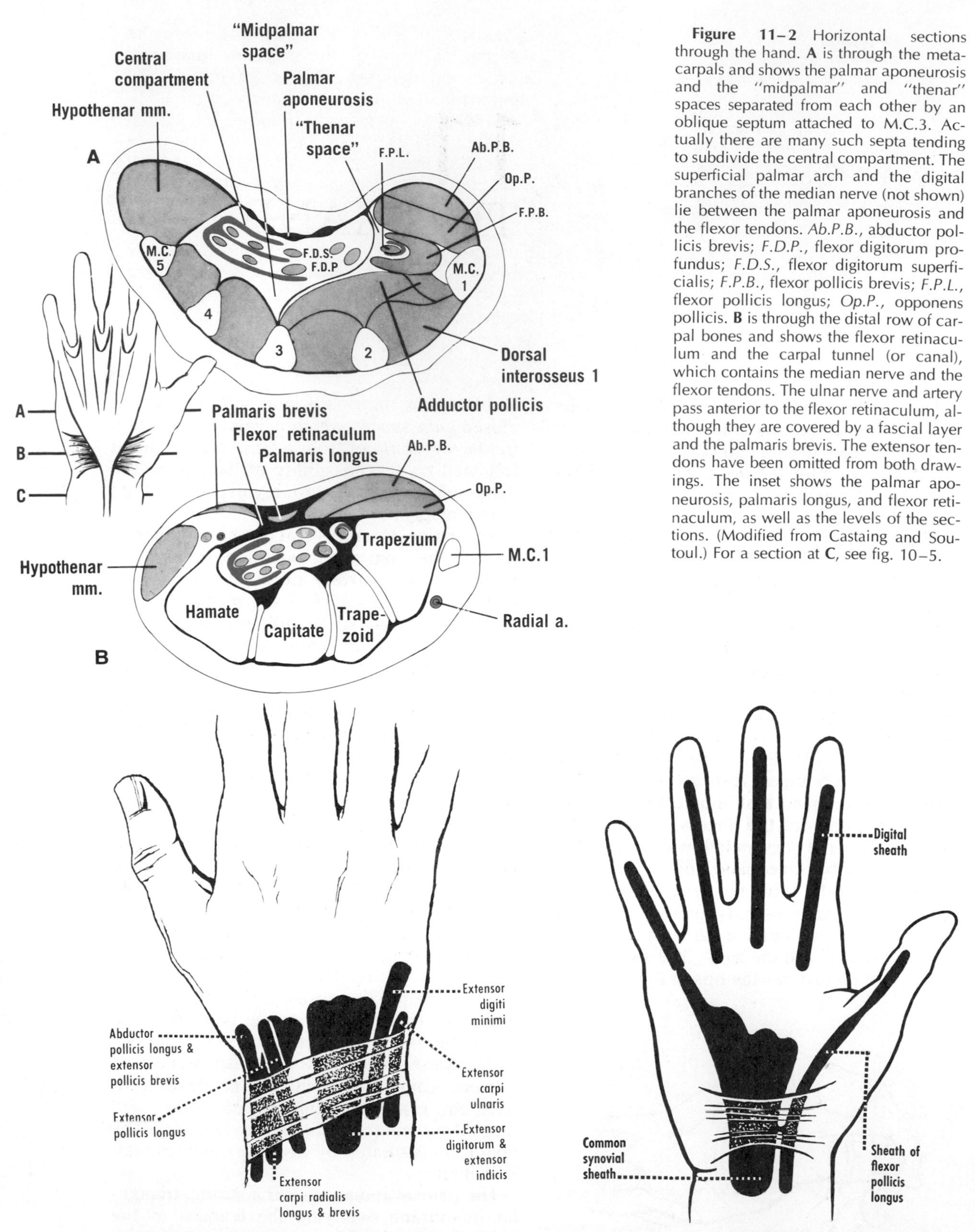

Figure 11–2 Horizontal sections through the hand. **A** is through the metacarpals and shows the palmar aponeurosis and the "midpalmar" and "thenar" spaces separated from each other by an oblique septum attached to M.C.3. Actually there are many such septa tending to subdivide the central compartment. The superficial palmar arch and the digital branches of the median nerve (not shown) lie between the palmar aponeurosis and the flexor tendons. *Ab.P.B.*, abductor pollicis brevis; *F.D.P.*, flexor digitorum profundus; *F.D.S.*, flexor digitorum superficialis; *F.P.B.*, flexor pollicis brevis; *F.P.L.*, flexor pollicis longus; *Op.P.*, opponens pollicis. **B** is through the distal row of carpal bones and shows the flexor retinaculum and the carpal tunnel (or canal), which contains the median nerve and the flexor tendons. The ulnar nerve and artery pass anterior to the flexor retinaculum, although they are covered by a fascial layer and the palmaris brevis. The extensor tendons have been omitted from both drawings. The inset shows the palmar aponeurosis, palmaris longus, and flexor retinaculum, as well as the levels of the sections. (Modified from Castaing and Soutoul.) For a section at **C**, see fig. 10–5.

Figure 11–3 The extensor and flexor tendons, and a common arrangement of their synovial sheaths. Note the extensor and flexor retinacula.

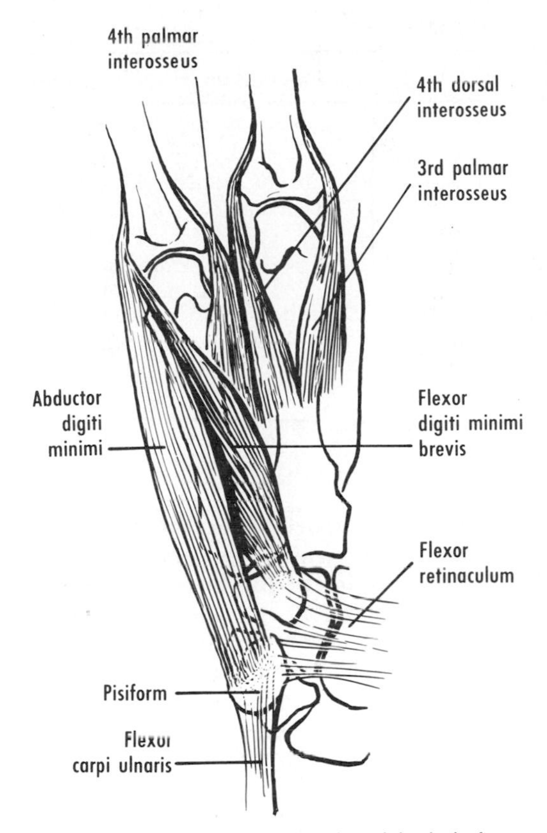

Figure 11–4 The intrinsic muscles of the little finger.

chored to the front of the flexor retinaculum, from which it can be distinguished by the longitudinal direction of its fibers. The aponeurosis continues distally as four slips (pretendinous bands), which overlie the flexor tendons of the medial four fingers. Flexion deformity (*Dupuytren's contracture*) of one or more fingers, especially the little and ring fingers, is commonly attributed to thickening and shortening of the palmar aponeurosis, although its causation is not really understood. Severe instances are usually treated by excision of the aponeurosis.

The flexor tendons leave the carpal canal and enter a central compartment in the palm (fig. 11–2). This zone contains the flexor tendons and their synovial sheaths, the lumbrical muscles, and the digital nerves and vessels. **The central compartment is bounded (1) in front by the deep aspect of the palmar aponeurosis, (2) behind by a fat pad on the interosseous fascia and by the fascia in front of the adductor pollicis muscle, and (3) on the sides by the fascia covering the thenar and hypothenar muscles. Uncontrolled infection (e.g., from a synovial sheath) may rupture into the central compartment and spread proximally into the forearm, in front of the pronator quadratus muscle and its fascia.** Although the central compartment is single proximally, it is subdivided distally by a large number of sagittal septa. In clinical writing, it is customary to select one of these septa (attached to the third metacarpal) for special consideration and to use it as the basis for a description of two fascial "spaces" of the palm. What are termed **the "midpalmar space" medially and the "thenar space" laterally are said to be situated behind the flexor tendons and are regarded as important sites for the accumulation of pus. Recent anatomical investigations, however, do not support the generally accepted clinical anatomy of the palm.***

The flexor tendons, being retained in place by fascial retinacula, are invested by *synovial sheaths,* which facilitate gliding (fig. 11–3). **Three synovial sheaths are found on the front of the wrist: (1) a common synovial flexor sheath envelops all the superficialis and profundus tendons, (2) a sheath surrounds the tendon of the flexor pollicis longus, and (3) the flexor carpi radialis tendon has a short sheath. Untreated infection of the synovial sheaths is liable to impair hand function. The common and pollical sheaths are frequently referred to in clinical writing as the ulnar and radial bursae, respectively.** These two sheaths project proximally a short distance above the flexor retinaculum, and they usually communicate with each other in the carpal canal. Distally, the *pollical sheath* extends almost to its insertion, and that part of the *common sheath* for the little finger usually does likewise. The ring, middle, and index fingers are provided with *digital sheaths* that usually extend proximally no further than the necks of their metacarpal bones, leaving a gap of 1 to 3 cm below the common sheath. Hence **infection of the synovial sheaths of the thumb or little finger may spread readily into the palm and even into the forearm.**

MUSCLES OF HAND

The muscles intrinsic to the hand are those of the thumb (thenar muscles) and of the little finger (hypothenar muscles), the palmar and dorsal interossei, and the lumbricals. They are supplied by the median and ulnar nerves,

* See, for example, F. Bojsen-Møller and L. Schmidt, J. Anat., *117*:55, 1974.

TABLE 11–1 THENAR MUSCLES

<table>
<tr><th>Muscle</th><th>Origin</th><th>Insertion</th><th>Innervation</th><th>Action</th></tr>
<tr><td>Abductor pollicis brevis</td><td rowspan="3">Flexor retinaculum & tubercle of trapezium</td><td rowspan="2">Lateral sesamoid & base of proximal phalanx</td><td rowspan="3">Recurrent branch of median</td><td>Abducts thumb</td></tr>
<tr><td>Flexor pollicis brevis</td><td>Flexes thumb</td></tr>
<tr><td>Opponens pollicis</td><td>Lateral side of 1st metacarpal</td><td>Rotates 1st metacarpal medially</td></tr>
<tr><td>Adductor pollicis</td><td>Capitate, trapezoid, base of 2nd metacarpal (oblique head), & front of 3rd metacarpal (transverse head)</td><td>Medial sesamoid & base of proximal phalanx</td><td>Deep branch of ulnar</td><td>Adducts thumb</td></tr>
</table>

which carry fibers of the T1 segment of the spinal cord. These muscles are of great functional importance, particularly those associated with the thumb. Their individual innervations and actions are significant in evaluating peripheral nerve injuries.

Thenar Muscles (table 11–1). The short muscles of the thumb (fig. 11–5) are the abductor pollicis brevis, flexor pollicis brevis, opponens pollicis, adductor pollicis, and the first palmar and dorsal interossei. The short abductor forms the lateral side of the thenar eminence, and the flexor (with which it is often fused) forms the medial side, whereas the opponens is deep. **The three muscles are supplied by the important recurrent branch of the median nerve, which lies very superficially at the distal border of the flexor retinaculum** (see figs. 10–2 and 11–9). The flexor is generally said to possess a variously defined deep head, e.g., arising in common with the oblique head of the adductor and supplied by the ulnar nerve. The adductor is more deeply placed and has two heads of origin. Two muscles, the abductor and flexor brevis, are supplied by the median nerve and are inserted in common on the lateral side of the proximal phalanx. Two other muscles, the adductor and first palmar interosseus, are supplied by the ulnar nerve and are inserted in common on the medial side of the proximal phalanx.

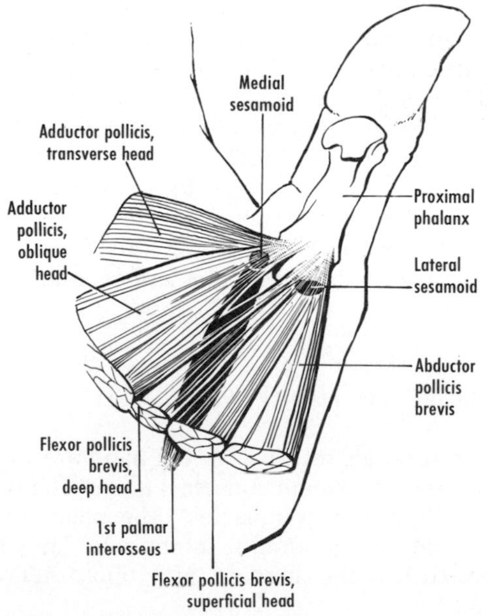

Figure 11–5 The intrinsic muscles of the thumb. The opponens is not shown.

Hypothenar Muscles (table 11–2). The short muscles of the little finger (see fig. 11–4) are the abductor digiti minimi, flexor digiti minimi brevis, and opponens digiti minimi. They are supplied by the ulnar nerve. The opponens is deep to the others and is not well named in that the fifth finger cannot be opposed. The hypothenar muscles are covered proximally by subcutaneous, transverse fibers termed the *palmaris brevis*.

Lumbricals and Interossei (table 11–3). The lumbricals and interossei (figs. 11–4 and 11–6) are a dozen small, important muscles that are inserted mainly into the extensor expansion (fig. 11–7). The lumbricals are associated with the tendons of the flexor digitorum profundus in the palm. The palmar interossei lie on the palmar surfaces of the metacarpals, whereas the dorsal interossei are more truly interosseous, i.e., between the bones. Each group of four muscles is numbered from one to four, from lateral to medial. It should be noted that the word *interosseous* is an adjective, whereas *interosseus* (plural: interossei) is a noun.

TABLE 11–2 HYPOTHENAR MUSCLES

Muscle	Origin	Insertion	Innervation	Action
Abductor digiti minimi	Pisiform	Medial side of proximal phalanx	Deep branch of ulnar	Abducts little finger
Flexor digiti minimi brevis	Hook of hamate	Medial side of proximal phalanx	Deep branch of ulnar	Flexes little finger
Opponens digiti minimi	Hook of hamate	Front of 5th metacarpal	Deep branch of ulnar	Draws 5th metacarpal forward
Palmaris brevis	Medial border of palmar aponeurosis	Skin on medial side of hand	Superficial branch of ulnar	Deepens hollow of palm & protects ulnar nerve & artery

TABLE 11–3 LUMBRICALS AND INTEROSSEI

Muscle	Origin	Insertion	Innervation	Action
Lumbricals 1 & 2	Lateral two flexor profundus tendons	Lateral sides of extensor expansions of fingers 2–5	Digital branches of median	Extend middle & distal phalanges
Lumbricals 3 & 4	Medial three flexor profundus tendons (two heads each)	Lateral sides of extensor expansions of fingers 2–5	Deep branch of ulnar	Extend middle & distal phalanges
Palmar interossei 1–4	Shafts of metacarpals 1, 2, 4, & 5	Extensor expansions of fingers 1, 2, 4, & 5	Deep branch of ulnar	Flex proximal phalanges & adduct fingers
Dorsal interossei 1–4	Sides of metacarpals 1–5 (two heads each)	Extensor expansions & proximal phalanges of fingers 2-4	Deep branch of ulnar	Flex proximal phalanges & abduct fingers

Each of the first two lumbricals arises by a single head from the lateral two tendons of the deep flexor, whereas each of the last two lumbricals arises by two heads from the adjacent sides of the medial three tendons. Each lumbrical has usually the same innervation as its deep tendon, i.e., the first two from the median and the last two from the ulnar. The lumbricals are inserted into the lateral (radial) sides of the extensor expansions of fingers 2 to 5, and they are chiefly extensors at the interphalangeal joints. Variations in attachments are common.

The four palmar interossei adduct the fingers toward an axis through the middle finger. Each arises from the metacarpal shaft of the digit that it adducts (see fig. 11–6). They are inserted into the extensor expansions.

The four dorsal interossei abduct the fingers away from an axis through the middle finger. Each arises by two heads from the adjacent sides of the metacarpal bones. They are inserted into the extensor expansions and, in certain fingers (see fig. 11–7B), into the bases of the proximal phalanges. The first dorsal interosseus can readily be seen and felt *in vivo* during abduction of the index finger against resistance.

The *extensor expansion,* or *dorsal aponeurosis,* is a fibrous sheet on the back of each finger, and it contains a hood of transverse fibers (fig. 11–7). The expansion is penetrated by the extensor tendon, which then divides into three slips, a central slip to the base of the middle phalanx and two collateral bands, which, fused with expansions from the interossei and lumbricals, unite and proceed to the base of the distal phalanx. The lumbricals and palmar interossei are inserted into the hood, whereas the dorsal interossei are generally inserted into both the hood and the proximal phalanx deep to the hood. The arrangement of the expansion deep to the hood is very complicated.

The precise details of lumbrical and interosseous movement *in vivo* and their correlation with the actions of the long flexors and extensors of the digits are still not clear.

In combination with dorsiflexion at the wrist, the interossei and lumbricals produce the "Z-position" (fig. 11–8), i.e., flexion at the metacarpophalangeal joints by the interossei and extension at the interphalangeal joints by the lumbricals. It should be noted that, in section of the ulnar nerve, with consequent paralysis of the interossei and medial two lum-

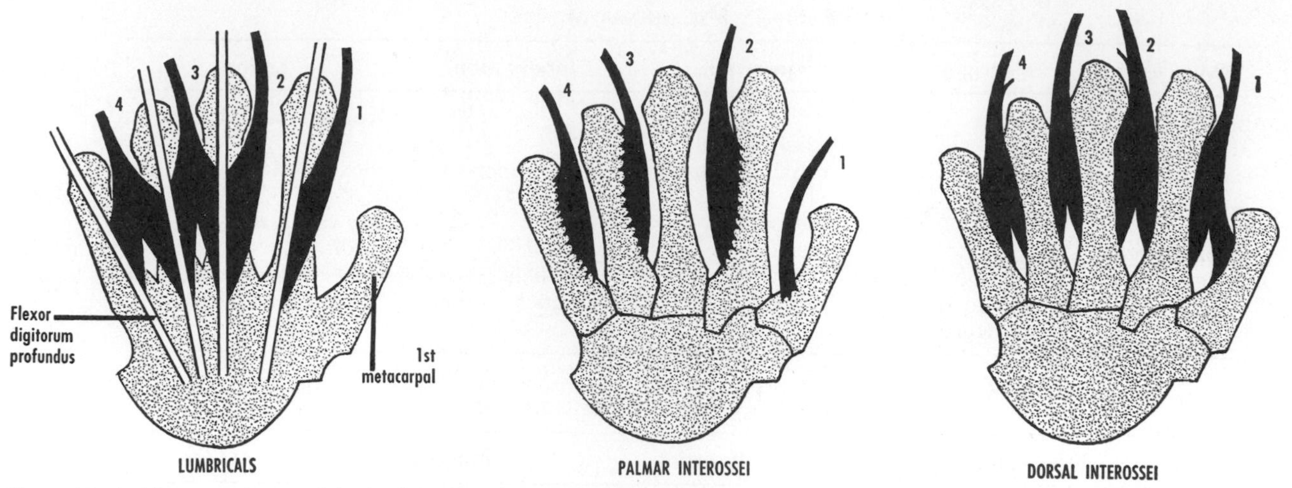

Figure 11–6 The arrangement of the lumbricals and interossei. See also fig. 11–7.

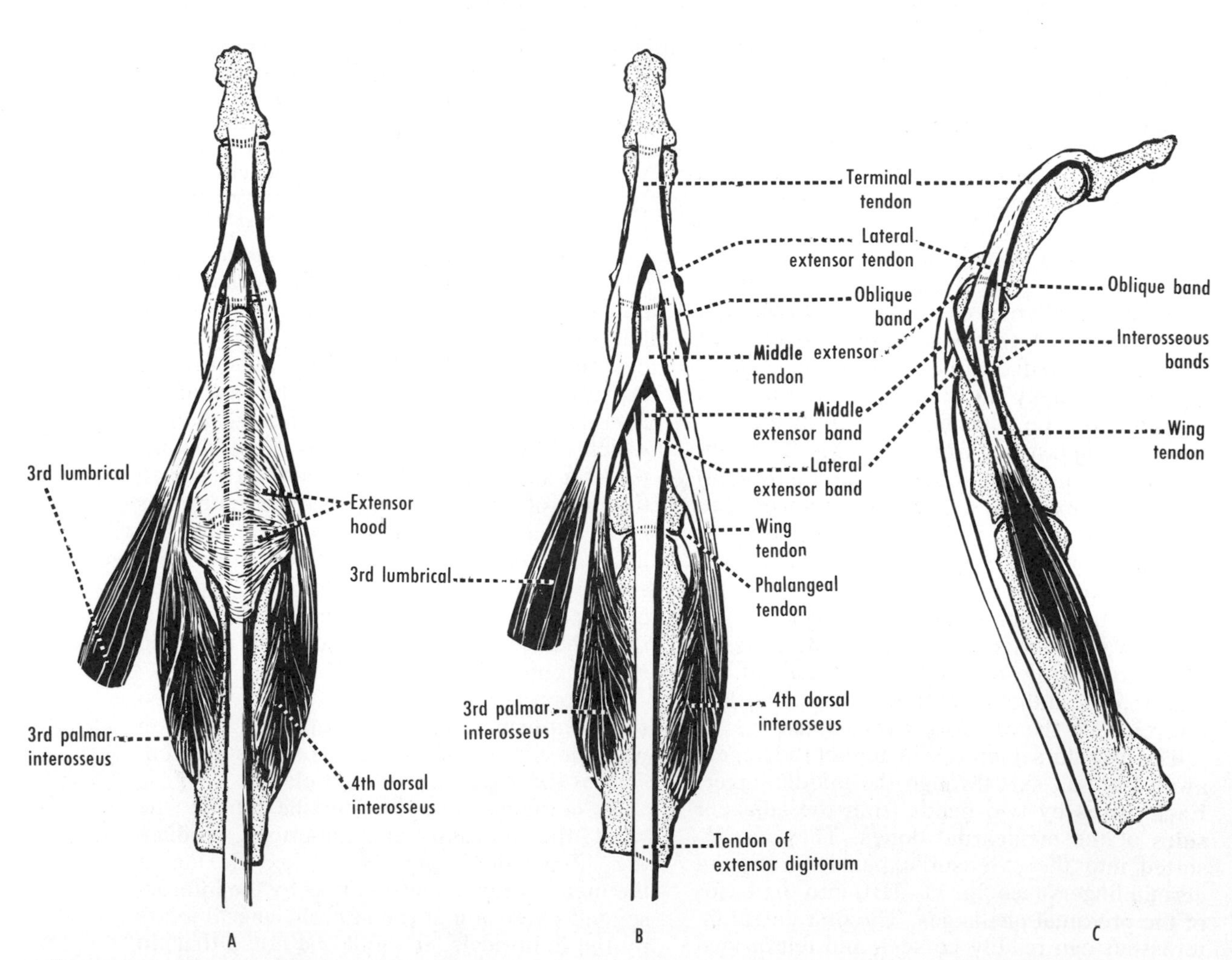

Figure 11–7 The extensor aponeurosis of the fourth finger. **A**, Dorsal view showing the extensor hood. **B**, Dorsal view with the hood removed. **C**, Medial view. (Based on Landsmeer.)

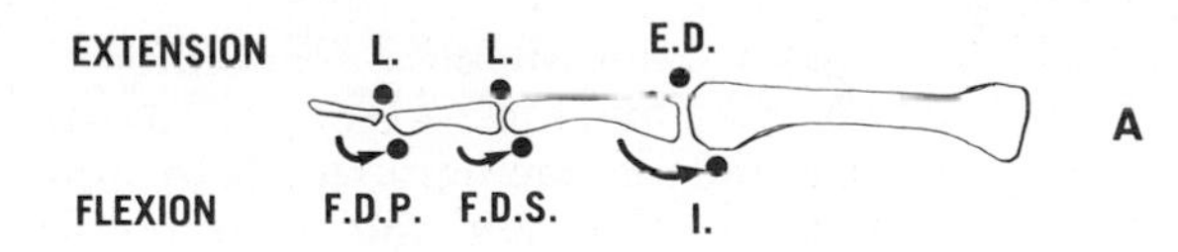

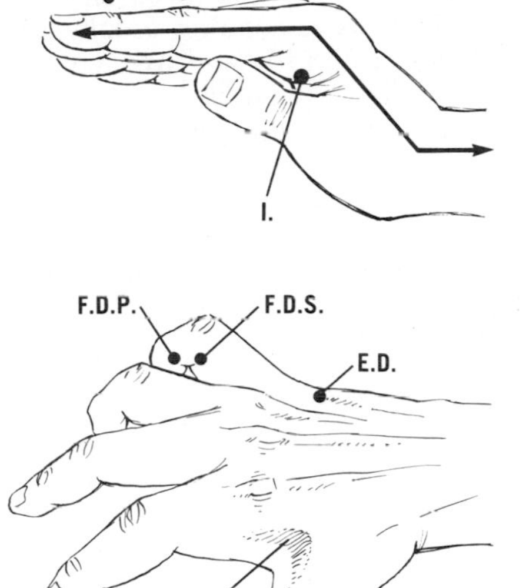

Figure 11–8 A, Diagram to show the chief muscles producing extension and flexion of the fingers. *E.D.*, extensor digitorum; *F.D.P.*, flexor digitorum profundus; *F.D.S.*, flexor digitorum superficialis; *I.*, interosseus; *L.*, lumbrical. **B**, The Z-position of the hand, produced by the interossei and lumbricals. **C**, Claw hand, due to paralysis of the interossei and lumbricals (following section of the ulnar nerve). The unopposed extensor and flexors produce the opposite of the Z-position.

bricals, the proximal phalanges are hyperextended by the unopposed long extensors and the middle and distal phalanges are hyperflexed (less so in fingers 2 and 3) by the unopposed long flexors. The result is a "claw hand" (fig. 11–8), which is the opposite of the Z-position.

NERVES OF HAND

The hand is supplied by the median, ulnar, and radial nerves (fig. 11–9). **The motor fibers to the intrinsic muscles of the hand, which are carried by the median and ulnar nerves, are derived from the T1 segment of the spinal cord.**

At the end of its descent along the front of the forearm, **the median nerve is found midway between the styloid processes of the radius and ulna, medial to the flexor carpi radialis tendon and lateral to the palmaris longus tendon when this is present** (see fig. 10–2). The nerve enters the hand by passing through

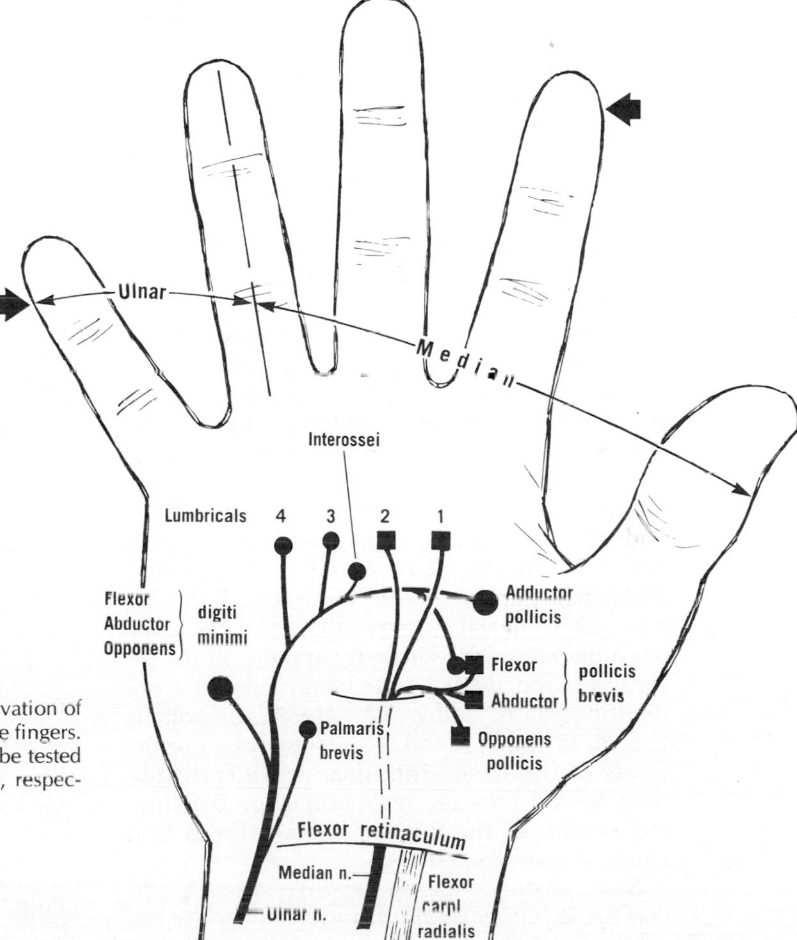

Figure 11–9 Diagram to show the motor innervation of the hand. Cutaneous innervation is indicated on the fingers. Sensation in the ulnar and median territories may be tested on the fifth and second fingers (*at broad arrows*), respectively.

the carpal canal, behind the flexor retinaculum and in front of the flexor tendons to the index finger. It spreads into its terminal branches under cover of the palmar aponeurosis and the superficial palmar arch. **An important recurrent branch emerges almost immediately below the flexor retinaculum, is dangerously superficial, and supplies the abductor and flexor pollicis brevis and the opponens** (see fig. 10–2). The median nerve gives *palmar digital nerves* to three and a half fingers. These supply also the first two or three lumbricals and send branches dorsally to supply the backs of the distal portions of the fingers. The median nerve (see fig. 11–2) may be compressed in the carpal canal (*carpal tunnel syndrome*), resulting in weakness of the thenar muscles and sensory loss in the distribution of the digital branches.

The median nerve has a complicated arrangement of fibers, and it characteristically supplies bones, joints, ligaments, the interosseous membrane, and blood vessels. **Injuries to the median nerve may be followed by severe, chronic sensory and trophic disturbances.**

In the middle of the forearm, the *ulnar nerve* gives off a *dorsal branch,* which descends between the ulna and the flexor carpi ulnaris and reaches the medial side of the hand. It supplies *dorsal digital nerves* to usually two and a half or one and a half fingers. **The ulnar nerve enters the hand anterior to the flexor retinaculum (hence outside the carpal canal) and lateral to the pisiform.** The ulnar artery lies on its lateral side, and both are sometimes covered by a superficial part of the retinaculum. The nerve then divides into superficial and deep branches. The *superficial branch* supplies the palmaris brevis and gives *palmar digital nerves* to one and a half fingers. The *deep branch* of the ulnar nerve passes between the abductor and flexor digiti minimi, both of which it supplies. It innervates the opponens also, then curves around the hook of the hamate and laterally across the palm with the deep palmar arch. There it supplies all the interossei, the medial two lumbricals, the adductor pollicis, and usually the flexor pollicis brevis, in which it ends. Many fibers go to the joints of the hand. **The ulnar nerve is distributed to the muscles responsible for the finer movements of the fingers; hence injuries to it cause severe disabilities.**

Some distance above the wrist, the *superficial branch of the radial nerve* leaves the lateral side of the radial artery, passes deep to the brachioradialis, and supplies *dorsal digital nerves* to two and a half fingers (see fig. 10–4).

In summary, **the median nerve tends to supply the thenar muscles, and the ulnar nerve most of the other short muscles of the hand. The dividing line between the two distributions, however, is variable, and either nerve may invade the territory of the other.**

The ulnar nerve supplies usually one and a half fingers anteriorly and two and a half fingers posteriorly. The other fingers are supplied by the median anteriorly and by the radial posteriorly. The median nerve, however, invades the posterior aspect of several fingers distally. Variations are common, and the ulnar distribution varies inversely with that of the median anteriorly and with that of the radial posteriorly.

ARTERIES OF HAND (fig. 11–10)

The *radial artery* leaves the forearm by curving dorsally to enter the anatomical snuffbox (see fig. 10–4). It then gains the palm between the heads of the first dorsal interosseus. Next it turns medially between the heads of the adductor pollicis and anastomoses with

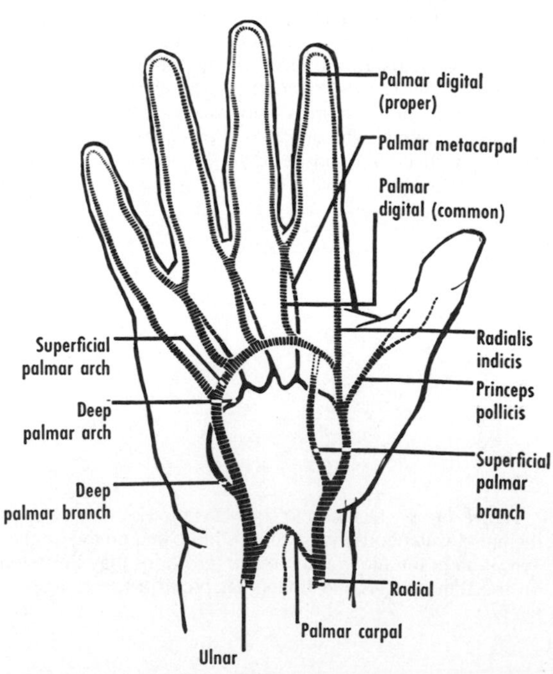

Figure 11–10 The arteries of the front of the hand. The superficial palmar arch is sometimes completed on the lateral side by the superficial palmar branch of the radial artery. The anastomosis is shown by dotted lines.

the deep branch of the ulnar artery to form the deep palmar arch.

In the forearm, the radial artery gives an inconstant *superficial palmar branch,* which joins the superficial palmar arch. Palmar and dorsal *carpal branches* from the radial artery form networks with corresponding vessels from the ulnar artery and give rise to *dorsal metacarpal* and *digital arteries*. Other digital vessels (see fig. 10–2) from the radial artery include those to the thumb (*princeps pollicis*) and radial side of the index finger (*radialis indicis*). The *deep palmar arch,* which lies on the interossei, is formed by the termination of the radial artery and the deep branch of the ulnar artery. It gives several *palmar metacarpal arteries*.

The *ulnar artery* enters the hand in front of the flexor retinaculum, on the lateral side of the pisiform. It then divides into the superficial palmar arch and a *deep palmar branch*. The latter joins the radial artery to form the deep palmar arch.

The *superficial palmar arch* (figs. 10–2 and 11–10), which is the main termination of the ulnar artery, is completed on the radial side in a variable manner, e.g., by the radialis indicis. The arch lies on the flexor tendons and the branches of the median nerve, under cover of the palmar aponeurosis. It gives off several *palmar digital arteries*. **The convexity of the superficial palmar arch is approximately at the level of a line drawn across the hand from the distal border of the extended thumb.** The anastomoses in the hand are extensive, and severe hemorrhage may occur from wounds of the palmar arches.

JOINTS OF WRIST AND HAND

The *distal radio-ulnar joint* is a pivot articulation between the head of the ulna and the ulnar notch of the radius. The *articular disc* (fig. 11–11) is a strong, triangular, fibrous or fibrocartilaginous plate. Its base is anchored to the radius, and its apex to the root of the styloid process of the ulna. The joint cavity lies between the disc and the head of the ulna and extends upward (as the *recessus sacciformis*) between the two bones of the forearm. The disc takes part in the wrist joint but excludes the ulna from that articulation. The movements at the distal radio-ulnar joint are supination and pronation.

The radiocarpal, or wrist, joint is an ellipsoidal articulation formed by the radius and articular disc and the scaphoid, lunate, and triquetrum (fig. 11–11). **The position of the wrist joint is indicated approximately by a line connecting the styloid processes of the radius and ulna.** The capsule is reinforced by *palmar, dorsal, radial collateral,* and *ulnar collateral ligaments*. The synovial membrane does not cover the articular disc. The cavity commonly communicates with those of some of the intercarpal joints but usually not with that of the distal radio-ulnar articulation.

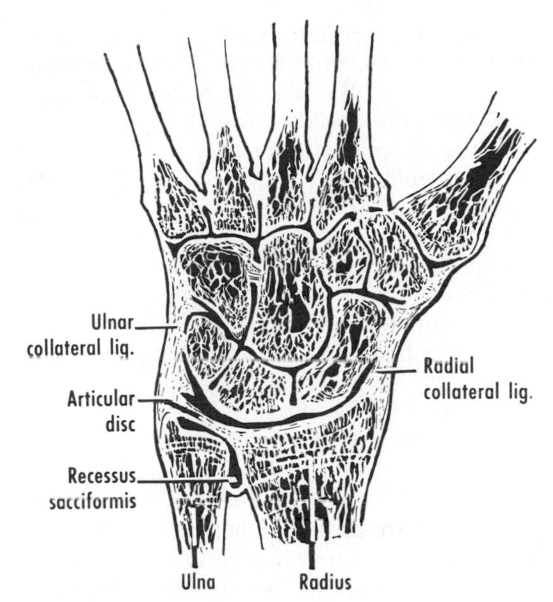

Figure 11–11 Schematic coronal section of the carpus as if the carpus were flat. Note the radiocarpal, intercarpal, and carpometacarpal joints.

Among the intercarpal joints, the most important is the *midcarpal joint,* formed between the proximal and distal rows of the carpus, exclusive of the pisiform (fig. 11–11). The three proximal bones form a socket for the hamate and head of the capitate (ellipsoidal joint), and the scaphoid articulates also with the trapezium and trapezoid (plane joint). The *pisotriquetral joint* is a small synovial articulation.

The radiocarpal and (medial part of the) midcarpal joints are ellipsoidal; hence their movements are basically flexion and extension, abduction and adduction, and combinations of these. The movements are produced by the extrinsic muscles of the hand: the chief flexors are the flexor carpi ulnaris and flexor carpi radialis; the extensors are the extensor carpi radialis longus and brevis and the extensor carpi ulnaris; the chief abductors are the flexor carpi radialis and extensor carpi radialis longus and brevis; the adductors are the extensor and flexor carpi ulnaris.

The medial four *carpometacarpal joints*

(with which intermetacarpal joints are associated) are plane articulations. **The carpometacarpal joint of the thumb is an important, freely movable, saddle articulation with reciprocally shaped surfaces on the trapezium and first metacarpal** (fig. 11–11). The loose capsule is reinforced by special ligaments. The joint may be involved in (Bennett's) fracture-dislocation of the base of the first metacarpal.

The *metacarpophalangeal joints* are ellipsoidal, and the *interphalangeal* joints are hinge articulations, but their ligamentous arrangements are similar. Each capsule is strengthened by a *collateral ligament* on each side, and the two ligaments fuse in front to form a dense, fibrous or fibrocartilaginous pad termed the *palmar ligament*. The four medial metacarpals are held together by a deep *transverse metacarpal ligament* related in front to the lumbrical tendons and behind to those of the interossei.

Movements of the Hand. In addition to free movement, the fingers (including the thumb) may be held forcibly against the palm (power grip) to transmit force to an object and to conform to its size and shape, as in holding a ball. Objects can also be rotated by fine movements (precision handling) of the fingers (including the thumb), as in winding a watch, or translated toward or away from the palm, as in threading a needle and pulling the thread through it. Furthermore, objects can be compressed (pinched) between the thumb and index (or index and middle) finger, as in holding a pen while writing. These mechanisms are of great importance.

In power grip, the extrinsic muscles provide the major force, the interossei are used as flexors and rotators at the metacarpophalangeal joints, and the thenar muscles are generally active. In precision handling, specific extrinsic muscles, the interossei and lumbricals, and the thenar triad (abductor and flexor brevis and opponens) are important. In pinching, compression is provided chiefly by the extrinsic muscles, assisted by the thenar musculature and the interossei (fig. 11–12).

The thumb is set at an angle to the plane of the palm; hence its movements are described differently from those of the other fingers. Forward movement of the thumb as a whole, away from the palm, is called abduction (fig. 11–13); the converse movement is called adduction. With the hand on a table, palm up, abduction points the thumb toward the ceiling. A medial movement of the thumb in the plane of the palm is termed flexion; the converse is termed extension. **Opposition is the movement whereby the palmar aspect of the thumb touches the palmar aspect of another finger of the same hand.** (When the contact is not by their palmar aspects, the movement is merely apposition.) The converse of opposition is reposition. These movements occur at all the joints of the thumb, although opposition takes place mainly at the carpometacarpal joint.

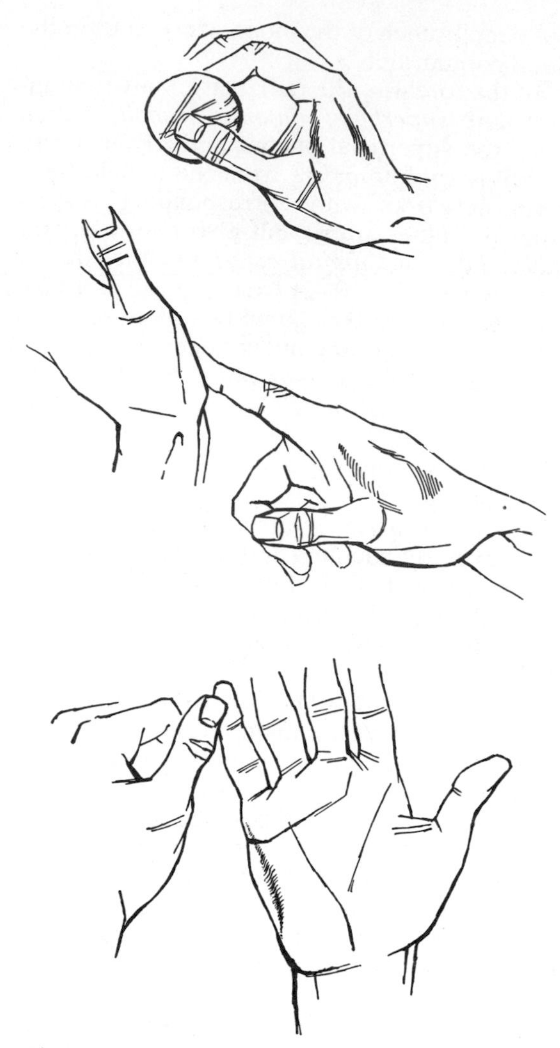

Figure 11–12 Testing finger movements. **Upper,** The use of a pinch to test the first dorsal interosseus and adductor pollicis. Adduction by the other interossei can be tested by having the subject hold a piece of paper between his fingers as the examiner tries to pull it away. **Middle,** The muscles of the thumb can be checked by palpating them as movements are made against resistance. Note the tenseness of the first dorsal interosseus of the examiner. **Lower,** the hypothenar muscles can be tested by abducting (or flexing) against resistance.

In the other fingers, flexion and extension occur at the metacarpophalangeal and interphalangeal joints (figs. 11–14 and 11–15), and an important rotatory component (produced by the interossei through abduction and adduction) takes place at the metacarpophalangeal joints (fig. 11–14).

The "position of rest," resulting from merely intrinsic muscular tone, is assumed by the inactive hand (see fig. 11–1). **When immobilization is necessary, the hand should be placed in the position of rest.**

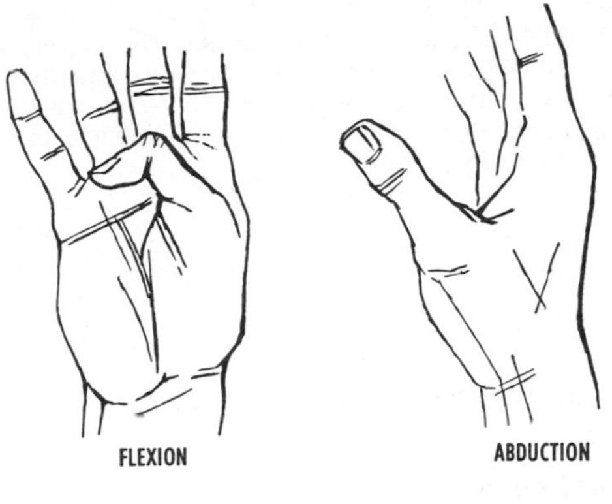

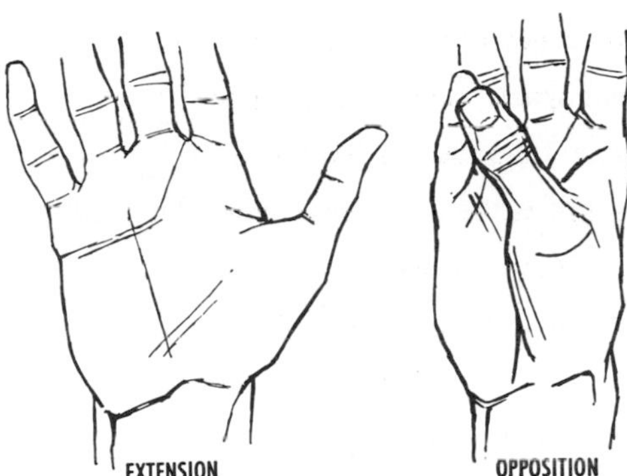

Figure 11–13 Movements of the thumb.

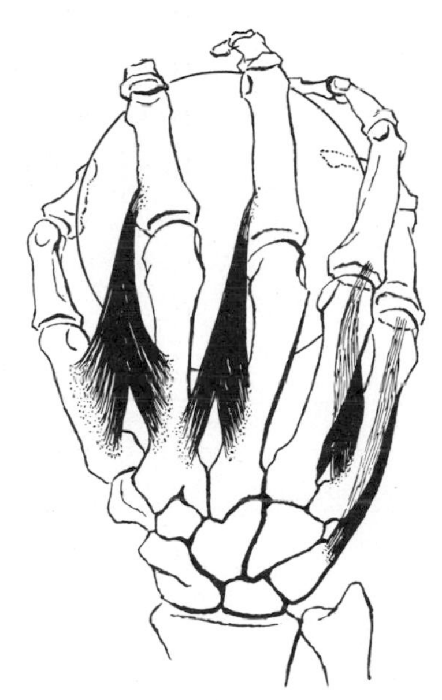

Figure 11–14 Diagram showing a power grip with rotation at metacarpophalangeal joints while squeezing a ball. The phalangeal insertions of the dorsal interossei and abductor digiti minimi are shown. (Based on Landsmeer.)

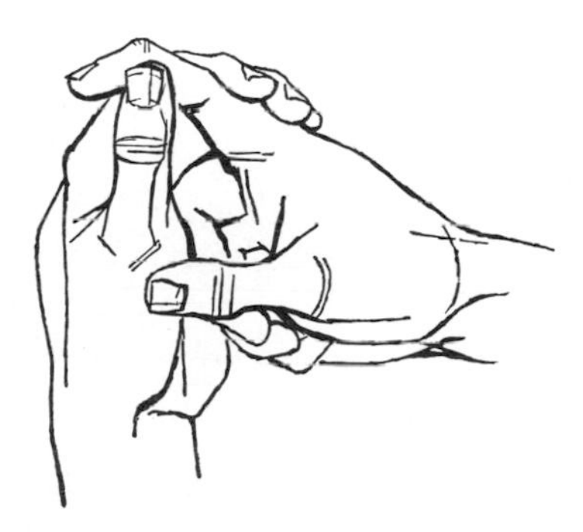

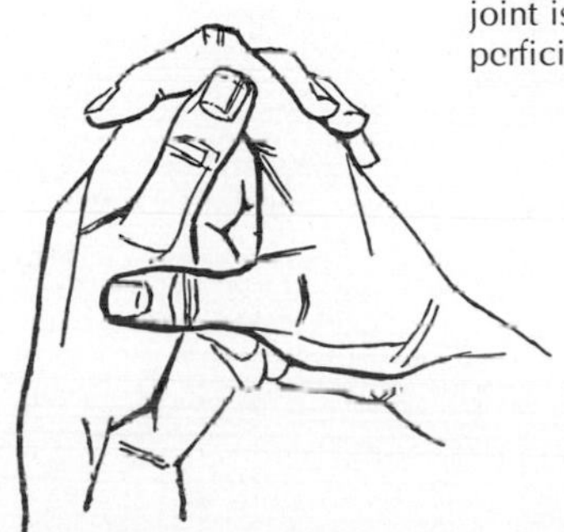

Figure 11–15 Testing finger movements. **Upper,** If the metacarpophalangeal and proximal interphalangeal joints are stabilized as shown, flexion at the distal interphalangeal joint (flexor profundus) can be tested. **Lower,** If the metacarpophalangeal joint is stabilized as shown, flexion at the proximal interphalangeal joint (flexor superficialis) can be tested.

ADDITIONAL READING

Jones, F. W., *The Principles of Anatomy as Seen in the Hand,* 2nd ed., Baillière, Tindall, and Cox, London, 1942. A very readable classic.

Kaplan, E. B., *Functional and Surgical Anatomy of the Hand,* 2nd ed., Lippincott, Philadelphia, 1965. A well-known text.

Landsmeer, J. M. F., *Atlas of Anatomy of the Hand,* Churchill Livingstone, Edinburgh, 1976. A very advanced study with numerous photomicrographs.

Tubiana, R. (ed.), *The Hand,* vol. 1, W. B. Saunders Company, Philadelphia, 1981. A detailed account including much anatomy.

QUESTIONS

11–1 When the hand has to be immobilized, in which position should it be placed?

11–2 What is the clinical importance of the carpal tunnel?

11–3 What is the clinical significance of the palmar aponeurosis?

11–4 What are the "midpalmar space" and "thenar space"?

11–5 What are the "ulnar bursa" and the "radial bursa"?

11–6 Which spinal segment supplies the small muscles of the hand?

11–7 What is the innervation of the muscles of the thumb?

11–8 What are the chief effects of section of the median nerve at the wrist?

11–9 Which muscles produce the "Z-position" of the hand?

11–10 Apart from claw hand, what motor deficits would be expected in ulnar nerve palsy?

11–11 Which nerves would be tested by examining for sensation distally along the lateral side of the index finger and the medial side of the little finger?

11–12 Why is a fracture of the base of the first metacarpal serious?

Part 3

THE LOWER LIMB

The lower limb consists of four major parts: a girdle formed by the hip bones, the thigh, the leg, and the foot. It is specialized for the support of weight, adaptation to gravity, and locomotion. In descriptions of the lower limb, it is customary to include regions that are transitional between the limb and the trunk, especially the gluteal and inguinal regions.

The following Latin words (given with their English equivalents) are the basis of many anatomical terms, e.g., femoral artery, sural nerve, calcaneus, and extensor hallucis: *Membrum* (limb), *inguen* (groin), *natis* or *clunis* (buttock; *gloutos* is a corresponding Greek word), *coxa* (hip; *ischion* is a corresponding Greek word), *femur* (thigh), *genu* (knee), *crus* (leg), *sura* (calf), *talus* (ankle), *pes* or *pedis* (foot), *calx* (heel), *planta* (sole), *digiti pedis* (toes), and *hallux* or *hallucis* (big toe).

The limb buds appear in the embryo at four weeks, the upper about two days before the lower. At first the limbs show a similar arrangement, with pre-axial and postaxial borders at seven weeks (see figs. 8–10 and 15–11). During fetal life, changes (generally termed the "rotation of the limbs") occur so that walking on the soles (plantigrade) will become possible later on. The big toes are then medial. Each limb possesses a girdle and a subsequent skeletal segment (humerus in the upper limb, femur in the lower limb). More distally are several pre-axial elements: radius (tibia); lunate and scaphoid (talus and navicular); trapezium, trapezoid, and capitate (cuneiforms); and metacarpals (metatarsals) 1 to 3. Postaxial components include the ulna (fibula), triquetrum (calcaneus), hamate (cuboid), and metacarpals (metatarsals) 4 and 5.

ADDITIONAL READING

See the list provided for the upper limb, p. 37. Particular attention is directed to *Frazer's Anatomy of the Human Skeleton*, an important classic, and Haymaker and Woodhall's *Peripheral Nerve Injuries*, a mine of information.

12
THE BONES OF THE LOWER LIMB

HIP BONE

The hip bones (which have been named—they are no longer "innominate"!) meet in front at the pubic symphysis. Together with the sacrum behind, they form a ring termed the *bony pelvis*. Each hip bone consists of an ilium, an ischium, and a pubis, all three of which in the adult are fused at the acetabulum to form a single bone.

The *hip bone* (figs. 12–1 to 12–9 and 31–1 to 31–3) connects the trunk to the lower limb by extending from the sacrum to the femur. The terminology of the hip bone is based on the anatomical position, in which the articular surface of the pubic symphysis is in a sagittal plane and **the pubic tubercle and anterior superior iliac spine are in the same coronal plane.** The internal (or visceral) aspect of the body of the pubis then faces almost directly upward so that the urinary bladder rests on it.

Ilium. The ilium consists of a body and an ala (or wing), which are not visibly demarcated from each other on the lateral aspect. The *body* of the ilium is fused with the ischium and pubis and forms about two fifths of the acetabulum. The *iliac crest,* which is the upper, sinuous border of the bone, is readily palpable *in vivo*. It ends in front at the *anterior superior iliac spine,* to which the inguinal ligament is attached, and behind at the *posterior superior iliac spine*. **The anterior superior iliac spine is an important landmark, which can be palpated (and sometimes seen) by tracing the iliac crest forward or the inguinal ligament upward. Clinically, the lower limb is measured from the anterior superior iliac spine to the tip of the medial malleolus of the tibia. The measuring tape is kept along the medial side of the patella. The posterior superior iliac spine is usually marked by a dimple in the skin. A line connecting the**

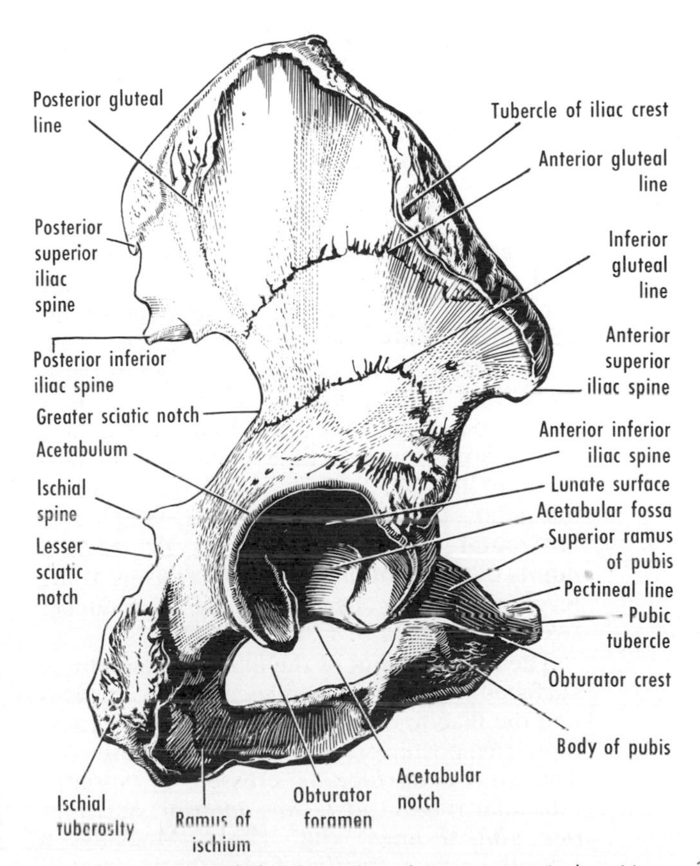

Figure 12–1 The right hip bone, lateral view, anatomical position.

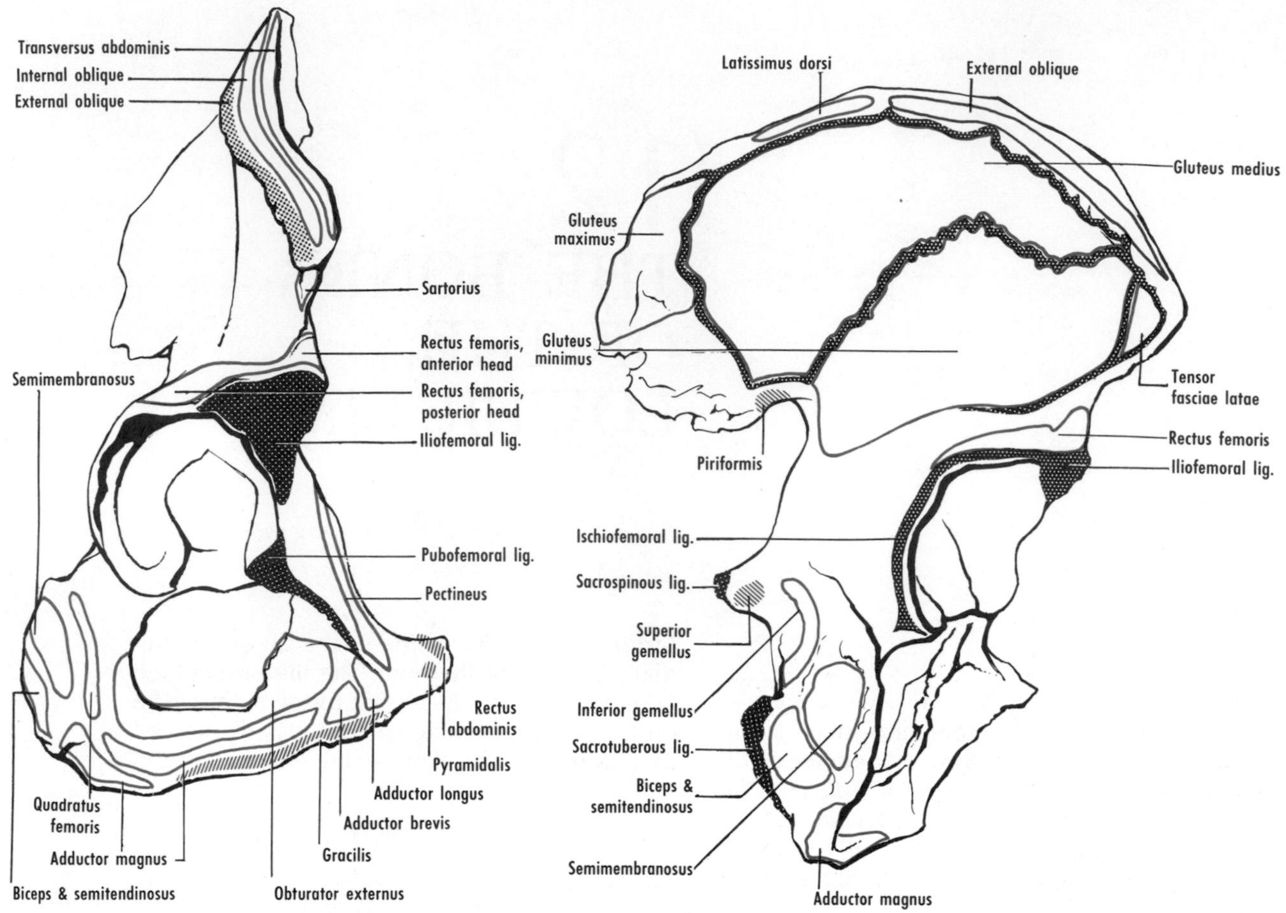

Figure 12–2 The right hip bone, inferolateral view, muscular and ligamentous attachments.

Figure 12–3 The right hip bone, posterolateral view, muscular and ligamentous attachments. The origin of the tensor fasciae latae often extends more posteriorly.

right and left dimples is at the level of the second sacral vertebra (S.V.2) and indicates approximately the level of the middle of the sacro-iliac joints. Most of the iliac crest presents *outer* and *inner lips* and a rough intermediate line. The *tubercle* of the crest is a projection of the outer lip about 5 cm behind the anterior superior iliac spine. The highest point of the iliac crest is slightly behind its midpoint. **The supracristal plane, which is a horizontal plane that connects the highest points of the right and left iliac crests, is at the level of L.V.4,** generally at the interval between the spines of L.V.3 and 4.

The *ala* (or wing) of the ilium presents three surfaces (the gluteal and sacropelvic surfaces and the iliac fossa) separated by three borders (anterior, posterior, and medial, respectively). The *gluteal surface* is crossed by variable muscular ridges (*posterior, anterior,* and *inferior gluteal lines*) (fig. 12–1). Medially, a smooth surface, the *iliac fossa,* forms a part of the lateral wall of the greater (or "false") pelvis (fig. 12–4). Behind the iliac fossa, a rough *sacropelvic surface* includes the *iliac tuberosity,* which is a very rough, ligamentous area, and, below it, the *auricular surface* for articulation with the sacrum to form the sacro-iliac joint. The *anterior border* begins at the anterior superior spine and includes the anterior inferior iliac spine, situated above the acetabulum. Below the anterior inferior spine and in front of the acetabulum, the junction of the ilium and pubis forms a diffuse swelling, the *iliopubic eminence.* The *posterior border* begins at the posterior superior spine and includes the posterior inferior iliac spine; it then runs forward and forms the *greater sciatic notch.* The *medial border* includes the anterior edge of the auricular surface. Below, it forms a rounded ridge, the *arcuate line,* which continues to the iliopubic eminence.

Ischium. The ischium, which forms the postero-inferior part of the hip bone, consists of a body and a ramus. The *body* is fused with the ilium and pubis and forms about two fifths of the acetabulum. The lower end of the body forms a rough impression known as the *tuber,* or *ischial tuberosity,* to which the hamstrings are attached (figs. 12–1 to 12–5). It is covered by a bursa, which may become enlarged ("weaver's bottom"). **The ischial tuberosity is obscured by the gluteus maximus when the hip is extended but is palpable when the thigh is flexed. The body weight rests on the ischial tuberosities in the sitting position.** The body of the ischium presents three surfaces: one facing the thigh (*femoral*), another related to the ischiorectal fossa (*pelvic*), and still another (*dorsal*), which is continuous with the gluteal surface of the ilium. Above the ischial tuberosity, the *lesser sciatic notch* leads to the *ischial spine,* which is at the lower limit of the greater sciatic notch. These notches are converted into foramina by the *sacrospinous* and *sacrotuberous ligaments.* The (formerly inferior) *ramus* of the ischium extends medially from the body and tuberosity and joins the inferior ramus of the pubis. The conjoined rami of the ischium and pubis complete the *obturator foramen.* They have two surfaces: one faces the thigh and the other the pelvis and perineum.

Pubis. The pubis consists of a body and two rami. The *bodies* of the two sides meet in the median plane at the *pubic symphysis.* The medial, or *symphysial, surface* is rough and covered *in vivo* by cartilage. The body also presents a *pelvic surface,* which faces upward and supports the bladder, and a *femoral surface,* which is roughened for muscular attachments. Anteriorly, a ridge termed the *pubic crest* ends laterally in the *pubic tubercle.* **The pubic tubercle is an important landmark in the lower part of the abdominal wall, about 3 cm from the median plane** (see fig. 25–7). **It can be found by tracing the tendon of the adductor longus upward. The pubic tubercle is crossed by the spermatic cord and is a guide to the superficial inguinal ring, femoral ring, and sa-**

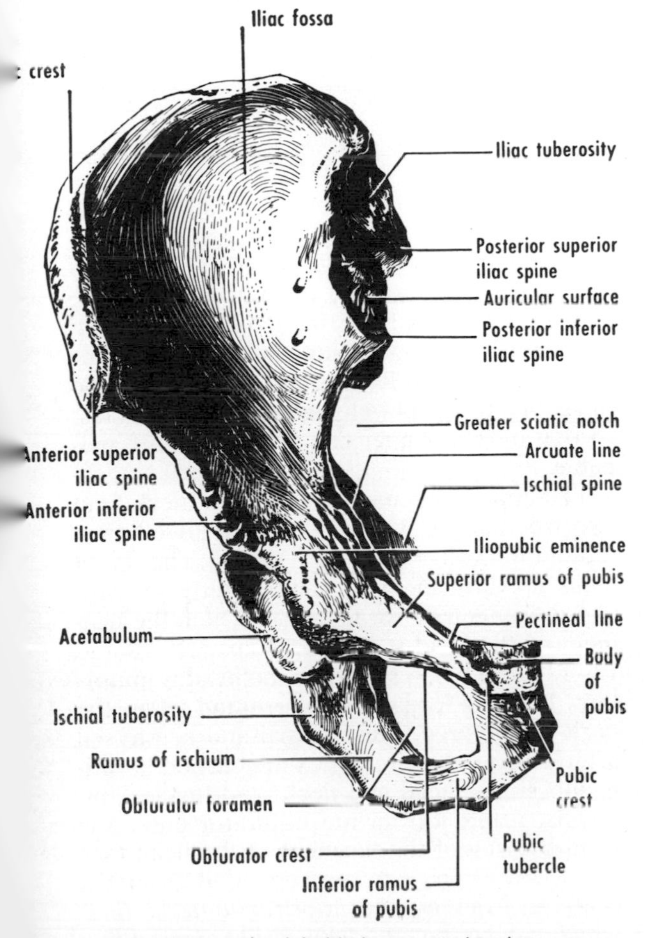

Figure 12–4 The right hip bone, anterior view, anatomical position.

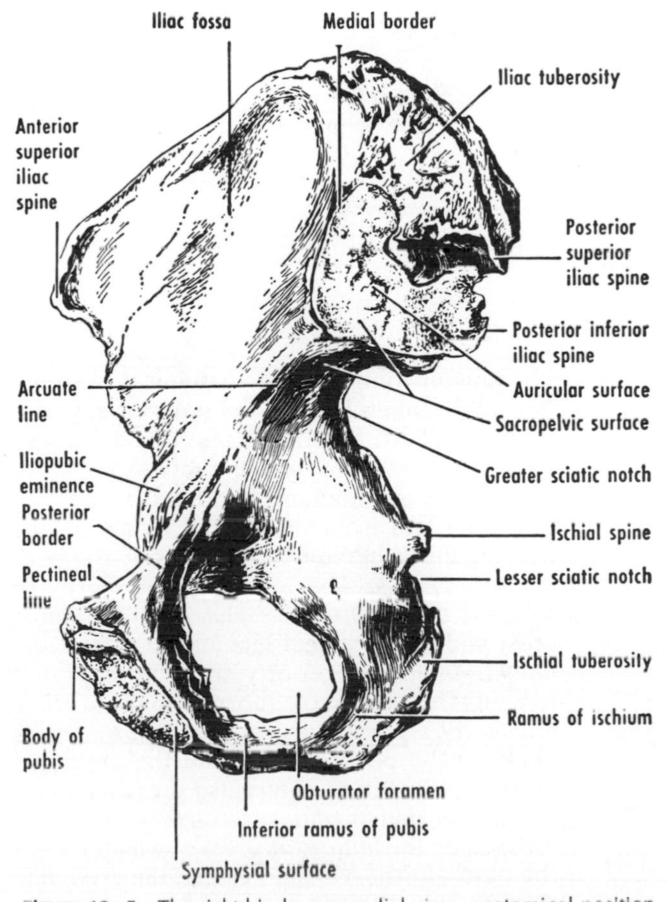

Figure 12–5 The right hip bone, medial view, anatomical position.

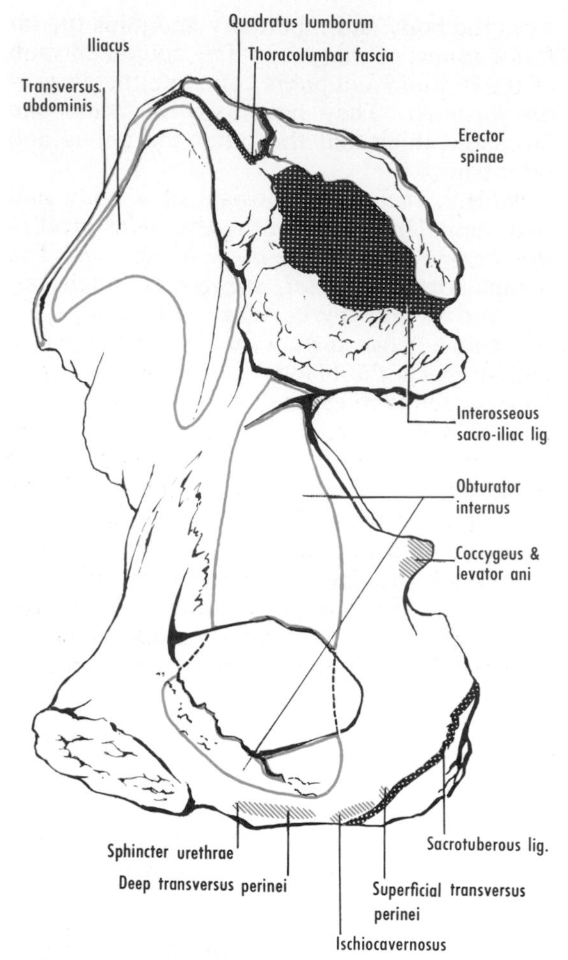

Figure 12–6 The right hip bone, medial view, muscular origins and ligamentous attachments.

phenous opening. In the anatomical position, the pubic tubercles and the anterior superior iliac spines are all in the same coronal plane. The *superior ramus* of the pubis extends upward to the acetabulum, where it is fused with the ilium and ischium, and forms about one fifth of the acetabulum. Anteriorly, the *pecten*, or *pectineal line*, extends from the pubic tubercle to the *iliopubic eminence*. The pubic crest and the pectineal line form a part of the *linea terminalis*. Inferiorly, the *obturator crest* extends from the pubic tubercle to the *acetabular notch*. A triangular *pectineal surface* lies between the pectineal line and the obturator crest. The superior ramus also presents a *pelvic surface* and an *obturator surface*, the latter crossed by the *obturator groove*, which lodges the obturator nerve and vessels; the groove is converted by the obturator membrane into the *obturator canal*. The *inferior ramus* of the pubis joins the ramus of the ischium.

Acetabulum and Obturator Foramen. The acetabulum (resembling a Roman vinegar cup; cf. acetic acid) is formed by the ilium, ischium, and pubis (see figs. 12–1 and 15–12). It looks downward, forward, and laterally and forms a socket for the head of the femur. Its horseshoe-shaped, articular surface is known as its *lunate surface*, whereas its rough, nonarticular floor is the *acetabular fossa*. The rim, which is deficient below at the acetabular notch, gives attachment to the *acetabular labrum*. The obturator foramen is bounded by the ischium and the pubis and their rami. Except at the obturator groove, the foramen is closed by the *obturator membrane*, which is attached along its margin.

Ossification. The three parts of the hip bone begin to ossify during the fetal period. By late childhood, the three primary centers are separated in the acetabulum by the Y-shaped triradiate cartilage (figs. 12–8*C* and 12–9). Secondary centers appear chiefly in that cartilage and also in the iliac crest, anterior inferior iliac spine, ischial tuberosity, and pubic symphysis.

FEMUR

The femur, or thigh bone (figs. 12–7, 12–8, and 12–10 to 12–18), is the longest and heaviest bone in the body. Its length varies from one fourth to one third of that of the body; hence stature can be estimated from it. When a subject is in the standing position, the femur transmits weight from the hip bone to the tibia. The femur is well covered with muscles, so that only its upper and lower ends are palpable.

The upper end consists of a head, neck, and two trochanters. Its trabeculae are well seen radiographically (see fig. 12–7). The *head* faces upward, medially, and slightly forward and presents a *pit* (or *fovea*) to which the ligament of the head is attached (figs. 12–10 and 12–12). **The blood supply to the head is important, because it may be interrupted when the neck is fractured.** Metaphysial and epiphysial arteries from the circumflex are carried in retinacula to the head and neck, and further epiphysial arteries from the obturator enter the head by way of the ligament of the head (see fig. 15–15). The *neck* is separated from the shaft in front by the *intertrochanteric line*, which can be traced downward to a *spiral line* that joins the medial lip of the linea aspera (fig.

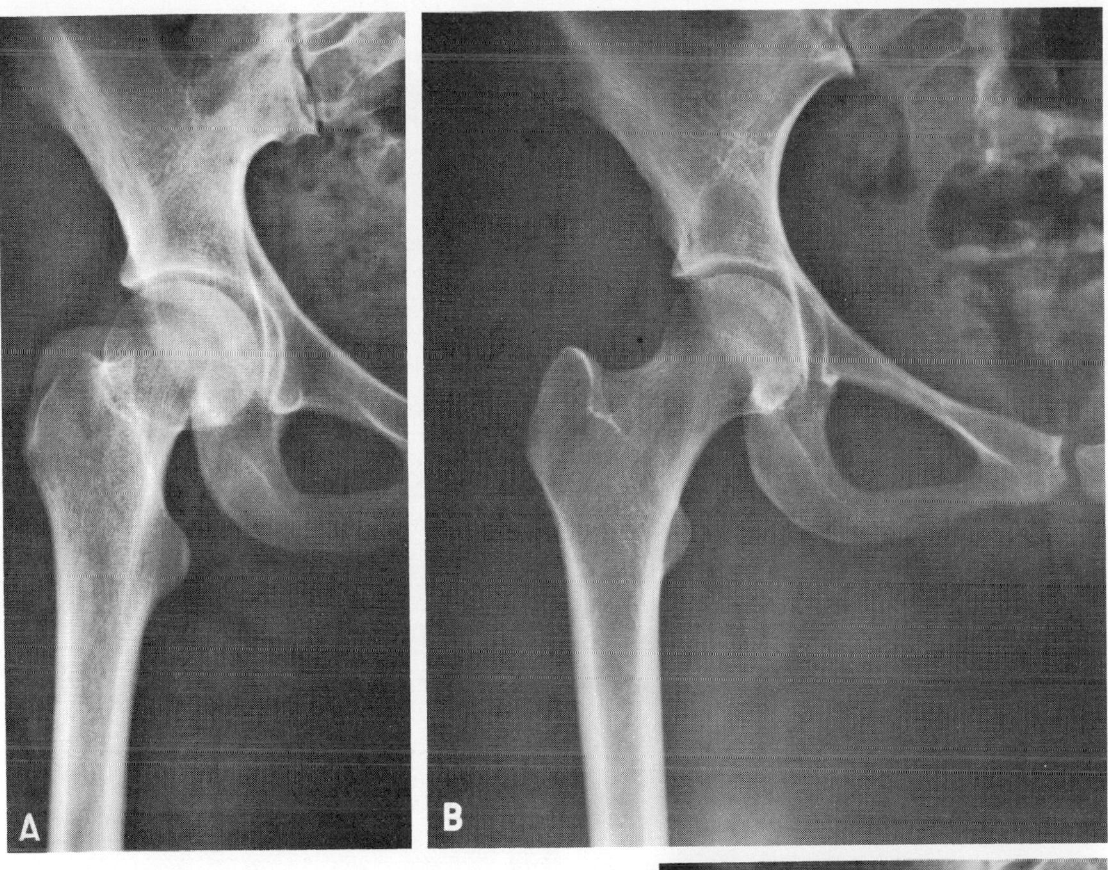

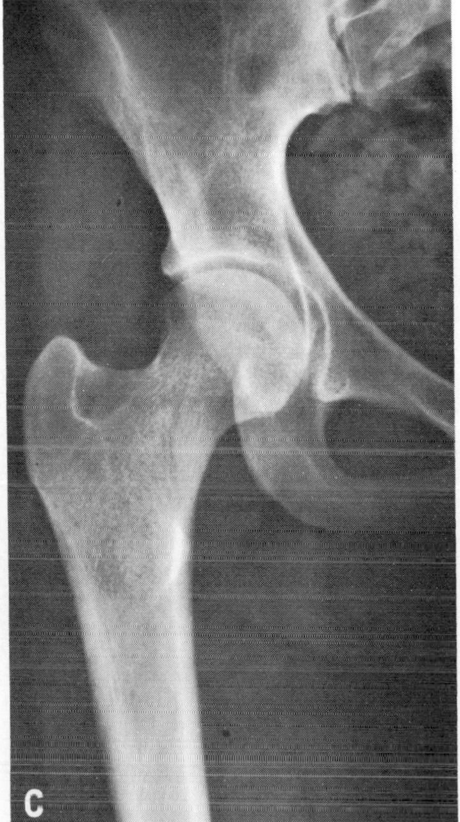

Figure 12–7 The hip in various positions. **A**, Maximum lateral rotation. Note the foreshortening of the neck of the femur. The lesser trochanter is clearly visible. **B**, anatomical position. Although the neck shows well, it is still slightly foreshortened, and a slight degree of medial rotation would be necessary to show it correctly. **C**, Maximum medial rotation. Note that the lesser trochanter is now overlapped completely by the shaft of the femur.

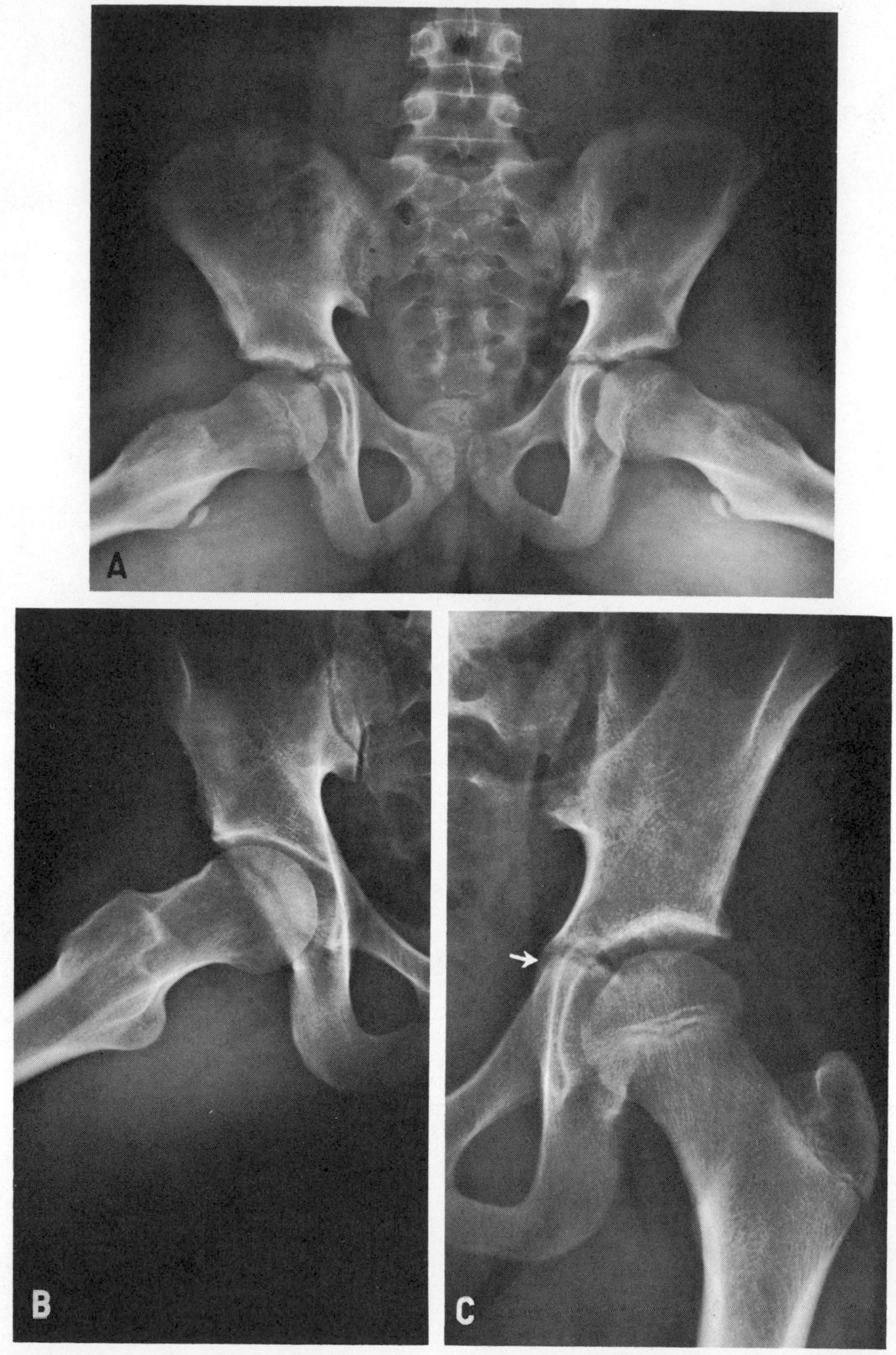

Figure 12–8 **A**, Child's hips, abducted. Note the epiphysis for the lesser trochanter on each side. **B**, Hip of an adult, abducted. Note the greater trochanter (*above*) and lesser trochanter (*below*). **C**, Child's hip. Note the site of the triradiate cartilage (*arrow*) and the epiphyses for the head and greater trochanter of the femur.

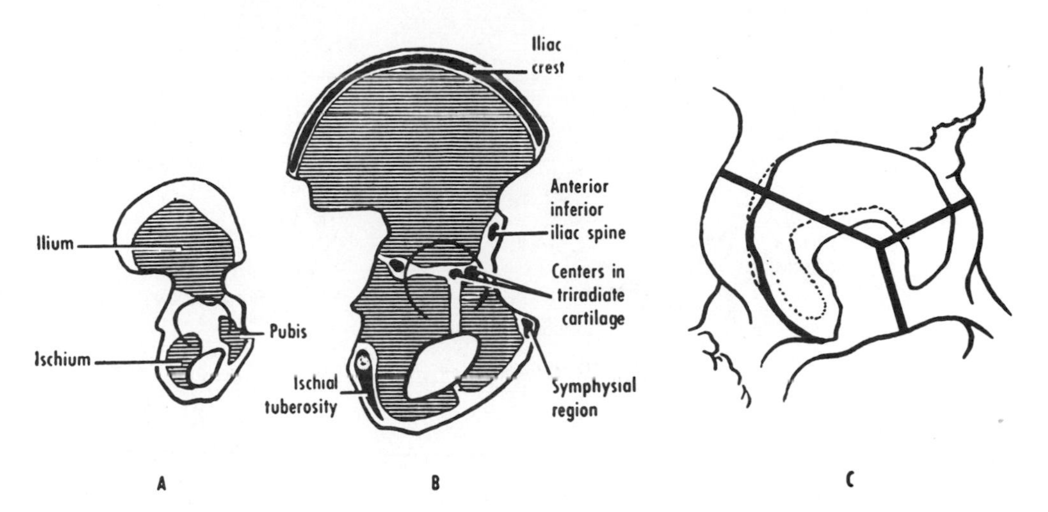

Figure 12–9 **A,** The hip bone at birth. Lined regions indicate bony areas in the ilium, ischium, and pubis. The rest is cartilage. **B,** The hip bone at puberty, showing the increase in ossification and secondary centers (*black*). **C,** Adult acetabulum. The lines indicate the site of fusion of the ilium, ischium, and pubis.

12–13). Posteriorly, about two thirds of the neck is intracapsular (fig. 12–18). The plane of the neck, followed medially, usually lies in front of that of the condyles at the lower end (*anteversion* of head of femur), and the two planes form an angle (fig. 12–16). **This angle of femoral torsion is about 15 degrees in the adult but is approximately twice as great in infancy. The degree of anteversion may be altered in pathological conditions, and its determination may be important in diagnosis and treatment.** In addition, the long axes of the neck and shaft make an *angle of inclination* (fig. 12–17), which may be altered by any pathological condition that weakens the neck of the femur. **When the angle of inclination is diminished, the condition is known as coxa vara; when it is increased, as coxa valga. The neck is a common site of fracture in older people, with danger of avascular necrosis of the head. The greater trochanter is situated laterally and can be palpated on the lateral side of the thigh, a handsbreadth or more below the iliac crest, most easily with the hip abducted. When a subject is in the erect position, the greater trochanters are in the same horizontal plane as the pubic tubercles, the heads of the femora (and hence the hip joints), and the coccyx.** Posteriorly, the *greater trochanter* can be traced downward into the *intertrochanteric crest,* which presents an elevation termed the *quadrate tubercle* (fig. 12–13). A depression medial to the greater trochanter posteriorly is the *trochanteric fossa* (fig. 12–12). The *lesser trochanter* projects medially at the junction of the neck with the shaft.

The *shaft* of the femur, which is convex forward, presents anterior, medial, and lateral surfaces. In the middle third, the prominent posterior border is known as the *linea aspera* (figs. 12–13 and 12–14). It has *medial* and *lateral lips* and an intermediate area that broadens into a posterior surface in the upper and lower thirds of the shaft. Above, the medial lip is continuous with the spiral line and the lateral lip with the *gluteal tuberosity*. Between the continuations of the lips, the *pectineal line* extends upward to the back of the lesser trochanter. Below, the medial lip is continuous with the *medial supracondylar line,* which is interrupted to allow the passage of the femoral artery and ends in the adductor tubercle. The lateral lip is continuous with the *lateral supracondylar line,* which descends to the lateral epicondyle. Posteriorly and below, the posterior, or *popliteal, surface* lies between the two supracondylar lines.

The lower end of the femur consists of two *condyles,* which are continuous in front but separated below and behind by the *intercondylar fossa*. In front, the condyles form the *patellar surface,* which comprises a wider lateral and a narrower medial part (fig. 12–10); these articulate with corresponding facets on the patella. Medially, the most prominent part of the medial condyle is the *medial epicondyle*. The *adductor tubercle,* a small prominence on the uppermost portion of the medial condyle, can be found by tracing the tendon of the adductor magnus downward. Laterally, the lateral condyle presents the *lateral epicondyle,* near which are the origins of the lateral head of the gastrocnemius and the popliteus.

The obliquity of the femur results in an obtuse, lateral angle at the knee. Its exaggeration is termed *genu valgum* (knock-knee). Genu valgum may tend to encourage lateral dislocation of the patella.

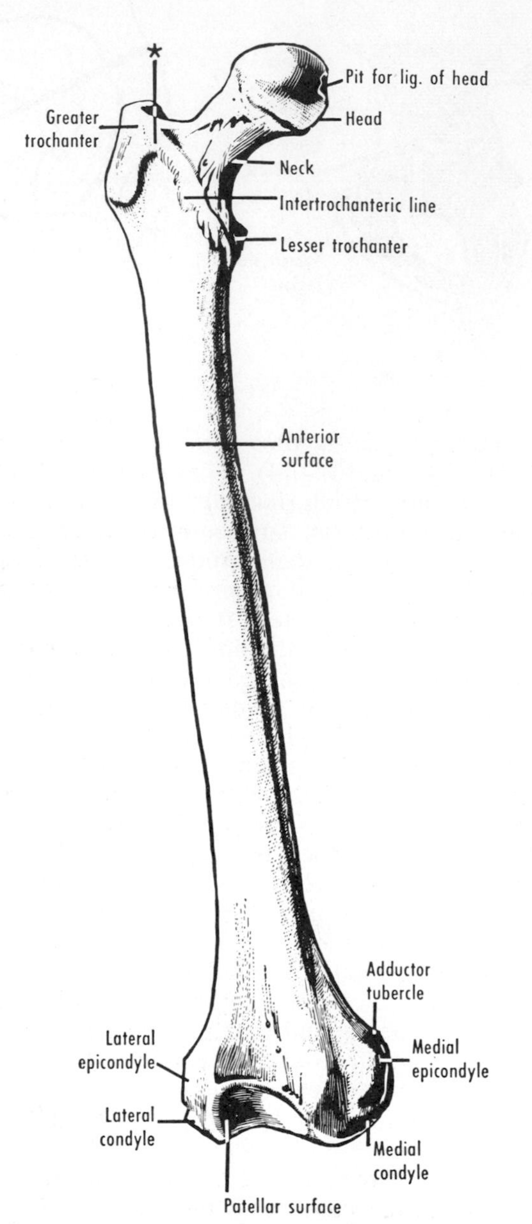

Figure 12–10 The right femur, anterior view, anatomical position. The asterisk indicates the cervical tubercle.

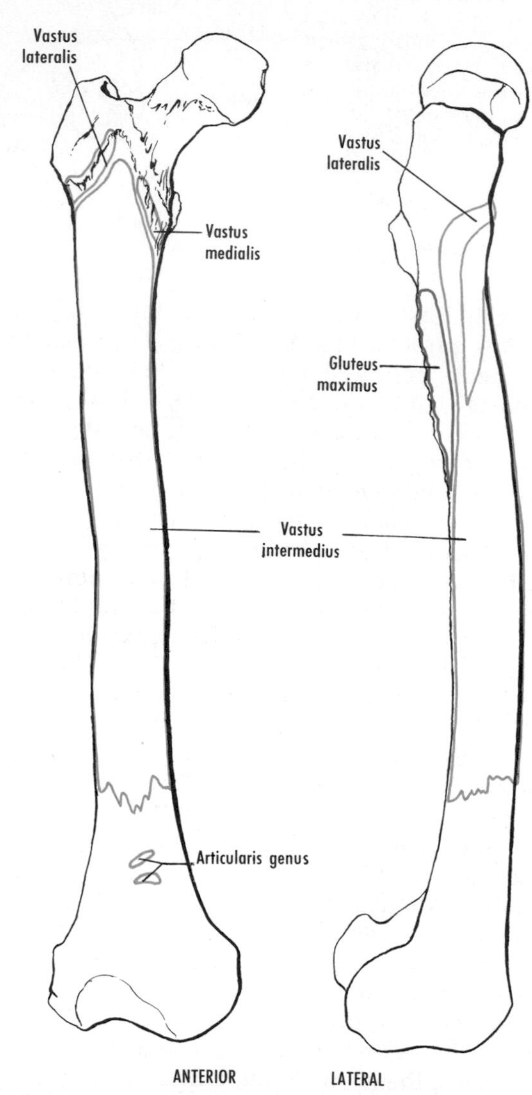

Figure 12–11 The right femur, showing muscular attachments. The lower part of the origin of the vastus intermedius fuses with that of the vastus lateralis.

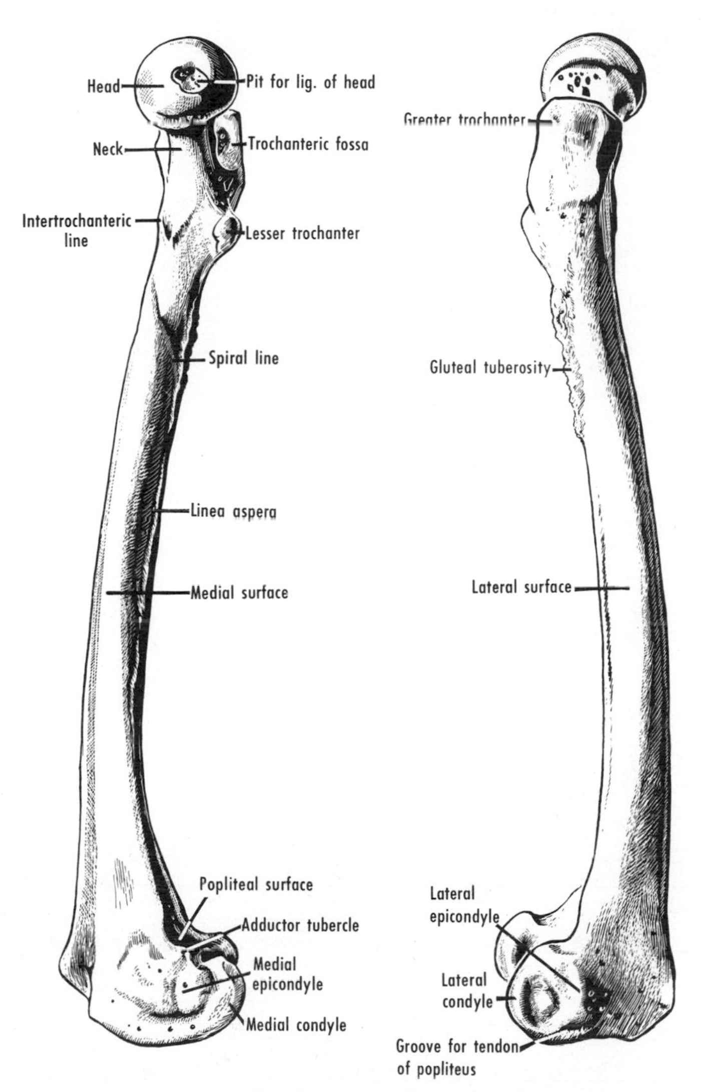

Figure 12–12 The right femur, medial and lateral views.

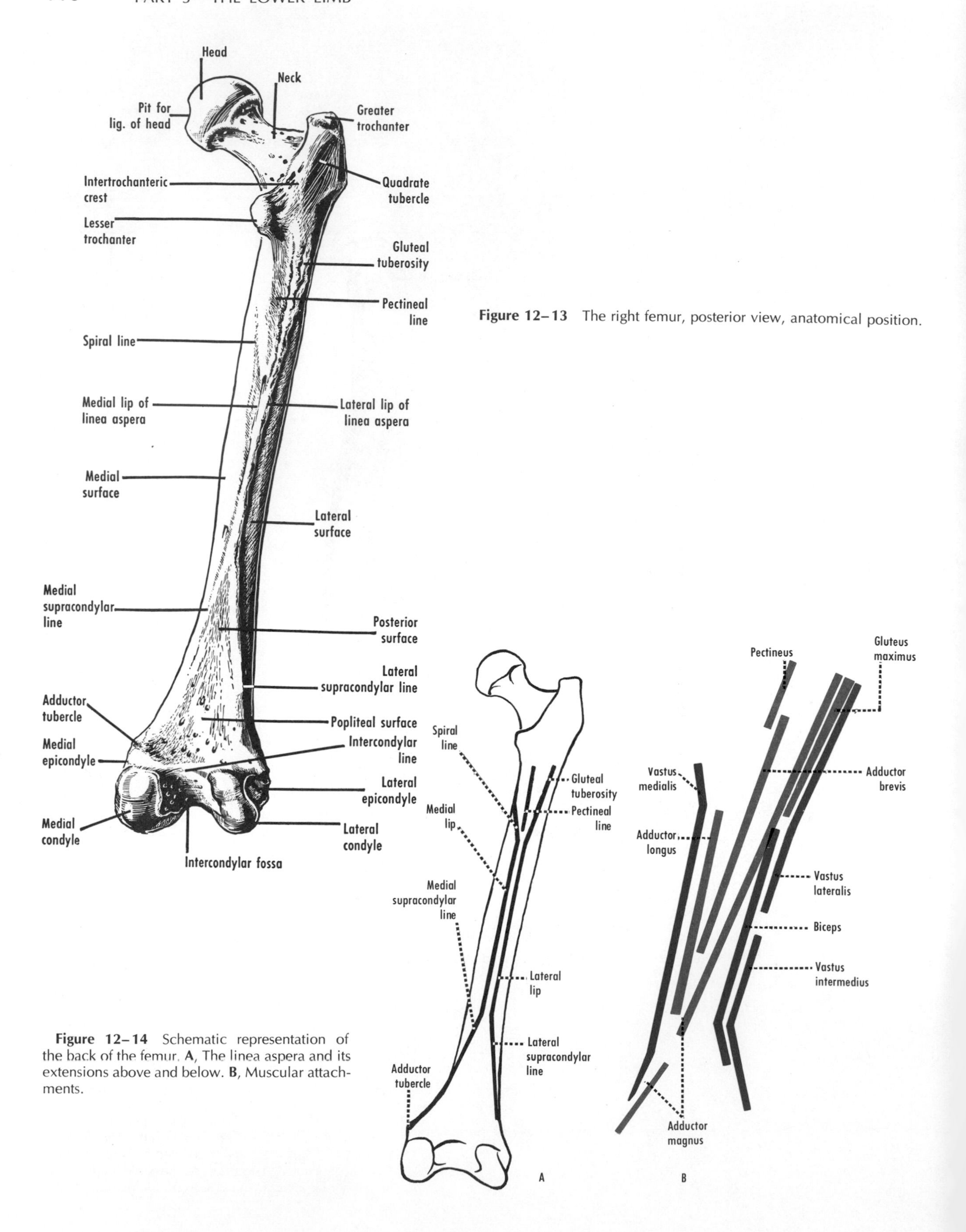

Figure 12–13 The right femur, posterior view, anatomical position.

Figure 12–14 Schematic representation of the back of the femur. **A**, The linea aspera and its extensions above and below. **B**, Muscular attachments.

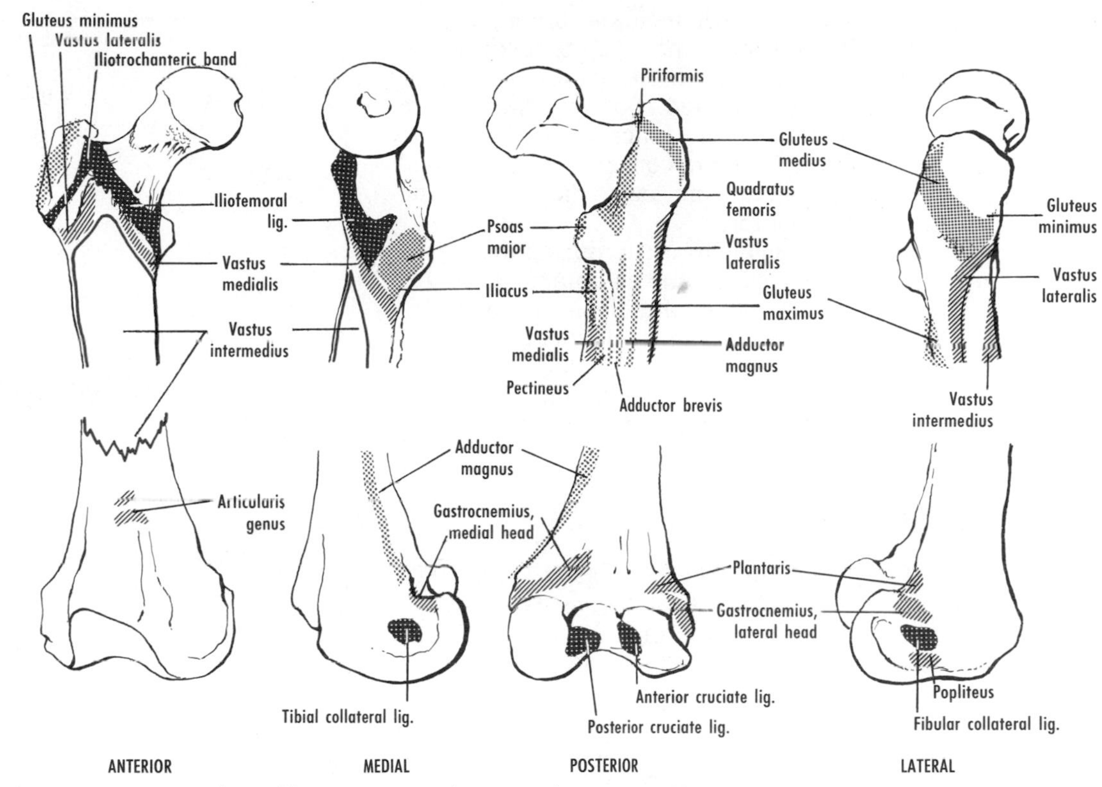

Figure 12–15 Muscular and ligamentous attachments to the upper and lower ends of the right femur. **Anterior,** The fascia that encloses the tensor fasciae latae meets at the front of that muscle, turns around the front edge of the gluteus minimus, and fuses with the rectus femoris and iliofemoral ligament at the ilium and with the tendon of the gluteus minimus below at the greater trochanter. This fascial strip constitutes the iliotrochanteric band. **Medial,** The attachment of the iliofemoral ligament turns upward above the lesser trochanter and constitutes the femoral attachment of the pubofemoral ligament. **Posterior,** See also fig. 12–14 for details. **Lateral,** The gluteus medius is inserted along an oblique line on the lateral aspect of the greater trochanter, continuous in front and below with the gluteus minimus (a bursa intervenes) and above and behind with the piriformis.

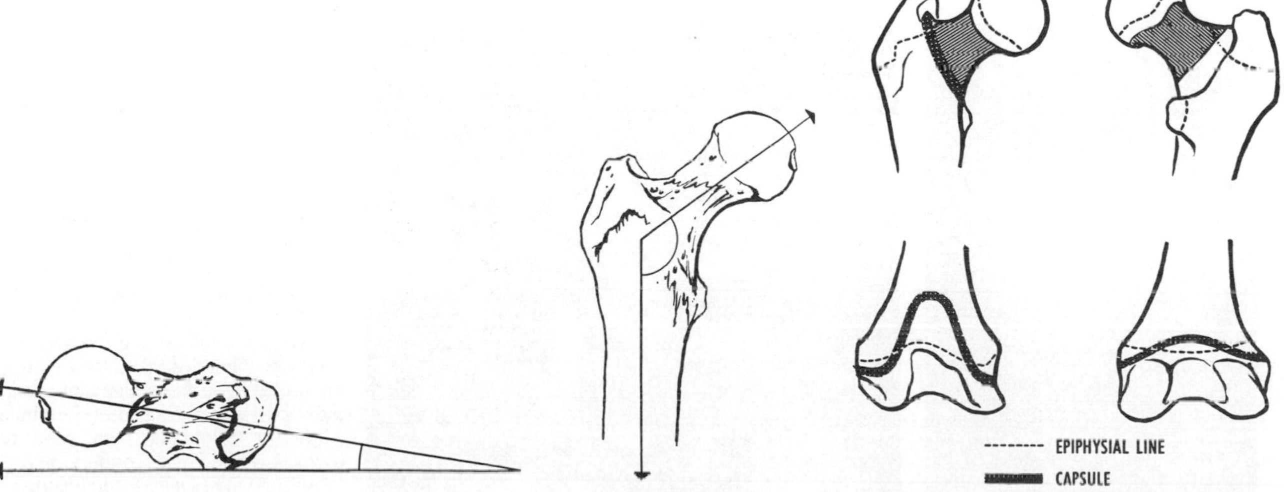

Figure 12–16 Anteversion of the head of the right femur as viewed from above. The angle of torsion is the angle between the long axis of the head (*upper arrow*) and the horizontal axis of the condyles (*lower arrow*).

Figure 12–17 The angle of inclination, which averages about 125 degrees in adults.

Figure 12–18 The upper and lower ends of the femur, showing the usual position of the epiphysial lines and the usual line of attachment of the joint capsule. The posterior part of the neck is covered by a reflection of synovial membrane but has little if any capsular attachment. (Based on Mainland.)

To determine whether shortening of the limb is in the head or neck of the femur (or in both), the relative positions of certain anatomical points are verified. For example, a line (Nélaton's) from the anterior superior iliac spine to the maximum convexity of the ischial tuberosity should normally lie above the greater trochanter.

The shaft begins to ossify during the eighth postovulatory week, and an epiphysial center is usually present in the distal end at birth. Centers appear for the head, the greater trochanter, and the lesser trochanter during infancy, childhood, and late childhood, respectively (see figs. 12–8 and 15–15).

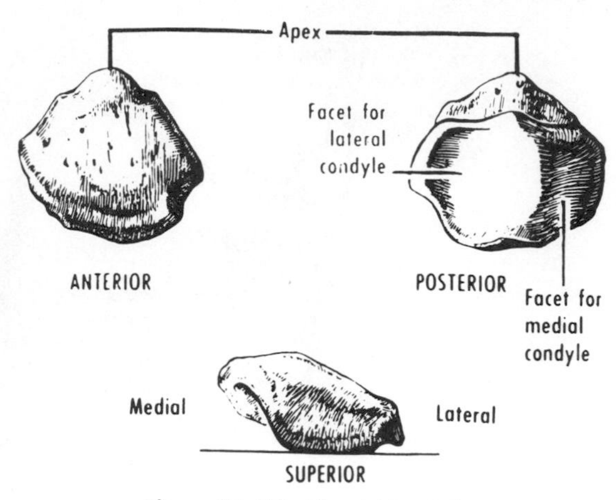

Figure 12–19 The right patella.

PATELLA

The patella, or knee cap (figs. 12–19 to 12–21), is a triangular sesamoid bone embedded in the tendon of insertion of the quadriceps femoris muscle. The superior border of the patella is the *base* of the triangle, and lateral and medial borders descend to converge at the *apex*. The patella can be moved from side to side when the quadriceps is relaxed. A part of the quadriceps tendon covers the *anterior surface* of the bone and is continued, as the *ligamentum patellae,* to the tuberosity of the tibia. The patella articulates behind with the patellar surface of the condyles of the femur. The *articular surface* of the patella comprises a larger, lateral facet and a smaller, medial one. Lateral dislocation of the patella is resisted by the shape of the lateral condyle of the femur and by the medial pull of the vastus medialis. Excision of the patella results in practically no functional deficiency. The patella ossifies from several centers, which appear during childhood.

TIBIA

The tibia, or shin bone (figs. 12–20 to 12–27), measures about one fourth to one fifth of

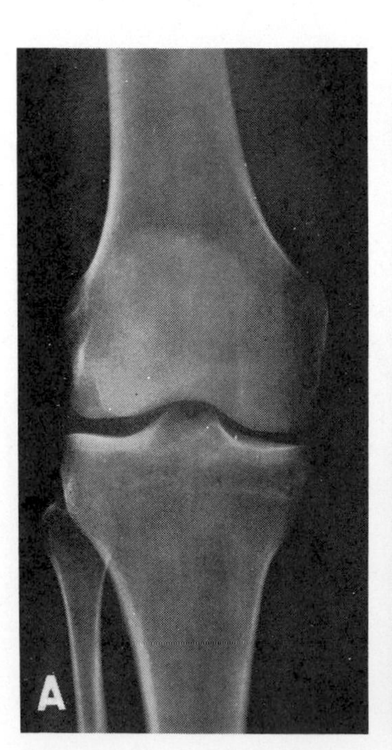

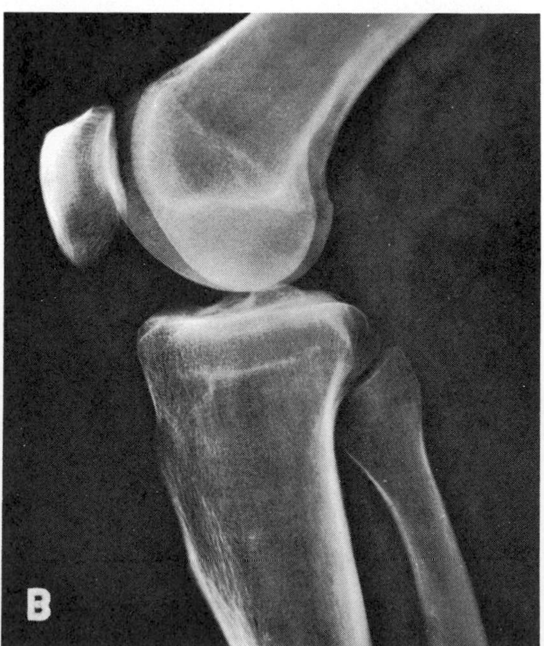

Figure 12–20 **A**, Anteroposterior view of the knee. Note the obliquity of the femur, outline of the patella, radiolucent interval occupied by the menisci and articular cartilage, and intercondylar eminence of the tibia (showing the lateral and medial intercondylar tubercles). **B**, Lateral view of the flexed knee. Note the patella, condyles of the femur, head of the fibula, (superior) tibiofibular joint, and tuberosity of the tibia. (Courtesy of V. C. Johnson, M.D., Detroit, Michigan.)

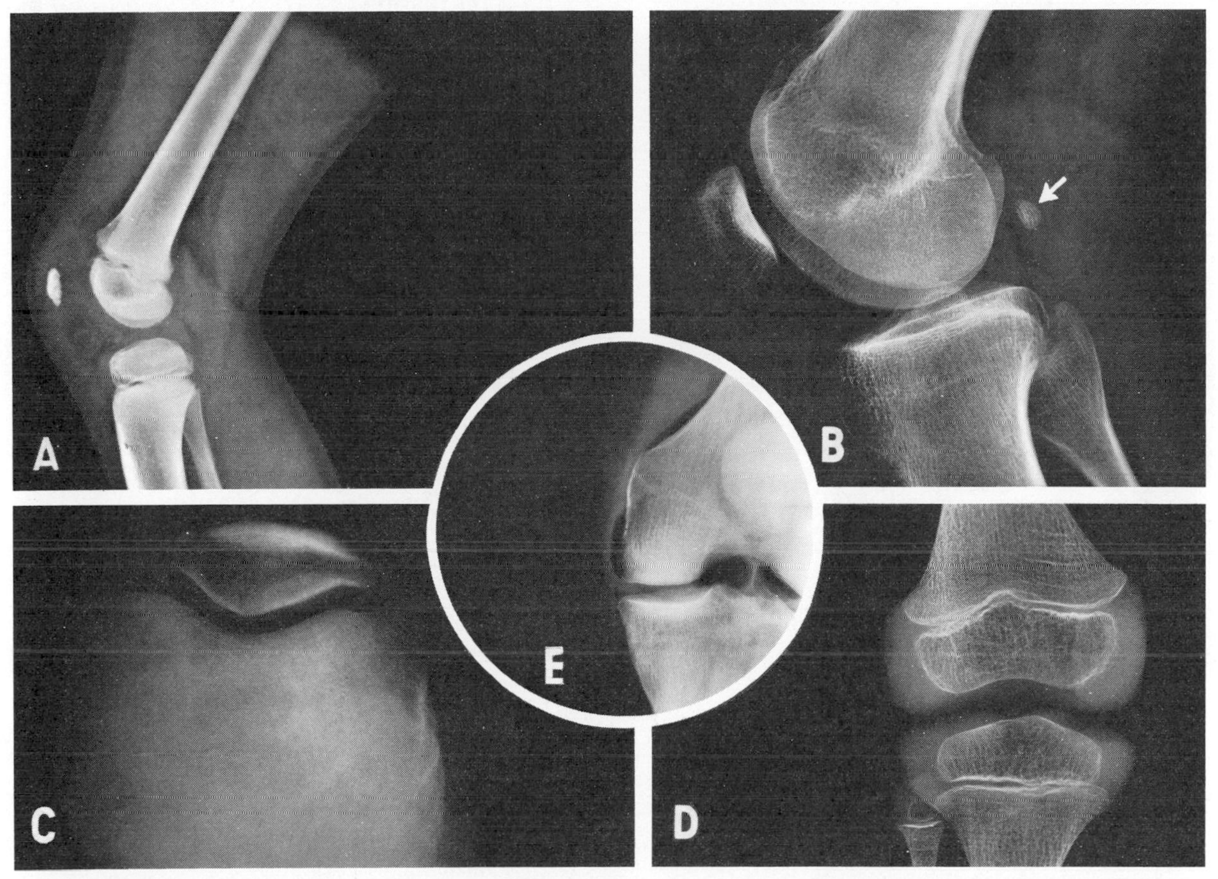

Figure 12–21 A, Child's knee, lateral view. Note the epiphyses for the lower end of the femur and the upper end of the tibia. The patella has begun to ossify, and the fat deep to the ligamentum patellae is visible as a radiolucent area. **B**, Lateral radiograph of the knee, showing fabella (*arrow*). **C**, Radiograph of the flexed knee. Note the radiological joint space between the femur and patella. The lateral condyle of the femur is that on the right-hand side of the illustration. **D**, Radiograph of dried bones of a 5-year old boy. Note the outline of the cartilage. **E**, Pneumoarthrogram of the knee produced by injecting air into the joint cavity. Note the medial meniscus and the cruciate ligaments. (**A** courtesy of V. C. Johnson, M.D., Detroit, Michigan. **E** courtesy of Sir Thomas Lodge, Sheffield, England.)

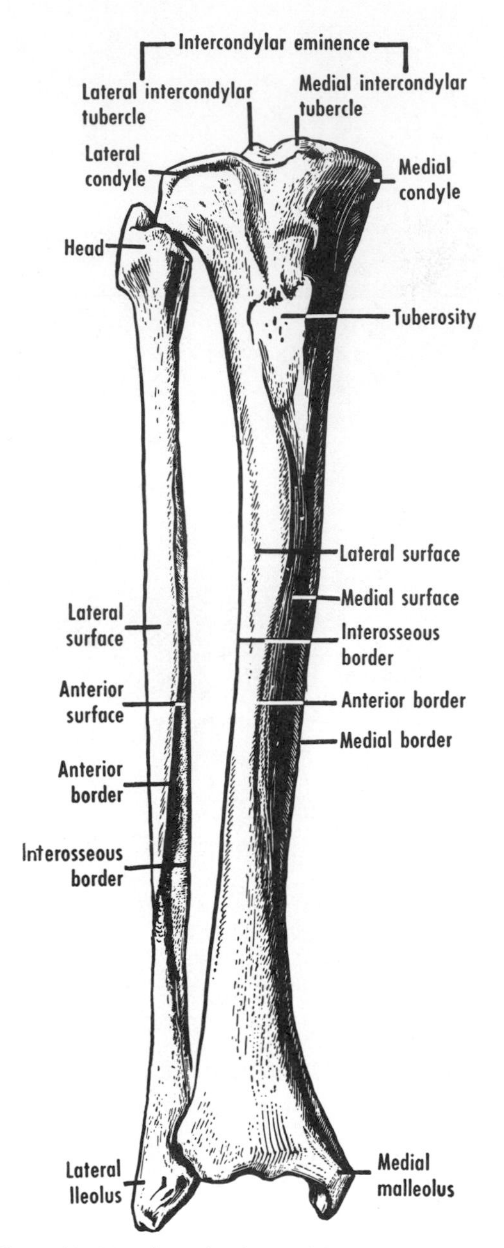

Figure 12–22 The right tibia and fibula, anterior view.

the length of the body. It can be palpated on the anterior and medial sides of the leg. When a subject is in the standing position, the tibia transmits the weight from the femur to the foot.

The upper end is expanded for articulation with the lower end of the femur. It consists of *medial* and *lateral condyles,* and a *tuberosity* is found anteriorly at the junction with the shaft. The upper end of each condyle articulates with the corresponding femoral condyle. Irregularities of the upper surface between the two tibial condyles form a series of *intercondylar areas,* an *eminence,* and *tubercles.* The lower aspect of the lateral condyle presents posteriorly a circular facet for the head of the fibula. **In the position of kneeling, the body rests on the lower part of the tuberosity, the ligamentum patellae, the front of the tibial condyles, and the patella.**

When viewed from above, the shaft of the tibia appears twisted (as if the upper end were rotated more medially than the lower). The *angle of tibial torsion* (usually 15 to 20 degrees) is that between a horizontal line through the condyles and one through the malleoli. The shaft of the tibia presents medial,

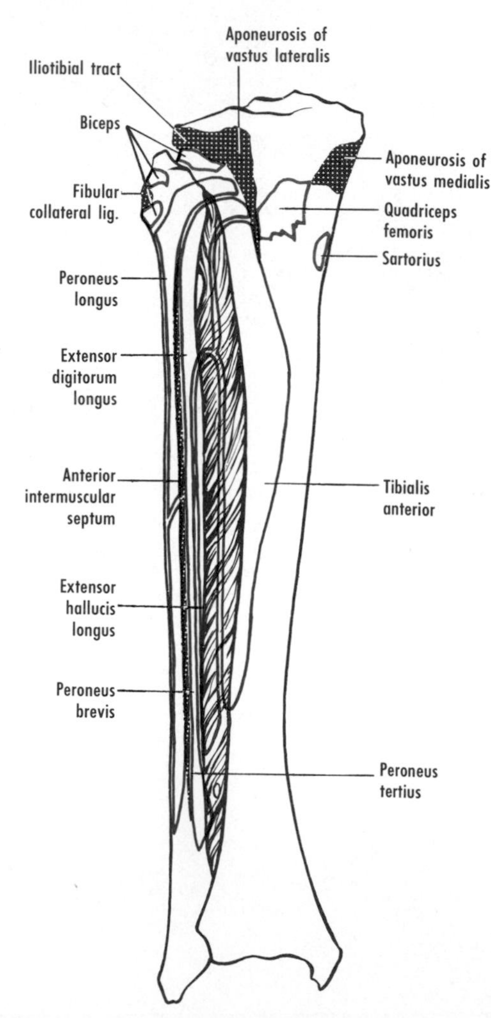

Figure 12–23 The right tibia and fibula, anterior aspect, muscular and ligamentous attachments. Note that the peroneus longus and extensor digitorum longus arise from the tibia as well as from the fibula.

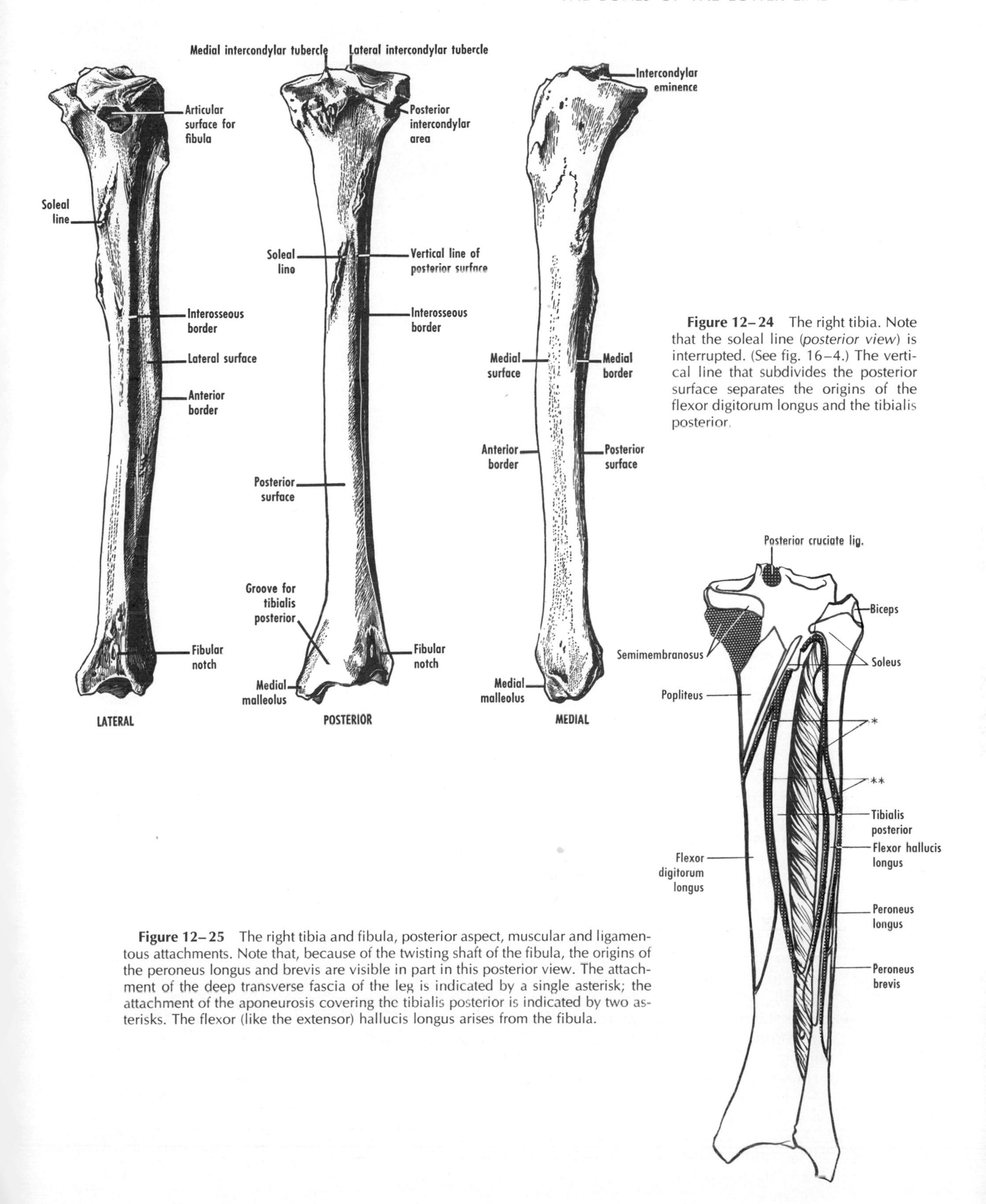

Figure 12–24 The right tibia. Note that the soleal line (*posterior view*) is interrupted. (See fig. 16–4.) The vertical line that subdivides the posterior surface separates the origins of the flexor digitorum longus and the tibialis posterior.

Figure 12–25 The right tibia and fibula, posterior aspect, muscular and ligamentous attachments. Note that, because of the twisting shaft of the fibula, the origins of the peroneus longus and brevis are visible in part in this posterior view. The attachment of the deep transverse fascia of the leg is indicated by a single asterisk; the attachment of the aponeurosis covering the tibialis posterior is indicated by two asterisks. The flexor (like the extensor) hallucis longus arises from the fibula.

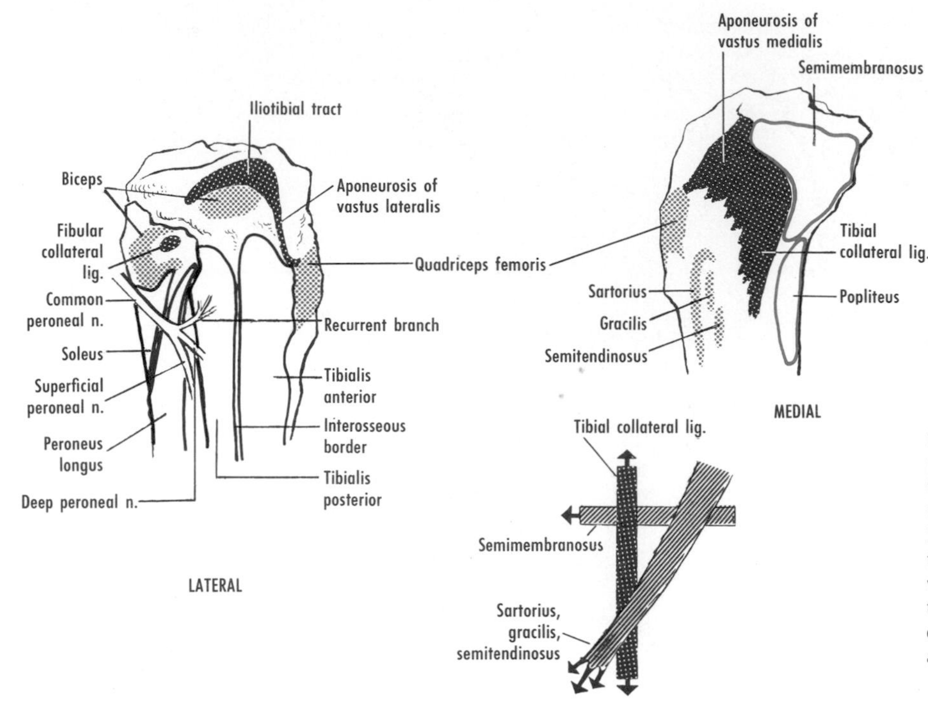

Figure 12–26 Muscular and ligamentous attachments of the upper ends of the right tibia and fibula. In the lateral view, note that the interosseous border of the tibia separates the attachments of the tibialis anterior and the tibialis posterior. The origin of the peroneus longus and extensor digitorum longus from the tibia is not shown. (See fig. 12–23.) **Lower right,** Schematic representation of the relations of the tibial collateral ligament, which crosses the semimembranosus tendon but which is deep to the tendons of the gracilis, semitendinosus, and sartorius (the pes anserinus).

lateral, and posterior surfaces and anterior, interosseous, and medial borders. The *medial surface* is readily palpable. The *posterior surface,* which is crossed above by a rough ridge termed the *soleal line,* may be considered to have medial and lateral parts, and these may be separated by an indistinct vertical ridge. The *anterior border,* or crest, forms the easily palpable "shin" and descends from the lateral side of the tuberosity to the front of the medial malleolus. The *interosseous* (or *lateral*) *border* gives attachment to the interosseous membrane.

The lower end of the tibia presents: (1) an anterior surface; (2) a lateral surface, which ends in the *fibular notch* (for the lower end of the fibula); (3) a posterior surface (grooved by the tibialis posterior and flexor digitorum longus tendons); (4) a medial surface, which runs onto the distal prolongation of the tibia known as the *medial malleolus;* and (5) an inferior surface, which articulates with the talus. It should be noted that the talus also articulates with the lateral surface of the medial malleolus.

The shaft begins to ossify during the eighth postovulatory week, and an epiphysial center

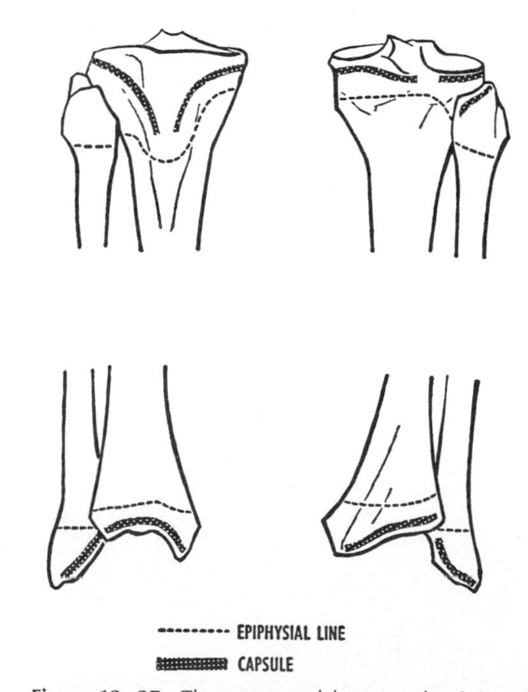

Figure 12–27 The upper and lower ends of the right tibia and fibula, showing the usual position of the epiphysial lines and the usual line of attachment of the joint capsule. (Based on Mainland.)

is usually present in the upper end at birth. Centers appear for the lower end during infancy (see fig. 12–30), commonly for the tuberosity (which ossifies mainly by downward growth from the upper epiphysis) and sometimes for the tip of the medial malleolus.

FIBULA

The fibula (figs. 12–20 to 12–23 and 12–25 to 12–31) is the slender, lateral bone of the leg. It does not bear weight. *Peroneal* is synonymous with *fibular*. The fibula articulates with the tibia above and with the talus below and is anchored in between to the tibia by the interosseous membrane. The upper and lower ends of the bone are palpable, but muscles cover its middle portion.

The upper end, or *head,* articulates with the lower aspect of the lateral condyle of the tibia posteriorly. **The head of the fibula is readily palpable by tracing the biceps tendon downward** (see figs. 12–23 and 12–25). It is on the

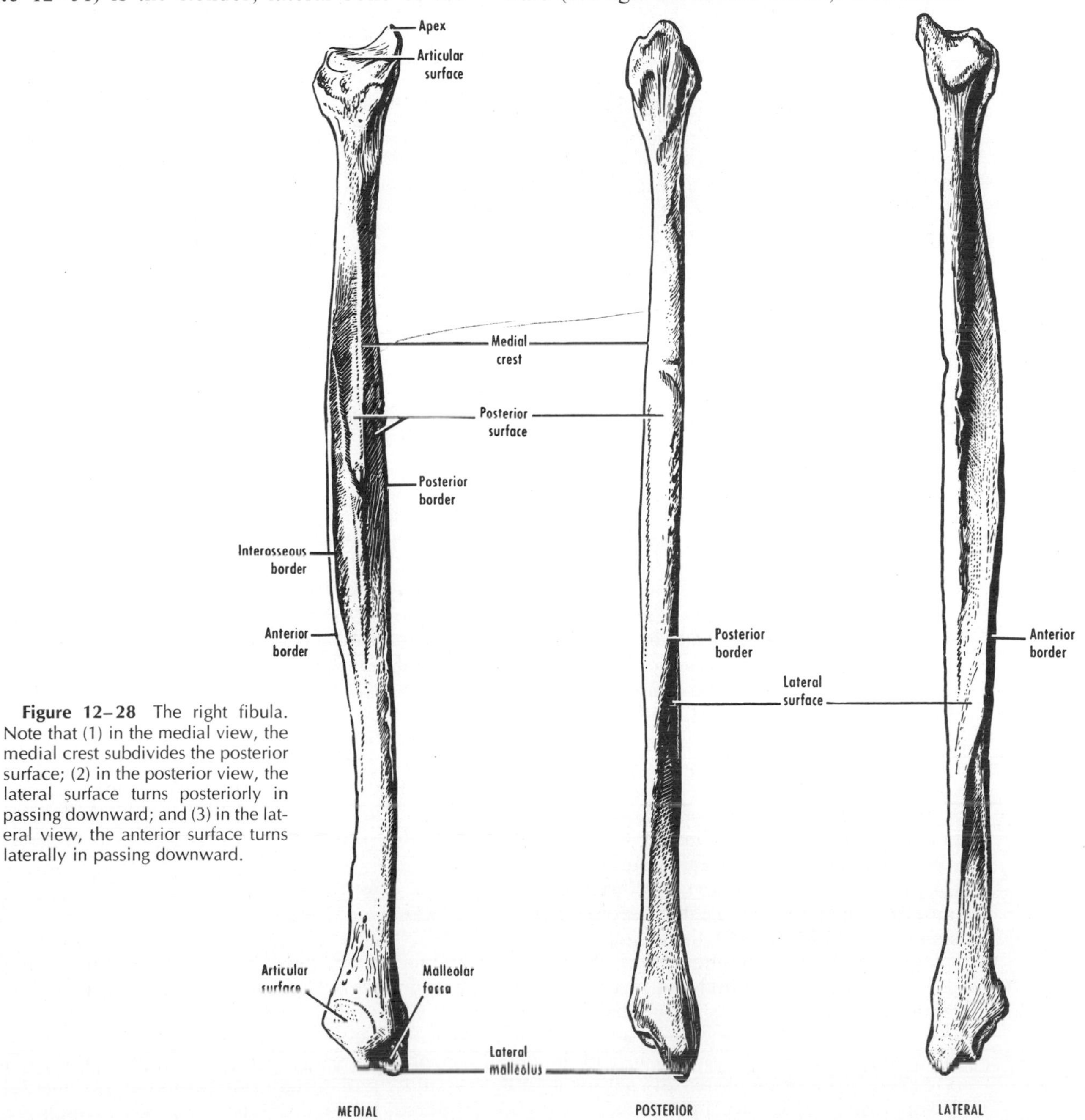

Figure 12–28 The right fibula. Note that (1) in the medial view, the medial crest subdivides the posterior surface; (2) in the posterior view, the lateral surface turns posteriorly in passing downward; and (3) in the lateral view, the anterior surface turns laterally in passing downward.

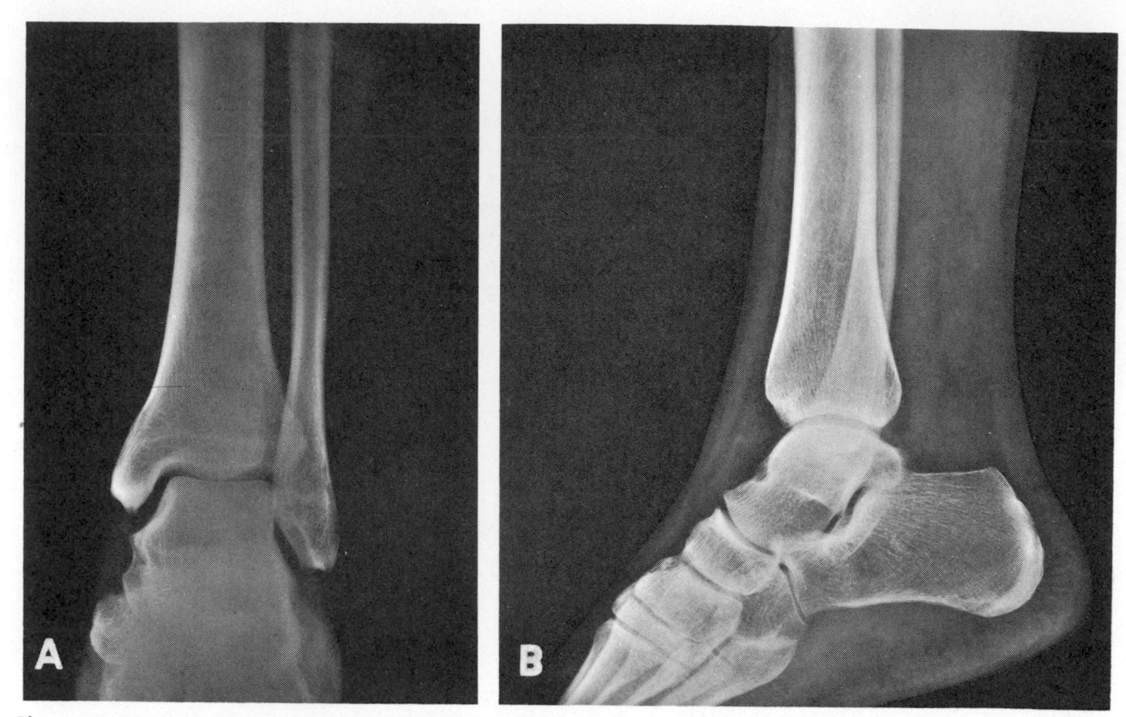

Figure 12–29 **A**, Anteroposterior view of the left ankle. Note the medial and lateral malleoli (and their different levels) and the trochlea of the talus. **B**, Lateral view of the ankle. Note the line of the talotibial part of the joint and the outlines of the talus, navicular, and calcaneus. (**B** courtesy of V. C. Johnson, M.D., Detroit, Michigan.)

same level as the tuberosity of the tibia. The head is prolonged upward into an *apex* (or *styloid process*) posterolaterally. **The common peroneal nerve winds from behind the head and onto the lateral aspect of the "neck," where it can be rolled between a finger and the bone** (see fig. 12–26). The *shaft* of the fibula is arched forward in such a way that the plane of the interosseous membrane is almost sagittal, except below, where it becomes coronal (as is the interosseous membrane of the forearm throughout its extent). **The seeming torsion of the fibula and the changing relationships of the fibula and tibia must be kept in mind in order to understand the topography of the leg.** The surfaces and borders of the shaft vary considerably (as does their nomenclature: the Birmingham Revision is followed here). When well-developed, at least three surfaces (anterior, lateral, and posterior) and three borders (anterior, posterior, and interosseous) can be distinguished. Moreover, the posterior surface is subdivided by a prominent ridge, the *medial crest,* as a result of which the medial portion of the posterior surface is frequently described as a separate, medial surface. In terms of muscular attachments, the anterior, lateral, and posterior surfaces are extensor, peroneal, and flexor, respectively (see fig. 16–1).

The lower end of the fibula, or the lateral malleolus, is more prominent and more posterior and extends about 1 cm more distally than the medial malleolus. It articulates with the tibia and with the lateral surface of the talus; the talus fits between the two malleoli. Posteromedially, a *malleolar fossa* gives attachment to ligaments. Posteriorly, a groove on the lateral malleolus is occupied by the peroneal tendons. The classic (Pott's) fracture at the ankle involves the lower end of the fibula.

The shaft begins to ossify at about the junction of the embryonic and fetal periods. Centers appear for the lower end during infancy and for the upper end during childhood. **It should be noted that the lower epiphysial line of the fibula is in line with the upper surface of the talus in the ankle joint** (figs. 12–27 and 12–30).

TARSUS

The tarsus (figs. 12–30 to 12–35) usually comprises seven bones, one of which, the talus, articulates with the bones of the leg. The tarsus is convex superiorly and concave inferiorly. The tarsal bones are the talus, navicular, and three cuneiforms on the medial side, and the calcaneus and cuboid, which are

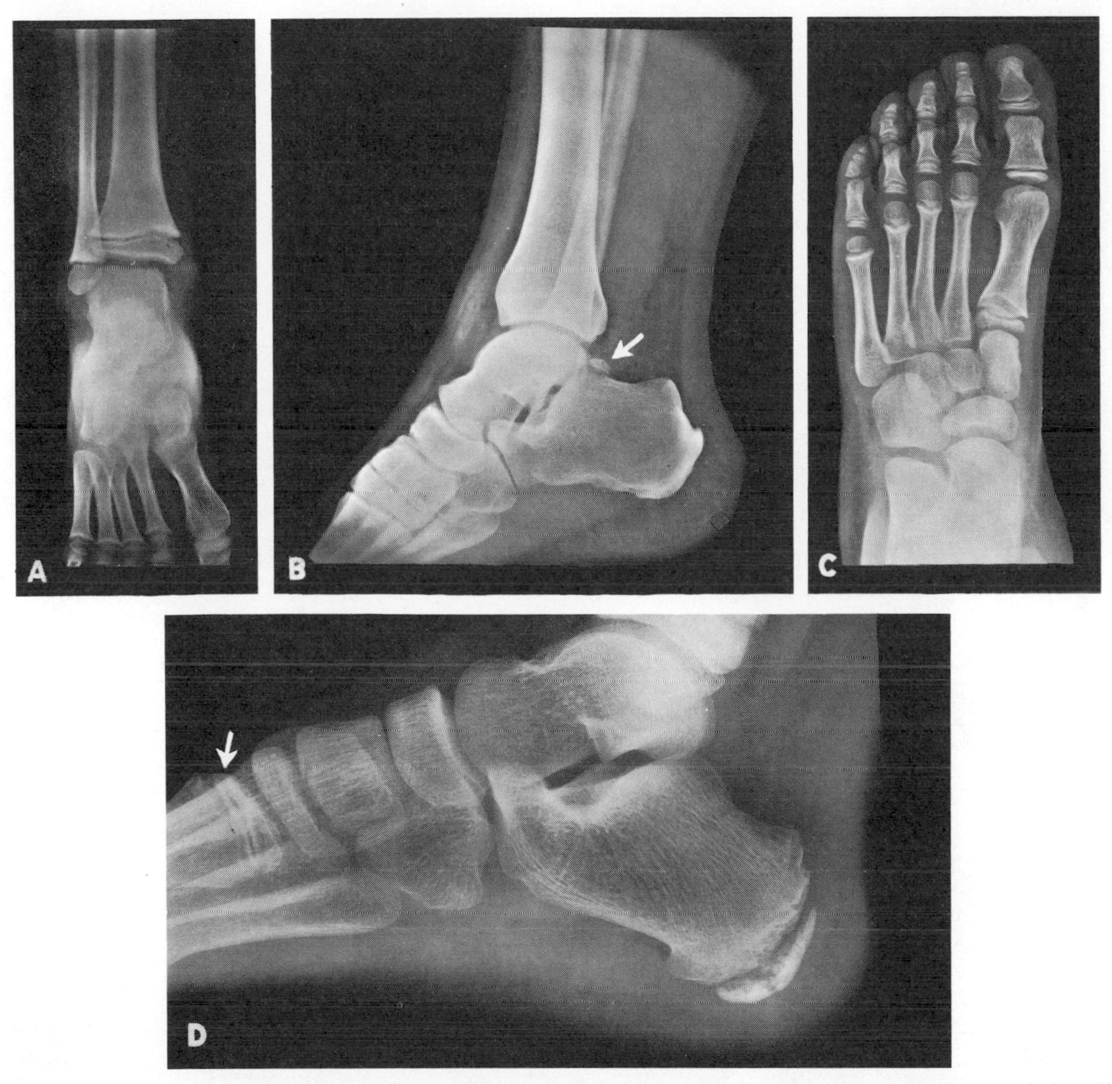

Figure 12–30 The ankle and foot. **A**, Child's ankle. Note the epiphyses for the lower ends of the fibula and tibia. The epiphysial line of the fibula is in line with the ankle joint. **B**, Lateral view of an adult ankle. Note the os trigonum (*arrow*) at the back of the talus. **C**, Child's foot. Note the epiphyses for the metatarsals and phalanges. Note also the irregularity in the ossification of the phalanges of the fifth toe. **D**, Lateral view of child's foot. Note the epiphyses for the base of the first metatarsal (*arrow*) and for the calcaneus (**A**, **B**, and **C** courtesy of V. C. Johnson, M.D., Detroit, Michigan. **D** courtesy of George L. Sackett, M.D., Painesville, Ohio.)

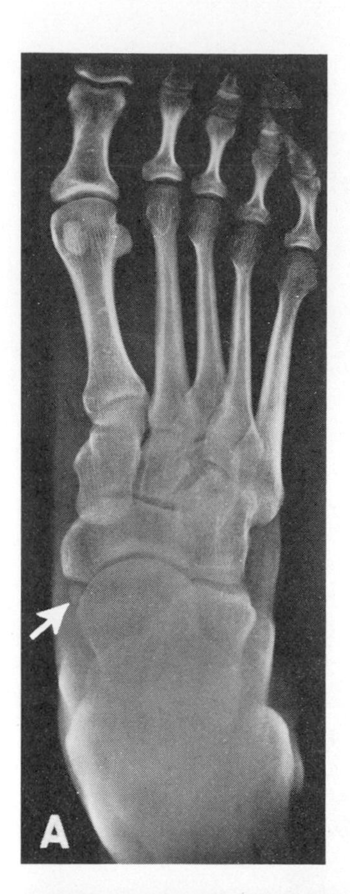

Figure 12–31 **A**, Dorsoplantar view of the foot. At least a portion of all seven tarsal bones can be identified. Note the os tibiale externum (*arrow*) near the tuberosity of the navicular. Note also the sesamoid bones below the head of the first metatarsal. **B**, Lateral view of the foot. Note the navicular and its tuberosity (overlapped by the head of the talus), the cuboid, a peroneal sesamoid, and the tuberosity of the fifth metatarsal. The numerals 1, 2, and 3 indicate the lines of the first, second, and third cuneometatarsal joints, respectively. **C**, Oblique view of the foot. Note the region where the calcaneus may meet the navicular. A peroneal sesamoid (*arrow*) is visible near the tuberosity of the cuboid. (Courtesy of V. C. Johnson, M.D., Detroit, Michigan.)

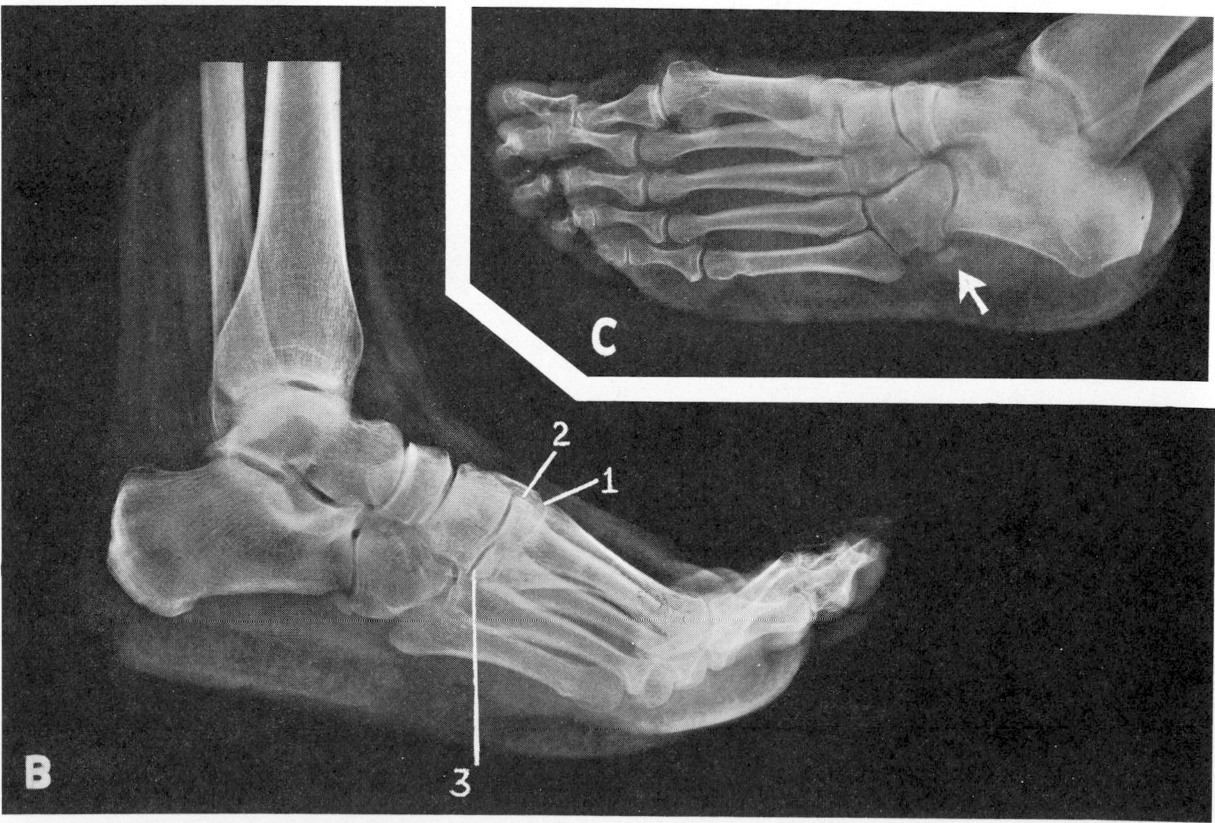

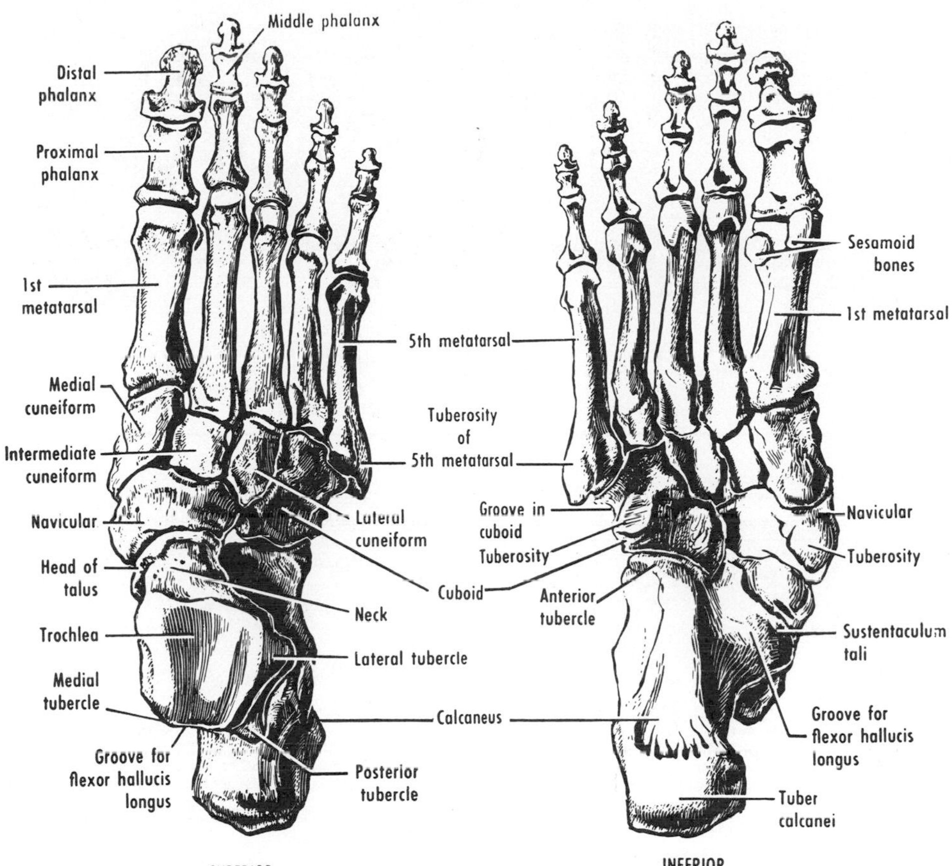

Figure 12–32 Bones of the right foot.

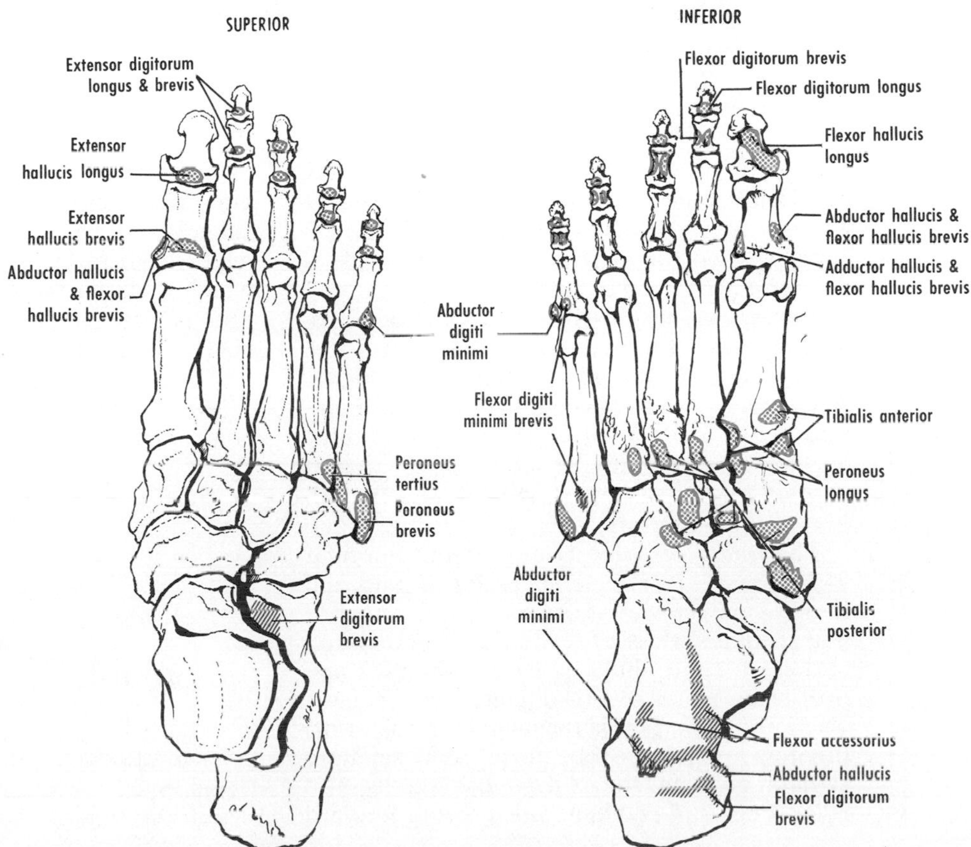

Figure 12–33 Bones of the right foot, muscular and ligamentous attachments. The attachments of the interossei are omitted.

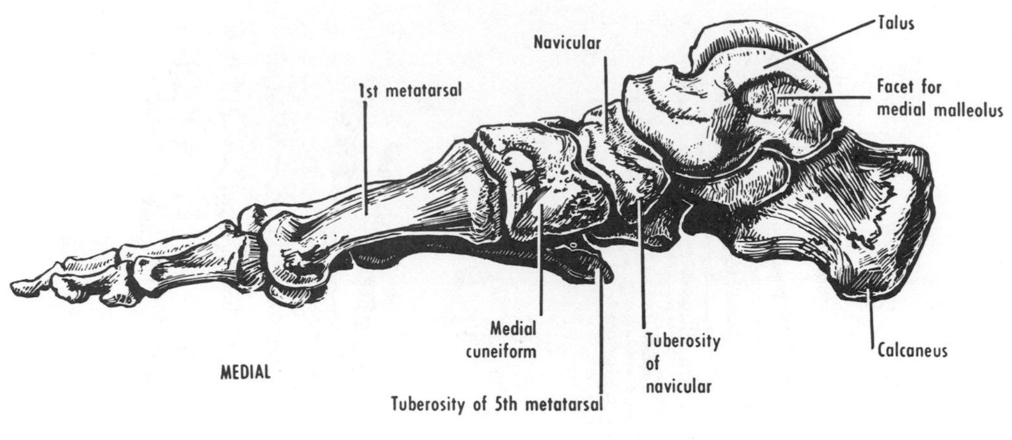

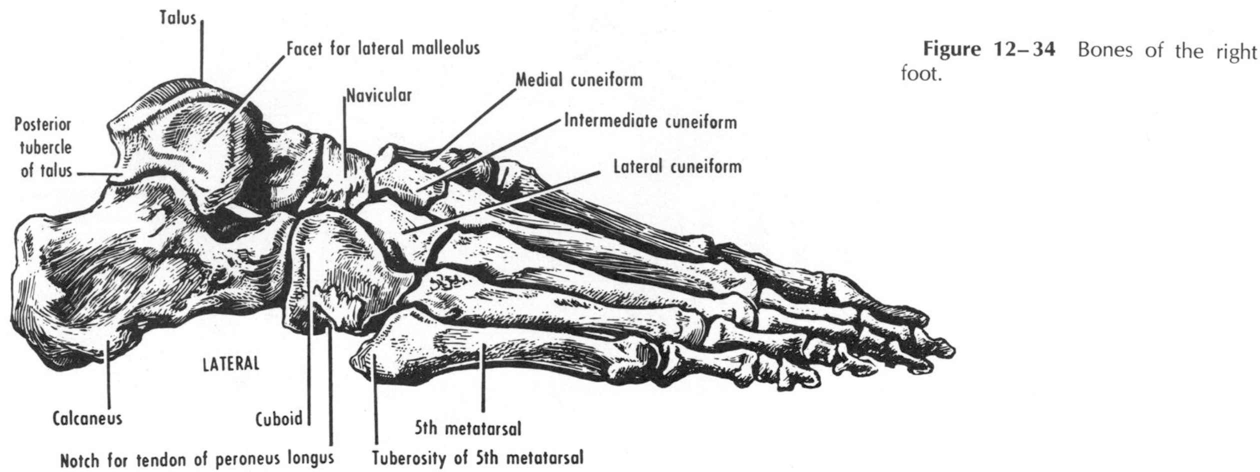

Figure 12–34 Bones of the right foot.

more laterally placed. Accessory ossicles may be found, e.g., a peroneal sesamoid, the *os tibiale externum* (near the tuberosity of the navicular), and the *os trigonum* (at the back of the talus; see fig. 12–30*B*).

The navicular, cuneiforms, and cuboid, together with the five metatarsals, form the transverse arch of the foot. Medially, a longitudinal arch is established by the calcaneus, talus, navicular, cuneiforms, and the first three metatarsals. Laterally, a longitudinal arch is formed by the calcaneus, cuboid, and the lateral two metatarsals.

The *talus*, or ankle bone, has no muscular attachments. Its *body* is its posterior part, and it presents the *trochlea*, which has superior and medial surfaces for the tibia and a lateral surface for the fibula. Various tubercles and processes have received conflicting names. The most important is the lateral (or posterior) tubercle, which may be found as a separate skeletal element, the *os trigonum* (see fig. 12–30*B*). The *neck* and head are directed forward and medially. A deep depression under the lateral side of the neck and above the calcaneus is known as the *sinus tarsi*. It becomes narrow medially to form the *tarsal canal*. The *head*, which rests on the sustentaculum of the calcaneus, articulates with the navicular and the plantar calcaneonavicular ligament. Inferiorly, the talus presents three facets for the calcaneus (fig. 12–36). Small portions of the talus may be palpable, especially in children.

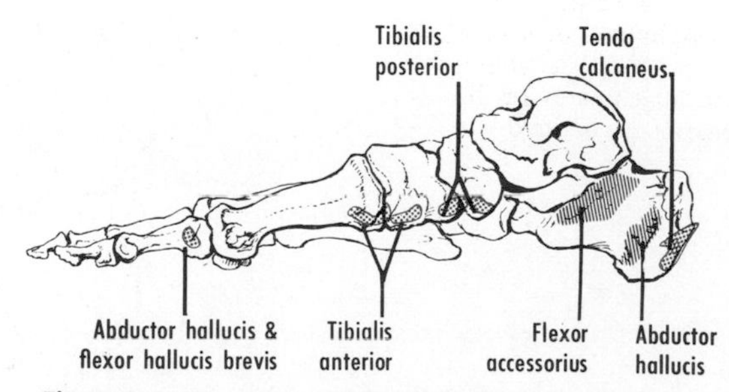

Figure 12–35 Bones of the right foot, medial aspect, muscular and tendinous attachments.

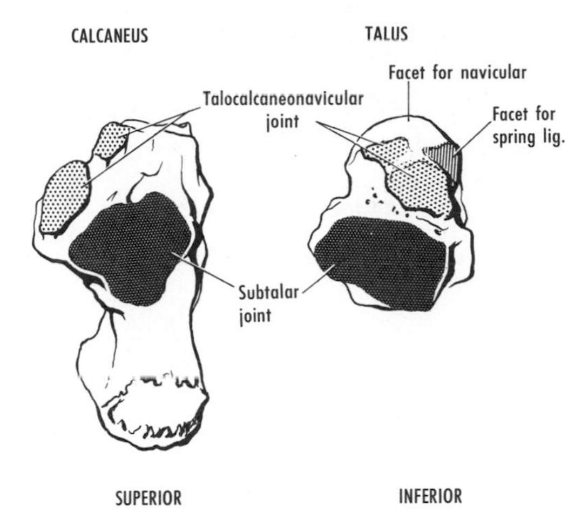

Figure 12–36 The right calcaneous and talus, showing corresponding articular facets. Cf. fig. 17–7.

The *navicular* lies between the talus behind and the three cuneiforms in front. Its medial projection, **the tuberosity of the navicular, can be palpated below and in front of the medial malleolus and in front of the sustentaculum tali. The back of the tuberosity may be taken as the medial end of the transverse tarsal joint. (The lateral end is about halfway between the lateral malleolus and the tuberosity of the fifth metatarsal, which is palpable.)** The tuberosity provides the main insertion for the tibialis posterior tendon.

The three *cuneiform bones,* so called because they are wedge-shaped, lie on the navicular behind and the first three metatarsals in front. The medial and lateral cuneiforms project farther forward than does the intermediate one, leaving a gap that is occupied by the second metatarsal. The line of the tarsometatarsal joints is thus irregular. The medial cuneiform can be identified by tracing the tibialis anterior tendon forward to it in the dorsiflexed foot.

The *calcaneus,* or *os calcis,* transmits much of the weight of the body from the talus to the ground. Its trabeculae are well seen radiographically (see fig. 12–30*D*). The front portion of the upper aspect is prolonged medially: **this projection, the sustentaculum tali, can be felt *in vivo* immediately below the medial malleolus.** The upper surface of the sustentaculum has a facet for the talus, and its lower surface is grooved by the flexor hallucis longus tendon. The posterior part of the calcaneus is known as the *tuber calcanei.* It gives attachment posteriorly to the tendo calcaneus and below to some of the short muscles of the sole as well as to the plantar aponeurosis. The calcaneus articulates with the cuboid in front. The front ends of the talus and calcaneus are more or less flush, and they form the transverse tarsal joint. Various tubercles and processes of the calcaneus have received confusing names.

The *cuboid* articulates behind with the calcaneus, in front with the fourth and fifth metatarsals, and medially with the lateral cuneiform and sometimes the navicular. The peroneus longus tendon lies in a notch laterally and may occupy a groove on the lower surface.

Ossification *(see fig. 12–30).* The calcaneus and talus begin to ossify during fetal life, and the cuboid shortly before birth. Epiphysial centers appear for the tuber calcanei and the posterior tubercle of the talus during childhood.

METATARSUS

The tarsus is connected to the phalanges by five metatarsal bones, referred to collectively as the metatarsus. They are numbered from 1 to 5, from the big to the little toe. They are longer and thinner than the metacarpals.

Each metatarsal is technically a long bone, consisting of a base proximally, a shaft, and a head distally. Each bone has characteristic features, e.g., the first (which carries more weight) is short and thick. The base of the fifth presents a *tuberosity,* which projects backward and laterally and is palpable *in vivo.*

The shaft of each metatarsal begins to ossify during fetal life, and centers appear postnatally in the heads of the four lateral bones and in the base of the first metatarsal (see fig. 12–30*C*). The first metatarsal, like the first metacarpal, may have a center for its head as well as its base. The tuberosity of the fifth frequently shows a separate center.

PHALANGES

The great toe (and frequently the little toe) has two phalanges, whereas each of the other toes has three. They are designated proximal, middle, and distal. Each phalanx is technically a long bone, consisting of a base proximally, a shaft, and a head distally. Although the phalanges of the foot are shaped differently from those of the hand, their basic arrangement is similar; e.g., each distal phalanx ends distally in a tuberosity. The middle and distal phalanges of the little toe are often fused. The phalanges usually begin to ossify during fetal life, and centers appear postnatally in the bases of most of them (see fig. 12–30*C*).

Sesamoid bones are found related to the lower aspect of some of the metatarsophalangeal and interphalangeal joints. Two below the head of the first metatarsal are almost constant.

ADDITIONAL READING

Frazer's Anatomy of the Human Skeleton, 6th ed., rev. by A. S. Breathnach, Churchill, London, 1965. A detailed, regional synthesis of skeletal and muscular anatomy.

QUESTIONS

12–1 How is the hip bone oriented in the anatomical position?

12–2 On what does the body weight rest in the sitting position?

12–3 To which features is the pubic tubercle a guide?

12–4 Compare the acetabulum with the glenoid cavity of the scapula.

12–5 Why is the blood supply to the head of the femur important clinically?

12–6 On which structures does one kneel?

12–7 How are (a) the adductor tubercle of the femur and (b) the head of the fibula found?

12–8 What is the relationship of the lateral malleolus to the medial?

12–9 What is the classic fracture at the ankle?

12–10 Do any tarsals show ossification at birth?

13

VESSELS AND LYMPHATIC DRAINAGE OF THE LOWER LIMB

VENOUS DRAINAGE

The chief superficial veins are the great and small saphenous veins. The *great* (or *long*) *saphenous vein* (fig. 13–1) begins on the medial side of the dorsal venous network of the foot. **It ascends in front of the medial malleolus, where it can be found for a transfusion.** It crosses the medial surface of the tibia obliquely, passes behind the medial condyles of the tibia and femur, and then ascends along the medial side of the thigh. In the femoral triangle, it pierces the cribriform fascia (at the saphenous opening) and the femoral sheath and ends in the femoral vein. **A varicose vein is one that has permanently lost its valvular efficiency, a condition that is not uncommon in the great saphenous vein, which is frequently tortuous and dilated when a subject is in the erect position. The saphenous vein receives many tributaries, including the superficial epigastric vein. Communications occur between the superficial epigastric vein and lateral thoracic veins (by way of the thoraco-epigastric vein), i.e., between the femoral and axillary veins. These communications are important, and they become enlarged in the event of obstruction of either the superior or inferior vena cava.**

The *small* (or *short*) *saphenous vein* begins on the lateral side of the dorsal venous network of the foot. It passes behind the lateral malleolus and up the back of the leg. It then passes between the heads of the gastrocnemius, pierces the fascia of the popliteal fossa, and ends variably in the popliteal, great saphenous, or some muscular veins (fig. 13–1).

The chief deep veins are the *femoral* and the *popliteal veins* (figs. 13–1 and 13–2). The deep veins begin on the plantar aspect of the foot and accompany the anterior and posterior tibial arteries and the peroneal artery. Many valves are present. **Most of the blood from the lower limb is returned by way of the deep veins, and alternate routes exist even when the femoral vein is ligated.** The superficial veins can be obliterated without seriously affecting the circulation, provided that the deep veins remain intact.

The superficial and deep veins are connected by perforating veins (figs. 13–1 and 13–3). An important series is found in the leg, where valves direct blood from superficial to deep veins. **Muscular action, combined with the arrangement of the valves, is important in returning blood from the lower limb, and blood flow is much reduced during quiet standing.**

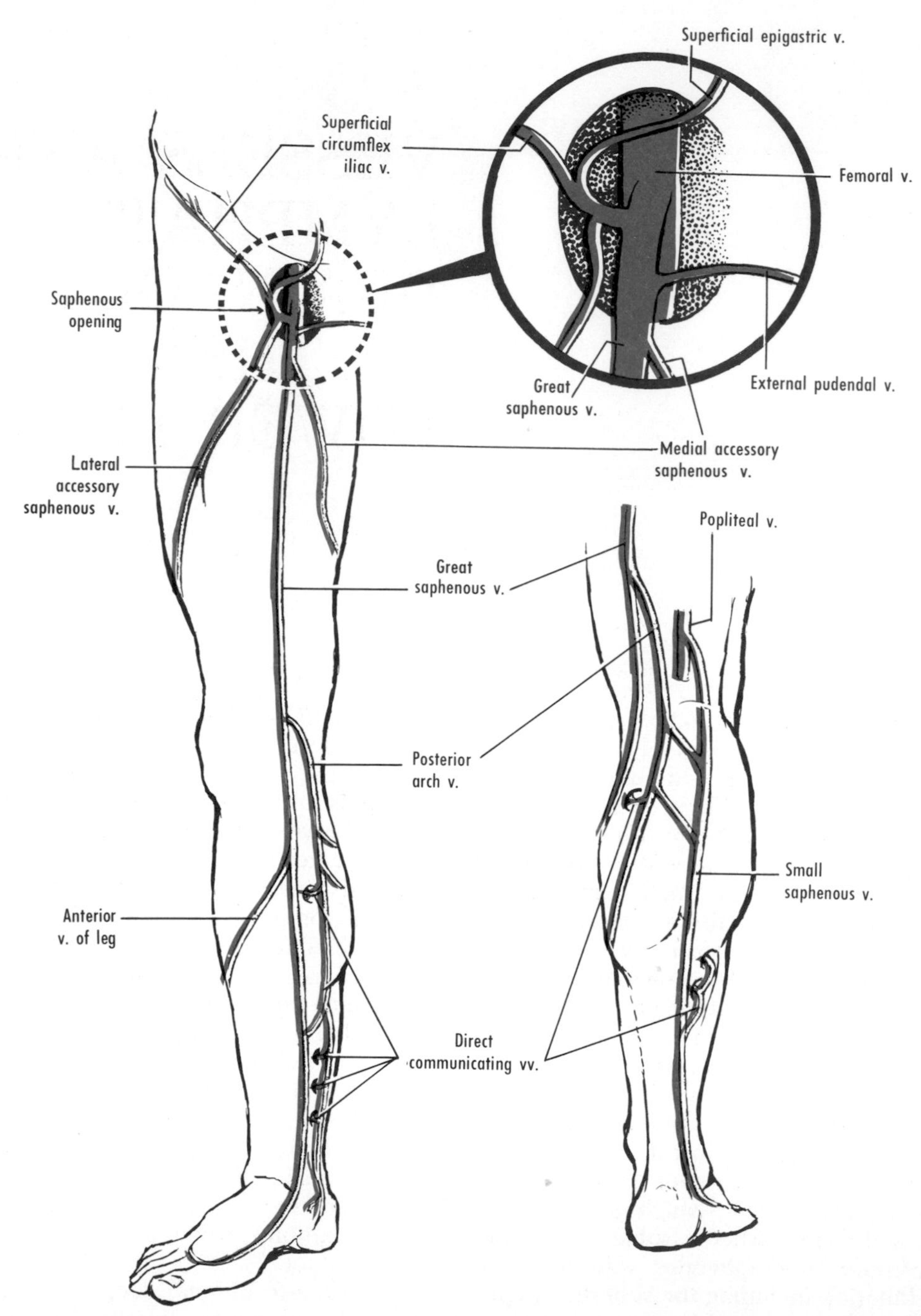

Figure 13–1 A simplified representation of the superficial veins of the lower limb. The major tributaries and also the major communicating veins above the ankle are shown according to Dodd and Cockett. Details of the veins of the foot have been omitted.

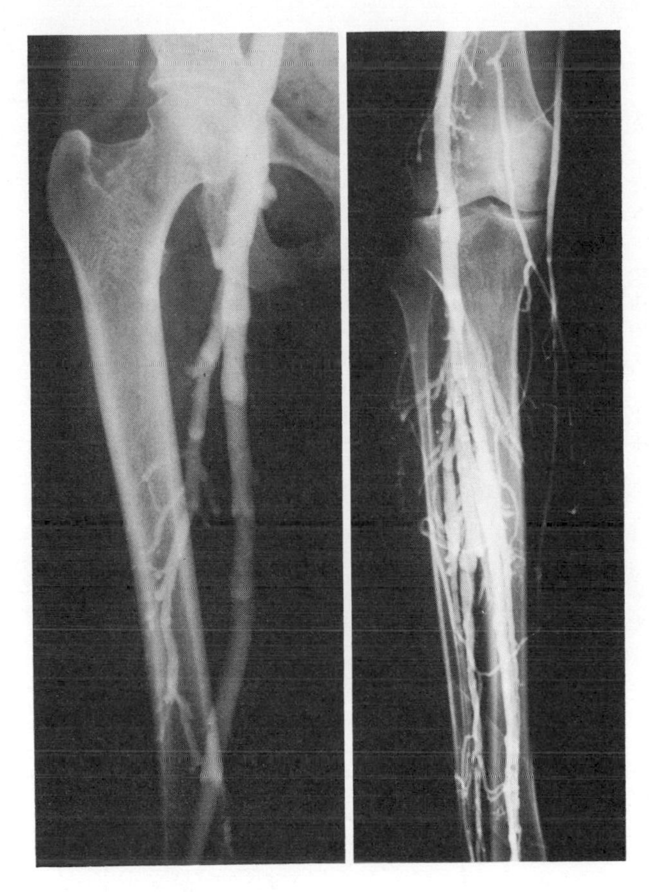

Figure 13–2 Normal venograms showing (*right*) the deep veins of the leg, the popliteal vein, the great saphenous vein, and (*left*) the femoral and deep femoral veins. Note the slight bulges at the valves, in some of which the cusps can be distinguished. (Courtesy of G. M. Stevens, M.D., Palo Alto Medical Clinic, Palo Alto, California.)

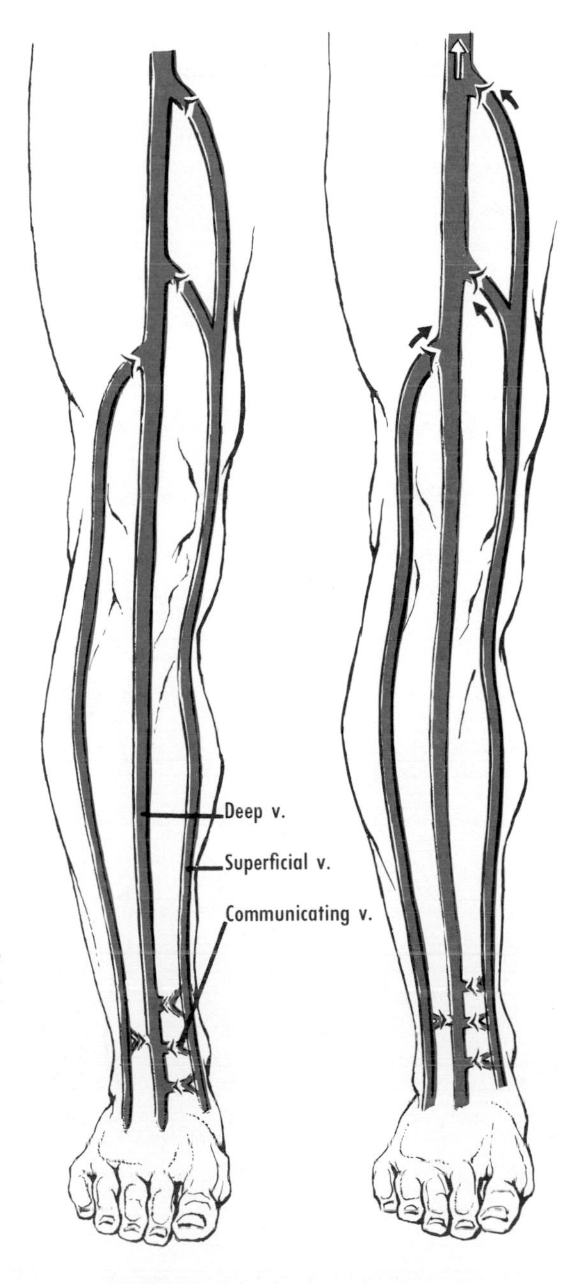

Figure 13–3 Schematic representation of venous circulation. The changes in venous pressure during exercise are such that nearly all blood returns by way of the deep veins, which receive blood from superficial veins by way of the communicating veins above the ankle.

LYMPHATIC DRAINAGE

The superficial lymphatic vessels of the lower limb are arranged as medial trunks, which end in the inguinal nodes, and as a lateral set, which either joins the medial trunks above the knee or ends in the popliteal nodes (fig. 13–4). The *popliteal nodes* receive the deep trunks that accompany the tibial blood vessels and the lateral group of superficial trunks. Their efferents accompany the femoral blood vessels to the deep inguinal nodes. The *inguinal nodes* are mostly subcutaneous and are frequently palpable *in vivo*. The superficial inguinal nodes lie both longitudinally along the great saphenous vein and transversely below the inguinal ligament, whereas the deep inguinal nodes (into which the superficial drain) lie medial to the femoral vein. The subdivision into superficial and deep groups, however, is unimportant. **When the inguinal nodes are found to be enlarged, their territory of drainage—the trunk below the level of the umbilicus (including perineum, external genitalia, and anus) as well as the entire lower limb—should be examined.** The efferents of the inguinal nodes proceed to the external iliac nodes and drain ultimately into the lumbar (aortic) nodes.

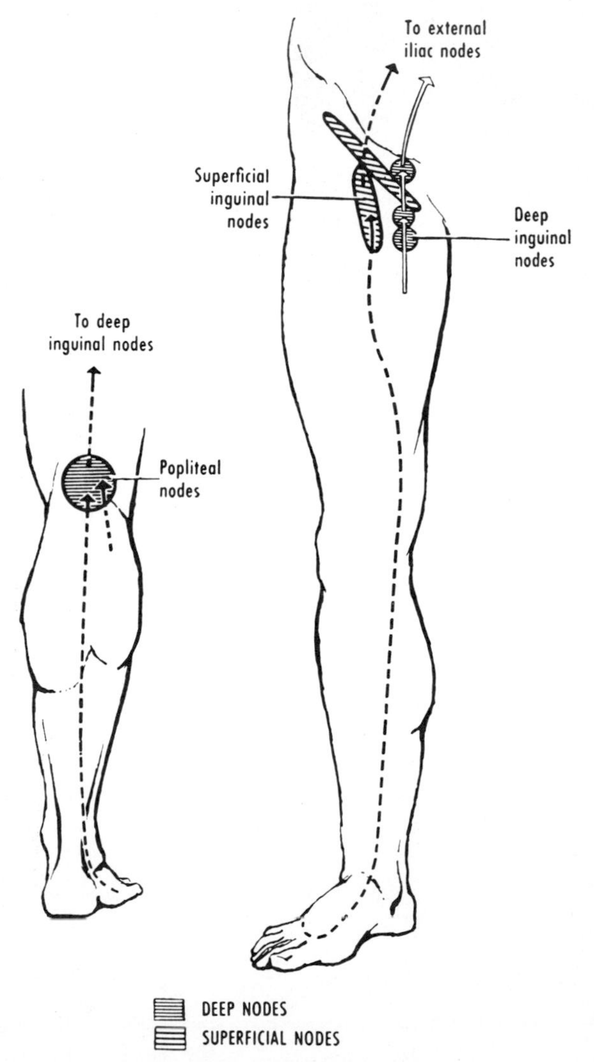

Figure 13–4 Schematic representation of the lymphatics of the lower limbs. The two lower arrows in the figure on the left indicate that the popliteal nodes receive both superficial and deep lymphatic vessels.

ARTERIES

The main artery carrying blood to the lower limb is named successively external iliac, femoral, and popliteal. The arrangement of this vessel and its branches is summarized in figure 13–5. The *external iliac artery* passes deep to the inguinal ligament and is then termed the *femoral artery*. In the lower third of the thigh, as it passes backward (through the adductor opening), its name is changed to *popliteal*. Below the knee joint (at the lower border of

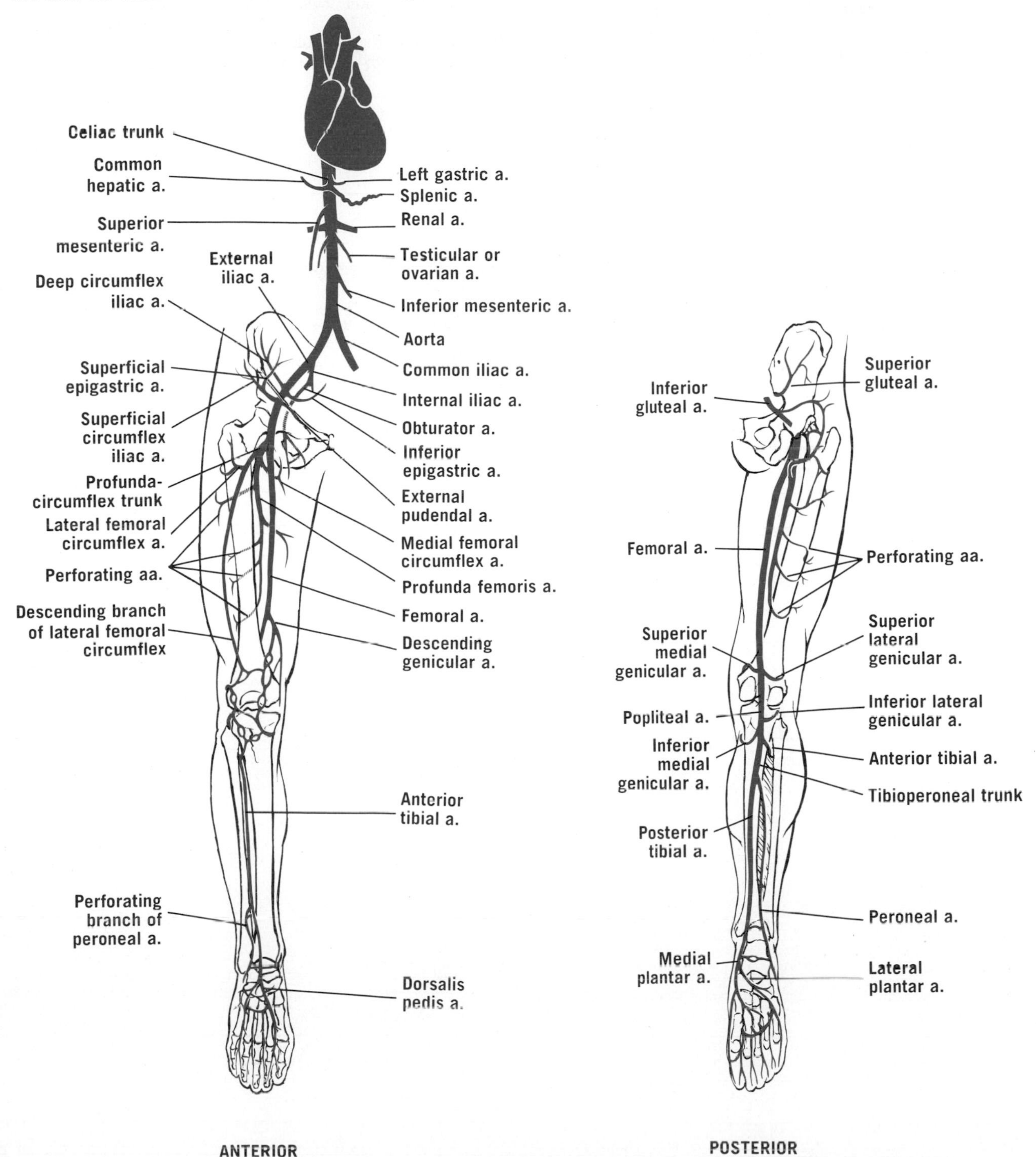

Figure 13–5 The arteries of the lower limb.

the popliteus), the popliteal artery either divides into the anterior and posterior tibial arteries or gives off the *anterior tibial artery* and continues as a common (tibioperoneal) trunk, which divides shortly thereafter into the *posterior tibial* and *peroneal arteries*. In front of the ankle, the anterior tibial artery continues as the *dorsalis pedis artery* (see fig. 17–4*C*). The posterior tibial divides into *medial* and *lateral plantar arteries,* and the lateral plantar artery forms the plantar arch.

ADDITIONAL READING

Dodd, H., and Cockett, F. B., *The Pathology and Surgery of the Veins of the Lower Limb,* 2nd ed., Churchill Livingstone, Edinburgh, 1976. Chapter 3 is devoted to the surgical anatomy of the veins.

QUESTIONS

13–1 Where are frequent sites for varicose veins?

13–2 What is the clinical importance of thoraco-epigastric venous connections?

13–3 What should be examined when inguinal lymph nodes are found to be enlarged?

13–4 What is meant by collateral circulation?

13–5 When a catheter is inserted in the thigh, how does it reach the vessels of the neck for cerebral angiography?

14
THE GLUTEAL REGION

The skin of the buttock is supplied by a number of small nerves derived ultimately from the region of the twelfth thoracic nerve to the third sacral nerve (T.N.12 to S.N.3). A bursa is found over the greater trochanter, and the strong gluteal aponeurosis over the gluteus medius continues downward as the iliotibial tract of the fascia lata. **The superolateral quadrant of the buttock is relatively free of nerves and vessels and is frequently used for intramuscular injections in order to avoid the sciatic nerve and other important structures. An alternative site is over the gluteus medius in a triangular area bounded by the anterior superior iliac spine, the tubercle of the iliac crest, and the greater trochanter** (fig. 14–1*B*).

MUSCLES OF GLUTEAL REGION
(table 14–1)

The glutei maximus, medius, and minimus, from superficial to deep, form the bulk of the buttock (fig. 14–1). They are supplied by the gluteal nerves and vessels, which reach them through the greater sciatic foramen (Figs. 14–2 and 14–3). The *gluteus maximus*, a large muscle with numerous attachments, is a powerful extensor of the thigh or of the trunk upon the fixed lower limbs. Surprisingly, however, it is not important posturally, is relaxed when one is standing, and is little used in walking. It is employed in running, climbing, and rising from a sitting or stooped position. It also regulates flexion at the hip on sitting down (paradoxical action). The *glutei medius* and *minimus* abduct the thigh and rotate it medially. **During walking, the glutei medius and minimus of the grounded limb abduct the pelvis, i.e., tilt it so that the swinging limb can clear the ground** (see fig. 18–2). **Paralysis results in a characteristic lurching gait.** A series of small lateral rotators of the thigh is found largely

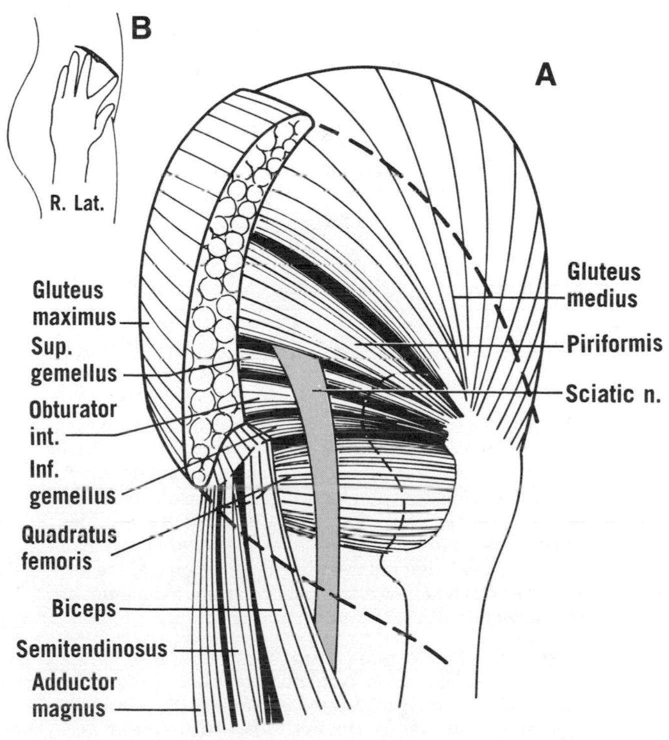

Figure 14–1 The gluteal region, posterior aspect. **A**, The deep relations of the gluteus maximus. The sciatic nerve usually emerges below the piriformis and lies on a succession of lateral rotators. **B**, Intramuscular injection. An intragluteal injection may be made safely in the area between the second and third fingers.

TABLE 14–1 MUSCLES OF GLUTEAL REGION

Muscle	Origin	Insertion	Innervation	Action
Gluteus maximus	Ilium behind posterior gluteal line, dorsal surface of sacrum, & gluteal aponeurosis	Iliotibial tract & gluteal tuberosity of femur	Inferior gluteal	Extends thigh
Gluteus medius	Ilium between anterior & posterior gluteal lines	Greater trochanter of femur (medius & minimus)	Superior gluteal (medius, minimus & tensor fasciae latae)	Abduct & rotate thigh medially (medius & minimus)
Gluteus minimus	Ilium between anterior & inferior gluteal lines			
Tensor fasciae latae	Anterior superior iliac spine & iliac crest	Iliotibial tract		Flexes & rotates thigh medially
Piriformis	Pelvic surface of sacrum	Greater trochanter	S.N.1 & 2	Rotate thigh laterally (piriformis to obturator externus)
Superior gemellus	Ischial spine	Obturator internus tendon	N. to obturator internus	
Obturator internus	Pelvic surface of obturator membrane & adjacent bone	Greater trochanter	Sacral plexus	
Inferior gemellus	Ischial tuberosity	Obturator internus tendon	N. to quadratus femoris	
Quadratus femoris	Ischial tuberosity	Intertrochanteric crest of femur	Sacral plexus	
Obturator externus	External surface of obturator membrane & adjacent bone	Trochanteric fossa of femur	Obturator	

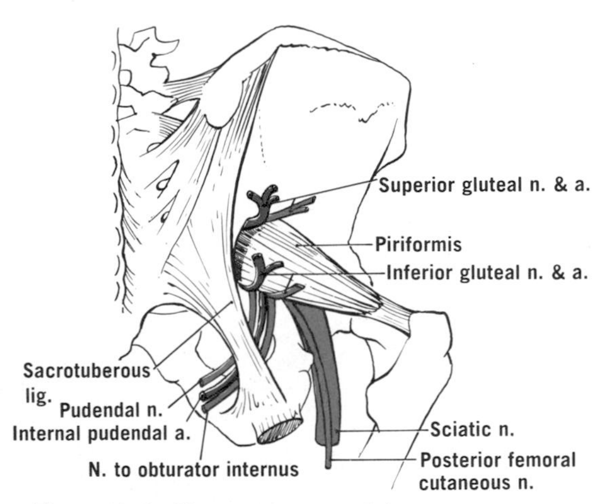

Figure 14–2 The arrangement of the structures emerging from the greater sciatic foramen. The foramen gives exit to the pirformis, to seven nerves (sciatic, posterior femoral cutaneous, superior gluteal, inferior gluteal, pudendal, nerve to the obturator internus, and nerve to the quadratus femoris), and to three groups of vessels (internal pudendal, superior gluteal, and inferior gluteal). The nerve to the quadratus femoris and the veins that accompany the arteries are not shown.

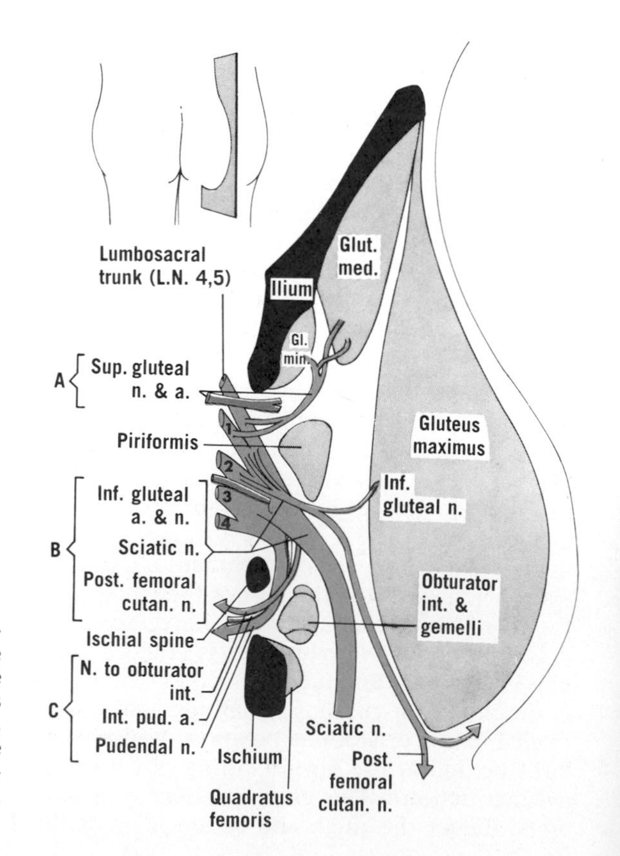

Figure 14–3 Oblique vertical section (*see inset*) to show relations in the gluteal region. Note that (*A*) a nerve and artery leave the greater sciatic notch above the piriformis, (*B*) seven nerves (the nerve to the quadratus femoris is not shown) and two arteries emerge below the piriformis (see the list in the legend of fig. 14–2), and (*C*) two nerves and an artery regain the pelvis through the lesser sciatic foramen. The sciatic nerve lies successively on the ischium, gemelli and obturator internus, and quadratus femoris. (Modified from Castaing and Soutoul.)

under cover of the gluteus maximus. One of them, the *piriformis*, emerges from the greater sciatic foramen, and the sciatic nerve appears at its lower border (figs. 14–1 to 14–3).

NERVES AND VESSELS OF GLUTEAL REGION

Several important nerves from the sacral plexus either supply or traverse the gluteal region. The *superior gluteal nerve* (L4 and 5 and S1) (figs. 14–2 and 14–3) passes backward through the greater sciatic foramen above the piriformis and supplies the glutei medius and minimus and the tensor fasciae latae. The *inferior gluteal nerve* (L5 and S1 and 2) (figs. 14–2 and 14–3) traverses the greater sciatic foramen below the piriformis and supplies the gluteus maximus. The pudendal nerve is merely passing through to re-enter the pelvis via the lesser sciatic foramen (fig. 14–3). The *posterior femoral cutaneous nerve* (S1 to 3) (figs. 14–2 and 14–3) passes backward through the greater sciatic foramen below the piriformis, descends deep to the gluteus maximus and down the middle of the back of the thigh, pierces the fascia, and reaches the calf. **The largest nerve in the body, the sciatic nerve, consists of two parts, tibial and peroneal, which are bound together and then separate at a variable level into two nerves.** A branch of the sacral plexus, the sciatic nerve (L4 to S3) (figs. 14–1 to 14–3 and 15–8) traverses the greater sciatic foramen, below the piriformis. The peroneal component, however, may pierce the piriformis (or even emerge above that muscle), and it then remains separate. The sciatic nerve descends under cover of the gluteus maximus and then along the middle of the back of the thigh. In terms of surface anatomy, the sciatic nerve leaves the pelvis at about a third or more of the way along a line from the posterior superior iliac spine to the

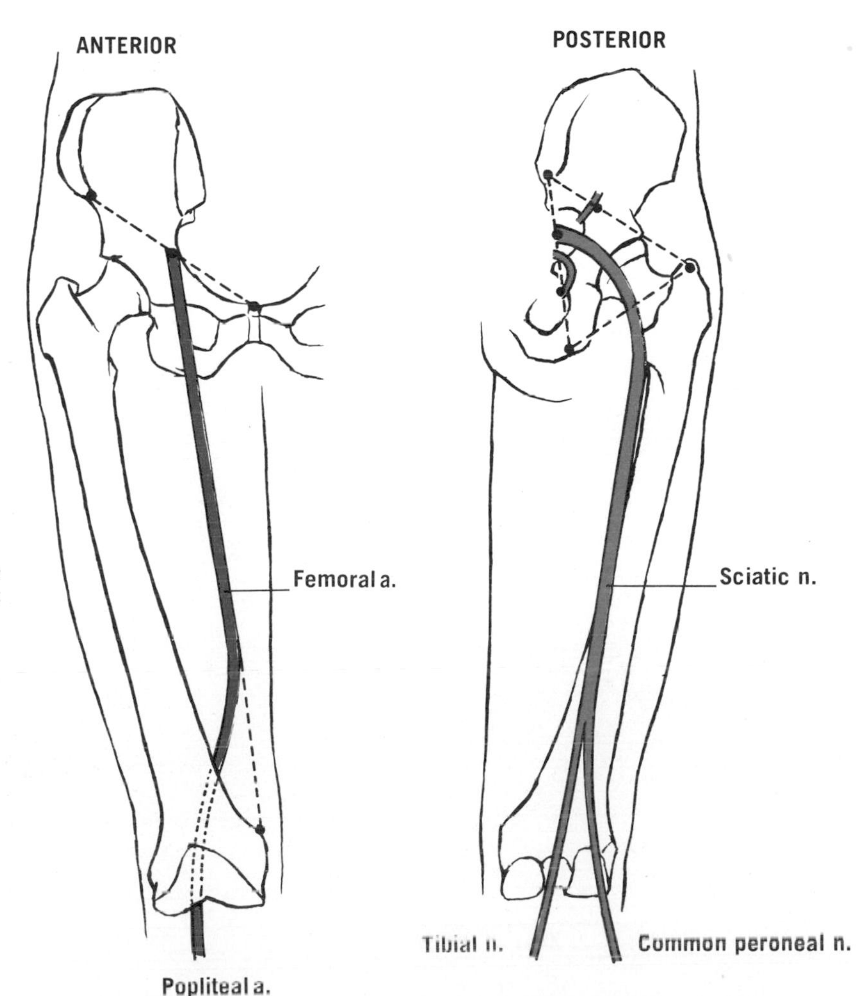

Figure 14–4 The surface anatomy of the femoral artery, gluteal nerves, and sciatic nerve.

ischial tuberosity (fig. 14–4). It then descends about halfway between the ischial tuberosity and the greater trochanter. The words *sciatic* and *ischium* are related.

The gluteal arteries (see figs. 13–5 and 14–2) are variable, but they arise, directly or indirectly, from the internal iliac artery. The *superior gluteal artery,* the largest branch of the internal iliac artery, accompanies the corresponding nerve and, under cover of the gluteus maximus, divides into branches that supply the gluteal muscles. The *inferior gluteal artery,* another branch of the internal iliac artery, accompanies the corresponding nerve, and, under cover of the gluteus maximus, lies medial to the sciatic nerve, to which it gives a companion artery. The *gluteal veins* accompany the arteries and drain into the internal iliac vein. They can return the blood from the lower limb even when the femoral vein is ligated.

The exit of the superior gluteal nerve and artery from the pelvis is indicated by the superior point of trisection of a line from the posterior superior iliac spine to the upper end of the greater trochanter (fig. 14–4). The exit of the inferior gluteal nerve and artery is indicated by the inferior point of trisection of a line from the posterior superior iliac spine to the ischial tuberosity.

QUESTIONS

14–1 Where is an intramuscular injection usually given?

14–2 What is the role of the glutei medius and minimus in walking?

14–3 From which spinal segments is the sciatic nerve derived?

14–4 Which lateral rotator of the thigh has a variable relationship to the sciatic nerve?

15
THE THIGH AND KNEE

FASCIA OF THIGH

The deeper portion of the subcutaneous tissue covers the saphenous opening (where it is termed the cribriform fascia) and fuses with the fascia lata below and parallel to the inguinal ligament, thereby preventing the fluid that is deep to the subcutaneous tissue of the abdomen from entering the thigh. The fascia of the thigh, known as the *fascia lata,* is attached to the hip bone, and it sends lateral and medial intermuscular septa inward to the femur (fig. 15–2). That part of the fascia lata overlying the vastus lateralis is the *iliotibial tract* (figs. 15–1 and 15–2). It is attached to the iliac crest above, where it forms the gluteal aponeurosis, and it receives the insertions of the gluteus maximus and tensor fasciae latae. This tripartite arrangement (which resembles the deltoid in the upper limb) is important in maintaining posture and in locomotion.

The *saphenous opening* (formerly known as the *fossa ovalis*) (figs. 13–1 and 15–3) is a gap in the fascia lata below and lateral to the pubic tubercle and overlying the femoral vein. It transmits the great saphenous vein, which joins the femoral vein. The fascia forms a falciform margin lateral to the great saphenous vein. The saphenous opening is covered by the *cribriform fascia,* which is pierced by the saphenous vein and some tributaries.

The uppermost parts of the femoral artery and vein lie in a vascular compartment behind the inguinal ligament and between the iliopsoas and pectineus. The femoral nerve, together with the iliopsoas, lies more laterally. The femoral artery and vein, together with the more medially placed femoral canal, are enclosed in a fascial funnel known as the femoral sheath (fig. 15–4). The sheath is formed in front by the fascia transversalis (pierced by the femoral branch of the genitofemoral nerve and the great saphenous vein) and behind by the fascia iliaca. After a few centimeters, the sheath tapers inferiorly and fuses with the vascular coats. The *femoral canal,* situated in front of the pectineus, contains fat and a few lymphatic vessels. Its upper end or base, termed the *femoral ring,* is closed by extraperitoneal tissue known as the *femoral septum.* The medial boundary of the ring is sharp and ligamentous. The femoral canal is important surgically: **a femoral hernia is the protrusion of extraperitoneal tissue, with or without an abdominal viscus, through the femoral ring. It may pass down the femoral canal and through the femoral sheath and saphenous opening.* Clinically, the neck of a femoral hernia is found immediately inferolateral to the pubic tubercle** (see fig. 25–7). (An inguinal hernia is more medial.) The hernial sac is formed by parietal peritoneum, but the external coverings of the hernia are frequently fused.

The femoral triangle (fig. 15–5), **which contains the femoral nerve and vessels, is situated in the upper third of the front of the thigh. It is**

* Based on his extensive experience in hernial surgery, W. J. Lytle disagrees with the traditional account and maintains that a femoral hernia *must* pass through two openings, namely the femoral ring *and* the *lower* end of the femoral canal (the "femoral hernial orifice") bounded medially by the curved edge of the lacunar ligament (Ann. R. Coll. Surg. Engl., *21*:244–262, 1957).

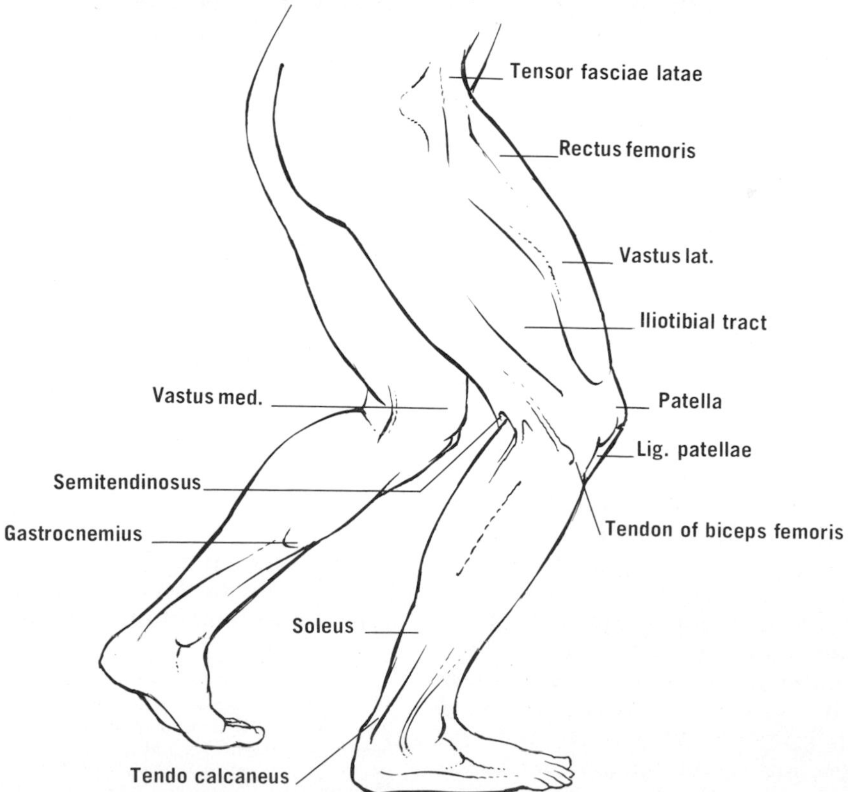

Figure 15–1 Surface landmarks of the lower limb. (From Royce, J., *Surface Anatomy,* Davis, Philadelphia, 1965.)

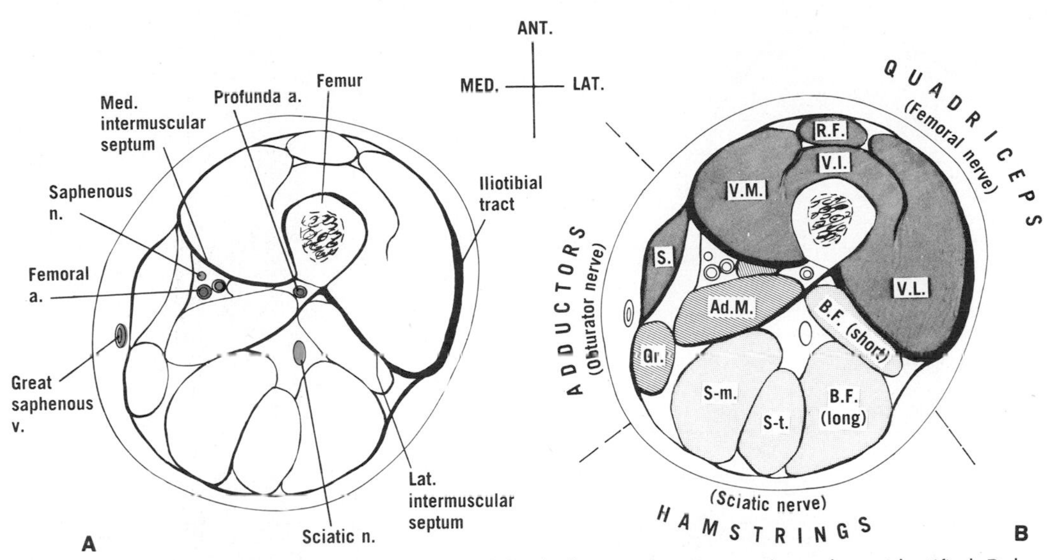

Figure 15–2 Horizontal section through the middle of the thigh. In **A**, the nerves and vessels are identified. **B** shows the quadriceps (supplied by the femoral nerve) anteriorly, the adductors (supplied mostly by the obturator nerve) medially, and the hamstrings (supplied mostly by the sciatic nerve) posteriorly. The adductor longus is shown immediately in front of the adductor magnus. The sartorius (*S*) has descended in a spiral from the anterior group and hence is supplied by the femoral nerve. The femoral vessels are situated subsartorially in the adductor canal. Although not shown here, the adductor magnus, in addition to its adductor part (supplied by the obturator nerve), has an extensor component supplied by the sciatic nerve. *Ad.M.*, adductor magnus; *B.F.*, biceps femoris; *Gr.*, gracilis; *R.F.*, rectus femoris; *S.*, sartorius; *S-m.*, semimebranosus; *S-t.*, semitendinosus; *V.I.*, vastus intermedius; *V.L.*, vastus lateralis; *V.M.*, vastus medialis.

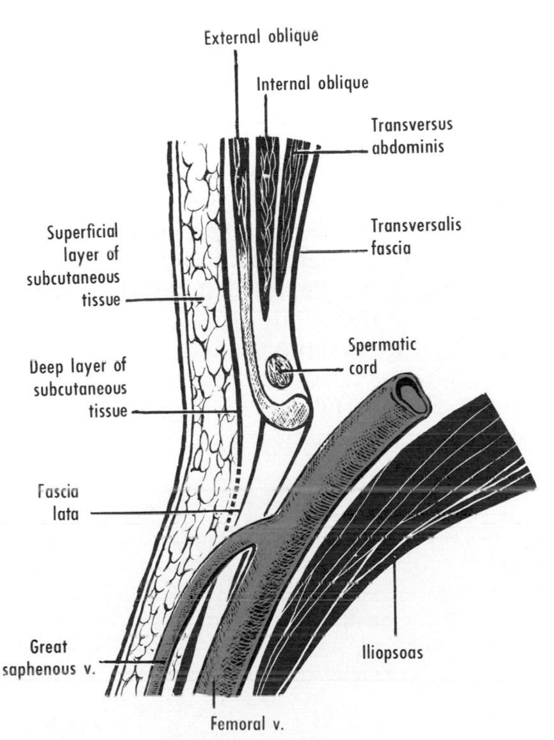

Figure 15–3 Schematic representation of the anterior abdominal wall, inguinal ligament, and saphenous opening in a sagittal plane. The saphenous opening is in the fascia lata, and it transmits the great saphenous vein. It is covered by the cribriform fascia (*interrupted line*).

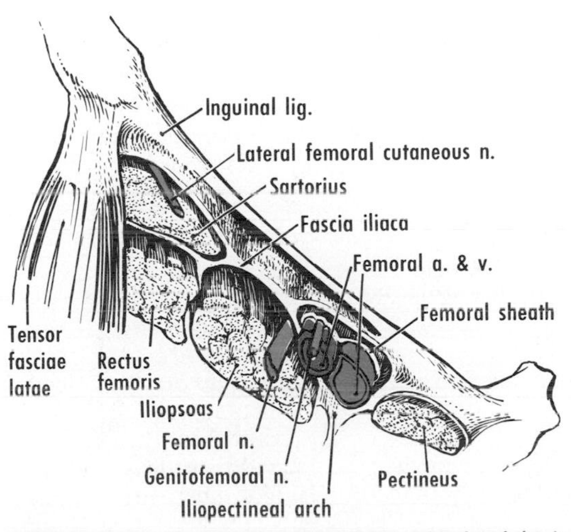

Figure 15–4 The structures that descend behind the inguinal ligament.

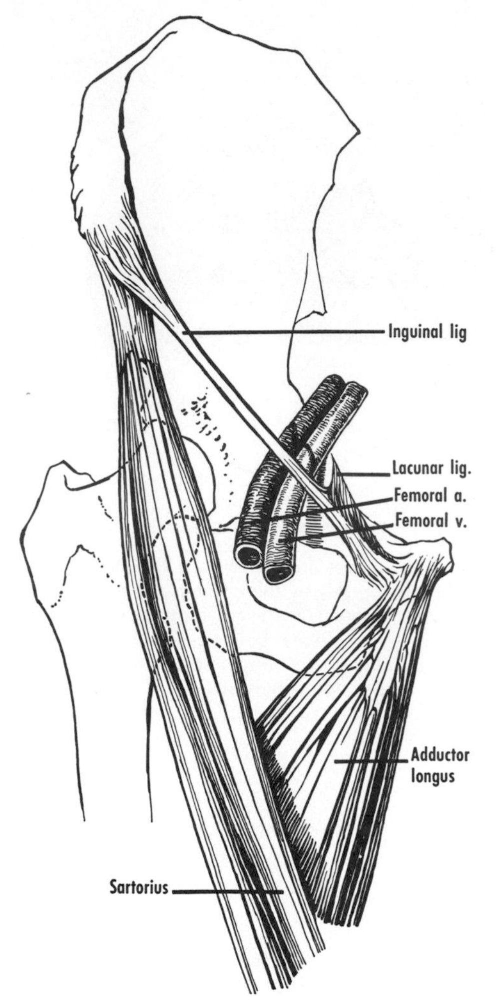

Figure 15–5 The femoral triangle: sartorius, adductor longus, and inguinal ligament.

bounded laterally by the medial border of the sartorius, medially by the medial (Continental authors use lateral) border of the adductor longus, and above by the inguinal ligament. Its roof is formed by the fascia lata and the cribriform fascia. Its floor is formed by the iliopsoas, pectineus, and adductor longus (fig. 15–6).

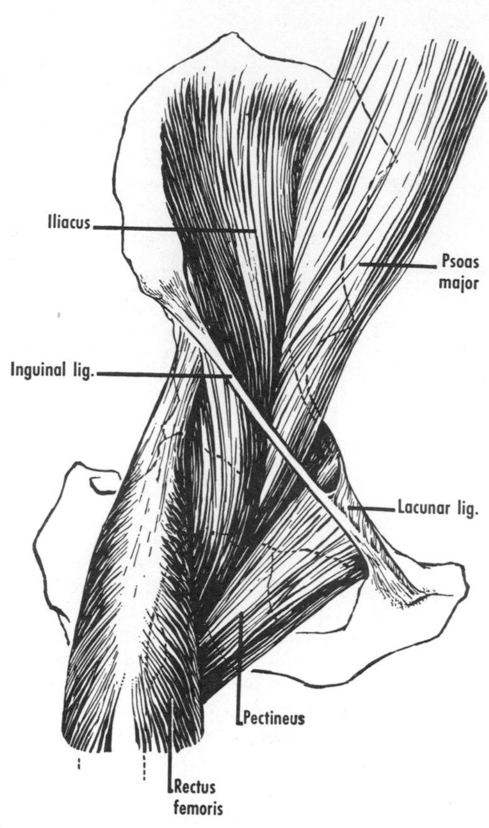

Figure 15–6 The floor of the femoral triangle. The adductor longus, which also forms a part of the floor, is not shown.

The adductor (or subsartorial) canal, which contains the femoral vessels and the saphenous nerve, is situated in the middle third of the medial part of the thigh. It is bounded laterally by the vastus medialis, medially by the adductor longus (and frequently magnus), and superficially by the sartorius and subsartorial fascia. The adductor canal was described in the eighteenth century by John Hunter, who ligated the femoral artery in it to treat popliteal aneurysm.

TABLE 15–1 MUSCLES OF BACK OF THIGH

Muscle	Origin	Insertion	Innervation	Action
Biceps femoris	Ischial tuberosity (long head); Lateral lip of linea aspera (short head)	Head of fibula	Sciatic (tibial part); Sciatic (peroneal part)	Extend thigh & flex leg
Semitendinosus	Ischial tuberosity	Medial surface of tibia	Sciatic (tibial part)	Extend thigh & flex leg
Semimembranosus	Ischial tuberosity	Medial condyle, medial border, & soleal line of tibia	Sciatic (tibial part)	Extend thigh & flex leg

BACK OF THIGH (table 15–1)

The muscles on the back of the thigh (biceps, semitendinosus, and semimembranosus) are known collectively as the *hamstrings* (see fig. 15–2). They are the main extensors of the thigh and flexors of the leg, especially during walking. Mostly they arise from the ischial tuberosity, cross two joints, have fascial and ligamentous as well as bony insertions, and are supplied by the tibial part of the sciatic nerve. The *biceps* and *semitendinosus* tendons, which are palpable and visible, form, respectively, the lateral and medial boundaries of the popliteal fossa (see fig. 15–9). The lower half of the semitendinosus is tendinous; hence its name. The tendons of origin and insertion of the *semimembranosus* together make up almost half of the length of the muscle.

The *sciatic nerve* (see figs. 14–1 to 14–3 and 15–8) descends under cover of the gluteus maximus and passes down the middle of the thigh, where it is crossed from behind by the long head of the biceps. **Separation of the sciatic into the tibial and common peroneal nerves may occur at any level in the gluteal region or thigh but usually takes place in the lower third of the thigh.** Most of the branches arise from the medial side. Tibial twigs supply the semitendinosus, semimembranosus, long head of biceps, and adductor magnus. Common peroneal twigs supply the short head of biceps.

Section of the sciatic nerve causes paralysis of the hamstrings (compensated for by the sartorius) and the muscles of the leg and foot, with loss of movement below the knee. Sensibility below the knee is lost except medially (saphenous nerve). In incomplete lesions of the sciatic nerve, the peroneal component is usually more severely damaged than the tibial one.

MEDIAL SIDE OF THIGH (table 15–2)

The muscles on the medial side of the thigh (pectineus, adductor longus, brevis, magnus, and gracilis) are mostly adductors of the thigh, supplied by the obturator nerve (see figs. 15–2 and 15–7). The three named adductors are used in all movements in which the thighs are pressed together. The *adductor magnus,* however, consists of an adductor part anterosuperiorly and an extensor part medially. The *adductor longus* is the medial boundary of the femoral triangle. Ossification may occur in its tendon ("rider's bone"). The *adductor brevis* is sandwiched between the anterior and posterior branches of the obturator nerve. The lower end of the adductor magnus allows the passage of the femoral vessels into the popliteal fossa.

The *obturator nerve* (L[2], 3, 4, [5]) (see figs. 15–7 and 30–7) arises from the lumbar plexus in the substance of the psoas major. It emerges at the medial margin of the psoas and accompanies the obturator vessels to the obturator groove, where it divides into anterior and posterior branches. These pass through the obturator foramen to reach the thigh, where they are separated by the adductor brevis. The anterior branch, which lies behind the pectineus and adductor longus, supplies these two (not always the pectineus) as well as the gracilis, adductor brevis, and skin on the medial side of the thigh. The posterior branch pierces and supplies the obturator externus, and gives twigs to the adductor magnus (and sometimes brevis). An accessory obturator nerve (L3, 4 or L2, 3), when present, communicates with the anterior branch of the obturator and supplies the pectineus.

The *obturator artery* (see fig. 13–5), a branch of the internal iliac, sends anterior and posterior branches around the margin of the

TABLE 15–2 Muscles of Medial Side of Thigh

Muscle	Origin	Insertion	Innervation	Action
Pectineus	Pectineal line of pubis	Pectineal line of femur	Femoral, obturator, or both	Adduct thigh
Adductor longus	Body of pubis	Medial lip of linea aspera	Obturator	Adduct thigh
Adductor brevis	Body & inferior ramus of pubis	Pectineal line & linea aspera	Obturator	Adduct thigh
Adductor magnus	Ischiopubic ramus	Linea aspera	Obturator	Adduct thigh
	Ischial tuberosity	Adductor tubercle	Sciatic (tibial part)	Extends thigh
Gracilis	Body & inferior ramus of pubis	Medial surface of tibia	Obturator	Adducts thigh

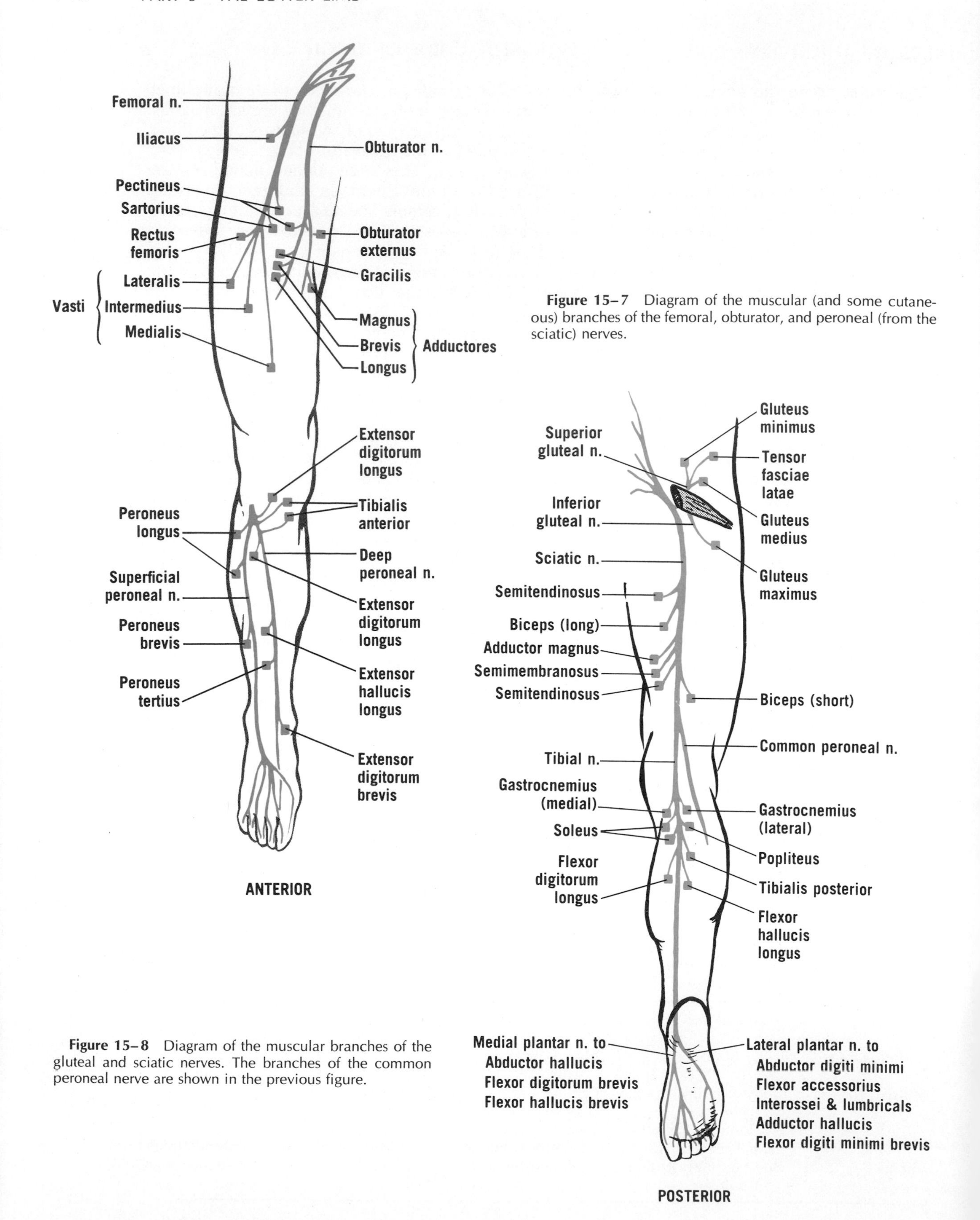

Figure 15–7 Diagram of the muscular (and some cutaneous) branches of the femoral, obturator, and peroneal (from the sciatic) nerves.

Figure 15–8 Diagram of the muscular branches of the gluteal and sciatic nerves. The branches of the common peroneal nerve are shown in the previous figure.

obturator foramen. The posterior branch gives off an acetabular branch, which passes through the acetabular notch and is an important vascular source for the head of the femur.

FRONT OF THIGH (table 15–3)

The chief muscles of the front of the thigh are the iliopsoas, quadriceps femoris, and sartorius. The *iliopsoas* (see figs. 15–4 and 15–6) consists of a lateral part, the iliacus, and a medial portion, the psoas major. It arises from the iliac fossa and the lumbar vertebrae, enters the thigh behind the inguinal ligament, and is inserted into the lesser trochanter of the femur. The iliopsoas is the chief flexor of the thigh and, when the thigh is fixed, of the trunk. It advances the limb during walking and is a postural muscle, being active during standing. It can also bend the vertebral column to one side. A psoas abscess originating in the spine may descend behind the inguinal ligament and simulate a femoral hernia.

The *quadriceps femoris* comprises the rectus femoris, which extends from the hip bone to the tibia, and three vasti (lateralis, medialis, and intermedius), which pass from the femur to the tibia (see fig. 15–2). The tendon of insertion of the quadriceps contains a large sesamoid bone, the patella, and its terminal portion is known as the *ligamentum patellae*. The *rectus femoris*, which has been termed the "kicking muscle," aids the iliopsoas. The *vastus medialis*, which superficially appears to be fused with the intermedius, usually forms a characteristic medial bulge in the lower part of the thigh (see fig. 15–1). The quadriceps extends the leg and is important in climbing, running, jumping, rising from sitting, and walking up and down stairs.

The knee jerk is elicited by tapping the ligamentum patellae, which causes sudden stretching of the quadriceps (and its contained neuromuscular spindles) and thence a rapid contraction of the muscle. The reflex center is in the L3 segment of the spinal cord.

The *sartorius* forms the lateral boundary of the femoral triangle (see fig. 15–5) and covers the adductor canal, thereby serving as a surgical guide to the femoral artery. It flexes the thigh and leg and assists in producing the crossed-leg position of tailors; hence its name (cf. sartorial).

The *femoral nerve* (L2 to 4) (see figs. 15–7 and 30–7), the largest branch of the lumbar plexus, descends between the iliacus and psoas major and enters the thigh behind the middle of the inguinal ligament. The femoral nerve, arising external to the fascia iliaca, remains outside the femoral sheath. In the femoral triangle, lateral to the femoral artery, it

TABLE 15–3 MUSCLES OF FRONT OF THIGH

Muscle	Origin	Insertion	Innervation	Action
Iliopsoas				
Iliacus	Iliac fossa	Tendon of psoas major	Femoral	Flexes thigh; flexes vertebral column laterally
Psoas major	Lumbar vertebrae & discs	Lesser trochanter	Lumbar plexus (L2, 3)	
Quadriceps femoris			Femoral	Extends leg
Rectus femoris	Anterior inferior iliac spine & rim of acetabulum	Patella & tuberosity of tibia		Also flexes thigh
Vastus lateralis	Lateral lip of linea aspera & bone above	Patella & lateral condyle of tibia		
Vastus medialis	Intertrochanteric & spiral lines	Patella & medial condyle of tibia		
Vastus intermedius	Anterior & lateral surfaces of femur	Tendon of rectus & other vasti		
Articularis genus	Front of femur	Capsule of knee joint	Femoral	Elevates synovial membrane during extension
Sartorius	Anterior superior iliac spine	Medial surface of tibia	Femoral	Flexes thigh & leg

breaks up into terminal branches. The femoral nerve supplies the iliacus, pectineus, and hip joint. The terminal branches include *intermediate* and *medial cutaneous nerves,* muscular twigs to the sartorius, quadriceps femoris, and articularis genus, and the saphenous nerve. The *saphenous nerve,* which may be regarded as the termination of the femoral nerve, descends with the femoral vessels through the femoral triangle and adductor canal and then becomes cutaneous. In the leg, it accompanies the great saphenous vein and ends on the medial side of the foot. An *accessory femoral nerve* from the lumbar plexus is not uncommon. The saphenous and other cutaneous nerves form *subsartorial* (beneath the sartorius) and *patellar plexuses.* The *lateral femoral cutaneous nerve,* a branch of the femoral nerve or of the lumbar plexus, supplies skin on the anterolateral aspect of the thigh and is liable to compression and paresthesia (*meralgia*).

The *femoral artery,* the continuation of the external iliac artery, enters the femoral triangle by passing behind the inguinal ligament (see fig. 15–5). It then enters the adductor canal and, passing backward between the adductor magnus and the femur, changes its name to popliteal (see figs. 13–5 and 15–10). The femoral artery can be represented by the upper two thirds of a line from the midinguinal point (midpoint between the anterior superior iliac spine and the pubic symphysis) to the adductor tubercle of the femur (see fig. 14–4). Its pulsations can be felt when the thigh is flexed, abducted, and rotated laterally. **The femoral artery can be compressed by pressing directly backward at the midinguinal point.** The femoro-popliteal axis supplies the muscles of the calf, and obstruction interferes with their blood supply on exertion (*intermittent claudication*). Proximally, the femoral artery gives off the *superficial epigastric* (which proceeds toward the umbilicus), *superficial circumflex iliac* (which runs toward the anterior superior iliac spine), and *superficial* and *deep external pudendal arteries* (to the inguinal and pudendal regions). The most important branch of the femoral artery is the *profunda femoris artery* (see fig. 15–15), which arises in the femoral triangle and gives lateral and medial circumflex arteries (one or both of which may arise directly from the femoral artery. The profunda descends along the medial side of the femur, gives origin to about three *perforating arteries* (which supply nearby muscles), and ends by passing through the adductor magnus as the last (fourth) perforating artery. The perforating arteries form an extensive anastomosis, and the first meets transverse branches of the circumflex and also the inferior gluteal artery (*cruciate anastomosis*). The *circumflex arteries* supply the head and neck of the femur. Distally, the femoral artery gives origin to the *descending genicular artery,* which supplies the knee joint. The *femoral vein,* which may be double below, accompanies the artery and finally lies medial to it in the femoral triangle. It receives the great saphenous vein and becomes the *external iliac vein.*

POPLITEAL FOSSA

The popliteal fossa (figs. 15–9 and 15–10) is a diamond-shaped area at the back of the knee. Its upper boundaries are the biceps laterally, and the semitendinosus and semimembranosus medially. Its lower boundaries are the lateral and medial heads of the gastrocnemius. The fascial roof is stretched on extension. The floor is formed, from above downward, by the popliteal surface of the femur, the oblique popliteal ligament (an expansion of the semimembranosus tendon), and fascia overlying the popliteus. **The popliteal fossa contains the common peroneal and tibial nerves, popliteal vessels, small saphenous vein, lymph nodes, bursae, and fat.**

The *common peroneal nerve* (formerly known as the lateral popliteal nerve) (L4 to S2) (fig. 15–10) arises from the sciatic nerve. It follows closely the medial edge of the biceps (fig. 15–9). **The common peroneal nerve winds around the neck of the fibula, where it can be felt and where it is liable to injury.** Under cover of the peroneus longus, it divides into the superficial and deep peroneal nerves (see fig. 15–7). The common peroneal nerve, while still a part of the sciatic nerve, supplies the short head of the biceps. In the popliteal fossa, it gives branches to the knee joint and to skin (*lateral sural cutaneous nerve*) and a communication to the medial sural nerve. It sometimes supplies the peroneus longus, tibialis anterior, and extensor digitorum longus. **Injury to the common peroneal nerve results in loss of eversion and of dorsiflexion of the foot (foot-drop)** and in a sensory loss on the lateral side of the leg and on the dorsum of the foot.

The *tibial nerve* (formerly known as the medial popliteal nerve) (L4 to S3) (fig. 15–9) arises from the sciatic nerve. Under cover of the gastrocnemius, it lies on the popliteus,

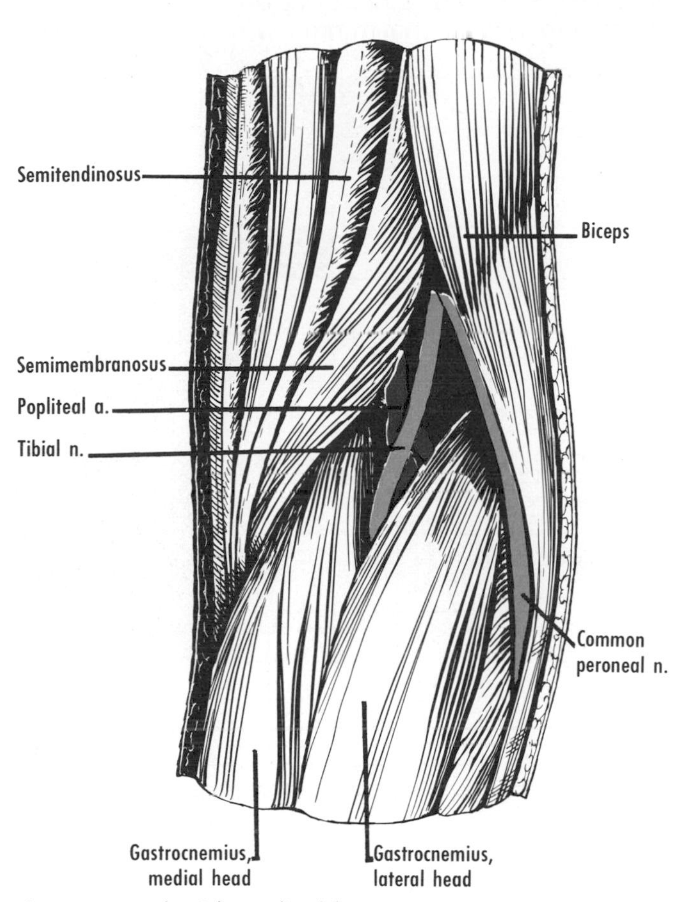

Figure 15–9 The right popliteal fossa.

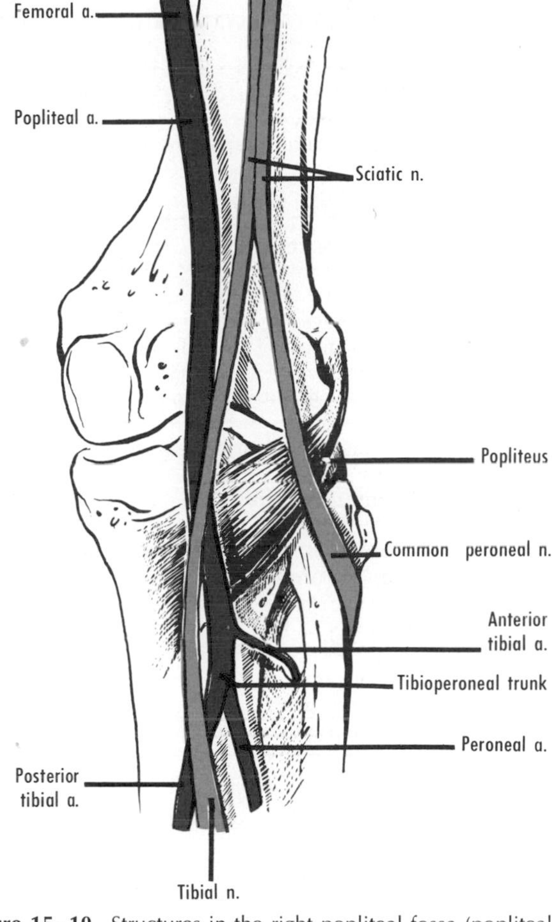

Figure 15–10 Structures in the right popliteal fossa (popliteal veins omitted). Note the relationship of the tibial nerve first to the popliteal and then to the posterior tibial artery. Note also that the anterior tibial artery passes laterally, not anteriorly, through an arch in the interosseous membrane.

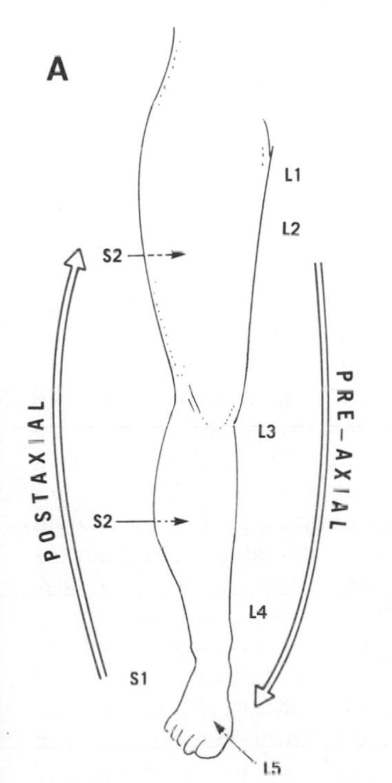

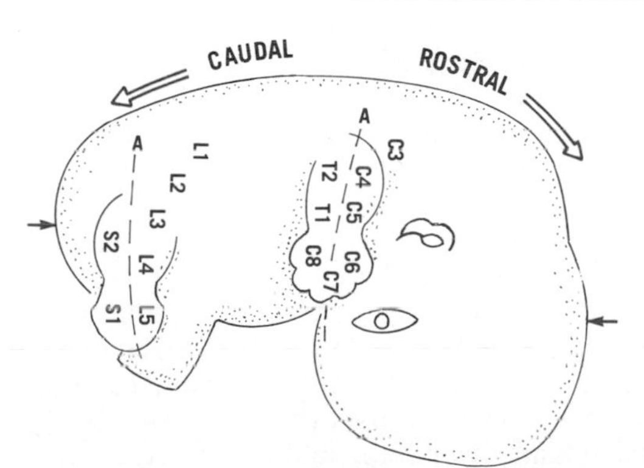

Figure 15–11 **A**, The basic arrangement of the segmental innervation (dermatomes) of the skin of the lower limb, which is supplied successively down the pre-axial border and then up the postaxial border. Note that the first toe is supplied by the fifth nerve and that the fifth toe is supplied by the first nerve. **B**, An embryo of 7 weeks (measuring 17 mm between the small arrows) showing the longitudinal axes (*A*) of the upper and lower limbs. The parts of the limbs rostral to the axes are termed postaxial. The sequence of dermatomes from pre-axial to postaxial is shown on each limb. Cf. fig. 8–10.

TABLE 15–4 SEGMENTAL INNERVATION OF MUSCLES OF LOWER LIMB*

Muscles	L2	3	4	5	S1	2
Psoas major, sartorius, pectineus, adductor longus	■	■				
Iliacus, quadriceps femoris, adductor brevis, gracilis	■	■	■			
Obturator externus		?	■			
Adductor magnus		■	■	■	■	
Tensor fasciae latae			■	■	?	
Glutei medii et minimi, plantaris, popliteus, muscles of front of leg			■	■	■	
Muscles of lateral side of leg			?	■	■	
Quadratus femoris, semimembranosus, tibialis posterior			?	■	■	?
Semitendinosus			?	■	■	■
Gluteus maximus, obturator internus, biceps (long head), flexor digitorum longus, flexor hallucis longus				■	■	■
Muscles of foot (mostly one or more of these nerves)				■	■	■
Piriformis				?	■	■
Gastrocnemius, soleus					■	■

* Adapted from various sources. It should be emphasized that, in the case of many muscles, some of these figures are uncertain.

and, at the lower border of the latter muscle, enters the leg by passing deep to the fibrous arch of the soleus. The tibial nerve, while still a part of the sciatic nerve, supplies the semitendinosus, semimembranosus, long head of the biceps, and adductor magnus. In the popliteal fossa, it gives branches to the knee joint and to the gastrocnemius, soleus, plantaris, popliteus, and tibialis posterior. The branch to the popliteus provides the *interosseous nerve* of the leg. The tibial nerve gives off the *medial sural cutaneous nerve,* which descends between the heads of the gastrocnemius and usually joins a communication from the common peroneal nerve to form the *sural nerve*. The sural nerve, which usually arises from both tibial and common peroneal components, lies on the tendo calcaneus, accompanies the small saphenous vein behind the lateral malleolus, and supplies the skin of the back of the leg and the lateral part of the foot, including the heel and at least the lateral side of the little toe. Sensation from the sole is important in posture and locomotion, and section of the tibial nerve results in a significant sensory loss on the sole and on the plantar aspects of the toes.

A simple scheme (fig. 15–11) shows the basic arrangement of the segmental innervation (dermatomes) of the skin of the lower limb, which is supplied successively down the pre-axial border and then up the postaxial border. Table 15–4 shows the segmental innervation of the muscles of the lower limb.

HIP JOINT

The hip joint is a ball-and-socket articulation between the acetabulum of the hip bone and the head of the femur (see figs. 12–7 and 12–8). The angle between the head and neck of the femur and the shaft may be abnormally diminished (*coxa vara*) or increased (*coxa valga*). More than half of the head of the femur is within the acetabulum, which is deepened by the fibrous or fibrocartilaginous *acetabular labrum* and completed below by the *transverse ligament* that bridges the acetabular notch (figs. 15–12 and 15–13). The capsule is attached to the margin of the acetabulum and to the intertrochanteric line of the femur (fig. 15–14). The capsule is thickened in front, to form the Y-shaped *iliofemoral ligament,* and

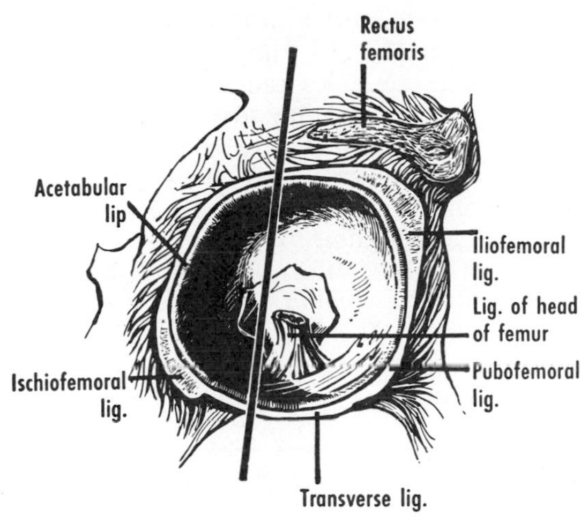

Figure 15–12 The acetabulum and capsule of the hip joint after removal of the femur. Note how the capsule varies in thickness. The line indicates the plane and position of the section in the next figure.

also below (*pubofemoral ligament*) and behind (*ischiofemoral ligament,* which encircles the neck of the femur as the *zona orbicularis*). **The back of the capsule is arranged so that the lateral one third to one half of the back of the neck of the femur is extracapsular** (see fig. 12–18). Capsular fibers attached to the femur tend to be reflected along the neck as retinacula that carry vessels to the head of the femur. The *ligament of the head* (formerly known as the *ligamentum teres*) extends from the acetabular notch and transverse ligament to a pit on the head of the femur (see fig. 12–12), and it transmits vessels (fig. 15–15). The hip joint may be approached surgically from in front, behind, or laterally, but it is surrounded by powerful muscles. The iliopsoas, pectineus, and femoral vessels are anterior relations. The hip joint may be tapped by a needle inserted anteriorly (halfway between the midinguinal point and the greater trochanter) or laterally

Figure 15–13 A coronal section of the hip joint in the plane indicated in the previous figure.

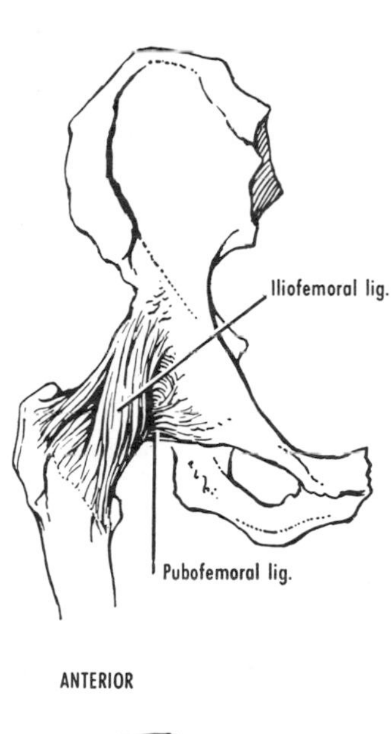

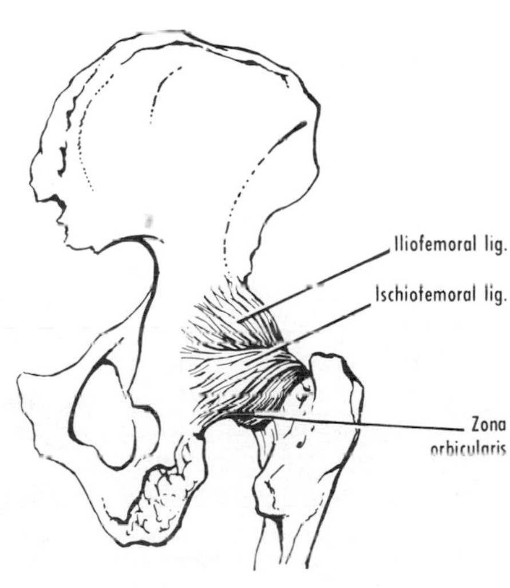

Figure 15–14 The capsule of the hip joint. Note that circular fibers form the zona orbicularis and that the neck of the femur is not completely covered posteriorly.

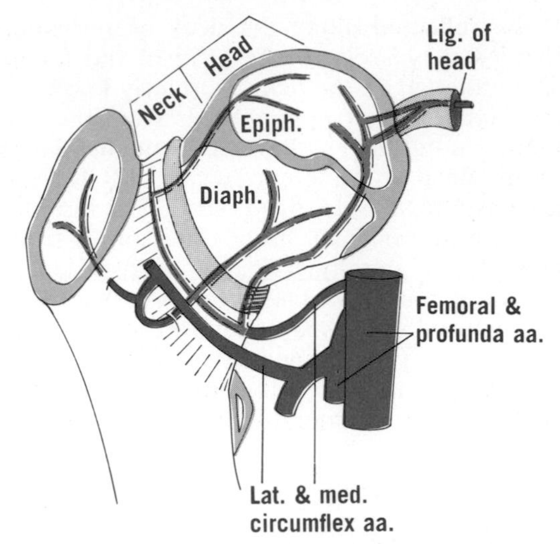

Figure 15–15 The blood supply to the head and neck of the femur. Apart from a small branch of the obturator artery that enters through the ligament of the head, most of the epiphysis of the head is supplied by diaphysial branches from the lateral and medial circumflex arteries of the femoral artery. These branches travel in retinacula, i.e., reflections of the capsule along the neck toward the head. The vessels may be damaged in fractures of the neck of the femur, which may result in avascular necrosis of the head of the bone. (Based on von Lanz and Wachsmuth.)

(above the greater trochanter). The hip joint is supplied by the femoral, sciatic, and obturator nerves, which also supply the knee joint. **Hip disease is an important cause of pain referred to the knee.**

The movements of the thigh at the hip joint are flexion and extension, abduction and adduction, and rotation and circumduction. The movements of the trunk at the hip joint are equally important, as when one lifts the trunk from the supine position. Flexion of the thigh is usually combined with flexion of the vertebral column. The chief flexor is the iliopsoas. The capsule becomes taut during extension. The chief extensors are the hamstrings. The abductors are the glutei medius and minimus, and the main adductors are the longus, brevis, and magnus. In rotation, the axis extends from the head of the femur to the medial condyle of the femur (not the long axis of the femur). The lateral rotators are the short muscles of the gluteal region. The chief medial rotators are the tensor fasciae latae and the glutei medius and minimus.

KNEE JOINT

The knee joint is a condylar articulation between the condyles of the femur, those of the tibia, and the patella (figs. 15–16 to 15–19). The articular surfaces are large, complicated, and incongruent. The angle between the vertical axes of the femur and tibia is exaggerated in knock-knee (*genu valgum*), whereas the knees appear more separated in bowlegs (*genu varum*). The capsule, thin and partly deficient, is attached to the margins of the condyles of the femur, to the patella and ligamentum patellae, and to the condyles of the tibia. The capsule is strengthened by *retinacula* derived from the vasti and by an expansion (*oblique popliteal ligament*) from the semimembranosus tendon.

A strong extracapsular ligament is present on each side (fig. 15–16). The *fibular collateral ligament* extends from the lateral epicondyle of the femur to the head of the fibula, and its stabilizing function is aided by the biceps

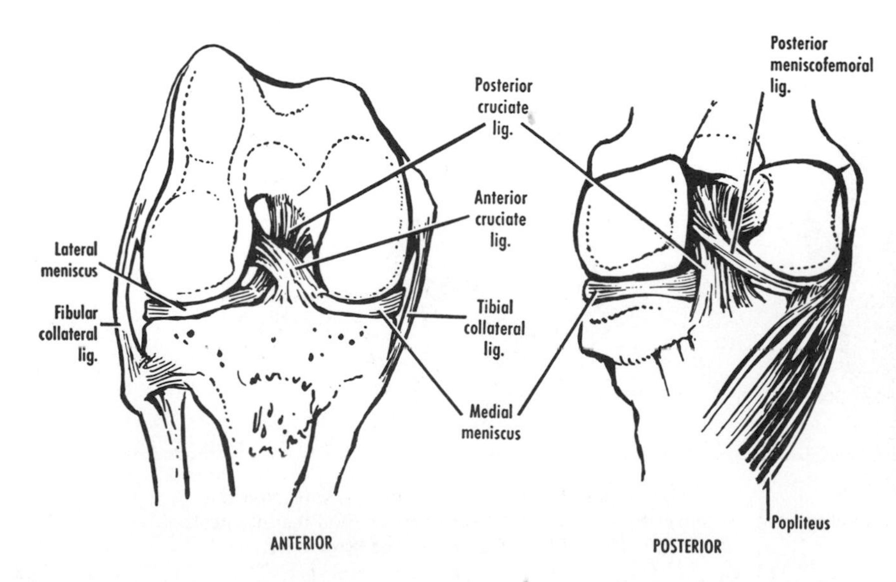

Figure 15–16 Anterior view of a flexed right knee joint in which the transverse ligament is absent. In the posterior view of the right knee joint, note that the popliteus arises in part from the lateral meniscus. The attachments of the collateral ligaments to the menisci are omitted.

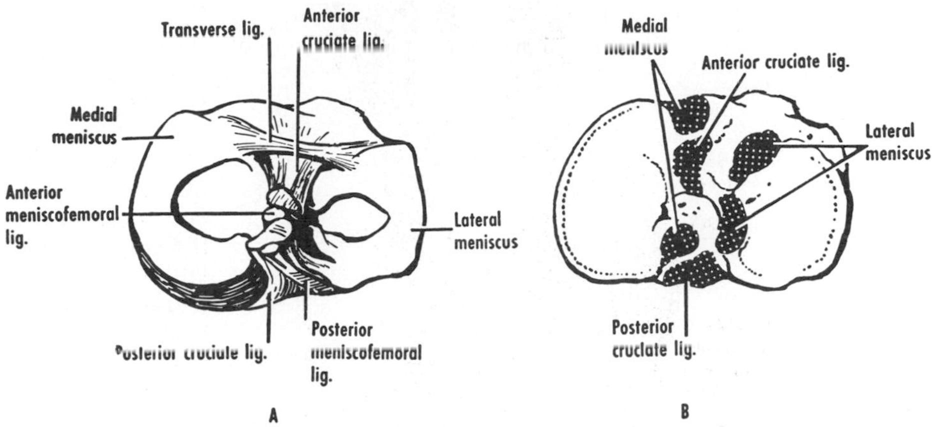

Figure 15–17 **A**, The cruciate ligaments and the menisci from above. Note the differences in the size and shape of the menisci. **B**, The right tibia from above, showing the attachments of the menisci and cruciate ligaments.

and popliteus tendons. The *tibial collateral ligament* extends from the medial epicondyle of the femur to the medial surface of the tibia, and it is attached to the medial meniscus.

The intra-articular ligaments are the cruciate ligaments and the menisci (fig. 15–16). The *anterior* and *posterior cruciate ligaments* limit, respectively, forward and backward slipping of the tibia on the femur. They are named for their tibial attachments, and they extend, respectively, in front of and behind the intercondylar eminence, from the proximity of the intercondylar fossa of the femur to the tibia (fig. 15–17). The ligaments cross each other; hence the name *cruciate*. You can "represent the ligaments by your lower limbs while standing, i.e., cross your right leg (right anterior cruciate ligament) in front of your left. Rotate your trunk to right and left" (Mainland). The *lateral* and *medial menisci* (or *semilunar cartilages*) are fibrous crescents that lie on the upper surface of the tibia. They act as cushions or shock absorbers and facilitate lubrication. Each meniscus is wedge-shaped in section, being thick externally and having a thin, free internal border. The ends, or horns, of the menisci are anchored to the tibia in front of and behind the intercondylar eminence, and they may be connected in front by a *transverse ligament* (see fig. 15–17). The lateral meniscus is almost circular, is anchored to the popliteus tendon (which probably pulls it backward in flexion), and is freer to move. **The medial meniscus is C-shaped, is anchored to the tibial collateral ligament, and is much more frequently torn by twisting injuries of the flexed knee.**

The synovial membrane is extensive, and it lines the infrapatellar fat pad between the patella and tibia. About a dozen bursae are situated near the knee (fig. 15–19). The most important is the **suprapatellar bursa or pouch, an extension of the joint cavity several centimeters above the patella, between the quadriceps and the front of the femur. Hence, when the cavity of the knee joint is distended with fluid, e.g., after trauma, the swelling presents itself above and at the sides of the patella. Other bursae that sometimes communicate with the joint cavity are related to the popliteus, medial head of the gastrocnemius (this bursa may form a cyst), semimembranosus, and lateral head of the gastrocnemius. A subcutaneous prepatellar bursa is situated between the skin and the lower part of the patella and may become inflamed ("housemaid's knee"). A subcutaneous infrapatellar bursa is found over the lower part of the tuberosity of**

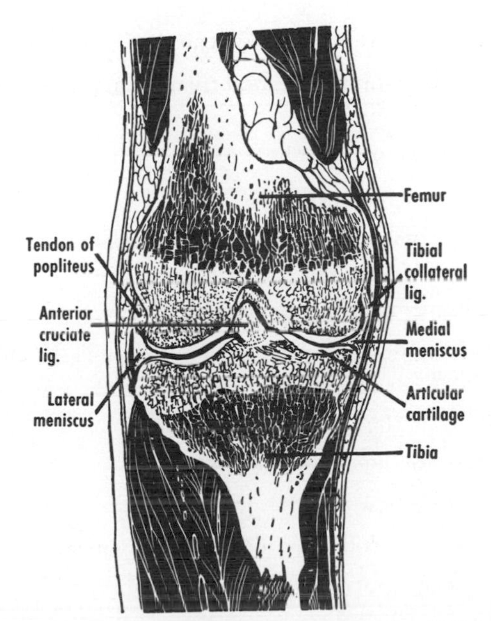

Figure 15–18 A coronal section of the knee joint.

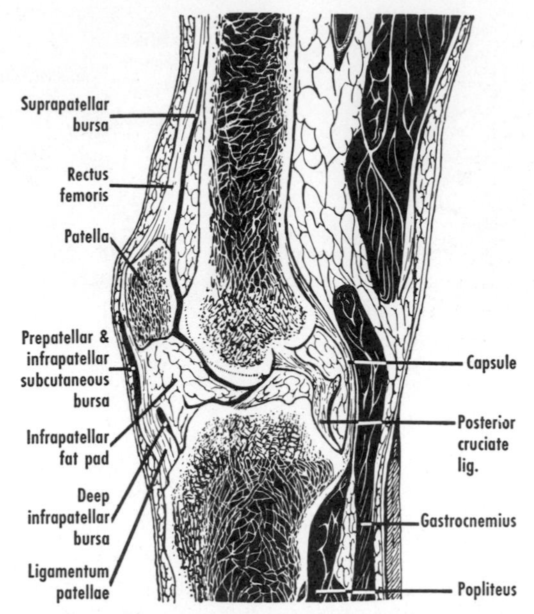

Figure 15–19 A sagittal section of the knee joint. The width of the articular and bursal cavities is exaggerated.

the tibia and may become inflamed ("clergyman's knee"). There is also a deep infrapatellar bursa between the ligamentum patellae and the tibia.

The knee joint is supplied by various nerves of the thigh and leg, e.g., the femoral, obturator, and sciatic nerves, and **pain may be referred to the knee from hip disease.**

The knee joint is very complicated and is best regarded as a condylar rather than a hinge joint, because the shapes and curvatures of the articular surfaces are such that hinge movements are combined with gliding, rolling, and rotation about a vertical axis. Flexion of the thigh at the knee is first accompanied by lateral rotation of the thigh (by the popliteus), and the femur then rolls backward on the tibia. Conversely, the last part of extension is accompanied by medial rotation of the thigh, and the ligaments are then taut and the joint is most stable. The quadriceps femoris extends the leg, and the hamstrings flex it. The biceps rotates the leg laterally, and the semitendinosus rotates it medially. The popliteus, acting from a fixed tibia, is believed to be significant in rotating the femur laterally. The muscles around the knee are very important in providing stability for the joint.

TIBIOFIBULAR JOINT

The tibiofibular joint is a small, plane articulation between the back of the lateral condyle of the tibia and the head of the fibula. The joint cavity may communicate with that of the knee.

QUESTIONS

15–1 What are the main compartments of the thigh?

15–2 What is the iliotibial tract?

15–3 What is the clinical importance of the femoral canal?

15–4 What are the chief contents of (a) the femoral triangle and (b) the adductor canal?

15–5 What is the chief action of the quadriceps?

15–6 Where is the reflex center for the knee jerk?

15–7 How far distally does the cutaneous territory of the saphenous nerve extend?

15–8 Which is the most important branch of the femoral artery?

15–9 What is the guide to the common peroneal nerve and what is the chief effect of injury to the nerve?

15–10 On what does the stability of the hip joint depend?

15–11 What is the innervation of the hip and knee joints?

15–12 Compare the knee with the elbow joint.

15–13 On what basis is the knee joint condylar and not a true hinge?

15–14 When is the knee joint "locked?"

15–15 Which bursae communicate with the cavity of the knee joint?

16
THE LEG

Inward extensions of the fascia of the leg form the *anterior* and *posterior intermuscular septa,* thereby giving rise to three compartments that allow of very little expansion: (1) the anterior, or extensor, compartment (muscles supplied by the common or deep peroneal nerve or by both; (2) lateral, or peroneal, compartment (muscles supplied by the superficial peroneal nerve); and (3) posterior, or flexor, compartment (muscles supplied by the tibial nerve) (fig. 16–1). The posterior compartment is subdivided by the deep transverse fascia of the leg, which runs between the medial border of the tibia and the posterior border of the fibula. The interosseous membrane, which connects the interosseous borders of the tibia and fibula, is almost sagittal in the upper part of the leg.

FRONT OF LEG (table 16–1)

The muscles of the front of the leg are the *tibialis anterior, extensor digitorum longus, peroneus tertius,* and *extensor hallucis longus* (figs. 16–1 and 17–4*C*). They arise from bone,

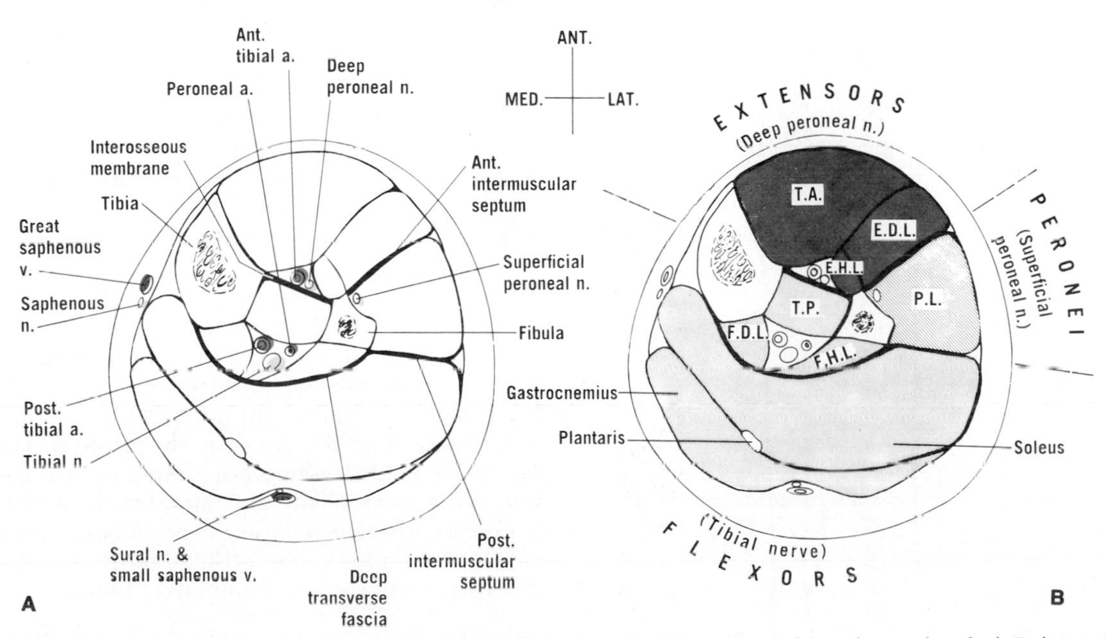

Figure 16–1 Horizontal section through the middle of the leg. In **A**, the nerves and vessels are identified. **B** shows the extensors (supplied by the deep peroneal nerve) anteriorly, the peronei (supplied by the superficial peroneal nerve) laterally, and the flexors (supplied by the tibial nerve) posteriorly. The posterior compartment contains superficial and deep muscles, separated by the deep transverse fascia of the leg. *E.D.L.*, extensor digitorum longus; *E.H.L.*, extensor hallucis longus; *F.D.L.*, flexor digitorum longus; *F.H.L.*, flexor hallucis longus; *P.L.*, peroneus longus; *T.A.*, tibialis anterior; *T.P.*, tibialis posterior.

TABLE 16–1 MUSCLES OF FRONT OF LEG

Muscle	Origin	Insertion	Innervation	Action
Tibialis anterior	Lateral condyle & surface of tibia	Medial cuneiform & base of 1st metatarsal	Deep & common peroneal	Dorsiflexes & inverts foot
Extensor digitorum longus	Lateral condyle of tibia & anterior surface of fibula	Middle & distal phalanges of toes 2-5	Deep (& common) peroneal	Extends metatarsophalangeal joints
Peroneus tertius	Anterior surface of fibula	Fascia, or base of 5th metatarsal	Deep peroneal	Aids in eversion
Extensor hallucis longus	Anterior surface of fibula	Distal phalanx of big toe	Deep peroneal	Extends big toe

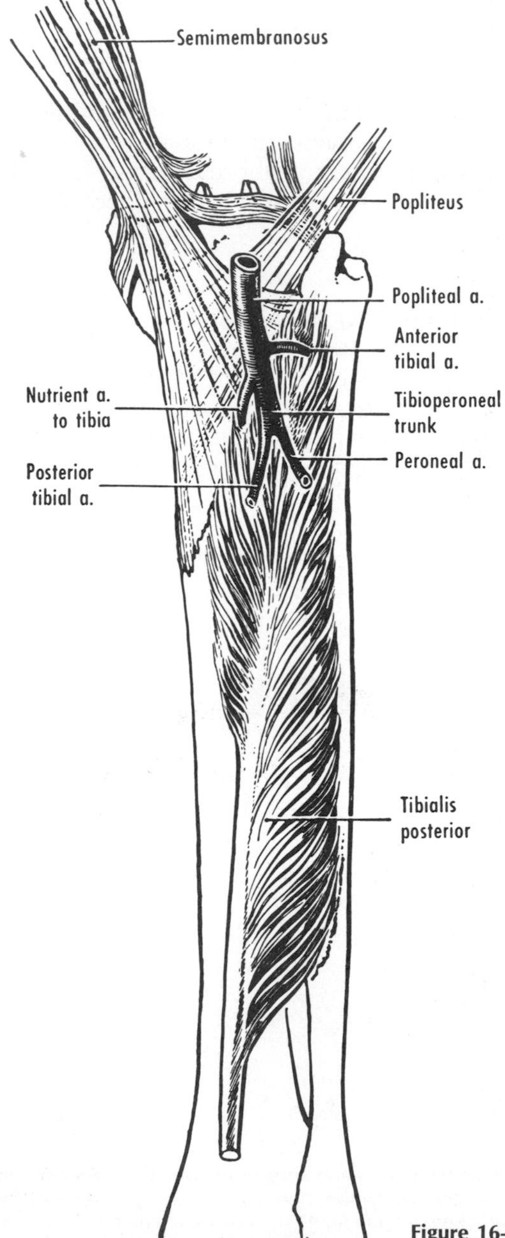

the strong investing fascia, and the interosseous membrane. **The muscles of the front of the leg are supplied by the common or deep peroneal nerve or by both, and they dorsiflex the foot.** The tendons of these muscles, surrounded by their synovial sheaths, are bound down by fascial thickenings known as the *superior* and *inferior extensor retinacula.*

The *deep peroneal nerve* (see fig. 15–7) (at one time known as the *anterior tibial nerve*) is one of the terminal branches of the common peroneal nerve. It continues around the neck of the fibula and descends on the interosseous membrane in company with the anterior tibial artery. It supplies the tibialis anterior, extensor digitorum longus, extensor hallucis longus, peroneus tertius, and extensor digitorum brevis, and it gives off dorsal digital nerves to the first two toes. Section of the deep peroneal nerve may result in footdrop and a "steppage" gait, owing to paralysis of the dorsiflexors of the foot and the extensors of the toes.

The *anterior tibial artery,* the smaller division of the popliteal artery, passes through the fibrous arch of the tibialis posterior and that of the interosseous membrane to meet its companion nerve (fig. 16–2). It descends on the interosseous membrane and, on the dorsum of the foot, is typically continued as the dorsalis pedis artery (see fig. 17–4*C*). **The pulsations of the anterior tibial artery are often palpable between the two malleoli and lateral to the extensor hallucis longus tendon.** Its branches supply adjacent muscles and contribute to the anastomoses around the knee and ankle joints.

Figure 16–2 The divisions of the popliteal artery and their relations to the tibialis posterior. (Based on Shellshear and Macintosh.)

LATERAL SIDE OF LEG (table 16–2)

The peroneus longus and brevis lie between the anterior and posterior intermuscular septa, and they arise from these septa, as well as from fascia and bone (figs. 16–1, 16–3, and 16–5). **The peroneus longus and brevis are supplied by the superficial peroneal nerve, and they evert the foot.** They have a common synovial sheath as they descend behind the lateral

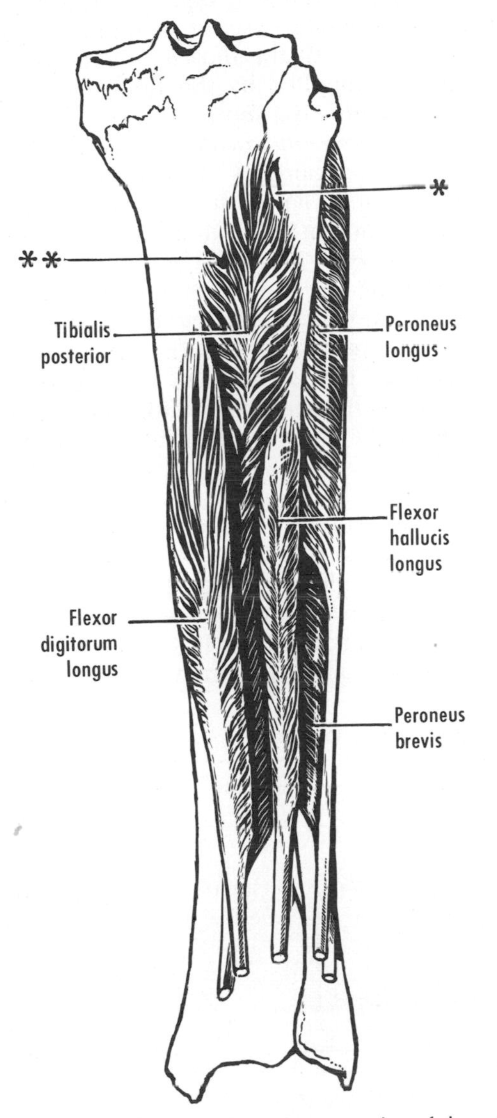

Figure 16–3 Origins of the deep muscles of the calf. Note the fibrous arch in the tibialis posterior for the anterior tibial vessels (*asterisk*). Note also the arch in the tibialis posterior for the nutrient artery to the tibia (*two asterisks*). (Based on Shellshear and Macintosh.)

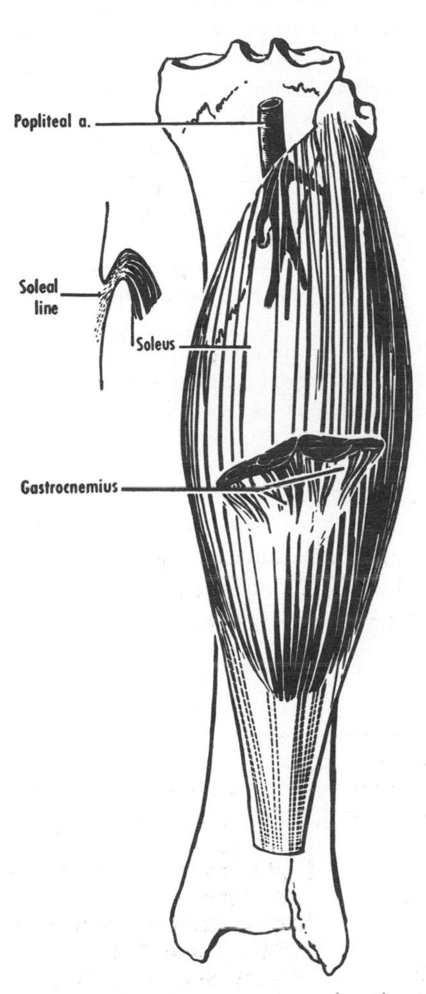

Figure 16–4 The soleus superimposed on the structures shown in fig. 16–2. Note that its apex is at the fibula. Note also from the inset that the soleus ascends from the soleal line and then turns downward. Thus, the upper border of the muscle is at a higher level than the soleal line. (Based on Shellshear and Macintosh.)

TABLE 16–2 MUSCLES OF LATERAL SIDE OF LEG

Muscle	Origin	Insertion	Innervation	Action
Peroneus longus	Lateral condyle of tibia & head & lateral surface of fibula	Medial cuneiform & base of 1st metatarsal	Superficial (& common) peroneal	Plantar-flexes & everts foot
Peroneus brevis	Lateral surface of fibula	Tuberosity of 5th metatarsal	Superficial peroneal	Everts foot

malleolus, and they are bound down by the *superior* and *inferior peroneal retinacula* (see fig. 17–4*B*). The *peroneus longus* traverses a notch in the cuboid and then contains either a sesamoid bone or a fibrocartilaginous thickening. The tendon crosses the sole of the foot to reach the medial cuneiform and the base of the first metatarsal opposite the more medial insertion of the tibialis anterior (see fig. 17–2*C*), with which it forms a stirrup. The *peroneus brevis* lies deep to the longus.

The *superficial peroneal nerve* (at one time known as the *musculocutaneous nerve*) is one of the terminal branches of the common peroneal nerve. It descends in front of the fibula (see fig. 15–7). It supplies the peroneus longus and brevis (and sometimes the extensor digitorum brevis) and provides cutaneous branches to usually all five toes. Section of the superficial peroneal nerve may result in impairment and loss of eversion.

BACK OF LEG (tables 16–3 and 16–4)

The superficial muscles of the back of the leg are the gastrocnemius and soleus (fig. 16–4), which together are sometimes termed the *triceps surae*, and the plantaris. The deep muscles are the popliteus, flexor digitorum longus, flexor hallucis longus, and tibialis posterior (figs. 16–1 to 16–3 and 16–5). The last three are separated from the superficial group by the deep transverse fascia of the leg and arise from the interosseous membrane as well as from bone. **All the muscles of the back of the leg are supplied by the tibial nerve, and they plantar-flex the foot.**

The *gastrocnemius* has two large heads that arise from the condyles of the femur (fig. 15–9). Bursae underlie the heads, and the lateral head may contain a sesamoid bone known as the *fabella* (see fig. 12–21*B*). At about the middle of the leg, the heads end in a common aponeurosis, which unites with the underlying tendon of the soleus to form the *tendo calcaneus* (fig. 16–5). Achilles is said to have been held by his heels when dipped in the river Styx to make him invulnerable, and, at the siege of Troy, he was mortally wounded by an arrow in his heel: for several centuries the tendo calcaneus was known as Achilles' tendon. The tendo calcaneus is inserted into the back of the calcaneus (see fig. 12–35). The *soleus* presents a tendinous arch between the fibula and tibia. The arch lies behind the popliteal vessels and tibial nerve. The triceps

TABLE 16–3 Superficial Muscles of Back of Leg

Muscle	Origin	Insertion	Innervation	Action
Gastrocnemius	Lateral aspect of lateral condyle of femur Popliteal surface & medial condyle of femur	Posterior aspect of calcaneus	Tibial	Plantar-flexes foot
Soleus	Head & posterior surface of fibula & soleal line of tibia	Posterior aspect of calcaneus	Tibial	Plantar-flexes foot
Plantaris	Popliteal surface of femur	Tendo calcaneus	Tibial	Assists gastrocnemius

TABLE 16–4 Deep Muscles of Back of Leg

Muscle	Origin	Insertion	Innervation	Action
Popliteus	Lateral aspect of lateral condyle of femur & lateral meniscus	Tibia above soleal line	Tibial	Rotates tibia medially or femur laterally
Flexor digitorum longus	Posterior surface of tibia	Distal phalanges of toes 2-5	Tibial	Flexes distal phalanges
Flexor hallucis longus	Posterior surface of fibula	Distal phalanx of big toe	Tibial	Flexes distal phalanx
Tibialis posterior	Posterior surface of fibula & soleal line of tibia	Tuberosity of navicular & all tarsals except talus	Tibial	Inverts foot

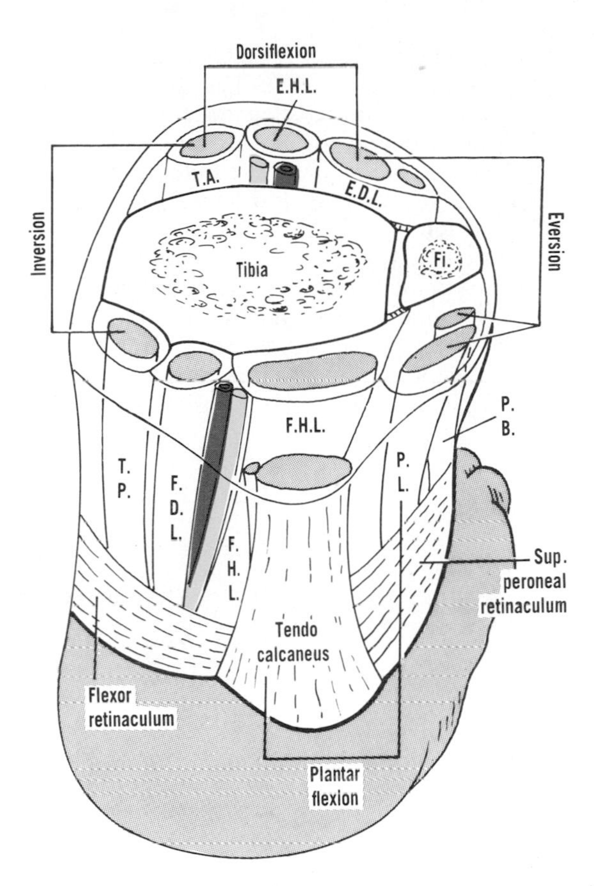

Figure 16–5 Horizontal section through the lower end of the tibia and fibula. The tendons of the muscles of the three compartments of the leg (see fig. 16–1) are shown, as are also the neurovascular bundles related to the hallucis and digitorum tendons (deep peroneal nerve and anterior tibial artery; posterior tibial artery and tibial nerve). The synovial sheaths (see fig. 17–1) are not represented. The plantaris is seen joining the medial side of the tendo calcaneus. Some important muscular functions are indicated. Dorsiflexion and plantar flexion take place at the ankle joint, whereas inversion and eversion occur at more distal joints. (See fig. 17–7.) (Modified from Castaing and Soutoul.)

surae plantar-flexes the foot and is an important muscle in both posture and locomotion. It is used in walking, running, jumping, and dancing. **The ankle jerk is a twitch of the triceps surae induced by tapping the tendo calcaneus. The reflex center is generally in the S1 segment of the spinal cord.**

The *popliteus* descends from the femur laterally to the tibia medially (see fig. 15–10). Hence it rotates the tibia medially or the femur laterally when the tibia is fixed. The popliteus arises also from the lateral meniscus, which it pulls backward at the beginning of flexion. The muscle is active in crouching, when it probably prevents the femur from sliding forward on the tibia. The *flexor digitorum longus* resembles the flexor digitorum profundus of the upper limb, in that each muscle has four tendons (omitting the pre-axial digit), gives origin to lumbricals, has fibrous and synovial sheaths, is anchored to the phalanges by vincula, and is inserted into the distal phalanges (see fig. 17–2*B*). Although it goes to the great toe, the *flexor hallucis longus* arises from the fibula. The *tibialis posterior,* more deeply placed, has an extensive origin from the interosseous membrane, fibula, and tibia. It is inserted into all the tarsal bones except the talus, but chiefly into the tuberosity of the navicular. The tibialis posterior is the principal inverter of the foot. The long flexors and tibialis posterior, together with their synovial sheaths and the posterior tibial vessels and tibial nerve, are limited behind the medial malleolus by an indistinct fascial thickening, the *flexor retinaculum* (see fig. 17–4*A*).

The *tibial nerve,* which lies on the popliteus and under cover of the gastrocnemius, passes in front of the tendinous arch of the soleus and descends on the deep muscles of the leg (see figs. 15–8 and 15–10). Deep to the flexor retinaculum, it divides into the medial and lateral plantar nerves. In addition to the gastrocnemius and plantaris, the tibial nerve supplies the soleus, flexor digitorum longus, flexor hallucis longus, tibialis posterior, and the skin of the heel and sole. Section of the tibial nerve is followed by sensory loss in the sole of the foot, which interferes with posture and locomotion. The extent of motor loss depends on the level of the lesion, but it may involve the muscles of the calf and the small muscles of the foot. Plantar flexion may be lost, resulting in a shuffling gait, and inversion (produced by the tibialis posterior) may be impaired.

The *posterior tibial artery,* the larger division of the popliteal artery, begins at the lower border of the popliteus (see figs. 16–2 and 16–4). It descends on the deep muscles of the leg and is covered by the soleus and gastrocnemius. Deep to the flexor retinaculum, it divides into the medial and lateral plantar arteries, which provide the chief blood supply to the foot (see fig. 17–4*A*). **The pulsations of the posterior tibial artery are often palpable between the medial malleolus and the tendo calcaneus.** The posterior tibial artery supplies adjacent muscles, a large nutrient artery to the tibia, branches to the anastomoses around the knee and ankle, and the peroneal artery. The *peroneal artery* (see fig. 16–2) arises from the posterior tibial artery below the lower border of the popliteus and descends along the medial crest of the fibula. It gives off a number of small branches, including its terminal calcanean ones. Its *perforating branch,* which reaches the front of the leg by piercing the interosseous membrane, anastomoses with branches of the dorsalis pedis. If the anterior tibial artery is small or absent, the peroneal is large and, by means of its perforating branch, may replace the dorsalis pedis.

QUESTIONS

16–1 What are the main compartments of the leg?

16–2 Which nerves accompany the arteries in the leg?

16–3 What are the chief (a) invertors and (b) evertors of the foot?

16–4 What is the triceps surae?

16–5 Where is the reflex center for the ankle jerk?

16–6 What is the action of the popliteus?

16–7 What are the chief effects of section of the tibial nerve?

17 THE ANKLE AND FOOT

The word *ankle* refers to the angle between the leg and the foot. The functions of the foot are support and locomotion, whereas the hand is a tactile and grasping organ. The toes are numbered from one to five, beginning with the big toe, or hallux. Thus, the pre-axial digit in either the hand or the foot is numbered one. The terms *abduction* and *adduction* of the toes are used with reference to an axis through the second toe. Thus, abduction of the big toe is a medial movement, away from the second toe. The tendons around the ankle (similar to those at the wrist) are bound down by retinacula (see fig. 17–4).

The fascia on the sole of the foot is a strong sheet termed the *plantar aponeurosis,* which acts as a mechanical tie. It extends forward from the tuber calcanei and divides into five processes, each of which is anchored at a metatarsophalangeal joint. Fascial "spaces" are situated above the plantar aponeurosis, and the big and little toes have special compartments. *Synovial sheaths* are found (1) in front of the ankle (fig. 17–1) for the (a) tibialis anterior, (b) extensor hallucis longus, and (c) extensor digitorum longus and peroneus tertius; (2) behind the medial malleolus for the (a) tibialis posterior, (b) flexor digitorum longus, and (c) flexor hallucis longus; and (3) behind the lateral malleolus for the peroneus longus and brevis. Some further sheaths are found in the sole and in relation to the toes.

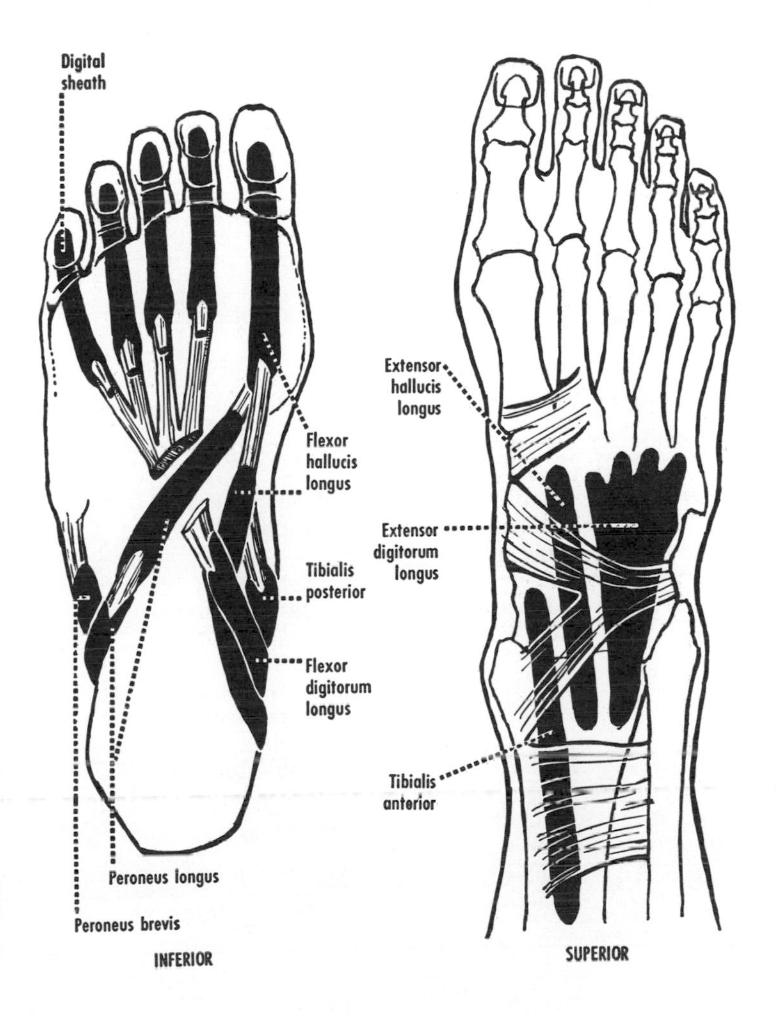

Figure 17–1 The synovial tendon sheaths of the foot and ankle. The more distal and somewhat less constant sheaths of the extensor digitorum longus and brevis have been omitted.

MUSCLES OF FOOT (table 17–1)

The *extensor digitorum brevis* is the only muscle on the dorsum. It and the extensor hallucis longus can be felt and sometimes seen on dorsiflexing the proximal phalanges against resistance, and these actions are used to test the integrity of the fifth lumbar nerve (L.N.5).

The muscles of the sole are collectively important in posture and locomotion, and they provide strong support for the arches of the foot during movement. They may be considered in three groups: for the big toe, the central portion of the sole, and the little toe. They may also be considered, however, in four layers (table 17–1 and fig. 17–2).

The *flexor digitorum brevis* resembles the flexor digitorum superficialis of the upper limb, in that each muscle has four tendons (omitting the pre-axial digit), which are perforated by the long flexor tendons and then divide to be inserted into the sides of the middle phalanges.

Although classified as dorsal and plantar, both groups of *interossei* are actually more plantar. **The lumbricals and interossei are arranged in a manner basically similar to those of the hand but with reference to the axis of the**

TABLE 17–1 MUSCLES OF FOOT

Muscle	Origin	Insertion	Innervation	Action
Extensor digitorum brevis	Sinus tarsi	Long extensor tendons of toes 1–4	Deep peroneal	Aids in extending toes 1–4
Layer 1:				
Abductor hallucis	Tuber calcanei	Medial sesamoid & proximal phalanx	Medial plantar	Abducts & flexes big toe
Flexor digitorum brevis	Tuber calcanei	Middle phalanges of toes 2–5	Medial plantar	Flexes middle phalanges
Abductor digiti minimi	Tuber calcanei	Proximal phalanx & 5th metatarsal	Lateral plantar	Abducts & flexes little toe
Layer 2:				
Quadratus plantae	Tuber calcanei	Flexor digitorum longus tendon	Lateral plantar	Accessory flexor
Lumbricals	Long flexor tendons	Medial sides of proximal phalanges of toes 2–5	Lateral (1) & medial (2–4) plantar	Aid interossei
Layer 3:				
Flexor hallucis brevis	Tibialis posterior tendon	Medial & lateral sesamoids & proximal phalanx	Medial plantar	Flexes big toe
Adductor hallucis	Sheath of peroneus longus (oblique head) & deep transverse metatarsal ligament (transverse head)	Lateral sesamoid & proximal phalanx	Lateral plantar	Adducts & flexes big toe
Flexor digiti minimi brevis	Sheath of peroneus longus	Proximal phalanx & 5th metatarsal	Lateral plantar	Flexes little toe
Layer 4:				
Plantar interossei 1–3	Metatarsals 3–5	Medial sides of bases of proximal phalanges of toes 3–5	Lateral plantar	Flex & adduct proximal phalanges
Dorsal interossei 1–4	Metatarsals 1–5	Bases of proximal phalanges of toes 2–4	Lateral plantar	Flex & abduct proximal phalanges

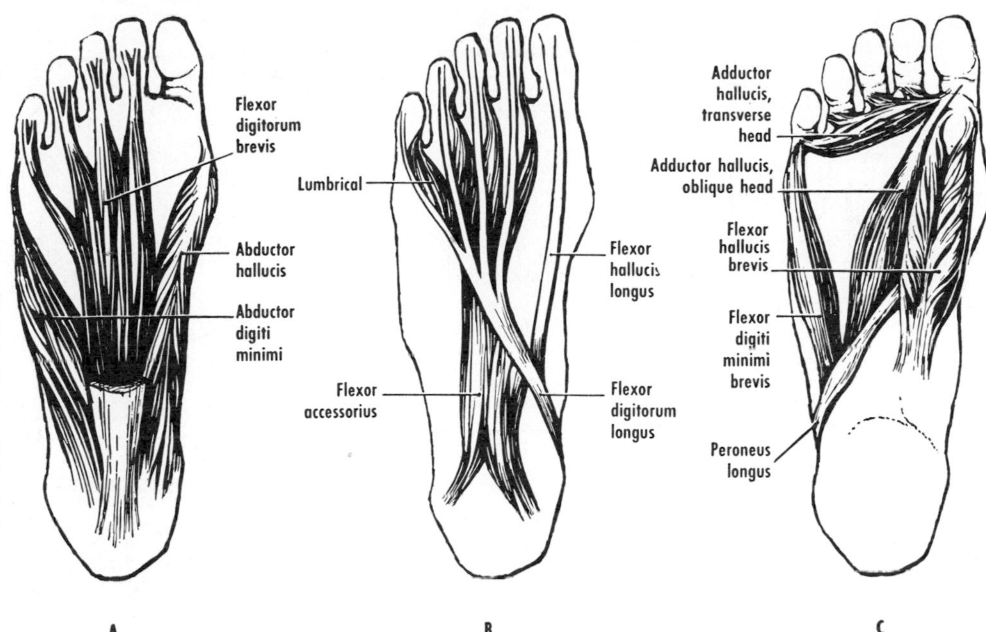

Figure 17–2 Muscles of the sole of the foot, shown in successive layers from below upward. **A,** The first, or most superficial, layer. **B,** The second layer. **C,** The third layer, including the peroneus longus, the insertion of which belongs to the fourth layer. See fig. 17–8 for the fourth layer.

second toe as compared with that of the third finger. There are, however, structural and functional differences; e.g., the interossei of the foot are not inserted into the extensor aponeuroses, and they probably strengthen the metatarsal arch by holding the metatarsals together.

The plantar skin reflex is a (plantar) flexion of the toes when the skin of the sole is stroked slowly along its lateral border. In infants before they walk and in patients with certain disorders of the motor pathways of the brain and spinal cord, however, similar stimulation of the sole results in a slow dorsiflexion of the big toe and a slight spreading of the other toes. This response is known as the Babinski sign.

NERVES OF FOOT

The *medial plantar nerve,* the larger terminal branch of the tibial nerve, arises behind the medial malleolus, deep to the flexor retinaculum and the abductor hallucis (see fig. 17–4*A*). It runs forward between the abductor hallucis and the flexor digitorum brevis and supplies these muscles (see fig. 15–2) as well as the skin on the medial side of the foot. It ends in *plantar digital nerves* that supply the flexor hallucis brevis, the first lumbrical, and the skin of the medial toes, including their nail beds. The medial plantar nerve is comparable to the median nerve in the hand.

The *lateral plantar nerve* arises behind the medial malleolus. It runs forward and laterally, deep to the flexor digitorum brevis, and divides into superficial and deep branches. It supplies the quadratus plantae, abductor digiti minimi, and the lateral side of the sole. The *superficial branch* supplies the flexor digiti minimi brevis and gives plantar digital nerves to the lateral toes. The *deep branch* turns medially and supplies the interossei, lumbricals 2 to 4, and the adductor hallucis. The lateral plantar nerve is comparable to the ulnar nerve in the hand.

VESSELS OF FOOT (fig. 17–3)

The *medial plantar artery,* one of the terminal branches of the posterior tibial artery, arises deep to the flexor retinaculum and the abductor hallucis. It runs forward with its companion nerve and gives digital branches to the medial toes (fig. 17–4*A*).

The *lateral plantar artery,* with its companion nerve, runs forward and laterally, deep to the flexor digitorum brevis. It then turns medially and forms the *plantar arch,* which lies between the third and fourth layers of the muscles of the sole. The arch gives off a series of *metatarsal* and *digital arteries.*

The *dorsalis pedis artery,* variable in size and course, is the continuation of the anterior tibial artery at a point midway between the

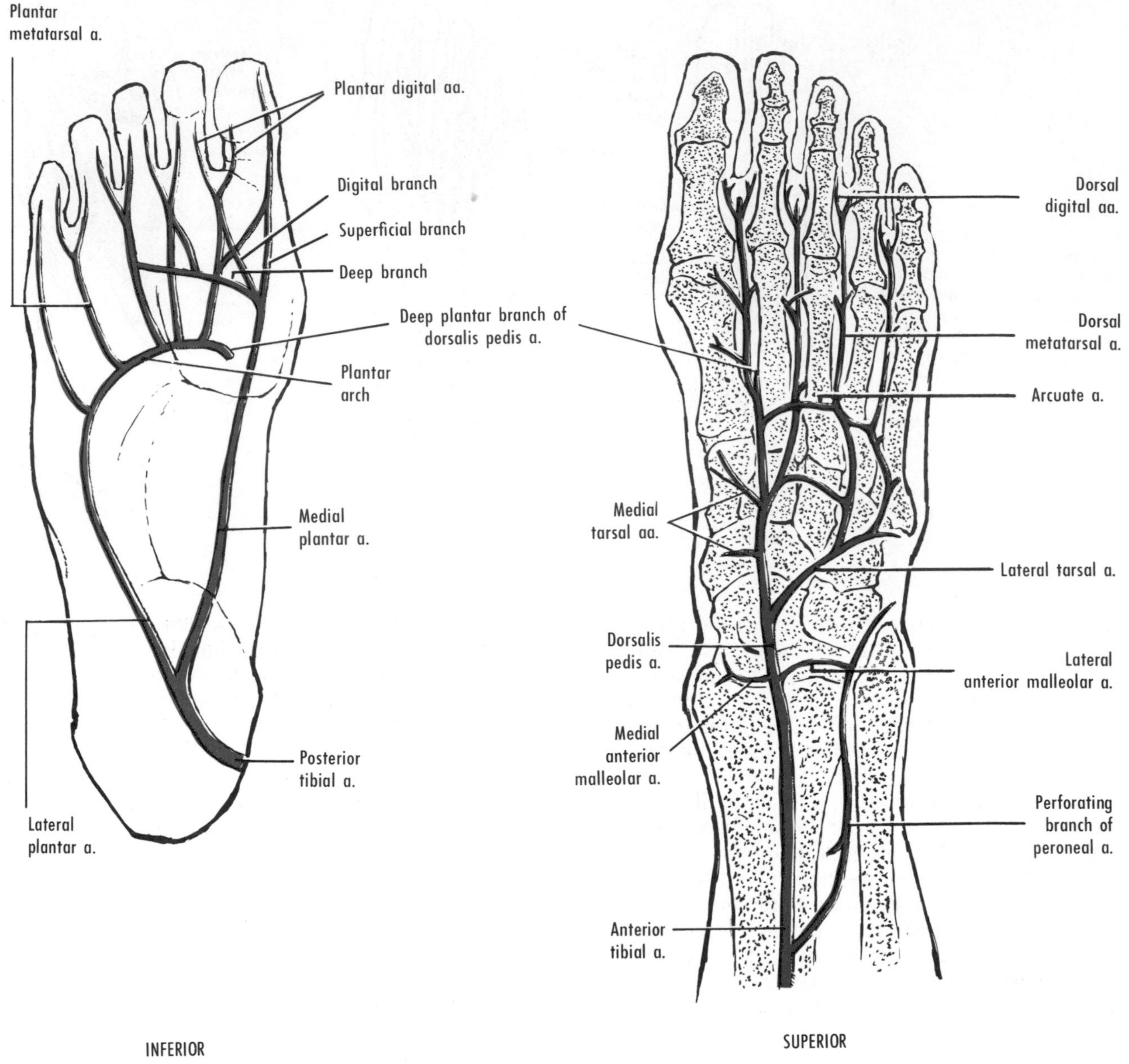

Figure 17–3 The arteries of the sole and dorsum of the foot.

malleoli (fig. 17–4*C*). It extends to the posterior end of the first intermetatarsal space. **The dorsalis pedis artery is important clinically in assessing peripheral circulation. Its pulsations should be sought, and can generally be felt, between the tendons of the extensor hallucis longus and extensor digitorum longus** (fig. 17–4*C*). The artery is crossed by the inferior extensor retinaculum and extensor hallucis brevis. It lies successively on the capsule of the ankle joint, the head of the talus, the navicular, and the intermediate cuneiform. Its branches form an arterial network on the dorsum of the foot. The tendon of the extensor hallucis longus crosses either the anterior tibial artery or the dorsalis pedis artery and comes to lie on the medial side of the latter. The dorsalis pedis artery ends in a *deep plan-*

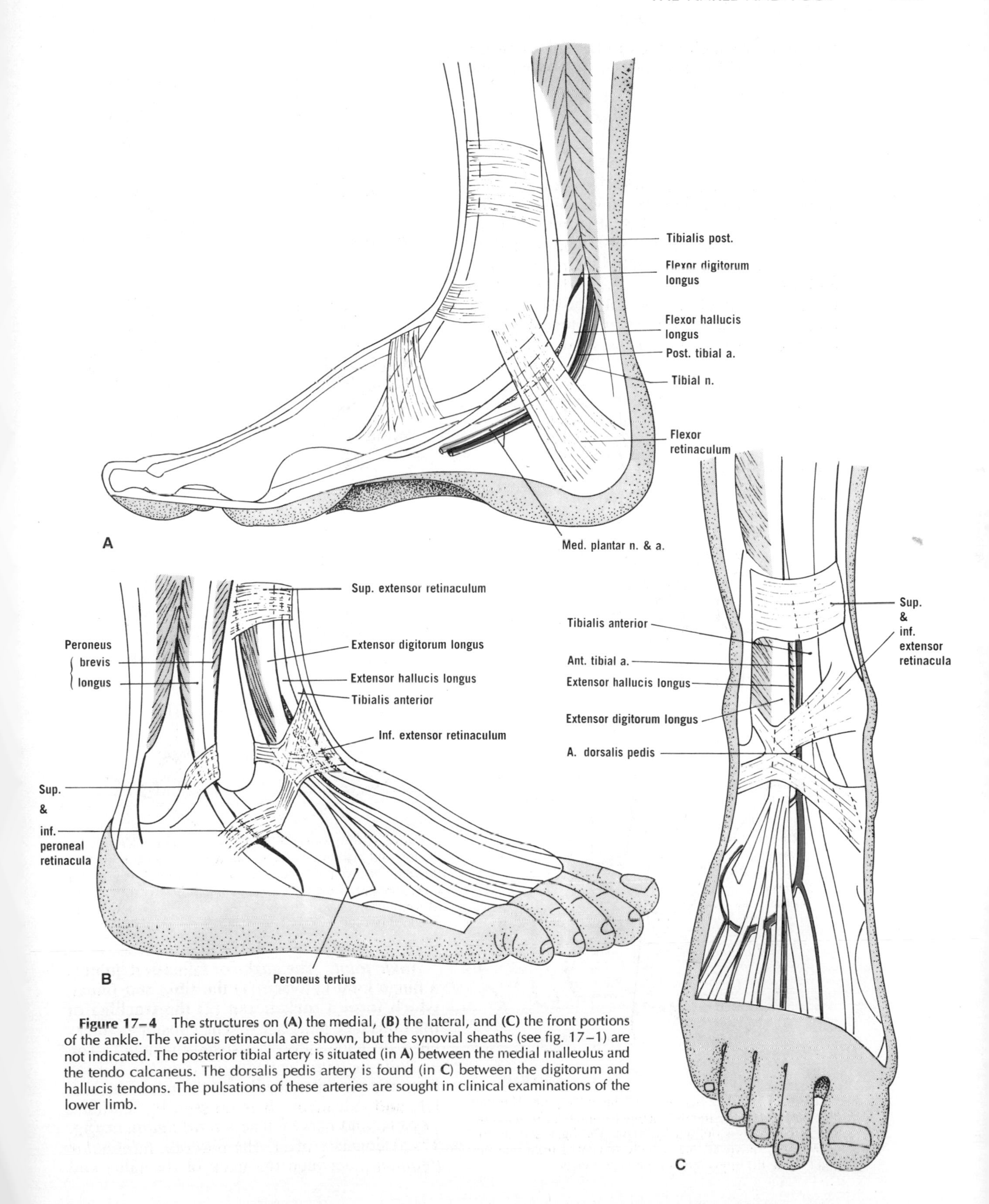

Figure 17–4 The structures on **(A)** the medial, **(B)** the lateral, and **(C)** the front portions of the ankle. The various retinacula are shown, but the synovial sheaths (see fig. 17–1) are not indicated. The posterior tibial artery is situated (in **A**) between the medial malleolus and the tendo calcaneus. The dorsalis pedis artery is found (in **C**) between the digitorum and hallucis tendons. The pulsations of these arteries are sought in clinical examinations of the lower limb.

tar branch, which passes to the sole between the heads of the first dorsal interosseus and completes the plantar arch.

JOINTS

Tibiofibular Syndesmosis. A strong fibrous union exists between the lower ends of the tibia and fibula. It consists of an interosseous ligament, strengthened in front and behind by *anterior* and *posterior tibiofibular ligaments,* and a *transverse ligament* from the malleolar fossa of the fibula to the back of the tibia. A recess of the ankle joint often extends upward into the lower portion of the syndesmosis.

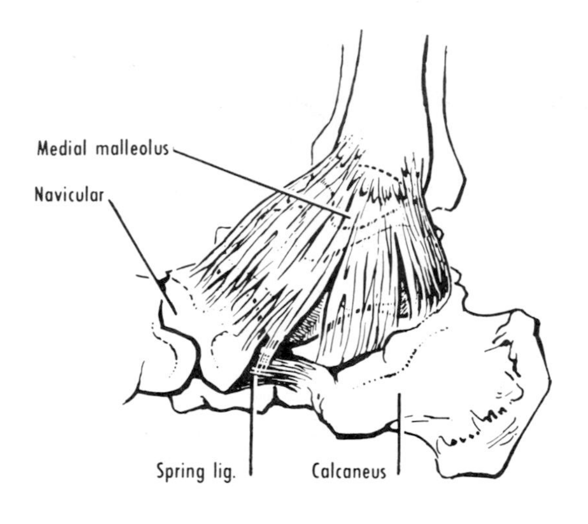

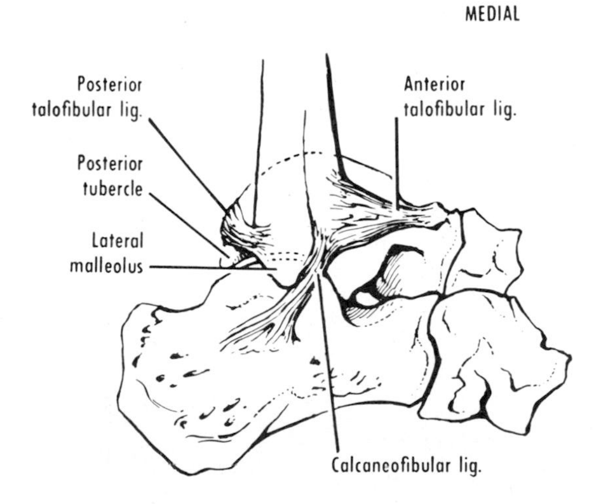

Figure 17–5 The ligaments of the ankle joint. The medial view shows the medial ligament, which forms a dense, almost continuous deltoid ligament. The ligaments on the lateral side, however, are usually separated from one another. Note the sinus tarsi in the lateral view.

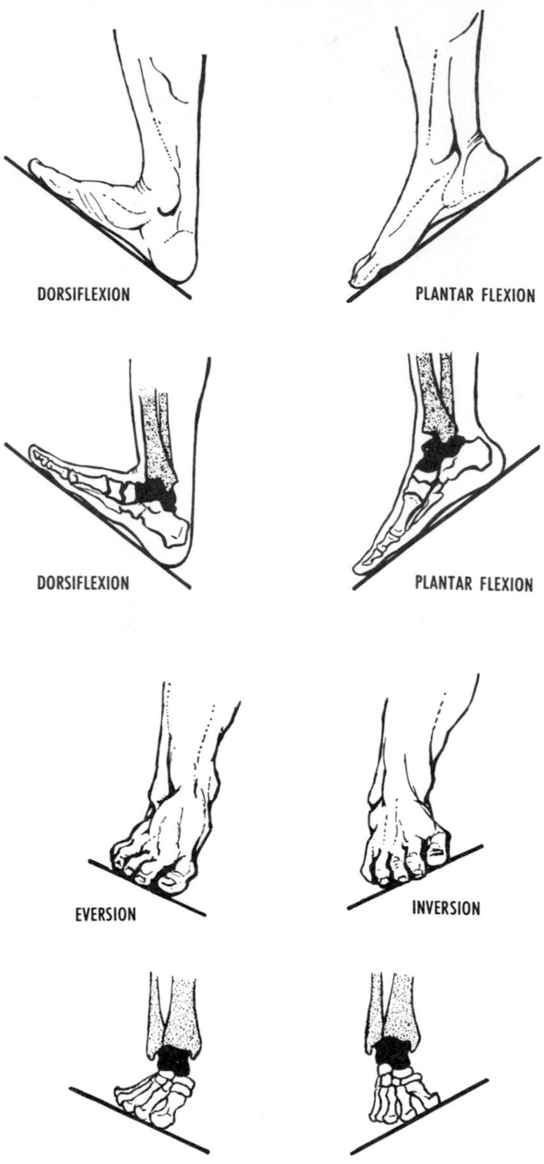

Figure 17–6 Movements of the foot and ankle. Dorsiflexion and plantar flexion are shown as in walking up and down hill. Movement occurs at the ankle joint. Eversion and inversion are shown as in standing sideways on a hill. Movement occurs at the tarsal joints, the talus remaining fixed. (Based on Mollier.)

Ankle Joint. **The ankle, or talocrural, joint is a hinge joint between (1) the tibia and fibula, which form a socket, and (2) the trochlea of the talus** (see fig. 12–29). The capsule is thickened on each side by a strong ligament. The *medial,* or *deltoid, ligament* (fig. 17–5) runs from the medial malleolus to the talus, navicular, and calcaneus. It is crossed by tendons, vessels, and nerves. The *lateral ligament* (fig. 17–5) consists of (1) the *anterior talofibular ligament,* between the neck of the talus and the lateral malleolus; (2) the *calcaneofibular ligament;* and (3) the *posterior talofibular ligament,* between the talus and the malleolar fossa. The medial and lateral ligaments prevent anterior and posterior slipping of the talus. They may be torn in injuries to the ankle, although, if they do not yield, one or both malleoli may be broken off in dislocations of the ankle joint. The shape of the surfaces at the ankle is such that, aided by the ligaments of the tibiofibular syndesmosis, the malleoli grip the talus tightly in dorsiflexion.

The ankle joint allows dorsiflexion and plantar flexion (fig. 17–6) around an axis that passes approximately through the malleoli. The range of movement varies. The triceps surae and peroneus longus plantar-flex the foot. The tibialis anterior and extensor digitorum longus dorsiflex the foot (see fig. 16–5). **It should be appreciated that, physiologically, plantar flexion of the foot and toes is an extensor response, whereas dorsiflexion of the foot and toes is a flexor response.**

Intertarsal Joints. The talus moves with the foot during dorsiflexion and plantar flexion. However, during inversion and eversion, which occur at *intertarsal joints,* the talus moves with the leg. **The most important intertarsal joints are the subtalar, the talocalcaneonavicular, and the calcaneocuboid. The last two constitute the transverse tarsal, or midtarsal, joint. The transverse tarsal joint can be represented by a line from the back of the tuberosity of the navicular to the midpoint between the lateral malleolus and the tuberosity of the fifth metatarsal.** The other intertarsal joints are the cuneocuboid, intercuneiform, and cuneonavicular, all of which are plane joints.

The *subtalar joint* (figs. 12–36 and 17–7) is a separate talocalcanean articulation lying behind the tarsal canal. The *talocalcaneonavicular joint,* a part of the *transverse tarsal joint,* lies in front of the tarsal canal. It resembles a ball-and-socket joint in that the head of the talus fits into a socket formed by the navicular in front, the calcaneus below, and the plantar calcaneonavicular ligament in between (fig. 17–7). This band, frequently termed the *spring ligament* (figs. 17–5 and 17–8), connects the sustentaculum tali with the navicular, and the tibialis posterior tendon lies immediately below it. The other part of the

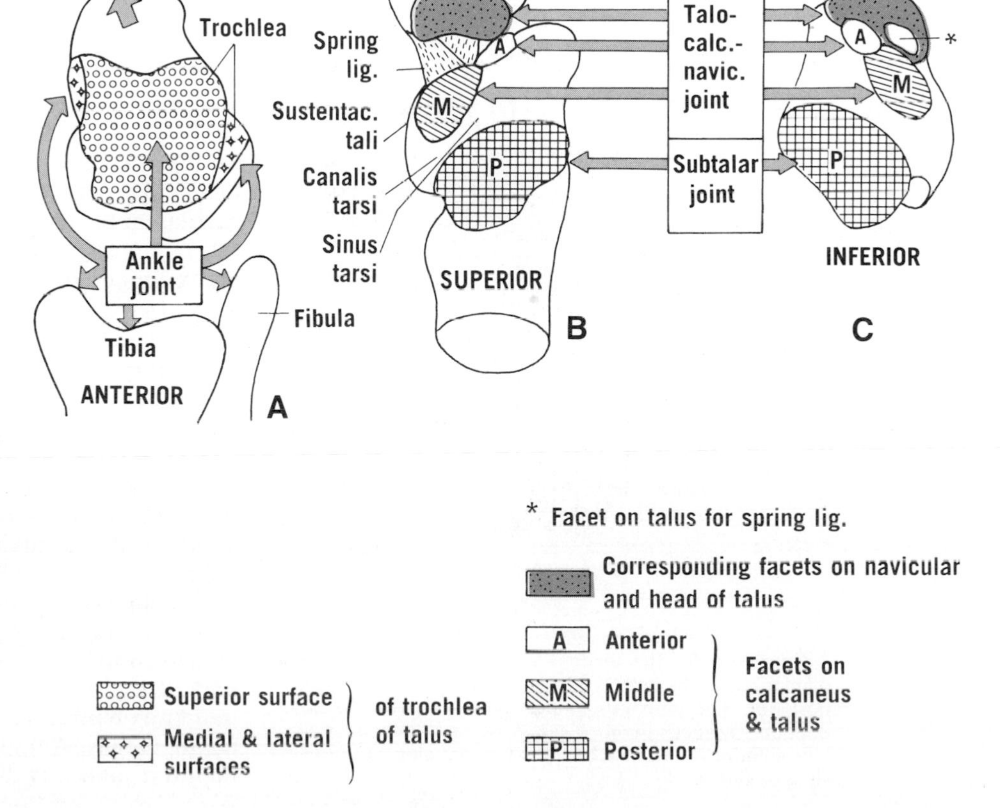

Figure 17–7 The facets of the ankle, subtalar, and talocalcaneonavicular joints. **A,** Diagram of the talus from above to show the three-surfaced trochlea that fits into the mortise formed by the tibia and fibula. **B,** Diagram of the calcaneus from above to show the posterior facet **(P)** for the subtalar joint, separated by the canalis and sinus tarsi from the middle **(M)** and anterior **(A)** facets of the talocalcaneonavicular joint. The socket of this latter joint is completed by the spring ligament and the concavity of the navicular. **C,** Diagram of the talus from below to show its corresponding facets for the subtalar and calcaneonavicular joints. Cf. fig. 12–36. A broad arrow in **A** emphasizes that the head of the talus is directed anteromedially.

transverse tarsal joint is the *calcaneocuboid,* which resembles a limited saddle joint. A strong *bifurcate ligament* extends from the floor of the sinus tarsi to the navicular and cuboid. (The sinus tarsi is the expanded anterolateral end of the *tarsal canal,* which runs obliquely between the talus and calcaneus.)

The foot may be disarticulated at the ankle joint (Syme's amputation) or at the transverse tarsal joint (Chopart's amputation).

The tension that develops during the support of body weight is taken up by strong ligaments on the plantar aspect of the tarsus (figs. 17–8 and 17–9). The *long plantar ligament* extends from the plantar aspect of the calcaneus to the tuberosity of the cuboid. The *short plantar ligament,* also calcaneocuboid, is more deeply placed.

The chief movements of the foot distal to the ankle joint are inversion and eversion. In inversion, the sole is directed medially. In eversion, it is turned so that it faces laterally (see fig. 17–6). The terms may be used also for the equivalent movements whereby the leg moves while the foot is fixed. **Inversion and eversion occur at mainly the subtalar and transverse tarsal joints.** The axes of movement at these articulations are situated obliquely with reference to the standard anatomical planes. Hence, each movement is a combination of two or more

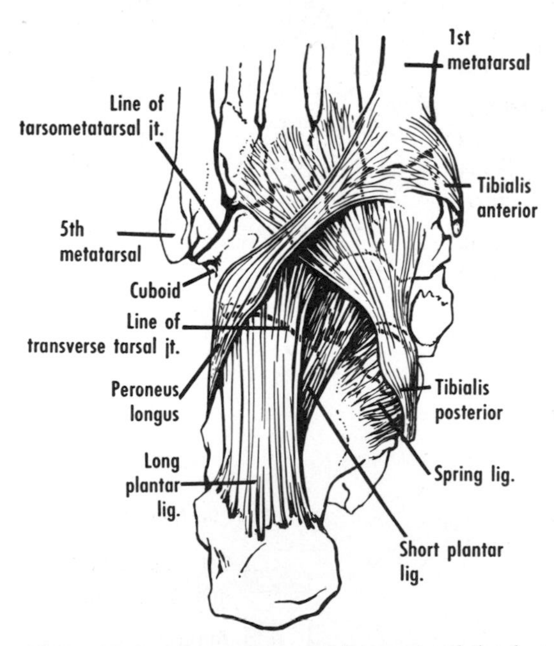

Figure 17–8 The tendons and ligaments of the foot, plantar aspect. Note the widespread insertion of the tibialis posterior. The peroneus longus tendon crosses the sole obliquely to reach the medial cuneiform, to which the tibialis anterior is also attached: the two muscles thus form a sling or stirrup.

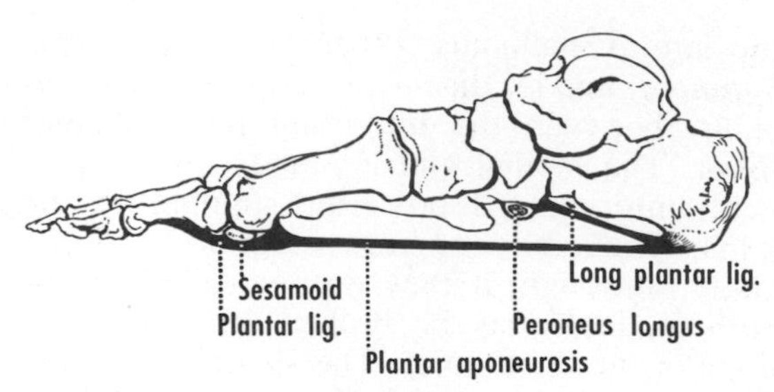

Figure 17–9 Schematic representation of the plantar aponeurosis and the long plantar ligament.

primary movements. Inversion comprises supination, adduction, and plantar flexion. Eversion involves pronation, abduction, and dorsiflexion. Usage is unfortunately variable, but supination and pronation of the foot generally refer to medial and lateral rotation about an anteroposterior axis. Abduction and adduction refer to movements of the front part of foot about a vertical axis. The tibialis posterior and anterior invert the foot. The peronei and extensor digitorum longus evert the foot (see fig. 16–5).

Remaining Joints. The *tarsometatarsal* and *intermetatarsal joints* are plane articulations that allow gliding. The medial cuneiform and first metatarsal have an independent joint cavity (see fig. 12–31*A*). The second metatarsal fits into a socket formed by the three cuneiforms (see fig. 12–32). The *metatarsophalangeal joints* are ellipsoid, and the *interphalangeal joints* are hinge, but the ligamentous arrangements of both are similar. Collateral ligaments are present, as are fibrous or fibrocartilaginous pads termed *plantar ligaments* (cf. palmar ligaments of fingers). The pads are interconnected by the *deep transverse metatarsal ligament,* which helps to hold the metatarsal heads together. The metatarsophalangeal joints allow flexion, extension, abduction, and adduction. The interphalangeal joints permit flexion and extension. The metatarsophalangeal joint of the big toe is specialized. Two grooves on the plantar aspect of the head of the first metatarsal articulate with the *sesamoids* that are embedded in the plantar ligament (fig. 17–9). The sesamoids are attached to the plantar aponeurosis and anchored to the phalanx. **The sesamoids of the big toe take the weight of the body, especially during the latter part of the stance phase of walking. The sesamoid mechanism is deranged in bunions and in hallux valgus.** A bunion is a swelling medial to the joint, and it is due to bursal thickening. In *hallux valgus* the big toe is displaced laterally because of angulation at the metatarsophalangeal joint.

ARCHES AND FLAT FEET

The arches of the foot (figs. 17–9 to 17–12) are the longitudinal and the transverse. The *longitudinal arch* is formed medially by the calcaneus, talus, navicular, cuneiforms, and first to third metatarsals and laterally by the calcaneus, cuboid, and fourth and fifth metatarsals. The *transverse,* or *metatarsal, arch* is formed by the navicular, cuneiforms, cuboid, and first to fifth metatarsals. These osseous arches depend on the mechanical arrangement of the bones, but they are supported by ligaments and, during movement, by muscles, especially the invertors and evertors. The arches develop during fetal life, but they are masked by a fat pad, which makes the sole convex in the newborn. The medial arch can be recognized in the footprints of most adults, but the extent of contact between the sole and the ground does not necessarily indicate precisely the height of the bony arches. The term "flatfoot" (*pes planus*) is used for several conditions, including a simple depression of the longitudinal arch, which in many individuals is not pathological (fig. 17–10*B*). The converse is *pes cavus,* in which the longitudinal arch is very high (fig. 17–10*C*). The term "clubfoot" (*talipes*) is used for a foot that appears twisted out of shape or position. The commonest variety of congenital clubfoot comprises plantar flexion, supination, and adduction (*talipes equinovarus*).

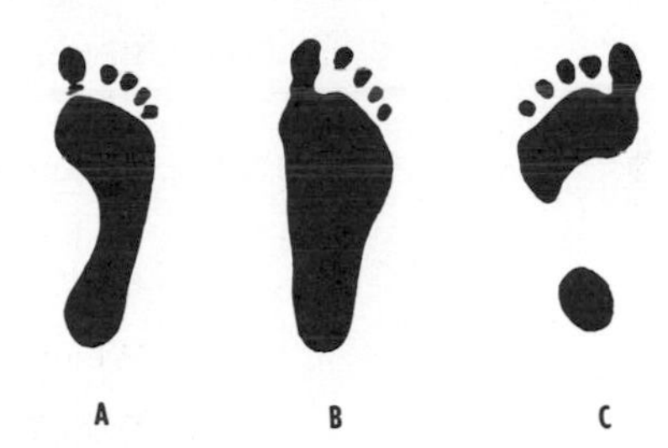

Figure 17–10 Footprints. **A,** Normal. **B,** Flatfoot. **C,** High longitudinal arch.

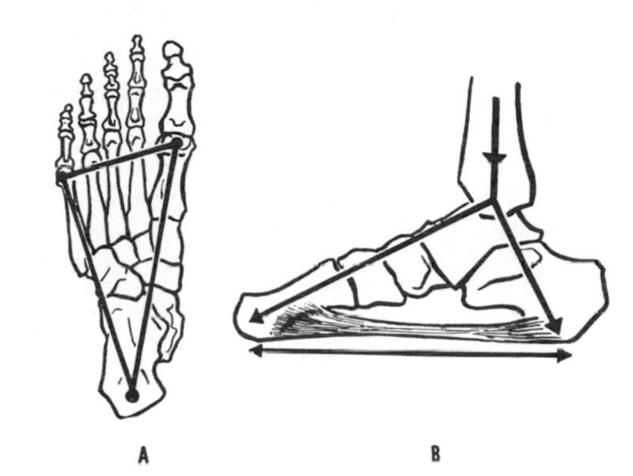

Figure 17–11 **A,** The three main points of weight bearing in the foot. **B,** The medial part of the longitudinal arch. The arrows indicate the distribution of weight that tends to flatten the arch. The connection between the calcaneus and the metatarsal represents schematically the ligamentous support of the arch. (Based on Mollier.)

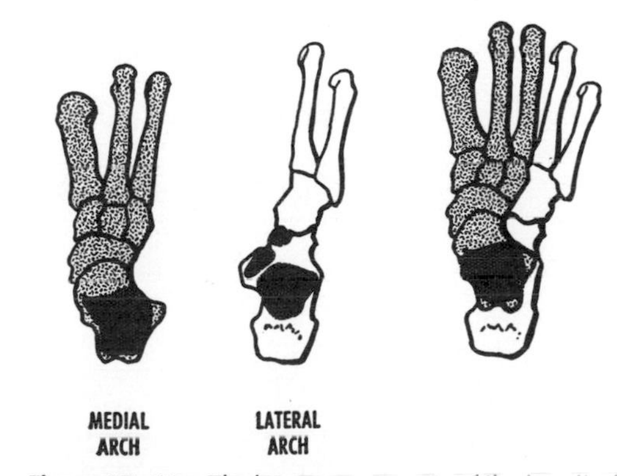

Figure 17–12 The bony components of the longitudinal arches. The third figure shows both arches.

ADDITIONAL READING

Inman, V. T., *The Joints of the Ankle,* Williams & Wilkins, Baltimore, 1976. Biomechanical studies of the ankle and subtalar joints.

Jones, F. W., *Structure and Function as Seen in the Foot,* 2nd ed., Baillière, Tindall, and Cox, London, 1949. A very readable classic.

QUESTIONS

17–1 What is the plantar reflex?

17–2 What is the Babinski sign?

17–3 How far distally do (a) the femoral, (b) the obturator, and (c) the sciatic cutaneous territories extend?

17–4 In terms of dermatomes, which spinal nerves supply the hand and the foot?

17–5 Which vessels are used in seeking a pulse at the ankle and foot?

17–6 Which structures are situated between the medial malleolus and the heel?

17–7 Which structures cross the front of the ankle joint?

17–8 Which structures are situated behind the lateral malleolus?

17–9 What is the deltoid ligament of the ankle joint?

17–10 What is the clinical importance of the lateral ligament of the ankle joint?

17–11 Which are the most important intertarsal joints and which important movements occur at them?

17–12 Which is the most frequent type of clubfoot?

17–13 What is the significance of the adjectives *varus* and *valgus?*

18
POSTURE AND LOCOMOTION

POSTURE

Standing involves a series of relatively immobile attitudes separated by brief intervals of movement during which swaying occurs. When a subject is in the easy standing position, few muscles of the back and lower limbs are active during the immobile periods. The position of the line of gravity, which is determined by the distribution of body weight, is important in determining the degree of muscular activity involved in maintaining all phases of posture. The line of gravity extends upward through the junctions of the curves of the vertebral column and downward behind the hip joints but in front of the knee and ankle joints (fig. 18–1). When a subject is in the easy standing position, the hip and knee joints are extended and are in their most stable positions. Because the line of gravity passes behind the hip joint and in front of the knee joint, the weight of the body tends to hyperextend these articulations. This is resisted by the iliofemoral ligament and by the ligamentous apparatus of the knee and the ligamentous action of the hamstrings. Similarly, the weight of the body tends to cause forward sway (dorsiflexion) at the ankle joint, and this is resisted by contraction of the calf muscles. Lateral stability in standing depends chiefly on the fascia lata, iliotibial tract, fibular collateral ligament of the knee, and tibialis anterior.

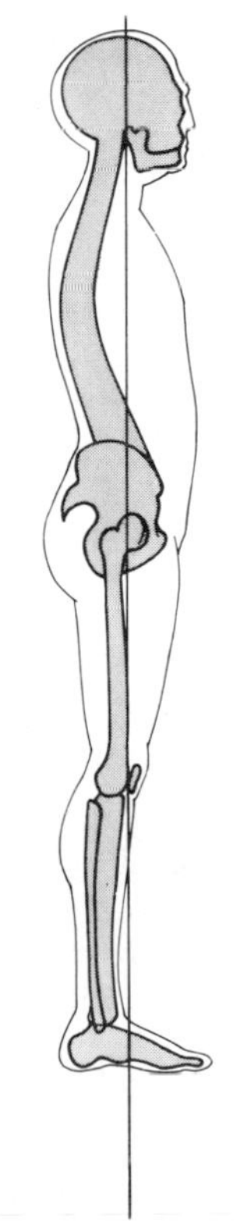

Figure 18–1 The line of gravity passes between the mastoid processes, in front of the shoulder joints, through or behind the hip joints, and in front of the knee and ankle joints. (Based on Carlsöö and on Basmajian.)

LOCOMOTION

Locomotion is very complicated, is laboriously acquired, and becomes almost entirely automatic. Disturbances of gait are important signs in many disorders of the central nervous system.

When a subject is walking on level ground, the movements of the lower limbs may be divided into "swing" and "stance" phases. The swing phase occurs when the limb is off the ground, and the stance phase when it is in contact with the ground and is bearing weight. A cycle of walking is the period from the heel-

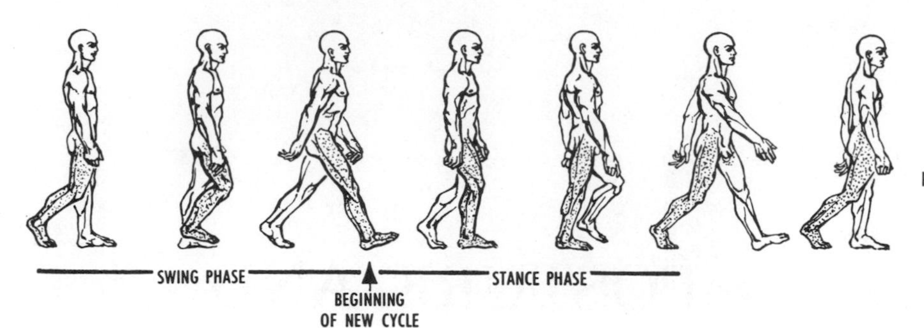

Figure 18–2 The swing and stance phases of the right lower limb.

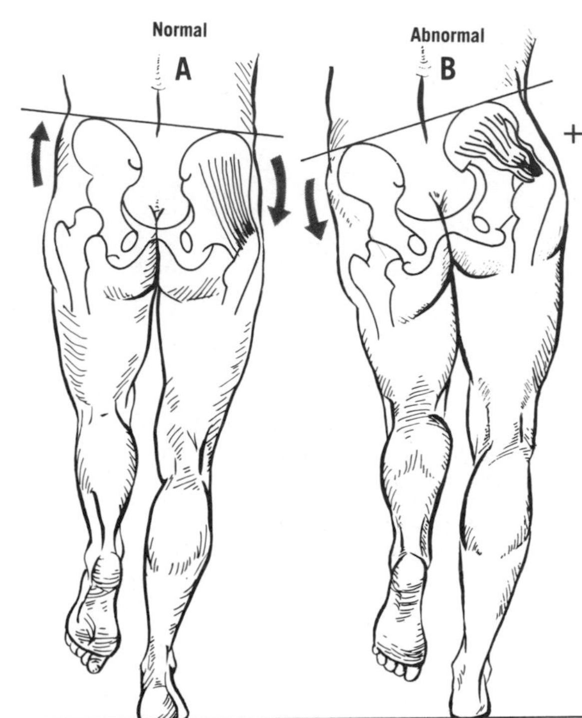

strike of one foot to the next heel-strike of the same foot.

The center of gravity moves upward and downward twice during each cycle, as is indicated by the bobbing up and down of the head. That is, the body is lifted as each limb is extended during its stance phase (fig. 18–2). There is also a side-to-side movement. The basic movements involved are (1) flexion and extension at the hip, knee, and ankle joints and at the front part of the foot; (2) abduction and adduction, chiefly at the hip joint (fig. 18–3); and (3) rotation, mainly at the hip and knee joints.

Figure 18–3 Abduction at the hip joint, viewed from behind. **A,** Normally, when weight is borne on one (e.g., the right) limb (during the stance phase of walking), the pelvis tends to sag on the free, or swing (left), side (because of gravity). This is counteracted by abduction of the hip on the stance (right) side, chiefly by strong contraction of the (right) gluteus medius, which acts on the pelvis from a fixed femur. **B,** When abduction of the (right) hip is interfered with on the supported side (positive Trendelenburg's sign), e.g., by dislocation, fracture, or paralysis, the pelvis sags (*arrow*) on the unsupported side. (Based on von Lanz and Wachsmuth.)

ADDITIONAL READING

Basmajian, J. V., *Muscles Alive,* 4th ed., Williams & Wilkins, Baltimore, 1978. Electromyographic findings, including chapters on posture and locomotion.

Carlsöö, S., *How Man Moves,* trans. by W. P. Michael, Heinemann, London, 1972. Kinesiological studies, including discussions of standing, sitting, and gait.

Inman, V. T., Ralston, H. J., and Todd, F., *Human Walking,* Williams & Wilkins, Baltimore, 1981. Technical account, including kinematics, kinetics, and muscles.

QUESTION

18–1 Where is the line of gravity in relation to the hip, knee, and ankle joints?

Part 4

THE THORAX

The thorax contains the heart, lungs, and other important structures within a skeletal framework that also protects some of the abdominal organs. The skeleton consists of the thoracic vertebrae and intervertebral discs, the ribs and costal cartilages, and the sternum (see fig. 19–1). The *thoracic cavity* communicates with the front of the neck by the *superior thoracic aperture,* or *thoracic inlet,* which is bounded by the upper margin of the first thoracic vertebra (T.V.1) behind, the first pair of ribs and their cartilages at the sides, and the upper border of the manubrium sterni in front. The inlet slopes downward and forward with the obliquity of the first ribs. It is occupied on each side by the apices of the lungs and pleurae and by the neurovascular bundles for the upper limbs. More medially, it is occupied by the vessels of the head and neck and by viscera. The thoracic cavity communicates with the abdomen by the *inferior thoracic aperture,* or *thoracic outlet,* which is closed by the diaphragm. The outlet is bounded by the twelfth thoracic vertebra (T.V.12), the twelfth pair of ribs, the free edges of the lower six pairs of costal cartilages, and the xiphosternal joint. Costal cartilages 7 to 10 unite and form the costal margin medially. The right and left costal margins meet at the *infrasternal,* or *subcostal, angle,* the apex of which is the xiphosternal joint. The xiphoid process descends into the infrasternal angle. The *epigastric fossa,* or "pit of the stomach," is a slight depression in front of the xiphoid process.

At birth the thorax is nearly circular in section, but between infancy and puberty it gradually becomes more elliptical until, in the adult, it is shallower from front to back than from side to side. The ratio between the anteroposterior and transverse diameters is the *thoracic index.* The shape of the chest varies from broad (hypersthenic individuals) with a wide infrasternal angle to narrow (asthenic individuals) with a narrow infrasternal angle.

ADDITIONAL READING

Edwards, E. A., Malone, P. D., and Collins, J. J., *Operative Anatomy of the Thorax,* Lea & Febiger, Philadelphia, 1972. Anatomical drawings of operative approaches used in thoracic surgery.

Kubik, S., *Surgical Anatomy of the Thorax,* ed. by J. E. Healey, W. B. Saunders Company, Philadelphia, 1970. Color photographs and key drawings of dissections of the thorax.

19
THE SKELETON OF THE THORAX

The skeleton of the thorax includes the sternum, ribs and costal cartilages, and thoracic vertebrae and intervertebral discs (fig. 19–1).

STERNUM

The sternum, or breast bone, is a flat bone that, from above downward, consists of three parts: manubrium, body, and xiphoid process (figs. 19–1 and 19–2). It is useful to remember that **the manubrium and body are approximately on the level of T.V.3, 4 and T.V.5 to 10, respectively** (see figs. 20–8 and 21–1). **Its accessibility and the thinness of its compacta have made the sternum a common site for puncture by a needle to obtain marrow for study.**

The upper border of the *manubrium* is marked by the *jugular notch,* which is easily palpable and is usually at the level of T.V.3. On each side of this is a notch for the medial end of the clavicle. The first costal cartilage is attached to the side of the manubrium. Below, the manubrium articulates with the *body* of the sternum at the *sternal angle,* which is marked by a palpable (and sometimes visible) transverse ridge about 5 cm below the jugular notch. **The sternal angle is an important bony landmark at the level of T.V.4 or 5. It indicates not only the manubriosternal junction but also the level of the second costal cartilages; hence it is a reference point in counting ribs.** Rarely, however, the sternal angle is at the level of the third costal cartilages. The *manubriosternal junction* (i.e., the union of the manubrium with the body of the sternum) is usually fibrocartilaginous, but it may become ossified.

The body of the sternum, about twice as long as the manubrium, is notched on each side to receive costal cartilages 2 to 7. Transverse ridges may indicate its development from several pieces. The *xiphosternal joint* (i.e., the union of the body of the sternum with the xiphoid process) is usually fibrocartilagi-

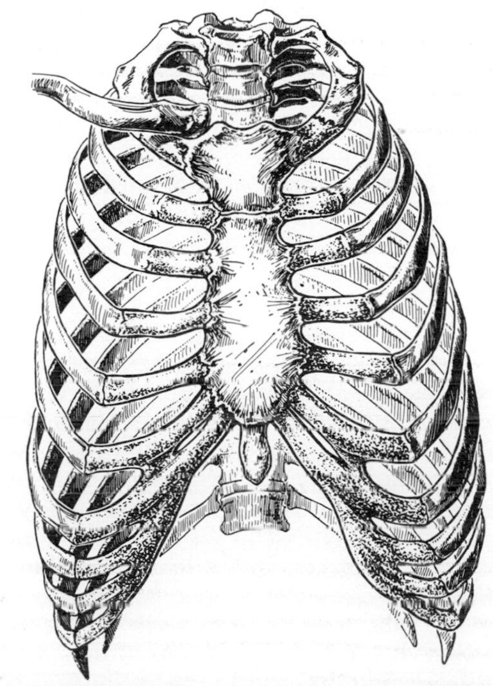

Figure 19–1 The bones of the thorax. Note that the upper two and a half and the lower two and a half thoracic vertebrae are visible.

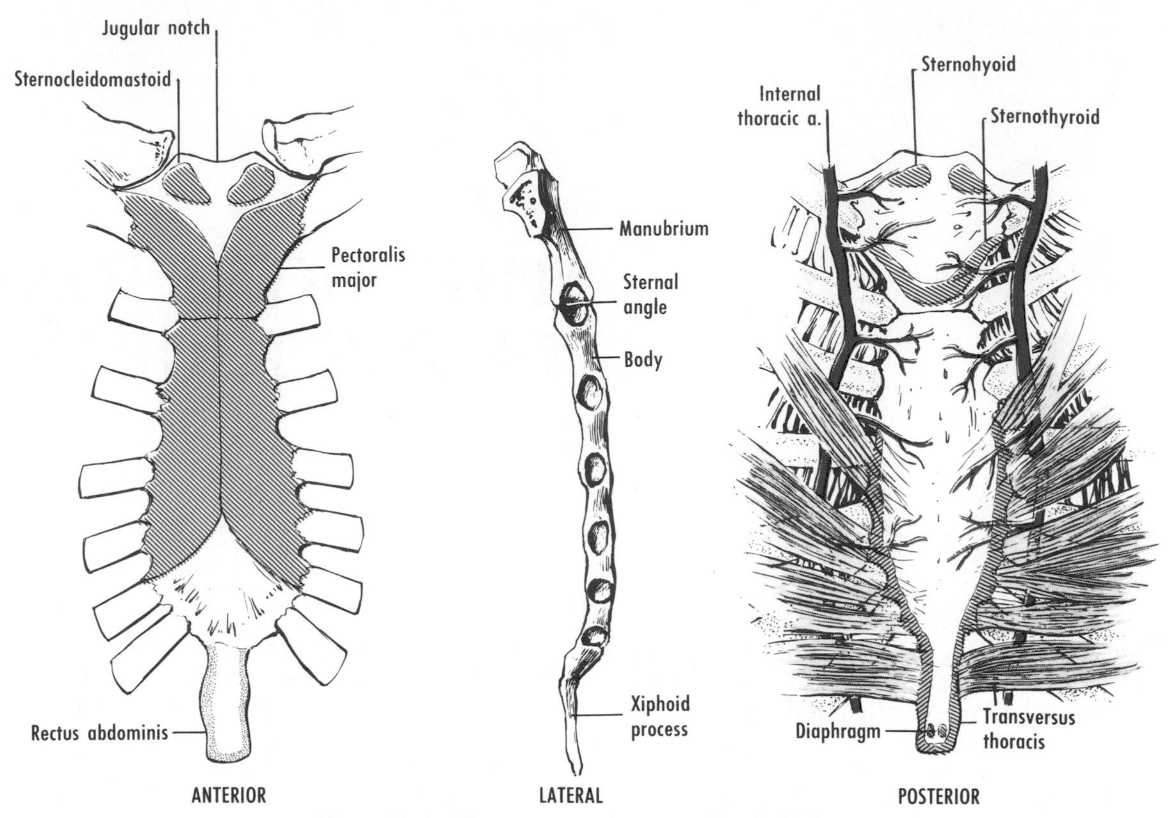

Figure 19–2 The sternum and its muscular attachments.

nous, but it too may become ossified. It is at the apex of the infrasternal angle and is usually at the level of T.V.10 or 11.

The *xiphoid process* is a small and variable piece of hyaline cartilage that contains a bony core. It lies in the epigastric fossa, or "pit of the stomach."

The sternum arises bilaterally in the embryo, and it later ossifies from a variable number of bilateral and median centers.

RIBS

There are usually 12 ribs on each side of the body (fig. 19–3). They are elongated yet flattened bones that curve downward and forward from the thoracic vertebrae. The ribs, as well as the costal cartilages, increase in length from the first to the seventh, and their obliquity increases from the first to the ninth. Generally, **ribs 1 to 7 are connected to the sternum by their costal cartilages and are called true ribs, whereas ribs 8 to 12 are termed false ribs. Usually, ribs 8 to 10, by means of their costal cartilages, join the costal cartilage immediately above, whereas ribs 11 and 12, which are free, are known as floating ribs.** A supernumerary rib may be found in either the cervical or lumbar region. In thoracic surgery, a portion of a rib can be shelled out of its periosteum, which later allows regeneration of the bone.

Ribs 3 to 9 have certain features in common and are known as *typical ribs* (figs. 19–4 and 19–5). Each has a head, neck, and shaft. The *head* presents an articular surface for the corresponding (n) and suprajacent ($n - 1$) vertebrae. The junction of the neck and shaft is marked by a *tubercle,* which articulates with the transverse process of the corresponding (n) vertebra. The *shaft,* which is curved and twisted, presents an *angle* posteriorly, which indicates the lateral extent of the erector spinae and is the weakest part of the rib. The curvature of the rib is such that a person lying on his back is supported by the spinous processes and the angles of the ribs. The concave, inner surface of the shaft is marked inferiorly by the *costal groove,* which gives attachment to the internal intercostal muscle and shelter to the intercostal vein, artery, and nerve from above downward. The ribs ossify from a primary center for the shaft and secondary centers for the head and tubercle.

The *first rib* (fig. 19–6) helps to bound the thoracic inlet. It is very short, and its broad,

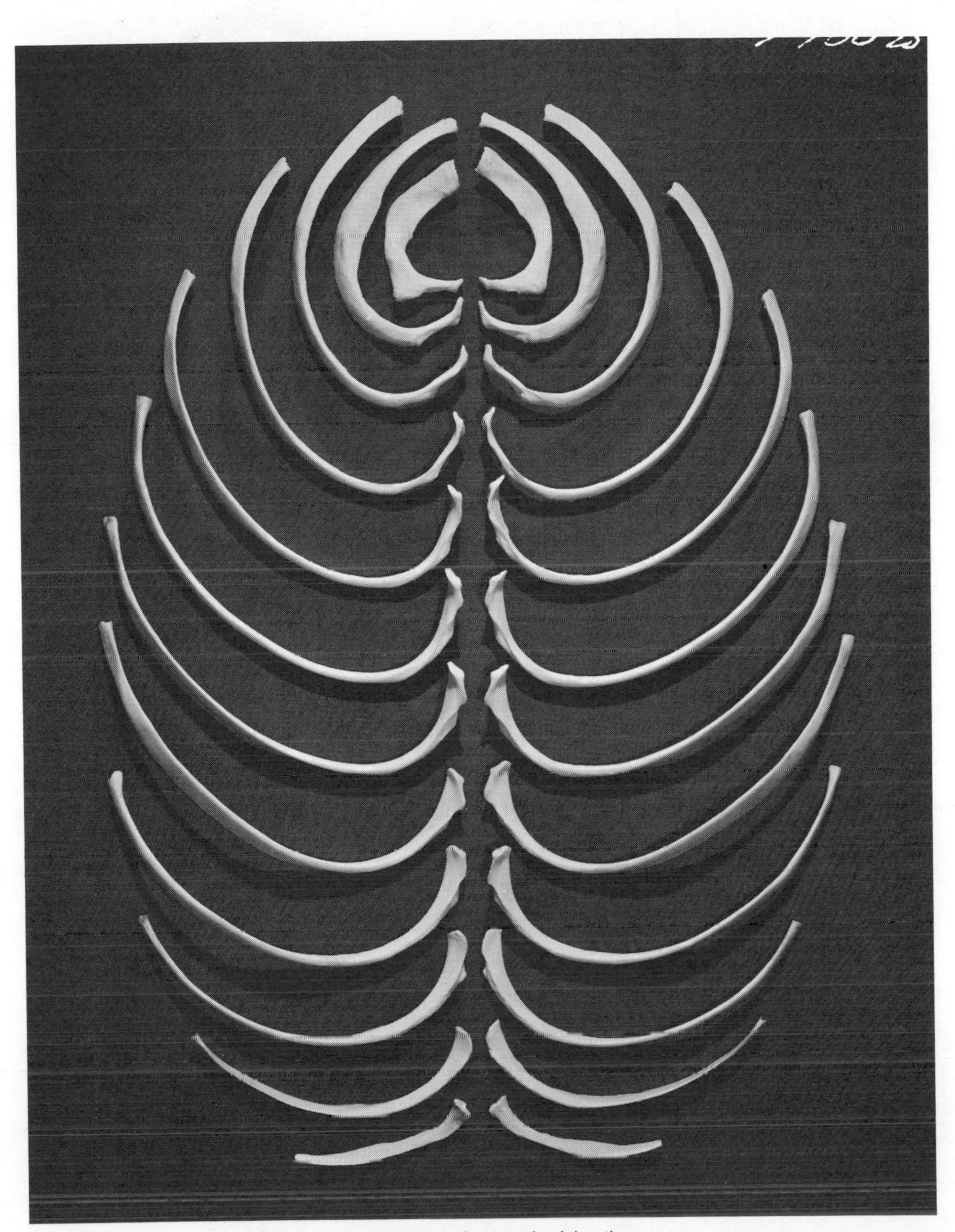

Figure 19–3 Photograph of the ribs.

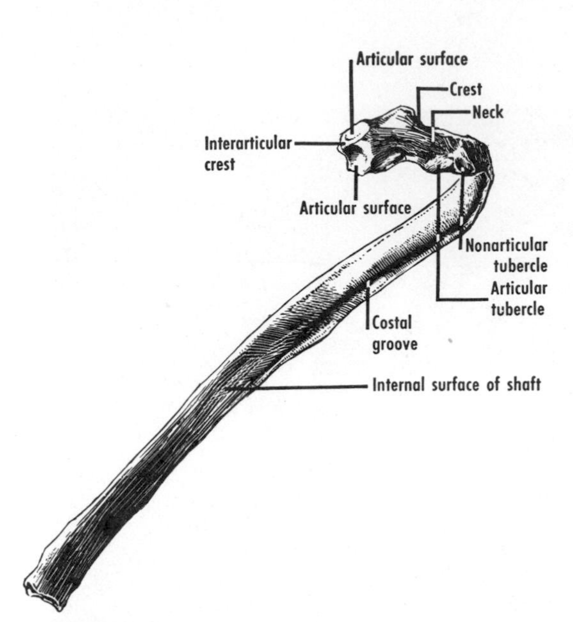

Figure 19–4 The internal aspect of the right seventh rib. Note the slope downward and forward and the twist of the shaft.

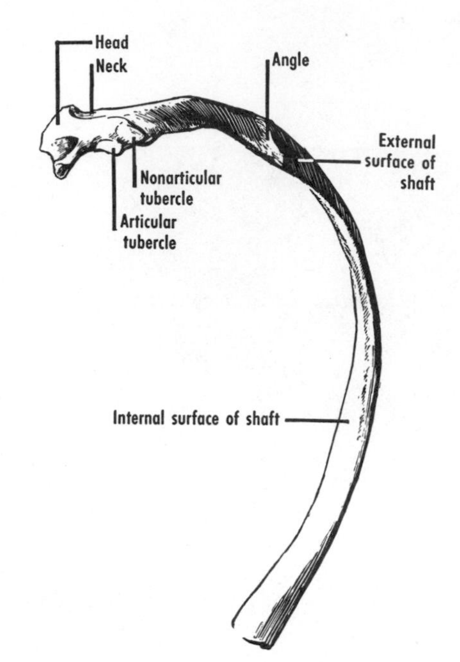

Figure 19–5 The right seventh rib from below and behind.

flat surfaces face upward and downward. The head articulates with T.V.1, and the neck lies behind the apex of the lung. The upper surface may present a *groove for the subclavian artery* and the lower trunk of the brachial plexus, in front of which is the tubercle for the scalenus anterior muscle internally. Further anteriorly is a shallow *groove for the subclavian vein.* The first rib is difficult to palpate *in vivo,* but the first intercostal space can be identified immediately below the clavicle. The *second rib,* which is much longer than the first, is curved but not twisted. It articulates with T.V.1 and 2 and presents a tuberosity for the serratus anterior muscle.

Rib 10 usually articulates with T.V.10 only. *Rib 11,* which articulates with T.V.11 only, has an indistinct tubercle, angle, and costal groove. *Rib 12,* which articulates with T.V.12 only, is small, slender, and variable in length.

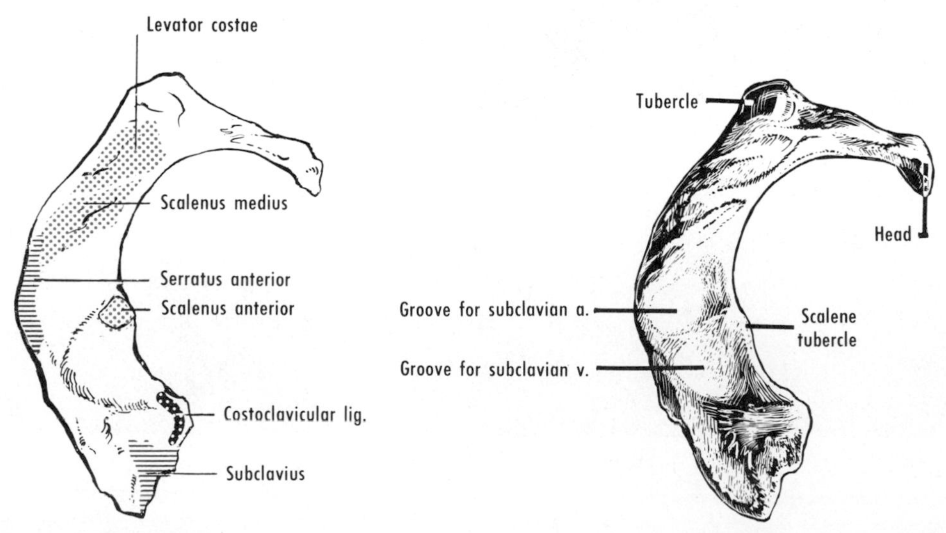

Figure 19–6 The first rib from above, showing grooves for vessels as well as muscular and ligamentous attachments. The scalenus minimus (not shown) is also inserted into the first rib.

The differences in length have to be kept in mind in surgical approaches to the kidney.

The costal cartilages are bars of hyaline cartilage, which later may become ossified. They fit into depressions in the anterior ends of the ribs, and the upper seven or eight articulate with the sternum. The costal cartilages impart resiliency to the chest wall. They may become partly ossified later in life.

THORACIC VERTEBRAE

The thoracic vertebrae are described with the back. The superior costal facet of a typical thoracic vertebra (n), together with the intervertebral disc and the inferior costal facet of the suprajacent vertebra ($n - 1$), forms a socket for the head of the corresponding rib (n).

QUESTIONS

19–1 What is the vertebral level of the manubrium sterni?

19–2 What lies on the upper surface of the first rib?

19–3 How are ribs counted?

20 THE THORACIC WALL AND MEDIASTINUM

THORACIC WALL

MUSCLES

The muscles of the thoracic and abdominal walls are in general arranged in external, middle, and internal layers. In the thorax (figs. 20–1 and 20–2), these are the (1) external intercostal muscles, (2) internal intercostal muscles, and (3) innermost intercostal muscles, subcostal muscles, and transversus thoracis. The internal layer and the thoracic skeleton are separated from the costal pleura by the *endothoracic fascia*. The diaphragm separates the thoracic and abdominal viscera.

The *external intercostal muscles* are attached to the lower margins of ribs 1 to 11. Their fibers pass downward and forward to the upper margin of the rib below. In front, at the costochondral junctions, the external intercostal muscles are replaced by the *external intercostal membranes* (fig. 20–2). The muscles are supplied by the corresponding intercostal or thoraco-abdominal nerves. They elevate the ribs and hence are considered to be muscles of inspiration. They are assisted posteriorly by the *levatores costarum,* which run from the transverse processes to the backs of the subjacent ribs and are supplied by dorsal rami.

The *internal intercostal muscles* are attached to the lower margins of the ribs and costal cartilages and to the floor of the costal groove. Their fibers pass downward and backward to the upper margin of the rib and costal cartilage below. At the back, at the angles of the ribs, the internal intercostal muscles are

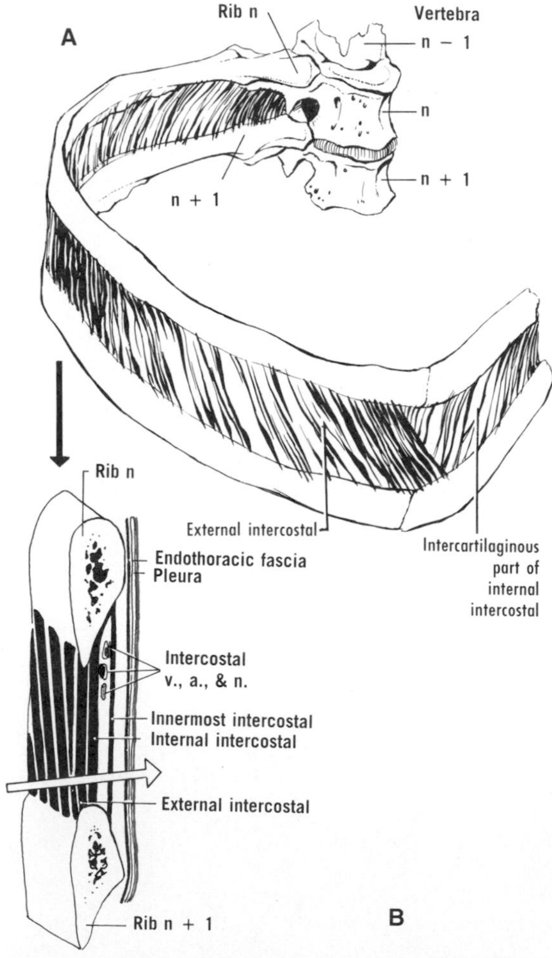

Figure 20–1 The intercostal muscles. **A** shows the direction of fibers of the external and internal intercostal muscles. **B** shows a vertical section through an intercostal space. The white arrow represents the path of a needle in pleural aspiration, avoiding the intercostal vessels and nerve.

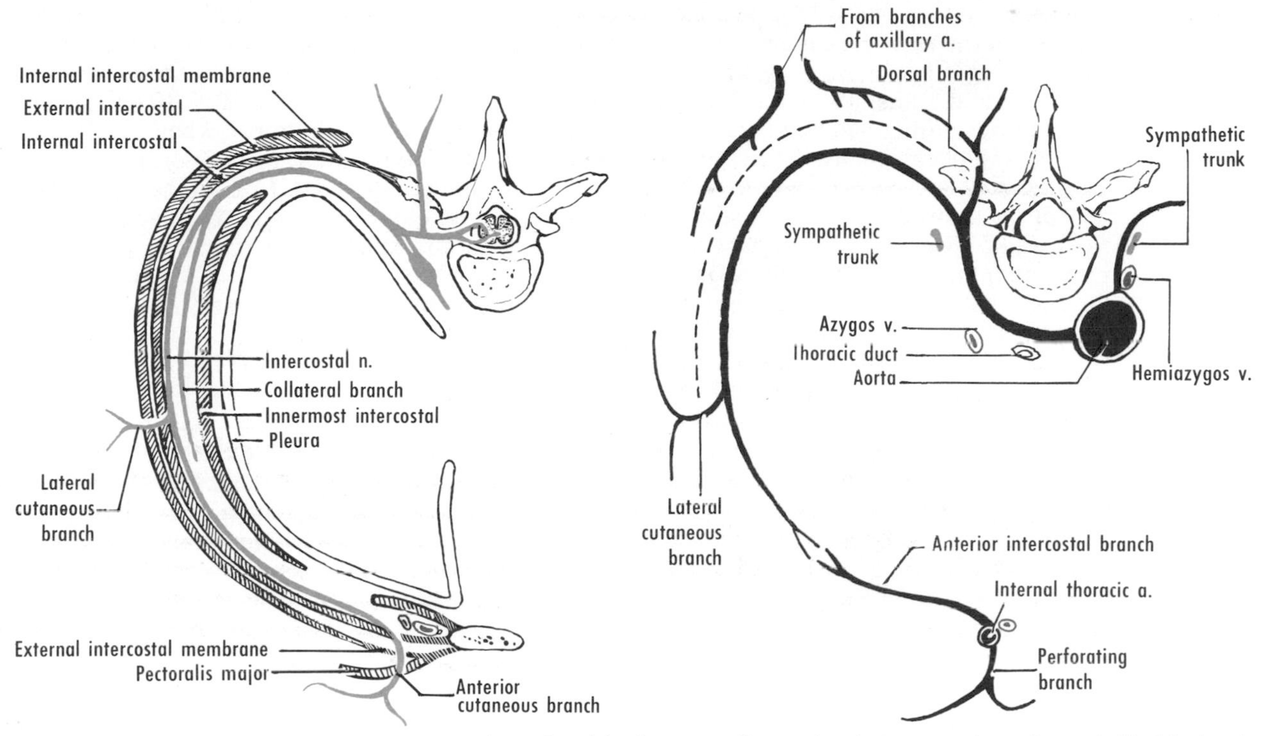

Figure 20–2 The nerves, arteries, and muscles of the thoracic wall. Note that the intercostal vessels pass behind the longitudinally disposed structures of the posterior mediastinum. The thickness of the intercostal muscles is exaggerated.

replaced by the *internal intercostal membranes* (fig. 20–2). The muscles are supplied by the corresponding intercostal or thoraco-abdominal nerves. For the most part, they are muscles of expiration.

The *innermost intercostal muscles* may be regarded as those parts of the internal intercostal muscles that are internal to the intercostal vessels and nerves. Their action is unknown. The *subcostal muscles,* which are quite variable, arise from the ribs posteriorly and are inserted into the second or third rib below. They probably elevate the ribs. The *transversus thoracis* (or *sternocostalis*) (see fig. 19–2) arises from the back of the xiphoid process and body of the sternum and is inserted into the back of several costal cartilages. It appears to be expiratory in function. All these muscles are supplied by the corresponding intercostal or thoraco-abdominal nerves.

The diaphragm, although not essential, is the most important muscle of respiration. It separates the thoracic and abdominal viscera. Three of its parts (sternal, costal, and lumbar) are inserted into the central tendon, a trifoliate structure that lies immediately below the heart. The *sternal part* consists of slips from the back of the xiphoid process, which (*in vivo*) descend to the central tendon. On each side, a small gap known as the *sternocostal triangle* is present between the sternal and costal parts. It transmits the superior epigastric vessels and some lymphatics, and it may be the site of a diaphragmatic hernia. The *costal parts,* which form the right and left "domes," arise from the inner surfaces of the lower costal cartilages and ribs and interdigitate with the transversus abdominis. They are inserted into the central tendon anterolaterally. Each *lumbar* (or *vertebral*) *part* arises from (1) a lateral arcuate ligament over the quadratus lumborum, (2) a medial arcuate ligament over the psoas major, and (3) a crus from the upper lumbar vertebrae (see fig. 25–13*B*). Usually the *right crus* arises from the first to third (or fourth) lumbar vertebrae (L.V.1 to 3 or 1 to 4) and the *left* from L.V.1 to 2 or 1 to 3. The crura are united in front of the aorta by the *median arcuate ligament,* a fibrous arch that forms the aortic opening. The right crus splits around the esophagus (see figs. 21–2 and 25–13*B*), and part of it continues into the suspensory ligament of the duodenum. The left crus is smaller and more variable.

The diaphragm is involved in three major

openings (see fig. 25–13*B*). The *esophageal opening* in the right crus transmits the esophagus and vagi. The *aortic opening* lies behind the crura and transmits the aorta, frequently the thoracic duct and greater splanchnic nerves, and occasionally the azygos vein. The *foramen for the inferior vena cava*, in the right half of the central tendon, transmits the vena cava, right phrenic nerve, and lymphatic vessels. Other structures that pierce or are related to the diaphragm include the splanchnic nerves, sympathetic trunk, subcostal nerves and vessels, superior epigastric and musculophrenic vessels, and azygos and hemiazygos veins.

The portion of the costal part of the diaphragm that arises from ribs 11 and 12 is often separated from the lumbar part by an interval termed the vertebrocostal trigone (see fig. 25–13*B*). **Such a triangle is occupied by connective tissue that separates the pleura above from the suprarenal gland and upper pole of the kidney below.**

Variations in the degree of development of the muscular parts are not uncommon. **Congenital diaphragmatic hernia, whereby an abdominal organ may enter the thoracic cavity, may occur through the esophageal opening (hiatal hernia), through a gap in the costal part of the diaphragm (e.g., from a persistent pleuroperitoneal canal), or through the sternocostal triangle.** Diaphragmatic herniae usually have peritoneal sacs.

The diaphragm, together with the adjacent pleura and peritoneum, is supplied by the phrenic nerves (see fig. 24–6), each of which is distributed to one half of the diaphragm. Although the two halves usually contract synchronously, **paralysis of one half does not affect the other half, because each half of the diaphragm has a separate innervation. The diaphragm is under only limited voluntary control. Hiccups are sharp, spasmodic contractions of the diaphragm.** The peripheral portion of the diaphragm is supplied with sensory and vasomotor fibers from the thoraco-abdominal nerves.

The diaphragm descends when it contracts, and it draws the central tendon downward with it. The volume of the thorax is thereby increased, whereas intrathoracic pressure is decreased; the converse holds for the abdominal cavity. The decreased intrathoracic pressure and increased abdominal pressure that accompany descent of the diaphragm facilitate the return of blood to the heart.

When a subject is in the erect position and the midphase of respiration, the summit of the domes of the diaphragm is at about the same level as the apical region of the heart (fifth intercostal space or rib 6; T.V.10 or 11). The diaphragm is a little lower when a subject is in the supine position. Moreover, the right dome is commonly about 1 cm higher than the left. During quiet breathing, the diaphragm undergoes an excursion of about $^1/_2$ cm, whereas, during deep breathing, the excursion may be as much as 10 cm.

Thoracic Nerves

Each of the 12 thoracic (spinal) nerves gives off a meningeal branch, emerges from an intervertebral foramen, and divides into a dorsal and a ventral ramus. These rami contain motor fibers to muscles, sensory fibers from skin and deep tissues, and postganglionic sympathetic fibers to blood vessels, sweat glands, and arrectores pilorum (fig. 20–4).

The *dorsal rami* (figs. 20–2 and 20–4) pass backward and supply the bones, joints, muscles, and skin of the back. The *ventral rami* run forward and supply the serous membranes, muscles, and skin of the thoracic and abdominal walls. Each is connected to the sympathetic trunk by a variable number of rami communicantes (figs. 20–3 and 20–4). **Although the distribution of the ventral rami is**

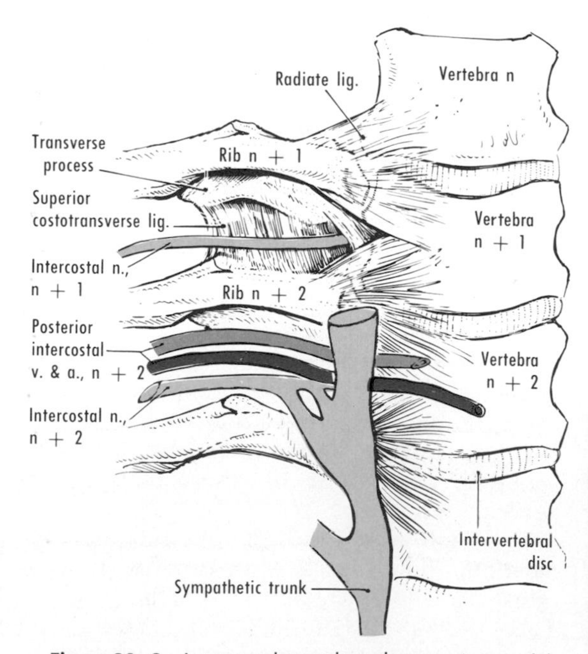

Figure 20–3 Intercostal vessels and nerve. A part of the sympathetic trunk is shown, including some rami communicantes.

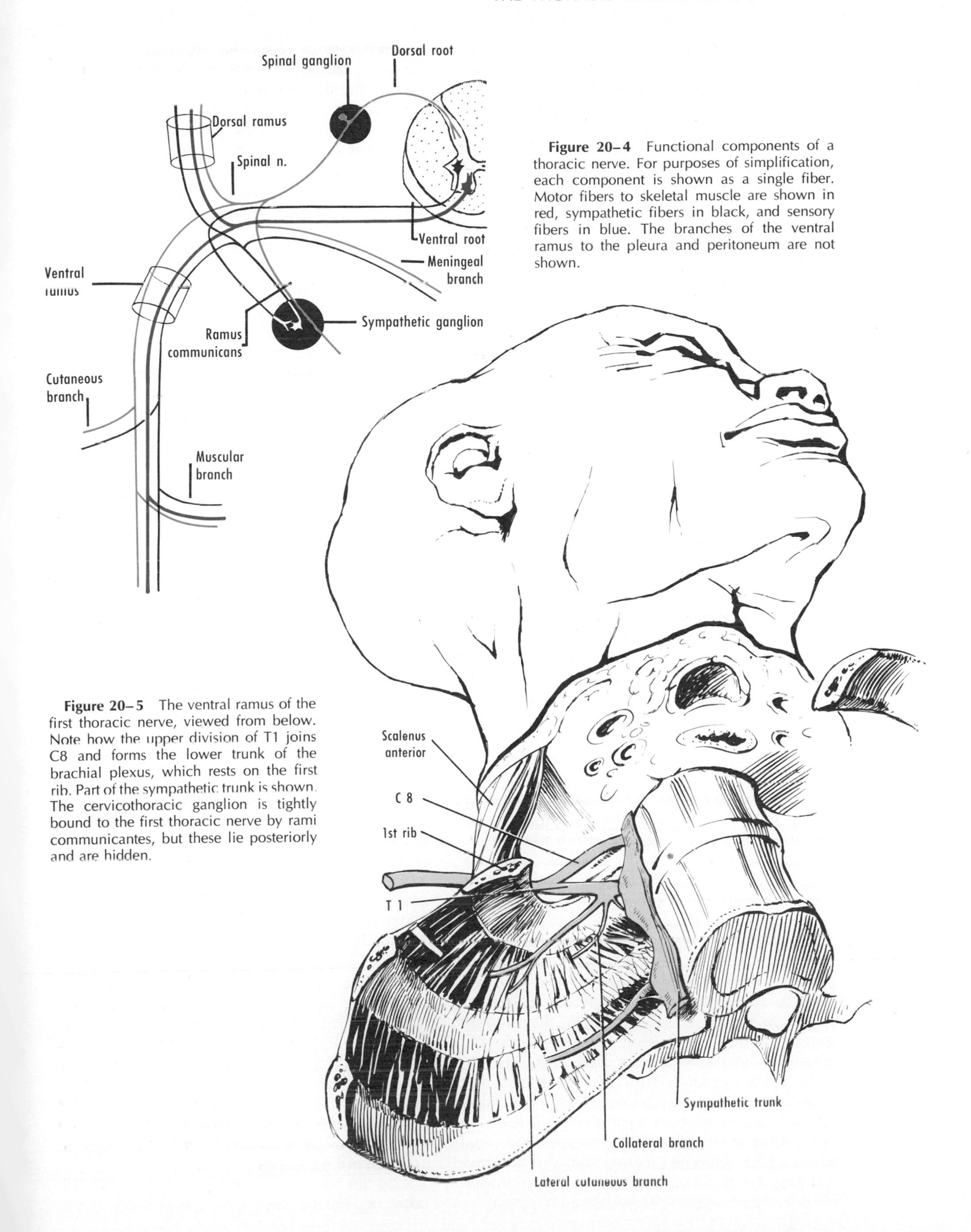

Figure 20–4 Functional components of a thoracic nerve. For purposes of simplification, each component is shown as a single fiber. Motor fibers to skeletal muscle are shown in red, sympathetic fibers in black, and sensory fibers in blue. The branches of the ventral ramus to the pleura and peritoneum are not shown.

Figure 20–5 The ventral ramus of the first thoracic nerve, viewed from below. Note how the upper division of T1 joins C8 and forms the lower trunk of the brachial plexus, which rests on the first rib. Part of the sympathetic trunk is shown. The cervicothoracic ganglion is tightly bound to the first thoracic nerve by rami communicantes, but these lie posteriorly and are hidden.

segmental, overlap of adjacent nerves is so great that section of three consecutive nerves would be necessary to produce complete anesthesia and paralysis within the middle one of the three intercostal spaces supplied.

The ventral rami of the first 11 thoracic (spinal) nerves are called intercostal nerves, whereas that of the twelfth is subcostal. Moreover, ventral rami 1 to 3 contribute to the upper limb as well as to the thoracic wall, and ventral rami 7 to 12 are thoraco-abdominal in their distribution. Intercostal nerves can be "blocked" posteriorly by a local anesthetic, e.g., for pain after fracture of a rib.

Intercostal nerves 4 to 6 are "typical" (figs. 20–1 to 20–4 in that they supply only the thoracic wall and its associated muscles (intercostal, subcostal, serratus posterior superior, and transversus thoracis). Each passes below the neck of the corresponding rib and enters the costal groove. In its course forward, it lies first on the pleura and endothoracic fascia, then between the innermost and internal intercostal muscles, and finally on the transversus thoracis and internal thoracic vessels. At the front of the intercostal space, it comes forward through the internal intercostal muscle, external intercostal membrane, and pectoralis major, to be distributed as the *anterior cutaneous branch* to the front of the chest. Each intercostal nerve gives off a *collateral branch* to the lower part of the intercostal space and a *lateral cutaneous branch* to the side of the chest. In addition to being distributed to muscle and skin, branches are given to the parietal pleura, mammary gland, and periosteum of the ribs.

The *first thoracic nerve* divides into an upper part, which joins the brachial plexus, and a lower part, which becomes the first intercostal nerve (fig. 20–5). The lateral cutaneous branches of intercostal nerves 1 to 3 contribute to the upper limb, that of the second being known as the *intercostobrachial nerve*.

Intercostal nerves 7 to 11 supply the abdominal as well as the thoracic wall; hence they may be termed *thoraco-abdominal* (see fig. 25–12). At the front of the intercostal spaces, they pass between the muscles of the abdominal wall and come to lie between the rectus abdominis and the posterior wall of its sheath. Here each nerve divides into branches that supply the rectus and the overlying skin. Their lateral cutaneous branches also contribute to the abdominal wall. The thoraco-abdominal nerves give branches to thoracic and abdominal muscles and skin and sensory twigs to the pleura, diaphragm, and peritoneum.

The ventral ramus of thoracic nerve 12 (T.N.12) is known as the *subcostal nerve*. It enters the abdomen behind the lateral arcuate ligament, crosses behind the kidney, penetrates the muscles of the abdominal wall, enters the rectus sheath, and becomes cutaneous (see fig. 25–12).

Blood Vessels and Lymphatic Drainage

The thoracic wall is supplied by branches of (1) the subclavian artery (internal thoracic and highest intercostal arteries), (2) the axillary artery, and (3) the aorta (posterior intercostal and subcostal arteries).

The *internal thoracic* (or *internal mammary*) *artery* (fig. 19–2) arises usually from the first part of the subclavian artery. It descends behind the sternomastoid, clavicle, and subclavian and internal jugular veins. It is crossed by the phrenic nerve, and it lies on the pleura behind. It then descends behind the upper six costal cartilages, immediately lateral to the sternum, and in front of the pleura. It gives branches to the intercostal spaces, pleura, pericardium, and breast. At the sixth intercostal space, it divides into the superior epigastric and musculophrenic arteries. The *superior epigastric artery* traverses the sternocostal triangle of the diaphragm, descends between the rectus abdominis and the posterior layer of its sheath, and anastomoses with the inferior epigastric artery. The *musculophrenic artery*, more laterally placed, supplies several intercostal spaces, pierces the diaphragm, and anastomoses with the deep circumflex iliac artery.

Posterior intercostal arteries 1 and 2 arise from the highest intercostal artery, which is a branch of the costocervical trunk of the subclavian artery. Posterior intercostal arteries 3 to 11 arise from the back of the aorta (figs. 20–2 and 20–3). The right arteries are longer because they have to cross the vertebral column. They are behind the pleura, azygos venous system, and sympathetic trunk. Each artery enters the costal groove, runs forward between the vein and nerve ("V.A.N.") (between the innermost and internal intercostal muscles), and anastomoses with branches of the internal thoracic or musculophrenic arteries. A lateral cutaneous branch accompanies the corresponding nerve. The two *subcostal arteries* are in series with the intercostal arteries, and they enter the abdomen with the corresponding nerves.

The anastomoses between the internal thoracic, posterior intercostal, and inferior epigastric arteries provide an important collateral circulation in obstruction of the aorta, e.g., from coarctation. In such instances, the enlarged intercostal arteries in the costal grooves may erode the bone and show radiographically as notching of the ribs.

The *parietal lymph nodes* of the thorax are the parasternal, phrenic, and intercostal. The *parasternal nodes,* situated along the upper part of the internal thoracic artery, receive lymphatics from the medial part of the breast, the intercostal spaces, the costal pleura, and the diaphragm and drain into the bronchomediastinal trunk. **The parasternal nodes allow the spread of carcinoma of the breast to the lungs and mediastinum and, by way of the diaphragm, even downward to the liver.** The *phrenic nodes* are situated on the thoracic surface of the diaphragm. They receive lymphatics from the pericardium, diaphragm, and liver and drain into the parasternal nodes. *Intercostal nodes* are found at the vertebral end of the intercostal spaces.

JOINTS

The joints of the thorax occur between (1) vertebrae, (2) ribs and vertebrae, (3) ribs and costal cartilages, (4) costal cartilages, (5) costal cartilages and the sternum, and (6) the parts of the sternum.

The *costovertebral joints* (figs. 20–3 and 20–6), synovial in type, are those of the heads of the ribs and the costotransverse joints. The head (n) of a typical rib (ribs 2 to 9) articulates with the inferior and superior costal facets of two adjacent vertebral bodies ($n - 1$ and n) and the intervening intervertebral disc. The heads of ribs 1 and 10 to 12 articulate with only one vertebra each. The tubercle of a typical rib forms a *costotransverse joint* with the costal facet on the transverse process of the corresponding vertebra. These joints are absent in ribs 11 and 12.

The *costochondral joints* are hyaline cartilaginous joints between the ends of the costal cartilages and ribs. The *interchondral joints,* generally synovial, are between costal cartilages 5 to 8 or 5 to 9.

The *sternocostal* (or *sternochondral*) *joints,* often considered synovial, are between costal cartilages 1 to 7 and the lateral margin of the sternum. The *manubriosternal joint* is fibrocartilaginous but may become ossified. The *xiphosternal joint* is cartilaginous but later becomes ossified.

MOVEMENTS OF THORACIC CAGE

The numerous thoracic joints are subject to continual movement, and any disorder that reduces their mobility hampers respiration.

In general, the ribs move around two axes. Movement at costovertebral joints 2 to 6 about a side-to-side axis results in raising and lowering the sternal end of the rib, the "pump-handle" movement (fig. 20–7). The downward slope of the rib ensures that, in elevation, the sternum moves upward and forward, increasing the anteroposterior diameter of the thorax. Movement at costovertebral joints 7 to 10 about an anteroposterior axis results in raising and lowering the middle of the rib, the "bucket-handle" movement (fig. 20–7). In elevation, this increases the transverse diameter of the thorax.

A movement of only a few millimeters is sufficient to increase the volume of the thoracic cage by the usual volume of air that enters and leaves the lungs during quiet

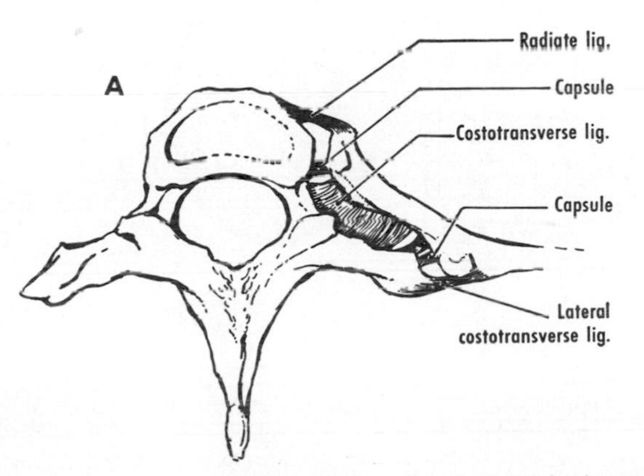

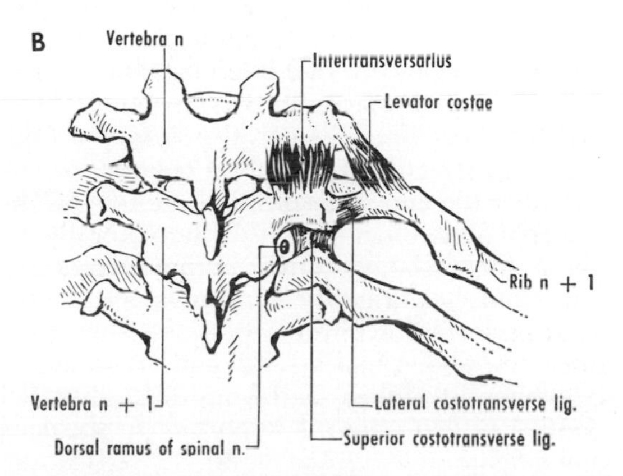

Figure 20–6 The costovertebral joints viewed from (**A**) above and (**B**) behind.

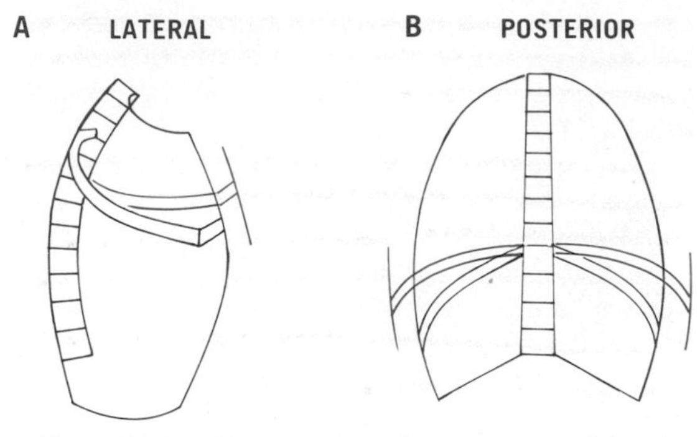

Figure 20–7 Diagram of certain movements of the ribs. In **A,** when the upper ribs are elevated, the anteroposterior diameter of the thorax is increased ("pump-handle" movement). In **B,** the lower ribs move laterally when they are elevated, and the transverse diameter of the thorax is increased ("bucket-handle" movement).

breathing. In deep breathing the excursions are greater. The descent of the diaphragm increases the height of the thoracic cavity and hence increases the volume of the thorax. However, in unilateral or even bilateral paralysis of the diaphragm, there may be no significant disability. Diaphragmatic breathing is often called abdominal (as distinct from thoracic) breathing.

The diaphragm, which increases the volume of the thoracic cavity, is the most important muscle of respiration. The mechanical actions of the intercostal muscles are not fully understood, but the external intercostal muscles (and intercartilaginous parts of the internal intercostal muscles) probably elevate the ribs and thus are inspiratory in function. The interosseous portions of the internal intercostal muscles probably depress the ribs and are thus expiratory in function.

In the inspiratory phase of quiet breathing, the diaphragm, "parasternal" intercostal muscles, and external intercostal muscles posteriorly are active, and, in some people, so are the scalene muscles. Expiration is mainly passive, because of the elasticity of the lungs, but the interosseous internal intercostal muscles of the lower interspaces also contribute. When breathing becomes deeper, the sternomastoids and extensors of the vertebral column are active near the end of inspiration. Moreover, the external abdominal muscles anterolaterally become increasingly active during expiration. These muscles draw the ribs down and are the most important expiratory muscles. They compress the abdominal viscera and are active in coughing, straining, and vomiting. Muscular control of expiration is important in speaking and singing.

MEDIASTINUM

The term *mediastinum* refers to a septum, and in the past it was applied to the partition between the right and left pleurae. However, **the mediastinum is now defined as the interval between the two pleural sacs. It is commonly considered to comprise a superior mediastinum, above the level of the pericardium, and three lower divisions: anterior, middle, and posterior.** Figure 20–8 shows the arrangement *in vivo*.

The *anterior mediastinum,* between the sternum and pericardium, contains the thymus. The *middle mediastinum* contains the pericardium, heart, and the main bronchi and other structures of the roots of the lungs. The *posterior mediastinum,* behind the pericardium, contains the esophagus and thoracic aorta. The *superior mediastinum* contains the

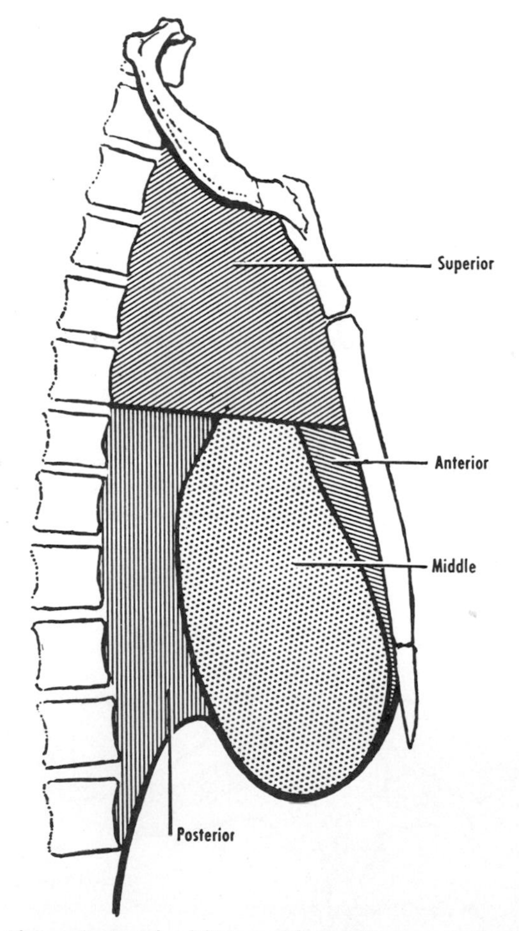

Figure 20–8 The divisions of the mediastinum. It should be stressed that the boundaries shown here are those *in vivo,* with the subject erect. See also figs. 22–4 and 22–5.

thymus, great vessels related to the heart, the trachea, and the esophagus. The mediastinum contains various groups of visceral lymph nodes. The various structures in the mediastinum are surrounded and supported by loose connective tissue often infiltrated with fat. The mediastinum can be inspected transcutaneously by means of a tubular instrument inserted through a small incision above the sternum. This procedure, termed *mediastinoscopy,* can also be used for biopsies and surgical interventions.

ADDITIONAL READING

Campbell, E. J. M., Agostoni, E., and Davis, J. N., *The Respiratory Muscles,* 2nd ed., Lloyd-Luke, London, 1970. A detailed account of respiratory mechanics and neural control.

QUESTIONS

20–1 What is the action of the diaphragm in respiration?

20–2 What is the vertebrocostal trigone?

20–3 Where may a congenital diaphragmatic hernia be found?

20–4 Which nerves are (a) intercostal, (b) subcostal, (c) thoraco-abdominal?

20–5 What does notching of ribs seen radiographically suggest?

20–6 What is the mediastinum?

21
THE ESOPHAGUS, TRACHEA, AND MAIN BRONCHI

THORACIC PART OF ESOPHAGUS

The esophagus, or gullet (figs. 21–1 to 21–3), which has cervical, thoracic, and abdominal parts, extends from the lower end of the pharynx (C.V.6) to the cardiac opening of the stomach (T.V.11 or 12). When a subject is in the erect position, it is about 25 to 30 cm long. **The esophagus is a median structure that lies first behind the trachea and then behind the left atrium.** It begins to deviate to the left below the left main bronchus. In the posterior mediastinum it is related to the vertebral column as a string is related to a bow. Hence there is a (retrocardiac) space between it and the vertebrae, which is visible radiographically in oblique and lateral views.

The esophagus presents constrictions (1) at its commencement, (2) frequently where it is crossed by the left main bronchus, and (3) commonly where it traverses the diaphragm. The impressions of adjacent structures, and their alterations in disease, can be seen radiographically after a thick barium paste is swallowed (fig. 21–3).

The esophagus conducts food and liquid and can be replaced successfully by a non-muscular tube. **The esophagus is distensible and can accommodate almost anything that can be swallowed,** e.g., a denture. The muscular layer is striated above (supplied by the vagi) and smooth below (supplied by parasympathetic, or vagal, and sympathetic fibers).

The process of swallowing may be watched fluoroscopically. A thin barium paste or liquid is "shot down" to the cardiac orifice, whereas a thick paste or a bolus of food travels more slowly.

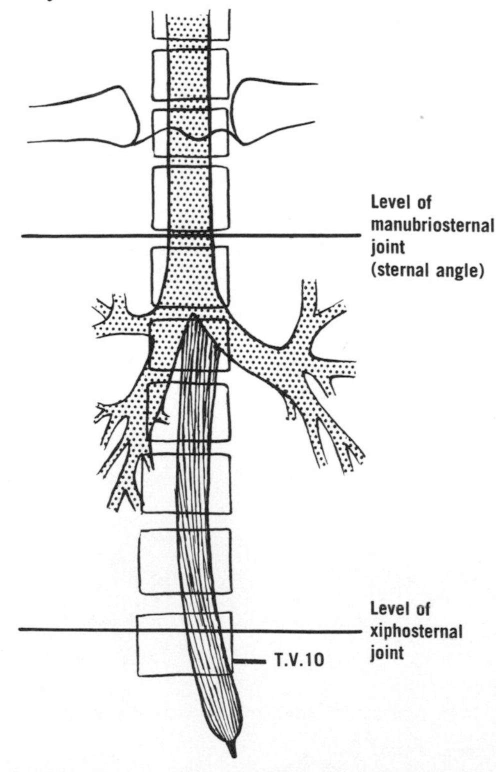

Figure 21–1 The trachea and esophagus in relation to vertebral and sternal levels in a subject in the erect position.

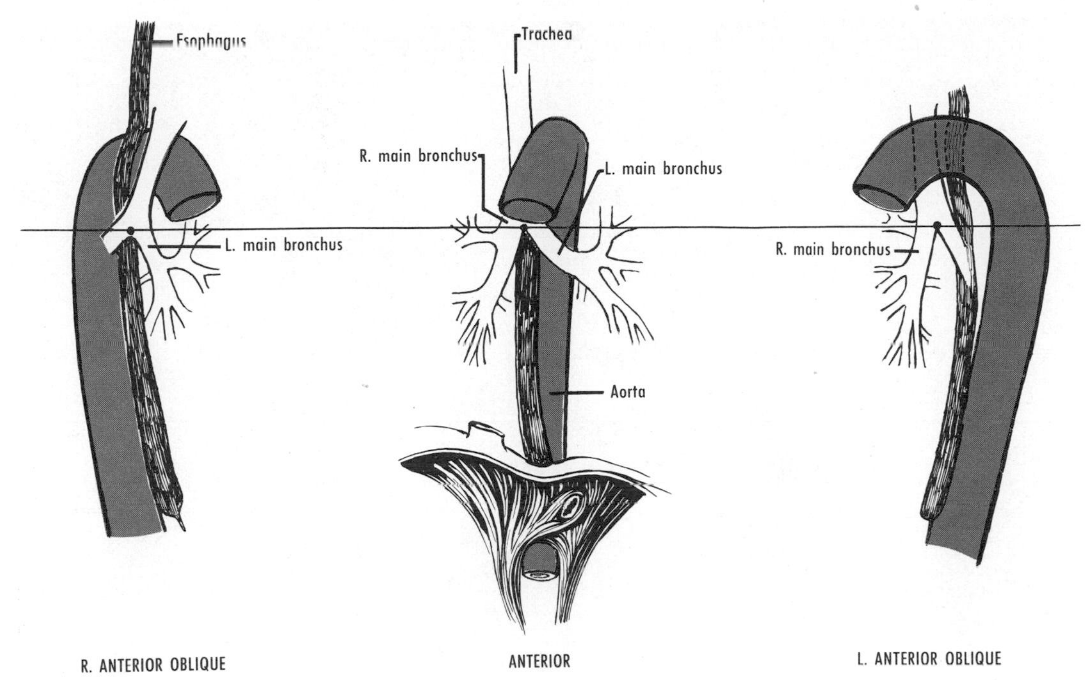

Figure 21–2 The relations of the trachea, bronchi, esophagus, and aorta to one another. In the right anterior oblique view, the right lobar and segmental bronchi are omitted because they are not clearly visible in radiographs of this view. For similar reasons, the left lobar and segmental bronchi are omitted from the left anterior oblique view. The horizontal line indicates the level of the carina.

The esophagus is supplied by arteries in the neck (inferior thyroid arteries), thorax (bronchial arteries, direct branches of the aorta, and phrenic arteries), and abdomen (left gastric artery). **Veins of the lower part of the esophagus communicate with the left gastric vein, thereby forming an important portal-systemic anastomosis.** Portal obstruction (e.g., in the liver) causes these channels to enlarge, and their varicosities may produce hemorrhage.

Pain fibers from the esophagus accompany the sympathetic system. A vague, deep-seated, esophageal pain may be felt behind the sternum or in the epigastrium, and it resembles that arising from the stomach or heart ("heartburn").

In *esophagoscopy,* measurements are taken from the upper incisor teeth to indicate the beginning of the esophagus (18 cm), the point at which it is crossed by the left bronchus (28 cm), and its termination (43 cm).

TRACHEA

The trachea, or windpipe (figs. 21–1 and 21–2), which has cervical and thoracic parts, extends from the lower end of the larynx (C.V.6) to its point of bifurcation (T.V.5 to 7). It is about 9 to 15 cm in length. The trachea descends in front of the esophagus, enters the superior mediastinum, and divides into right and left main bronchi. **The trachea is a median structure but, near its lower end, deviates slightly to the right, because of which the left main bronchus crosses in front of the esophagus.** Owing to the translucency of the air within it, the trachea is usually visible above the arch of the aorta in radiographs.

The trachea has 15 to 20 C-shaped bars of hyaline cartilage that prevent it from collapsing. Longitudinal elastic fibers enable the trachea to stretch and descend with the roots of the lungs during inspiration. **When a subject is in the erect position, the trachea divides at the level of T.V.5 to 7.** The *carina* is the upward-directed ridge seen internally at the bifurcation and is a landmark during bronchoscopy.

The arch of the aorta is at first in front of the trachea and then on its left side immediately above the left main bronchus. Other close relations include the brachiocephalic and left common carotid arteries. The trachea is supplied mainly by the inferior thyroid arteries. Its smooth muscle is supplied by parasympathetic and sympathetic fibers, and pain fibers are carried by the vagi.

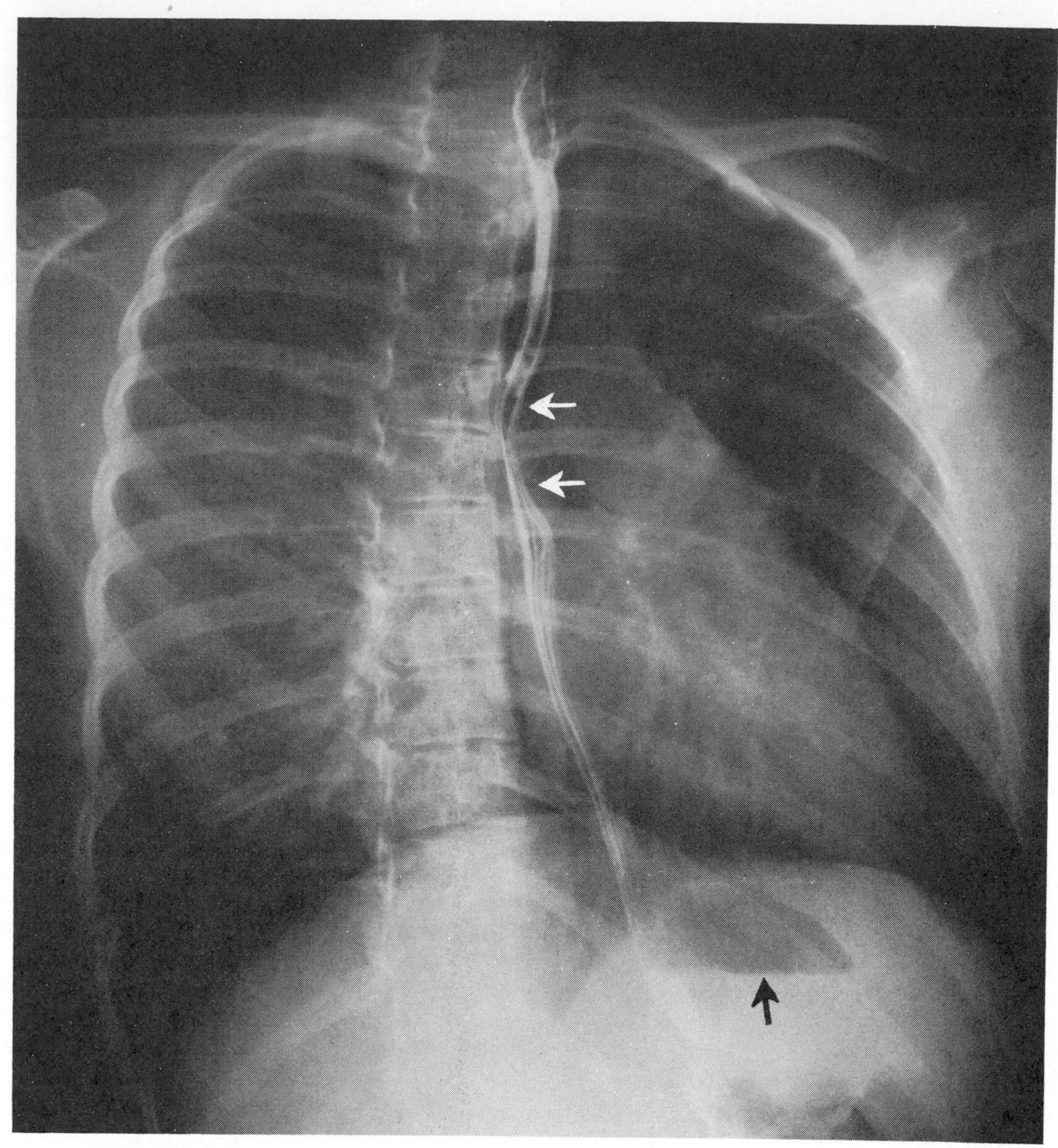

Figure 21–3 Oblique (R.A.O.) view of thorax. The esophagus is shown coated with barium paste. Note the vertical folds of the mucosa and the indentations produced by the arch of the aorta (*upper white arrow*) and by the left main bronchus (*lower white arrow*). Note also the right and left domes of the diaphragm (on the left and right sides of the illustration, respectively) and the fluid level in the stomach (*black arrow*). (From Bassett, D.L., *A Stereoscopic Atlas of Human Anatomy,* Sawyer's, Portland, Oregon, 1958. Copyright 1958, Sawyer's Inc.)

MAIN BRONCHI

Each main bronchus extends from the tracheal bifurcation to the hilus of the corresponding lung. The *right main bronchus* may be considered as comprising (1) an upper (eparterial) part, from which the segmental bronchi for the upper lobe arise, and (2) a lower part, from which the segmental bronchi for the middle and lower lobes emerge (fig. 21–4). The *left main bronchus* divides into two lobar bronchi, one each for the upper and lower lobes. The upper lobar bronchus may be considered as having (1) an upper division and (2) a lower, or lingular, division.

The right main bronchus, about 2½ cm in length, is shorter, wider, and more nearly vertical than the left. Because it is in almost a direct

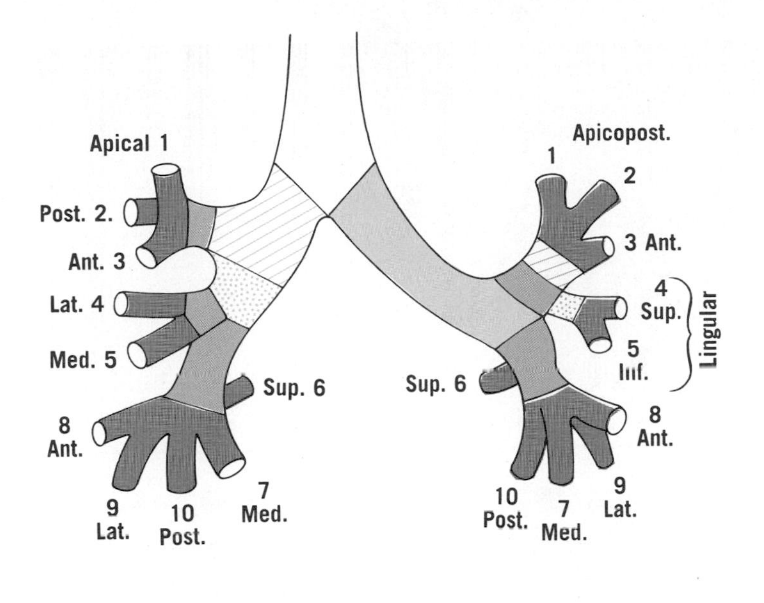

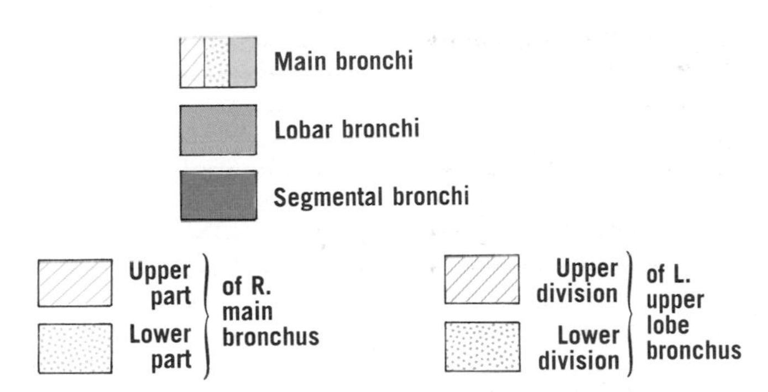

Figure 21–4 The main, lobar, and segmental bronchi. See also table 22–1.

line with the trachea, foreign objects traversing the trachea are more likely to enter the right main bronchus. The left main bronchus, 5 cm or more in length, crosses in front of the esophagus (fig. 21–2), which it indents. Both bronchi have cartilaginous rings that are replaced by separated plates at the roots of the lungs. The bronchi are supplied by the bronchial arteries and veins, and their innervation is similar to that of the trachea.

ADDITIONAL READING

Terracol, J., and Sweet, R. H., *Diseases of the Esophagus*, W. B. Saunders Company, Philadelphia, 1958. A general reference with an extensive bibliography.

QUESTIONS

21–1 What lies in front of the esophagus in the lower part of the thorax?

21–2 Where does the trachea divide?

21–3 How does the right main bronchus differ from the left?

22
THE PLEURAE AND LUNGS

PLEURA

The two lungs and their pleural sacs are situated in the thoracic cavity (figs. 22–1 and 22–2). The pleura is a thin, glistening, slippery serous membrane, inflammation of which is called *pleurisy*. The pleura lines the thoracic wall and diaphragm, where it is known as the *parietal pleura*. It is reflected onto the lung, where it is called the *visceral* (or *pulmonary*) *pleura*. The visceral pleura covers the lung and dips into its fissures. The facing surfaces of the parietal and visceral pleurae slide smoothly against each other during respiration. The contact between the parietal and visceral pleurae depends on the atmospheric pressure (1) on the outside of the chest wall and (2) inside the alveoli (which are connected to the exterior by the bronchial tree). On the other hand, the two pleural layers tend to be separated by the elasticity of (1) the thoracic wall (directed outward) and (2) the lungs (stretched by inspiration). The *pleural cavity*, which is the potential space between the two layers, contains merely a thin film of fluid. Air in the pleural cavity (*pneumothorax*) results in collapse of the lung. The pleura is supplied by adjacent arteries and nerves and has numerous lymphatics. Irritation of the parietal pleura causes pain referred to the thoraco-abdominal wall (intercostal nerves) or to the shoulder (phrenic nerve).

The parietal pleura presents costal, mediastinal, and diaphragmatic parts and a cupola (fig. 22–3). The *costal pleura* is separated from the sternum, costal cartilages, ribs, and muscles by a loose connective tissue termed *endothoracic fascia*, which provides a natural cleavage plane for surgical separation of the pleura from the thoracic wall. In front, the costal pleura turns sharply onto the mediastinum, and the underlying portion of the pleural cavity is called the *costomediastinal recess*. Below, the costal pleura is continuous with the diaphragmatic pleura, and the underlying cavity is termed the *costodiaphragmatic recess*. **In the adult *in vivo*, the anterior borders of the right and left pleurae probably meet at or near the median plane during a part of their course.** The left anterior border sometimes diverges to leave a part of the pericardium uncovered (bare area). Behind, the pleura crosses the twelfth rib. At the root of the lung, the *mediastinal pleura* turns laterally, enclosing the structures at the root and becoming continuous with the visceral pleura. This reflection projects downward as a tapering double fold called the *pulmonary ligament*. The mediastinal pleura is adherent to the pericardium except where the phrenic nerve descends between them. Above the arch of the aorta, the right and left pleurae approach each other behind the esophagus. The *diaphragmatic pleura* covers most of the diaphragm except the central tendon. The *cupola* (or *cervical pleura*) is the continuation of the costal and mediastinal parts of the pleura over the apex of the lung. The cupola is strengthened by a thickening of the endothoracic fascia termed the *suprapleural membrane*, which is attached to the inner margin of the first rib and the transverse process of the seventh cervical vertebra (C.V.7). Some muscular fibers (*scalenus minimus*) may be inserted into the membrane. **Because of the slope of the first rib, the cupola of the pleura and the apex of the lung**

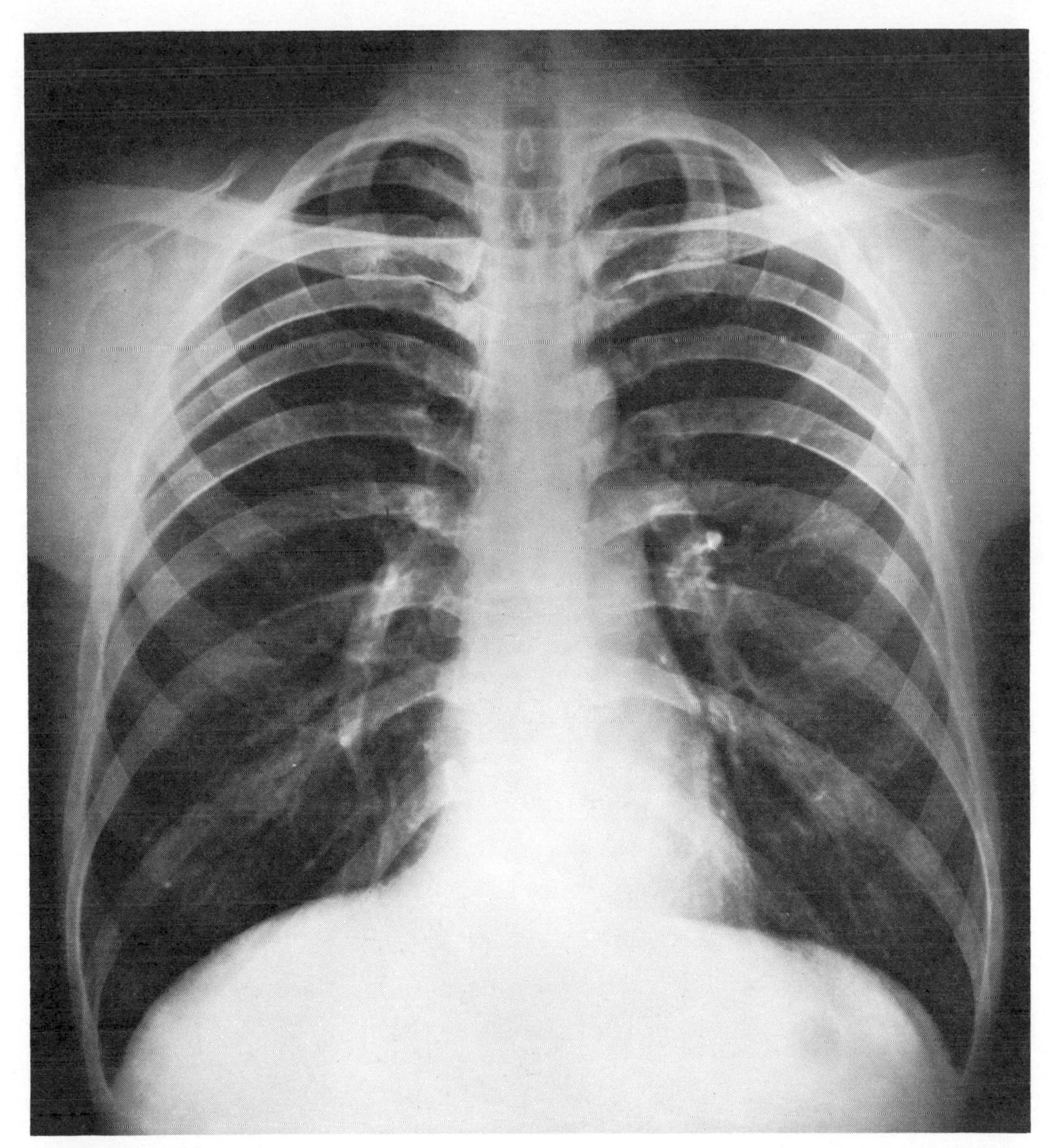

Figure 22–1 Thorax of an adult. Note the clavicles, ribs, diaphragm, cardiovascular shadow (including the right atrium, aortic knuckle, and left ventricle), trachea, and lungs (including vascular markings). Large descending branches of the pulmonary arteries are visible on each side of the heart.

project upward into the neck, behind the sternomastoid, and hence may be injured in wounds of the neck. Their summit is 2 to 3 cm or more above the level of the medial third of the clavicle. Behind the cupola are found the sympathetic trunk and first thoracic nerve.

The *anterior border* of the pleura extends downward from the cupola, passing behind the sternoclavicular joint, then to the middle of the sternal angle, and next to approximately the level of the xiphosternal joint. The *inferior border* of the pleura extends laterally from the xiphosternal joint, crosses rib 8 in the midclavicular line and rib 10 in the midaxillary line, and then proceeds toward the spine of T.V.12 (see fig. 22–2). Considerable individual variation occurs, but generally the pleura extends two fingerbreadths below the lung.

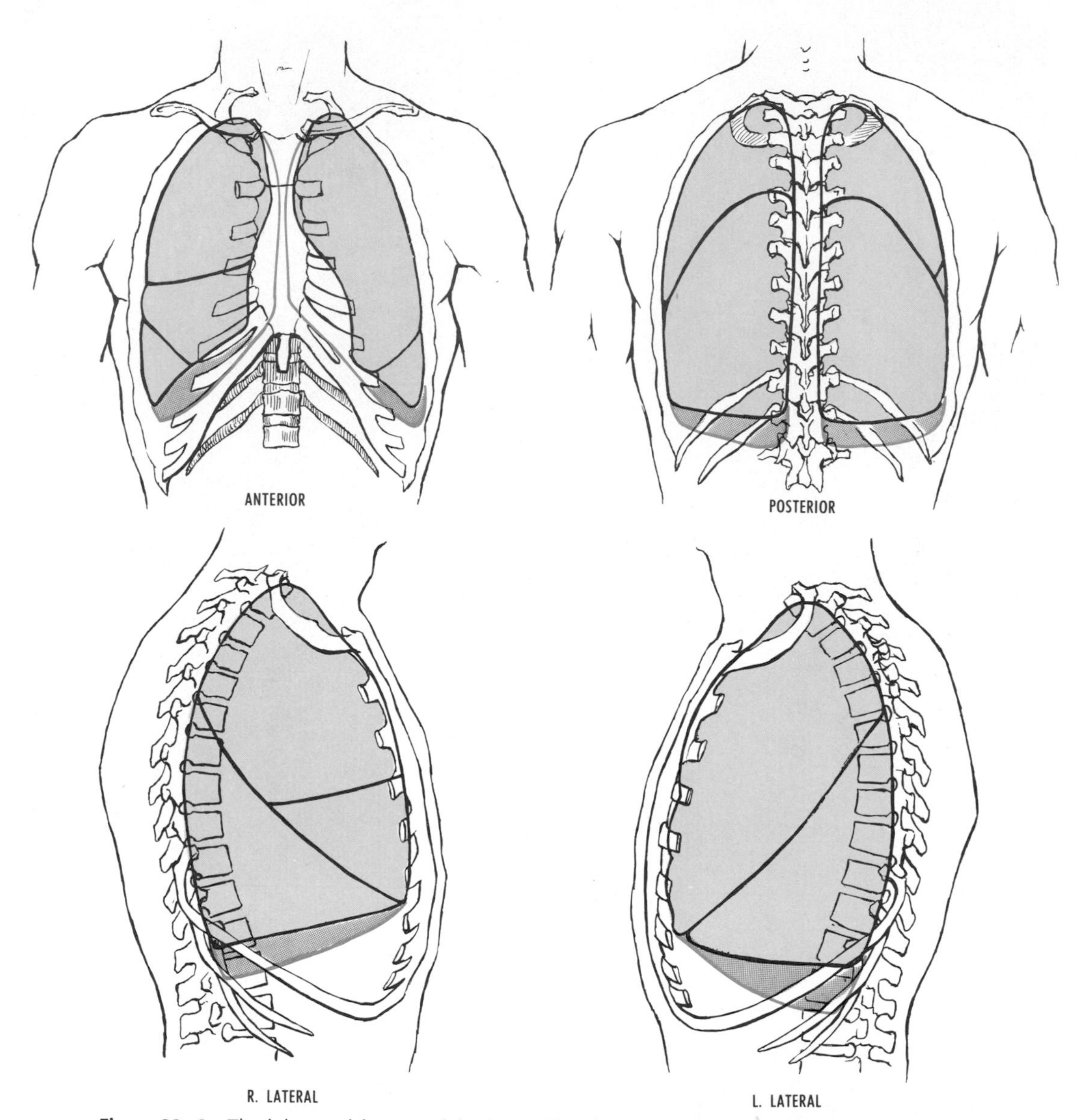

Figure 22–2 The lobes and fissures of the lungs. The pleurae are shown in blue. (Based on Brock.)

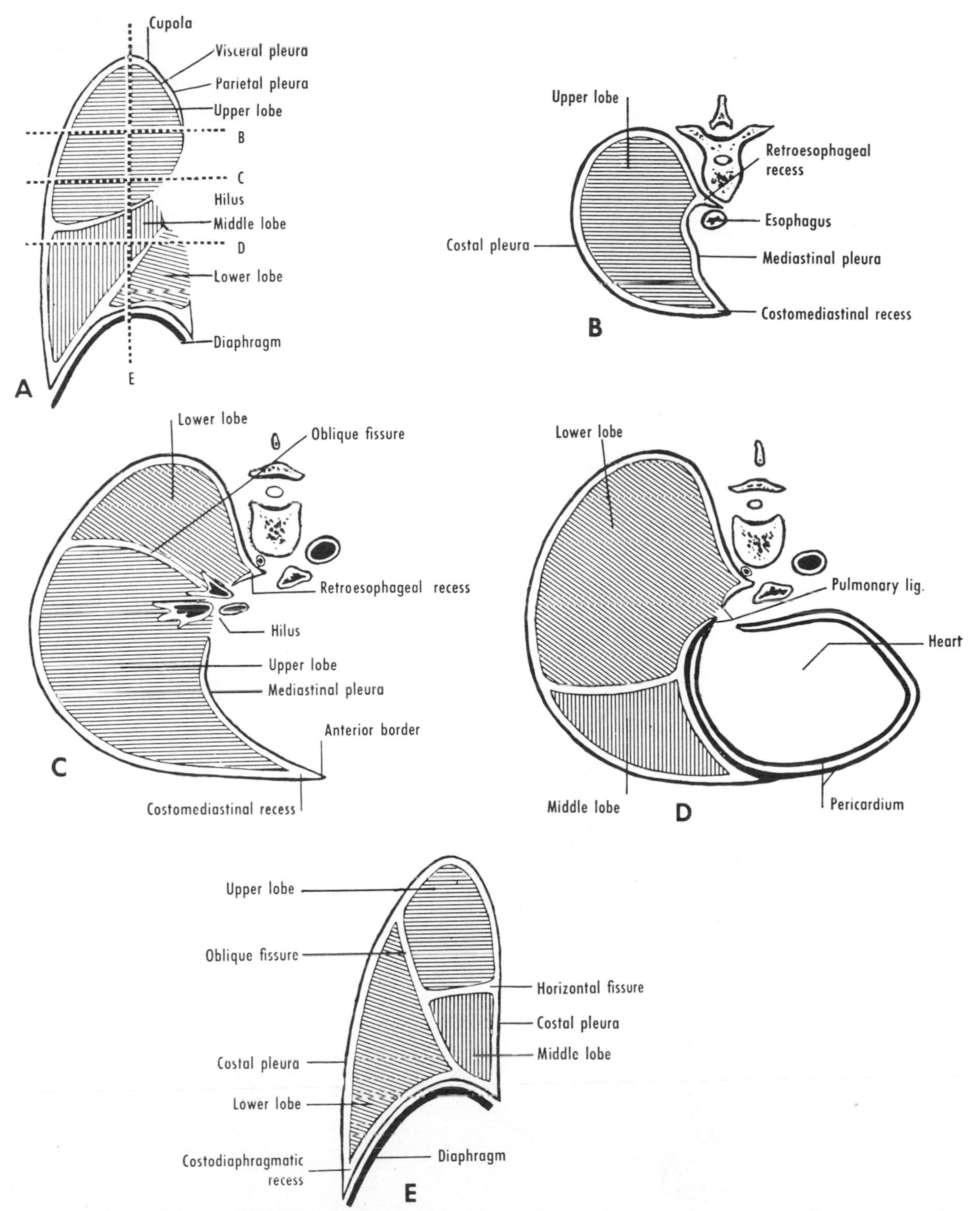

Figure 22–3 Diagrams of pleural reflections. **A,** Coronal section of the right lung and pleura. Lines *B* to *E* indicate the respective planes and levels of sections shown in diagrams **B** to **E. B,** Upper horizontal section. Note the costomediastinal and retro-esophageal recesses. **C,** Middle horizontal section. Note that the anterior border of the pleura forms the edge of the costomediastinal recess and that the oblique fissure reaches almost to the hilus. **D,** Lower horizontal section, showing also relationships to the pericardium. Note that the pulmonary ligament is formed by the double reflection of the pleura below the hilus of the lung. The mediastinal pleura is adherent to the fibrous pericardium, except where the phrenic nerve descends between them (not shown). **E,** Sagittal section.

LUNGS

The lungs are the essential organs of respiration. The Latin word *pulmo,* lung, gives rise to the adjective *pulmonary*. The corresponding Greek word provides *pneumonia,* inflammation of the lungs. Each lung is attached by its root and pulmonary ligament to the heart and trachea but is otherwise free in the thoracic cavity. The lungs are light, soft, spongy, and elastic, and, because they contain air, they float in water. If an infant has not drawn a breath, however, the lungs will not float. (From two thirds of the way through prenatal life, a fetus is viable, i.e., has sufficient pulmonary development to live *ex utero.*) The air in the lungs renders them translucent to x-rays. The surface of an adult lung is usually mottled, and it presents dark gray or bluish patches caused by inhalation of atmospheric dust.

The main bronchus enters the hilus and subdivides within the substance of the lung to form the "bronchial tree." The tubes of the tree carry air to the alveoli, where respiratory exchanges with the blood occur. The bronchial tree elongates on deep inspiration.

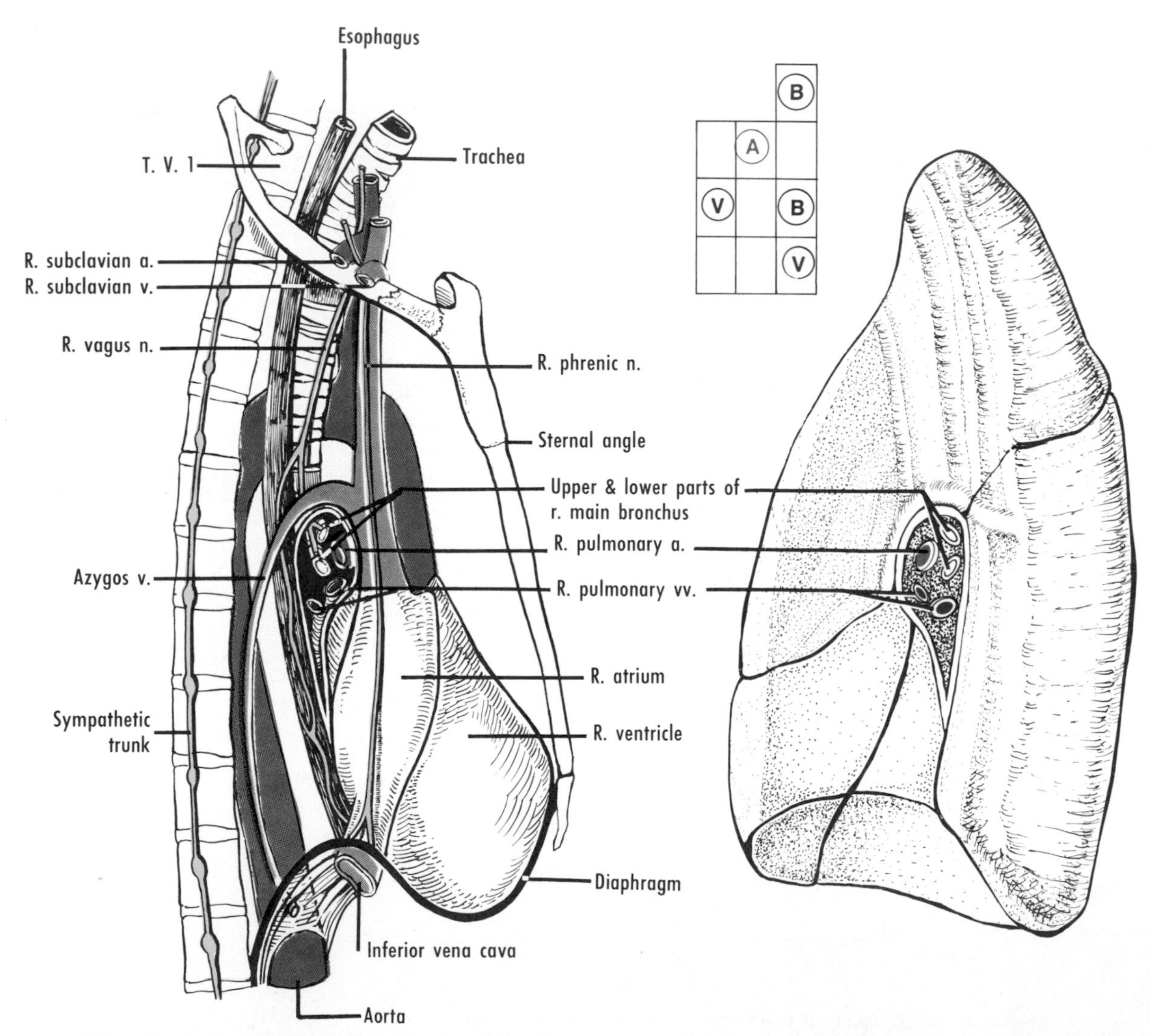

Figure 22–4 Mirror-image views of the right lung and mediastinal structures. The line of reflection of the parietal to the visceral pleura is shown as a white line around the hilus, prolonged below as the two-layered pulmonary ligament. Impressions produced by mediastinal structures have been indicated on the medial surface of the lung as they might appear in a hardened lung. (Based on Mainland and on Mainland and Gordon.) The inset shows the relationships at the hilus, including the eparterial (upper lobe) bronchus. Note that in this and in similar drawings, because the pulmonary veins contain oxygenated blood, they are shown in red; correspondingly, the pulmonary arteries are shown in blue.

The right lung, which is heavier than the left, is also shorter (the right dome of the diaphragm being higher) and wider (the heart bulging more to the left).

Surfaces and Borders

Each lung presents an apex, three surfaces (costal, medial, and diaphragmatic), and three borders (anterior, inferior, and posterior). **The right lung is divided into upper, middle, and lower lobes by oblique and horizontal fissures, whereas the left lung has usually only upper and lower lobes, separated by an oblique fissure.**

The bronchi and pulmonary vessels, which extend from the trachea and heart, respectively, collectively form the *root of the lung.* The part of the medial surface where these structures enter the lung is known as the *hilus* (figs. 22–4 and 22–5).

The *apex* is rounded and follows the cupola of the pleura. The *costal surface,* which is re-

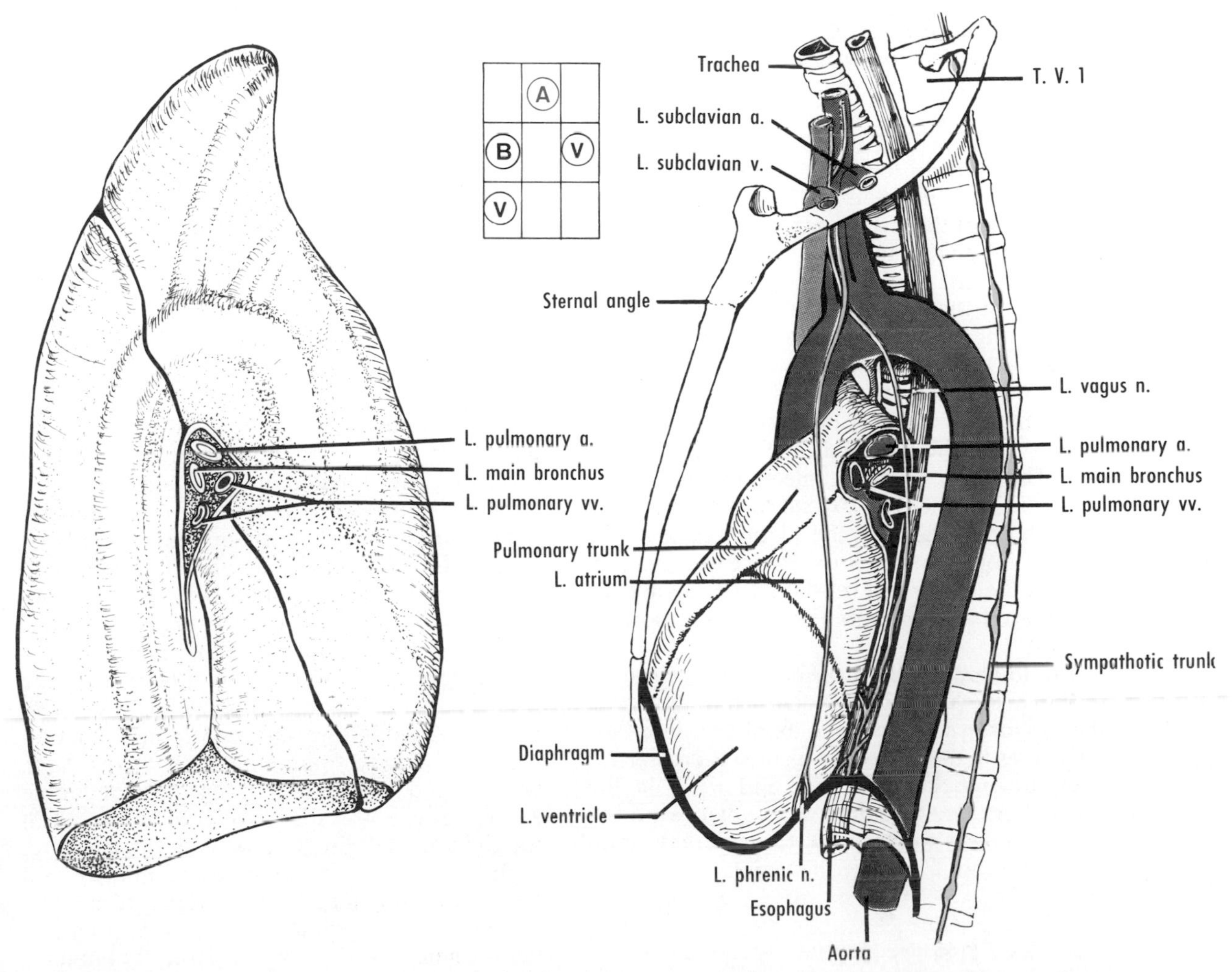

Figure 22–5 Mirror-image views of the left lung and mediastinal structures, similar to the previous figure. The inset shows the relationships at the hilus.

lated to the sternum, costal cartilages, and ribs, joins the medial surface at the anterior and posterior borders and the diaphragmatic surface at the inferior border. The *medial surface* is related behind to the sides of the bodies of the vertebrae. In front, the medial surface is related to the superior, middle, and posterior parts of the mediastinum and includes the hilus. The *diaphragmatic surface,* or *base,* corresponds to the dome of the diaphragm, which separates the lung from the liver (on the right side) or the stomach, spleen, and sometimes liver and left colic flexure (on the left side).

Rarely, the azygos vein, instead of arching over the hilus of the right lung, arches over the upper lobe so that it isolates a medial part of the lung, called the *lobule of the azygos vein.* Parietal pleura (*meso-azygos*) extends into the fissure, and the condition may be visible radiographically.

The anterior border of the lung corresponds to that of the pleura, although it is uncertain whether the costomediastinal recess of the pleura is completely filled by the lung during quiet breathing, as it is in deep inspiration. The anterior border of the left lung probably deviates more to the left (cardiac notch) than does that of the pleura. The portion of the upper lobe of the left lung that lies between the cardiac notch and the oblique fissure is known as the *lingula,* and it corresponds to the middle lobe of the right lung. The inferior border of the lung occupies the costodiaphragmatic recess of the pleura, although it is too thin to be demonstrated by percussion during quiet breathing. **For surgical purposes, it is safest to assume that the lung and pleura are coextensive. The liver, stomach, spleen, colon, kidney, and peritoneal cavity extend to a higher level than the periphery of the diaphragm and the inferior border of the lung. Hence any perforation of the lower intercostal spaces should be considered an abdominal as well as a thoracic wound.**

The lowermost limit of the lung that can be outlined by percussion extends laterally from the xiphosternal joint and about two intercostal spaces higher than the pleura. It crosses rib 6 in the midclavicular line and rib 8 in the midaxillary line and then proceeds toward the spinous process of T.V.10. Considerable individual variation occurs, however.

Lobes and Fissures (see fig. 22–2)

The right lung is divided into upper, middle, and lower lobes by an oblique and a horizontal fissure. The left lung is divided into upper and lower lobes by an oblique fissure. The *oblique fissure* follows approximately the line of rib 6 as far as the inferior border of the lung. When the arm is abducted and the hand placed on the back of the head, the medial border of the scapula indicates approximately the oblique fissure. The *horizontal fissure* begins at the oblique fissure near the midaxillary line (of the right side), at about the level of rib 6. It extends forward to the anterior border at the level of costal cartilage 4. It may be incomplete or even absent.

Root (figs. 22–4 and 22–5)

The root of the lung consists of the structures entering and emerging at the hilus. It connects the medial surface of each lung to the heart and trachea. It is surrounded by pleura, which is prolonged below as the pulmonary ligament. The roots of the lungs descend on deep inspiration. The chief structures in the root are the bronchi and pulmonary vessels. Also included are nerves, bronchial vessels, and lymphatics and nodes. The heart and great vessels are anterior to the trachea and main bronchi, and this relationship is maintained in the root of the lung, where the order, from in front backward, is veins, artery, and bronchus, with the artery above the veins.

Bronchopulmonary Segments (fig. 22–6)

The main bronchus divides into lobar bronchi, each of which divides into third-order bronchi. **The portion of lung supplied by a third-order bronchus is known as a bronchopulmonary segment. A given segment may be located by radiography or bronchoscopy. Pulmonary disorders may be localized in a bronchopulmonary segment, and surgical removal of a segment is feasible.** The segments are separated from each other by connective tissue septa, which are continuous with the visceral pleura. Although variations are not uncommon, the bronchopulmonary segments have been named and numbered (fig. 22–6 and table 22–1). There are slight differences between the right and left lungs: briefly, in the left lung, segments 1 and 2 are generally combined, and commonly segments 7 and 8 are also. The branches of the pulmonary artery accompany the bronchi but are more variable. Pulmonary veins do not accompany the bronchi, but run between the segments; hence they are guides to intersegmental planes.

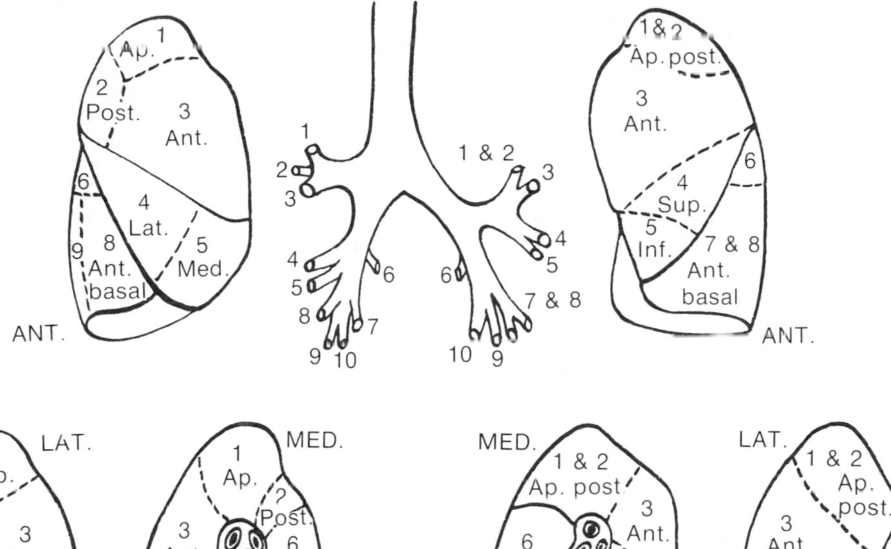

Figure 22–6 The segmental bronchi and bronchopulmonary segments. See also figs. 21–4 and 22–8 and table 22–1.

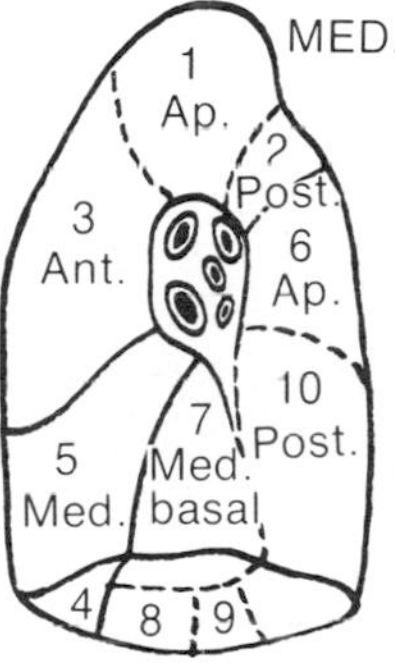

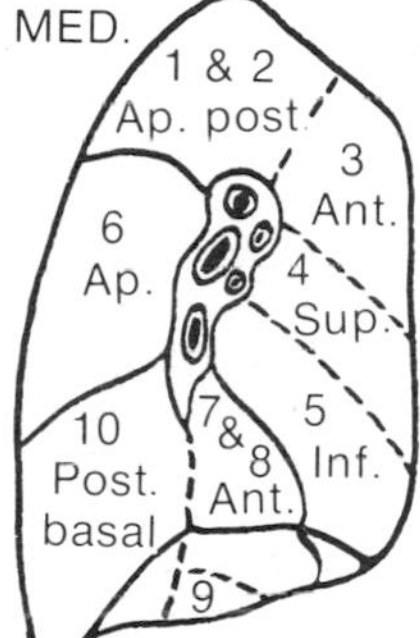

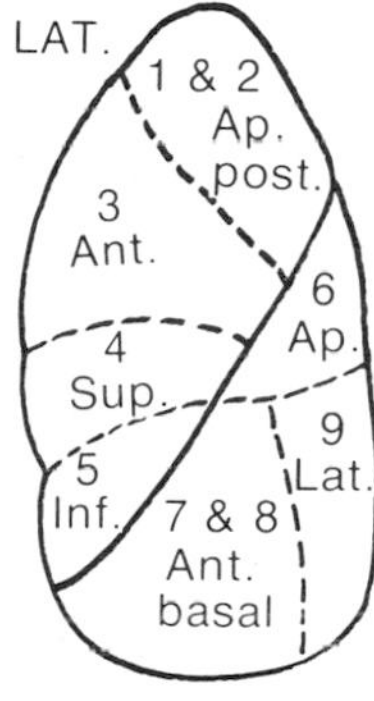

TABLE 22–1 Bronchopulmonary Segments

Right Lung	Left Lung
Superior Lobe	*Superior Lobe*
1. Apical	1 & 2. Apicoposterior
2. Posterior	
3. Anterior	3. Anterior
Middle Lobe	
4. Lateral	4. Superior lingular
5. Medial	5. Inferior lingular
Inferior Lobe	*Inferior Lobe*
6. Apical (superior)	6. Apical (superior)
7. Medial basal (cardiac)	7 & 8. Anterior basal (Medial basal [cardiac] is independent in about 1/3 of instances)
8. Anterior basal	
9. Lateral basal	9. Lateral basal
10. Posterior basal	10. Posterior basal

Blood Supply, Lymphatic Drainage, and Innervation

Blood to be oxygenated is carried by the pulmonary arteries, whereas the tissue of the bronchial tree and alveoli is nourished by the bronchial arteries. The branches of the *pulmonary arteries* within the lungs accompany the bronchi and end in capillary networks in the alveoli. The arteries at the hilus are visible radiographically and form a pattern that extends into the lung. The *pulmonary veins* collect oxygenated blood from the lung and deoxygenated blood from the bronchi and visceral pleura. Pulmonary veins are intersegmental in location. Usually four pulmonary veins enter the left atrium.

Bronchial arteries, usually one on the right and two on the left, arise commonly from the aorta, but variations are frequent. They supply oxygenated blood to the non-respiratory tissues of the lungs, including the visceral pleura. *Bronchial veins* carry deoxygenated blood from the first few bronchial divisions to the azygos system.

Carbon particles in the superficial lymphatics give the lung a grayish and mottled appearance. *Superficial* and *deep lymphatic vessels* drain toward the hilus and end in *pulmonary* and *bronchopulmonary nodes*. These in turn drain into the *tracheobronchial nodes*.

The *anterior* and *posterior pulmonary plexuses* around the root of the lung are formed by branches of the vagi and sympathetic trunks. Parasympathetic fibers (of vagal origin) supply the smooth muscle and glands of the bronchial tree. Spasm of the bronchial musculature occurs in asthma and can be relieved by epinephrine. Sympathetic fibers supply blood vessels and are probably inhibitory to the bronchi. Afferent fibers (vagal) from the visceral pleura and bronchi are concerned with the reflex control of respiration. Irritation of endings in the bronchial mucosa provokes coughing.

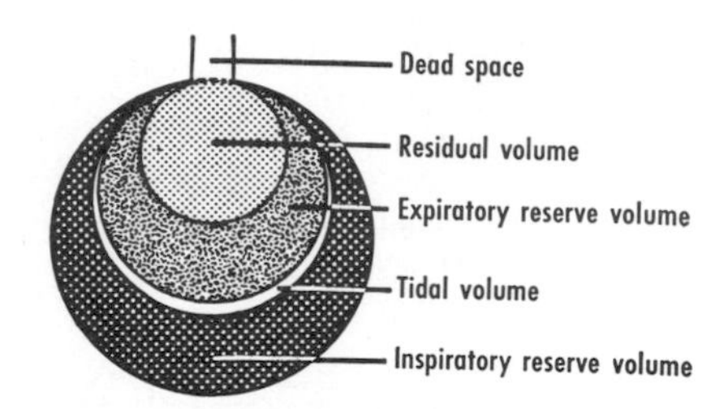

Figure 22–7 The primary lung volumes. The outermost line represents the greatest size to which the lung can expand. (Based on Comroe *et al.*)

PHYSICAL AND RADIOLOGICAL EXAMINATIONS

The classic methods of physical examination of the living body, in order, are inspection, palpation, percussion, and auscultation. Similar points on the two sides of the chest should be compared systematically. Details of such examinations are available in special works on physical diagnosis. *Inspection* includes examination of the scapulae, clavicles, ribs, sternal

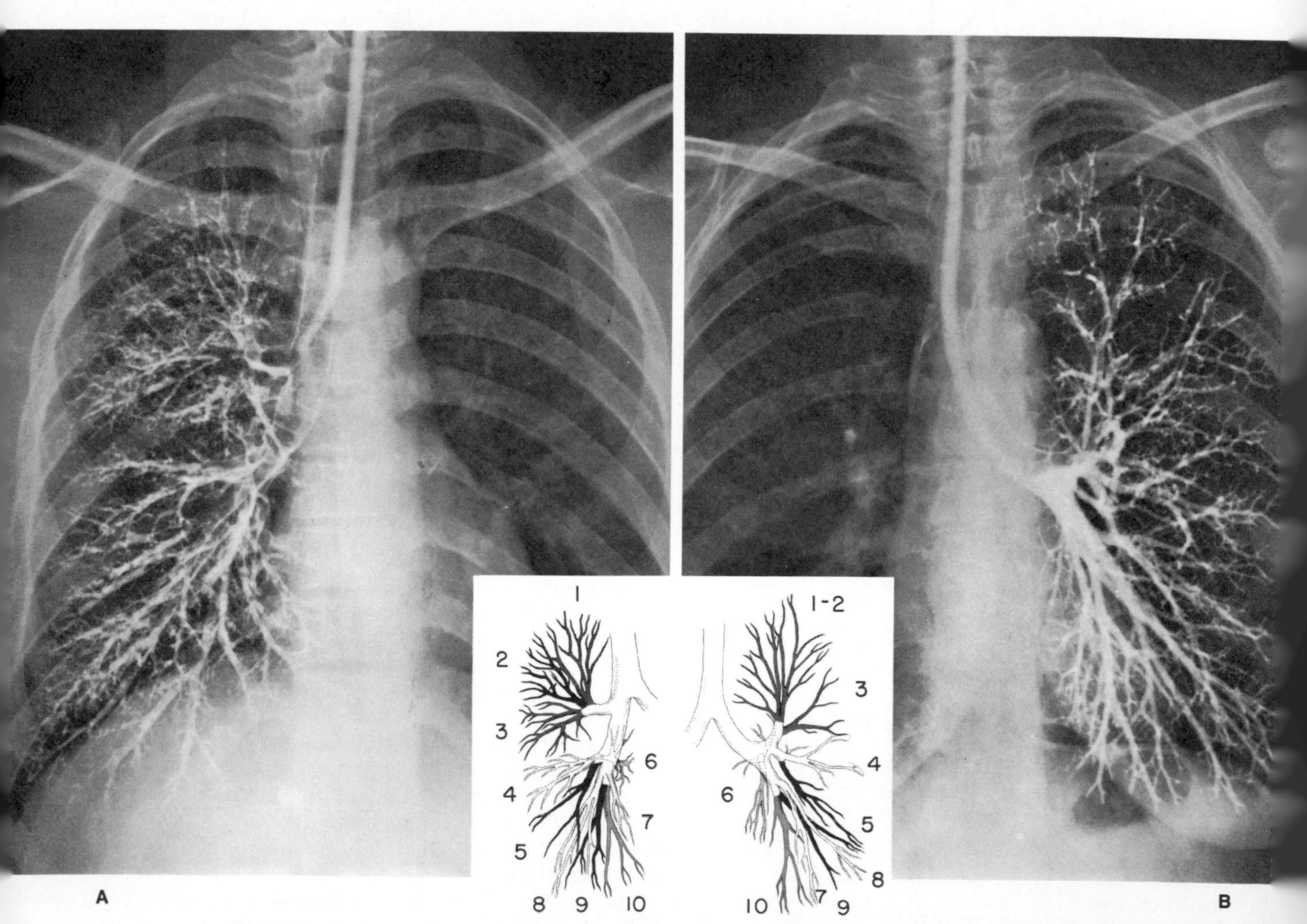

Figure 22–8 Postero-anterior bronchograms. **A,** Right lung. **B,** Left lung. (From *Medical Radiography and Photography*, courtesy of J. Stauffer Lehman, M.D., and J. Antrim Crellin, M.D., Philadelphia, Pennsylvania.) For terminology see table 22–1.

angle, subcostal angle, expansion of the chest, and movement of the abdominal wall. The rate of respiration is normally 11 to 14 per minute, i.e., about one fourth of the pulse rate. It is higher in children, and the neonatal rate is two to three times as high as the adult rate. *Palpation* involves feeling the trachea and also the vibrations ("vocal fremitus") of the chest wall. *Percussion* yields a note termed "resonance" over air-containing organs such as the lungs. *Auscultation* of the breath sounds is undertaken with a stethoscope, as is also the detection of vibrations ("vocal resonance") of the chest wall.

The chief x-ray methods used in the examination of the chest are fluoroscopy, radiography, tomography, and bronchography.

The most frequently employed view in radiography of the chest is the *anterior* view, i.e., *postero-anterior projection,* in which the x-ray tube is behind the erect patient and the film is vertically in front of the chest. Other views are indicated in figure 23–15. The bronchi and pulmonary tissue are radiolucent, but the branches of the pulmonary arteries form a visible pattern. The dense hilar shadows are produced by a combination of vascular, bronchial, lymphatic, and connective tissue components. The bronchial tree can be outlined by a contrast medium injected by way of an intratracheal catheter introduced through the mouth or nose. This procedure is termed *bronchography* (fig. 22–8).

ADDITIONAL READING

Boyden, E. A., *Segmental Anatomy of the Lungs,* Blakiston (McGraw-Hill), New York, 1955. A classic on the patterns of the segmental bronchi and related pulmonary vessels.

Hayek, H., von, *Die menschliche Lunge,* 2nd ed., Springer, Berlin, 1970. The first edition (1960) is available in English (*The Human Lung*).

Nagaishi, C., et al., *Functional Anatomy and Histology of the Lung,* University Park, Baltimore, 1972. A mine of pulmonary information.

QUESTIONS

22–1 What are the serous membranes?

22–2 What is pneumothorax?

22–3 What is the cupola?

22–4 How does the right lung differ from the left?

22–5 What is the lowermost limit of the lung and pleura?

22–6 Why is a perforation of the lower intercostal spaces to be considered an abdominal as well as a thoracic wound?

22–7 What is a bronchopulmonary segment?

22–8 What is the main constituent of the hilar shadows seen radiographically?

22–9 What are the classic methods of physical examination?

22–10 Which is the most frequently employed view in radiography of the chest?

23 THE PERICARDIUM AND HEART

PERICARDIUM

The heart is enclosed in a fibroserous sac termed the *pericardium* (figs. 23–1 and 23–2), which occupies the middle mediastinum (see fig. 20–8). The pericardium and its fluid lubricate the moving surfaces of the heart.

The *fibrous pericardium* is the outermost layer, and it is firmly bound to the central tendon of the diaphragm. Extrapericardial fat, which may be visible radiographically, is often found in the angles between the pericardium and diaphragm on each side. The pericardium is attached to the sternum (by the *sternopericardial ligaments*) and is adherent to the mediastinal pleura except where the two are separated by the phrenic nerves.

TABLE 23–1 LAYERS OF PERICARDIUM AND HEART

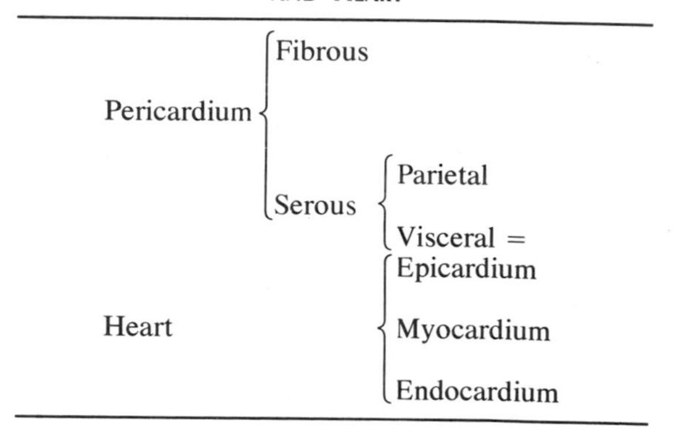

Pericardium	Fibrous		
	Serous	Parietal	
		Visceral =	Epicardium
Heart			Epicardium
			Myocardium
			Endocardium

The *serous pericardium* (fig. 23–1) is a closed sac, the parietal layer of which lines the inner surface of the fibrous pericardium and is reflected onto the heart as the visceral layer, or *epicardium*. The potential space between the parietal and visceral layers contains a thin film of fluid and is known as the *pericardial cavity*.

Because of the manner in which the reflections of the pericardium occur during development at the arterial and venous ends of the heart, when the pericardial sac is opened from the front in the adult **it is possible to pass a finger behind the aorta and pulmonary trunk (arterial end of heart) and in front of the left atrium and superior vena cava (venous end of**

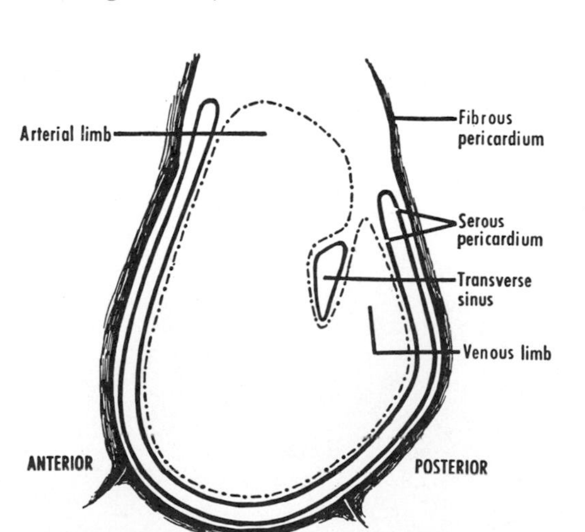

Figure 23–1 Schematic sagittal section through the heart and pericardium. Note how the serous layer of the pericardium is reflected onto the heart and forms a double layer.

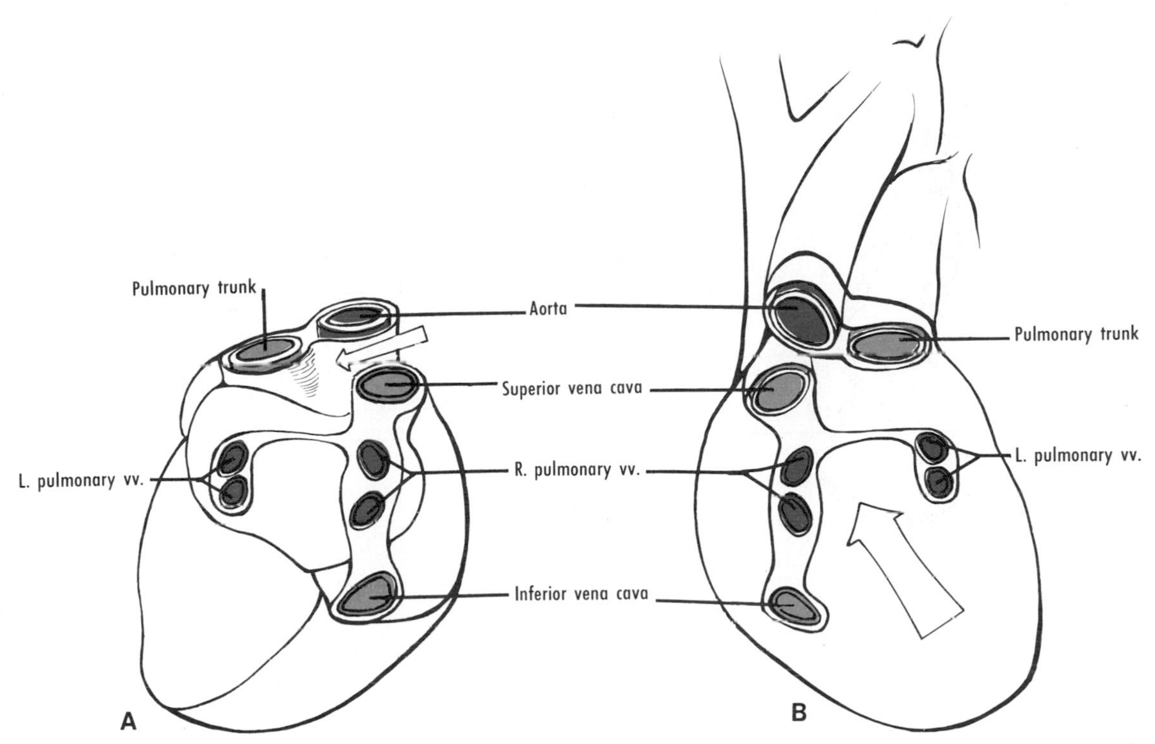

Figure 23–2 Mirror images of the pericardial reflections. **A,** The reflections onto the heart, viewed from behind. The arrow indicates the transverse sinus of the pericardium. **B,** The heart removed, and the posterior part of the pericardium viewed from in front. The reflection at the veins forms an irregular continuous line that begins at the inferior vena cava, extends up to the lower right pulmonary vein, and turns to the left across the left atrium to the left pulmonary veins. The irregular space thereby bounded is the oblique sinus of the pericardium (*arrow*).

heart). This passage is termed the *transverse sinus of the pericardium* (figs. 23–1 and 23–2). The reflection at the venous end is complicated and occurs as an inverted U along the pulmonary veins and inferior vena cava: the recess is termed the *oblique sinus of the pericardium*.

The pericardium is supplied by adjacent arteries, e.g., the internal thoracic, and by the phrenic nerves, which contain vasomotor and sensory fibers. Pericardial pain is felt diffusely behind the sternum but may radiate.

HEART

The heart is situated in the middle mediastinum and is divided into right and left halves by an obliquely placed, longitudinal septum. Each half consists of an *atrium,* which receives blood from the veins, and a *ventricle,* which propels the blood along the arteries. The heart is more in the left half of the thorax than in the right (see figs. 23–4 and 23–21). The adjective *cardiac* is derived from the Greek *kardia,* heart. The cardiac wall consists, from outside inward, of epicardium (visceral pericardium), myocardium, and endocardium.

Course of Circulation

The superior and inferior venae cavae and the intrinsic veins of the heart discharge venous blood into the right atrium (fig. 23–3). The blood then enters the right ventricle, from which it is ejected into the pulmonary trunk, which, by way of the right and left pulmonary arteries, sends blood to the lungs. The pulmonary veins return blood to the left atrium. The blood then enters the left ventricle and is ejected into the aorta. Four important valves are encountered: the right and left *atrioventricular,* known as the *tricuspid* and *mitral valves,* respectively, and the right and left *semilunar,* known as the *pulmonary* and *aortic valves,* respectively (see fig. 23–7). Insufficient closure results in an incompetent valve and reflux of blood.

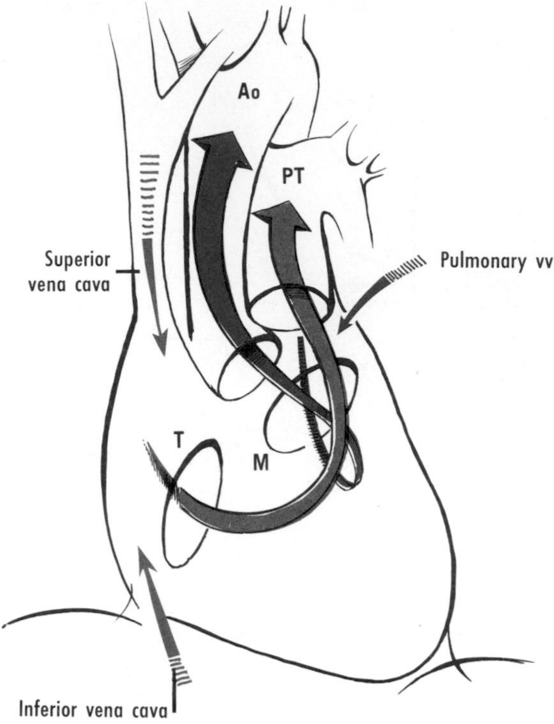

Figure 23–3 The circulation of blood through the chambers of the heart. Note that blood flows almost horizontally forward from the right atrium to reach the outflowing part (conus arteriosus, or infundibulum) of the right ventricle. *M*, mitral orifice; *T*, tricuspid orifice.

Size and Position

The size of the heart is frequently described as that of the person's fist. The surface area of the heart, as determined from anterior radiograms, is used as an index of size. The maximum transverse diameter is the sum of the greatest distances of the right and left sides of the cardiac silhouette from the median plane.

When a subject is in the erect position, the heart (fig. 23–4) is lower than is generally illustrated. Radiographic data are frequently not in agreement with findings obtained by percussion, and **locating the apex beat is probably a better guide to the left border than is percussion.**

The so-called *apex* of the heart is often rounded and ill-defined radiographically. When recognizable, it is usually at the level of costal cartilage 6, below and medial to the point at which the apex beat can be felt. The so-called *apex beat,* which is an impulse imparted by the heart, can usually be felt on the front of the left side of the chest. The site of this point of maximum pulsation is generally in intercostal space 4 or 5, about 6 or 7 cm from the median plane. The apex beat is produced by a complex movement of the left ventricle during contraction. It is a fairly reliable guide to the left border of the heart, but in some people it is felt outside the cardiac area.

The atria, which form the base of the heart, lie behind the ventricles, and the right chambers contribute more to the front of the heart than do the left. The planes of the atrioventricular orifices are more vertical than horizontal; hence the blood flows almost horizontally forward from the atria to the ventricles (see fig. 23–3). **The longitudinal axis of the heart, which extends from the base to the apex, is directed forward, downward, and to the left.** The axis is frequently largely vertical, but it may be more horizontal (transverse heart), especially in infancy, obesity, or pregnancy. The size, shape, and position of the heart vary from one individual to another, and also from time to time in the same individual (see fig. 23–17). Tall, slender people are more likely to have "vertical" hearts, whereas heavy-set, stocky individuals are more apt to have "transverse" hearts. **The position and movements of the diaphragm are the most important factors in determining the position of the heart.** During deep inspiration, the heart descends and becomes narrower and more "vertical."

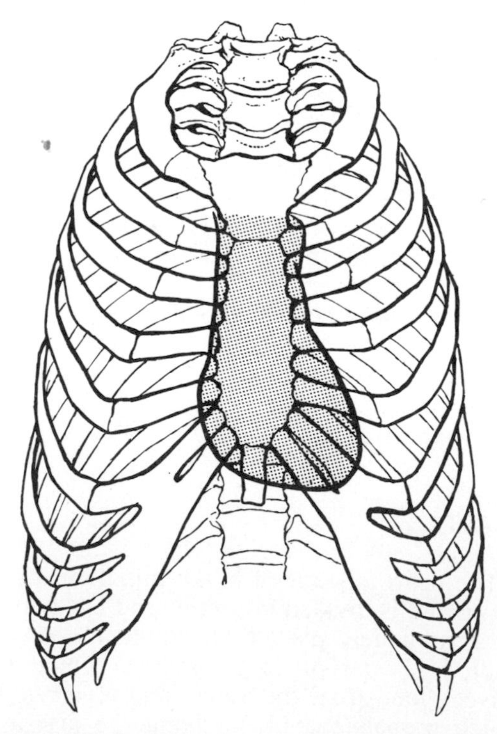

Figure 23–4 The cardiovascular shadow (of a "vertical" heart) in relation to the bony cage. Depending on posture and phase of respiration, the lower margin of the heart may be at a still lower level, as much as 5 cm below the xiphosternal joint (see figs. 22–4 and 22–5). See fig. 23–16 for the composition of the cardiovascular shadow.

When a subject is in the supine position, the heart moves upward and backward. In the newborn and infant, the heart is relatively large, globular, "transverse," and higher than in the adult.

External Features (fig. 23–5)

The heart is usually said to have an apex, a base, and three surfaces: sternocostal, pulmonary (or left), and diaphragmatic. Borders are

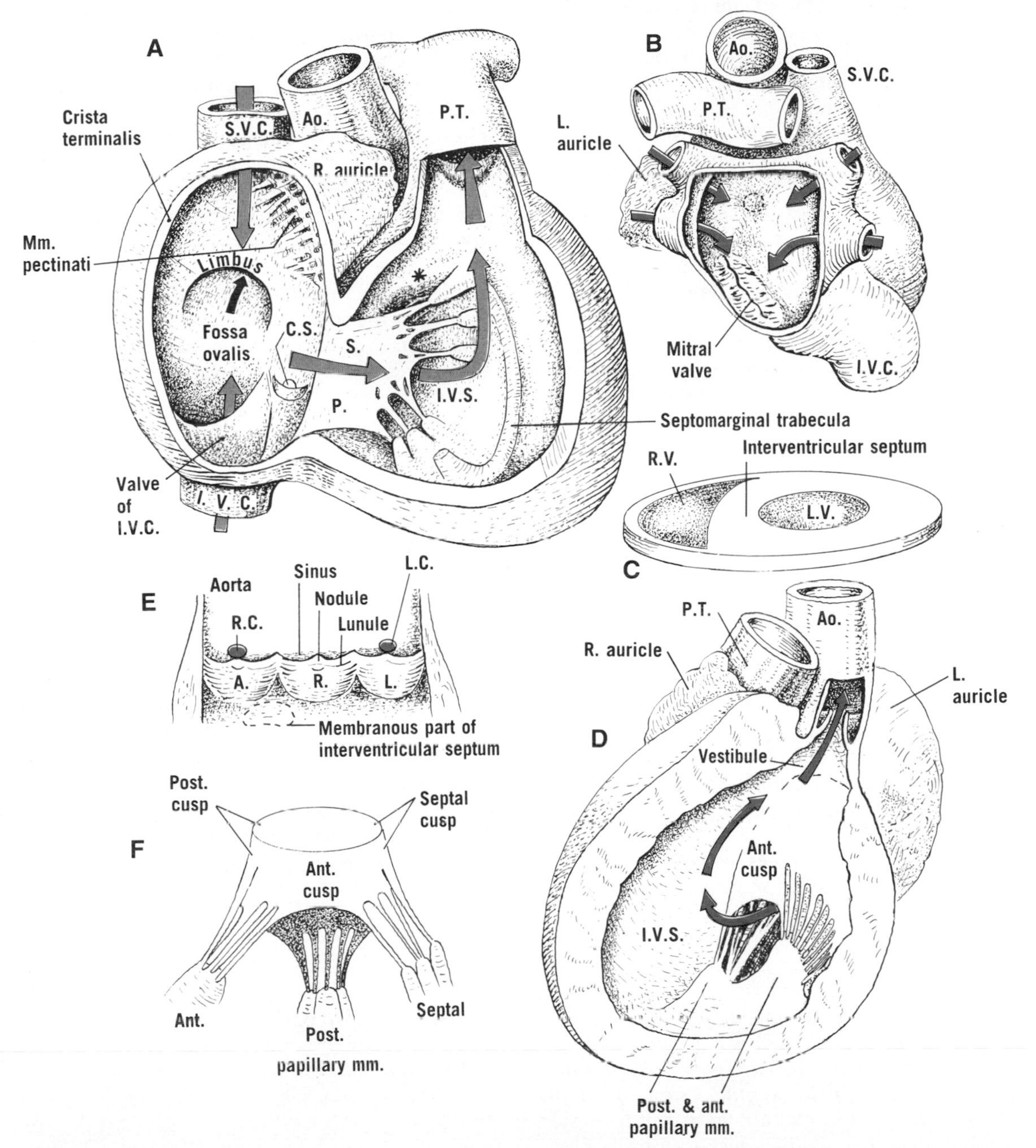

Figure 23–5 The external and internal anatomy of the heart. **A,** Right atrium and right ventricle, showing the tricuspid orifice. The arrows indicate the circulation of the blood. An arrow in the fossa ovalis represents "probe patency" of the foramen ovale. The membranous part of the interventricular septum lies mostly under cover of the septal cusp. The asterisk indicates the supraventricular crest. The septomarginal trabecula is frequently called the moderator band. **B,** Interior of the left atrium, dorsal view. The openings of the pulmonary veins (variable in number) and the left side of the fossa ovalis can be seen. **C,** Cross section through the ventricles, showing the much thicker wall of the left ventricle and the curved shape of the interventricular septum. **D,** Left ventricle, showing the mitral valve. **E,** Aortic valve, spread out. **F,** Tricuspid valve, showing the papillary muscles, chordae tendineae, and cusps (or leaflets). *A.*, anterior cusp of aortic valve; *Ao.*, aorta; *C.S.*, opening of coronary sinus; *I.V.C.*, inferior vena cava; *I.V.S.*, interventricular septum; *L.*, left cusp of aortic valve; *L.C.*, opening of left coronary artery; *L.V.*, left ventricle; *P.*, posterior cusp of tricuspid valve; *P.T.*, *pulmonary trunk;* *R.*, right cusp of aortic valve; *R.C.*, opening of right coronary artery; *R.V.*, right ventricle; *S.*, septal cusp of tricuspid valve; *S.V.C.*, superior vena cava. (Based on Testut and other sources.)

indefinite. The *base,* formed by the atria, is directed backward, and it lies behind the ventricles. Each *atrium* continues forward, on each side of the aorta and pulmonary trunk, as an ear-shaped appendage termed the *auricle.* (In clinical usage, an atrium is sometimes still called an auricle and an auricle an auricular appendage.) The right atrium may show a slight groove from the front of the superior caval orifice to the right side of the inferior caval opening. It indicates a muscular band interiorly, the *crista terminalis,* the upper part of which is occupied by the sinu-atrial node.

The atria and ventricles are separated behind by an *atrioventricular* (or *coronary*) *groove,* which lodges the coronary sinus and portions of the coronary arteries (see fig. 23–9).

The sternocostal surface of the heart is formed mainly by the right ventricle, which is prolonged upward as the conus arteriosus, or infundibulum. A shallow *anterior interventricular groove* superiorly lodges the interventricular branch of the left coronary artery. A *posterior interventricular groove* inferiorly (on the diaphragmatic surface), lodges most often the interventricular branch of the right coronary artery. The diaphragmatic surface is formed by both ventricles, whereas the left surface is mainly the left ventricle.

Internal Features of Atria (fig. 23–5)

The inner surfaces of both auricles present muscular ridges termed *musculi pectinati,* whereas the inner aspects of the atria are mostly smooth.

The *right atrium* receives the openings of the superior and inferior venae cavae, which are separated by a muscular wall called the *intervenous tubercle.* The *valve of the inferior vena cava* is a variable semilunar fold in front of the orifice. The *valve of the coronary sinus* is also variable. The walls of all four chambers show a number of small foramina for the intrinsic veins of the heart (*venae cordis minimae*). The *interatrial septum,* seen from the right side, presents an ovoid depression termed the *fossa ovalis,* which is bounded by a fold called the *limbus fossae ovalis.* The upper part of the fossa may be separated from the limbus by a variable aperture, the *foramen ovale,* which is the persistence of a fetal interatrial opening. The right *atrioventricular orifice* is guarded by the tricuspid valve, which usually can admit three fingers.

The *left atrium* is prolonged on each side as pouches for the openings of the pulmonary veins. The fossa ovalis appears as a translucent area in the interatrial septum. The upper edge of this region is called the *valve of the foramen ovale.* The left *atrioventricular orifice* is guarded by the mitral valve, which usually can admit two fingers.

The right atrium is usually on the same side of the body as the eparterial (upper lobe) bronchus, the suprahepatic part of the inferior vena cava, and the liver (hepatocavo-atrial concordance). In some anomalies of alignment, a morphological right atrium may be present on the left side of the heart. Thus, in transposition of the viscera (*situs inversus*), the morphologically right cardiac chambers are generally on the left and the morphologically left chambers, including the apex, are frequently on the right.

Internal Features of Ventricles (figs. 23–5 to 23–7)

The ventricular part of the heart presents tricuspid and pulmonary openings on the right and mitral and aortic openings on the left. The inner surfaces of the ventricles are mostly irregular, because of the projection of muscular bundles known as *trabeculae carneae.* Three kinds occur: (1) "ridges" on the ventricular surface; (2) "bridges," free in the middle; and (3) "pillars," termed *papillary muscles,* the bases of which are attached to the ventricular wall. The apices are continued as fine strands, the *chordae tendineae,* which are anchored to the cusps of the atrioventricular valves and prevent eversion of those valves (see fig. 23–5).

Each *atrioventricular valve* has *cusps* that are attached to a *fibrous ring* around the opening (figs. 23–5 and 23–7). The atrial surfaces of the cusps are smooth, whereas the ventricular surfaces are rough to allow for attachment of the chordae tendineae.

The *semilunar valves* are situated at the roots of the pulmonary trunk and aorta (see fig. 23–5). Each has three *cusps,* and the free edge of each cusp has a small, fibrous thickening, the *nodule,* on either side of which is a thin crescentic area termed a *lunule.* When the valve is closed, the nodules and lunules are in apposition. The spaces between the cusps and the walls of the vessels "distally" are the pulmonary and aortic *sinuses,* respectively.

The terminology of the cusps of the semilu-

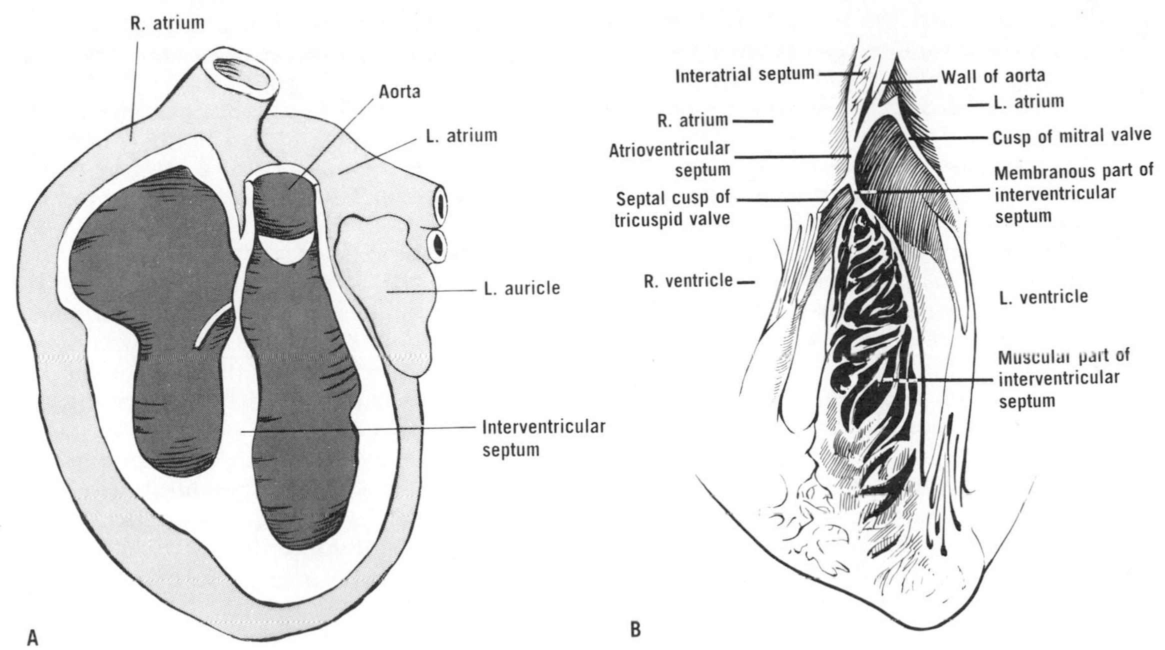

Figure 23–6 **A,** Longitudinal section of the heart to show the interventricular septum. **B,** Higher magnification to show the atrioventricular and interventricular septa.

nar valves is important in comparing congenital defects in which the great vessels are misaligned, out of position, or transposed. Nevertheless, the terminology is confusing because the cusps are frequently named (as in the *Nomina anatomica*) from the excised organ instead of being based (as in this book) on the relationships *in situ* (fig. 23–7).

The cardiac valves are not visible radiographically, unless calcified, but they can be delimited during angiocardiography. The aortic valve is situated in the middle of the cardiac shadow, as seen from the front, and the mitral valve is a little lower and to the left (cf. fig. 23–18).

The right ventricle is usually characterized by (1) a tricuspid valve, (2) a trabeculated septal surface, and (3) a supraventricular crest and conus arteriosus (infundibulum). (In some anomalies of alignment, a morphological right ventricle may be present on the left side of the heart.)

The right ventricle lies in front of the right atrium, the plane of the tricuspid opening is

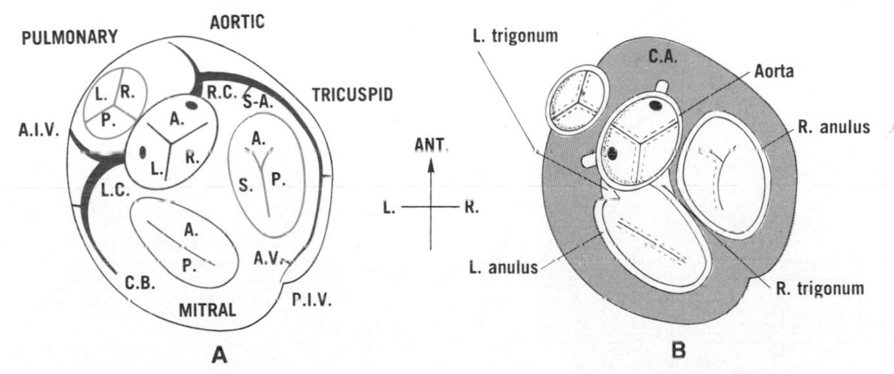

Figure 23–7 **A,** Diagram of the cardiac valves and their cusps *in situ*, as seen from above. The coronary arteries are also shown. *A.*, anterior cusp; *A.I.V.*, anterior interventricular branch; *A.V.*, branch to atrioventricular node; *C.B.*, circumflex branch; *L.*, left cusp; *L.C.*, left coronary artery; *P.*, posterior cusp; *P.I.V.*, posterior interventricular branch; *R.*, right cusp; *R.C.*, right coronary artery; *S.*, septal cusp; *S.-A.*, branch to sinu-atrial node. Cf. fig. 23–9. **B,** Cardiac skeleton seen from above. The arrangement of the anuli and trigona shown is based on J. Zimmerman, J. Alb. Einstein Med. Cent., 7:77, 1959. *C.A.*, roof of conus arteriosus (or infundibulum).

nearly vertical, and the blood flows horizontally from the atrium to the ventricle (see fig. 23–3). The *tricuspid (right atrioventricular) valve* has three cusps: anterior, posterior, and septal (fig. 23–7). The associated papillary muscles are classified also as anterior, posterior, and septal. The *septomarginal trabecula* (or *moderator band*) is a bridge-type trabecula that extends from the interventricular septum to the base of the anterior papillary muscle. Its importance is in the fact that it contains Purkinje fibers from the right limb of the atrioventricular bundle (cf. fig. 23–8).

The cavity of the right ventricle is U-shaped, but with the U on its side: ⊂ . The lower limb of the U, which receives blood from the right atrium, is the venous, or inflowing, part of the ventricle. The limbs of the U are separated by a muscular ridge termed the *supraventricular crest*. The upper limb, or *conus arteriosus* (or *infundibulum*), is the arterial, or outflowing, part of the ventricle, and it ends in the pulmonary trunk. The walls of the conus are usually smooth. The junction of the conus and pulmonary trunk contains dense fibrous tissue that encircles the pulmonary valve and is continuous with the cardiac skeleton.

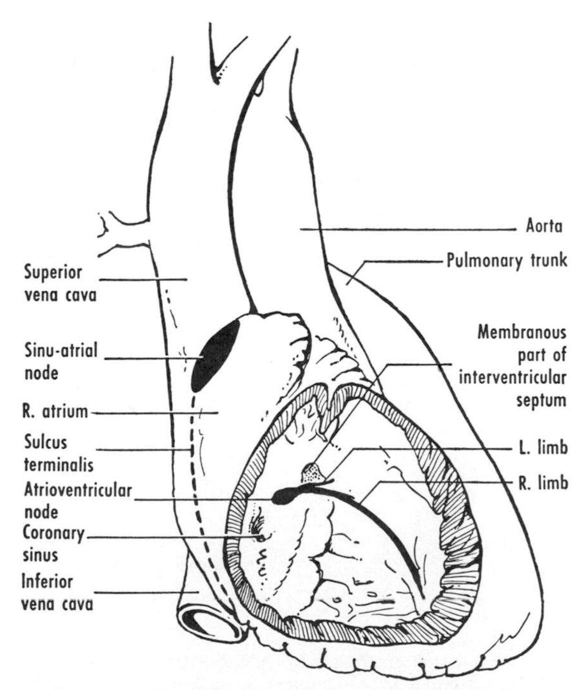

Figure 23–8 The conducting system, showing the positions of the sinu-atrial and atrioventricular nodes. The atrium and right ventricle are opened, and the interventricular septum is exposed.

The *pulmonary valve in situ* has two cusps (right and left) in front and one (posterior) behind (see figs. 23–7 and 23–11).

The left ventricle is usually characterized by (1) a bicuspid valve, (2) a smooth septal surface, and (3) the absence of a crest or conus. (In some anomalies of alignment, a morphological left ventricle may be present on the right side of the heart.)

Because the arterial pressure in the systemic circulation is much higher than in the pulmonary circulation, the left ventricle performs more work; hence **the wall of the left ventricle is usually more than twice as thick as that of the right.**

The left ventricle lies mostly in front of the left atrium, the plane of the mitral opening is nearly vertical, and blood flows obliquely forward from the atrium to the ventricle. The *mitral (left atrioventricular) valve* has two major cusps: anterior and posterior (see fig. 23–7). The associated papillary muscles are commonly the anterior and posterior.

The lower, or inflowing, part of the left ventricle receives blood from the left atrium (see fig. 23–19). The upper part, which has mainly fibrous walls, is known as the *aortic vestibule*, and it ends in the aorta. The junction of the vestibule and aorta contains dense fibrous tissue that encircles the aortic valve and is continuous with the cardiac skeleton.

The *aortic valve in situ* has generally one cusp (anterior) in front and two (right and left) behind (see figs. 23–7 and 23–11).

The *interventricular septum* (see fig. 23–6) is a strong, oblique partition consisting of muscular and membranous parts. One surface of the interventricular septum faces forward and to the right and bulges into the cavity of the right ventricle. The other surface looks backward and to the left and is concave toward the left ventricle. The *membranous part* of the septum is the small, thin, smooth, fibrous, upper portion. The atrioventricular bundle is situated at the lower border of the membranous part. The septal cusp of the tricuspid valve is usually attached to the right side of the membranous part so that the uppermost portion of the septum (between right atrium and left ventricle) is atrioventricular in position.

The *cardiac skeleton* (see fig. 23–7) consists of fibrous or fibrocartilaginous tissue that partially surrounds the atrioventricular and perhaps the semilunar openings and gives attachment to the valves and the muscular layers.

The *myocardium* is arranged in complicated

sheets and bands. Apart from the conducting system, the atrial and ventricular musculatures are separate. The ventricular musculature comprises superficial spiral and deep constrictor sheets, and the heart is twisted during systole like a cloth being wrung out.

Conducting System

The conducting system (fig. 23–8) consists of specialized (i.e., impulse-conducting) muscle fibers that connect certain "pacemaker" regions of the heart with ordinary "working" cardiac muscle fibers. The intrinsic, rhythmic contractions of cardiac muscle fibers are regulated by pacemakers, and the intrinsic rhythmicity of the pacemakers is regulated in turn by nerve impulses from vasomotor centers in the brain stem. If the conducting system between the atria and ventricles is destroyed (complete heart block), the atria and ventricles beat at different rates.

The conducting system comprises the sinuatrial (S.A.) node, the atrioventricular (A.V.) node, and the A.V. bundle, with its two limbs and the subendocardial plexus of Purkinje fibers. The impulse begins at the S.A. node, activates the atrial musculature, and is thereby conveyed to the A.V. node. Special bundles (internodal tracts) of atrial muscle fibers pass more or less directly from the S.A. to the A.V. node, but their functional significance is unclear.

The *S.A. node* is the usual pacemaker for the heart. It is situated anterolaterally at the junction of the superior vena cava and the right atrium, near the upper end of the sulcus terminalis and immediately beneath the epicardium. It consists of a network of specialized cardiac muscle fibers, which are continuous with the atrial muscle fibers. The node is supplied by autonomic fibers and by a branch of the right (sometimes the left) coronary artery.

The *A.V. node* is situated beneath the endocardium of the right atrium, in the interatrial septum immediately anterosuperior to the opening of the coronary sinus. It consists of a network of specialized cardiac muscle fibers, which are continuous with the atrial muscle fibers and with the A.V. bundle.

The *A.V. bundle* leaves the A.V. node and ascends to the membranous part of the interventricular septum. It then runs forward and divides into *right* and *left limbs,* or *crura,* which straddle the muscular part of the septum. The limbs proceed to the papillary muscles of their respective ventricles (the right limb traverses the septomarginal trabecula), and their fibers then ramify subendocardially as a plexus of Purkinje fibers. The bundle and crura are surrounded by a fibrous sheath.

Blood Supply and Innervation

The heart is supplied by the right and left coronary arteries (fig. 23–9), which usually arise from the anterior and left aortic sinuses, respectively (see figs. 23–7 and 23–11).

The *right coronary artery* emerges between the pulmonary trunk and right auricle. It usually supplies the S.A. node, conus arteriosus, right atrium, and right ventricle, and then winds to the back of the heart, where it anastomoses with the left coronary artery. A *posterior interventricular branch* supplies the A.V. node and portions of both ventricles.

The *left coronary artery* emerges between the pulmonary trunk and left auricle, gives off an *anterior interventricular branch* (which descends to the apical region), supplies the left atrium, and, as the *circumflex branch,* winds to the back of the heart, where it supplies the left ventricle and anastomoses with the right coronary artery. In many instances, the branch to the S.A. node arises from the circumflex branch (rather than from the right coronary artery), and the branch to the A.V. node may occasionally come from the left artery also. The anterior interventricular branch appears to be a direct continuation of the left coronary artery, and it provides the chief supply to the interventricular septum. It turns around the apical region and, on the back of the heart, anastomoses with the posterior interventricular branch of the right coronary artery.

Commonly, the heart is probably supplied equally by the two coronary arteries, but preponderance of one vessel may occur. Variations and anomalies are not uncommon. Thus, although the posterior interventricular branch arises from the right coronary in at least 90 per cent of hearts, it comes from the left coronary in approximately 10 per cent, in which case the left artery gives rise to both interventricular branches. These alternative arrangements are known as dominance of the right or of the left coronary artery, respectively.

The heart is drained (1) partly by small veins that empty directly into the cardiac chambers and (2) partly by veins that empty into the coronary sinus (fig. 23–9). The *coronary sinus* is

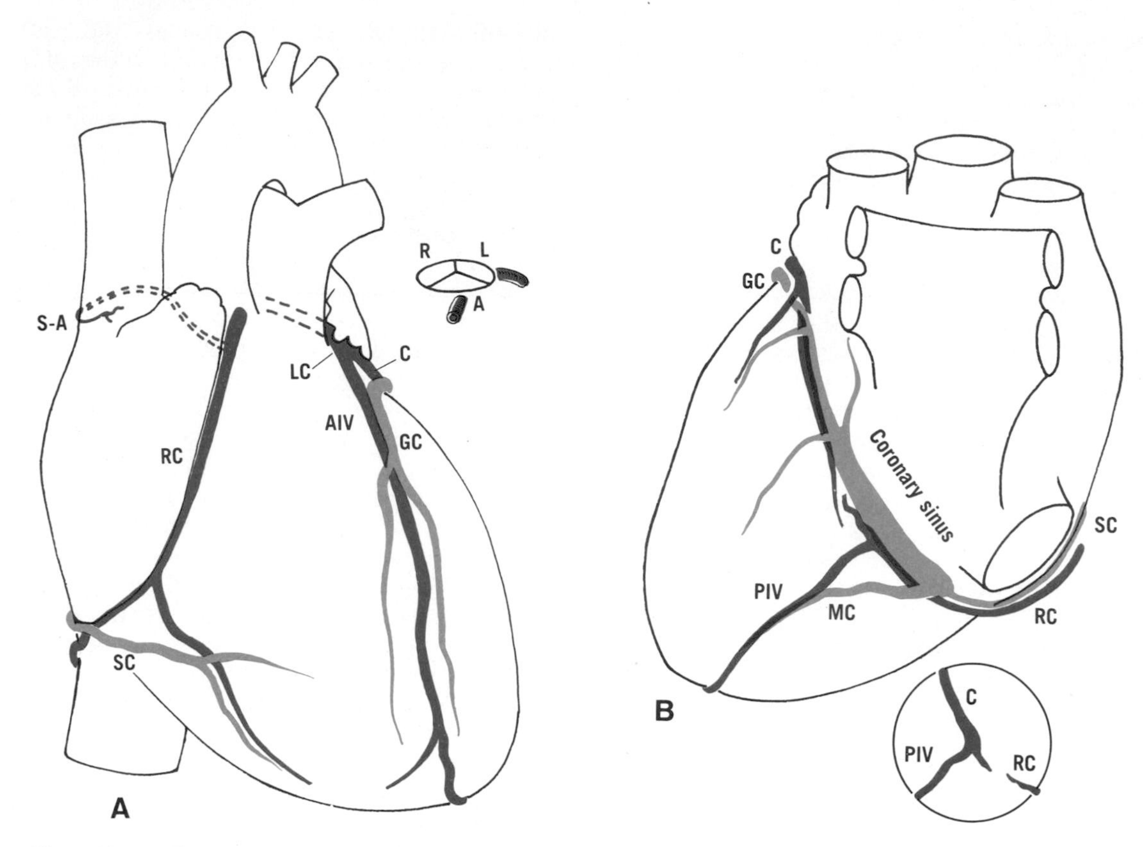

Figure 23–9 The coronary arteries and the veins that drain into the coronary sinus. The posterior interventricular branch (*PIV*), although usually a branch of the right coronary artery (*RC*), may arise from the circumflex branch (*C*) of the left coronary artery (*inset*). In **B,** the left marginal vein can be seen ascending to join the great cardiac vein. The posterior vein of the left ventricle ascends and the oblique vein of the left atrium descends to end in the coronary sinus. *AIV,* anterior interventricular branch; *C,* circumflex branch; *GC,* great cardiac vein; *LC,* left coronary artery; *MC,* middle cardiac vein; *PIV,* posterior interventricular branch; *RC,* right coronary artery; *S.-A,* branch to sinu-atrial node; *SC,* small cardiac vein. Cf. fig. 23–7.

a short, wide trunk that lies in the coronary groove between the left atrium and left ventricle. Its tributaries include the *small, middle,* and *great cardiac veins* and the *oblique vein of the left atrium* (which represents a left superior vena cava). The coronary sinus opens into the right atrium between the openings of the inferior vena cava and the tricuspid valve. Its right side is guarded by a valve.

The heart is supplied by autonomic nerve fibers and by sensory fibers of the sympathetic trunks and vagi (fig. 23–10). Many ganglion cells (mainly parasympathetic) and complicated sensory endings are found, especially in the atria and near the veins.

Preganglionic sympathetic fibers from the spinal cord (T1 to 6) synapse in cervical and thoracic ganglia, and postganglionic fibers travel in cervical and thoracic cardiac branches of the sympathetic trunk. Preganglionic parasympathetic fibers in the vagi travel in cervical and thoracic cardiac branches of the vagi to ganglion cells in the heart. The postganglionic fibers of both autonomic systems supply the S.A. and A.V. nodes and the coronary vessels. The arrangement of the various cardiac nerves is extremely variable.

Sensory fibers from complicated endings in the heart travel in the vagi. They are concerned with the reflex control of blood pressure, blood flow, and heart rate. Pain fibers from blood vessels in the heart proceed to the sympathetic trunks and enter the spinal cord (by dorsal roots T1 to 5). **Cardiac pain is usually referred to the left shoulder and medial side of the left upper limb (ulnar distribution), although it may be felt in the chest.**

CARDIAC CYCLE

The contraction of the heart is termed *systole;* its relaxation is termed *diastole*. When the ventricles are filled, they begin to con-

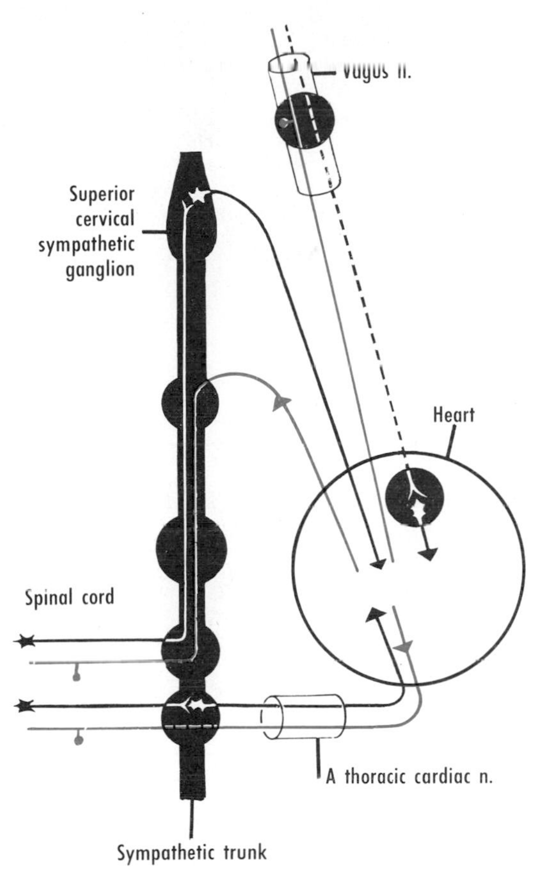

Figure 23–10 The sympathetic and afferent (*in blue*) fibers to the heart. For vagal fibers, see fig. 24–7.

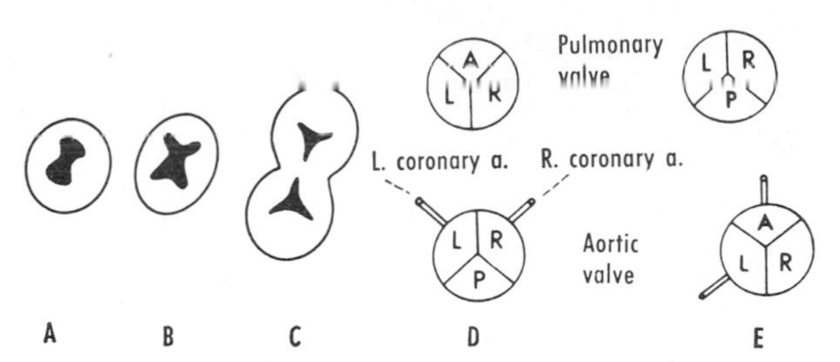

Figure 23–11 The partitioning of the truncus arteriosus, and the development of the semilunar valves. **A,** Truncus arteriosus. **B, C,** and **D,** An intermediate position during the rotation that the heart undergoes during development; the terminology used by some authors for naming the cusps is shown. **E,** The usual position of the valves of the adult heart *in situ;* the terminology used in this book is shown.

tract. The increased intraventricular pressure causes the A.V. valves to close, and the resultant vibrations are a major cause of the first heart sound. The A.V. valves are prevented from being pushed into the atria by the contraction of the papillary muscles. When intraventricular pressures surpass those in the aorta and pulmonary trunk, the semilunar valves open and blood is ejected into these arteries by the deep constrictor layer of myocardium. The highest pressure reached during the ejection phase is the *systolic blood pressure*. The closure of the semilunar valves is a major cause of the second heart sound. The ventricular musculature relaxes, and the intraventricular pressures become lower than those in the atria. The A.V. valves then open, and blood flows from atria to ventricles. The ventricles dilate as they fill (diastole), the arterial pressure reaches its minimum (*diastolic blood pressure*), and the atria contract. The electrical activity of the heart is recorded as an electrocardiogram (E.C.G. or E.K.G.). Diseased valves result in abnormal vibrations that may be heard as murmurs. If the heart stops beating, attempts can be made to start it by electrical stimulation, closed-chest cardiac massage, or open-chest cardiac massage.

Additional Features

The development of the heart (figs. 23–11 and 23–12), its congenital anomalies, and fetal circulation (fig. 23–13) are all important but complicated topics that should be studied in special books.

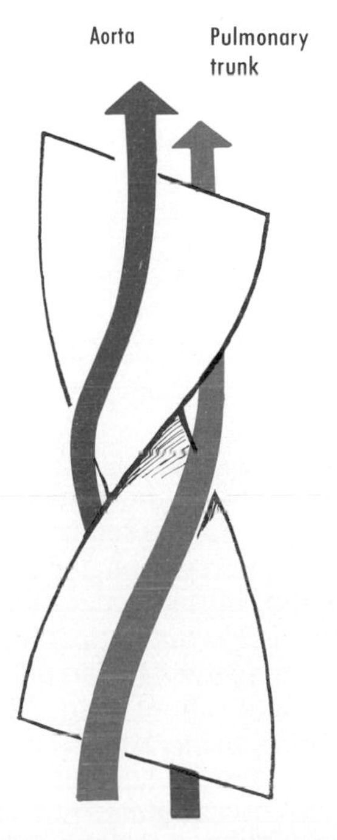

Figure 23–12 Scheme of the spiral septum in the truncus arteriosus and bulbus cordis.

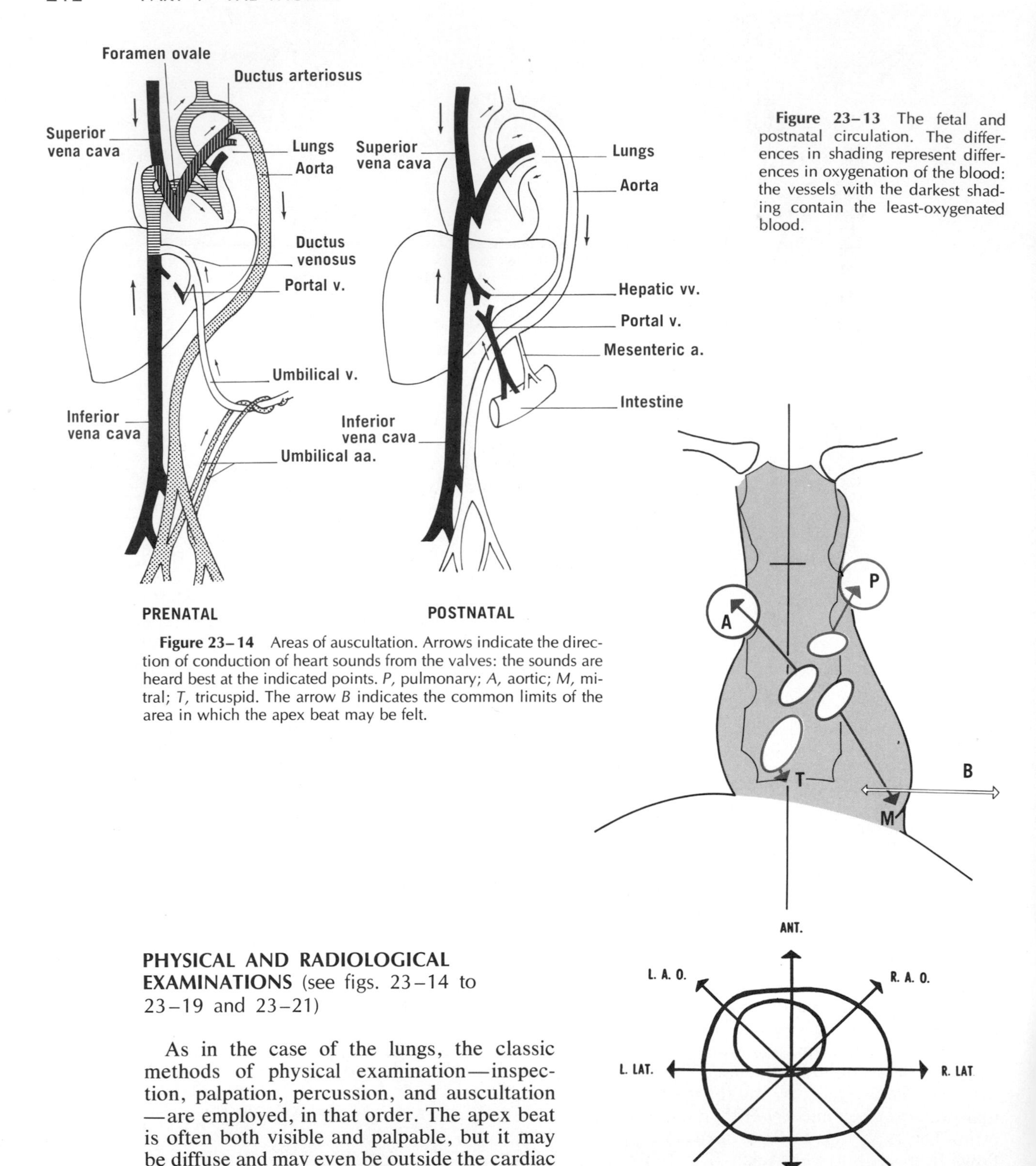

Figure 23–13 The fetal and postnatal circulation. The differences in shading represent differences in oxygenation of the blood: the vessels with the darkest shading contain the least-oxygenated blood.

Figure 23–14 Areas of auscultation. Arrows indicate the direction of conduction of heart sounds from the valves: the sounds are heard best at the indicated points. *P*, pulmonary; *A*, aortic; *M*, mitral; *T*, tricuspid. The arrow *B* indicates the common limits of the area in which the apex beat may be felt.

PHYSICAL AND RADIOLOGICAL EXAMINATIONS (see figs. 23–14 to 23–19 and 23–21)

As in the case of the lungs, the classic methods of physical examination—inspection, palpation, percussion, and auscultation—are employed, in that order. The apex beat is often both visible and palpable, but it may be diffuse and may even be outside the cardiac area. It is usually about 6 to 7 cm or more from the median plane and is generally in intercostal space 4 or 5. The heart rate is commonly 70 beats per minute, but considerable variation (50 to 90) occurs, and the neonatal rate is twice as fast. Percussion has limited

Figure 23–15 The usual views used in fluoroscopy and radiography are anterior, posterior, right anterior oblique (R.A.O.; fencing position), left anterior oblique (L.A.O.; boxing position), and sometimes right and left lateral. (Based on Zdansky.)

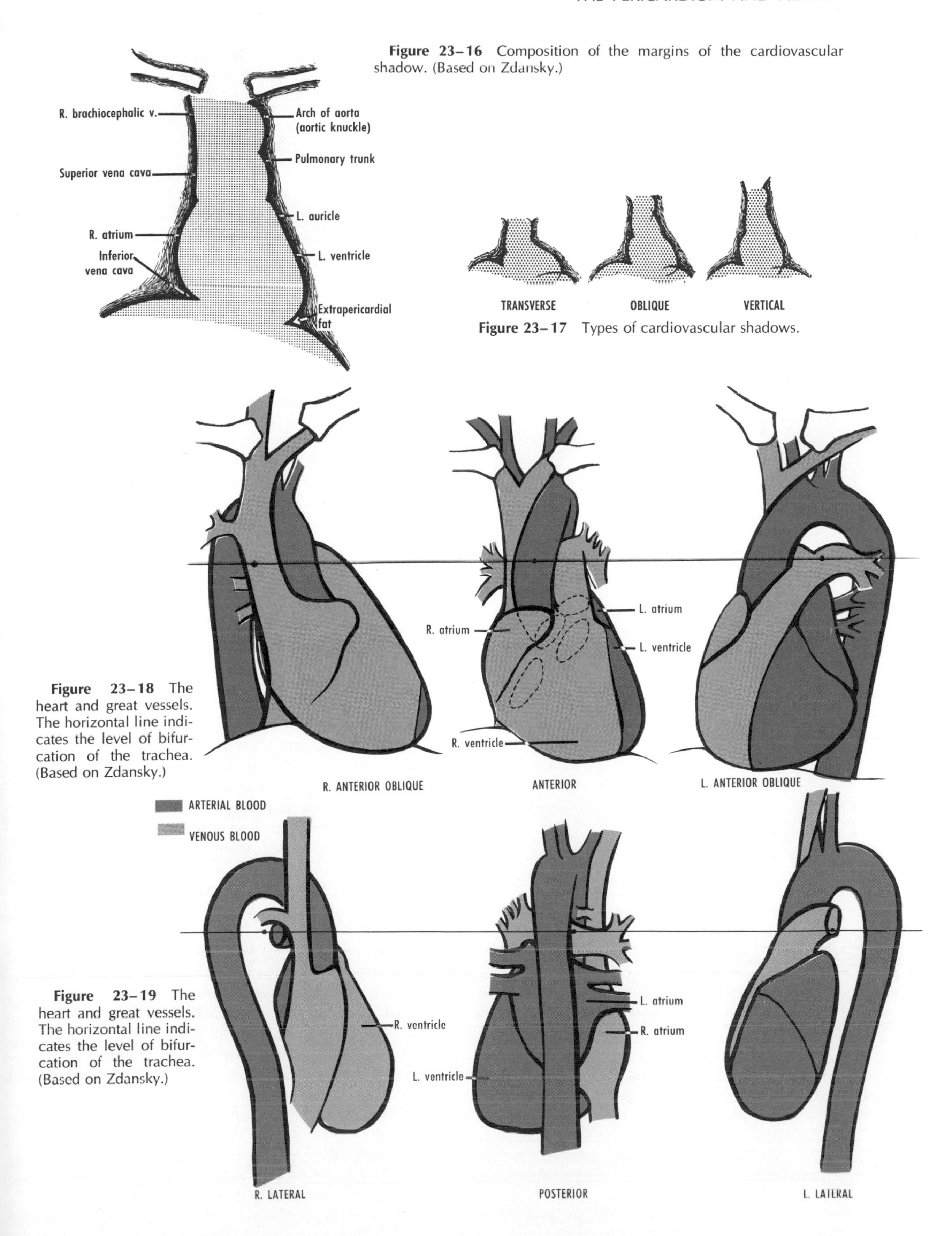

Figure 23–16 Composition of the margins of the cardiovascular shadow. (Based on Zdansky.)

Figure 23–17 Types of cardiovascular shadows.

Figure 23–18 The heart and great vessels. The horizontal line indicates the level of bifurcation of the trachea. (Based on Zdansky.)

Figure 23–19 The heart and great vessels. The horizontal line indicates the level of bifurcation of the trachea. (Based on Zdansky.)

Figure 23–20 The shape of the heart at maximum inspiration and maximum expiration.

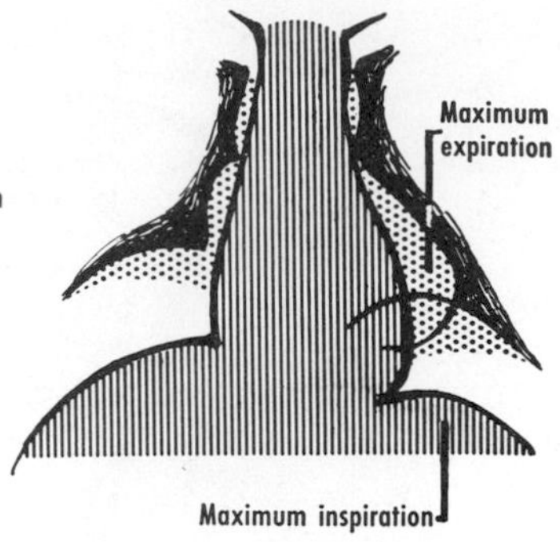

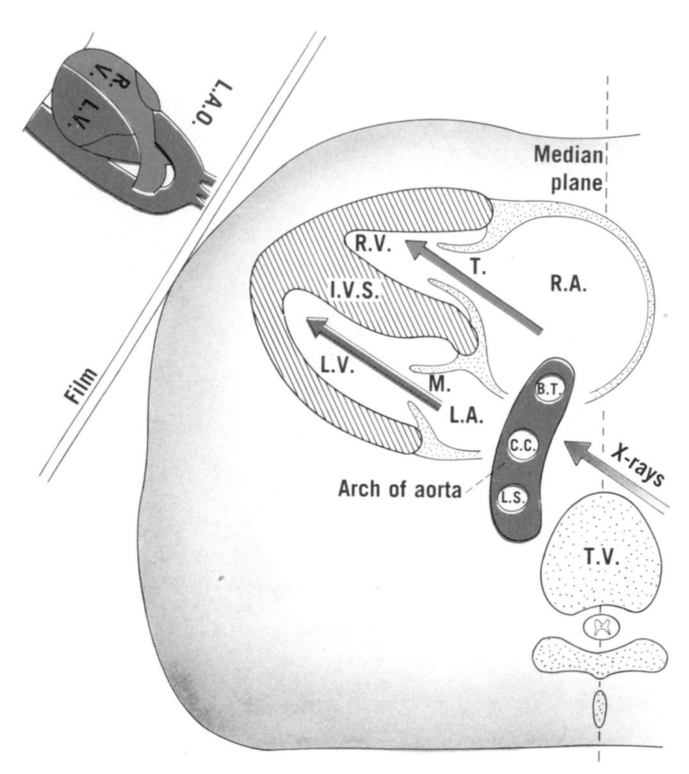

Figure 23–21 Horizontal section showing all four chambers of the heart. The left atrium is almost entirely on the back of the heart. The interventricular and interatrial septa are at approximately a 45-degree angle to the median plane. The "right heart" (*blue arrow*) lies in front of the "left heart" (*red arrow*). The outline of the arch of the aorta (with the origin of its three branches), as seen from above, has been included, although it would lie about three vertebrae higher than the plane of this section. The position of a film for a left anterior oblique (*L.A.O.*) projection is shown. In such a view (shown in miniature next to the film), the right and left portions of the heart would appear approximately equally separated along the line of the interventricular septum. Moreover, because the arch of the aorta passes almost directly backward, rather than transversely, its curve would appear well opened out. A valuable article on cardiac anatomy is that by R. Walmsley and H. Watson (Br. Heart J., *28*:435-447, 1966). *B.T.*, brachiocephalic trunk; *C.C.*, left common carotid artery; *I.V.S.*, interventricular septum; *L.A.*, left atrium; *L.S.*, left subclavian artery; *L.V.*, left ventricle; *M.*, mitral valve; *R.A.*, right atrium; *R.V.*, right ventricle; *T.*, tricuspid valve; *T.V.*, thoracic vertebra.

value in assessing the cardiac outline. In auscultation, the area of maximum intensity of the heart sounds for each valve corresponds not to the anatomical location of the respective valve but to the area where the cavity in which the valve lies is nearest the body surface, as far as possible from other valves, and "distal" to the valve with reference to the blood flow. **The areas of maximum audibility are: (1) for the pulmonary valve, over left intercostal space 2; (2) for the aortic valve, over right intercostal space 2; (3) for the mitral valve, over the apical region; and (4) for the tricuspid valve, over the lower part of the body of the sternum** (fig. 23–14).

The chief radiographic methods used in the examination of the heart and great vessels are fluoroscopy, radiography and teleradiography, and special methods such as angiocardiography, in which a radio-opaque medium is injected into a peripheral vein and followed through the heart. The commonest view is anterior (postero-anterior projection) (fig. 23–15). **The right border of the cardiovascular shadow is generally formed by the right brachiocephalic vein, superior vena cava, right atrium, and inferior vena cava. The left border is formed by the arch of the aorta (constituting a prominence known as the aortic knob, or knuckle), pulmonary trunk, left auricle, left ventricle, and extrapericardial fat** (fig. 23–16). The apical region is at the lower left part of the cardiac silhouette: if an apex is present, it is usually below the level of the diaphragmatic shadow. The cardiovascular shadow is generally oblique, but it may be transverse (especially in infancy, obesity, and pregnancy) or vertical (fig. 23–17). In a left anterior oblique (L.A.O.) view, the curve of the aorta is "opened" and the x-rays are frequently projected in the plane of the interventricular septum (figs. 23–18 and 23–21). Further views are shown in figure 23–19.

Cardiac position and configuration depend chiefly on the diaphragm, the position of which depends mainly on posture and respiration (fig. 23–20). When a subject is in an erect position, the heart lies at the level of T.V.7 to 10, and its lower border may be 5 cm below the level of the xiphosternal joint (see fig. 23–4). In recumbency, the heart rises about one vertebra. During inspiration, the heart appears more vertical (fig. 23–20), and the hili of the lungs are more readily seen.

CARDIOPULMONARY RESUSCITATION

Cardiopulmonary resuscitation (C.P.R.) involves the combined use of closed-chest manual heart compression and direct mouth-to-mouth breathing. The "ABCs" are to keep the *A*irway open and to see that *B*reathing is restored and *C*irculation is re-established. Details should be sought elsewhere (e.g., J. Am. Med. Assoc., *198:*372, 1966).

ADDITIONAL READING

Hurst, J. W. (ed.), *The Heart,* 4th ed., Blakiston (McGraw-Hill), New York, 1978. Contains several important chapters on anatomy.

James, T. N., *Anatomy of the Coronary Arteries,* Hoeber (Harper), New York, 1961. Provides details of injection-corrosion preparations.

McAlpine, W. A., *Heart and Coronary Arteries,* Springer, New York, 1975. Superb color photographs.

Milhiet, H., and Jager, P., *Anatomie et chirurgie du péricarde,* Masson, Paris, 1956. A detailed account of the pericardium.

Walmsley, T., "The Heart." Vol. 4, Part 3, of *Quain's Elements of Anatomy,* 11th ed., Longmans, Green, London, 1929. A classic.

QUESTIONS

23–1 What is the transverse sinus of the pericardium?

23–2 Where is the apex beat?

23–3 In which direction does blood flow from the atria to the ventricles?

23–4 What is the main component of the sternocostal surface of the heart?

23–5 What is the foramen ovale?

23–6 What is hepatocavo-atrial concordance?

23–7 What is the outflowing part of (a) the right ventricle and (b) the left ventricle termed?

23–8 Which important structure is situated at the lower border of the membranous part of the interventricular septum?

23–9 What is meant by dominance of the left coronary artery?

23–10 How does a left anterior oblique (L.A.O.) view show certain features to better advantage?

24
BLOOD VESSELS, LYMPHATIC DRAINAGE, AND NERVES OF THE THORAX

BLOOD VESSELS

PULMONARY CIRCULATION

The pulmonary trunk and arteries carry deoxygenated blood, but they are arteries in the sense that they transmit blood away from the heart at a relatively high pressure (20 to 30 mm of mercury) and in a pulsatile manner and in the sense that they have elastic walls like the aorta.

The *pulmonary trunk* extends from the conus arteriosus of the right ventricle to the concavity of the arch of the aorta, at the left of the ascending aorta, where it divides into the right and left pulmonary arteries. The point of division is approximately at the left side of the sternal angle. **The pulmonary arteries and their branches are largely responsible for the normal shadows seen radiographically in the roots and hili of the lungs.**

The *right pulmonary artery,* longer and wider than the left, passes below the arch of the aorta and enters the hilus of the lung (see fig. 22–4). The *left pulmonary artery* is connected to the arch of the aorta by the *ligamentum arteriosum* (fig. 24–1), which is the fibrous remains of a prenatal vessel, the *ductus arteriosus*.

A *pulmonary vein* arises in each lobe. The right upper and middle veins usually unite, so that four veins, upper and lower on each side, enter the left atrium.

SYSTEMIC CIRCULATION

The systemic supply of the thorax is derived mainly from branches of the *aorta,* the chief systemic artery of the body. The aorta is divided into the ascending aorta, arch of the aorta, and descending aorta. The part of the descending aorta in the thorax is called the thoracic aorta. The aorta is an elastic artery that withstands the systolic blood pressure and provides elastic recoil. The walls of the ascending aorta and arch contain pressoreceptors, which are connected to aortic depressor fibers in the vagi.

The *ascending aorta* begins in the root of the aorta, which presents the three aortic sinuses. Its branches are the right and left coronary arteries. It ascends to the level of the sternal angle.

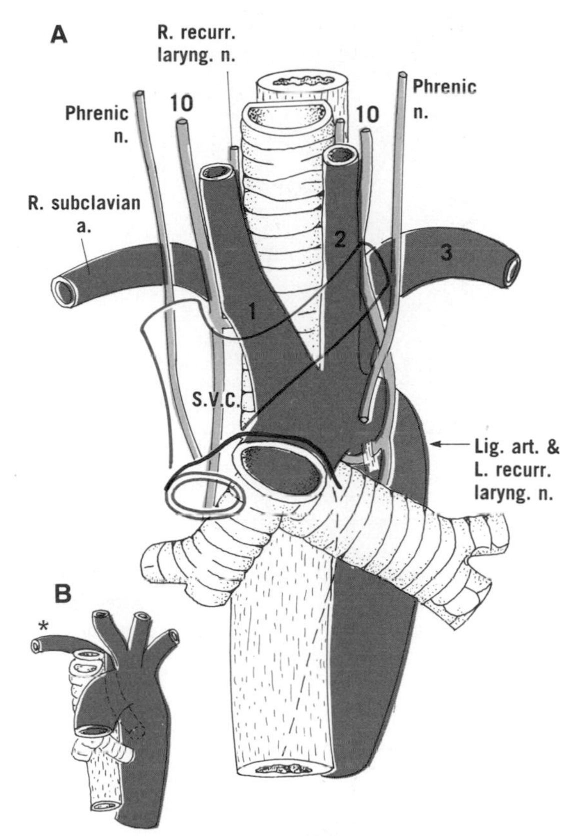

Figure 24–1 **A,** The three arteries (*1, 2, 3*) usually given off by the arch of the aorta. Also shown are the brachiocephalic and superior caval veins in outline (as if transparent) and the trachea, esophagus, and phrenic and vagus nerves. Note the different course of the recurrent laryngeal nerves on the two sides, the left nerve being related to the ligamentum arteriosum. Cf. fig. 50–17. **B,** An instance in which a retro-esophageal right subclavian artery (*asterisk*) arises as the last branch of the arch of the aorta.

The *arch of the aorta* (the term *aortic arch* is better reserved for special embryonic vessels) runs backward on the left side of the trachea and esophagus and above the left main bronchus (see figs. 22–5 and 24–1). **The arch of the aorta, in proceeding backward, occupies an almost sagittal plane in the superior mediastinum, behind the lower part of the manubrium sterni** (see fig. 23–21). It is visible radiographically as the aortic knob, or knuckle (see fig. 23–16). The arch is related below to the bifurcation of the pulmonary trunk and is connected to the left pulmonary artery by the ligamentum arteriosum. The left recurrent laryngeal nerve hooks below the arch. Above are the three main branches of the arch (crossed in front by the left brachiocephalic vein): the brachiocephalic trunk, left common carotid artery, and left subclavian artery. Immediately beyond the last-named branch, the aorta is slightly constricted (*isthmus*). A severe constriction (*coarctation* of the aorta) may occur here during development, in which case the adequacy of the collateral circulation depends on the relationship of the constriction to the opening of the ductus arteriosus (which connects the pulmonary trunk and aorta).

The *brachiocephalic trunk* divides behind the right sternoclavicular joint into the right subclavian and right common carotid arteries. The left common carotid and left subclavian arteries enter the neck behind the left sternoclavicular joint. Variations in the branches of the arch of the aorta may be encountered, e.g., a common origin of the brachiocephalic trunk and left common carotid artery. **Sometimes an aortic ring may encircle the trachea and esophagus and press on them. The right subclavian artery may arise from the thoracic aorta and be retro-esophageal** (fig. 24–1*B*), **which is frequently stated to cause dysphagia (difficulty in swallowing).**

The *thoracic aorta* descends in the posterior mediastinum and traverses the diaphragm to become the abdominal aorta. It begins at the left of the vertebral column, gradually reaches the front (where it lies behind the esophagus), and enters the abdomen in the median plane. The branches of the thoracic aorta are parietal and visceral. The parietal include several posterior intercostal arteries and the subcostal and some phrenic arteries. The visceral branches are the bronchial, esophageal, pericardial, and mediastinal.

Each *brachiocephalic vein* is formed by the union of the subclavian and internal jugular veins behind the corresponding sternoclavicular joint. The right vein descends vertically, whereas the left crosses obliquely in front of the branches of the arch of the aorta. Near the level of the sternal angle, the two brachiocephalic veins unite to form the *superior vena cava* (fig. 24–1). The superior vena cava descends on the right side of the ascending aorta, receives the azygos vein, and ends in the right atrium. Rarely, a left superior vena cava may persist: it comprises parts of the left brachiocephalic vein, left superior intercostal vein, oblique vein of the left atrium, and coronary sinus (fig. 24–2*B*).

The *azygos system* consists of veins on each side of the vertebral column, which drain the back as well as the walls of the thorax and abdomen. These veins are highly variable, but they end in the azygos, hemiazygos, and accessory hemiazygos veins (fig. 24–2). The *azygos vein* (Gk, *a zygon,* "unpaired") is formed by small vessels (such as the right sub-

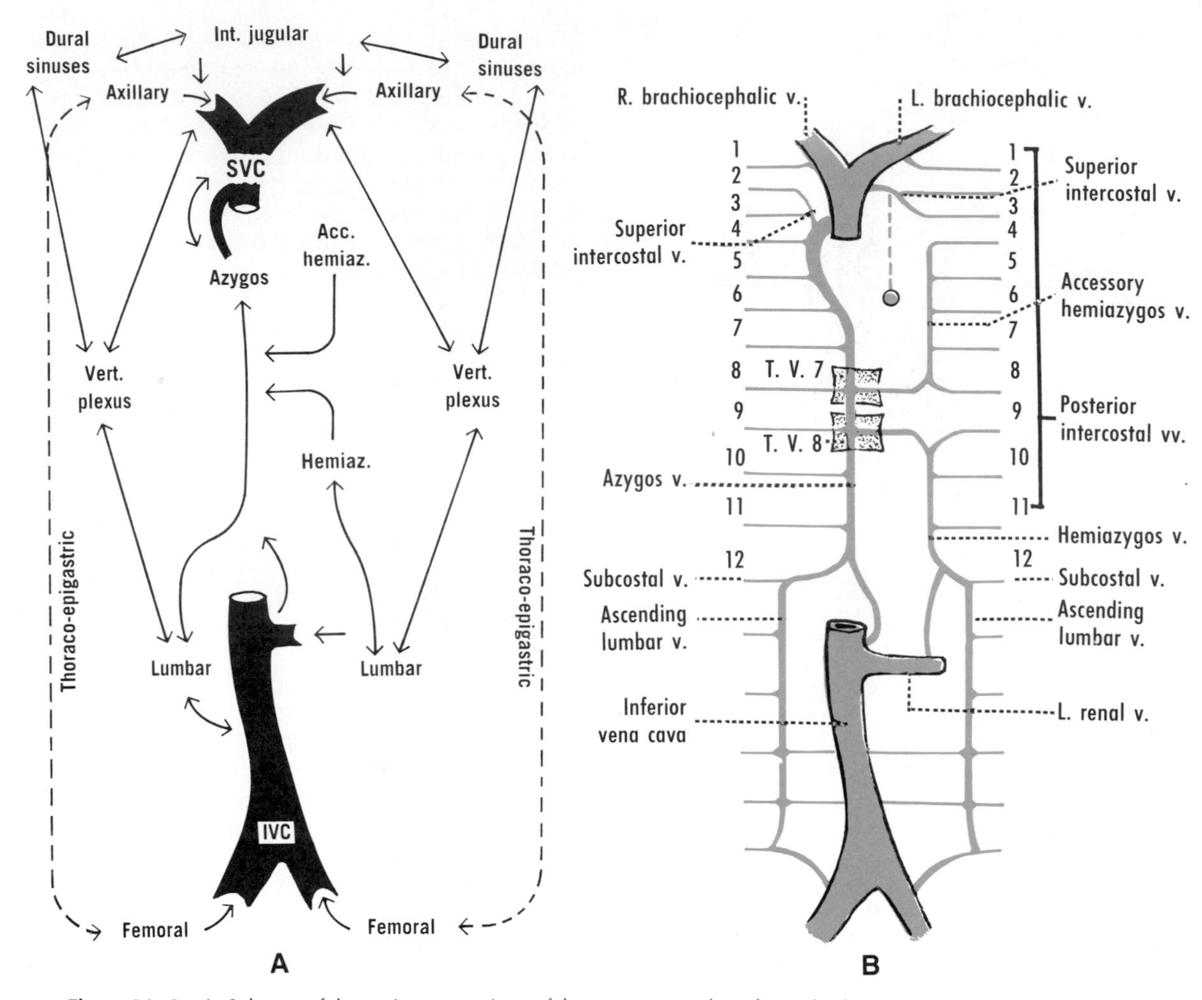

Figure 24–2 **A,** Scheme of the main connections of the azygos, caval, and vertebral systems of veins. Connections of the azygos and hemiazygos veins with the posterior intercostal veins also occur. **B,** The main veins of the thorax. An interrupted blue line indicates the course of a left superior vena cava (a rare anomaly) on its way to the coronary sinus.

costal and right ascending lumbar veins), and it ascends through the posterior and superior mediastina. It arches forward over the root of the right lung and ends in the superior vena cava. It receives the hemiazygos, accessory hemiazygos, and a number of posterior intercostal veins. The *hemiazygos* and *accessory hemiazygos veins* form a very variable arrangement on the left side. The hemiazygos vein arises in a manner comparable to that of the azygos. The accessory hemiazygos vein corresponds to the upper portion of the azygos vein.

The *vertebral venous system* consists of plexuses that drain the back, vertebrae, and structures in the vertebral canal. They communicate above with the intracranial veins and below with the portal system, and they empty into the vertebral, posterior interosseous, lumbar, and sacral veins. The veins in the vertebral plexus are valveless: blood may flow in either direction, and pressure in them is reflected in the cerebrospinal fluid. Reversed blood flow permits tumor cells to be transported from the breast, thorax, abdomen, or pelvis to the vertebrae, spinal cord, or brain. Veins of the thoracic wall (such as the *thoraco-epigastric veins,* which are superficially placed) connect the superior and inferior venae cavae and can provide a collateral circulation in obstruction of one of the venae cavae.

Extensive anastomoses among the caval, azygos, and vertebral systems provide multiple

routes for the return of blood to the heart. In effect, the azygos and vertebral systems bypass the caval system.

LYMPHATIC DRAINAGE

Lymph Nodes

The *parietal nodes* of the thorax are the parasternal, phrenic, and intercostal. The *visceral nodes* drain the lungs, pleurae, and mediastinum. The nodes in the roots and hili of the lungs are arranged in several groups: *pulmonary* along the larger bronchi, *bronchopulmonary* mainly at the hilus, and *tracheobronchial* near the bifurcation of the trachea (fig. 24–3). Lymph nodes in the roots of the lungs tend to be involved secondarily in infections, such as tuberculosis, and in tumors of the lungs and mediastinum. Their density may increase so that they become visible radiographically, especially if they become calcified. The tracheobronchial nodes drain into the *tracheal* (or *paratracheal*) *nodes*. *Mediastinal nodes* are scattered in the superior mediastinum, and they receive vessels from the thymus, pericardium, and heart. The efferents of the tracheal and mediastinal nodes form a bronchomediastinal trunk on each side of the trachea. There are also *posterior mediastinal nodes*, most of which drain directly to the thoracic duct.

Lymphatic Vessels

All of the lymphatic drainage of the thorax is directed toward the bronchomediastinal trunks, thoracic duct, and descending intercostal lymphatic trunks (fig. 24–3), but the actual lymphatic trunks themselves are highly variable.

The *thoracic duct* extends from the abdomen to the neck, where it ends in one of the large veins (figs. 24–3 and 24–4). It begins as either a plexus or a dilatation called the *cisterna chyli*, passes through or near the aortic opening of the diaphragm, and ascends in the posterior mediastinum between the aorta and the azygos vein. Next it crosses obliquely to the left, behind and then along the left side of the esophagus. Finally it passes behind the left subclavian artery, enters the neck (where it forms an arch above the level of the clavicle), and commonly ends in the left internal jugular vein (fig. 24–3). Variations are common. The thoracic duct receives the *left subclavian* and *jugular trunks* and often the *left bronchomediastinal trunk*.

Most of the lymph in the body reaches the venous system by way of the thoracic duct (figs. 24–4 and 24–5), but anastomoses are so extensive that no serious effects result if the thoracic duct is ligated.

On the right side, the *bronchomediastinal trunk* forms various combinations with the subclavian and jugular trunks. Rarely, all three unite to form a *right lymphatic duct*, which then empties directly into the junction of the internal jugular and subclavian veins.

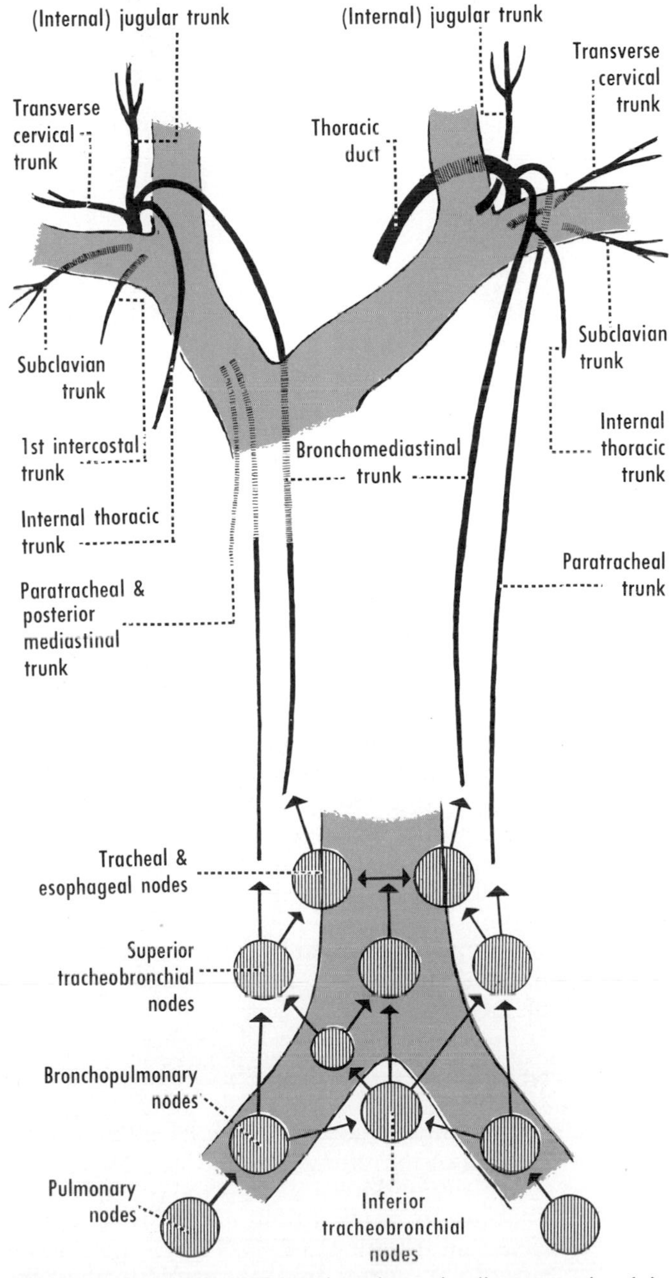

Figure 24–3 The visceral lymph nodes and collecting trunks of the thorax.

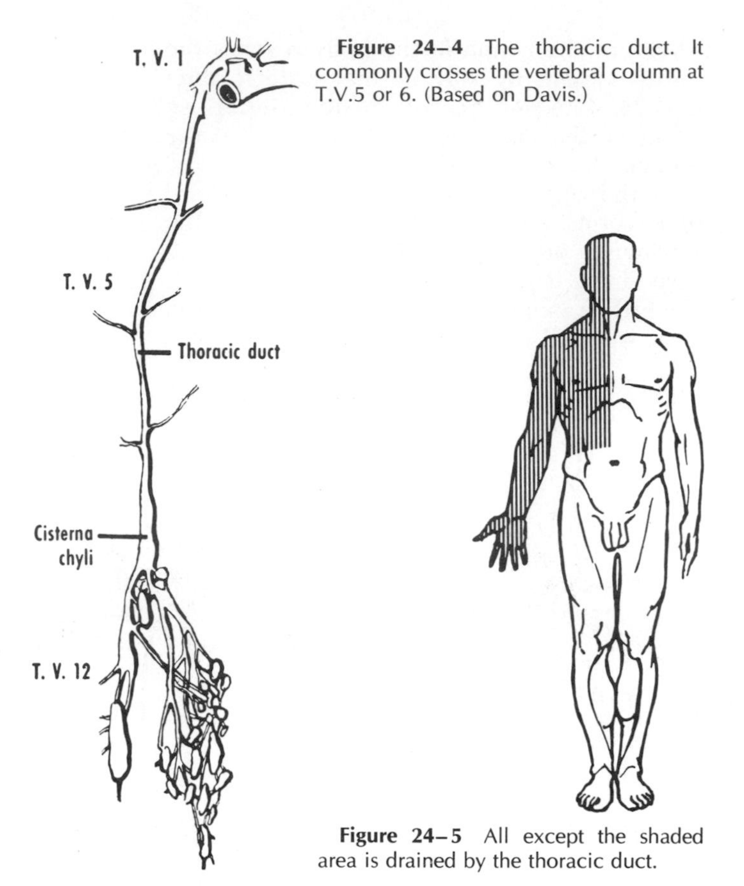

Figure 24–4 The thoracic duct. It commonly crosses the vertebral column at T.V.5 or 6. (Based on Davis.)

Figure 24–5 All except the shaded area is drained by the thoracic duct.

THYMUS

The thymus is partly in the neck and partly in the thorax. It comprises one to three lobes, each of which consists of numerous lobules containing lymphocytes, which are important in the development and maintenance of the immune system. The cervical part of the thymus lies on the front and sides of the trachea, whereas the thoracic part lies behind the upper portion of the sternum. The organ has a profuse blood supply and lymphatic drainage. The thymus reaches its greatest size at puberty and then begins to regress. Much of its substance is replaced by fat and fibrous tissue, but thymic tissue never disappears completely.

NERVES (figs. 24–1 and 24–6 to 24–8)

The thoracic nerves are described in the discussion of the thoracic wall.

The *phrenic nerves* supply the diaphragm. Each arises from usually C.N.(3), 4, and 5. (The contribution of C.N.5 to a phrenic nerve often arises from the nerve to the subclavius, and it may enter the thorax separately as the accessory phrenic nerve.) After crossing the front of the scalenus anterior in the neck, the phrenic nerves enter the thorax and pass in front of the roots of the lungs. Each gives branches to the pericardium and pleura and then divides into several branches, which pierce the diaphragm and supply that muscle, as well as part of the peritoneum, from below. The phrenic (fig. 24–6) nerves contain (1) motor fibers to the diaphragm, (2) pain fibers from the pericardium, pleura, and peritoneum, and (3) sympathetic vasomotor fibers. **Pain is usually referred to the skin (C.N.3 to 5) over the trapezius** and sometimes to the region of the ear (C.N.2, 3).

The *vagi* descend through the neck and enter the thorax, where they contribute to the pulmonary plexuses and then form the esophageal plexus. This is continued as *anterior* and *posterior vagal trunks,* which pass through the esophageal opening of the diaphragm. Each vagus has a recurrent laryngeal branch and a variable number of cardiac branches in the neck and thorax.

The *right recurrent laryngeal* nerve arises as the vagus crosses in front of the subclavian artery, hooks below that vessel, and ascends between the trachea and esophagus. The *left recurrent laryngeal* nerve arises as the vagus crosses the left side of the arch of the aorta, hooks below the arch (to the left of the ligamentum arteriosum), and then ascends on the right side of the arch between the trachea and esophagus. **The left recurrent laryngeal nerve is liable to damage from disorders of the aorta (e.g., aneurysms) or of the mediastinum (e.g., tumors), resulting in hoarseness.** It is believed that both recurrent laryngeal nerves owe their adult relationships to their embryonic arrangement caudal to the sixth aortic arches.

The vagi contain (1) parasympathetic fibers (e.g., to the heart), (2) sensory fibers (many of which are concerned with cardiovascular and pulmonary reflexes; others, in the mucosa of the bronchial tree, cause coughing), and (3) motor fibers to the pharynx and larynx (fig. 24–7), in the head and neck.

The *sympathetic trunks* (figs. 24–8 and 32–5) descend through the neck and enter the thorax, where they lie in front of the necks of the ribs (see figs. 20–2, 20–3, and 20–5). The thoracic part of each trunk has about a dozen ganglia, the first of which is often fused with the inferior cervical ganglion to form the *stellate ganglion*. The sympathetic trunks gain the abdomen by piercing the crura of the diaphragm or by passing behind the medial arcuate liga-

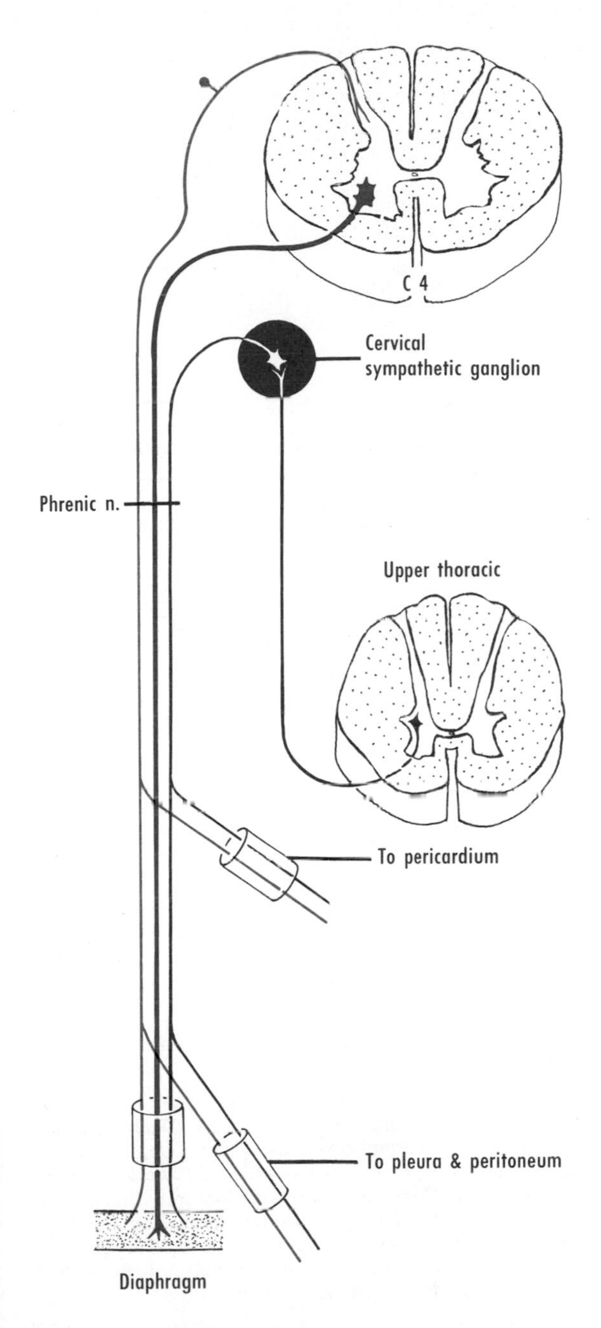

Figure 24–6 Functional components of the phrenic nerve. For purposes of simplification, each component is shown as a single fiber. Afferent (*blue*), efferent (*red*), and sympathetic (*black*) fibers are distinguished.

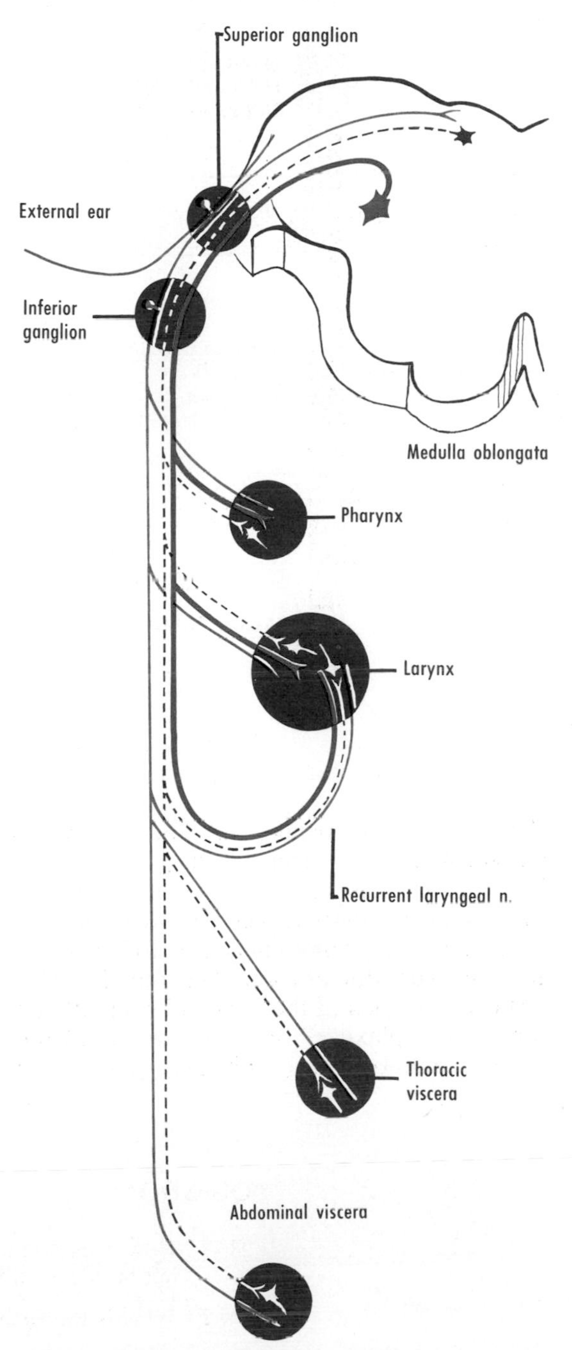

Figure 24–7 Functional components of the vagus nerve. For purposes of simplification, each component is shown as a single fiber. Afferent (*blue*), efferent (*red*), and parasympathetic (*interrupted*) fibers are distinguished. The distinction between accessory and vagal components is not shown.

ments. The trunks and ganglia are connected with the thoracic ventral rami by *rami communicantes,* which convey preganglionic and postganglionic fibers. Preganglionic fibers from segments T1 to 6 of the spinal cord supply the heart, coronary vessels, and bronchial tree. Apart from cardiac and pulmonary branches, the main visceral branches are the three *splanchnic nerves.* The *greater, lesser,* and *lowest splanchnic nerves* arise variably from the lower thoracic ganglia, pierce the crura of the diaphragm, and end in ganglia and plexuses (celiac and renal) in the abdomen.

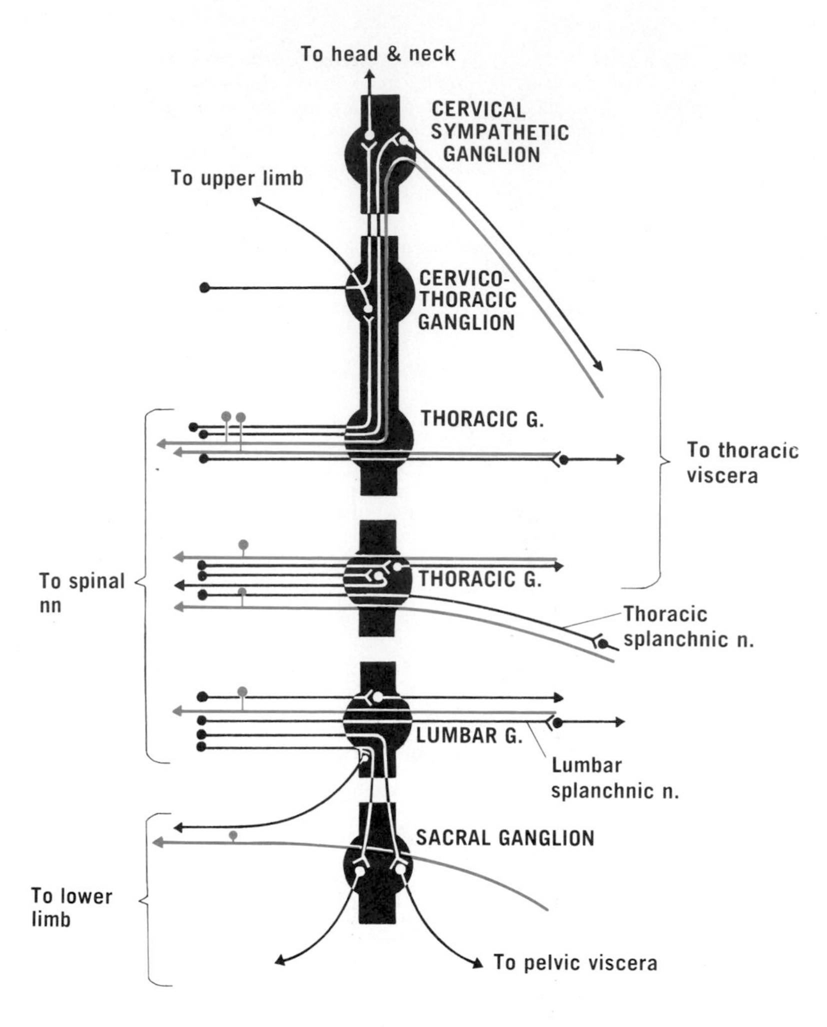

Figure 24–8 Functional components of the sympathetic trunk, showing the origin of preganglionic fibers and the sites of ganglion cells. Afferent (mostly pain) fibers from viscera are indicated in blue.

The sympathetic trunks and their branches contain pain fibers from the thoracic and abdominal viscera and from blood vessels. These sensory fibers traverse the sympathetic trunks and rami communicantes to reach the spinal nerves, dorsal roots, and spinal cord.

Many branches of the vagi and sympathetic trunks form plexuses in the thorax, e.g., the cardiac, pulmonary, esophageal, and aortic plexuses.

ADDITIONAL READING

Clemens, H. J., *Die Venensysteme der menschlichen Wirbelsäule,* de Gruyter, Berlin, 1961. Well-illustrated account of vertebral venous system.

Jacobsson, S.-I., *Clinical Anatomy and Pathology of the Thoracic Duct,* Almquist and Wiksell, Stockholm, 1972. An excellent, concise monograph on anatomy.

Pontes, A. de P., *Arterias supra-aorticas,* University of Brazil, Rio de Janeiro, 1963. A study of the arch of the aorta and variations in its branches.

QUESTIONS

24–1 How does (a) the right pulmonary artery differ from the left and (b) the left bronchus differ from the right?

24–2 What is the ligamentum arteriosum?

24–3 Which structure hooks below the arch of the aorta?

24–4 What is a retro-esophageal right subclavian artery?

24–5 Which regions of the body are drained by the thoracic duct?

24–6 What is the origin of the phrenic nerves?

Part 5

THE ABDOMEN

The trunk comprises the thorax, abdomen proper, pelvis, and back. (A portion of the back extends into the neck). The terms *abdomen* and *abdominal cavity* are frequently used to include the pelvis and pelvic cavity, respectively. The abdominal cavity proper is separated from the pelvic cavity below and behind by an arbitrary plane passing through the terminal lines of the bony pelvis. A considerable part of the abdominal cavity lies under cover of the thoracic bony cage. The abdominal cavity contains most of the organs of the digestive system, the spleen, the suprarenal glands, and the peritoneum. Certain organs may, at certain times, pass from the abdomen to the pelvis or vice versa.

A subdiaphragmatic thrust (applied quickly above the umbilicus) can force air from the lungs sufficiently to expel an obstructing object in the airway (Heimlich procedure).

EXAMINATION OF THE ABDOMEN

The classic methods of clinical examination are inspection, palpation, percussion, and auscultation. On *inspection,* the umbilicus, the rectus muscles, the respiratory movements, and the relative levels of the testes are noted. For *palpation,* the subject should be on a couch and the examiner's hands should be warm. The flat of the hand is used. The abdominal muscles must be relaxed. **On deep palpation, the following may sometimes be identified in normal subjects: pulsations of the abdominal aorta, the lumbar vertebrae, the lower pole of the right kidney, possibly the liver, and occasionally the spleen.** Stroking the skin with a sharp point may induce contraction of the abdominal muscles (superficial abdominal reflex). Stroking the medial side of the thigh may induce elevation of the testis (cremasteric reflex). The superficial inguinal ring is examined with the subject erect by invaginating the scrotum with the little finger. The body of the uterus can be palpated bimanually, i.e., through the anterior abdominal wall and *per vaginam.* On *percussion,* a tympanitic note is obtained over the alimentary canal, whereas dullness is found over the liver, the spleen, and a full bladder. *Auscultation* is used chiefly to hear fetal heart sounds during pregnancy and bowel sounds in assessing intestinal peristalsis.

RADIOLOGICAL ANATOMY

An anterior radiogram ("scout" or survey film) of the abdomen shows the lower ribs, lumbar vertebrae, ilia, and sacro-iliac joints (see fig. 25–2). The shadow of the liver is usually evident, and it demarcates the domes of the diaphragm. The spleen, psoas muscles, and kidneys are often identifiable. Air in the fundus of the stomach and gas bubbles in the large intestine can usually be seen.

In addition to routine fluoroscopy and radiography, contrast media are frequently employed, e.g., a barium suspension (orally or by enema) for the alimentary canal (see figs. 27–2, 27–3, and 27–9) and organic iodides (orally, intravenously, or by injection into the bladder and ureters) to demarcate the calices of the kidney (see fig. 29–3). Radio-isotopic scanning (e.g., of the liver), computerized axial tomography, and ultrasound are being used with increasing frequency.

25
ABDOMINAL WALLS

SURFACE LANDMARKS

The xiphosternal joint is at the apex of the infrasternal angle, which is formed by costal cartilage 7 on each side. The xiphoid process extends into the angle, and the slight depression of the anterior abdominal wall in front of it is the *epigastric fossa* (or "pit of the stomach"). The costal margin is formed by costal cartilages 7 to 10.

The whole of the iliac crest is usually palpable. Its highest part is situated somewhat posteriorly. The anterior superior iliac spine is frequently visible, and the posterior superior iliac spine is usually marked by a dimple. The pubic tubercle is about 2 to 3 cm lateral to the median plane.

The linea alba (a median furrow) and the linea semilunaris (lateral border of the rectus muscle) are usually evident in lean, muscular individuals on contraction of the abdominal wall. The umbilicus is usually between L.V.3 and L.V.5, but its position is highly variable.

The inguinal ligament is in the groin and extends from the anterior superior iliac spine to the pubic tubercle. The skin crease at the junction of the abdomen and thigh lies immediately below and parallel to the inguinal ligament. The deep inguinal ring is immediately above the midinguinal point, and the superficial inguinal ring is about 1 cm above and lateral to the pubic tubercle.

PLANES (fig. 25–1)

The two most important abdominal planes are the transpyloric and the supracristal. **The transpyloric plane is a horizontal plane halfway between the jugular notch of the sternum and the pubic symphysis, at the level of L.V.1. The pyloric part of the stomach may be, but is not necessarily, on the transpyloric plane. The supracristal plane is a horizontal plane through the highest points of the iliac crests, at the level of the spinous process of L.V.4.**

The *subcostal plane* is a horizontal plane through the lowest points of the costal margin. The *transtubercular plane* is a horizontal plane through the tubercles of the iliac crests. The right and left *lateral planes* are sagittal planes midway between the anterior superior iliac spine and the pubic symphysis, i.e., through the midinguinal point (slightly medial to the middle of the inguinal ligament).

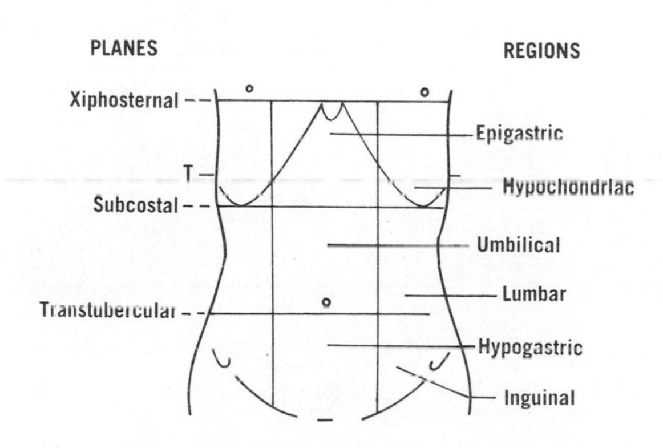

Figure 25–1 Chief planes and classic regions of the abdomen. The right and left lateral planes cross the costal margins. *T*, transpyloric plane.

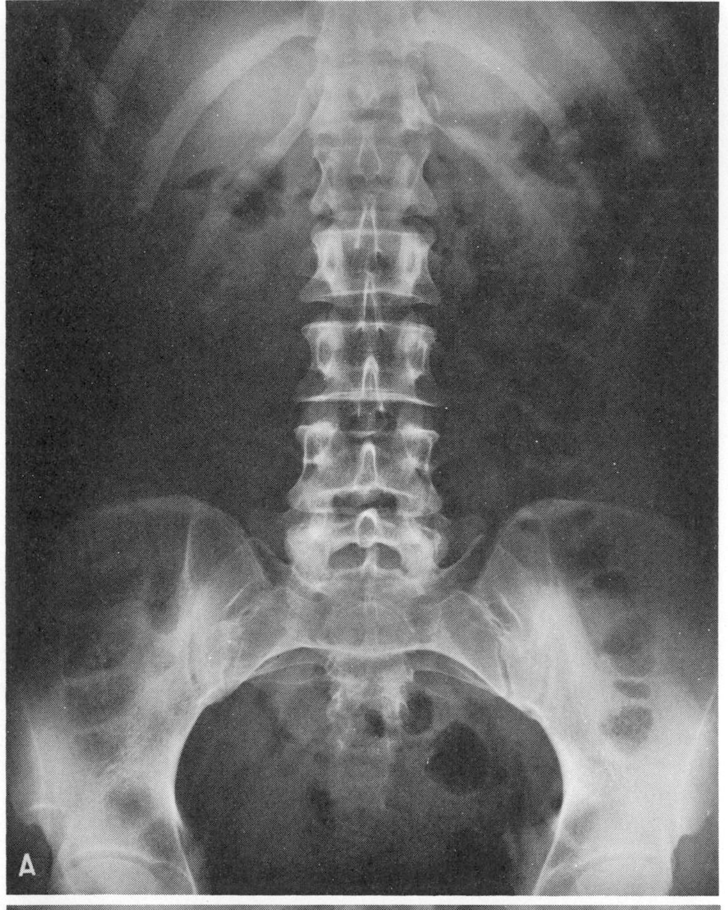

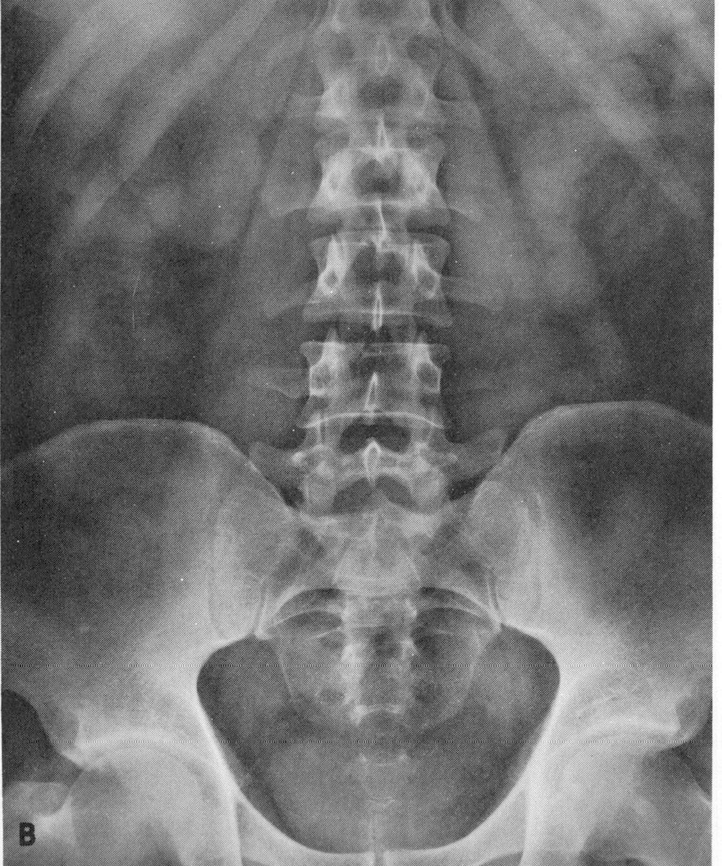

REGIONS

In examining and describing the abdomen, **it is customary to relate pains, swellings, or the positions of organs to one of nine regions,** as shown in figure 25–1. The transpyloric plane is sometimes used instead of the subcostal plane. Moreover, variations in the costal margins and their relationship to the iliac crests are not unusual, so that some prefer to divide the linea alba into thirds, whereas others prefer to **divide the abdomen merely into quadrants by using the median plane and the umbilicus.**

In front, the abdominal wall includes the rectus abdominis muscles, and the aponeuroses of the three muscles (the external and internal oblique and the transversus) that form much of the side of the abdomen. Behind, the lumbar vertebrae and intervertebral discs project forward (fig. 25–2). Most of the abdominal wall is layered in the following manner: (1) skin, (2) subcutaneous tissue, (3) muscles and fasciae (or bone), (4) extraperitoneal tissue, and (5) peritoneum. The superficial part of the subcutaneous tissue is fatty; the deep part is more membranous (i.e., collagenous) and is anchored to the inguinal ligament and fascia lata, thereby limiting the spread of urine into the thigh after rupture of the urethra below the urogenital diaphragm.

Figure 25–2 **A,** Survey, or "scout," film of the abdomen, with emphasis on the lumbar vertebrae. Note the twelfth ribs, the bodies and transverse and spinous processes of vertebrae (L.V.4 spinous process is on the supracristal plane), and gas in the large intestine, particularly in the descending colon. **B,** Survey film with emphasis on soft tissues. Note the kidneys, psoas major muscles, and urinary bladder. The twelfth ribs are much shorter than in **A.** Note the transverse processes of the lumbar vertebrae and the sacro-iliac joints, sacrum, and coccyx.

TABLE 25–1 MUSCLES OF THE ANTEROLATERAL ABDOMINAL WALL

Muscle	Origin	Insertion	Innervation	Action
External oblique	External surfaces of ribs 5-12	Linea alba, pubic tubercle, & iliac crest	Thoraco-abdominal & subcostal	Retain viscera, increase intra-abdominal pressure, and act in forced expiration
Internal oblique	Thoracolumbar fascia & iliac crest	Ribs 10-12, rectus sheath & linea alba, & pubis		
Transversus abdominis	Costal cartilages 7-12, thoracolumbar fascia, & iliac crest	Linea alba & pubis		
Rectus abdominis	Pubic crest & symphysis	Costal cartilages 5-7 and xiphoid process		Flexes trunk
Pyramidalis	Body of pubis	Linea alba	Subcostal	Tenses linea alba

ANTEROLATERAL ABDOMINAL WALL

MUSCLES (figs. 25–3 to 25–7)

In front are the rectus abdominis and pyramidalis; anterolaterally are the external and internal oblique muscles and the transversus (table 25–1).

The fibers of the rectus run vertically. In general, those of the external oblique muscle (cf. the external intercostal muscles) run downward and forward (as in inserting a hand in a pocket), those of the internal oblique muscle (cf. the internal intercostal muscles) go mostly upward and forward, and those of the transversus pass transversely.*

External Oblique Muscle. The aponeurosis of the *external oblique* muscle passes in front of the rectus abdominis. Its lower edge extends from the anterior superior iliac spine to the pubic tubercle and is known as the *inguinal ligament.* A part that continues horizontally backward is termed the *lacunar ligament.* A further extension laterally is called the *pectineal ligament* (fig. 25–7).†

Lateral to the pubic tubercle, the aponeurosis of the external oblique muscle divides into medial and lateral crura, which diverge to form the superficial inguinal ring. **The superficial inguinal ring may be found by pushing the**

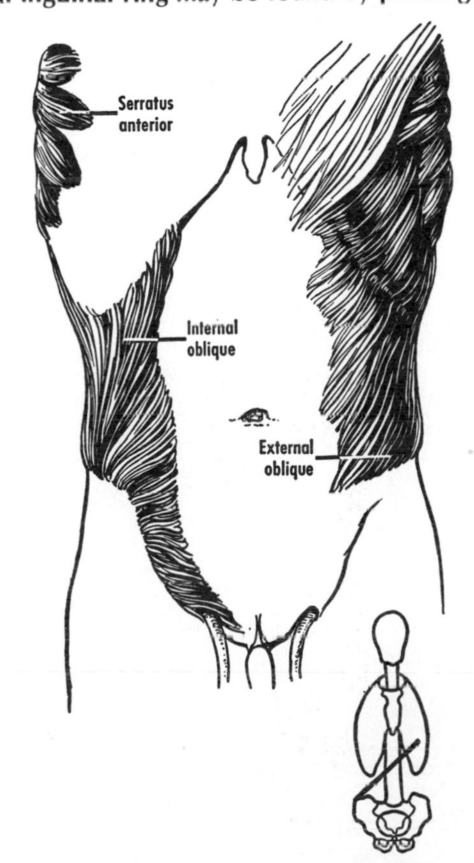

Figure 25–3 External and internal oblique muscles. The fibers of one external oblique muscle are approximately parallel to those of the opposite internal oblique muscle. The lower drawing shows the line of pull of the left external oblique and the right internal oblique muscles. These muscles, acting together, flex and rotate the trunk.

* A more complicated arrangement of the anterior abdominal wall has been proposed (N. N. Rizk, J. Anat., *131*:373-385, 1980). Each abdominal aponeurosis is said to be bilaminar, and the six layers of each side are oblique and cross the median plane in a common area of decussation, namely the linea alba.

† The traditional account of inguinal anatomy is not confirmed by recent authors. The inguinal "ligament," which shows no ligamentous thickening, is complicated (J. F. Doyle, J. Anat., *108*:297-304, 1971). W. J. Lytle maintains that the curved edge of the lacunar ligament is fixed around the medial wall of the femoral sheath and that the femoral canal is *not* bounded medially by the curved edge, is *not* a vertical funnel, and does *not* enter the thigh (J. Anat., *128*:581-594, 1979).

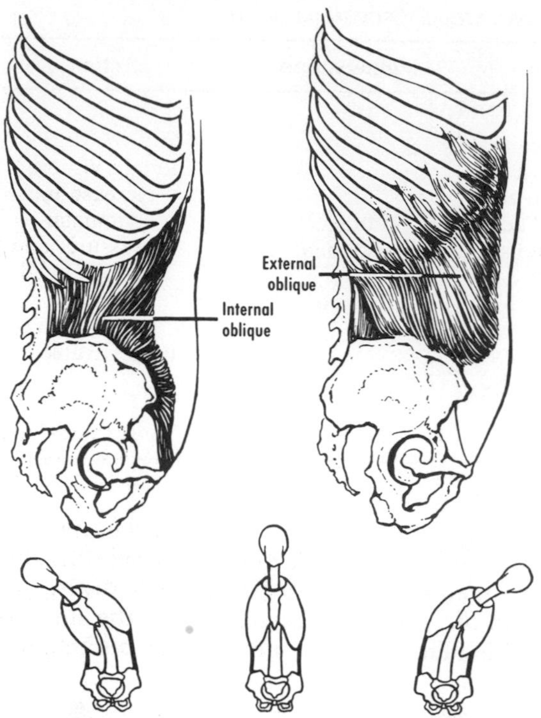

Figure 25–4 Internal and external oblique muscles. The lower drawings show the external and internal oblique muscles of one side acting together in bending the trunk toward that side.

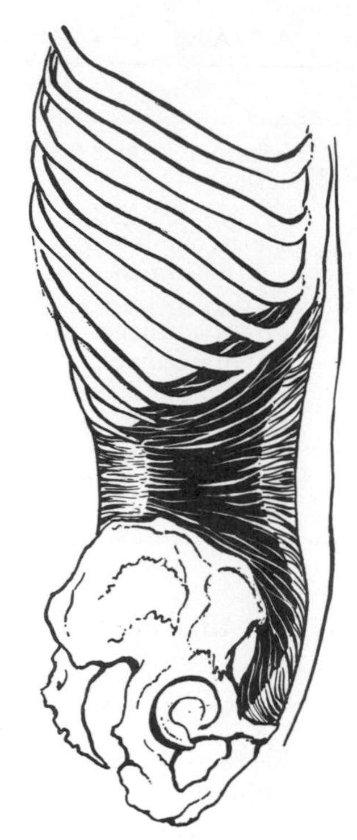

Figure 25–5 The transversus abdominis.

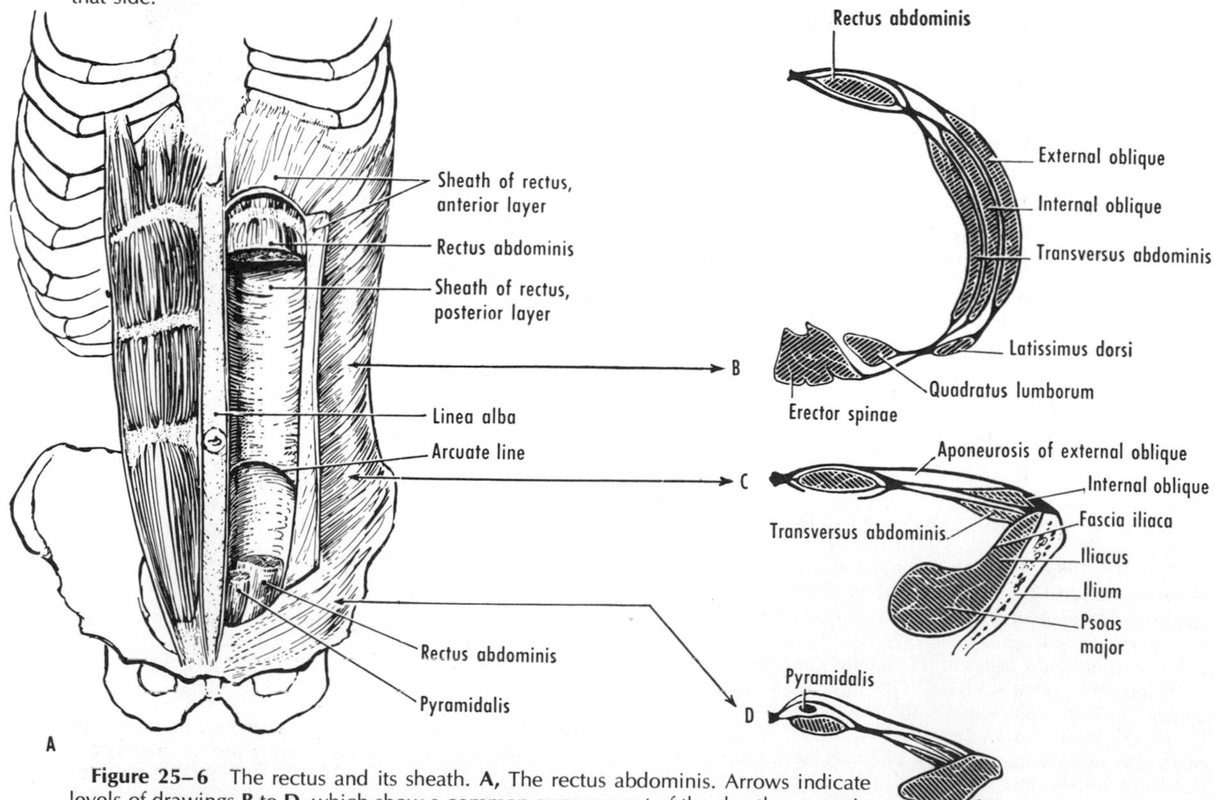

Figure 25–6 The rectus and its sheath. A, The rectus abdominis. Arrows indicate levels of drawings B to D, which show a common arrangement of the sheath as seen in horizontal section. The fascia transversalis is not shown separately from the transversus aponeurosis.

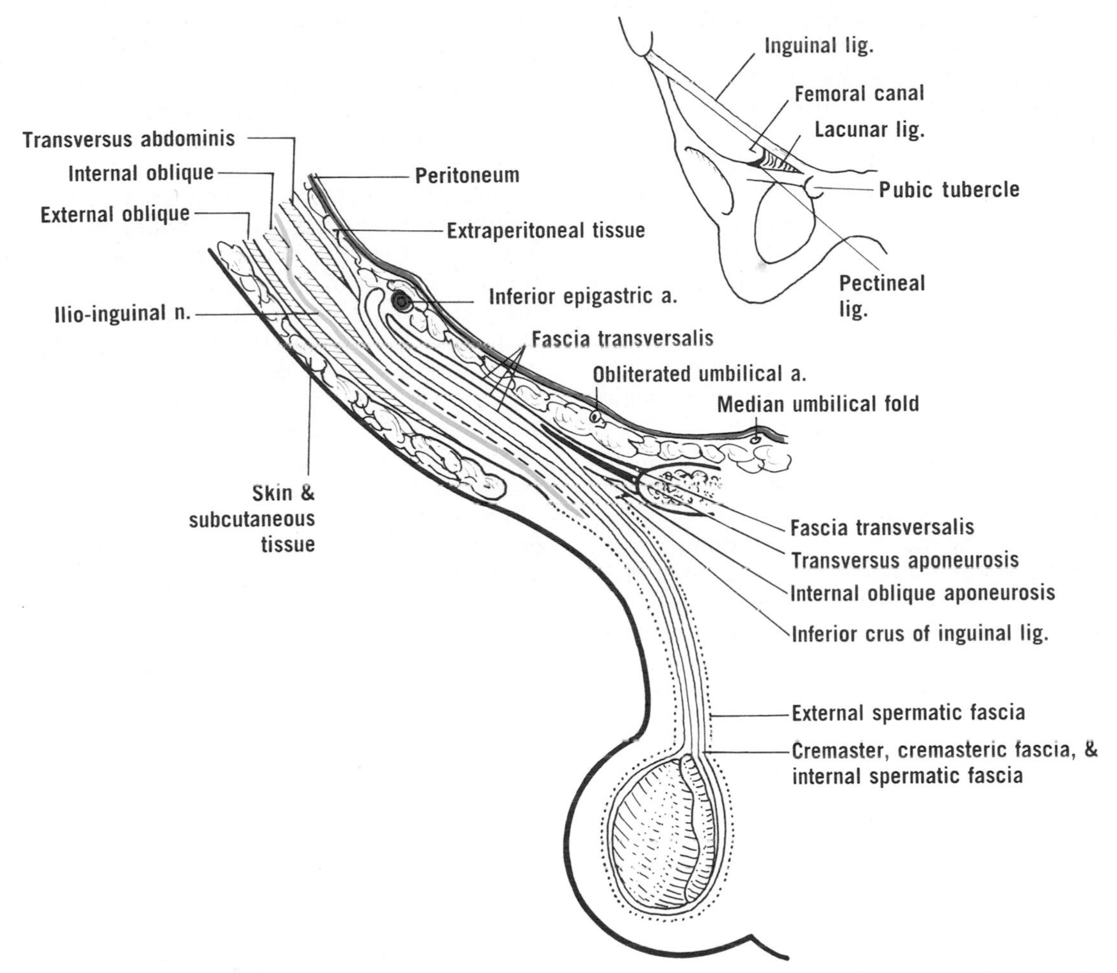

Figure 25–7 Inguinal ligament (*inset*), inguinal canal, and layers of scrotum. The main diagram combines two planes: that of the inguinal ligament and a sagittal plane through the scrotum. The separation of the layers is exaggerated. The fascia transversalis forms most of the posterior wall of the inguinal canal and is prolonged along the ductus deferens as the internal spermatic fascia.

scrotal skin upward along the spermatic cord to a point immediately above the pubic tubercle, and then passing the finger backward. The ring normally admits the tip of the little finger.

Internal Oblique Muscle. The aponeurosis of the *internal oblique* muscle divides into anterior and posterior layers, which pass, respectively, in front of and behind the rectus to reach the linea alba. The *linea alba* is the median, fibrous intersection of the aponeuroses between the xiphoid process and pubic symphysis. The division into anterior and posterior layers is absent below, where the aponeuroses of all three muscles pass in front of the rectus to gain the linea alba.

Transversus Abdominis. The fascia on the inner surface of the *transversus abdominis* serves as epimysium and is known as the *fascia transversalis*. It passes behind the rectus sheath and crosses the median plane. It is continuous with the general fascia of the abdomen to the extent that it is regarded by some as a part of the extraperitoneal connective tissue. Below, the medial portion of the fused internal oblique and transversus aponeuroses is termed the *conjoined tendon*.

Rectus Abdominis. The *rectus sheath* (fig. 25–6) is formed mainly by the aponeuroses of the internal oblique muscle and the transversus. These two aponeuroses meet at the lateral edge of the rectus along a curved line termed the *linea semilunaris,* which extends from costal cartilage 9 to the pubic tubercle and is often visible *in vivo*. Above, the transversus passes behind the rectus, whereas below, it passes in front. The crescentic lower border of the part behind the rectus is called the *arcuate line*. Its level is variable, but below it the rectus lies on the fascia transversalis posteriorly. The conjoined tendon (fused internal oblique and transversus aponeuroses) continues medially in front of the rectus sheath. In summary, the sheath of the rectus abdominis is variable but is classically described as consisting of anterior and posterior layers. Above the arcuate line, the anterior layer is formed by the aponeuroses of the internal and external oblique muscles; below the arcuate line, by all three aponeuroses. Above the arcuate line, the posterior layer is formed by the aponeuroses of the internal oblique muscle and the transversus; below the arcuate line, the sheath is deficient posteriorly. In paramedian abdominal incisions, when the sheath is incised, the rectus may be displaced laterally (or medially).

Actions. The muscles of the abdominal wall protect the viscera and help to maintain or to increase intra-abdominal pressure. They also move the trunk and help to maintain posture. The recti flex the trunk against resistance, and they can be tested by having a supine subject flex the trunk without using the arms. The obliqui and transversi increase intra-abdominal pressure and hence are important in respiration, defecation, micturition, parturition, and vomiting. The obliqui also aid in movements of the trunk.

Inguinal Canal. **The inguinal canal is an oblique passage through the abdominal wall. It is occupied by the spermatic cord or by the round ligament of the uterus,** and it contains the ilio-inguinal nerve. It is about 3 to 5 cm long (figs. 25–7 to 25–10 and table 25–2). The canal is potentially a weak area through which an inguinal hernia may occur. Many divergent viewpoints exist concerning the canal and its surgery, largely because of variations and a confusing terminology.

The *ductus deferens* hooks around the lateral side of the inferior epigastric artery and is joined by nerves and vessels embedded in extraperitoneal connective tissue to form the *spermatic cord.* Above the midinguinal point, the cord traverses the deep ring. **The deep inguinal ring is a slit-like opening in the fascia**

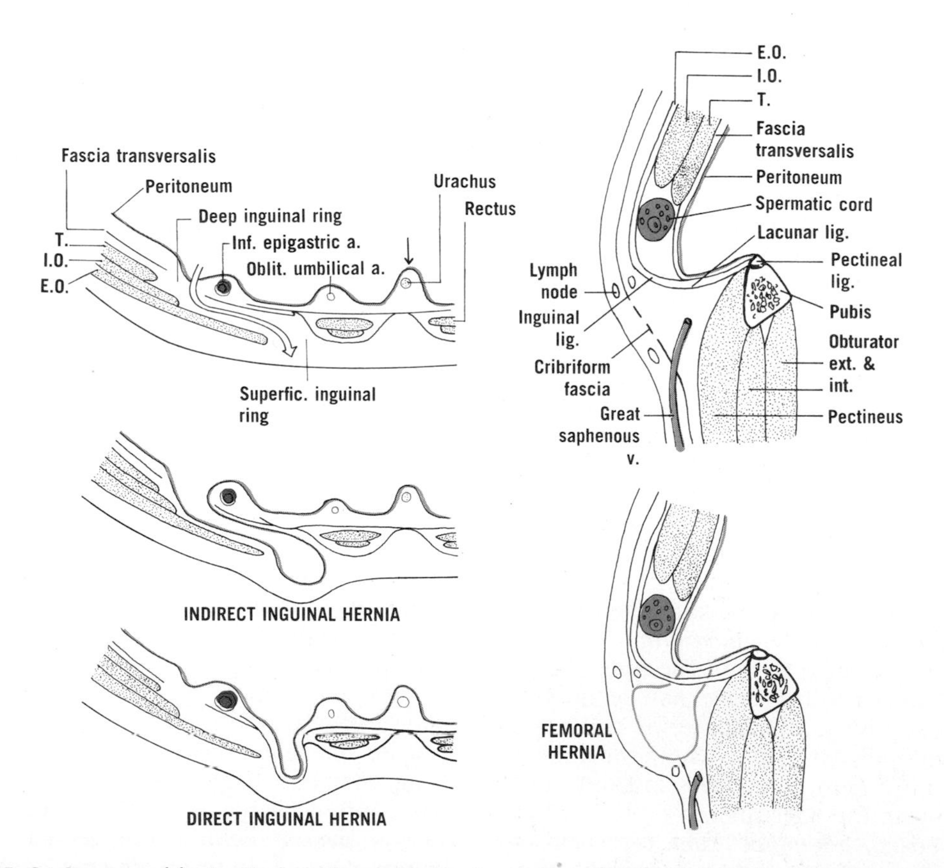

Figure 25–8 Structure of the inguinal and femoral regions in relation to hernia. The broad arrow represents the course of the spermatic cord. The small arrow indicates the median plane. (After Maisonnet and Coudane. The sagittal sections are based largely on Lytle.)

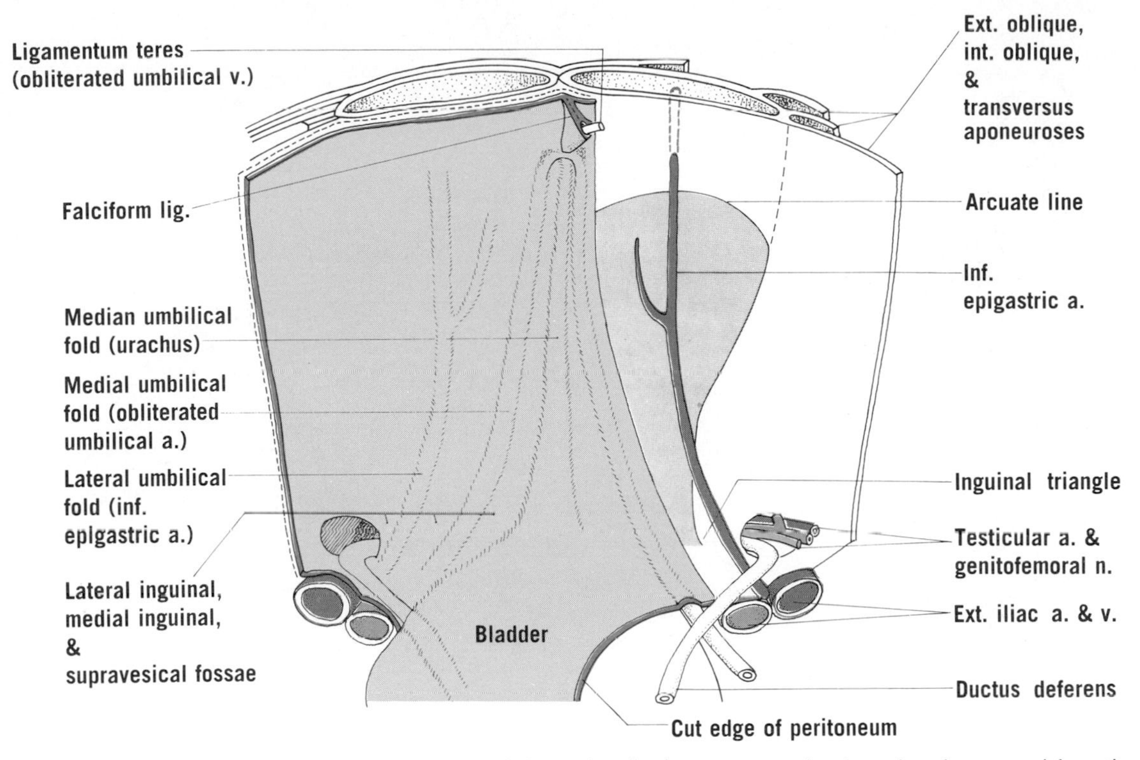

Figure 25–9 The posterior aspect of the anterior abdominal wall. The peritoneum has been largely removed from the right side. A direct inguinal hernia would enter the inguinal canal through its posterior wall, medial to the inferior epigastric artery.

transversalis. The cord then runs medially and downward in the inguinal canal and emerges through the superficial ring. As the spermatic cord traverses the muscular part of the internal oblique muscle, it acquires a covering of *cremaster muscle* and *cremasteric fascia*. **The superficial inguinal ring is a triangular opening in the aponeurosis of the external oblique muscle** (fig. 25–10*A*). In its course through the canal, the spermatic cord acquires sheaths from each of the layers of the abdominal wall. The posterior wall of the inguinal canal is formed by the fascia transversalis and the aponeurosis of the transversus. The anterior wall is formed by the aponeurosis of the external oblique muscle and, laterally, by internal oblique muscular fibers. Above are the arching fibers of the internal oblique muscle and the transversus. The floor is formed by the inguinal and lacunar ligaments. The inferior epigastric vessels lie behind the canal, immediately medial to the deep ring. The inferior epigastric vessels laterally, the lateral border of the rectus medially, and the inguinal ligament inferiorly form the *inguinal triangle* (fig. 25–9).

The chief protection of the inguinal canal is muscular. The muscles that increase intra-abdominal pressure and tend to force abdominal contents into the canal at the same time tend to narrow the canal and close the rings. For example, the deep ring moves laterally and upward, closing like a shutter and making the canal longer and more oblique.

TABLE 25–2 Corresponding Layers of Abdominal Wall and Scrotum

Abdominal Wall	Spermatic Cord and Scrotum
Skin	Skin
Subcutaneous tissue	Dartos
External oblique muscle	External spermatic fascia
Internal oblique muscle	Cremaster and cremasteric fascia
Transversus	—
Fascia transversalis	Internal spermatic fascia
Extraperitoneal tissue	—
Peritoneum	Tunica vaginalis

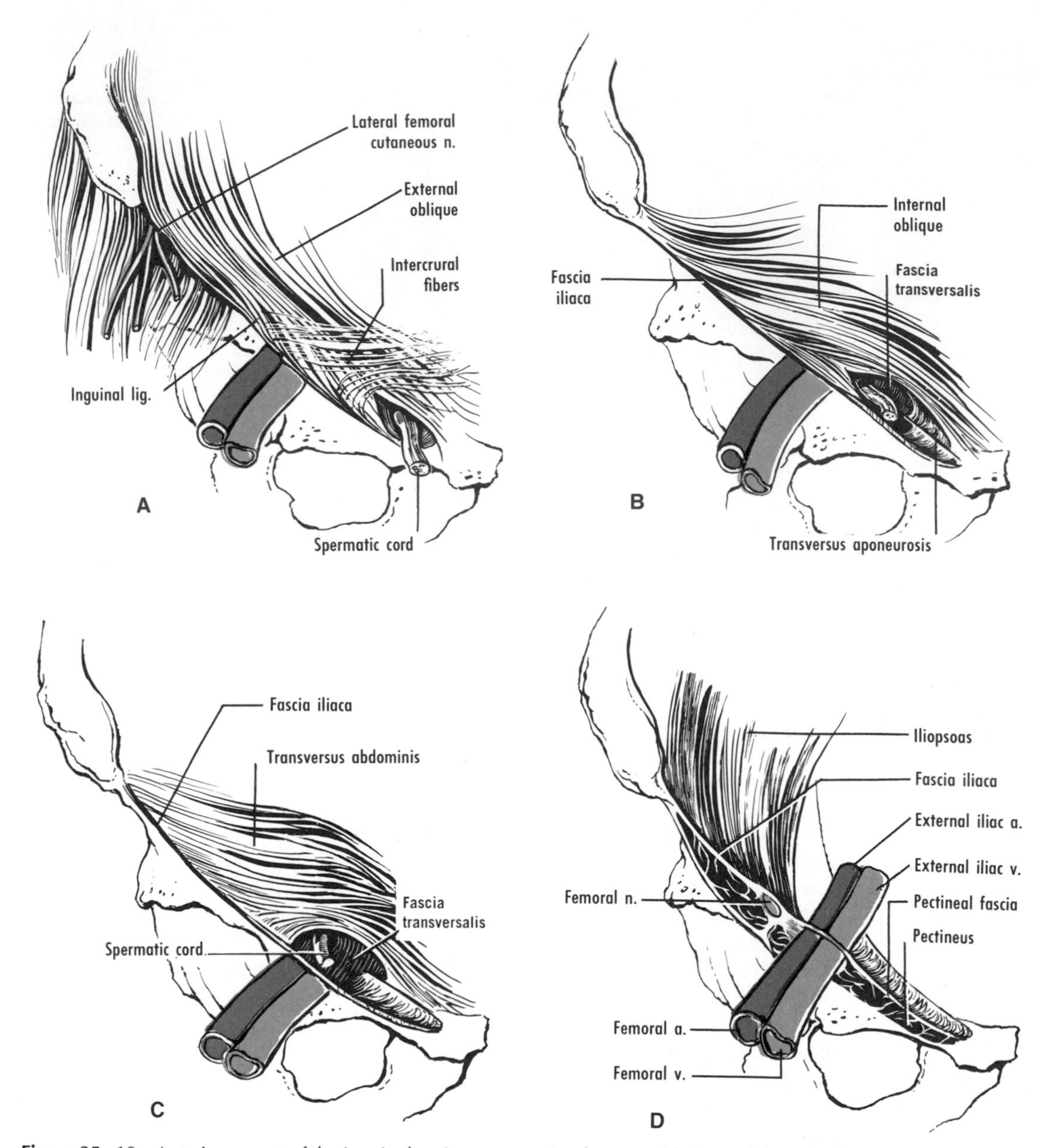

Figure 25–10 Anterior aspect of the inguinal region, progressing from outside inward, layer by layer. **A,** External oblique muscle and superficial inguinal ring. **B,** Internal oblique muscle extending medially to reach pubis. **C,** Transversus abdominis and fascia transversalis. **D,** Cut edges of iliopsoas and pectineus.

HERNIAE (figs. 25–8 and 25–11)

A hernia is an abnormal protrusion of a viscus. Herniae occur most commonly in the inguinal, femoral, and umbilical regions.

Inguinal herniae are either indirect or direct. In indirect, or oblique, inguinal hernia, abdominal contents enter the inguinal canal through the deep inguinal ring. Congenital factors, especially a partly or wholly patent processus vaginalis, are believed to be important. The layers are basically those of the spermatic cord. In direct inguinal hernia, abdominal contents enter the inguinal canal through its posterior wall, medial to the inferior epigastric artery, i.e., in the inguinal triangle. The hernia protrudes forward to (but rarely through) the superficial inguinal ring. Direct herniae are acquired and are due to weakness of the posterior wall. The sac is formed by peritoneum.

Femoral herniae (see fig. 25–8) are discussed with the thigh (Chapter 15).

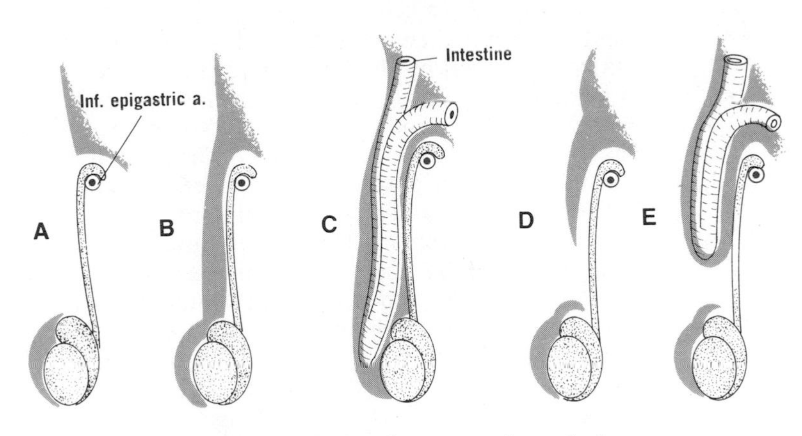

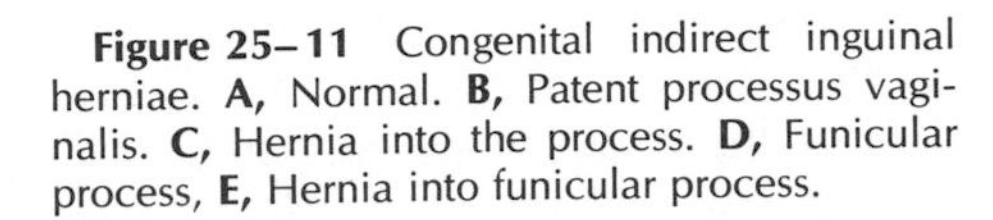

Figure 25–11 Congenital indirect inguinal herniae. **A,** Normal. **B,** Patent processus vaginalis. **C,** Hernia into the process. **D,** Funicular process, **E,** Hernia into funicular process.

Umbilical herniae are usually congenital and result from an incomplete closure of the abdominal wall. *Ventral herniae* may also occur through defects in the linea alba or along the linea semilunaris.

Blood Vessels And Lymphatic Drainage (see figs. 13–5, 24–2, and 32–1)

The cutaneous veins and lymphatic vessels drain in two directions from approximately the level of the umbilicus: (1) upward to the thoraco-epigastric and lateral thoracic veins (thereby providing collateral circulation in caval obstruction) and to the axillary nodes, respectively, and (2) downward to the great saphenous vein and superficial inguinal nodes, respectively. Subcutaneous veins near the umbilicus anastomose with the portal vein by way of branches along the ligamentum teres of the liver.

Apart from branches (the superficial epigastric and superficial circumflex iliac) of the femoral artery, the chief arteries of the abdominal wall are two above (the superior epigastric and musculophrenic) from the internal thoracic artery and two below (the inferior epigastric and deep circumflex iliac) from the external iliac artery. The *superior epigastric artery* enters the rectus sheath and descends behind that muscle. The *musculophrenic artery* courses along the costal margin. The *inferior epigastric artery* (see fig. 25–9), arising near the midinguinal point, ascends past the medial margin of the deep ring, where the ductus deferens hooks around its lateral side. As it proceeds toward the lateral edge of the rectus abdominis, it forms the lateral boundary of the inguinal triangle. Finally the artery ascends behind the rectus in a compartment of the rectus sheath. **The anastomoses between the superior and inferior epigastric arteries provide collateral circulation between the subclavian and external iliac arteries.** The deep circumflex iliac artery (see fig. 32–1) proceeds laterally between the transversus and the internal oblique muscles and reaches the anterior superior iliac spine.

Nerves (see figs. 25–12 and 30–7)

The abdominal wall is supplied by intercostal nerves 7 to 11 (the thoraco-abdominal nerves) and by the subcostal, iliohypogastric,

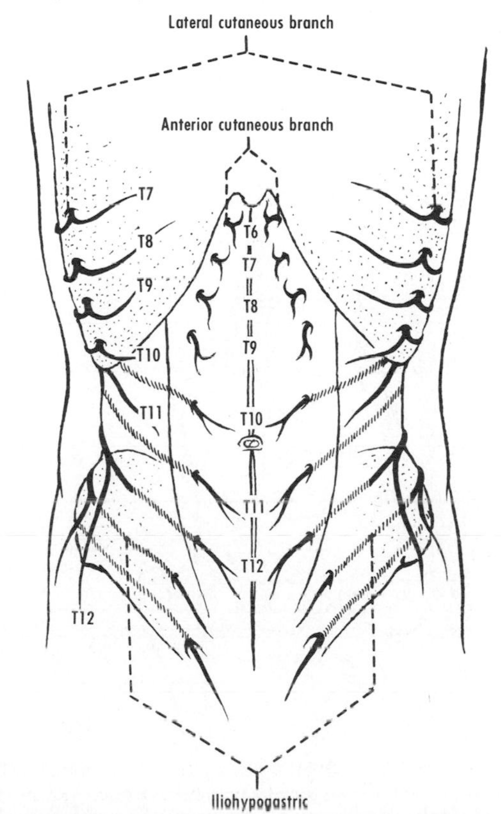

Figure 25–12 The cutaneous distribution of the thoraco-abdominal nerves.

and ilio-inguinal nerves (fig. 25–12). A band of skin is supplied by the lateral and anterior cutaneous branches of each of these nerves (except the ilio-inguinal, which may be regarded as the collateral branch of L.N.1). The overlap in distribution is such that section of a single nerve results in only diminished sensation in its area of supply. The lower intercostal nerves travel basically between the internal oblique muscle and the transversus abdominis. (Similarly in the thorax, the intercostal nerves run between the middle [internal intercostal] and deep [innermost intercostal] layers.)

Umbilicus

The umbilicus (*omphalos* in Greek), or navel, is a median depression some distance above the pubis. It indicates the site of attachment of the umbilical cord before birth, and, even in the adult, some constituents of the cord are recognizable on the inner aspect of the abdominal wall. All layers of the abdominal wall are fused at the umbilicus, and subcutaneous fat that accumulates around the margins causes the umbilicus to appear depressed.

Congenital anomalies may be (1) alimentary, e.g., persistence of the vitello-intestinal duct; (2) urachal, e.g., partial or complete patency; (3) vascular, e.g., a persistent omphalomesenteric vein; or (4) somatic, e.g., faulty development of the abdominal wall, including ventral herniae, such as *omphalocele* (or *exomphalos*), which involves a protrusion of intestine through a large defect at the umbilicus.

POSTERIOR ABDOMINAL WALL (see figs. 15–6, 25–6, 25–13, and 29–5)

The posterior abdominal wall is formed by the bodies and intervening discs of the lumbar vertebrae and by the iliopsoas, quadratus lumborum, ilium, and diaphragm (fig. 25–13). Against this background lie the aorta, inferior vena cava, kidneys, suprarenal glands, and ascending and descending colon.

The *iliopsoas* is the main flexor of the thigh and trunk. The iliacus arises from the iliac

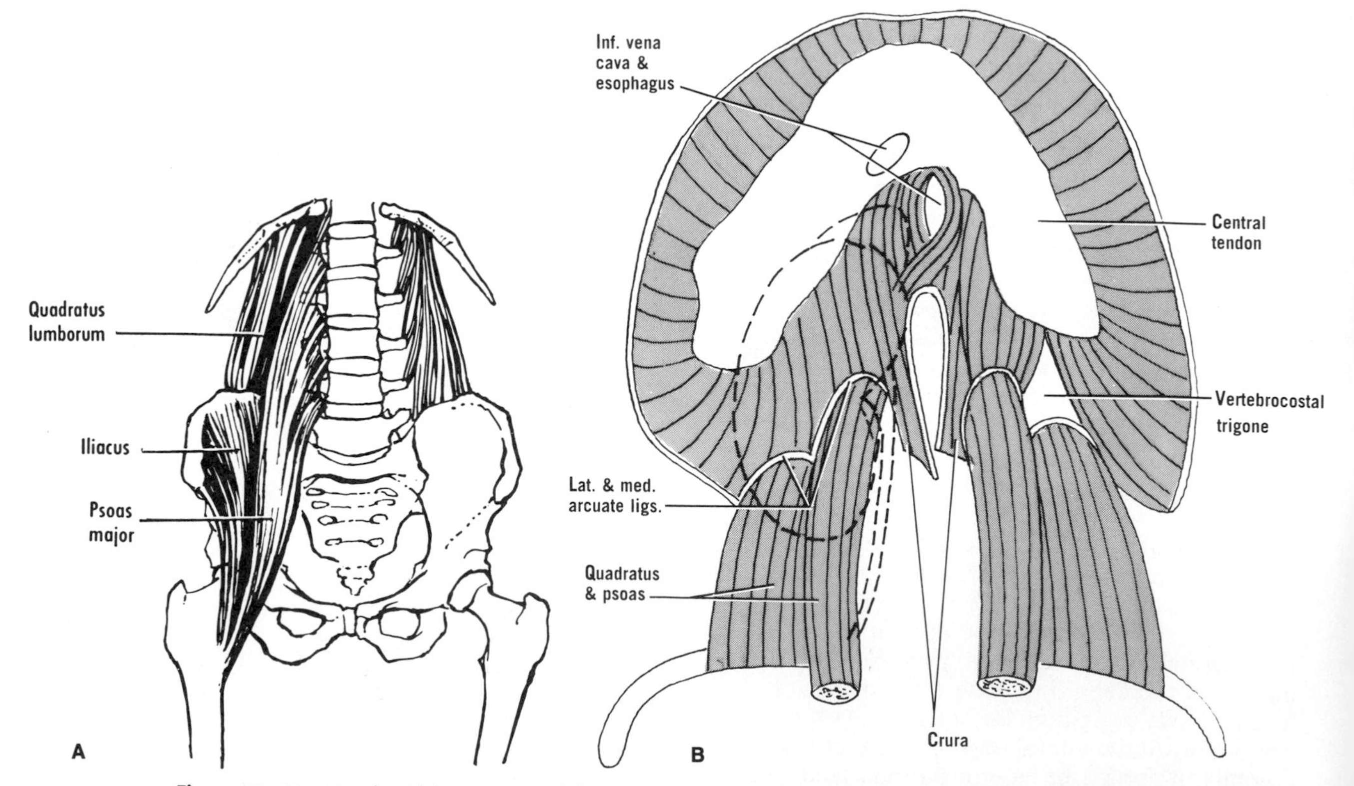

Figure 25–13 Muscles of the posterior abdominal wall. **A,** Quadratus lumborum and iliopsoas. In **B,** the diaphragm has been added and the arcuate ligaments have been indicated. The outline of the right suprarenal, kidney and ureter are shown. **Defects in the diaphragm (diaphragmatic herniae), many of which are congenital, are most frequent around the esophageal opening (hiatal herniae).** Other defects are commoner on the left than on the right side. In **B,** a congenital defect (*vertebrocostal trigone*) on the left side would result in close contact between the left kidney and pleura.

fossa, and the psoas major arises from the lumbar vertebrae; the combined muscle (iliopsoas) is inserted into the lesser trochanter of the femur. The psoas may be assisted by an anterior slip, the psoas minor, which arises from vertebral bodies (T.V.12 and L.V.1) and is inserted chiefly into the arcuate line. Fascial continuity occurs between the transversalis and thoracolumbar fasciae, psoas sheath, and fascia iliaca. The psoas sheath, attached to the lumbar transverse processes and bodies, allows the spread of infection (e.g., a tuberculous abscess from a vertebral body) into the thigh (psoas abscess). The quadratus lumborum ascends from the iliac crest to the last rib and is anchored medially to the lumbar transverse processes. It probably flexes the trunk laterally. The iliacus is supplied by the femoral nerve, and the psoas and quadratus are supplied by the lumbar plexus.

QUESTIONS

25–1 At which vertebral level is the transpyloric plane?

25–2 What is the direction of the fibers of the external oblique muscle?

25–3 What is the inguinal ligament?

25–4 What and where is the superficial inguinal ring?

25–5 What and where is the deep inguinal ring?

25–6 What provides the chief protection of the inguinal canal?

25–7 What are the boundaries of the inguinal triangle?

25–8 In which dermatome is the umbilicus?

25–9 What is exomphalos?

26 THE ABDOMINAL VISCERA AND PERITONEUM

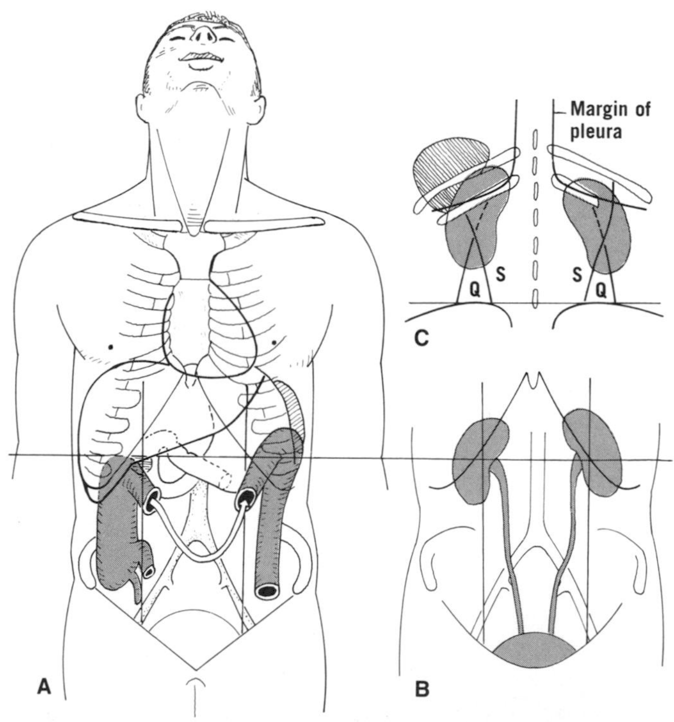

Figure 26–1 General relations of abdominal viscera. In **A,** most of the small intestine and transverse colon, as well as the sigmoid colon and rectum, has been removed. The liver and spleen are visible above the right and left flexures of the colon, respectively. The transpyloric and the right and left lateral planes are indicated. In **B,** the position of the kidneys and ureters is shown. In **C,** which is a posterior view, the kidneys and spleen are represented. The right twelfth rib is shorter here than the left. *Q* represents the portion of the quadratus lumborum not under cover of *S,* the erector spinae.

ABDOMINAL VISCERA

The viscera of the abdomen proper include the stomach, intestine, liver and biliary system, pancreas, spleen, kidneys, ureters, and suprarenal glands. Most of the stomach and intestine is anchored to the body wall by peritoneal mesentery, whereas the three paired glands (kidneys, suprarenals, and gonads before birth) lie retroperitoneally. The general relationships are shown in figures 26–1 to 26–4. The positions of the abdominal viscera vary with the individual and with gravity, posture, respiration, and degree of filling. Radiological studies have shown that "**the normal abdominal viscera have no fixed shapes and no fixed positions, and every description of them must be qualified by a statement of the conditions existing at the time of observation. Moreover, profound change may be caused not only by mechanical forces but also by mental influences**" (A. E. Barclay). Organs tend to sink when a subject is in the erect position, and the most mobile organs are those attached by mesenteries.

The following structures can sometimes be palpated *in vivo* in normal subjects: pulsations of the abdominal aorta, the lumbar vertebrae, the lower pole of the right kidney, possibly the liver, and occasionally the spleen. The body of the uterus can be palpated bimanually.

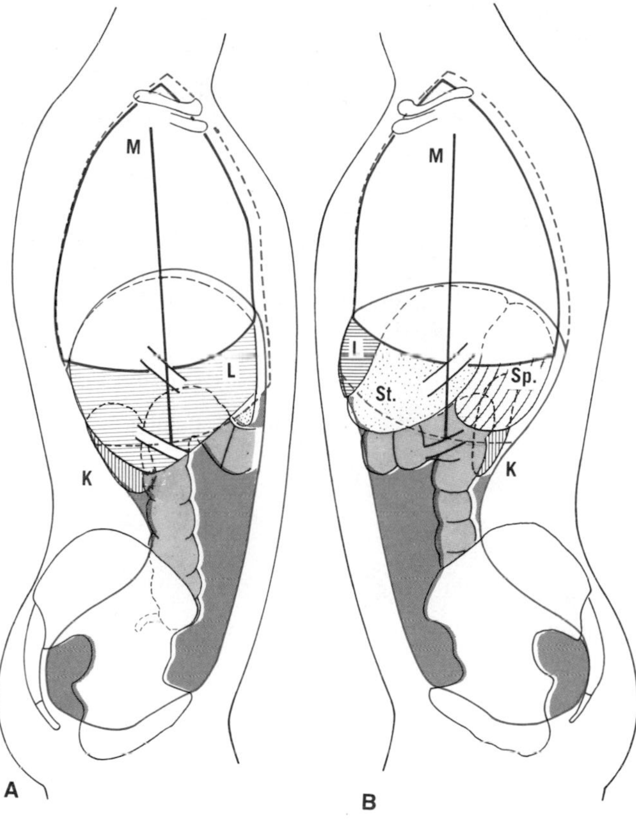

Figure 26–2 Right and left lateral aspects of the trunk, showing the topography of the viscera. The clavicles and parts of ribs 2, 8, and 10 are shown. In the thorax, the outlines of the pleurae (*interrupted line*) and lungs are provided. In the abdomen, the peritoneum and peritoneal cavity are shown. In **A,** the liver (*L*), kidney (*K*), right flexure, and ascending colon can be seen. In **B,** the liver (*L*), stomach (*St.*), spleen (*Sp.*), kidney (*K*), left flexure, and descending colon can be seen. *M*, midaxillary line. (After Pernkopf.)

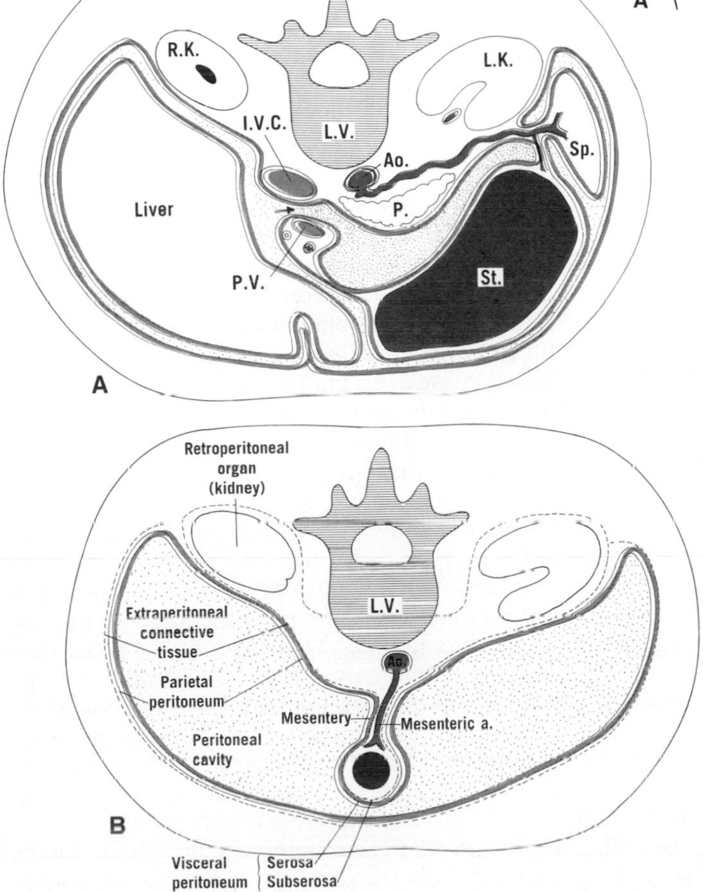

Figure 26–3 Horizontal sections through the abdomen. In **A,** the liver, right and left kidneys (*R.K., L.K.*), pancreas (*P.*), spleen (*Sp.*), and stomach (*St.*) are shown, as well as the aorta (*Ao.*) and splenic artery, inferior vena cava (*I.V.C.*), portal vein (*P.V.*), bile duct, and hepatic artery. The greater sac of the peritoneal cavity (around the liver, for example) can be traced through the epiploic foramen (*arrow*) into the lesser sac between the stomach and pancreas. *L.V.*, lumbar vertebra. (After Symington.) **B** represents the principle of the arrangement of the peritoneum and its mesenteries.

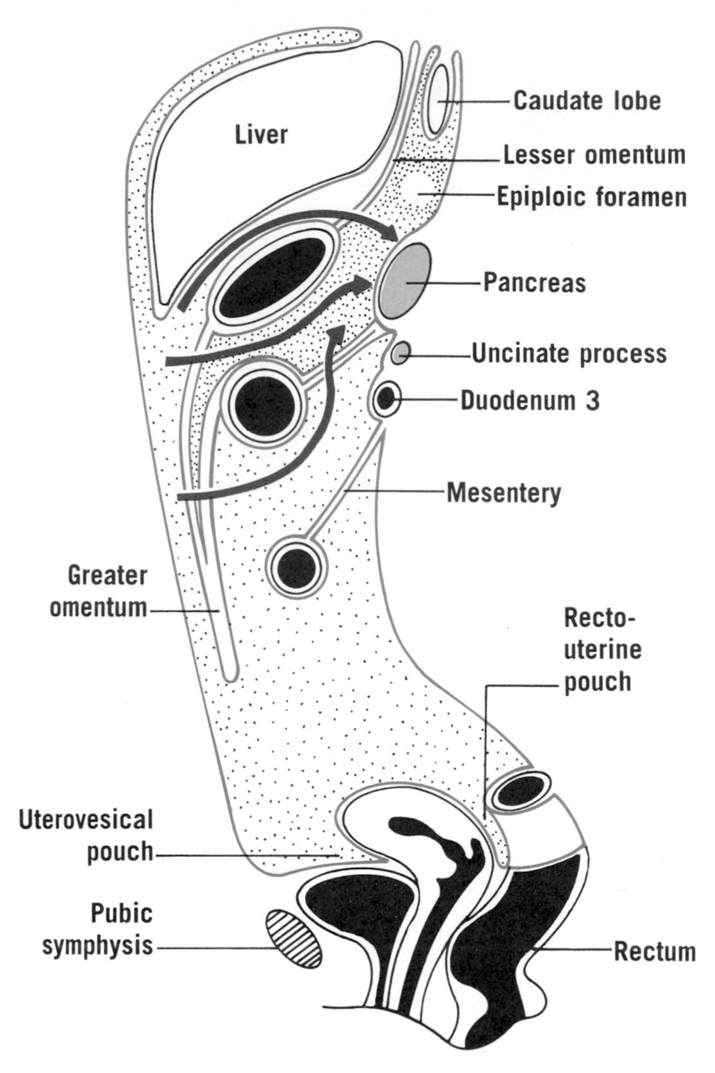

Figure 26–4 Median section through the abdomen to show the peritoneal cavity. The lesser sac is situated behind the stomach and in front of the pancreas. The caudate lobe invaginates the lesser sac. Below the liver, the stomach, transverse colon, and small intestine are seen in section. The greater omentum and transverse mesocolon are fused posteriorly. The arrows indicate surgical approaches to the lesser sac. The relations in the pelvis are shown also in fig. 31–9.

PERITONEUM

The peritoneum is a smooth, glistening, serous membrane that lines the abdominal wall as the *parietal peritoneum* and is reflected from the body wall to various organs, where, as *visceral peritoneum,* it forms an integral part as the outermost, or serosal, layer. **The pericardium, pleura, and peritoneum have a similar arrangement in the parietal and visceral layers, with a cavity between.** The extraperitoneal tissue external to the parietal peritoneum is carried with the reflections to the organs and becomes a part of the serosal layer. Organs, such as most of the intestine, that are almost completely invested by peritoneum are connected to the body wall by a *mesentery.* Other viscera, however, such as the kidneys, are *retroperitoneal;* i.e., they lie on the posterior abdominal wall and are covered by peritoneum only anteriorly.

The peritoneal cavity contains merely a thin film of fluid. **The peritoneal cavity is completely closed in the male, whereas, in the female, it communicates with the uterine tubes and hence indirectly with the exterior of the body.**

The peritoneum minimizes friction, resists infection, and stores fat. It allows free movement of the abdominal viscera. In response to injury or infection (*peritonitis*), it exudes fluid and cells and tends to wall off or localize infection.

The parietal peritoneum is supplied by nerves (e.g., phrenic and thoraco-abdominal) to the adjacent body wall, and most of it is very sensitive to pain. Painful stimuli to the central part of the diaphragmatic peritoneum are referred to the shoulder. The visceral peritoneum is insensitive.

General Arrangement of Peritoneum

The peritoneal reflection to the jejunum and ileum is termed *the* mesentery, whereas those to the colon are each known as a *mesocolon.* Some reflections are termed ligaments or folds, e.g., gastrohepatic ligament or recto-uterine fold. Most such ligaments contain blood vessels, and most folds are raised by underlying vessels in their free edges. A broad peritoneal sheet or reflection is termed an *omentum,* the Greek word for which is *epiploon.* The general arrangement of the peritoneum is shown in figures 26–2 to 26–5.

Greater Sac. An incision through the anterior abdominal wall and parietal peritoneum enters that part of the peritoneal cavity known as the greater sac. This extends from the diaphragm to the pelvic floor. Above the umbilicus, its anterior wall presents the *falciform ligament,* which contains the *ligamentum teres* (obliterated umbilical vein) in its free margin. Below the umbilicus, the anterior wall of the greater sac contains five folds: (1) the *median umbilical fold* (containing the urachus), which extends from the urinary bladder to the umbil-

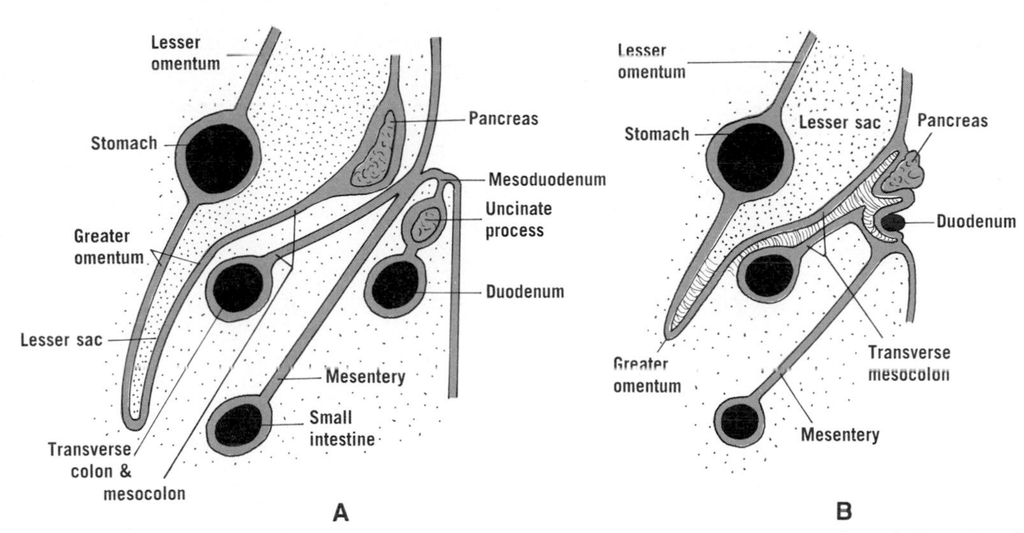

Figure 26–5 Sagittal schemes of the greater omentum and transverse mesocolon **(A)** prenatally and **(B)** in the adult. The numerous short lines in **B** represent sites of fusion.

icus; (2) two *medial umbilical folds* (each containing an obliterated umbilical artery), which extend from the sides of the bladder to the umbilicus; and (3) two *lateral umbilical folds* (each containing an inferior epigastric artery), which extend from the deep inguinal ring to the arcuate line. Three depressions on each side of the median plane are produced by the umbilical folds (see fig. 25–9).

The *greater omentum* is a prominent peritoneal fold that hangs down from the stomach in front of the transverse colon, to which it is attached. The greater omentum is a double fold that connects the stomach to the posterior abdominal wall (figs. 26–6 and 26–7), being derived from the dorsal mesogastrium in the embryo. The greater omentum is fused behind with, but is separable from, the transverse colon and mesocolon. If the greater omentum is turned up, the coils of the small intestine can be examined and *the mesentery* can be traced to its root. **The root of the mesentery extends from the duodenojejunal flexure (above and to the left) to the ileocolic junction (below and to the right)** (fig. 26–6). If the greater omentum is followed to the transverse colon, the *transverse mesocolon* can be traced back to the posterior abdominal wall (see fig. 26–5*B*). The transverse colon ends in right and left colic flexures, which lead to the ascending and descending colon, respectively. These have usually no mesocolon (i.e., they are retroperitoneal), although at least the ascending colon appears to be more mobile *in vivo* than is generally appreciated. The descending colon is continued into the sigmoid colon, which commonly has a mesocolon. The attachment of the sigmoid mesocolon may be shaped like an inverted V, the apex being at the pelvic brim, in front of the left ureter.

Traced upward, the falciform ligament is found to be reflected onto the diaphragmatic surface of the liver (see fig. 28–3*B*), thereby preventing a hand placed between the liver and the diaphragm from passing from one side to the other. More posteriorly, the peritoneal reflection from the diaphragm to the liver diverges and forms the upper layer of the *coronary ligament* on the right and the upper layer of the left *triangular ligament* on the left (see fig. 28–3*C*). The subdiaphragmatic portions of the greater sac on each side of the falciform ligament are known as the right and left *subphrenic spaces,* respectively, and are important clinically because an abscess may form in one of them. Below the coronary ligament, the recess of the greater sac between the liver and the right kidney is known as the *hepatorenal pouch* (fig. 26–6).

The fundus of the gallbladder is usually visible at the lower border of the liver. Traced upward, the gallbladder leads to the *lesser omentum,* which extends between the liver and (1) the stomach (*gastrohepatic ligament*), and (2) the duodenum (*hepatoduodenal ligament*). The free edge of the lesser omentum is its right border, which lies immediately in front of the opening into the lesser sac.

Lesser sac. The lesser sac is a large, irregular part of the peritoneal cavity that lies mostly behind the stomach and lesser omentum (see figs. 26–3 to 26–5). It communicates with the

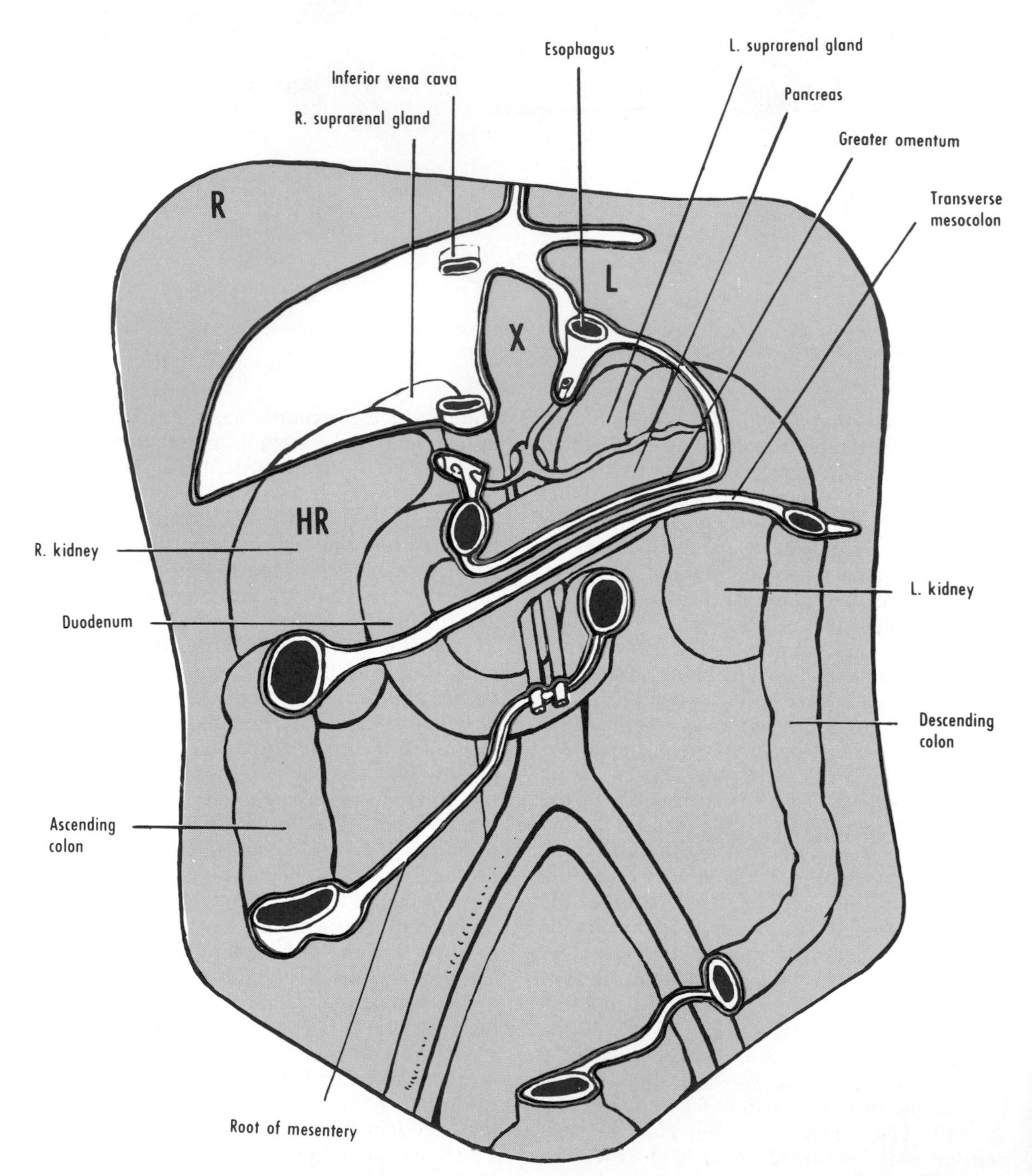

Figure 26–6 The attachments of the peritoneum to the posterior abdominal wall, viewed from in front. *HR*, the hepatorenal pouch. *R* and *L*, right and left subphrenic spaces. *X*, the superior recess of the lesser sac.

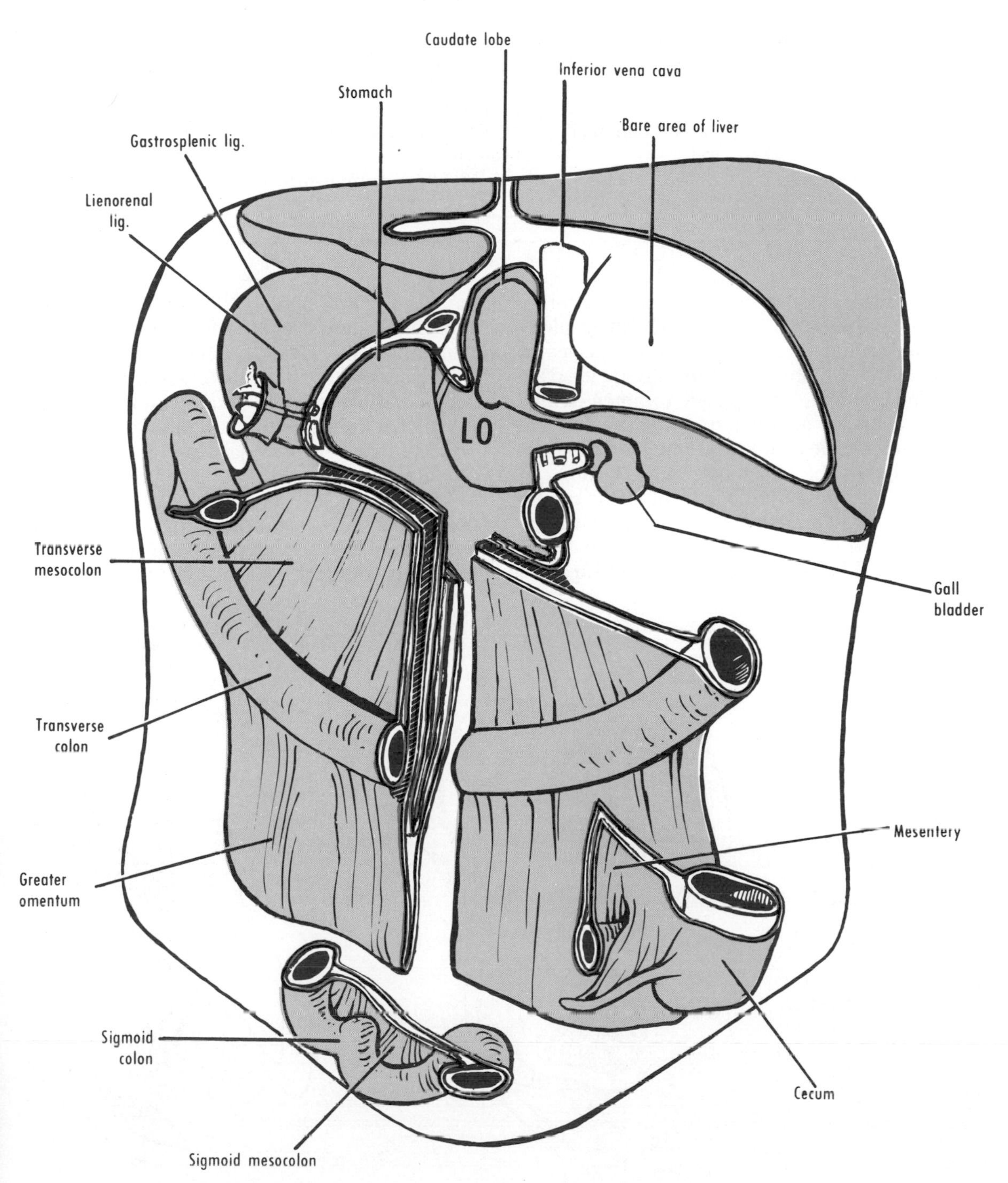

Figure 26–7 The attachments of the peritoneum to the abdominal viscera, viewed from behind. This is a mirror image of the previous view. The peritoneum on the back of the anterior abdominal wall has not been colored. A vertical segment of the transverse mesocolon and greater omentum has been removed. *LO*, the lesser omentum.

TABLE 26–1 Relationships of Clinically Important Subphrenic and Subhepatic Spaces

Space	Important Relationships				
	Anterior	*Posterior*	*Superior*	*Right*	*Left*
Right subphrenic	Anterior abdominal wall	Upper layer of coronary ligament	Diaphragm	Diaphragm	Falciform ligament
Left subphrenic	Anterior abdominal wall	Left triangular ligament	Diaphragm	Falciform ligament	Spleen
Right subhepatic (also termed hepatorenal pouch)	Visceral surface of right lobe of liver	Right kidney	Lower layer of coronary ligament	Diaphragm	Epiploic foramen

greater sac by the so-called *epiploic foramen,* which can be found by running a finger along the gallbladder to the free edge of the lesser omentum. Two fingers can usually be inserted into the opening. **The epiploic foramen, which is the aditus, or opening, from the greater into the lesser sac, lies immediately behind the free, right edge of the lesser omentum. A finger in the opening and a thumb in front of the omentum would catch the bile duct (at the right), the hepatic artery (at the left), and the portal vein (behind and between them). The inferior vena cava is situated behind the epiploic foramen, the liver is above, and the first part of the duodenum is below** (see fig. 26–3*A*).

The anterior wall of the lesser sac is formed by the peritoneum of (1) the lesser omentum, (2) the back of the stomach, and (3) the front portion (anterior two layers) of the greater omentum (see fig. 26–5*B*). The posterior wall of the lesser sac is formed by (1) the peritoneum that covers the diaphragm, pancreas, left kidney and suprarenal gland, and duodenum and (2) the back portion (posterior two

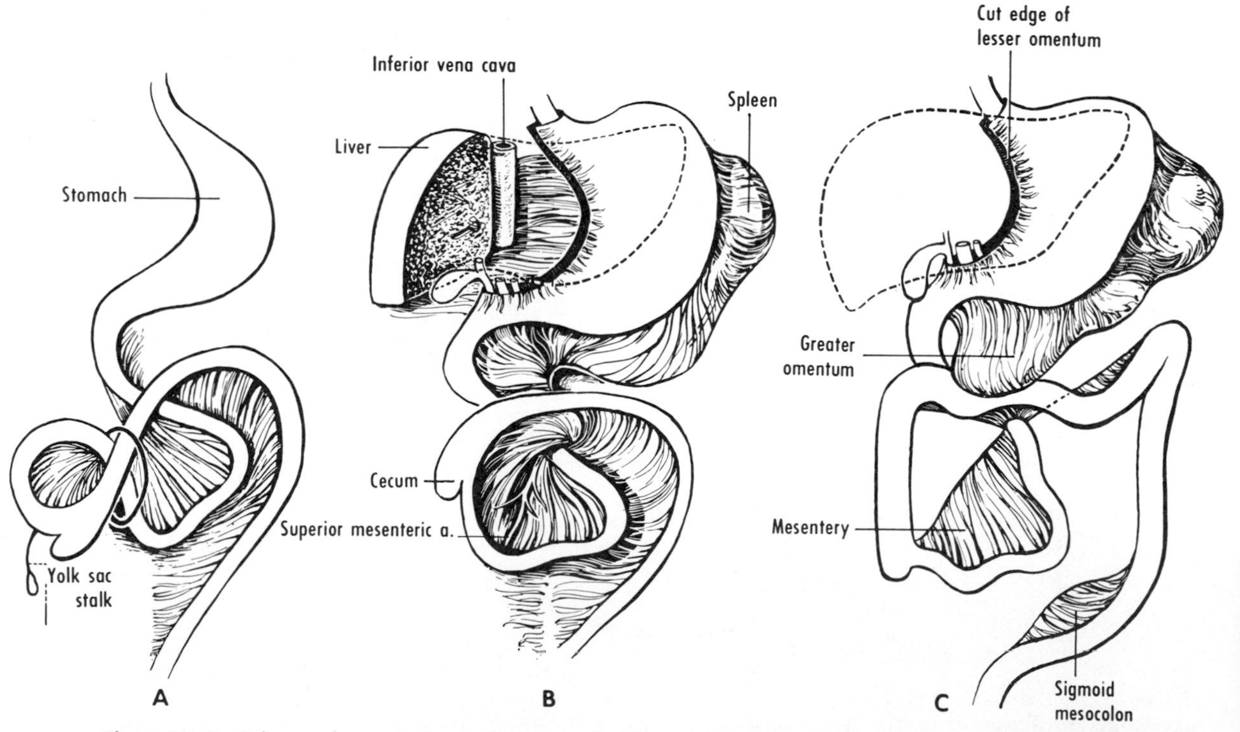

Figure 26–8 Scheme of mesenteric attachments and of the rotation of the gut. **A,** Early fetal period. Note that the colon crosses ventral to the duodenum. The small intestine is being reduced, that is, withdrawn into the abdominal cavity through the umbilical ring. **B,** Slightly later in the fetal period. The intestine is completely reduced. The axis of rotation of the gut is the superior mesenteric artery. **C,** The fixation of mesenteries in the adult. Note that every second part of the alimentary canal retains a mesentery.

layers) of the greater omentum (see fig. 26–5*B*). The lesser sac can be entered artificially by penetrating (1) the lesser omentum, (2) the greater omentum between the stomach and transverse colon, or (3) the fused greater omentum and transverse mesocolon (see fig. 26–4).

The lesser sac extends upward (as its superior recess) behind the liver, being invaginated by the caudate lobe (see fig. 26–4). Most of the lesser sac (the *omental bursa* of the embryo) lies behind the stomach and in the greater omentum.

A number of minor peritoneal folds, fossae, and recesses may be found, especially around the duodenum and cecum.

DEVELOPMENT OF PERITONEUM (figs. 26–5 and 26–8)

The stomach and intestine in the embryo are attached by a continuous mesentery to the posterior abdominal wall. This *dorsal mesentery* persists in only every second portion of the gastro-intestinal canal, namely: in the stomach as the greater omentum; not in most of the duodenum; in the jejunum and ileum as *the* mesentery; seemingly not in the ascending colon; in the transverse colon as the transverse mesocolon; usually not in the descending colon; in the sigmoid colon as the sigmoid mesocolon; and not in the rectum.

The embryonic *ventral mesentery,* which is confined to the upper part of the abdomen, is more complicated. It is usually stated that the liver grows into it, thereby dividing it into (1) a portion from the anterior abdominal wall to the liver, the falciform ligament, and (2) a part from the liver to the stomach and duodenum, the lesser omentum.

The complicated rotation and fixation of the intestine are summarized in figure 26–8. The vague term *malrotation* is used for various anomalies of these processes.

QUESTIONS

26–1 What are the serous membranes?

26–2 How does the peritoneal cavity communicate with the exterior of the body?

26–3 What is the relationship of the transverse colon to the greater omentum?

26–4 What is the extent of the root of the mesentery?

26–5 Where may a subphrenic abscess be found?

26–6 What are the attachments of the lesser omentum?

26–7 What is the lesser sac of the peritoneum and how may it be approached surgically?

26–8 What and where is the epiploic foramen?

26–9 What is malrotation of the intestine?

27
THE ESOPHAGUS, STOMACH, AND INTESTINE

The alimentary canal comprises the esophagus, stomach, and intestine. Details of the development of the canal from the embryonic gut, and also the complicated question of rotation of the gut, should be sought in books on embryology. Briefly, **the stomach and upper half of the duodenum are derived from the foregut and supplied by the celiac trunk. The lower half of the duodenum, the jejunum and ileum, cecum and appendix, ascending colon, and two thirds of the transverse colon are derived from the midgut and supplied by the superior mesenteric artery. The left third of the transverse colon, the descending colon, sigmoid colon, rectum, and the upper part of the anal canal are derived from the hindgut and supplied by the inferior mesenteric artery. Hence the developmental and vascular junctions are at (1) the middle of the duodenum, where the bile duct ends, and (2) the left part of the transverse colon.**

The esophagus is a conducting tube for food, whereas the stomach and intestine, together with their associated glands, are concerned with the digestion of food and the excretion of undigested material. The products of digestion traverse the gastro-intestinal mucosa to capillaries that ultimately form the portal vein. The portal vein breaks up into a second set of capillaries (sinusoids) in the liver, and these ultimately form hepatic veins.

The alimentary canal presents sphincteric mechanisms at junctional areas: pharyngo-esophageal, gastro-esophageal, pyloric, ileocolic, and anal. The sphincters are under neural and hormonal control, and they help to prevent regurgitation of contents.

ABDOMINAL PART OF ESOPHAGUS

After its cervical and thoracic course, the esophagus traverses the diaphragm and joins the stomach at the gastro-esophageal junction. The junction is a barrier to the reflux of contents from the stomach to the esophagus. **A complicated sphincteric segment extends several centimeters above the gastro-esophageal junction and is partly in the thorax, partly in the diaphragm, and partly in the abdomen.**

An important portal-systemic anastomosis takes place between the esophageal and left gastric veins. Stimulation of the lower end of the esophagus (e.g., by reflux of acid gastric contents) may cause pain ("heartburn") deep to the sternum or in the epigastrium. Diaphragmatic herniae occur most commonly through the esophageal opening and are termed *hiatal herniae*.

STOMACH

The stomach (Gk, *gaster,* "belly"; adjective *gastric* from Latin) presents a cardiac

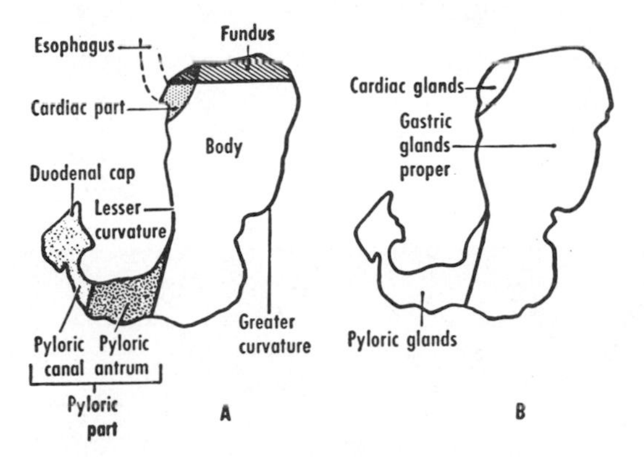

Figure 27–1 **A,** The parts of the stomach. **B,** The location of the glands.

part, fundus, body, and pyloric part (based on histological differences) (fig. 27–1*A*); greater and lesser curvatures; and cardiac and pyloric openings.

The esophagus enters the stomach at the *cardiac opening,* and the immediately adjacent portion of the stomach is termed its *cardiac part.* The *fundus* is the part of the stomach above the level of the cardiac opening. It usually contains swallowed air and hence is visible radiographically. The *body* of the stomach lies between the fundus and the pyloric part. The *pyloric part* comprises the *pyloric antrum* followed by the *pyloric canal.* The pyloric opening, or *pylorus,* between the stomach and duodenum, is surrounded by the *pyloric sphincter.* Congenital thickening of the sphincter is known as *congenital hypertrophic pyloric stenosis,* which produces vomiting, loss of weight, and a palpable lump in the infant's abdomen. The hypertrophied sphincter can be incised surgically. The gastroduodenal junction is often normally marked by a prominent prepyloric vein.

The *greater* and *lesser curvatures* extend between the cardiac and pyloric openings. The greater is on the left and is convex and longer; the lesser is on the right and is concave and shorter. The lesser curvature usually presents an *angular notch* (*incisura angularis*) at its most dependent point, and this is commonly visible in radiographs of the barium-filled stomach of a subject in the erect position. The stomach, which is sometimes J-shaped when empty, is very variable in shape, capacity, and position. The front of the organ faces the greater sac; the back limits the lesser sac. The stomach lies on a variable visceral bed that includes the diaphragm, pancreas, and transverse mesocolon. Posterior gastric ulcers may involve the pancreas and the splenic artery, resulting in severe pain and bleeding.

The cardiac orifice is the most fixed part of the stomach and may be indicated on the surface at the left costal margin. The fundus fits into the curve of the left dome of the diaphragm. The pyloric part is very mobile and is frequently below the transpyloric plane. Moreover, the greater curvature may even enter the true pelvis.

The stomach can be examined *in vivo* by radiography, usually after a barium meal (figs. 27–2 and 27–3), and directly through a tube passed down the esophagus (*gastroscopy*).

The food in the stomach is transformed into a liquid termed *chyme,* which, by rhythmic muscular contractions (*peristalsis*) of the pyloric part, is emptied into the duodenum. Reflux is prevented by the pyloric sphincter.

Peritoneal Relations. The liver is connected to the lesser curvature by the lesser omentum, the two layers of which surround the stomach and leave the greater curvature as the greater omentum (see fig. 26–5*B*). The stomach is covered entirely by peritoneum except for a small posterior "bare area" near the cardiac opening.

Blood Supply (fig. 27–4). The stomach is supplied by the celiac trunk: (1) the right gastric (from the hepatic) and left gastric arteries run along the lesser curvature; (2) the right gastro-epiploic (derived from the hepatic) and left gastro-epiploic and short gastric (from the splenic) arteries course along the greater curvature. The veins end directly or indirectly in the portal vein. The junctions between the left gastric and esophageal veins are important portal-systemic anastomoses.

Lymphatic drainage (fig. 27–5). Plexuses drain into regional nodes that accompany the arteries and end ultimately in the thoracic duct. Carcinoma can spread (1) to the liver, (2) to the pelvis by retroperitoneal lymphatics, and (3) to the rest of the body by veins and by the thoracic duct.

Innervation (fig. 27–6). The stomach receives innervation from several sources: (1) sympathetic fibers via the splanchnic nerves and celiac ganglion (synapse) supply blood vessels and musculature, (2) parasympathetic fibers from the medulla travel in the gastric branches of the vagi, and (3) sensory vagal fibers include those concerned with gastric secretion. The vagi are sometimes sectioned (*vagotomy*) in the treatment of chronic peptic ulcer to reduce hypersecretion and hypermotility of the stomach.

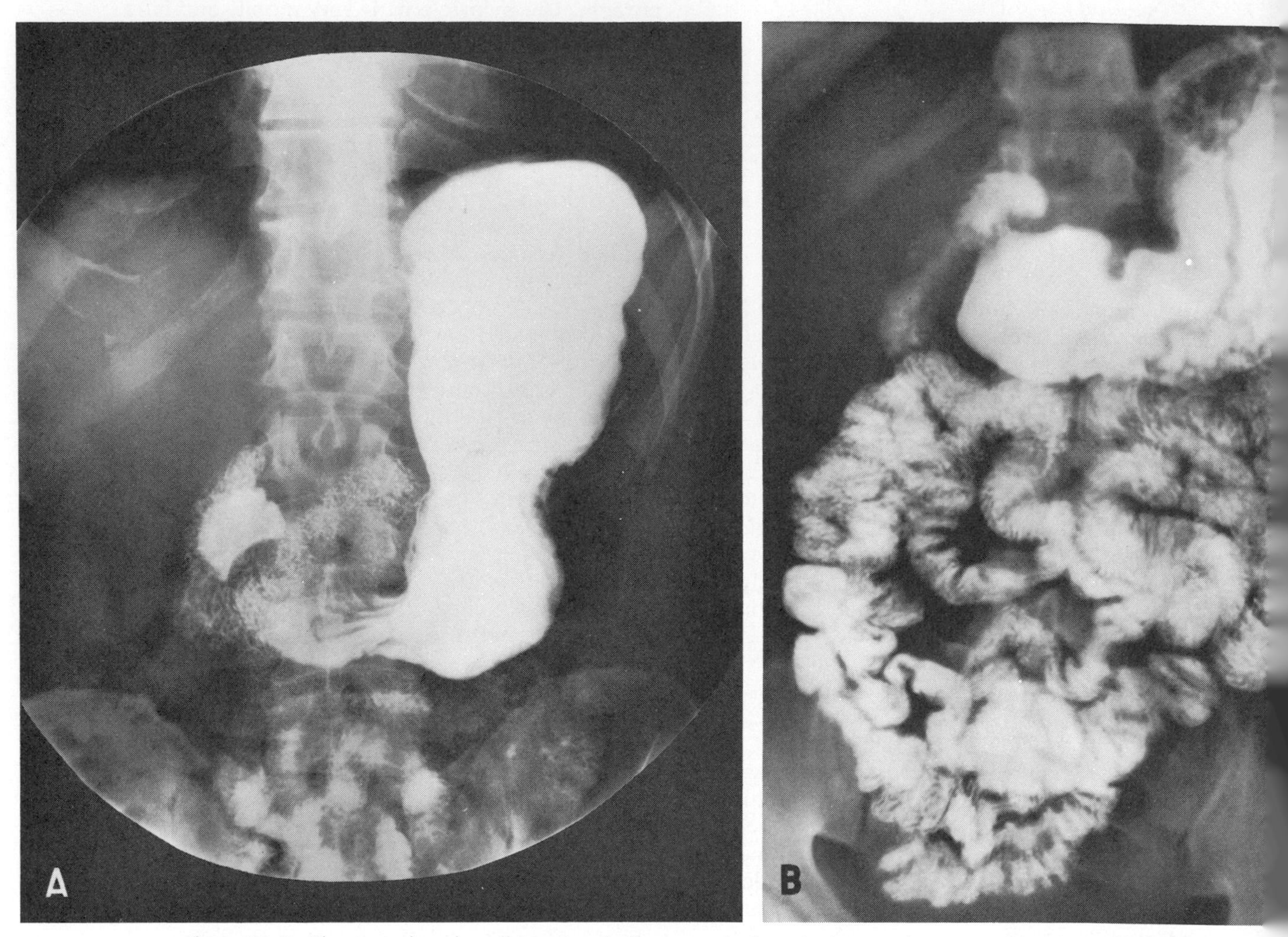

Figure 27–2 The stomach and small intestine. **A,** The stomach after a barium meal, prone position. Note the duodenal cap, the feathery pattern of the barium in the small intestine, the twelfth rib on the right side of the body, and the pyloric part of the stomach at the level of L.V.4. **B,** The small intestine 25 minutes after ingestion of a barium meal. Note the duodenal cap, the feathery pattern of the barium in the jejunum, and the ileum in the lower part of the photograph.

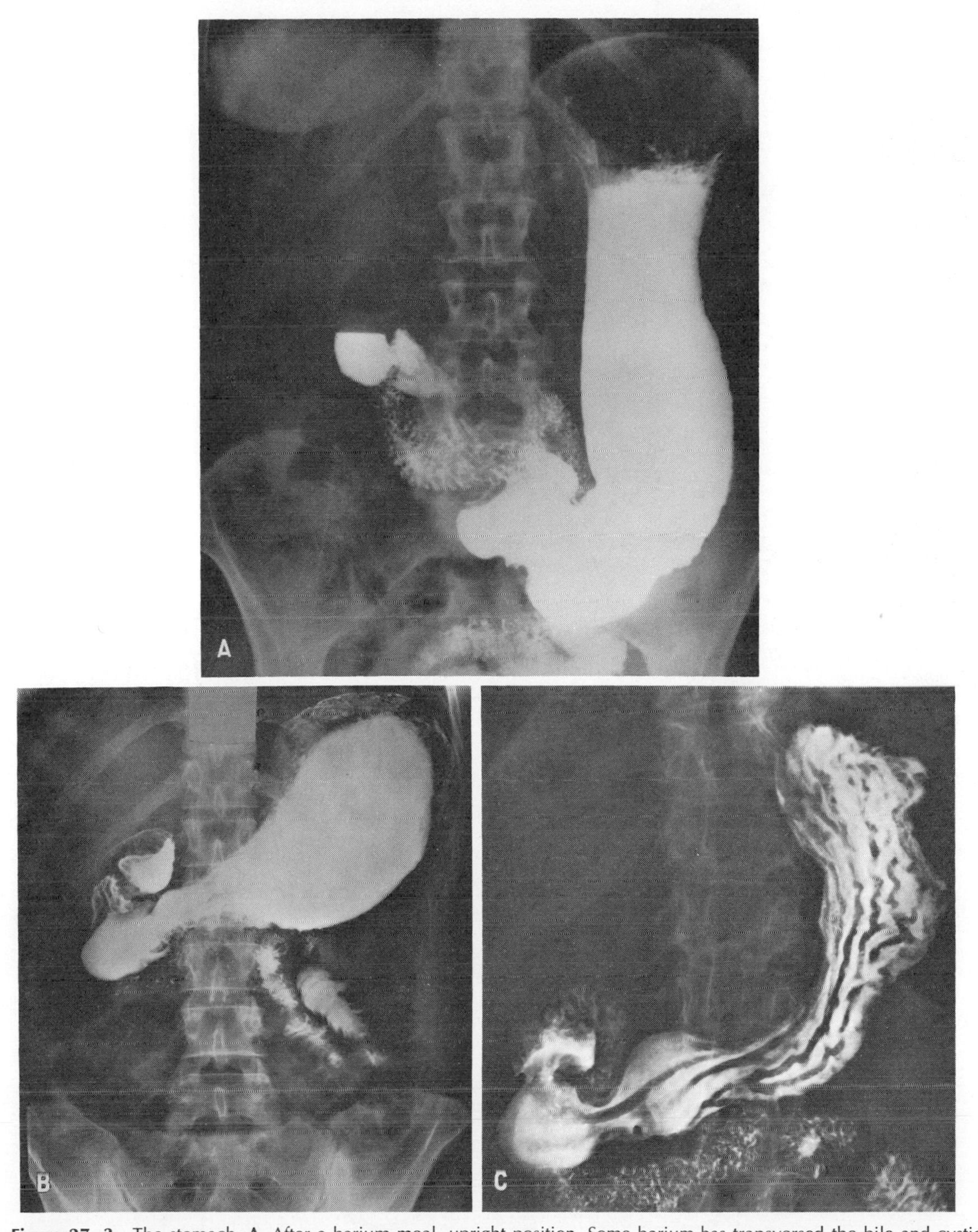

Figure 27–3 The stomach. **A,** After a barium meal, upright position. Some barium has transversed the bile and cystic ducts and is seen in the lower part of the gallbladder. Both the stomach and the gallbladder show a fluid level. The lower part of the stomach extends considerably below the supracristal plane here, and the fundus of the gallbladder is at the level of L.V.4. In **B,** note the horizontal position of the "steerhorn" stomach. In **C,** the stomach is coated with barium and shows mucosal folds. (**A,** Courtesy of A. J. Chilko, M.D., New York, New York.)

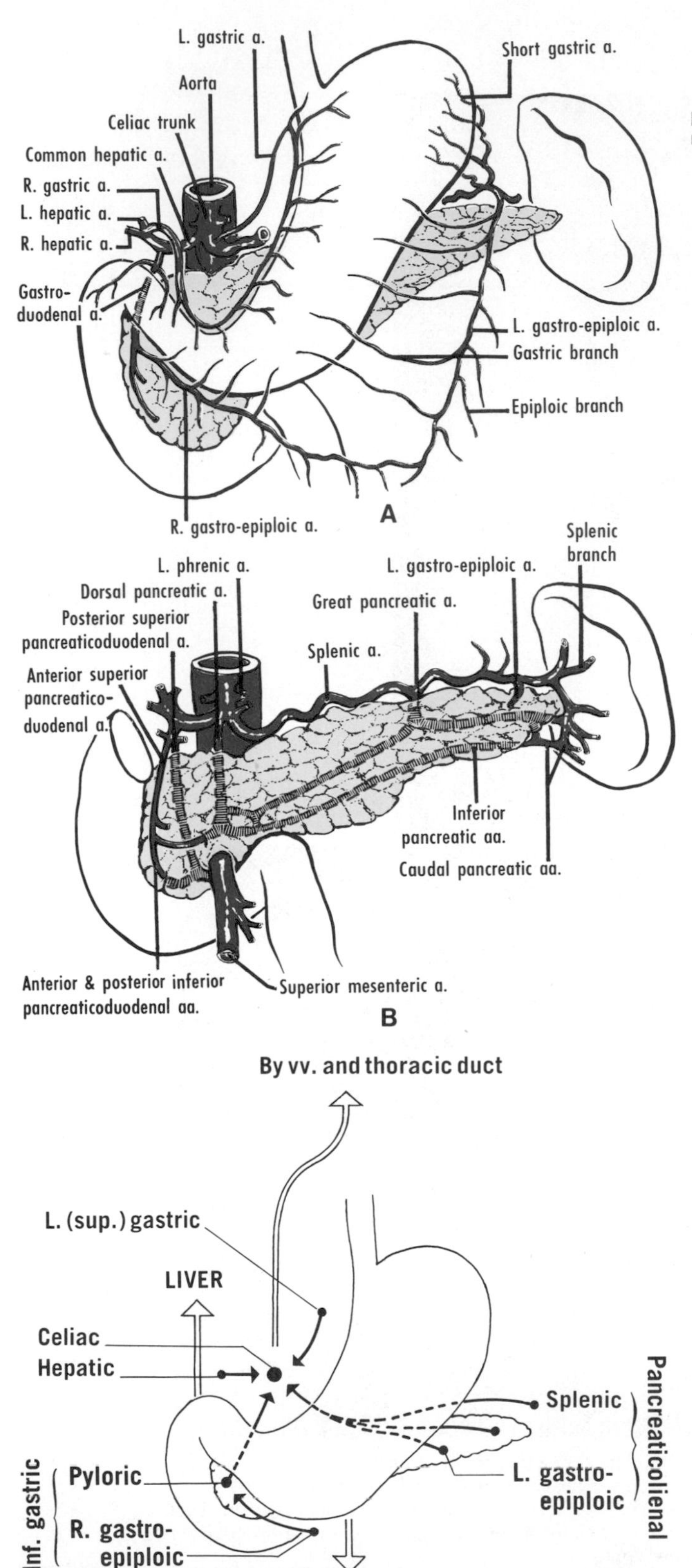

Figure 27–4 Arterial supply of the stomach, duodenum, pancreas, and spleen. The stomach and first part of the duodenum have been removed in **B.**

Figure 27–5 Lymphatic drainage of the stomach, pancreas, and spleen. Each dot represents one or more nodes. The white arrows show the three major directions of spread.

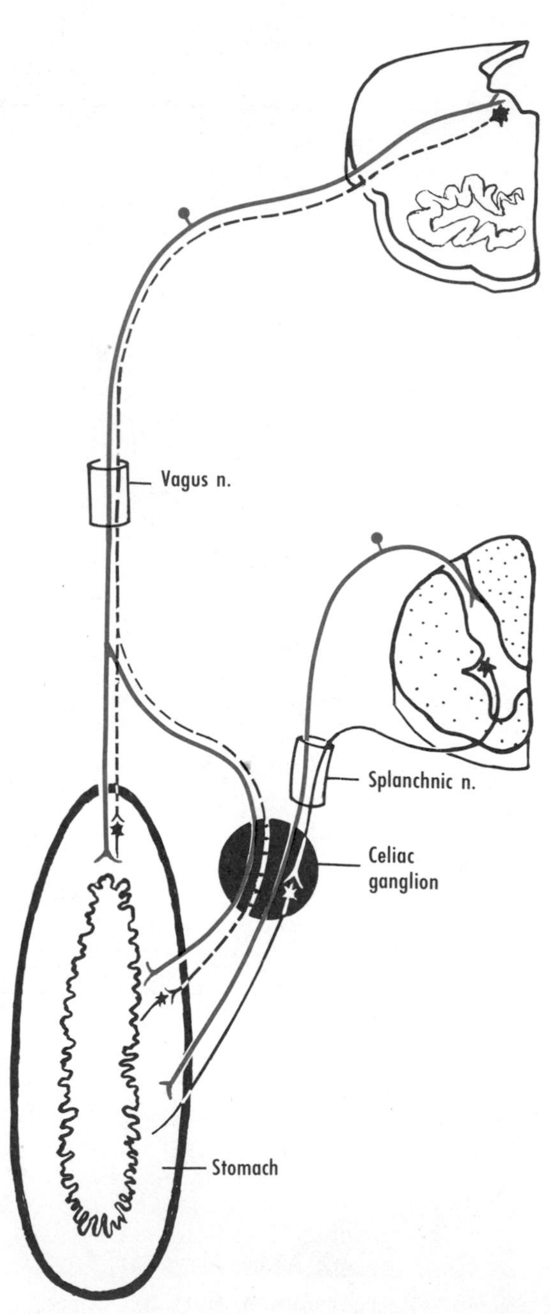

Figure 27–6 Functional components of the nerve supply of the stomach. For simplification, each component is shown as a single fiber. Sympathetic fibers are shown as continuous lines, parasympathetic fibers as interrupted lines, and sensory fibers in blue.

SMALL INTESTINE

The major part of digestion occurs in the small intestine, which extends from the pylorus to the ileocolic junction and includes the duodenum, jejunum and ileum. The Greek word *enteron,* meaning "gut," refers to the intestine (*enteritis* is inflammation of the intestine) and is used for its peritoneal attachment in the term *mesentery*. The small intestine, which is an indispensable organ, is about 7 m (varying from 5 to 8 m) in length. Its characteristic feathery appearance after a barium meal (see fig. 27–2) is due to permanent *circular folds* and villi. Intestinal movements in humans are poorly understood. The entrance of food into the stomach tends to cause the ileum to empty into the large intestine (*gastro-ileal reflex*).

DUODENUM (see figs. 27–4 and 28–8)

The duodenum, which is derived from both foregut and hindgut, is so called because it is approximately 12 (L., *duodenarius*) fingerbreadths in length. It is usually C-shaped, the concavity enclosing the head of the pancreas. It extends from the pylorus to the *duodenojejunal flexure*. Smooth muscle and elastic fibers from the stems of the superior mesenteric and celiac arteries to parts 3 and 4 of the duodenum constitute the *suspensory muscle* of the duodenum. The suspensory muscle opens out the angle at the duodenojejunal flexure and thereby facilitates the passage of the contents.* The duodenum is described in four parts (see fig. 28–8).

The *first part* runs to the right and backward close to the vertebral column. In front are the liver and gallbladder; behind are the bile duct, portal vein, and pancreas. **The beginning of the first portion is the free part; i.e., it is not attached to the posterior abdominal wall. It is mobile and follows the movements of the pyloric part of the stomach. It lacks circular folds, giving the appearance of a "duodenal cap" on radiography** (see fig. 27–2).

The *second part* descends in front of the right renal vessels and behind the transverse colon.

The *third part* runs horizontally to the left across the inferior vena cava and aorta. It is crossed in front by the superior mesenteric vessels and the root of the mesentery.

The *fourth part* ascends on the left of the aorta and then turns anteriorly as the duodenojejunal flexure.

Peritoneal Relations. **The duodenal cap, which is often at the level of L.V.2 *in vivo,* is attached to the liver by the hepatoduodenal part of the lesser omentum and is mobile.** The rest of the duodenum, however, is retroperitoneal and relatively fixed (see fig. 26–6).

Bile and Pancreatic Ducts. **The descending (second) part of the duodenum receives the bile, pancreatic, and accessory pancreatic ducts.** The bile and pancreatic ducts frequently unite and form a short *hepatopancreatic ampulla,* which opens into the *greater duodenal papilla,* a small projection in the interior of the posteromedial aspect of the concavity of the duodenum (see fig. 28–6). In other cases, the ducts meet but open separately on the papilla. Each duct usually has a sphincter, the sphincter of the bile duct frequently being termed the *choledochal sphincter*. A sphincter of the hepatopancreatic ampulla may also be present. The accessory pancreatic duct empties into the *lesser duodenal papilla,* which is situated on the anteromedial aspect of the descending part of the duodenum, about 2 cm above the greater papilla. It is frequently absent.

Blood Supply. The upper half of the duodenum is supplied by branches of the celiac trunk, the lower half by branches of the superior mesenteric artery (see fig. 27–4). The arteries approach the duodenum at its concavity. An incision along the right edge of the second part of the duodenum would mobilize the duodenum and the head of the pancreas without endangering their blood supply.

JEJUNUM AND ILEUM

The jejunum and ileum are the continuous coiled part of the small intestine. **In contrast to the ileum, the jejunum is shorter and typically emptier, more vascular (redder *in vivo*), and more thickly walled. Moreover, its mesentery shows translucent areas between the vessels, owing to the absence of fat. However, it is often difficult to distinguish the distal part of the jejunum from the proximal part of the ileum.**

Occasionally a remnant of the vitello-intestinal duct persists in the adult as a *diverticulum ilei,* which is situated near the ileocolic junction. It may contain gastric or pancreatic

* This description (I. Jit and S. S. Grewal, J. Anat., *123*:397–405, 1977) differs from the former idea that the suspensory muscle angulates the duodenojejunal flexure.

tissue, and inflammation of the diverticulum may simulate acute appendicitis.

Peritoneal Relations. The jejunum and ileum are suspended from the posterior abdominal wall by the mesentery (see fig. 26–4); they are highly mobile and occupy much of the abdominal and some of the pelvic cavity.

The root of the mesentery (see fig. 26–6), about 15 cm long, runs downward and to the right from the duodenojejunal flexure to the *ileocolic junction,* which is approximately at the level of the right sacro-iliac joint. Between the two layers of the mesentery are the branches of the superior mesenteric vessels, as well as nerves, lymph nodes and vessels, and some fat.

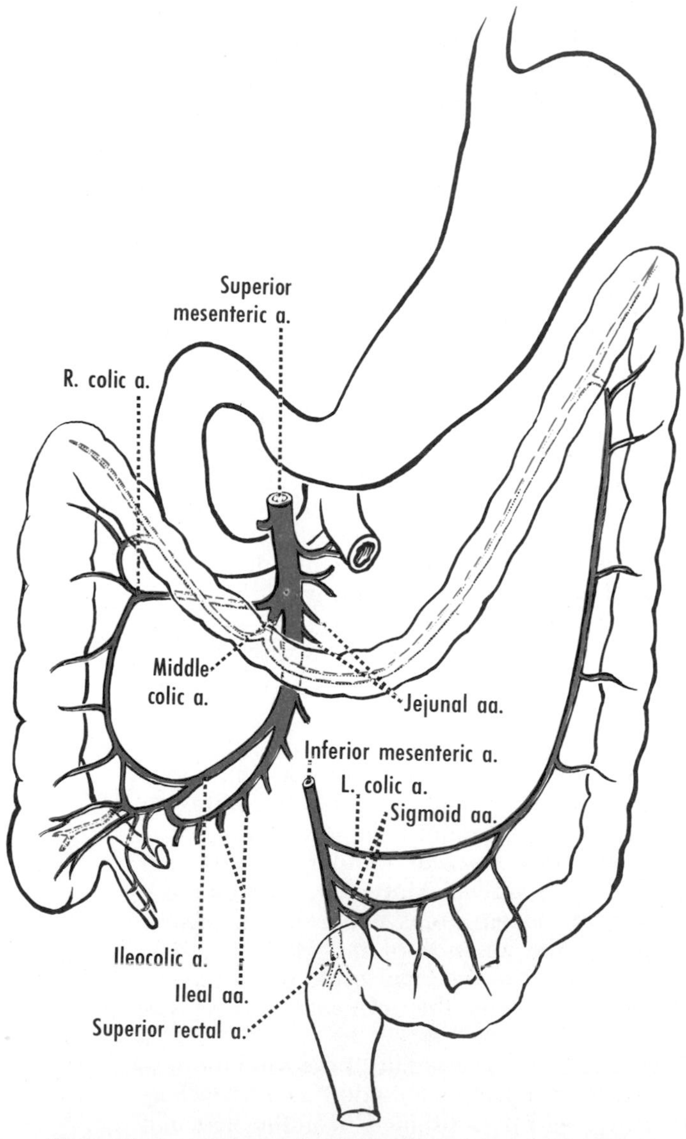

Figure 27–7 Arterial supply of the jejunum, ileum, and colon.

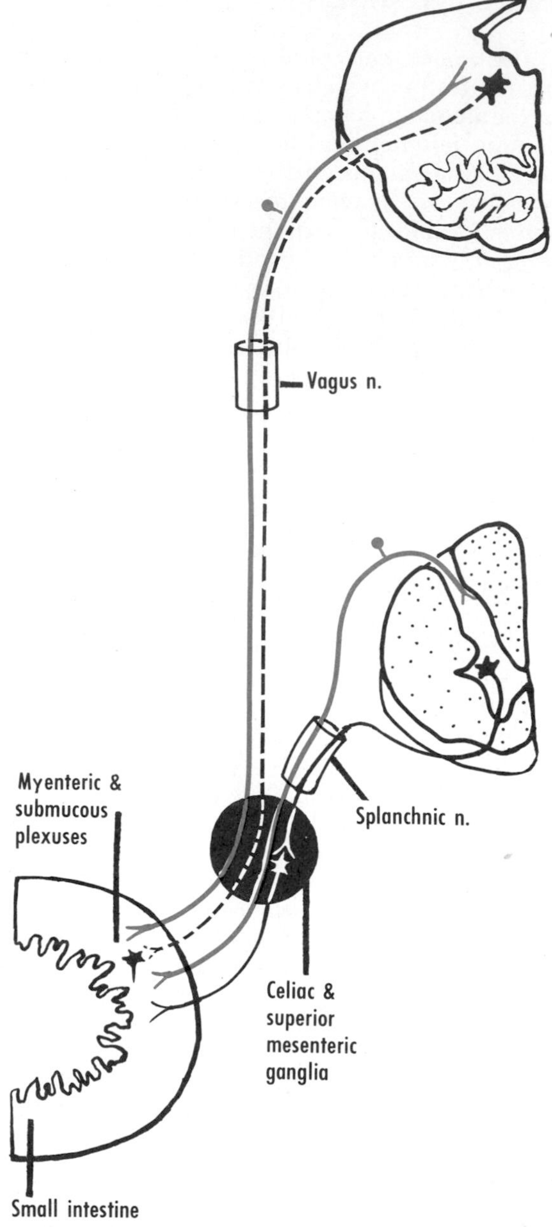

Figure 27–8 Functional components of the nerve supply of the small intestine. Sympathetic fibers are shown as continuous lines, parasympathetic fibers as interrupted lines, and sensory fibers in blue.

Blood Supply (fig. 27–7). The jejunum and ileum are supplied by the superior mesenteric artery through a series of arcades within the mesentery.

Lymphatic Drainage. After a fatty meal, the mesenteric lymphatics contain emulsified fat in a creamy lymph termed *chyle.* Lymph nodes and vessels are abundant in the mesentery. Carcinoma of the intestine can spread to the liver through the portal vein as well as through lymphatics.

Innervation (fig. 27–8). The small intestine is supplied by autonomic and sensory fibers from the celiac and superior mesenteric plexuses. The sensory fibers include those for pain, the intestine being sensitive to distension ("cramps").

LARGE INTESTINE (fig. 27–9)

The large intestine comprises the cecum and appendix; the ascending, transverse, descending, and sigmoid colon; and the rectum and anal canal (see fig. 27–7). Air and gas bubbles

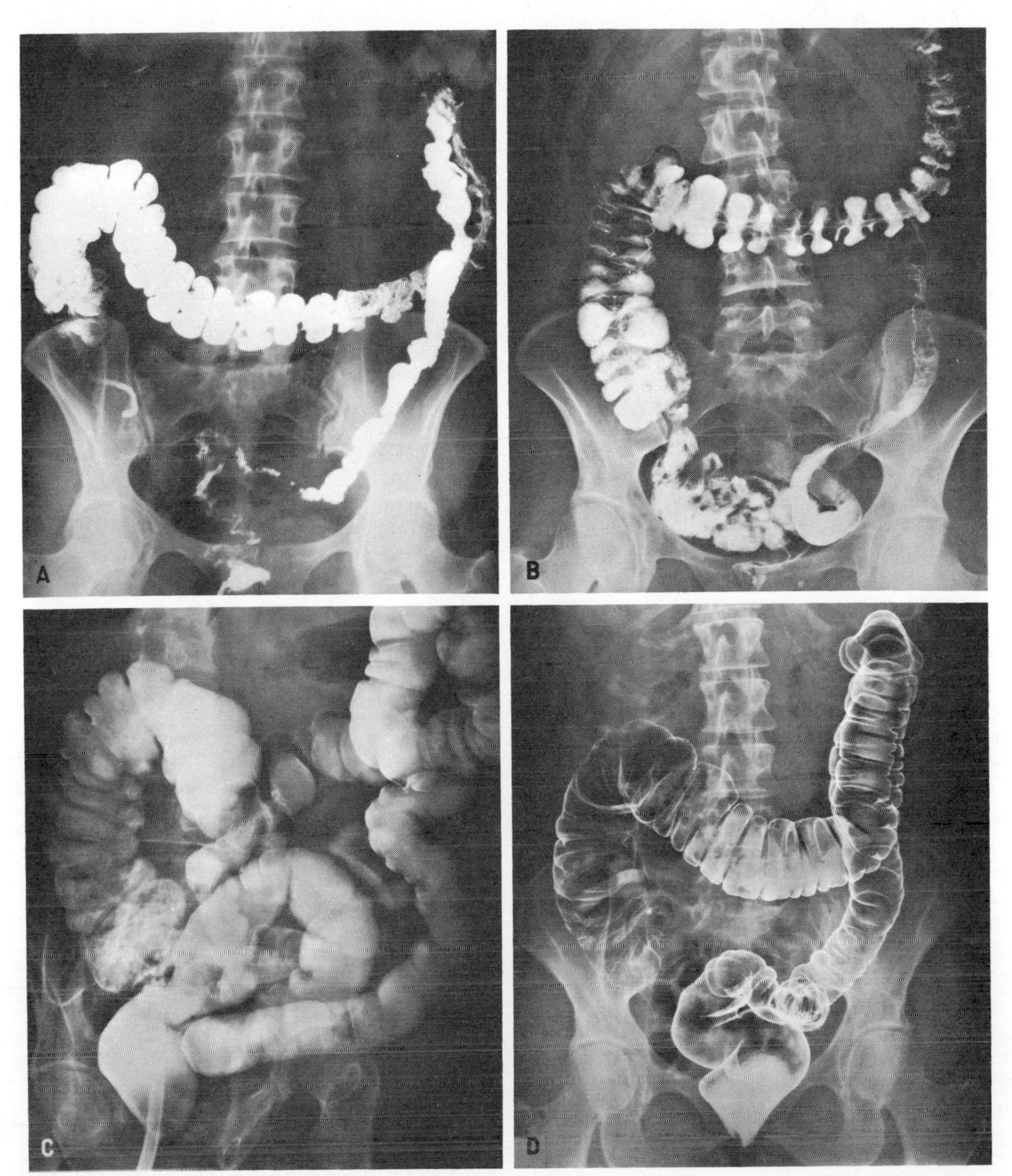

Figure 27–9 The large intestine shown by a barium enema. Note the different levels of the transverse colon. **A** shows the appendix. **B** shows the ileum and the pattern of the colon. **C,** Oblique view of the colon and rectum. **D,** The colon and rectum outlined by a double contrast enema. (**A,** Courtesy of Maurice C. Howard, M.D., Omaha, Nebraska. **C,** Courtesy of Robert A. Powers, M.D., Palo Alto, California. **D,** Courtesy of Eugene F. Ahern, M.D., Minneapolis, Minnesota. **C** and **D** are from *Medical Radiography and Photography.*)

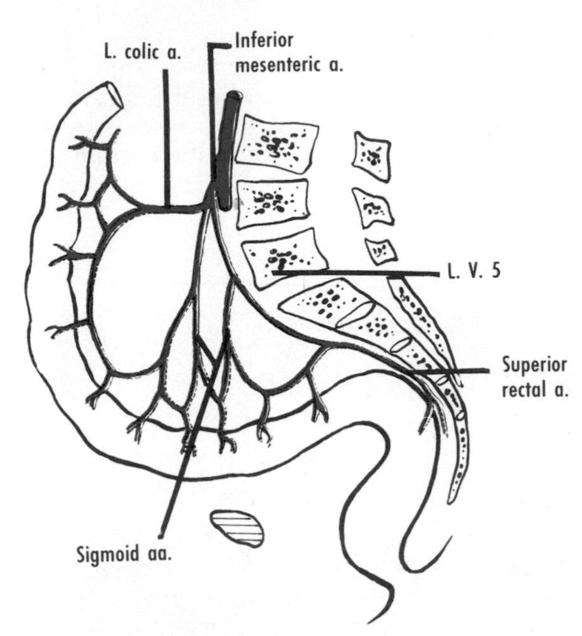

Figure 27–10 A common pattern of the inferior mesenteric artery as viewed in lateral perspective. The descending colon and sigmoid colon have been pulled forward.

are often seen in the large intestine radiographically. Most of the large intestine is characterized by a series of shifting sacculations termed *haustra* (see fig. 27–12). Multiple diverticula may develop in the colon (*diverticulosis*) and may become inflamed (*diverticulitis*).

The large intestine is concerned with the formation, transport, and evacuation of feces and is associated with the absorption of water and the secretion of mucus. The large intestine is distensible and mobile.

Blood Supply (figs. 27–7 and 27–10). The large intestine is supplied successively by the superior and inferior mesenteric arteries. The branches form a long marginal artery that usually extends from the cecum to the sigmoid colon. The venous drainage is mostly into the portal vein ultimately. The lymphatic drainage is similar to that of the small intestine.

Innervation (fig. 27–11). Autonomic and sensory fibers travel in continuations of the celiac and mesenteric plexuses that accompany the colic arteries. The parasympathetic supply to the sigmoid colon and to at least a part of the descending colon, however, is from

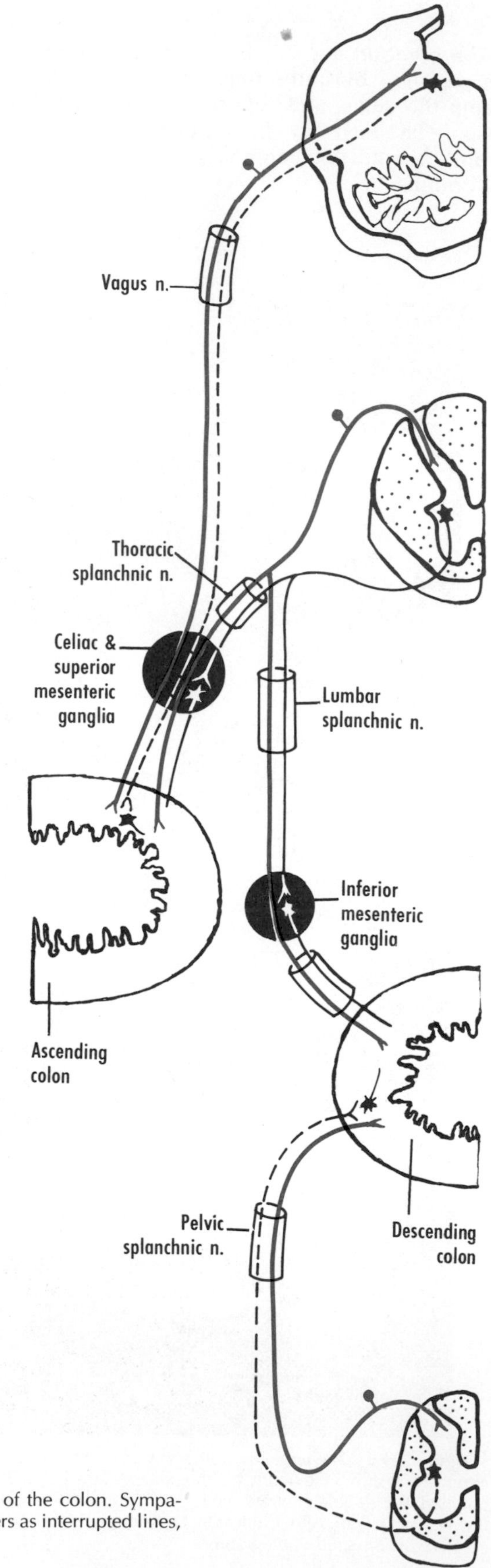

Figure 27–11 Functional components of the nerve supply of the colon. Sympathetic fibers are shown as continuous lines, parasympathetic fibers as interrupted lines, and sensory fibers in blue.

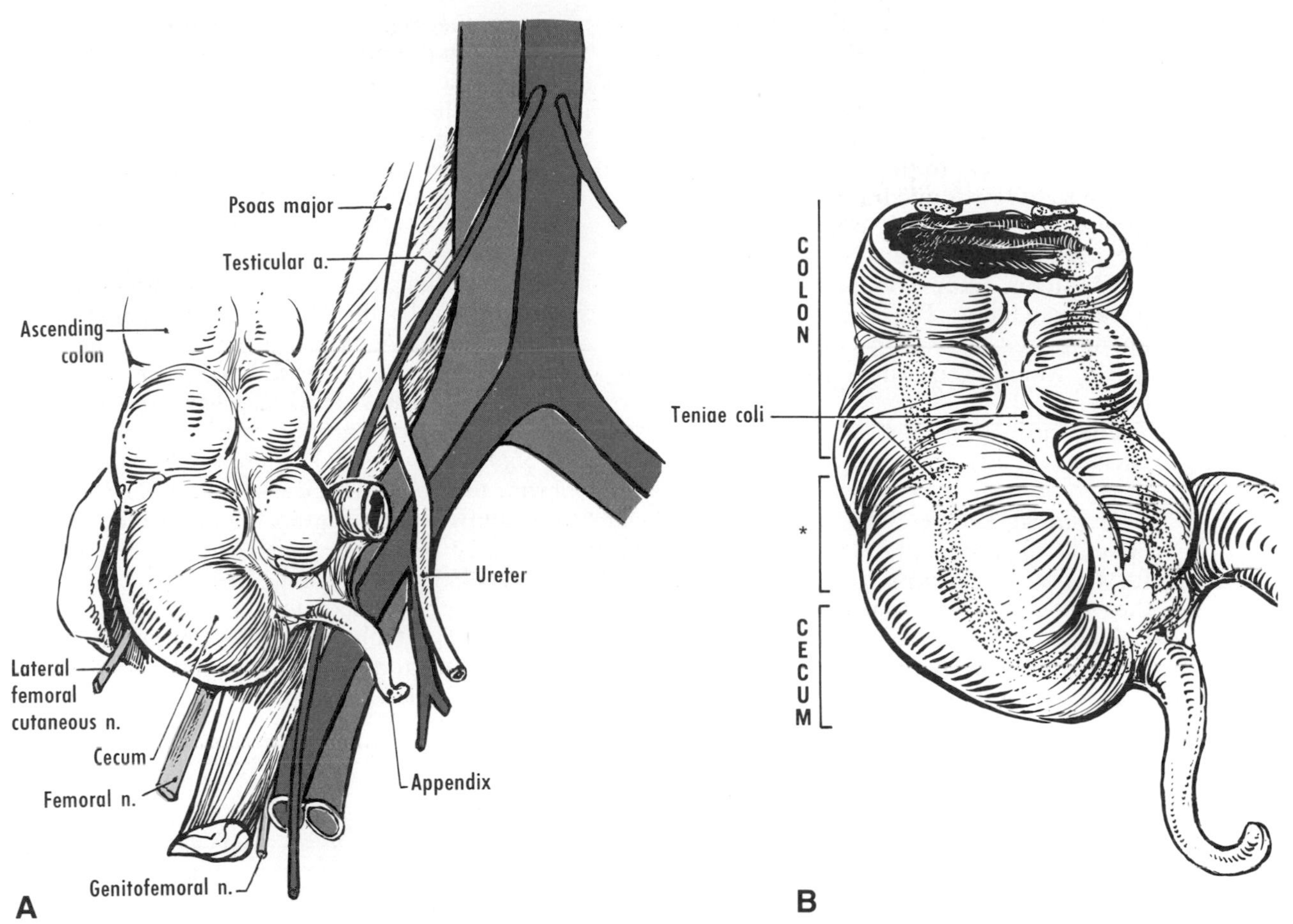

Figure 27–12 **A,** Common relations of the cecum and appendix. A free appendix may hang over the pelvic brim or be retrocecal. **B,** Positions of the teniae, which meet at the appendix. The region marked by the asterisk may be regarded as ileocolic or ileocecal.

the pelvic splanchnic nerves by way of the hypogastric nerves and the inferior hypogastric plexuses. Distension activates pain fibers in the splanchnic nerves.

Cecum. Although the cecum is frequently said to lie at as well as below the level at which the ileum joins the large intestine, it is considered in this book to lie below only, the term *ileocolic opening* being preferred to ileocecal (fig. 27–12*B*). The cecum lies in the right iliac fossa, may reach the pelvic brim, and is usually surrounded by peritoneum (see fig. 26–7).

The outer longitudinal layer of muscle of the cecum and colon is thickened as three bands known as *teniae coli* (fig. 27–12). The anterior tenia serves as a guide to the appendix.

Appendix (fig. 27–12). The vermiform appendix, often about 9 or 10 cm in length, arises typically from the posteromedial aspect of the cecum at the junction of the three teniae coli and about 1 or 2 cm below the ileum. Its mucosa is laden with lymphoid tissue.* The appendix usually possesses a peritoneal fold (its "mesentery" or *mesoappendix;* see fig. 26–7), which contains the appendicular artery (a branch of the ileocolic artery).

The appendix is variable in position and may be classified as (1) anterior, with ileal or pelvic (the commonest) positions, or (2) posterior, with subcecal, retrocecal, or retrocolic positions. Appendices may also be free or fixed. A free appendix may point in any direction, whereas a fixed appendix is usually either retrocecal or retrocolic. An inflamed appendix goes into spasm, becomes distended, and

* The vermiform appendix has a rich blood supply and "far from being a vestigial organ, it has actually developed progressively in primates" (G. B. D. Scott, J. Anat., *131:*549–563, 1980). Moreover, "the human appendix, at least in children, has the characteristics of a well-developed lymphoid organ, suggesting that it has important immunological functions" (P. Gorgollón, J. Anat., *126:*87–101, 1978).

causes pain referred to the epigastrium. Inflammation of the adjacent parietal peritoneum causes pain in the lower right quadrant of the abdomen, and the overlying muscles often show reflex spasm. **The point of maximum tenderness to pressure may be anywhere in the right lower quadrant.** Local pain may be minimal in retrocecal appendicitis, because the parietal peritoneum is not involved.

The ileum enters the large intestine usually posteromedially. The ileocolic (or ileocecal) opening presents two lips that form a so-called *valve,* which is of slight mechanical importance.

The surface of the colon presents small masses of fat, enclosed in peritoneum and termed *appendices epiploicae*.

Ascending Colon. The colon begins at the ileocolic junction and ascends in the right iliac fossa and on the posterior abdominal wall to the *right colic flexure,* which lies in front of the right kidney. An ascending mesocolon is said to be rare but perhaps may be frequent *in vivo*.

Transverse Colon. The part of the colon between the right and left flexures is derived from both the midgut and hindgut. Most of the transverse colon loops down, often below the level of the iliac crests and even into the true pelvis (see fig. 26–1). The *transverse mesocolon* is attached to the pancreas and is adherent to, but separable from, the greater omentum (see fig. 26–5*B*). The *left colic flexure* is usually higher, more acute, and less mobile than the right, and it may be anchored to the diaphragm (by the phrenicocolic ligament).

Descending Colon. From the left flexure, the colon descends, usually without a mesocolon, to the pelvic brim, where the sigmoid colon begins.

Sigmoid (or Pelvic) Colon. Below the pelvic brim, the colon acquires the *sigmoid mesocolon* and forms a variable loop, which, when full, may reach the epigastrium or lie in the pelvis. Feces are usually held in the sigmoid colon until immediately before defecation. The line of attachment of the sigmoid mesocolon may form an inverted V, with its apex in front of the left ureter.

QUESTIONS

27–1 Where are the junctions between the foregut and midgut and between the midgut and hindgut?

27–2 Provide an example of a sphincteric mechanism of the alimentary canal.

27–3 What is a hiatal hernia?

27–4 What is the origin of the terms *cardiac part, pylorus,* and *duodenum?*

27–5 What is congenital pyloric stenosis?

27–6 What is the "duodenal cap"?

27–7 How does the jejunum differ from the ileum?

27–8 What is the diverticulum ilei?

27–9 Devise a term for a surgical anastomosis (Gk, *stoma,* "mouth") between the stomach and jejunum.

27–10 What are the most frequent positions of the vermiform appendix?

27–11 What are the main differences between the small and large intestines?

28
THE LIVER, BILIARY PASSAGES, PANCREAS, AND SPLEEN

The ductal systems of the liver and pancreas develop in the embryo as outgrowths of the alimentary canal at the junction of the foregut and midgut, i.e., at the middle of the future duodenum.

LIVER

The liver (Gk, *hepar;* hence the adjective *hepatic*) is a large, soft, reddish organ and the largest gland in the body. **The exocrine secre-**

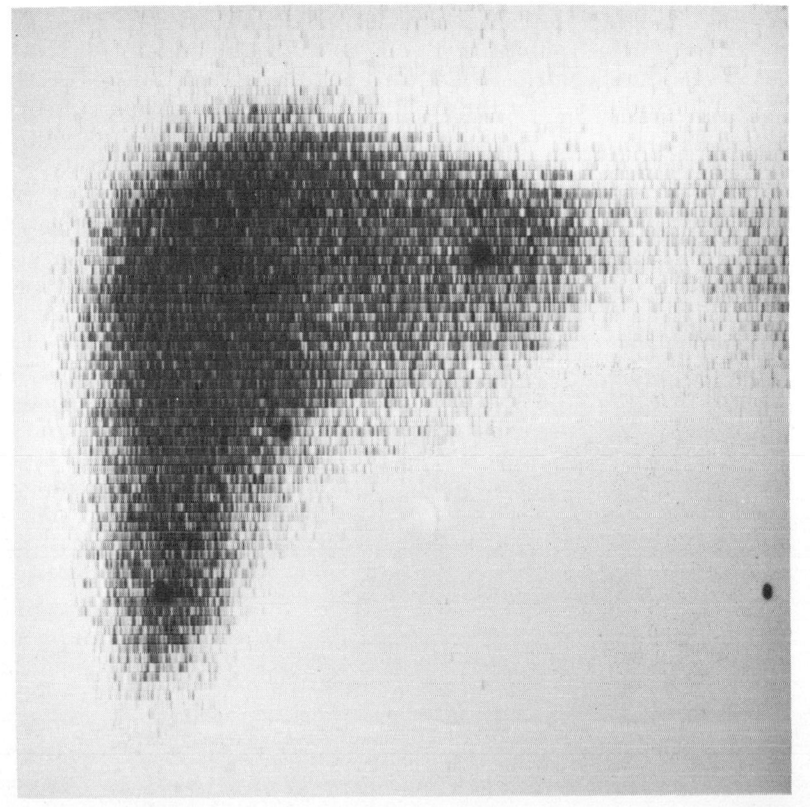

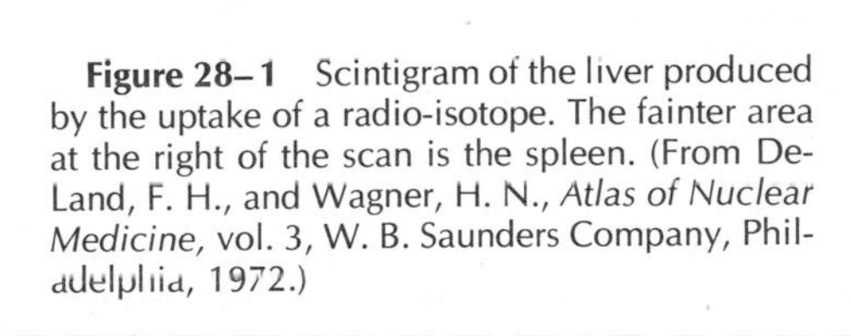

Figure 28–1 Scintigram of the liver produced by the uptake of a radio-isotope. The fainter area at the right of the scan is the spleen. (From DeLand, F. H., and Wagner, H. N., *Atlas of Nuclear Medicine,* vol. 3, W. B. Saunders Company, Philadelphia, 1972.)

tion of the liver is termed bile. Many products of the hepatic cells are discharged directly into the blood stream and may be considered the endocrine secretion of the liver.

The liver lies mostly under cover of the thoracic bony cage (see fig. 26–1) and is covered by the diaphragm. It can be demonstrated *in vivo* by the uptake of radio-active isotopes (fig. 28–1). The liver, which is relatively large at birth, presents diaphragmatic and visceral surfaces.

The *diaphragmatic surface,* smooth and convex, is separated in front and below from the visceral surface by the sharp *inferior border*. The *visceral surface* faces downward, backward, and to the left. It is related to the right colic flexure and right kidney and to the duodenum and stomach. The visceral surface presents *quadrate* and *caudate lobes* (see fig. 28–3*C*) marked off by an H-shaped series of grooves.

Porta Hepatis. **The cross-bar of the H is the porta hepatis, or hilus of the liver, which contains the hepatic ducts and the branches of the portal vein and hepatic artery.** The structures (vessels and ducts) at the porta constitute the hepatic pedicle and show many variations that are of surgical importance. The limbs of the H (see fig. 28–3*C*) are (1) the *fissure for the ligamentum teres,* which contains that ligament (obliterated left umbilical vein), (2) the *fissure for the ligamentum venosum,* which contains that ligament (obliterated ductus venosus), (3) the fossa for the gallbladder, which contains that organ, and (4) the sulcus for the vena cava, which lodges the inferior vena cava.

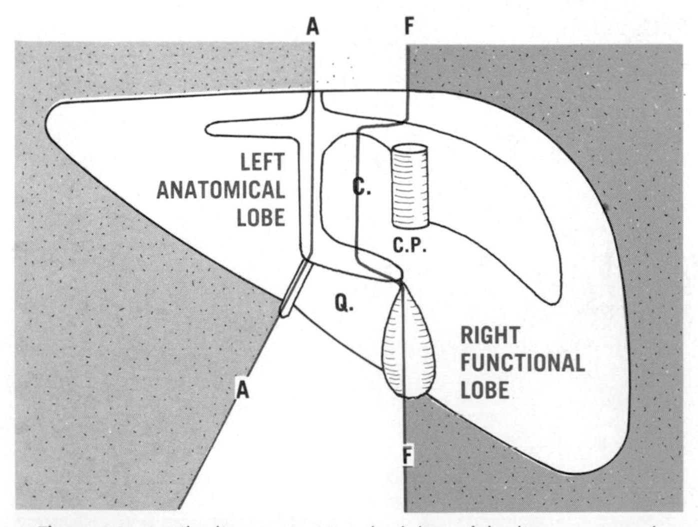

Figure 28–2 The lines separating the lobes of the liver, projected on the visceral surface. Line *A* separates the left and right anatomical lobes, the caudate (C.) and quadrate (*Q.*) lobes belonging to the right anatomical lobe. Line *F* separates the left and right functional lobes. *C.P.*, caudate process.

Lobes. **The liver can be divided into right and left anatomical lobes along the left-hand limb of the H** (fig. 28–2) and by the attachment of the falciform ligament on the diaphragmatic surface (fig. 28–3). The caudate and quadrate lobes are classified with the right anatomical lobe of the liver. However, in regard to the ductal and vascular distributions, **the right and left functional lobes would be separated by an irregular plane close to the right-hand limb of the H** (see fig. 28–2) and to the right of the falciform ligament. A number of *portal segments* have been described within these functional lobes (fig. 28–4). Hepatic segments, unlike bronchopulmonary segments, are not demarcated by connective tissue septa. The segmental anatomy of the liver is of diagnostic and surgical importance.

Peritoneal Relations (see fig. 28–3). The liver is nearly surrounded by peritoneum, which (as the ventral mesogastrium of the embryo) attaches it to the body wall (falciform ligament) and to the stomach (lesser omentum) (see fig. 26–4). The *falciform ligament* connects the anterior abdominal wall and diaphragm to the liver. The free edge of the falciform ligament meets the inferior border of the liver at a notch for the *ligamentum teres,* where that ligament (obliterated left umbilical vein) is conveyed to the porta. As the two layers of the falciform ligament reach the liver, the left layer becomes continuous above with the *left triangular ligament* and the right layer with the upper layer of the *coronary ligament*. The upper and lower layers of the coronary ligament, which meet at the right as the right triangular ligament, diverge toward the left and enclose the triangular "bare area" of the liver, which is in direct contact with the diaphragm. The junction of the left triangular and coronary ligaments leads to the lesser omentum, which has an L-shaped attachment to the liver. The vertical limb is at the fissure for the ligamentum venosum, and it extends to the lesser curvature of the stomach (*gastrohepatic ligament*). The horizontal limb of the L is at the porta hepatis and extends to the first part of the duodenum (*hepatoduodenal ligament*). The right free border of the lesser omentum contains the bile duct, portal vein, and hepatic artery.

Surface Anatomy. **The liver lies mostly under cover of the thoracic bony cage and is covered by the diaphragm. On the right side, it extends above the inferior border of the lung**

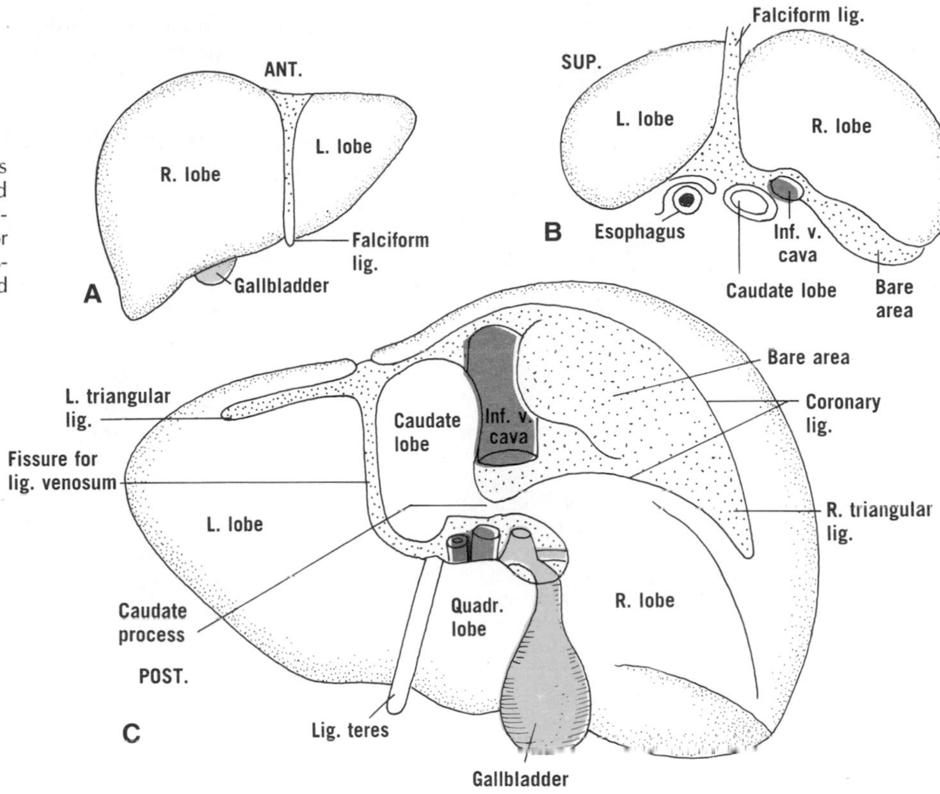

Figure 28–3 The liver and its peritoneal relations. The stippled areas represent surfaces not covered by peritoneum. **A,** Anterior view. **B,** Superior view. **C,** The visceral surface of the liver, viewed from behind.

and causes dullness there on percussion. The liver lies largely, and in some people entirely, to the right of the median plane. Its sloping, lower margin may reach to, or even be below, the right iliac crest. The radio-opacity of the liver is largely responsible for the outline of the diaphragm on radiography. The liver moves with respiration, and its position varies with the diaphragm and with body type. Hepatic tissue may be obtained for biopsy by puncture through a lower intercostal space.

Blood Supply (fig. 28–5). The liver has a double blood supply: from the hepatic artery and from the portal vein. Blood is returned to the vena cava by the hepatic veins.

Lymphatic Drainage. Subperitoneal networks drain to the internal thoracic nodes, and some vessels accompany the ligamentum teres to the umbilicus. Lymphatics accompany the blood vessels in the lesser omentum to reach the celiac nodes. Metastases can reach the liver from the thorax, breast, or any region drained by the portal vein.

Innervation. The hepatic plexus is derived

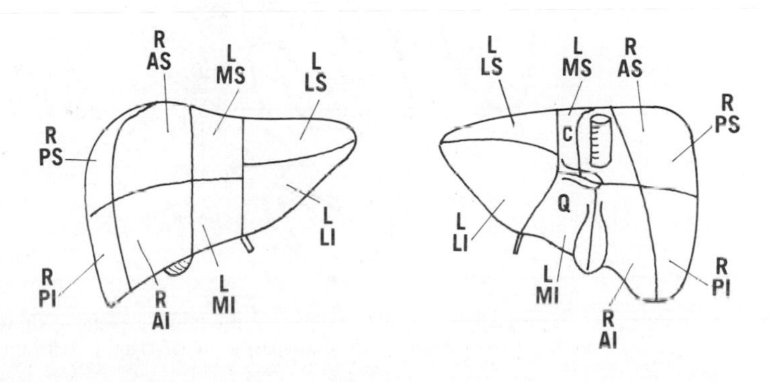

Figure 28–4 The vascular segments of the liver from in front and as projected onto the visceral surface. *AI,* antero-inferior; *AS,* anterosuperior; *L,* left; *LI,* latero-inferior; *LS,* laterosuperior; *MI,* medio-inferior; *MS,* mediosuperior; *PI,* postero-inferior; *PS,* posterosuperior; *R,* right.

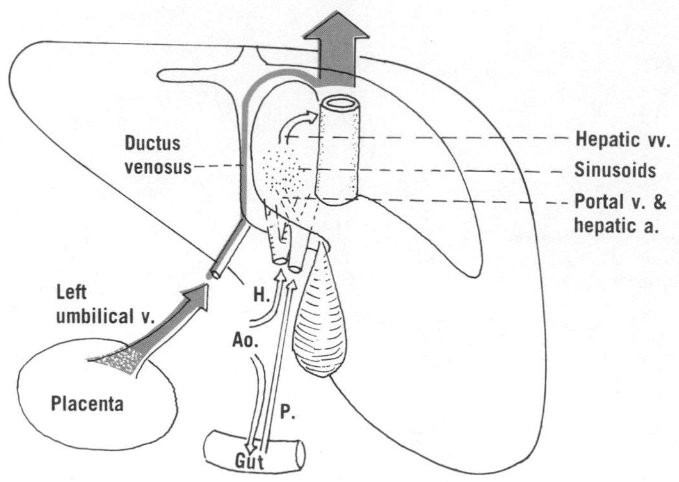

Figure 28–5 The blood supply of the liver and the portal circulation, projected onto the visceral surface. In prenatal life, oxygenated blood from the placenta is carried by the left umbilical vein and ductus venosus to the inferior vena cava and right atrium. Postnatally, part of this pathway becomes the ligamentum teres and the ligamentum venosum. The alimentary canal is supplied by blood from the aorta (*Ao.*) and is drained by the portal vein (*P.*). The portal blood and that from the hepatic artery (*H.*) provide a double blood supply to the liver. The blood is distributed by sinusoids within the liver and is drained by the hepatic veins into the inferior vena cava.

from the celiac plexus (see fig. 32–6). Pain fibers from the biliary passages are significant.

BILIARY PASSAGES (figs. 28–3 and 28–6 to 28–8)

The biliary passages are important surgically because of the frequency of inflammation and of gallstones. The passages are supplied chiefly by the cystic artery (a branch of the right hepatic artery). Pain fibers are carried by the splanchnic nerves, and pain from distension or spasm may be severe.

The right and left *hepatic ducts* emerge from the liver and unite to form the *common hepatic duct*. This receives the *cystic duct* from the gallbladder and becomes the *bile duct* (or *choledochal duct;* from Gk, *chole,* "bile"), which opens into the second part of the duodenum in common with or at least beside the pancreatic duct (fig. 28–6).

Gallbladder. The gallbladder stores and concentrates bile. It lies in a fossa on the visceral surface of the liver, covered below and at the sides by peritoneum. The gallbladder

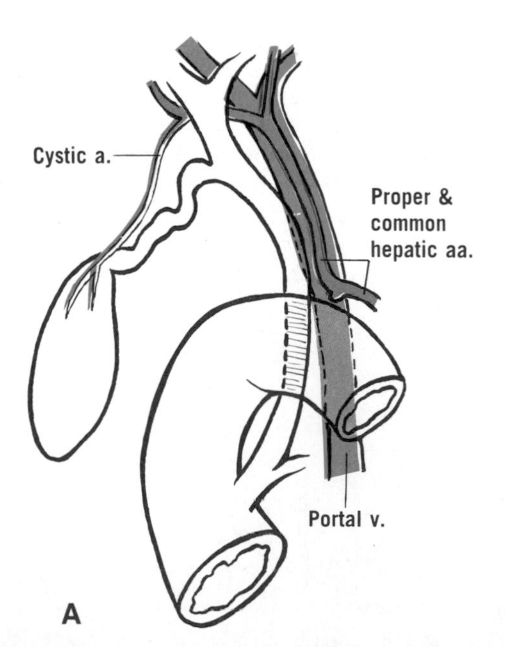

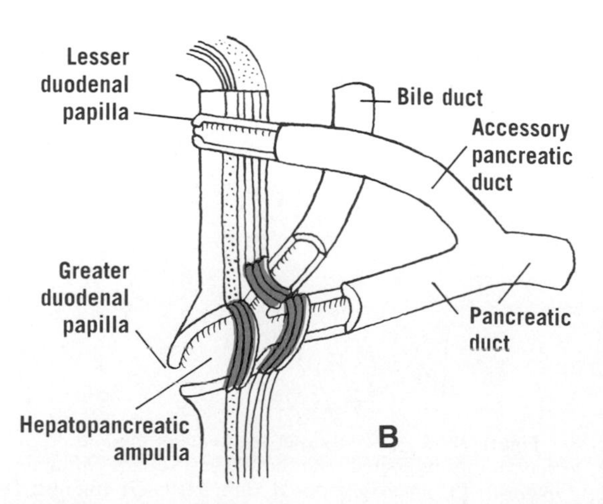

Figure 28–6 **A,** One pattern of the hepatic pedicle, showing structures important in cholecystectomy. **B,** An enlarged view to show the bile and pancreatic ducts, together with their sphincters (indicated in red).

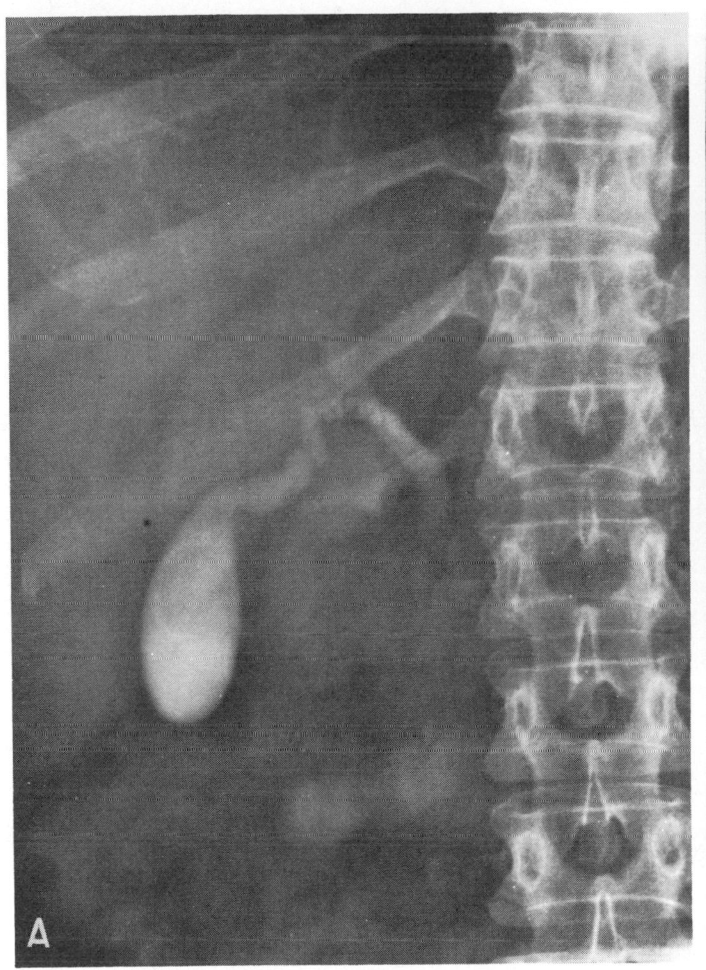

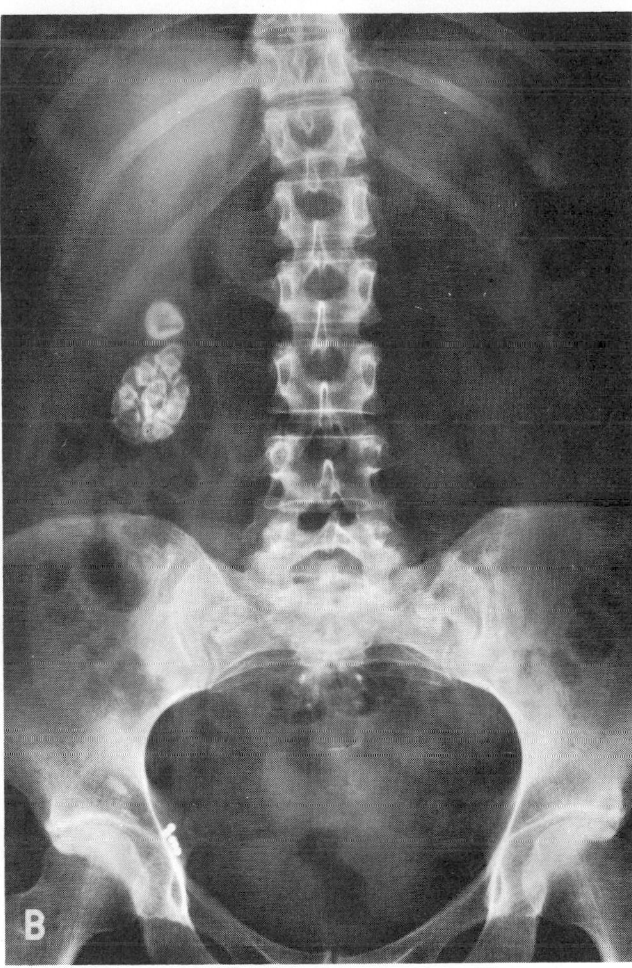

Figure 28–7 Gallbladder. **A,** A cholecystogram showing the right, left, and common hepatic ducts, cystic duct, gallbladder (the fundus is at the level of L.V.2 here), and bile duct. Cf. fig. 28–6*A*. **B,** The position of the gallbladder is shown by radio-opaque, multifaceted gallstones. The fundus is at the level of the upper portion of L.V.4 here. (**A,** Courtesy of John Pepe, M.D., Brooklyn, New York.)

comprises a *fundus* (at or below the lower border of the liver), *body*, and *neck*, which continues as the cystic duct. The region of the neck is frequently S shaped and may present an abnormal pouch, and its mucosa presents spiral folds. The gallbladder lies between the liver, the first or second part (or both) of the duodenum, and the anterior abdominal wall. **The gallbladder is sought in the angle between the right costal margin and the linea semilunaris, usually on the transpyloric plane, but it may be as low as the iliac crest.** The gallbladder can be made radio-opaque (*cholecystography*, fig. 28–7). Biliary constituents may be thrown out of solution and form gallstones (*cholelithiasis*).

Bile Duct. The bile duct (the qualification "common" is unnecessary) runs in the free edge of the lesser omentum, then behind the first part of the duodenum and through (or at least enfolded by) the head of the pancreas, ending in the second part of the duodenum. As the bile duct traverses the duodenal wall, its wall acquires the *choledochal sphincter* and its lumen becomes narrowed. It is usually united to the pancreatic duct, and the two ducts frequently empty into a common channel (*hepatopancreatic ampulla*), which enters the duodenum at the *greater duodenal papilla* (see fig. 28–6).

PANCREAS (fig. 28–8)

The pancreas, an exocrine and endocrine gland, comprises a head, neck, body, and tail.

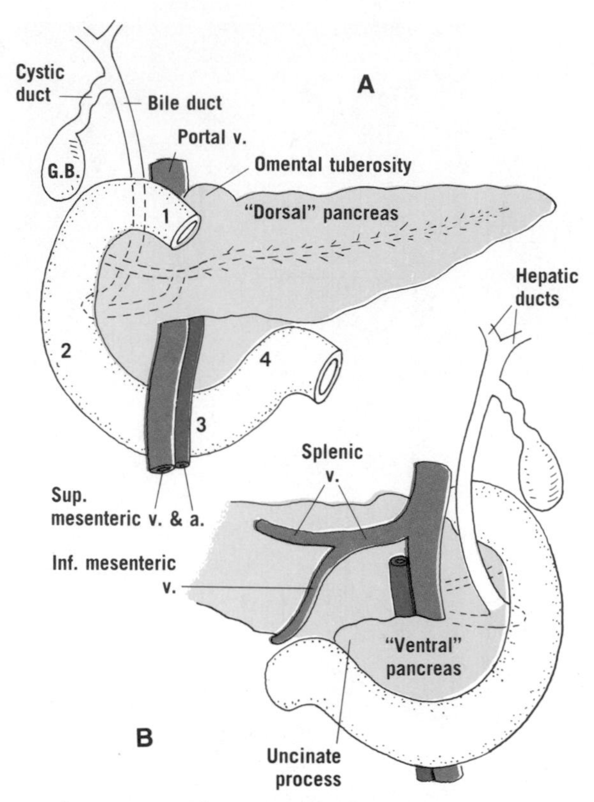

Figure 28–8 The pancreas and portal vein. **A,** Anterior view showing the head of the pancreas within the cavity of the duodenum. **B,** Posterior view showing the formation of the portal vein behind the neck of the pancreas. The pancreas develops by the fusion of a dorsal with a ventral component, hence the frequency of two ducts (main and accessory) that open into the duodenum. *1* to *4*, parts of the duodenum; *G.B.*, gallbladder.

The portal vein is formed behind the neck of the pancreas by the union of the superior mesenteric and splenic veins.

The *head* lies within the curve of the duodenum and usually surrounds the lower part of the bile duct. The head is prolonged below and to the left as the *uncinate process,* which hooks behind the superior mesenteric vessels. The *body* and *tail* extend to the left, across the vertebral column. The tail (within the lienorenal ligament) ends at the spleen. The body lies above the duodenojejunal flexure and presents anterior, inferior, and posterior surfaces.

The pancreas lies behind the stomach and in front of the inferior vena cava, aorta, left crus of the diaphragm, left suprarenal gland, and left kidney (see figs. 26–6 and 29–2). The splenic artery runs tortuously along the upper border of the pancreas, whereas the splenic vein lies behind the organ.

Aberrant pancreatic tissue may be found in the stomach, small intestine, gallbladder, or spleen. Rarely, the pancreas is annular, and the ring may constrict or obstruct the duodenum.

Peritoneal Relations. The pancreas (except for its tail) is retroperitoneal. The root of the transverse mesocolon is attached to the anterior border of the body (see fig. 26–6), and the anterior surface of the organ is related to the lesser sac, whereas the inferior surface is associated with the greater sac (see fig. 26–4).

Pancreatic Ducts. The pancreatic duct begins in the tail, runs to the right, and, meeting the bile duct, empties into the second part of the duodenum at the greater duodenal papilla. A part of the head is drained by an *accessory pancreatic duct,* which frequently empties separately into the duodenum higher up, at the *lesser duodenal papilla* (see fig. 28–6*B*).

The pancreas develops from the gut as dorsal and ventral diverticula that soon fuse, and their ductal systems also unite. The body of the adult organ is drained by the dorsal duct (the original termination of which may open on the lesser duodenal papilla), whereas the head is drained by both dorsal and ventral ducts (which, by their union, empty on the greater duodenal papilla).

Blood Supply, Lymphatic Drainage, and Innervation. The front and back of the head are supplied by the anterior and posterior superior pancreaticoduodenal arteries (from the hepatic via the gastroduodenal artery) and by the anterior and posterior inferior pancreaticoduodenal arteries (from the superior mesenteric) (see fig. 27–4). The remainder of the organ is supplied by branches of the splenic artery. The lymphatic drainage is to the adjacent nodes. Pain fibers are carried by the splanchnic nerves.

SPLEEN (see figs. 26–1 to 26–3, 26–7, 27–4, and 29–2)

The spleen (Gk, *splen;* L., *lien;* hence the adjectives *splenic* and *lienal*) is a soft, vascular organ surrounded by a fibrous capsule and belonging to the lymphatic system. It filters blood, removes iron from hemoglobin, produces lymphocytes and antibodies, and stores and releases concentrated blood. **The spleen lies against the diaphragm and ribs 9 to 11 on the left-hand side of the body. Usually it is palpable only when enlarged.**

It is convenient to speak of diaphragmatic and visceral surfaces, superior and inferior borders, and anterior and posterior ends of the spleen. The *diaphragmatic surface* is related to the costal part of the diaphragm. The *visceral surface* is related to the stomach, left colic flexure, and left kidney and presents a fissure, the *hilus,* for blood vessels. The superior border is notched.

Accessory splenic tissue may be found in any part of the abdominal cavity, but chiefly in the tail of the pancreas.

Peritoneal Relations. The spleen (except at the hilus) is surrounded by peritoneum. It is connected to the stomach (*gastrolienal,* or *gastrosplenic, ligament*) and to the left kidney (*phrenicolienal* and *lienorenal ligaments*). The lienorenal ligament (see fig. 26–7) transmits the splenic vessels and contains the tail of the pancreas.

Blood Supply. The spleen is supplied by the splenic artery, usually a branch of the celiac trunk. It runs tortuously to the left along the upper border of the pancreas (see fig. 27–4). The splenic vein runs to the right behind the body of the pancreas and unites with the superior mesenteric vein to form the portal vein (fig. 28–8).

QUESTIONS

28–1 How is the liver divided into right and left anatomical lobes?

28–2 How is the liver divided into right and left functional lobes?

28–3 What is the ligamentum teres?

28–4 How is the liver maintained in position?

28–5 What and where is the "bare area" of the liver?

28–6 What is contained in the right free border of the lesser omentum?

28–7 What is the Greek word for bile?

28–8 Where is the gallbladder in terms of surface anatomy?

28–9 What are the main sphincters found in the biliary region?

28–10 Where is the portal vein formed?

28–11 What is an annular pancreas?

28–12 What is the developmental basis of the main and accessory pancreatic ducts?

28–13 On which side is the spleen?

28–14 Is the spleen palpable?

28–15 Where may accessory splenic tissue commonly be found?

29 THE KIDNEYS, URETERS, AND SUPRARENAL GLANDS

KIDNEYS

The kidneys (L., *ren;* Gk, *nephros;* hence the adjectives *renal* and *nephric*) belong to the urinary system, maintain the ionic balance of the blood, and excrete waste products as urine. They are reddish-brown organs covered by a thin, glistening, fibromuscular capsule that normally can be stripped easily. Each kidney has anterior and posterior surfaces, upper and lower poles, and lateral and medial borders. The medial border is indented at the

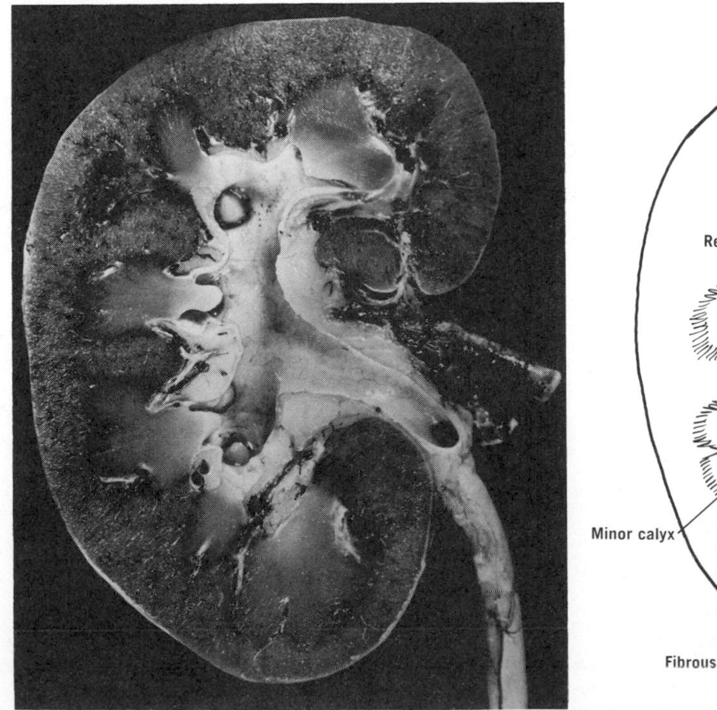

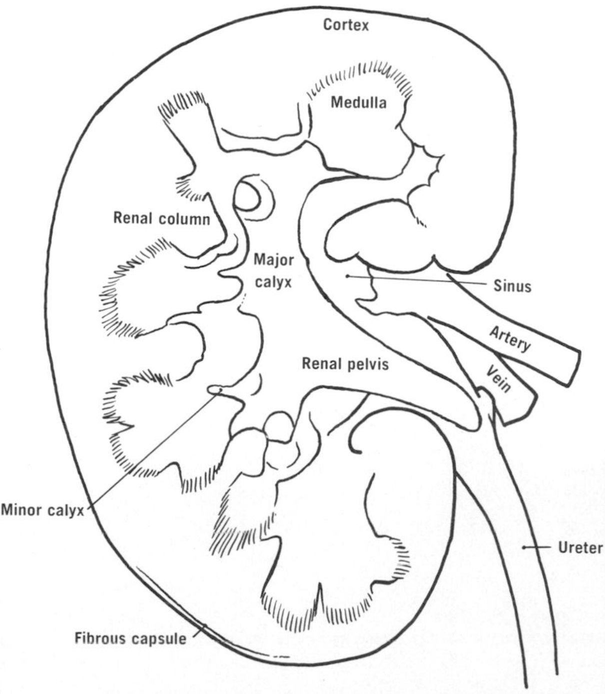

Figure 29–1 Coronal section of a kidney. (From Yokochi, C., *Photographic Anatomy of the Human Body,* Igaku Shoin, Ltd., Tokyo, 1971.)

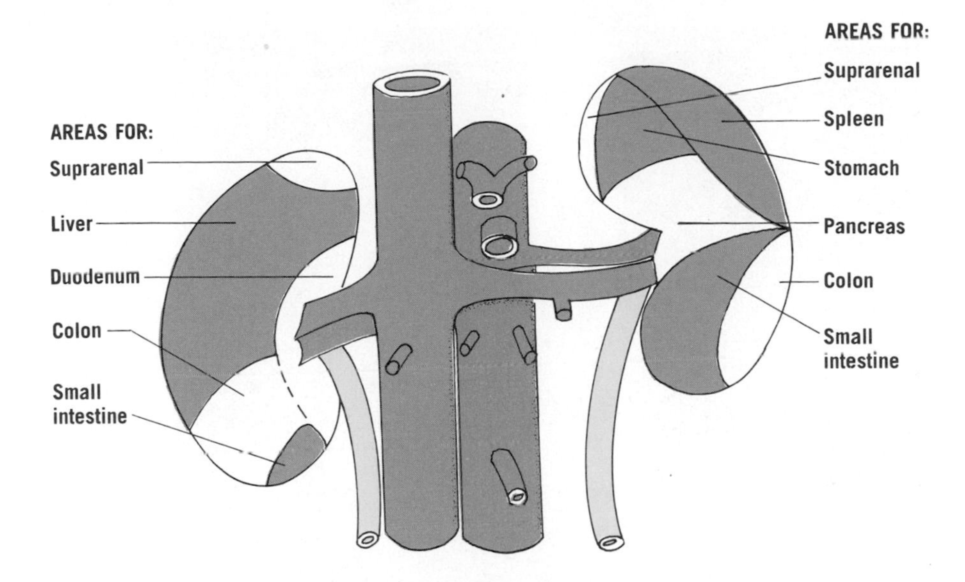

Figure 29–2 Anterior relations of the kidneys. The areas covered by peritoneum are shown in blue. In addition to the renal vessels, the origins of the celiac, superior mesenteric, gonadal (testicular or ovarian), and inferior mesenteric arteries are included, as are the terminations of the gonadal veins.

hilus, where blood vessels enter and leave, and there the ureter emerges. Each kidney is composed of a paler cortex and a darker medulla (fig. 29–1). **The kidneys lie obliquely along the vertebral column, against the psoas major muscles.**

Relations (fig. 29–2). The upper pole of the kidney is covered by the suprarenal gland. In front, the right kidney is related to the liver, duodenum, ascending colon or right colic flexure, and small intestine. The left is related to the spleen, stomach, pancreas, descending colon or left colic flexure, and small intestine. Posteriorly, the kidneys are related to rib 12 and the diaphragm, psoas major, quadratus lumborum, and transversus abdominis (see fig. 29–5).

The upper part of the kidney is usually separated by the diaphragm from the pleura and lung. In the vertebrocostal trigone, however, the kidney and pleura may be separated only by connective tissue (see fig. 25–13*B*).

PERITONEAL RELATIONS. The kidneys are retroperitoneal. Certain areas of each kidney are covered in front by peritoneum, whereas others are "bare" (fig. 29–2).

Surface Anatomy. **When a subject is in the erect position, the kidneys are opposite the first four lumbar vertebrae; they are one vertebra higher when the subject is recumbent. The right kidney is frequently a little lower than the left (probably because of the liver), and its lower pole may be palpable.** The levels alter with respiration as well as with posture.

Hilus. The hilus for the renal vessels and ureter is situated on the medial border and leads into a recess termed the *renal sinus*. The sinus contains the renal vessels and an expansion of the ureter termed the *pelvis*. Within the sinus, the ureteric pelvis usually divides into two or three short tubes, the *major calices*, each of which subdivides into 7 to 14 *minor calices* (fig. 29–3). Each minor calyx receives the openings of collecting tubules on papillae that project into the calices (see fig. 29–1).

Renal Pedicle. The ureter and renal vessels near the hilus form the pedicle, important variations of which are common. The renal vein (figs. 29–2 and 29–4) is in front, the ureter behind, and the arteries more or less between.

Renal Fascia (fig. 29–5). The kidney is enclosed in a condensation of the extraperitoneal tissue termed the *renal fascia*. Its anterior layer continues across the median plane, whereas the posterior layer merges with the prevertebral connective tissue. The two layers are fused strongly above and weakly below the suprarenal gland. *Perirenal* (or *perinephric*) *fat* lies between the fascia and the renal capsule. *Pararenal fat* is situated external to the renal fascia.

Blood Supply and Lymphatic Drainage. The renal arteries arise from the aorta, and the right one passes behind the inferior vena cava (see figs. 29–2 and 29–4). Renal segments based on the arterial distribution have been identified (fig. 29–6). The left renal vein, which is longer than the right, drains not only the kidney but also the suprarenal gland, gonad, diaphragm, and body wall (see fig. 29–4). Lymphatic drainage is into adjacent nodes and thence into lumbar nodes.

Innervation. Extensions of the celiac (aor-

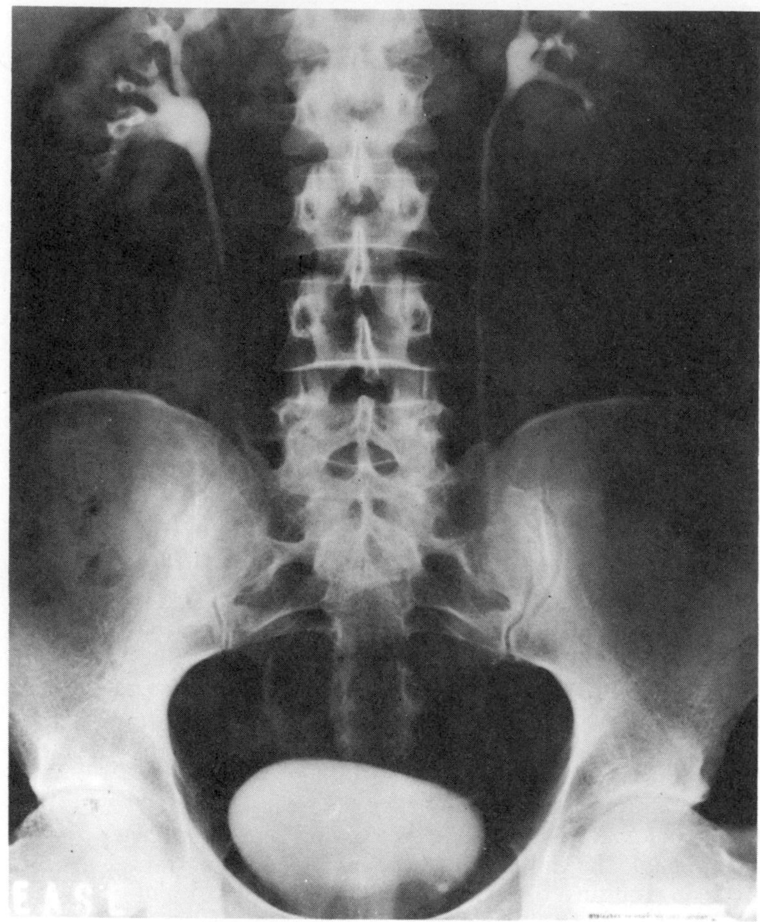

Figure 29–3 Intravenous pyelogram. Note the calices, some of which are seen from the side and others end-on, and the pelves of the ureters, which differ in shape and level. (Courtesy of Sir Thomas Lodge, Sheffield, England.)

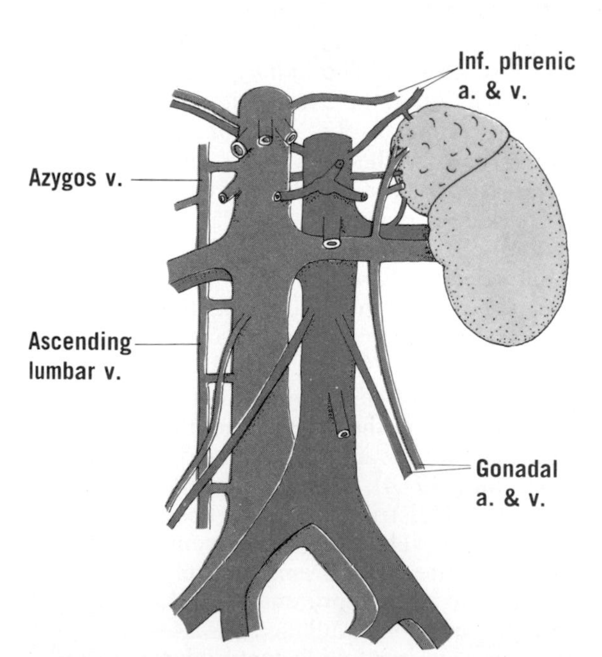

Figure 29–4 Renal and related vessels. A large area is drained by the left renal vein, which receives tributaries from the back, abdominal wall, diaphragm, suprarenal gland, and gonad.

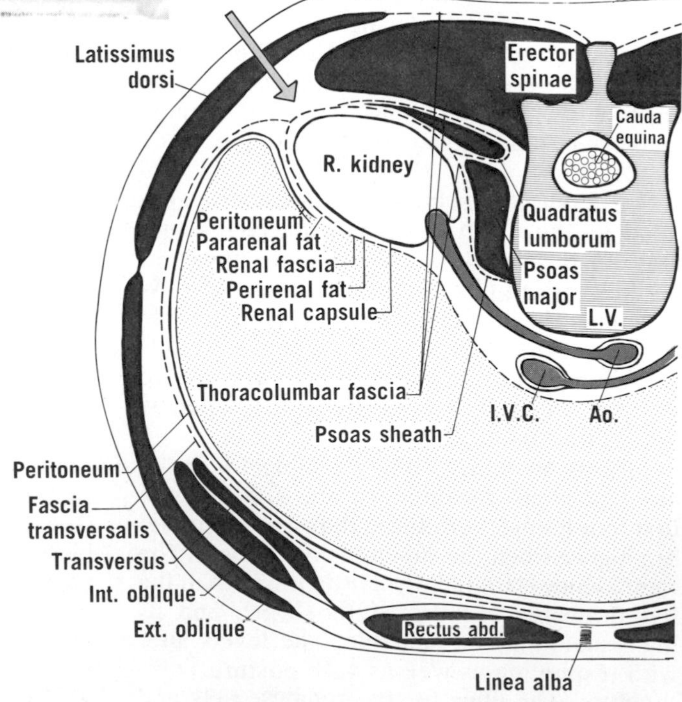

Figure 29–5 Horizontal section through the abdomen to show renal fascia, as described in the text. The arrow indicates the lumbar approach to the kidney. Other features shown include the rectus sheath. (Cf. fig. 25–6.) See also fig. 40–1. (After Symington.)

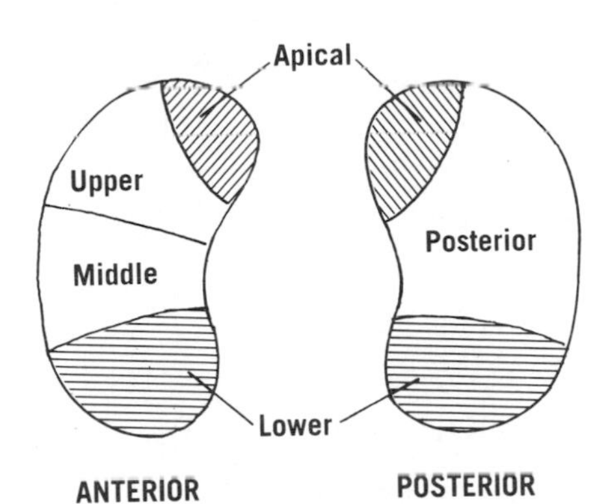

Figure 29–6 Arterial segments of right kidney. The left is similar. (After Fourman and Moffat.) Interpretation of the posterior segment as two units (upper dorsal and lower dorsal) would provide a total of six arterial segments. The relationship between these and the four to nineteen lobes of the kidney has been examined in detail by Inke (in Vidrio, E. A., and Galina, M. A., eds., *Advances in the Morphology of Cells and Tissues,* Liss, New York, 1981, pp. 71–78).

ticorenal) and intermesenteric plexuses accompany the renal arteries, and the splanchnic nerves supply branches that include pain fibers from the ureteric pelvis (see figs. 30–4 and 32–6).

Congenital Anomalies. Congenital anomalies of the kidneys include lobation (which is normally evident at birth), so-called accessory (e.g., polar) arteries, duplications (e.g., of ureters), ectopia and malrotation (e.g., pelvic kidney and horseshoe kidney), and cystic disease.

URETERS (see figs. 26–1, 29–1, and 29–3)

The ureter is a retroperitoneal, distensible muscular tube that connects the kidney with the bladder. In position, the upper half is abdominal, the lower half pelvic. The ureter commences as a dilatation, the *pelvis,* behind the renal vessels, and it descends on the psoas major. It crosses the common or the external iliac artery, courses along the lateral wall of the pelvis, and turns medially to reach the bladder. Near the ischial spine, the ureter turns downward, forward, and medially below the uterine vessels, about 2 cm from the cervix (where it may be endangered in hysterectomy).

The ureter may be constricted (1) at the narrowing of the ureteric pelvis, (2) where it crosses the pelvic inlet, and (3) during its course through the wall of the bladder. These are potential sites of obstruction.

The ureter is supplied by nearby arteries (renal, gonadal, and vesical) and from adjacent nervous plexuses (renal and hypogastric). Obstruction by a renal calculus (stone) causes acute distension and severe pain (renal colic). **Depending on the level of obstruction, the pain of renal (actually ureteric) colic may be referred to the lumbar or the hypogastric region or to the external genitalia.**

SUPRARENAL GLANDS

The suprarenal glands are paired endocrine organs situated (in most animals) adjacent to the kidneys (hence the terms adrenal and adrenaline) and, in the human, above the kidneys (hence suprarenal; cf. epinephrine).

Each suprarenal gland consists of two distinct endocrine organs, the *cortex* and the *medulla*. Some of their hormones are essential to life. Each suprarenal gland is surrounded by renal fascia and lies on the superomedial aspect of the front of the kidney (see fig. 29–2). The right gland is in contact with the bare area of the liver and projects behind the inferior vena cava (see fig. 30–3). The left gland, a little different in shape, is related in front to the lesser sac, the splenic artery, and the pancreas (see fig. 26–6). Both glands lie against the diaphragm.

Accessory cortical tissue may sometimes be found near the kidney or in the pelvis. The suprarenal medulla is part of the chromaffin system, other portions of which are found near sympathetic ganglia along the abdominal aorta and are termed *paraganglia* or *para-aortic bodies*.

Blood Supply (see fig. 29–4). The suprarenal glands are supplied by multiple and variable arteries (from the inferior phrenic and renal arteries and the aorta). The suprarenal vein emerges from a hilus and enters the inferior vena cava (right side) or the renal vein (left side).

Innervation. The celiac plexus and splanchnic nerves supply branches. Most of the fibers are preganglionic sympathetic and go directly to the cells of the medulla.

ADDITIONAL READING

Fourman, J., and Moffat, D. B., *The Blood Vessels of the Kidney*, Blackwell, Oxford, 1971. Renal vasculature in relation to function.

Graves, F. T., *The Arterial Anatomy of the Kidney*, Wright, Bristol, 1971. Anatomical studies as a basis for surgical technique.

QUESTIONS

29–1 Which level of the vertebral column would be crossed by a line joining the hili of the kidneys?

29–2 How may the kidney be approached surgically from behind?

29–3 Where would a perinephric abscess be situated?

29–4 List some congenital anomalies of the kidneys.

29–5 Where is a ureter likely to be obstructed?

29–6 To where is the pain of renal colic referred?

29–7 Behind which structures are the suprarenal glands found?

29–8 What are the cortical and chromaffin systems?

30 BLOOD VESSELS, LYMPHATIC DRAINAGE, AND NERVES OF THE ABDOMEN

BLOOD VESSELS

Apart from the anterolateral abdominal wall, the abdomen is supplied by branches of the abdominal aorta.

ABDOMINAL AORTA (see figs. 13–5, 30–1, 30–3, and 30–4)

The abdominal aorta begins at the aortic opening in the diaphragm (at about T.V.12) and descends in front of the vertebral bodies, at the left of the inferior vena cava (figs. 30–1 and 30–4). It ends (at about L.V.4 and slightly to the left) by dividing into the right and left *common iliac arteries*. The celiac plexus and ganglia are in front of it above, and the intermesenteric part of the aortic plexus is in front at a lower level. **The abdominal aorta may be compressed against the vertebral column by backward pressure on the anterior abdominal wall at the level of L.V.4, especially in children and thin adults.**

Parietal Branches. All of the parietal branches of the abdominal aorta are paired except the *median sacral artery*, which arises

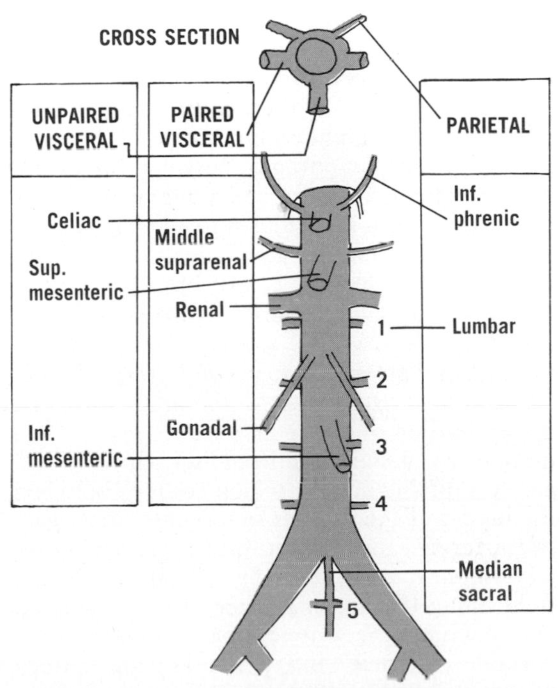

Figure 30–1 The branches of the aorta are arranged as unpaired visceral, paired visceral, and parietal.

near the bifurcation and descends to the coccygeal body. The *inferior phrenic arteries,* which frequently arise from the celiac trunk, supply the diaphragm and suprarenal glands. The *lumbar arteries* are small segmental branches that supply the muscles of the back and provide spinal branches.

Each common iliac artery runs downward and laterally and divides into the *external* and *internal iliac arteries*. The external iliac artery descends in the iliac fossa and passes behind the inguinal ligament to become the femoral artery. The aortic plexus is continued along the external iliac and femoral arteries. The external iliac artery gives off the *inferior epigastric* and *deep circumflex iliac arteries*. The internal iliac artery is described with the pelvis.

Visceral Branches. The middle suprarenal, renal, and gonadal arteries are paired. The *renal arteries,* which give off *inferior suprarenal arteries,* divide near or at the hili of the kidneys. The *gonadal arteries* arise below the renal arteries (the gonads develop near the kidneys) and are either testicular or ovarian in distribution. Each *testicular artery,* long and slender, descends on the psoas and reaches the deep inguinal ring. It then accompanies the ductus deferens and supplies the spermatic cord and testis. Each *ovarian artery* descends similarly and then enters the suspensory ligament of the ovary, runs medially in the mesovarium, and supplies the ovary.

The celiac trunk and superior and inferior mesenteric arteries are unpaired. **The celiac trunk is the artery to the caudal part of the foregut, i.e., as far as the middle of the duodenum.** It arises immediately below the aortic opening of the diaphragm, between the crura, and is embedded in the celiac plexus and ganglia. Commonly it divides into the left gastric, splenic, and common hepatic arteries. The *left gastric artery* (see fig. 27–4) runs to the left to reach the stomach, where it courses along the lesser curvature and anastomoses with the right gastric artery. The *splenic artery* (see fig. 27–4) runs tortuously to the left along the upper border of the pancreas. It gives off a number of pancreatic branches and the left gastro-epiploic artery (which courses between the layers of the greater omentum), short gastric arteries, and splenic branches. The *common hepatic artery* (see fig. 27–4) runs to the right along the upper border of the pancreas. On reaching the duodenum, it divides in a variable manner into the hepatic artery proper, the right gastric artery, and the gastroduodenal artery. The *hepatic artery proper* ascends in the free edge of the lesser omentum to the liver, where it divides into right and left branches, the former of which gives off the highly variable *cystic artery*. The *right gastric artery* courses along the lesser curvature and anastomoses with the left gastric artery. The *gastroduodenal artery* descends behind the first part of the duodenum, gives off the *posterior superior pancreaticoduodenal* artery, and divides into the *anterior superior pancreaticoduodenal* and *right gastro-epiploic arteries*. The last-named courses between the layers of the greater omentum and, like its left counterpart, gives rise to gastric and epiploic branches.

The superior mesenteric artery supplies the midgut, i.e., from the middle of the duodenum to the left part of the transverse colon. It arises from the front of the aorta below the origin of the celiac trunk. It emerges in front of the uncinate process of the pancreas and the third part of the duodenum (see fig. 27–7) to enter the root of the mesentery and gain the right iliac fossa. Its branches include the *inferior pancreaticoduodenal* artery (which divides into anterior and posterior components; see fig. 27–4); the *ileocolic, right colic,* and *middle colic arteries,* which contribute to the marginal artery and supply the terminal ileum, cecum, appendix, and much of the colon; and the *jejunal* and *ileal arteries* to the small intestine.

The inferior mesenteric artery supplies the hindgut, i.e., from the left part of the transverse colon. It arises from the aorta a little above its bifurcation and descends to the left on the psoas. It gives off the *left colic* and *sigmoid arteries* (see fig. 27–7), which contribute to the marginal artery. On crossing the pelvic inlet, the inferior mesenteric artery becomes the *superior rectal artery,* which passes between the layers of the sigmoid mesocolon to reach the rectum (see fig. 27–10).

Veins

The portal, caval, and vertebral systems require special consideration.

Portal System. **Venous blood from the stomach and intestine is collected by the portal vein and carried to the sinusoids of the liver and thence via the hepatic veins to the inferior vena cava** (see fig. 28–5).

The portal vein is formed behind the neck of the pancreas, usually by the union of the superior mesenteric and splenic veins (fig. 30–2). The portal vein enters the hepatoduodenal ligament (lesser omentum) and ascends to divide at the porta hepatis into right and left branches.

PORTAL-SYSTEMIC ANASTOMOSES. Because valves are insignificant or absent, **portal hypertension from obstruction of the portal vein readily causes enlargement of portal-systemic anastomoses and reversed flow of blood into systemic veins.** The important anastomoses are as follows:

1. Between the inferior mesenteric vein and the inferior vena cava and its tributaries.
2. Between gastric veins and the superior vena cava and its tributaries. Rupture of gastro-esophageal *varices* in portal hypertension may cause serious hemorrhage.
3. Between retroperitoneal veins and the caval system. Numerous, small retroperitoneal veins drain the "bare" (i.e., non-peritonealized) surfaces of the organs.
4. Between para-umbilical and subcutaneous veins. Para-umbilical veins in the falciform ligament connect the portal vein with subcutaneous vessels around the umbilicus, and these last drain into the epigastric veins and thence into the venae cavae.
5. Between (a) the superior and (b) the middle and inferior rectal veins. *Hemorrhoids* (piles) are varicose veins in the anal region.

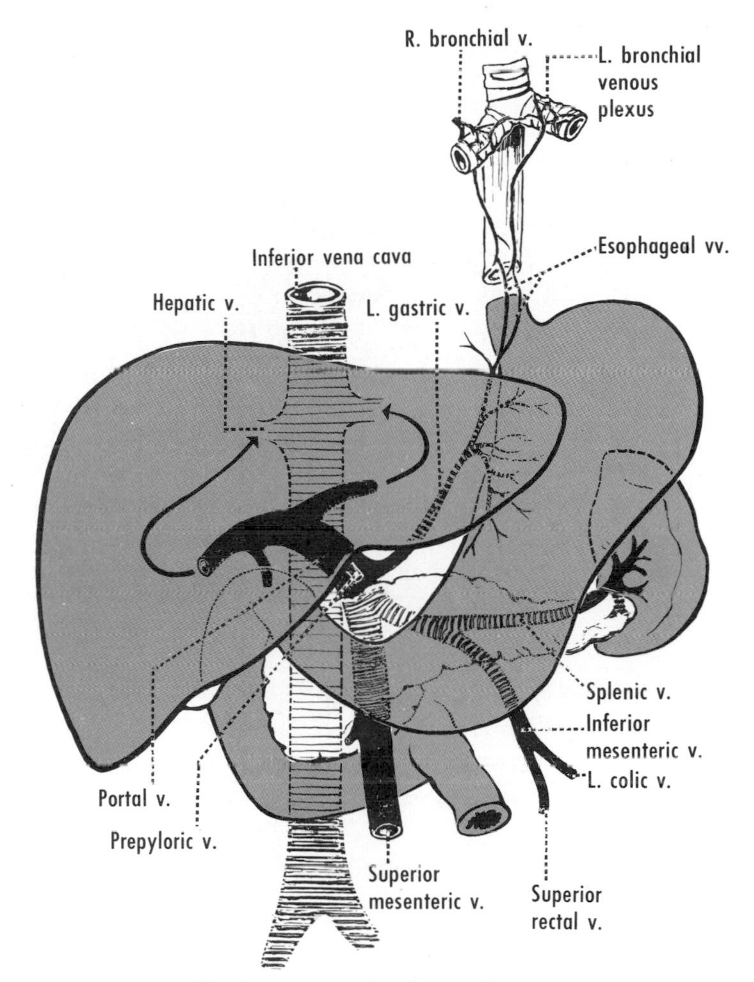

Figure 30–2 The portal vein and its major tributaries.

Inferior Vena Cava (see fig. 29–4). The inferior vena cava receives the blood from the lower limbs and much of the blood from the back and from the walls and contents of the abdomen and pelvis. It is formed by the union of the two common iliac veins at L.V.5, ascends at the right of the aorta, traverses the central tendon of the diaphragm, and empties into the right atrium. The usual tributaries are the common iliac, gonadal, renal, suprarenal, inferior phrenic, lumbar, and hepatic veins. **In some instances the vena cava crosses in front of (instead of behind) the ureter (pre-ureteric vena cava or postcaval ureter) and may cause ureteric obstruction.** Uncommonly, the embryonic left inferior vena cava may persist.

Vertebral Plexus. The main systemic channels have widespread valveless connections with the valveless vertebral plexus (see fig. 24–2). These are important in the spread of infections and tumors (e.g., from the prostate to the skeleton). During inspiration, venous return increases, and blood flows up into the vertebral plexus from the abdomen and out of the vertebral plexus into the thorax. The converse occurs during expiration. **Unlike the constant direction of flow in the main systemic venous channels, the direction of flow in the vertebral plexus varies according to the respiratory phase.** Furthermore, the flow of blood from the abdomen and pelvis into the vertebral plexus is accentuated by any increase in intra-abdominal pressure caused by coughing or straining.

LYMPHATIC DRAINAGE

Lumbar lymphatics ascend from the *iliac nodes* (see fig. 32–7) along the vertebral bodies and join the thoracic duct. *Lumbar (aortic) nodes* (fig. 30–3) overlie the transverse processes and the aorta. Bilateral connections are common. Nodes are also found scattered along the vessels supplying the abdominal organs.

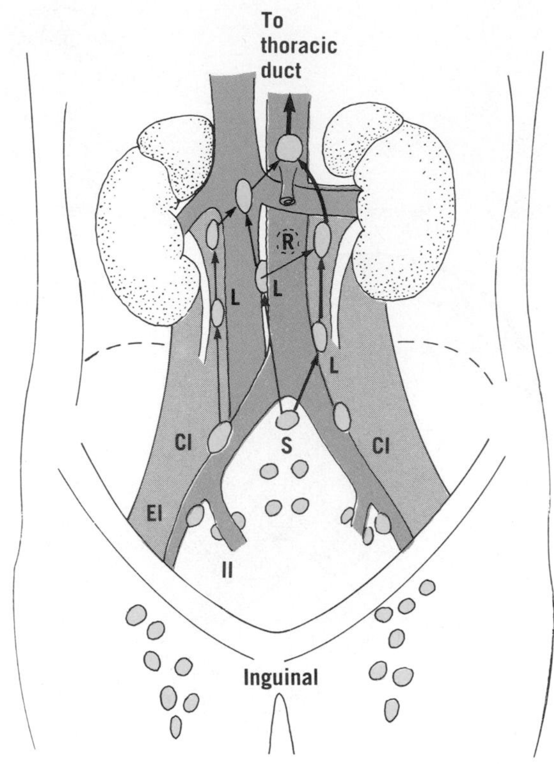

Figure 30–3 Lymphatic pathways and nodes of the posterior abdominal wall. *CI*, common iliac; *EI*, external iliac; *II*, internal iliac; *L*, lumbar; *R*, retro-aortic; *S*, sacral.

The *thoracic duct* begins as either a plexus or a dilatation termed the *cisterna chyli,* which receives a variable number of collecting trunks. The thoracic duct lies behind and on the right side of the aorta adjacent to the vertebral column and the right crus of the diaphragm. It passes through the aortic opening in the diaphragm and ascends in the thorax to the root of the neck, where it empties into the venous system (see figs. 24–4 and 24–5).

NERVES

Thoraco-abdominal Nerves. Intercostal nerves 7 to 11 supply the abdominal as well as the thoracic wall (see fig. 25–12).

Phrenic Nerves. The phrenic nerves, which contain motor, sensory, and sympathetic fibers (see fig. 24–6), pierce the diaphragm and supply it (mostly from below), as well as giving branches to the peritoneum.

Vagi. *Anterior* and *posterior vagal trunks* from the esophageal plexus supply the front and back of the stomach, respectively. Each trunk contains fibers from both right and left vagi. The trunks provide hepatic, gastric, and celiac branches. Vagal fibers enter the celiac and superior mesenteric plexuses and are distributed to the derivatives of the foregut and

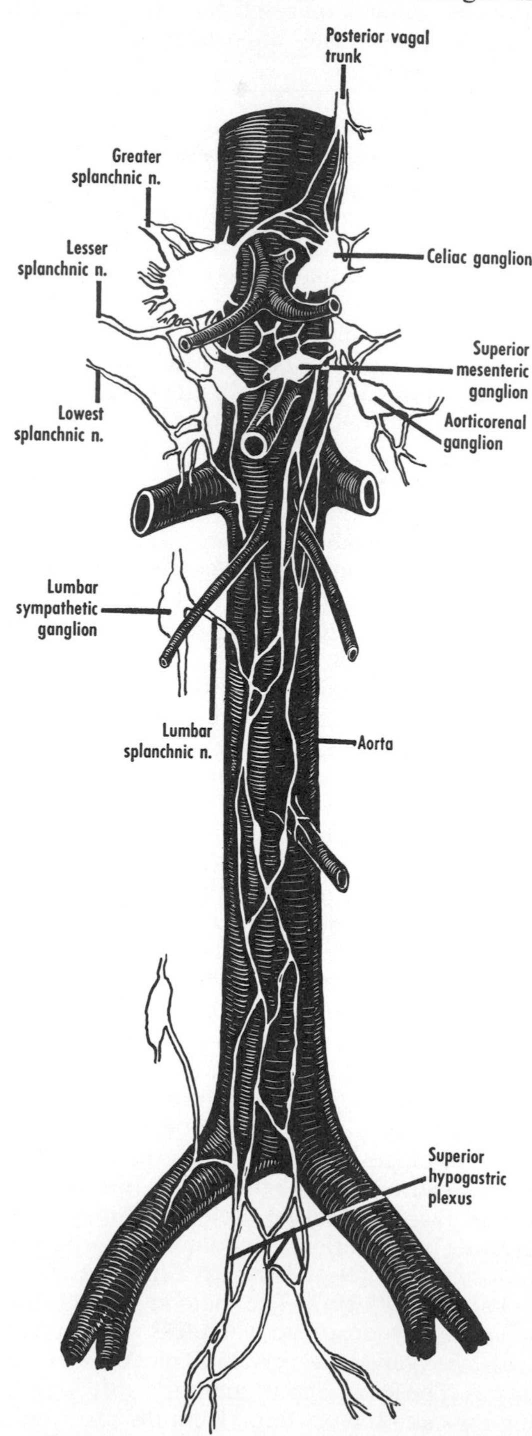

Figure 30–4 The prevertebral plexus and ganglia. See also fig. 32–5 for the sympathetic trunks and ganglia.

midgut: stomach, liver, pancreas, and intestine as far as the left colic flexure. (The remainder of the large intestine receives parasympathetic fibers from the pelvic splanchnic nerves).

Thoracic Splanchnic Nerves. The *greater, lesser,* and *lowest splanchnic nerves* arise from the thoracic part of the sympathetic trunk and carry much of the sympathetic and sensory supply of the abdominal viscera. The nerves pierce the diaphragm and reach the celiac and aorticorenal ganglia (fig. 30–4).

Sympathetic Trunks and Ganglia (see figs. 32–5 and 32–6). The sympathetic trunks pierce the diaphragm (or pass behind the medial arcuate ligaments) and descend on the vertebral column. The right trunk lies behind the inferior vena cava, the left one beside the aorta. Each trunk commonly presents three to five lumbar ganglia, each of which is connected by rami communicantes to the ventral rami of the spinal nerves. The trunk and ganglia give rise to several *lumbar splanchnic nerves,* which join the celiac, intermesenteric, and superior hypogastric plexuses. Both sympathetic trunks continue into the pelvis in front of the sacrum.

Autonomic Plexuses. **A dense prevertebral plexus is formed in the abdomen by the splanchnic nerves, branches from both vagi, and masses of ganglion cells.** The prevertebral plexus, which lies in front of the aorta, extends along that vessel and its branches (fig. 30–4) and contains preganglionic and postganglionic sympathetic, preganglionic parasympathetic, and sensory fibers. Parts of the plexus are named after their associated arteries. The *celiac* (or "solar") *plexus,* situated at the origin of the celiac trunk, contains paired *celiac ganglia* (fig. 30–5), which distribute branches along the arteries (e.g., to the liver and stomach). *Aorticorenal ganglia* are found near the origin of the renal arteries. Continuing fibers constitute the *aortic plexus,* parts of which are named the *superior mesenteric, intermesenteric,* and *inferior mesenteric plexuses.* The aortic plexus, which continues as the *superior hypogastric plexus,* receives many branches from the lumbar splanchnic nerves. Filaments descend from the aortic plexus along the iliac and femoral arteries.

Lumbar Plexus. The ventral rami of the lumbar spinal nerves enter the psoas major muscle, where they combine in a variable manner to form the lumbar plexus (figs. 30–6 and 30–7). Within the muscle, the rami are connected to the lumbar sympathetic trunk by

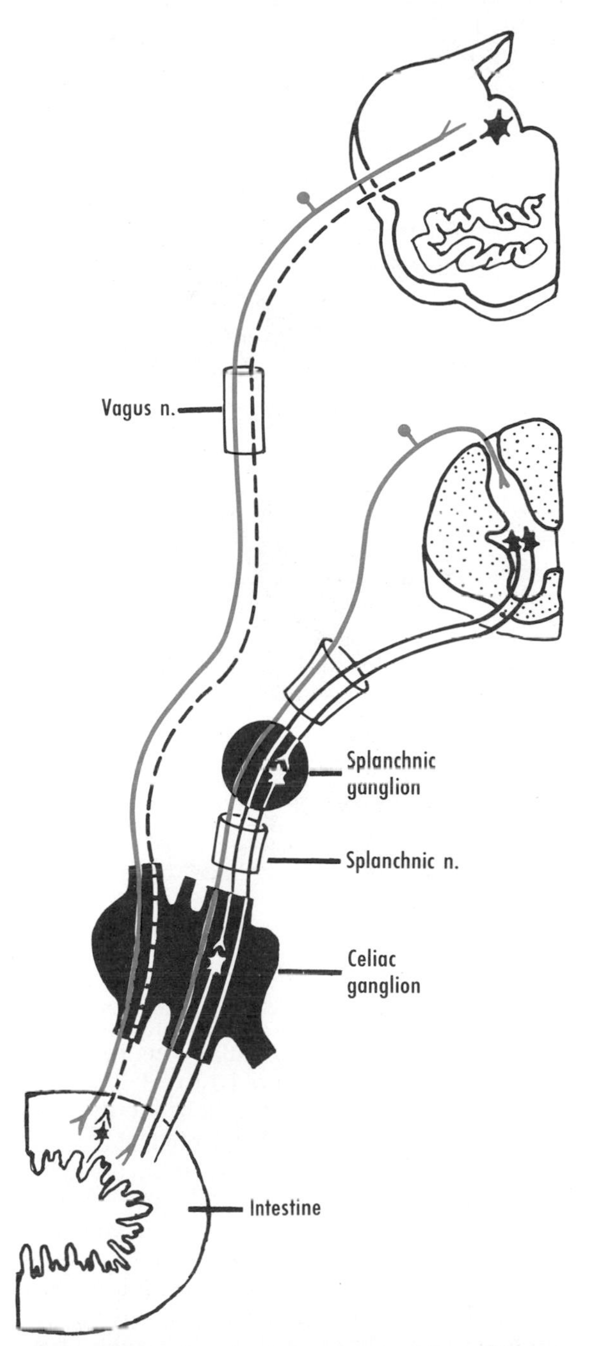

Figure 30–5 Functional components of the celiac ganglia and plexus. Sympathetic fibers are shown as continuous lines, parasympathetic fibers as interrupted lines, and sensory fibers in blue.

rami communicantes. **The term "lumbosacral plexus" is used for the lumbar plexus proper (which is formed by L2 to 4) and the sacral plexus (which is formed by L4 to S4).** The fourth lumbar nerve bifurcates (*nervus furcalis*) to go to both the lumbar and sacral plexuses (fig. 30–6). The rami that supply the

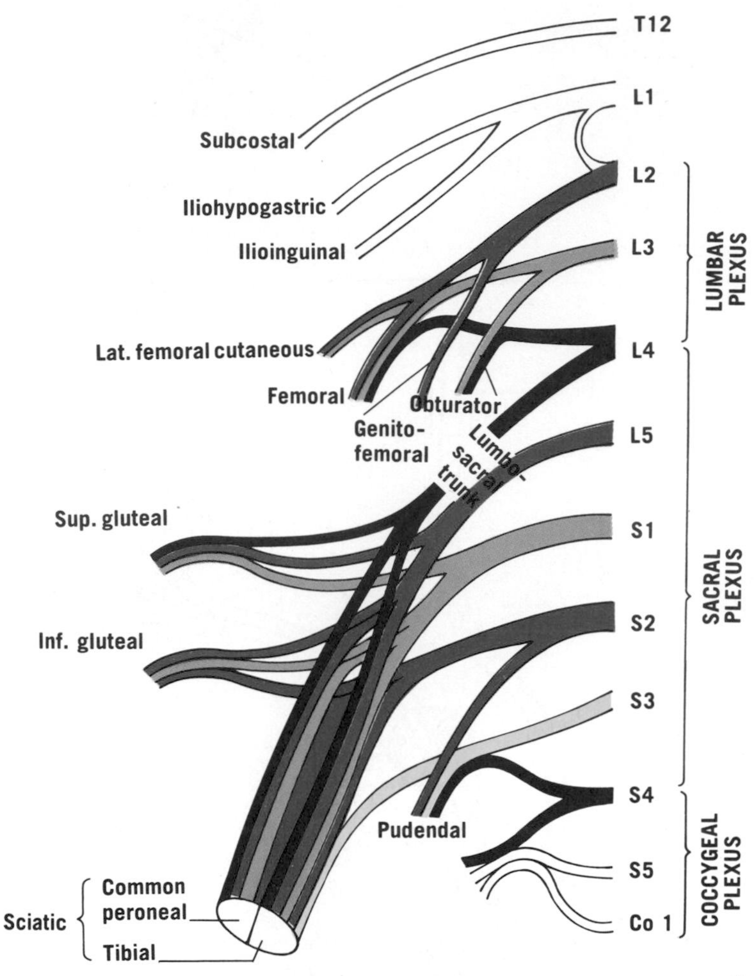

Figure 30–6 Simplified scheme of lumbosacral and coccygeal plexuses. See table 15–4 for the ultimate distribution of rami to muscles. (Based partly on Seddon.)

lower limbs (exclusive of cutaneous branches of T12 and L1) extend from L1 to S3. The lumbar plexus supplies direct branches to the quadratus lumborum and psoas muscles.

The first lumbar nerve resembles an intercostal nerve in giving off a collateral branch, the *ilio-inguinal nerve,* and then continuing as the *iliohypogastric nerve,* which has a lateral cutaneous branch (fig. 30–8). The iliohypogastric nerve runs through the muscles of the anterolateral abdominal wall to reach the skin. The ilio-inguinal nerve accompanies the spermatic cord or round ligament through the inguinal canal, emerges from the superficial ring, and becomes cutaneous.

The *lateral femoral cutaneous nerve* (chiefly from L2), which is frequently bound

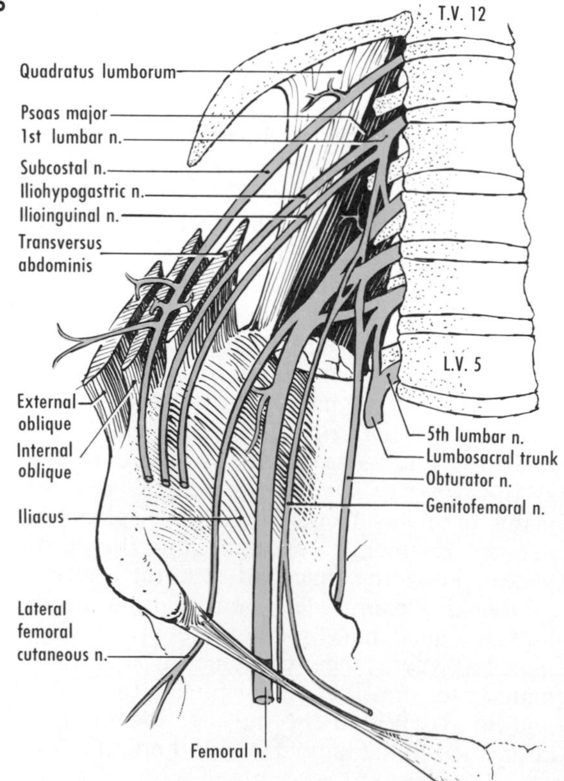

Figure 30–7 The lumbar plexus in relation to the muscular layers of the abdominal wall. The lateral cutaneous branch of the iliohypogastric nerve (fig. 30–8) is not shown. (After Pitres and Testut.)

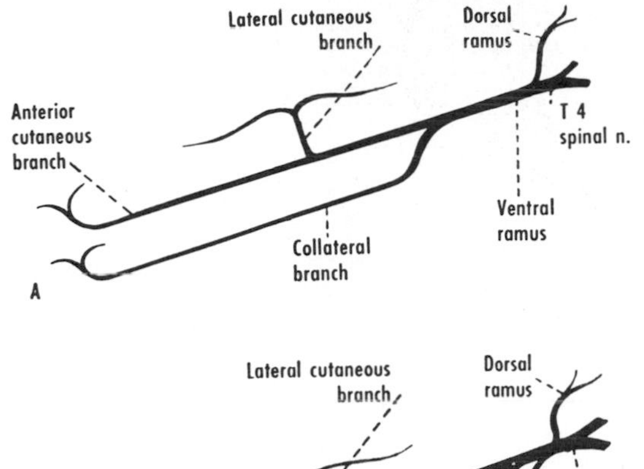

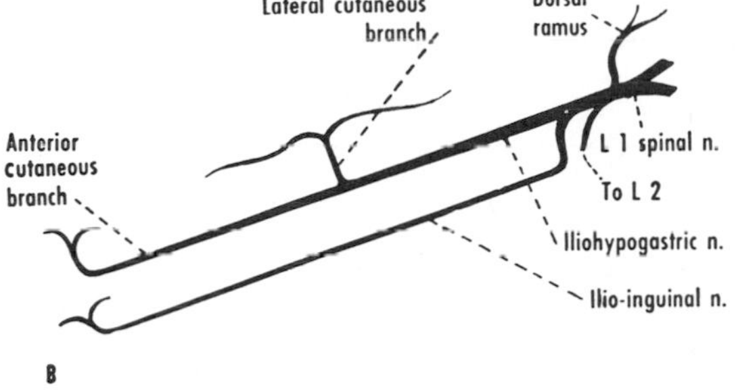

Figure 30–8 Comparison of an intercostal nerve **(A)** with the first lumbar nerve **(B)**. (After Davies.)

with the femoral nerve, enters the thigh behind the inguinal ligament.

The *femoral nerve* (L2 to 4) emerges from the lateral side of the psoas major, descends between the psoas and iliacus, and enters the thigh behind the inguinal ligament. It supplies the iliacus, quadriceps femoris, pectineus, and sartorius muscles, the skin of the medial side of the thigh and leg, and the hip and knee joints.

The *genitofemoral nerve* (chiefly L2) divides into (1) a *genital branch,* which enters the inguinal canal through the deep ring and supplies the cremaster and scrotum (or labium majus), and (2) a *femoral branch,* which enters the femoral sheath and supplies the skin of the femoral triangle.

The *obturator nerve* (L2 to 4) emerges from the medial side of the psoas and enters the thigh through the obturator foramen. It supplies the adductor muscles and gracilis, the skin of the medial side of the thigh, and the hip and knee joints. The *accessory obturator nerve* (L3, 4), when present, enters the thigh deep to the pectineus, which it supplies.

ADDITIONAL READING

Michels, N. A., *Blood Supply and Anatomy of the Upper Abdominal Organs,* Lippincott, Philadelphia, 1955. A descriptive atlas of patterns and variations for surgeons.

QUESTIONS

30–1 Where does the aorta end?

30–2 Which arteries supply (a) the caudal part of the foregut, (b) the midgut, (c) the hindgut?

30–3 Name two important sites of portal-systemic anastomosis.

30–4 Where does the inferior vena cava begin?

30–5 What is the relationship of the inferior vena cava to the right ureter?

30–6 How far distally do vagal fibers extend?

30–7 From which spinal segments is the lumbosacral plexus derived?

30–8 What is the distribution of the iliohypogastric, ilio-inguinal, and genitofemoral nerves?

Part 6

THE PELVIS

The word *pelvis* means "basin" in Latin, and in a number of other languages (e.g., French and German) the same word is used for both.

The lesser (or "true") pelvis, as distinct from the abdomen proper (which includes the greater pelvis) is the subject of this section. It is that part of the trunk below the pelvic inlet. Its funnel-shaped cavity extends backward and downward from the abdominal cavity proper. The part of the abdominal cavity that lies between the iliac fossae is known as the greater (or "false") pelvis.

The thorax, abdomen, and pelvis are (together with the back) parts of the trunk. The thoracic and abdominal cavities are separated from each other by the diaphragm. The abdominal cavity proper and the true pelvic cavity are continuous across the plane of the pelvic inlet, or brim (see fig. 31–3). The combined abdominopelvic cavity varies in shape as it is traced downward.* Above, it appears kidney shaped in horizontal sections, because of the forward projection of the vertebral bodies (T.V.12 and L.V.1). Lower down, the shape is modified by the psoas muscles and becomes crescentic (L.V.2 to L.V.5; see figs. 26–3*B* and 29–5). In the sacral region, the transverse diameter diminishes (as the iliac fossae disappear), whereas the anteroposterior diameter increases (because of the backward slope of the sacrum; see figs. 26–2 and 31–3).

Pelvic structure and function are particularly important in obstetrics and gynecology. An interesting pictorial history has been compiled by H. Speert, *Iconographia gyniatrica,* Davis, Philadelphia, 1973.

---

* J. Symington in *Quain's Elements of Anatomy,* 11th ed., vol. 2, part 2, Longmans, Green, London, 1914, has provided instructive horizontal sections (fig. 68), sagittal sections (fig. 69), and models of the contained viscera (figs. 73 to 75).

ADDITIONAL READING

Francis, C. C., *The Human Pelvis,* Mosby, St. Louis, 1952. A brief account.

Smout, C. F. V., and Jacoby, F., *Gynaecological and Obstetrical Anatomy and Functional Histology,* 3rd ed., Arnold, London, 1953. A good account, longer than the above.

Uhlenhuth, E., *Problems in the Anatomy of the Pelvis. An Atlas,* Lippincott, Philadelphia, 1953. Discussions and illustrations of selected topics, e.g., the retrovesical space.

31
THE BONES, JOINTS, AND WALLS OF THE PELVIS

BONY PELVIS

The bony pelvis is formed by the hip bones in front and at the sides and by the sacrum and coccyx behind (figs. 31–1 and 31–2). **When a subject is in the anatomical position, the anterior superior iliac spines and the pubic tubercles are in the same coronal plane. The pelvic surface of the body of the pubis, on which the bladder rests, faces more upward than backward. The pelvic surface of the sacrum faces more downward than forward** (fig. 31–3).

The (lesser) pelvis has an inlet, a cavity, and an outlet, each of which has three main diameters: anteroposterior (or conjugate) (fig. 31–3), oblique, and transverse (see fig. 31–1).

The *pelvic inlet*, or *brim* (*upper pelvic aperture*), is indicated by the *lineae terminales*, the

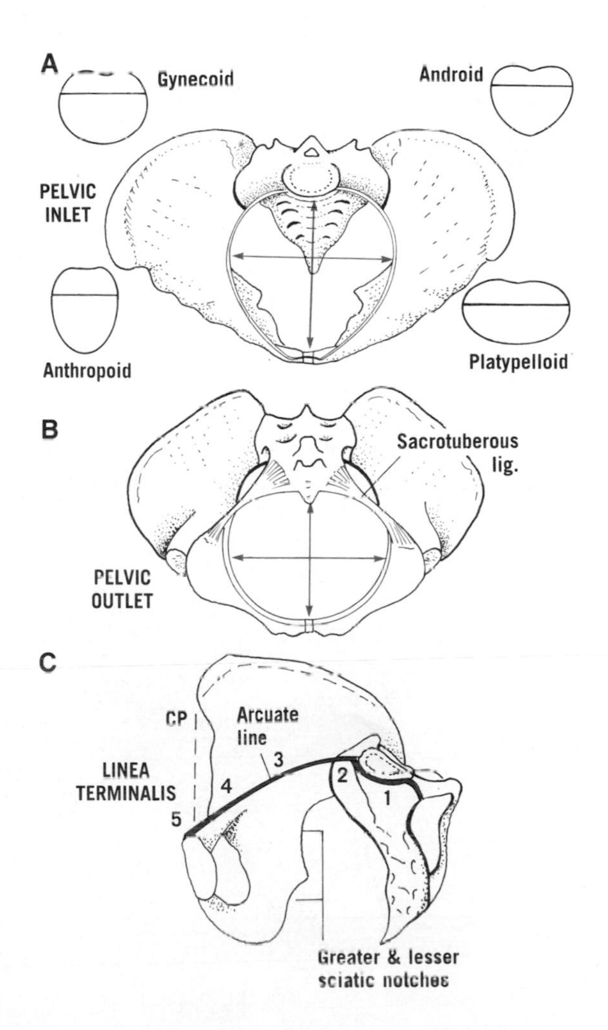

Figure 31–1 Female pelvis. **A,** View from above, showing inlet and anteroposterior (conjugate) and transverse diameters and surrounded by drawings of the four main types of female pelves. **B,** View from below, showing outlet and anteroposterior (conjugate) and transverse diameters. **C,** The pelvic cavity with the left hip bone removed. The anterior superior iliac spines and the pubic tubercles are in the same coronal plane (*CP*). The linea terminalis comprises the (*1*) promontory, (*2*) ala of the sacrum, (*3*) medial border of the ilium (arcuate line), (*4*) pectineal line, and (*5*) pubic crest. (After Smout and Jacoby.)

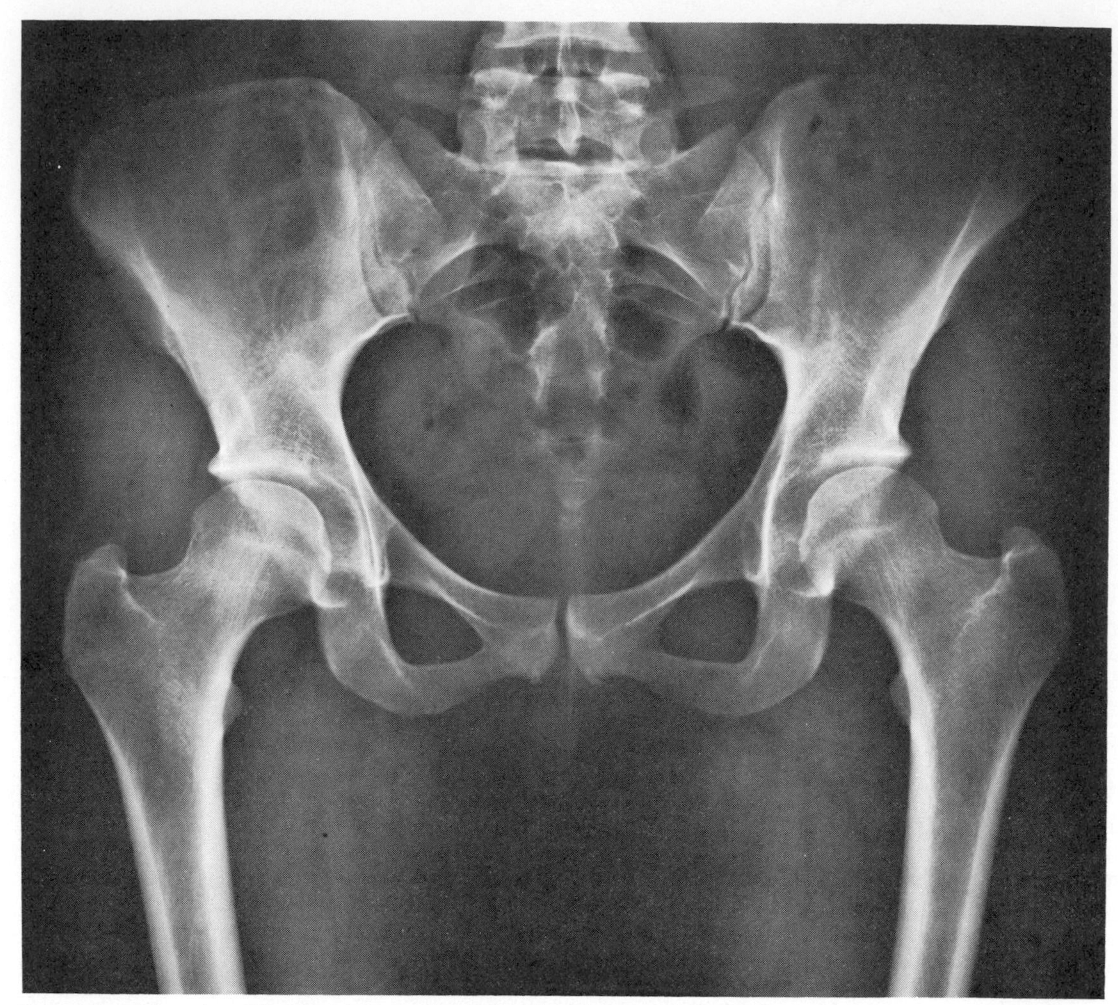

Figure 31–2 Female pelvis. Note the sacro-iliac joints, the subpubic angle, and the continuous curvature of the margin of the obturator foramen and the neck of the femur (Shenton's line).

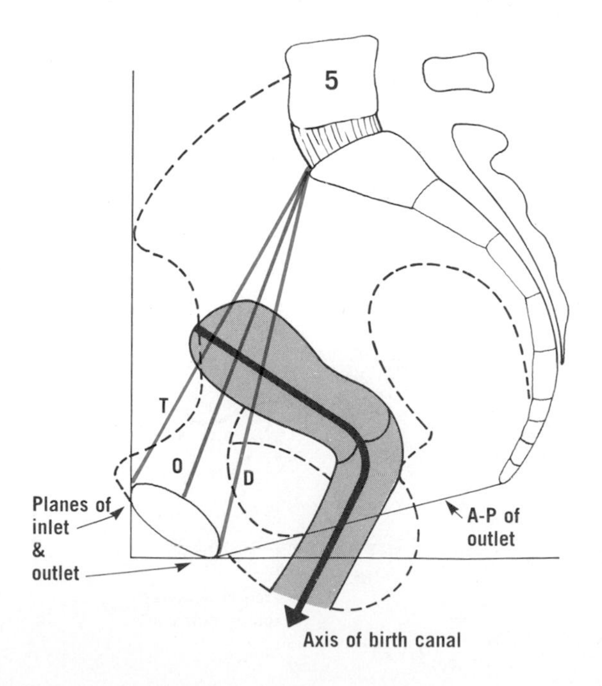

Figure 31–3 Median section of the pelvis, showing the planes of the inlet and outlet. The true (*T*), obstetrical (*O*), and diagonal (*D*) conjugate diameters are indicated. The axis of the birth canal, that is, the path taken by the fetal head in its passage through the pelvic cavity, can be seen to turn at the uterovaginal angle. *5*, fifth lumbar vertebra.

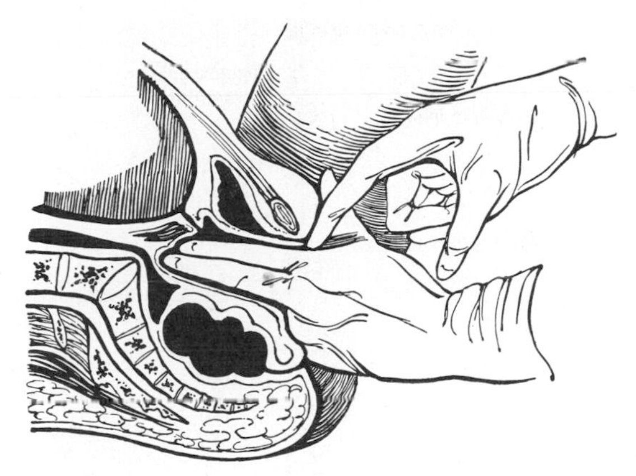

Figure 31–4 Measurement of the diagonal conjugate diameter by the middle finger. The indicated length on the index finger gives the true conjugate, because the index is about 1½ cm shorter than the middle finger. (After Smout and Jacoby.)

iliac parts of which are the *arcuate lines* (see fig. 31–1). The inlet is at about half a right angle to the horizontal. The anteroposterior (or true) conjugate diameter extends from the upper margin of the pubic symphysis to the middle of the sacral promontory. The *obstetrical conjugate diameter,* which is measured from the back of the pubic symphysis (fig. 31–3), is the shortest diameter through which the fetal head must pass in its course through the inlet. The *diagonal conjugate diameter,* between the lower margin of the pubic symphysis and the sacral promontory (fig. 31–3), is measured *per vaginam* (fig. 31–4). Inability to palpate the sacral promontory suggests that the conjugate diameter of the inlet is adequate for parturition, whereas palpation indicates a contracted pelvis.

The *pelvic cavity* extends backward and downward from the inlet to the outlet. It curves with the sacrum and coccyx, and hence is longer behind than in front (see fig. 31–3).

The *pelvic outlet* (*lower pelvic aperture*) extends from the pubic symphysis to the tip of the coccyx (its anteroposterior, or conjugate, diameter; see fig. 31–3) and, from side to side, between the ischial tuberosities; hence it is diamond shaped. The outlet is at a slight angle (10 to 15 degrees) to the horizontal.

The pubic arch formed by the conjoined rami of the pubes and ischia has its apex at the symphysis, where it forms the *subpubic angle* (fig. 31–5).

The path taken through the pelvic cavity by the fetal head is known as the axis of the birth canal (see fig. 31–3). The axis intersects the inlet at a right angle, turns forward at the uterovaginal angle (level of ischial spines), and follows the axis of the vagina. During parturition, the fetal head (usually the *suboccipitobregmatic diameter*) occupies successively the inlet (transverse diameter), cavity (oblique diameter, and outlet (anteroposterior diameter) (fig. 31–6).

Classification of Pelves. Although pelves can be arranged by the measurements of their diameters, it is usual in obstetrics and radiology to **classify pelves according to the shape of the pelvic inlet. Four main types are recognized: (1) gynecoid, a rounded inlet; (2) android, a heart-shaped inlet; (3) anthropoid, a long, narrow, oval inlet; and (4) platypelloid, an ovoid inlet with its long axis transverse, like a flat bowl** (see fig. 31–1). A female pelvis may belong to any of the four types (only about half are gynecoid), but intermediate types are more frequent. The female pelvis tends to have thinner and lighter bones with less prom-

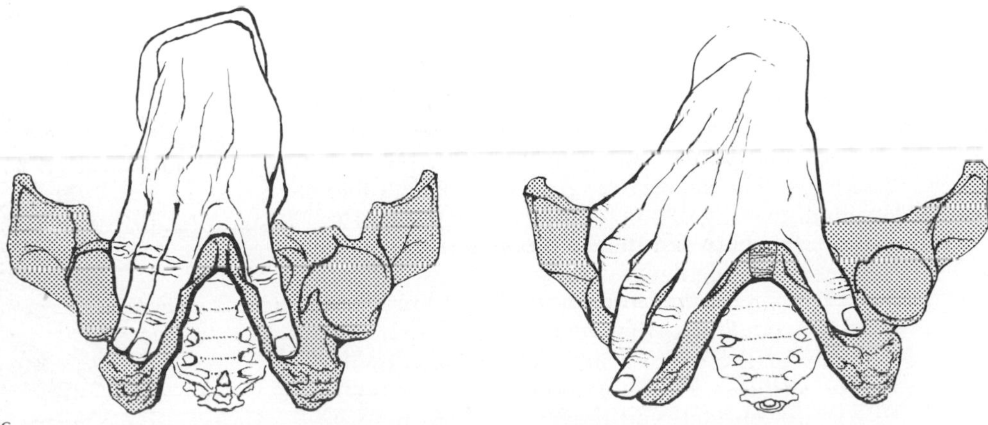

Figure 31–5 The subpubic angle is nearly a right angle in the female and about 60 degrees in the male.

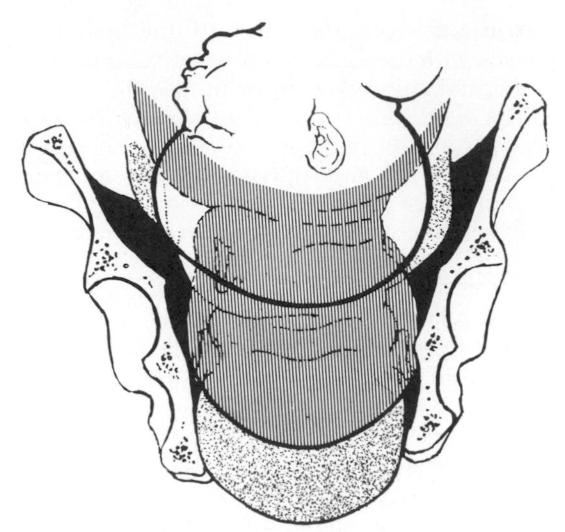

Figure 31–6 The bony pelvis and fetal head. Note how the head turns as it occupies first the inlet, then the cavity, and finally the outlet. (Based on Smout and Jacoby, after Bumm.)

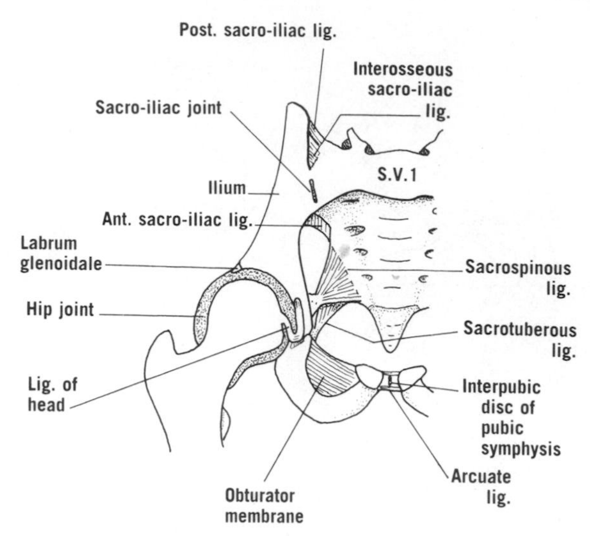

Figure 31–7 The sacro-iliac and hip joints and the pubic symphysis, as seen in an oblique section through the first sacral vertebra. (After Quain.)

inent muscular markings, and the subpubic angle is less acute (approximating a right angle; see fig. 31–5). Pelvic diameters and shape can be determined *in vivo* by radiographic pelvimetry.

ADDITIONAL READING

Steer, C. M., *Moloy's Evaluation of the Pelvis in Obstetrics*, 3rd ed., Plenum, New York, 1975. Readable and useful.

JOINTS OF PELVIS

The joints of the pelvis include the lumbosacral, sacrococcygeal, and sacro-iliac, and the pubic symphysis. Associated ligaments include the sacrotuberous, sacrospinous, and iliolumbar.

The *lumbosacral joint* is that between L.V.5 and the sacrum. It includes an intervertebral disc and joints between the articular processes. The *sacrococcygeal joint,* which may undergo bony fusion, consists of an intervertebral disc between the sacrum and coccyx and accessory ligaments.

The *sacro-iliac joints* (fig. 31–7) are synovial articulations between the auricular surfaces of the sacrum and ilium on each side. The surfaces may be smooth and flat or reciprocally curved and irregular. The joint is strengthened posteriorly by *interosseous* and dorsal *sacro-iliac ligaments*. The weight of the body is transmitted through the sacrum and ilia to the femora during standing and to the ischial tuberosities in sitting.

The *pubic symphysis* (fig. 31–7) is a cartilaginous joint between the bodies of the pubic bones in the median plane. The symphysial surfaces, each covered by hyaline cartilage, are united by an *interpubic disc* of fibrocarti-

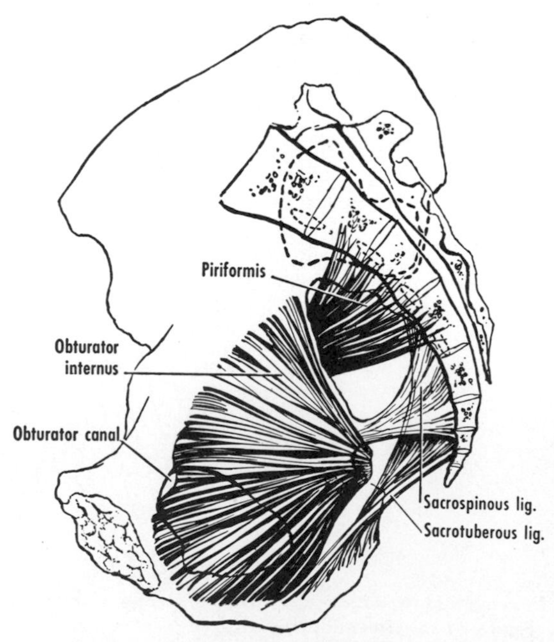

Figure 31–8 The muscles and ligaments of the lateral pelvic wall, pelvic aspect. (After Shellshear and Macintosh.)

lage, which may present a cleft. The ligaments around the joint become relaxed during pregnancy.

The *sacrotuberous ligament* (fig. 31–8) extends from the dorsal surface of the sacrum (as well as from the ilium and coccyx) to the ischial tuberosity. The *sacrospinous ligament* (fig. 31–8) extends from the lateral margin of the sacrum (and coccyx) to the ischial spine. The sacrotuberous ligament converts the sciatic notches into foramina, which are separated from each other by the sacrospinous ligament (see figs. 14–2 and 31–8). The *greater sciatic foramen* transmits the piriformis muscle, superior and inferior gluteal vessels and nerves, (internal) pudendal vessels and nerve, sciatic and posterior femoral cutaneous nerves, and the nerves to the obturator internus and quadratus femoris muscles. The *lesser sciatic foramen* transmits the obturator internus tendon, the nerve to the obturator internus, and the (internal) pudendal vessels and nerve (see fig. 14–3).

WALLS OF PELVIS

The wall of the pelvic cavity includes (1) superficial muscles, such as the glutei; (2) the hip bones, the sacrum and coccyx, and their associated ligaments; and (3) deep muscles, blood vessels, nerves, and peritoneum. For descriptive purposes, the pelvic wall can be subdivided into two lateral walls, a posterior wall, and a floor.

Each lateral wall (see figs. 31–8, 32–2 and 35–1), limited by the hip bone below the linea terminalis, is lined by the obturator internus muscle, medial to which are the obturator nerve and vessels and other branches of the internal iliac artery. Rarely, the intestine may protrude through the obturator canal (*obturator hernia*) and lie under cover of the pectineus. The lateral wall of the pelvis is crossed behind by the ureter and in front by the round ligament or the ductus deferens. The ovary lies in a slight depression on the lateral wall. The lateral and posterior walls are separated by the sacrotuberous and sacrospinous ligaments and by the greater and lesser sciatic foramina.

The posterior wall, formed by the sacrum and coccyx, is lined laterally by the piriformis and coccygeus muscles. The lumbosacral trunk and sacral plexus are situated in front of the piriformis. In the median plane is the median sacral artery (from the aorta), which ends in a vascular mass, the coccygeal body or glomus.

The pelvic floor (a term variously defined)

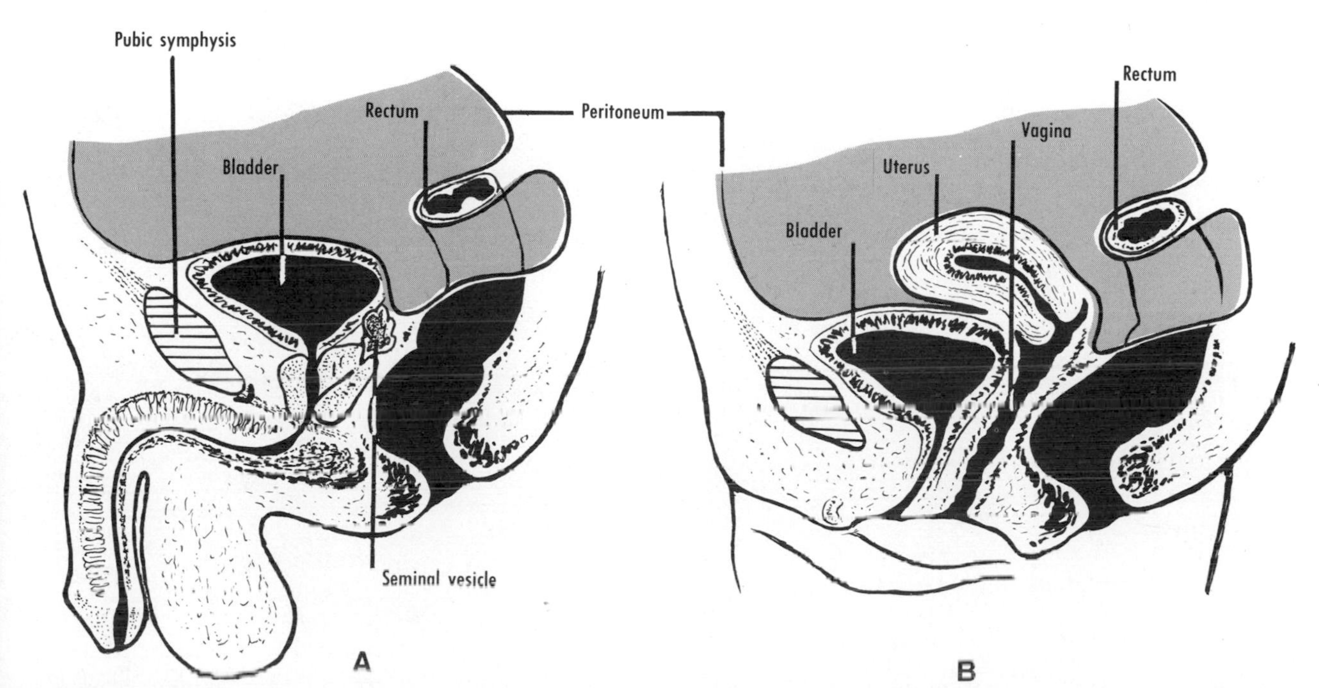

Figure 31–9 The peritoneal reflections from the pelvic viscera. Note the rectovesical, uterovesical, recto-uterine, and rectovaginal pouches.

may conveniently be considered as the main structures that support the abdominal and pelvic viscera, i.e., the peritoneum and the pelvic and urogenital diaphragms. The peritoneum descends to be reflected from the front of the rectum to the bladder (rectovesical pouch in the male) or to the uterus and vagina (recto-uterine and rectovaginal pouches in the female) (figs. 31–9 and 35–4). The pouches are bounded laterally by peritoneal elevations that are frequently termed *sacrogenital folds*. In front, the peritoneum is reflected from the uterus to the bladder (uterovesical pouch). The blood vessels and neural plexuses to the viscera (as well as the ureter and ductus deferens) are situated in the connective tissue between the peritoneum and the pelvic diaphragm. Localized thickenings of this extraperitoneal tissue form ligaments. The pelvic floor has two median openings, one for the rectum and the other for the urethra (and vagina).

QUESTIONS

31–1 What is the boundary between the lesser and greater pelves?

31–2 In the anatomical position, in which direction does the visceral (pelvic) surface of the body of the pubis face?

31–3 What is the subpubic angle?

31–4 What is the diagonal conjugate diameter?

31–5 Of which type is the sacro-iliac joint?

31–6 What is the structure of the pubic symphysis?

31–7 What are the chief supports for the pelvic viscera?

31–8 How is the rectovesical pouch formed?

32 BLOOD VESSELS, NERVES, AND LYMPHATIC DRAINAGE OF THE PELVIS

BLOOD VESSELS

The internal iliac (hypogastric) artery supplies most of the blood to the pelvis (figs. 32–1 and 32–2). It arises from the common iliac artery in front of the sacro-iliac joint. Although the internal iliac artery is often described as ending in anterior and posterior divisions, its various branches arise in a variable manner. They may be divided into parietal and visceral branches.

Parietal branches. The parietal branches of the internal iliac artery include the iliolumbar, lateral sacral, obturator, superior and inferior gluteal, and internal pudendal arteries. The *iliolumbar artery* supplies bone and muscle in the iliac fossa. The *lateral sacral arteries,* which give off spinal branches, supply the sacrum and coccyx. The *obturator artery,* which is crossed by the ureter, descends and traverses the obturator foramen. It divides into anterior and posterior branches, which encircle the margin of the obturator foramen; an *acetabular branch* supplies the ligament of the head of the femur. The obturator artery may arise from the inferior epigastric artery (see fig. 25–9) and, if it passes medial to the femoral ring, it is liable to damage during an operation for femoral hernia. The *superior* and *inferior gluteal arteries* pass backward between the sacral nerves and leave the pelvis through the greater sciatic foramen, running superior and inferior to the piriformis, respectively.

The *internal pudendal artery* (fig. 32–3) descends to the greater sciatic foramen, through which it leaves the pelvis. It then crosses the back of the ischial spine and enters the perineum through the lesser sciatic foramen. Next, accompanied by the branches of the pudendal nerve, it traverses the pudendal canal in the lateral wall of the ischiorectal fossa. Finally, it pierces the urogenital diaphragm, traverses the deep perineal pouch, and divides into the deep and dorsal arteries of the penis (or clitoris). The branches of the internal pudendal artery include the inferior rectal artery and vessels that supply the scrotum (or labia), perineum, bulb of the penis (or vestibule), and urethra.

Visceral branches. These include the umbilical and superior and inferior vesical arteries, the uterine artery (or the artery of the ductus deferens), and the vaginal and middle

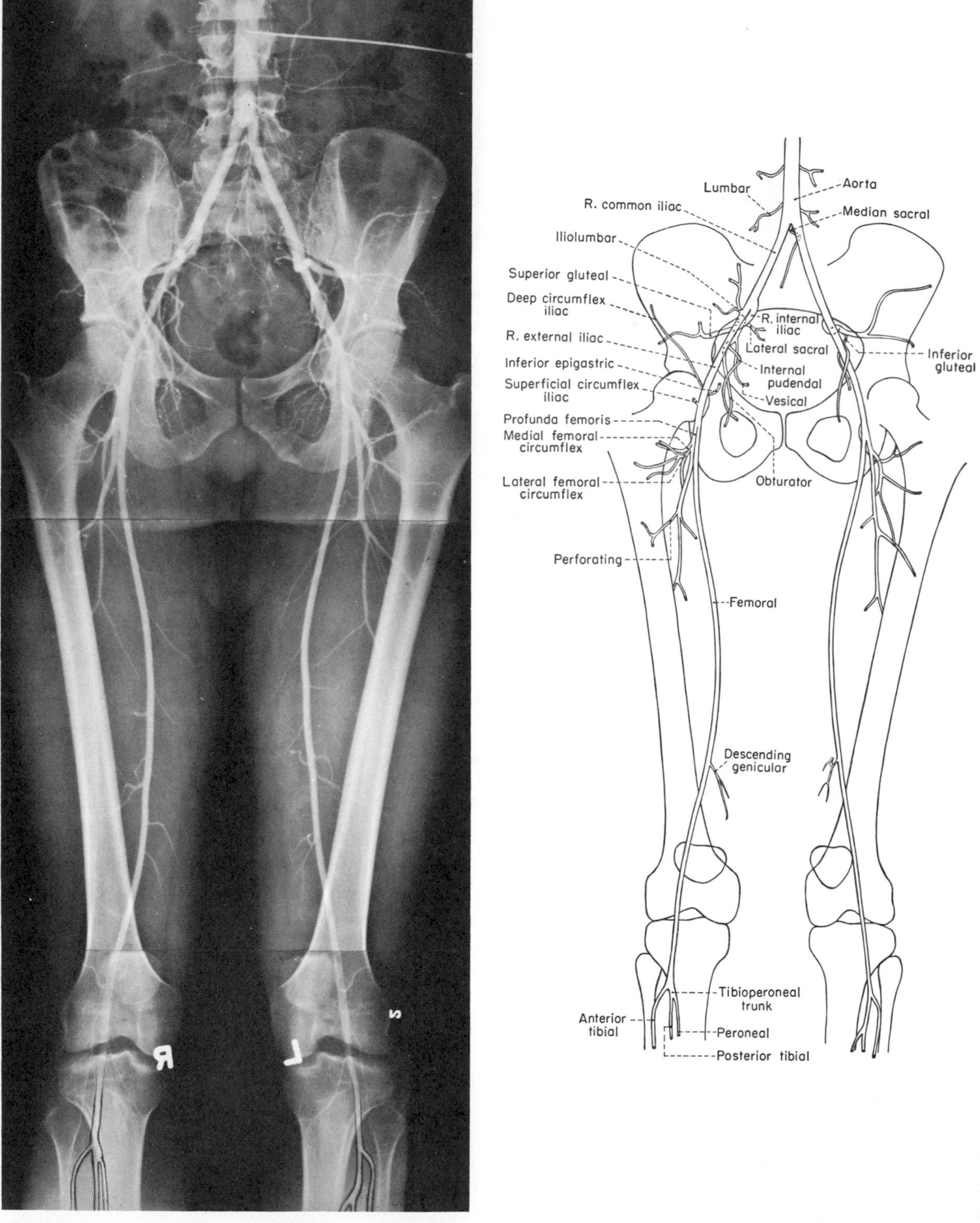

Figure 32–1 Lumbar aortogram. (From Abrams, H. L., ed., *Angiography,* Little, Brown, Boston, 1961. Courtesy of S. M. Rogoff, M.D.)

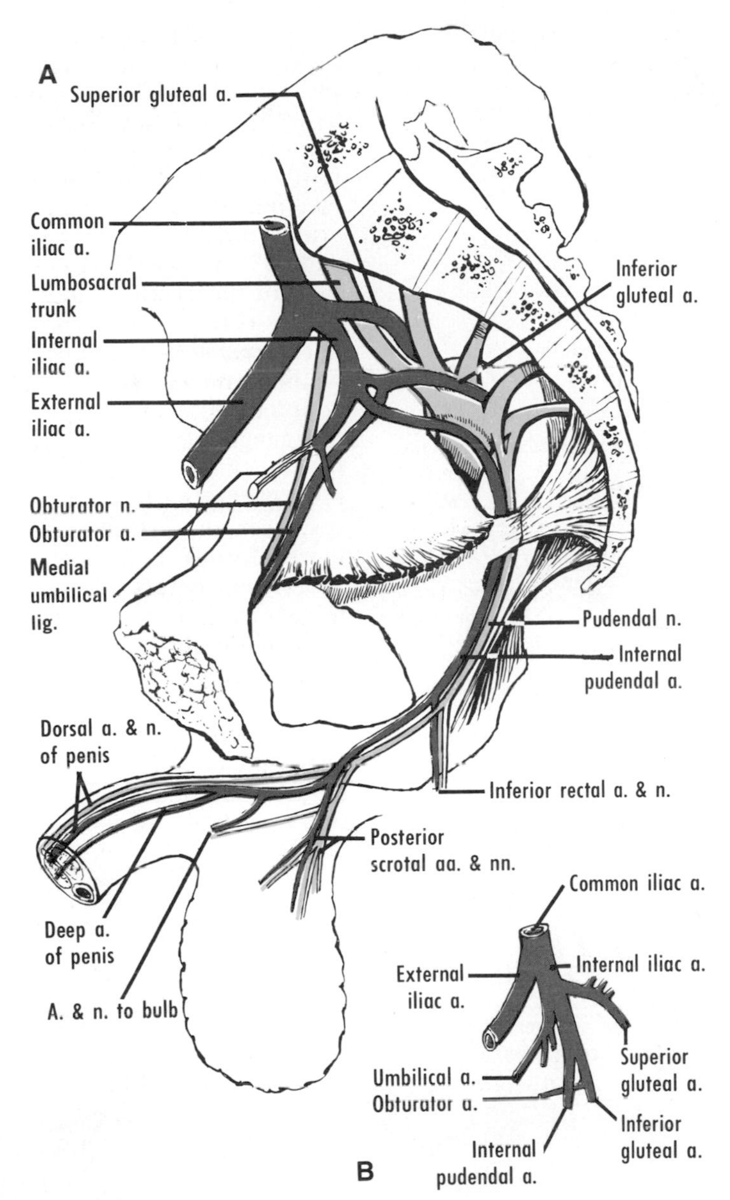

Figure 32–2 **A,** The sacral plexus and branches of the internal iliac artery, medial aspect. **B,** The most frequent pattern of branches of the internal iliac artery.

rectal arteries. The *umbilical arteries* return deoxygenated blood in the fetus from the aorta to the placenta. The distal part of each umbilical artery becomes the medial umbilical ligament, whereas the proximal part remains patent and gives rise to some of the next branches. The *superior* and *inferior vesical arteries* supply the bladder, and the *middle rectal artery* supplies the rectum.

The *uterine artery* (see fig. 35–9), comparable to the artery of the ductus deferens, may arise separately or in common with other branches of the internal iliac artery. It descends to the lower part of the broad ligament, where it lies near the lateral fornix of the vagina. It crosses in front of the ureter and ascends along the body of the uterus, between the layers of the broad ligament. Near the uterine tube, it turns laterally and anastomoses with the ovarian artery. Its branches supply the vagina, uterus, and tube. The *vaginal artery* (see fig. 35–9) arises from the uterine

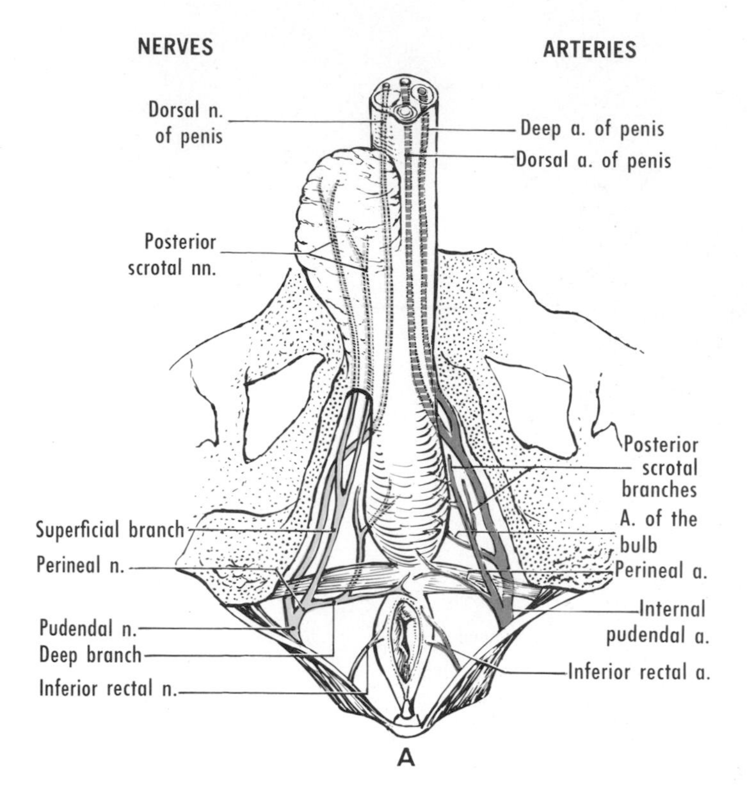

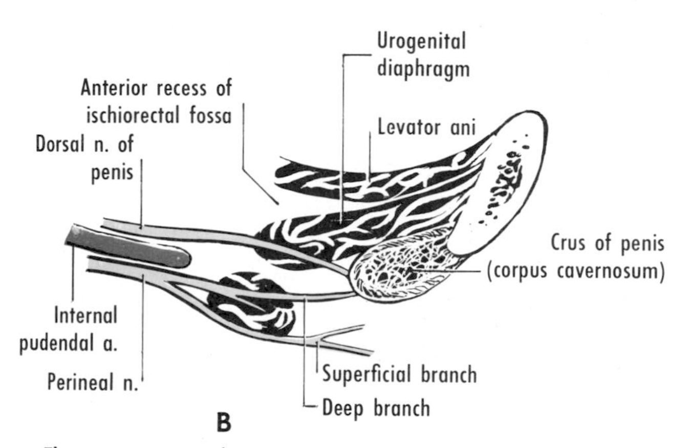

Figure 32–3 **A,** The internal pudendal artery and pudendal nerve. **B,** An almost sagittal section through the lateral part of the perineum.

artery or from the internal iliac artery and gives branches to the front and back of the vagina.

The *internal iliac* (*hypogastric*) *vein* lies behind its artery, and its tributaries correspond mostly with the branches of the artery.

NERVES

The pelvis is innervated chiefly by the sacral and coccygeal spinal nerves and by the pelvic part of the autonomic nervous system.

Sacral and Coccygeal Plexuses. **The sacral plexus** (fig. 32–4), **which lies in front of the piriformis, supplies the buttock and lower limb as well as structures belonging to the pelvis. It is formed by the lumbosacral trunk, the ventral rami of S1 to 3, and the upper division of S4** (see fig. 30–6). The rami are connected to sacral sympathetic ganglia by rami communicantes. The gluteal vessels pass between the rami of the plexus. The largest branch of the plexus is the sciatic nerve.

The pudendal nerve (S2 to 4) supplies most of the perineum (see figs. 32–2 and 32–3). It contains motor, sensory (pain and reflex), and postganglionic sympathetic fibers, and it may be "blocked" medial to the ischial tuberosity, e.g., during parturition. The pudendal nerve traverses the greater sciatic foramen below the piriformis, crosses the back of the ischial spine, and enters the perineum through the lesser sciatic foramen. It traverses the pudendal canal in the lateral wall of the ischiorectal fossa, gives off the *inferior rectal nerve,* and divides into the *perineal nerve* and the *dorsal nerve of the penis* (or *clitoris*). The perineal nerve divides into a deep branch to perineal muscles and a superficial branch to the scrotum (or labium majus).

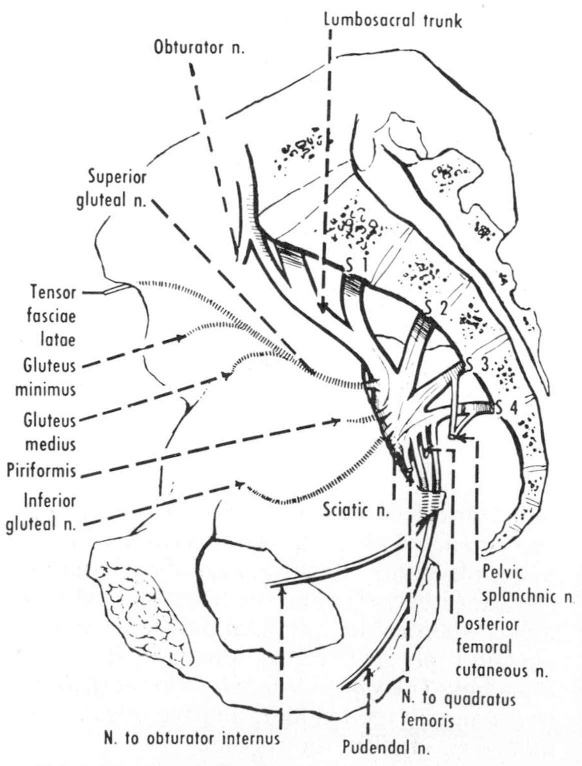

Figure 32–4 The sacral plexus and its main branches. (After Pitres.)

The *pelvic splanchnic nerves* (S2 to 4) contain parasympathetic preganglionic and sensory fibers. They help to form the inferior hypogastric plexus and thereby supply the sigmoid colon.

The *coccygeal plexus* (S4, 5; Co1) supplies the skin over the coccyx.

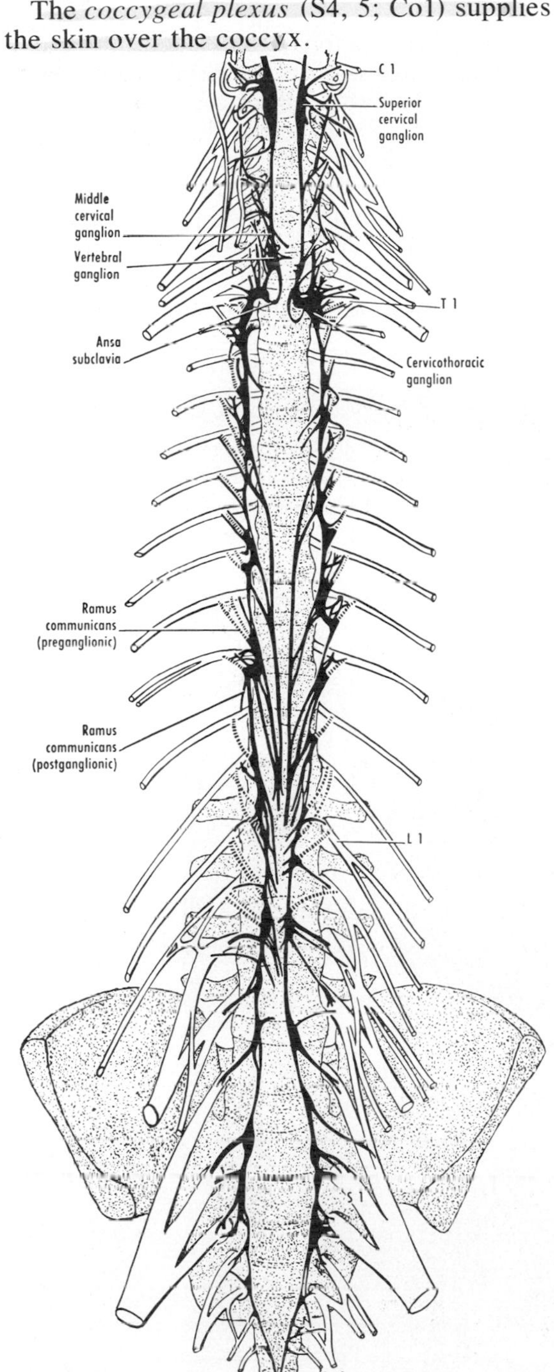

Figure 32–5 The sympathetic trunks. Preganglionic rami communicantes are shown as interrupted lines, postganglionic rami as solid black lines. (After Pick and Sheehan.)

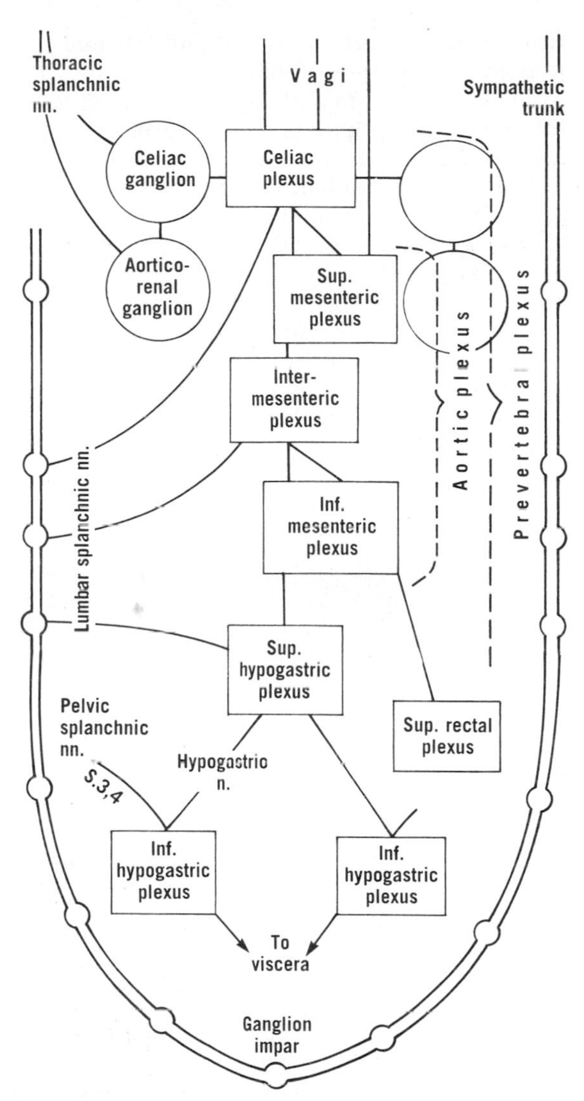

Figure 32–6 Highly simplified scheme of the sympathetic trunks and ganglia and of the autonomic plexuses in the abdomen and pelvis.

Pelvic Part of Autonomic Nervous System. **Sympathetic fibers reach the pelvis by downward continuations of the sympathetic trunks and of the aortic plexus** (fig. 32–5).

In front of the sacrum, the sympathetic trunks consist largely of preganglionic fibers and present three or four ganglia each. The two trunks meet in front of the coccyx, where they may form an enlargement termed the *ganglion impar*. The ganglia on the trunk are connected with the spinal nerves by rami communicantes, which transmit postganglionic fibers chiefly to the lower limbs and perineum.

The aortic plexus is continued as the *superior hypogastric plexus* (or "presacral nerve"; see figs. 30–4 and 32–6), which divides in front of the sacrum into right and left inferior hypogastric nerves. These descend on the sides of the rectum and vagina and unite with the pelvic splanchnic nerves to form the right and left *inferior hypogastric plexuses* (fig. 32–6), which give branches to the pelvic viscera (e.g., the rectum, bladder, and uterus). The inferior hypogastric plexuses contain: (1) postganglionic sympathetic fibers; (2) preganglionic parasympathetic fibers, which arise from the sacral part of the spinal cord, travel in the pelvic splanchnic nerves to the inferior hypogastric plexus, and supply the descending and sigmoid colon and the pelvic viscera; and (3) sensory fibers, including pain fibers (many of which travel in the lumbar splanchnic nerves) and reflex fibers from the bladder (which ascend in the pelvic splanchnic nerves).

LYMPHATIC DRAINAGE

Most of the lymphatic vessels from the pelvis drain into groups of nodes associated with the iliac arteries and their branches (see figs. 30–3 and 32–7). External iliac lymph nodes receive vessels from the inguinal nodes, external genitalia, vagina, and cervix; they drain into the common iliac nodes. Internal iliac and sacral lymph nodes receive afferents from all the pelvic viscera (e.g., cervix, prostate, and rectum) and from the perineum, buttock, and

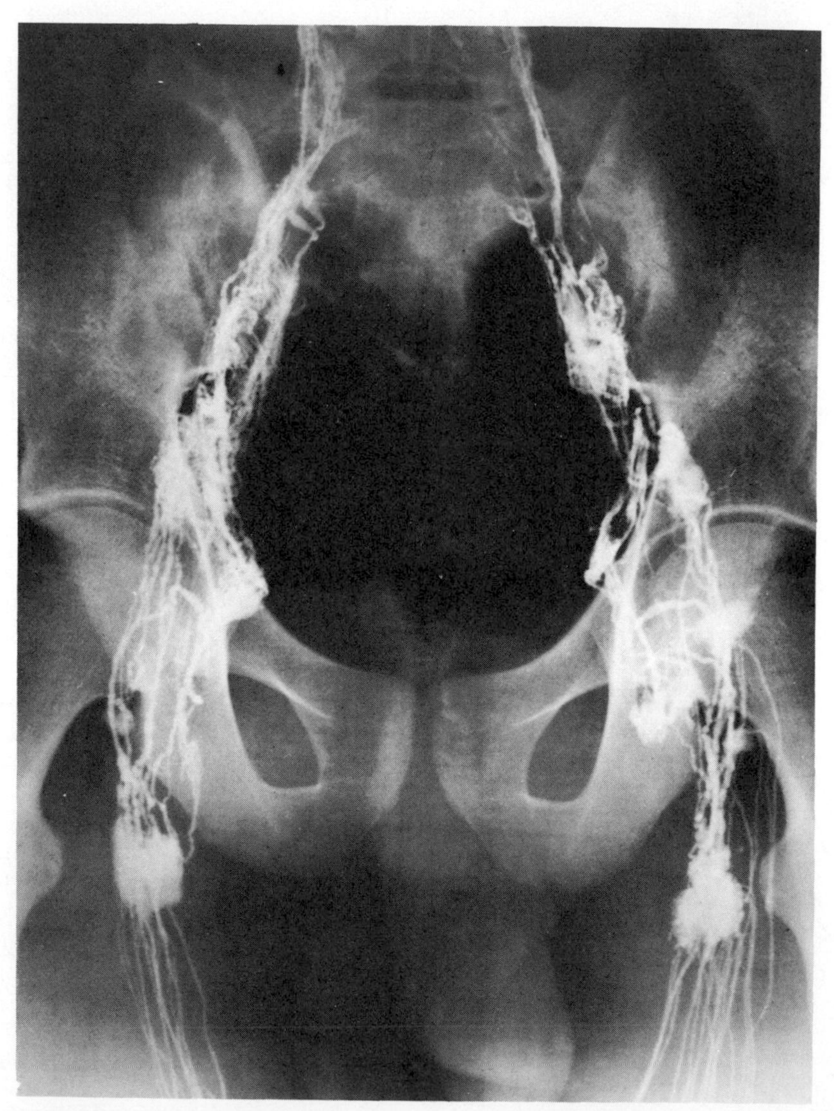

Figure 32–7 Iliac lymphogram, showing inguinal as well as iliac nodes. Large efferent vessels are visible. (From Kinmouth, J. B., *The Lymphatics,* Arnold, London, 1972.)

thigh; they drain into the common iliac nodes. Common iliac lymph nodes drain the two preceding groups and send their efferents to the lumbar group of aortic nodes, which also receives the afferents of the testis and ovary.

The cervix drains chiefly into the external and internal iliac nodes, the body of the uterus mainly into the external iliac and lumbar nodes. The prostate drains principally into the internal iliac nodes, and the bladder into the external iliac. The upper part of the rectum drains into the inferior mesenteric nodes, the lower part (together with the upper part of the anal canal) into the internal iliac nodes. **The lower part of the anal canal, as also the external genitalia, drains into the inguinal nodes.**

QUESTIONS

32–1 Which artery supplies most of the pelvis?

32–2 What is the course of the umbilical artery?

32–3 Which nerve supplies most of the perineum?

32–4 What are the pelvic splanchnic nerves?

32–5 What is the lymphatic drainage of the colon, rectum, and anal canal?

33
THE URETER, BLADDER, AND URETHRA

URETER (see figs. 26–1, 27–12, 29–1 to 29–3, and 33–2)

The upper half of the ureter is in the abdomen proper; the lower half is in the pelvis. The ureter descends retroperitoneally on the lateral pelvic wall. At the level of the ischial spine, it turns forward and medially. In the male, the ureter lies in the sacrogenital fold and is crossed medially by the ductus deferens. In the female, the ureter is at first related to the posterior border of the ovary; it then lies in the uterosacral ligament and is crossed anteriorly by the uterine artery. It passes about 2 cm lateral to the cervix uteri (and hence may be endangered in hysterectomy) and courses in front of the lateral border of the vagina. A ureteric stone at this level may even be palpable *per vaginam.* On entering the back of the bladder, the ureter is embedded for about 2 cm in the wall of that organ. There the ureteric lumen is narrowest, and the muscular coats of the ureter and bladder are continuous. The slit-like opening of the ureter on the trigone may be catheterized through a cystoscope, and a radio-opaque medium may be injected (*ascending,* or *retrograde, pyelography*). Alternatively, the ureters may be examined after injection of a radio-opaque medium into the blood stream (*descending,* or *intravenous, pyelography*).

URINARY BLADDER

The bladder (L., *vesica;* hence the adjective *vesical*) varies in size, shape, and position with the amount of contained urine (figs. 33–1, 33–2, and 38–2). **The empty bladder *in vivo* lies almost entirely within the pelvis and rests on the pubis and pelvic floor. As the organ fills, it ascends into the abdomen proper and may reach the level of the umbilicus. In infancy, however, even the empty bladder lies mostly within the abdomen proper.**

Relations. The empty bladder is less rounded and is commonly said to have four surfaces: superior, right and left inferolateral, and posterior (or base) (fig. 33–2). The superior surface and the upper part of the base are covered by peritoneum. As the bladder fills and ascends, the peritoneum is lifted off the abdominal wall; hence the reflection becomes higher. **The peritoneal relations are important in rupture of the bladder, which may result in either intraperitoneal or extraperitoneal extravasation of urine** (see fig. 33–1). Behind, the peritoneum forms the rectovesical (or uterovesical) pouch (see fig. 31–9). The *superior surface* is related to the intestine and the body of the uterus. The *inferolateral surfaces* are related to the retropubic space, which contains veins and a pad of fat. The *base* faces backward and downward and is related to the seminal vesicles, ductus deferentes, and rectum or to the vagina and supravaginal cervix.

The apex of the bladder is connected to the umbilicus by the median umbilical ligament, which is a remnant of the urachus. The bladder is connected to the umbilicus also by the right and left medial umbilical ligaments, which are the obliterated umbilical arteries (see fig. 25–9). The main part of the bladder is

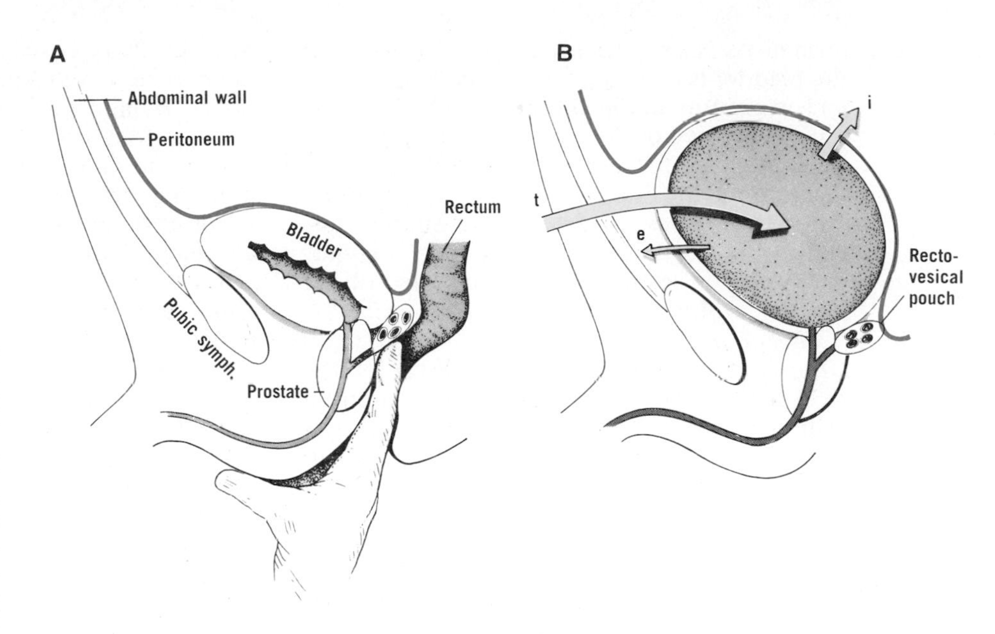

Figure 33–1 The urinary bladder empty (**A**) and full (**B**). The peritoneum is stripped away from the anterior abdominal wall as the bladder fills. Hence access to a full bladder may be gained extraperitoneally (*t*, trocar; in **B**). Rupture may be intraperitoneal (*i*) or extraperitoneal (*e*). In **A**, the prostate is being palpated *per rectum*. (Based on Testut and Latarjet.)

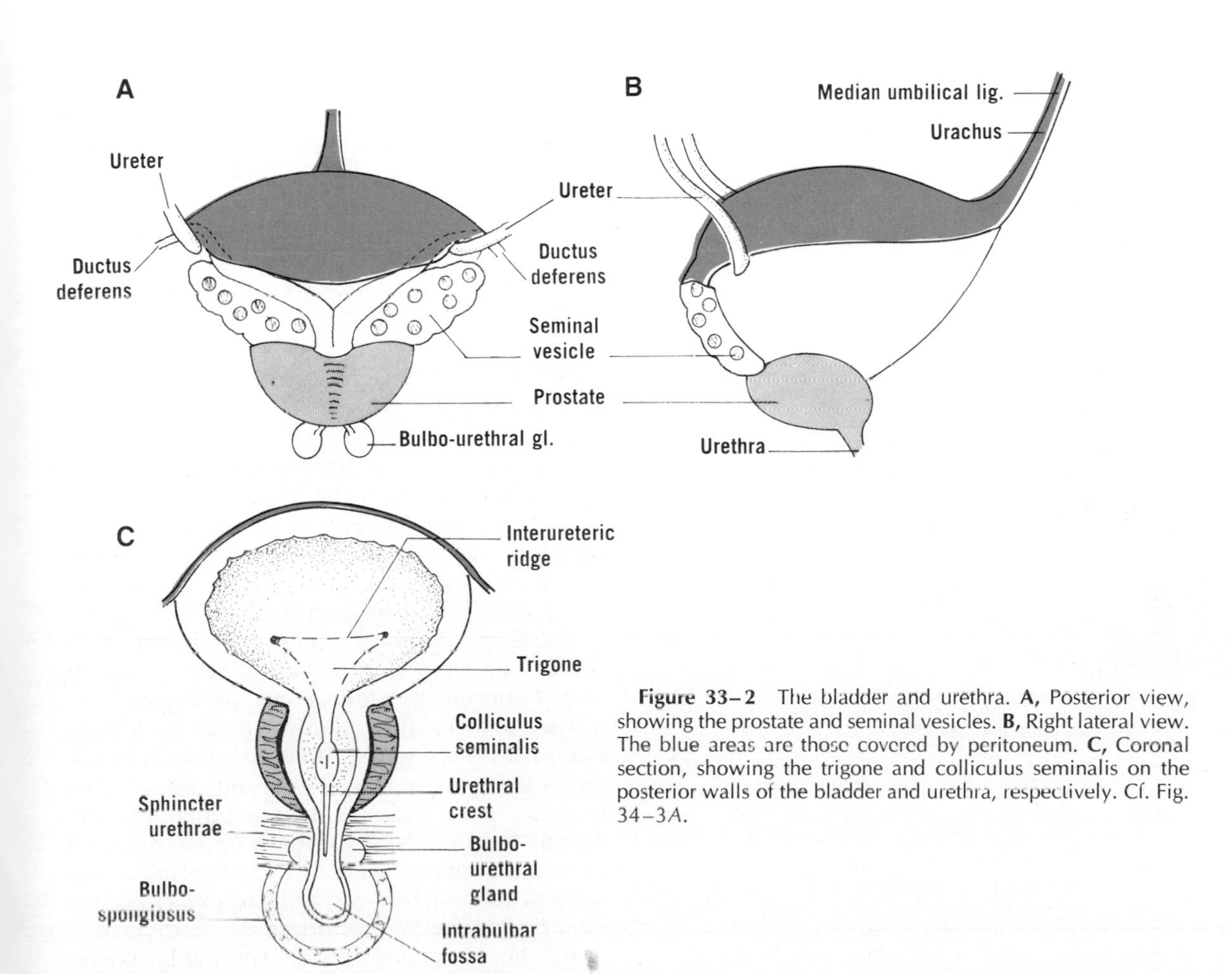

Figure 33–2 The bladder and urethra. **A,** Posterior view, showing the prostate and seminal vesicles. **B,** Right lateral view. The blue areas are those covered by peritoneum. **C,** Coronal section, showing the trigone and colliculus seminalis on the posterior walls of the bladder and urethra, respectively. Cf. Fig. 34–3*A*.

termed its *body*. The lowest part, or *neck*, of the bladder is attached to the pelvic diaphragm and is continuous in the male with the prostate. This region is anchored by localized thickenings of the superior fascia of the pelvic diaphragm: the *medial* and *lateral puboprostatic* (or *pubovesical*) *ligaments*. Another fascial support, the *lateral ligament*, extends backward on each side from the base of the bladder to the sacrogenital fold.

Interior (*figs. 33–2 and 34–1*). **The triangular area formed between the orifices of the right and left ureters and the internal urethral orifice is termed the trigone.** Its mucosa is smooth, flat, red to pink, and firmly attached.

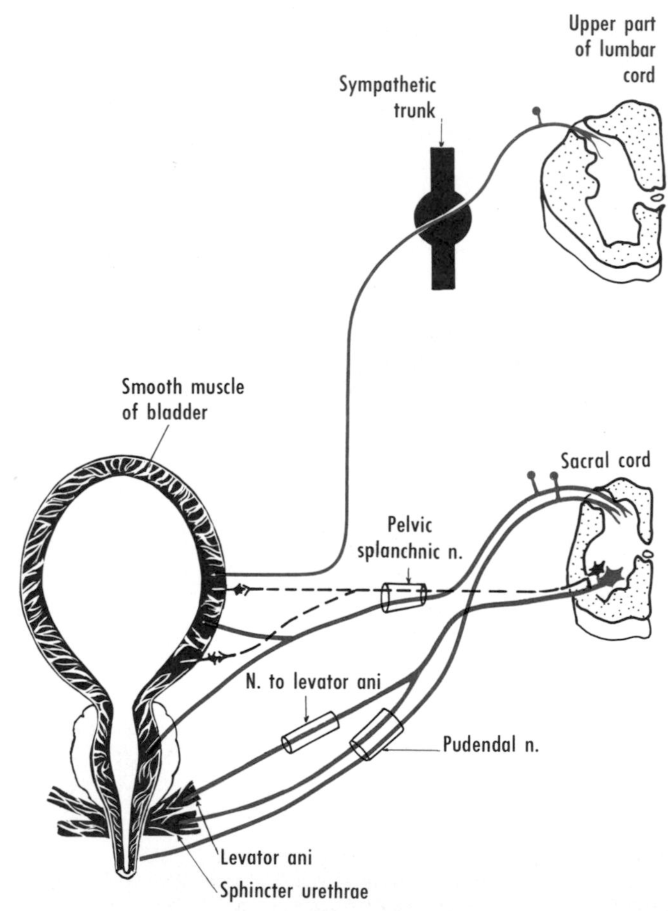

Figure 33–3 Nerve supply of the urinary bladder and related structures. The preganglionic parasympathetic fibers (*interrupted lines*) to the smooth muscle (detrusor) of the bladder synapse with ganglion cells in the wall of the organ. Most of the afferent fibers (*blue*) from the bladder and urethra course in the pelvic splanchnic nerves. A few pain fibers from the bladder ascend in the hypogastric plexuses to the upper lumbar part of the spinal cord. The sympathetic supply is not shown. Pain fibers from the urethra course in the pelvic splanchnic and pudendal nerves. The red lines represent motor fibers to the levator ani and sphincter urethrae.

Muscle fibers between the two ureteric orifices raise a fold known as the *interureteric ridge*. Behind the internal urethral orifice, a median fold (the *uvula*) formed by muscle fibers, the middle lobe of the prostate, or by both may develop with increasing age. The interior of the bladder may be viewed *in vivo* by an electrically lit instrument termed a cystoscope. Access to the bladder may be gained by either urethral catheterization or suprapubic puncture (see fig. 33–1). A calculus may be removed through a suprapubic incision (*lithotomy*). Perineal lithotomy ("cutting for stone") is a very ancient operation.

Blood Supply and Lymphatic Drainage. The bladder is supplied mainly by the superior and inferior vesical arteries, which arise directly or indirectly from the internal iliac artery. The veins drain into the internal iliac vein. The lymphatic vessels go to the various iliac nodes.

Innervation (*fig. 33–3*). The bladder is supplied by branches of the vesical and prostatic plexuses, which are extensions of the inferior hypogastric plexuses. The branches include (1) parasympathetic motor fibers to the *detrusor* (i.e., the muscular coat); (2) sensory fibers that are stimulated by stretching, causing a sensation of fullness and activating reflexes; and (3) sympathetic fibers to blood vessels.

Micturition (or urination) is preceded by contraction of the diaphragm and abdominal wall. The neck of the bladder descends, the detrusor contracts reflexly, and urine is expelled from the bladder.

URETHRA (see figs. 33–2, 34–1, 34–3, and 35–1)

The urethra is a fibromuscular tube that conducts urine from the bladder (and semen from the ductus deferens) to the exterior. It begins at the neck of the bladder, traverses the pelvic and urogenital diaphragms, and ends at the external urethral orifice.

The female urethra, about 4 cm in length, is fused with the anterior wall of the vagina. It ends between the clitoris and the vagina.

The male urethra, about 20 cm in length, comprises three parts: prostatic, membranous, and spongy (see figs. 34–1 and 38–4). The *prostatic part*, which is the most dilatable, descends through the prostate. Its posterior wall presents a median ridge, the *urethral crest*, the summit of which is termed the *colliculus seminalis* (or the *verumontanum*). A diverticulum, the *prostatic utricle* (probably corresponding to portions of the uterus and vagina) opens on the colliculus, as do the ejaculatory ducts. The prostatic ducts open into a groove, the *prostatic sinus,* on each side of the urethral crest. The *membranous part* descends from the apex of the prostate to the bulb of the penis and is surrounded by the sphincter urethrae. **The lowermost part of the membranous urethra is liable to rupture or to penetration by a catheter.** The *spongy part* lies in the corpus spongiosum and traverses the bulb, body, and glans of the penis. It is slightly dilated near its origin (*intrabulbar fossa*) and termination (*navicular fossa*). The two *bulbourethral glands* (which are situated bilaterally in the sphincter urethrae and behind the membranous urethra) open into the proximal portion of the spongy urethra. The external urethral orifice is the narrowest portion of the entire urethra. The interior of the urethra may be viewed *in vivo* by an electrically lit instrument termed a urethroscope.

QUESTIONS

33–1 How does the distended bladder differ from the empty organ?

33–2 How does the bladder of a child differ from that of an adult?

33–3 How does the trigone differ from the rest of the bladder?

33–4 What is lithotomy?

33–5 Which is the narrowest portion of the urethra?

33–6 Which structures open into the urethra?

34 MALE GENITALIA

The male genital organs comprise the testes and epididymides, ductus deferentes, seminal vesicles and ejaculatory ducts, prostate, bulbo-urethral glands, and penis (fig. 34–1). Spermatozoa, formed in the testes and stored in the epididymides, are contained in the semen, which is secreted by the testes and epididymides, seminal vesicles, prostate, and bulbo-urethral glands. The spermatozoa, on leaving the epididymides, pass through the ductus deferentes and ejaculatory ducts to reach the urethra. The complete pathway is as follows (figs. 34–1 and 34–2): (1) convoluted seminiferous tubules, (2) straight seminiferous tubules, (3) rete testis, (4) efferent ductules of testis, (5) lobules (or cones) of epididymis, (6) duct of epididymis, (7) ductus deferens, (8) ejaculatory duct, (9) prostatic urethra, (10) membranous urethra, and (11) spongy urethra.

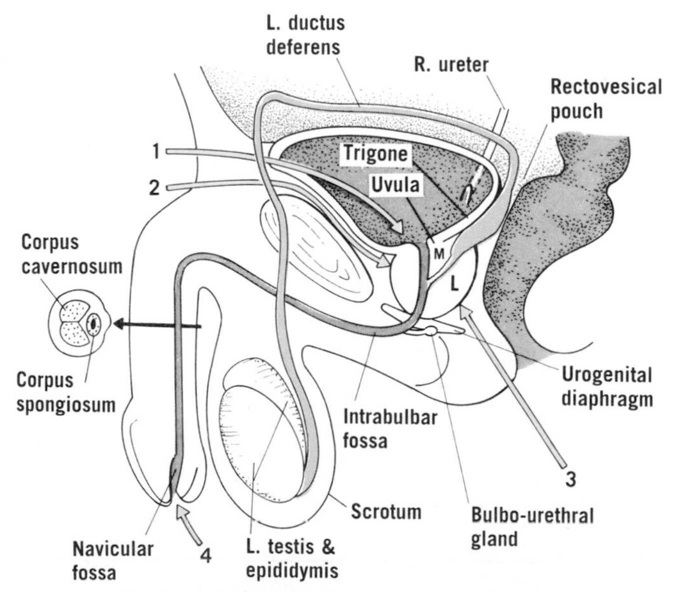

Figure 34–1 Male reproductive system. The approaches to the prostate are (*1*) transvesical, (*2*) retropubic, (*3*) perineal, and (*4*) urethral. *M*, middle lobe of the prostate (which may produce a uvulva); *L*, median groove between the lateral lobes of the prostate.

TESTIS AND EPIDIDYMIS (figs. 34–1 and 34–2)

The testis is an ovoid gland that produces spermatozoa and secretes steroid hormones. In late fetal life, the testes descend from the abdomen through the inguinal canal and so reach the scrotum. The left testis is usually lower than the right. Failure of descent of a testis is termed *cryptorchism* (Gk, "hidden testis"). A testis may descend to an anomalous site, e.g., the perineum (*ectopia testis*). The testis presents upper and lower ends, medial and lateral surfaces, and anterior and posterior margins.

The testis is covered by the *tunica vaginalis,* which is derived prenatally from the peritoneum. Beneath the visceral layer, a connective tissue coat, the *tunica albuginea,* sends fibrous *septa* into the interior, and these converge posteriorly to form the *mediastinum testis. Efferent ductules* connect the testis with the head of the epididymis.

The epididymis (fig. 34–2) is applied to the posterior margin of the testis and comprises a *head, body,* and *tail.* It should be noted that the plural is *epididymides.* A recess of the tunica vaginalis, termed the *sinus of the epididymis,* extends between the body of the epididymis and the lateral surface of the testis (fig. 34–2*A* and *C*).

The efferent ductules of the testis form *lobules* (or *cones*) in the head of the epididymis, and these drain into the *duct of the epididymis,* which descends through the body and tail of the organ and becomes the ductus deferens. Small embryonic remnants may be found: the *appendix testis* at the upper end and the *appendix of the epididymis* on the head of that organ.

Blood Supply and Lymphatic Drainage. The testis and epididymis are supplied by the testicular artery (fig. 34–2*C*), and their veins drain into the *pampiniform plexus,* which forms the bulk of the spermatic cord. **The veins of the pampiniform plexus often become varicose, a condition termed varicocele.** Lymphatics accompany the testicular vessels and drain into the lumbar (aortic) nodes.

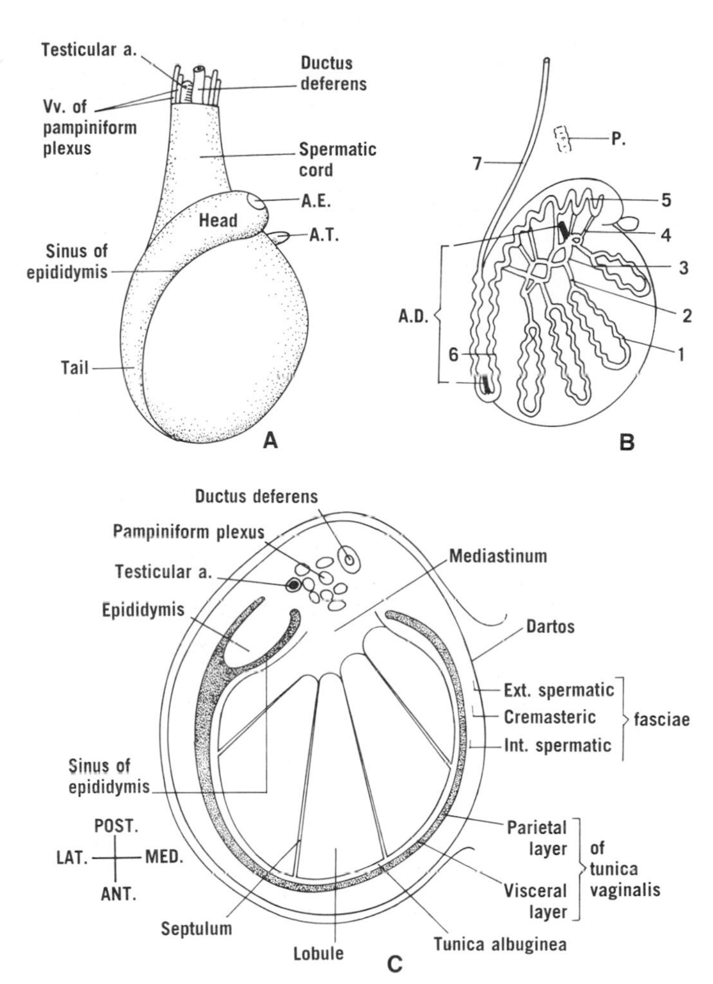

Figure 34–2 The testis and epididymis. **A,** Right testis, lateral aspect. **B,** Ductal system. (*See text.*) **C,** Horizontal section, showing the tunica vaginalis. The numbers 1 to 7 refer to the parts listed in the text.

DUCTUS DEFERENS AND SPERMATIC CORD (figs. 33–2, 34–1, and 34–2)

A ductus (formerly called *vas;* hence *vasectomy*) deferens (fig. 34–2) continues the duct of the epididymis to the ejaculatory duct. The plural is *ductus deferentes* (*vasa deferentia*). The ductus deferens extends from the tail along the medial side of the epididymis and becomes surrounded by the pampiniform plexus of veins as it becomes incorporated into the spermatic cord, where it can be felt *in vivo* as a firm cord. After its passage through the inguinal canal, it leaves the spermatic cord and turns around the inferior epigastric artery. Next it enters the pelvis, crosses the ureter, gains the back of the bladder, and becomes expanded as the *ampulla*. Finally, it joins the duct of the seminal vesicle to form the ejaculatory duct. Embryonic remnants include superior and inferior *aberrant ductules* and the *paradidymis* (fig. 34–2*D*).

The *seminal vesicles* (fig. 34–3) are two saccular pouches on the back of the bladder (see fig. 33–2). They produce much of the semen. They are related to the rectum behind, and their uppermost parts are covered by peritoneum. Each vesicle consists of a coiled tube that ends below as a duct and joins the ductus deferens to form the ejaculatory duct. The two *ejaculatory ducts* penetrate the prostate and open on the colliculus seminalis into the prostatic urethra.

The spermatic cord extends from the deep inguinal ring, where the ductus deferens begins to acquire its coverings, to the posterior border of the testis. It consists of the ductus deferens and associated arteries (including the testicular artery), nerves, and lymphatics; the pampiniform plexus of veins; and remnants of the *processus vaginalis peritonei.* The coverings derived from the fasciae associated with the three oblique muscles of the abdomen (see fig. 25–7) are the external spermatic, cremasteric, and internal spermatic fasciae, respectively. The cremasteric fascia contains bundles of skeletal-type muscle known collectively as the *cremaster* and supplied by the genital branch of the genitofemoral nerve. Contraction results in elevation of the testis and can often be produced by gently stroking the skin of the medial side of the thigh (*cremasteric reflex*).

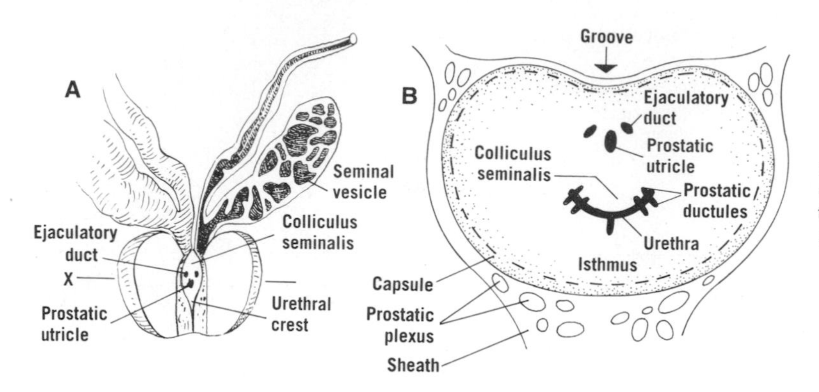

Figure 34–3 The prostate. **A,** Coronal scheme, showing the seminal vesicles and colliculus seminalis. Cf. fig. 33–2C. The depression on each side of the urethral crest is termed the prostatic sinus. **B,** Horizontal section.

Deep to these layers, the testis and epididymis are covered by a double serous layer, the *tunica vaginalis testis,* derived prenatally from the processus vaginalis of the peritoneum (see fig. 25–11). **The potential cavity between the parietal and visceral layers of some part of the processus vaginalis (usually the tunica) may become distended with fluid, a condition termed hydrocele.** Although the commonest type of hydrocele is in the tunica around the testis, a remnant of the processus may form an encysted hydrocele of the cord.

PROSTATE (fig. 34–3)

The prostate is a fibromuscular, pelvic organ surrounding the male urethra and containing glands that contribute to the semen. It is situated behind the pubic symphysis and in front of the rectum, through which it can be felt *in vivo.* The prostate presents a base, an apex, and anterior, posterior, and two inferolateral surfaces. The *base* is continuous with the bladder, although it is separated by a slight groove. The *apex* is the lowermost part. The *anterior surface* is narrow, and the *inferolateral surfaces* are related to the superior fascia of the pelvic diaphragm. The *posterior surface* is triangular and presents a median groove (see figs. 33–2*A* and 34–3*B*). **The normal prostate can be palpated *per rectum* as an elastic swelling with a median groove ending above in a notch.** The upper part of the posterior surface is covered by the seminal vesicles and the ampullae of the ductus deferentes.

The *prostatic glands* within the organ open chiefly by *ductules* into the prostatic sinuses of the urethra. The main glands, situated laterally and posteriorly, are those involved in carcinoma.

The prostate increases rapidly in size at puberty. During the fifth decade, it either begins to atrophy or undergoes benign hypertrophy.

Access to the prostate may be gained by one of the following routes: transvesical, retropubic, perineal, or urethral (see fig. 34–1).

Lobes. Although the fetal prostate consists of several lobes, these are not distinguishable in the adult.* Nevertheless, right and left lobes, united in front by a muscular *isthmus,* are often postulated. The portion of the prostate that extends forward from the upper part of the posterior surface and lies between the ejaculatory ducts and the urethra is commonly known as the *middle* (or *median*) *lobe.* **Enlargement of the so-called middle lobe may accentuate the uvula of the bladder, which then acts as a valve over the internal urethral orifice, thereby blocking the passage of urine.**

Prostatic Sheath. The superior fascia of the pelvic diaphragm forms the sheath or fascia of the prostate and is continued over the bladder (see fig. 38–2). The sheath, which is attached to the pubis by ligamentous and smooth muscle fibers, is separated from the *capsule* of the prostate anteriorly and laterally by the *prostatic venous plexus.* Beneath the capsule, the compressed outer zone of an enlarged prostate may form a ''false capsule.''

Blood Supply and Lymphatic Drainage. The

* In a study by J. E. McNeal (J. Urol., *107*: 1008-1016, 1972) it was emphasized that (1) the prostatic lobes of the fetus do not persist in the adult; (2) the urethral glands, which form a central zone, are not really a part of the prostate; (3) the prostatic urethra comprises an upper portion distinguished by the urethral glands and a ''preprostatic sphincter;'' (4) ''benign prostatic hypertrophy'' is really benign peri-urethral hyperplasia, being confined to the central zone and upper portion of the prostatic urethra; (5) the preprostatic sphincter is important in maintaining continence; and (6) the peripheral zone is the site of carcinoma.

prostate is supplied mainly by the inferior vesical artery of the internal iliac artery. The prostatic venous plexus drains into the internal iliac veins and communicates with the vertebral plexus, thereby allowing neoplastic spread to bones. The lymphatics end mostly in the internal iliac nodes, although some end in the external iliac nodes.

QUESTIONS

34–1 What is meant by descent of the testis?

34–2 What is the tunica vaginalis testis?

34–3 List some embryonic remnants related to the testis, epididymis, and ductus deferens.

34–4 What is a varicocele?

34–5 What is the spermatic cord?

34–6 What is the key feature of the normal prostate as palpated *per rectum?*

34–7 What are the surgical approaches to the prostate?

34–8 Where is the middle (or median) lobe of the prostate?

34–9 Where is the uvula of the bladder?

34–10 What are the coverings of the prostate?

35
FEMALE GENITALIA

The female genital organs comprise the ovaries, uterine tubes, uterus, vagina, and external genitalia (fig. 35–1). The vagina is situated partly in the pelvic cavity and partly in the perineum. The internal organs can be examined *in vivo* by an electrically lit tubular instrument inserted into the peritoneal cavity (*laparoscopy*).

OVARY (figs. 35–1, 35–2, and 35–8)

An ovary is an ovoid gland that produces oocytes and secretes steroid hormones. It is commonly situated on the lateral wall of the pelvis (typically in the angle between the external iliac vein and the ureter), where it can be palpated bimanually. The ovary presents

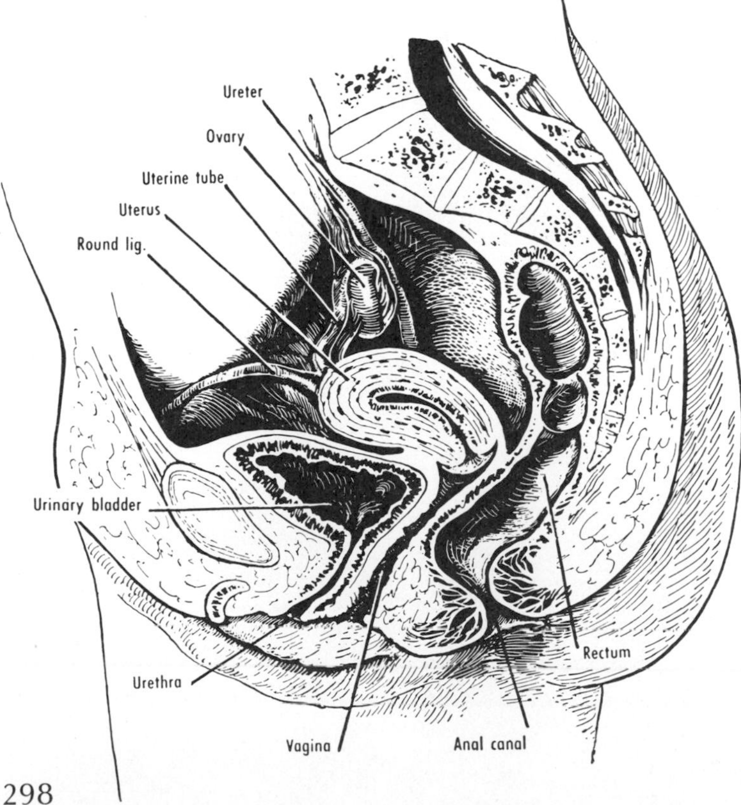

Figure 35–1 Median section of the female pelvis, including a medial view of the right lateral pelvic wall. (Modified from Appleton, Hamilton, and Tchaperoff and from Shellshear and Macintosh.)

tubal and uterine ends, medial and lateral surfaces, and mesovarian and free borders. The upper, or *tubal, end,* closely related to the uterine tube, is attached to the suspensory ligament of the ovary. The lower, or *uterine, end* is attached to the ovarian ligament. The *medial surface* is related to the uterine tube and the ileum. The *lateral surface* is in contact with the parietal peritoneum that lines the side wall of the pelvis. The anterior, or *mesovarian, border* is attached to the mesovarium, and it provides the *hilus* of the ovary. The posterior, or *free, border* is related to the uterine tube and ureter.

Ligaments. The ovary is anchored to the back of the broad ligament by a double fold of peritoneum, the *mesovarium,* which is continuous with the so-called germinal epithelium around the ovary. The *suspensory ligament of the ovary* (or *infundibulopelvic ligament*) ascends to become lost in the connective tissue of the pelvis. The ovarian artery descends in the suspensory ligament and, by way of the broad ligament and mesovarium, gains the hilus of the ovary. The *ovarian ligament* proceeds to the body of the uterus, immediately behind the opening of the uterine tube.

UTERINE TUBES (figs. 35–2, 35–3, and 35–8)

The uterine tubes are paired conduits between the ovaries and the uterus. The uterine tube transmits oocytes from the ovaries and spermatozoa from the uterus. It is the usual site of fertilization, and it conveys the early embryo to the uterine cavity.

The uterine tubes develop as outgrowths of the peritoneal cavity; they maintain this continuity and thereby allow communication between the peritoneal cavity and the exterior of the body. The Greek word *salpinx,* meaning "tube," is used in such compounds as *mesosalpinx.*

Each uterine tube is situated in the upper, free border and between the layers of the broad ligament. **The uterine tube is subdivided into four parts, from lateral to medial: the infundibulum, ampulla, isthmus, and uterine part.** The *infundibulum,* which is closely related to the ovary, presents the *abdominal opening of the uterine tube,* by which the tube is in communication with the peritoneal cavity. Oocytes pass from the ovary through the abdominal opening and along the uterine tube. The *fimbriae* are irregular fringes that project

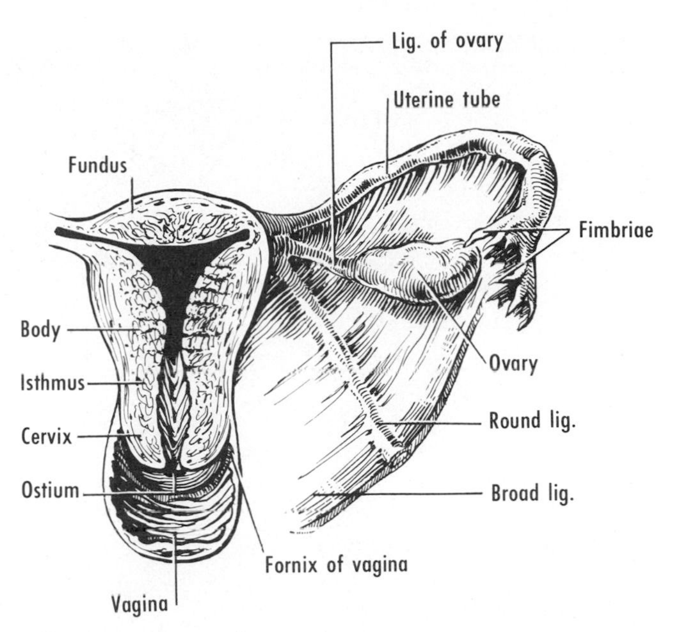

Figure 35–2 Female reproductive organs, posterior view.

from the margin of the infundibulum, and one (*ovarian fimbria*) may be longer than the others. The *ampulla,* the longest and widest part, continues gradually into the *isthmus.* The *uterine part,* which lies in the wall of the uterus, presents the *uterine opening of the uterine tube.*

Patency of the uterine tubes can be demonstrated radiographically (*hysterosalpingography*) by the injection of a radio-opaque medium into the uterus (fig. 35–3).

UTERUS (figs. 35–1, 35–2, 35–4, and 35–5)

The uterus is a muscular organ in the lining of which the embryo becomes implanted and in which the embryo and fetus develop. The uterine cavity receives the openings of the uterine tubes, and the uterine cavity and vagina (the "birth canal") allow the exit of the fetus at birth (fig. 35–6). The Greek words *hystera* and *metra* are used in such compounds as *hysterectomy* and *endometrium.* The uterus has three layers: a mucosa (the *endometrium*), a muscular coat (the *myometrium*), and a serosa (the *perimetrium*).

Parts. The nulliparous uterus resembles an inverted pear and consists of two main parts: the body and the cervix. The *body* is twice as long as the cervix, whereas the converse is true in the newborn. The body includes the *fundus,* which is the portion that lies above

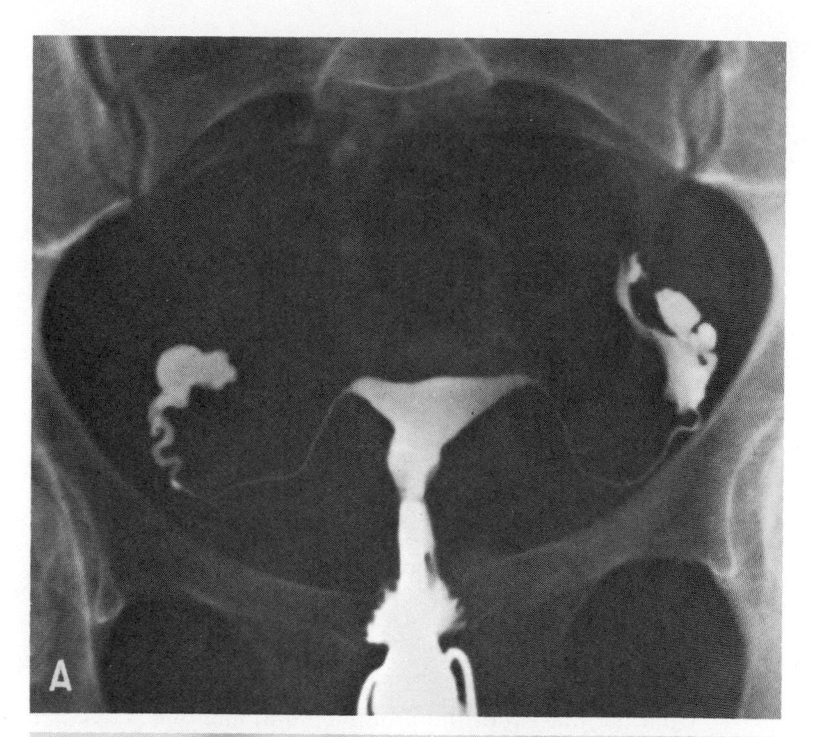

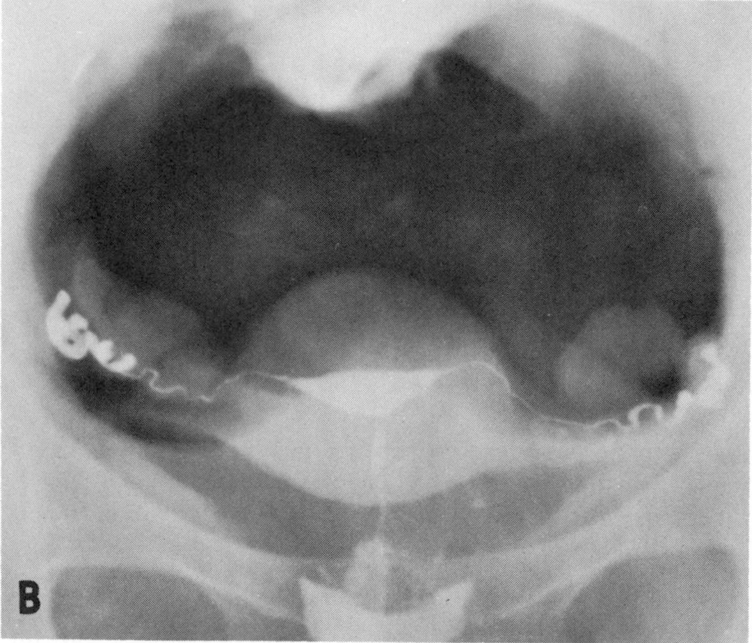

Figure 35–3 Hysterosalpingograms, showing the uterus and uterine tubes. In **A,** note the cavity of the uterus, uterine tubes, and the bilateral "spill" of the radio-opaque medium into the peritoneal cavity, thereby demonstrating the patency of the tubes. **B,** View from in front and above. Note the slit-like shape of the uterine cavity in this view. Note also the thickness of the uterine wall. (**A,** Courtesy of Sir Thomas Lodge. **B,** Courtesy of Robert A. Arens, M.D., Chicago, Illinois.)

and in front of the openings of the uterine tubes. The body is usually bent forward onto the bladder (anteflexion, see fig. 35–4), being separated by the uterovesical pouch. Above and behind, the body is separated from the sigmoid colon by the recto-uterine pouch (see fig. 35–4), which generally contains coils of ileum. Right and left margins are anchored to the broad ligaments. **The junctional region between the body and cervix is referred to as the isthmus: during pregnancy, it is taken up into the body of the uterus as the "lower uterine segment."** The cavity of the isthmus was formerly called the "internal os." The *cervix* extends downward and backward and usually forms approximately a right angle with the vagina (the angle of anteversion, see fig. 35–4). As the bladder fills, the uterus tends to be-

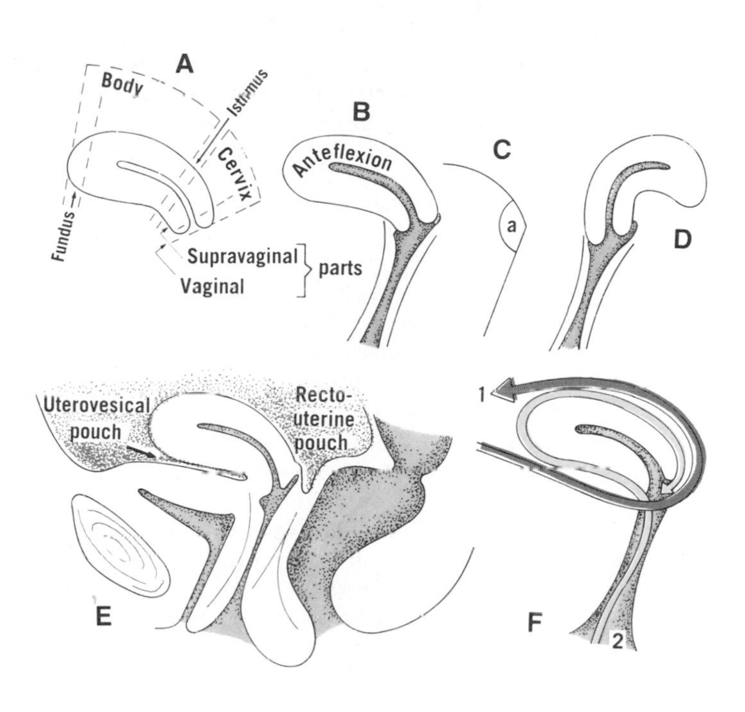

Figure 35–4 The uterus. **A** shows the parts of the organ. **B** shows the normal position of anteflexion and anteversion. **C** shows the angle (*a*) of anteversion. **D** represents a retroverted uterus. **E** shows the uterovesical and recto-uterine pouches. **F** demonstrates the principle of (*1*) abdominal and (*2*) vaginal hysterectomy (*arrows*).

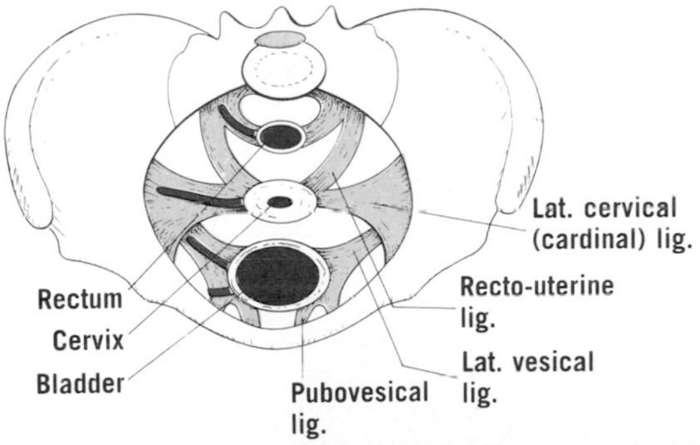

Lat. cervical (cardinal) lig.
Recto-uterine lig.
Lat. vesical lig.
Pubovesical lig.
Rectum
Cervix
Bladder

Figure 35–5 Horizontal section of the pelvic viscera, showing the ligaments of the uterus. The arteries shown are, from posterior to anterior, the middle rectal, uterine, and inferior and superior vesical.

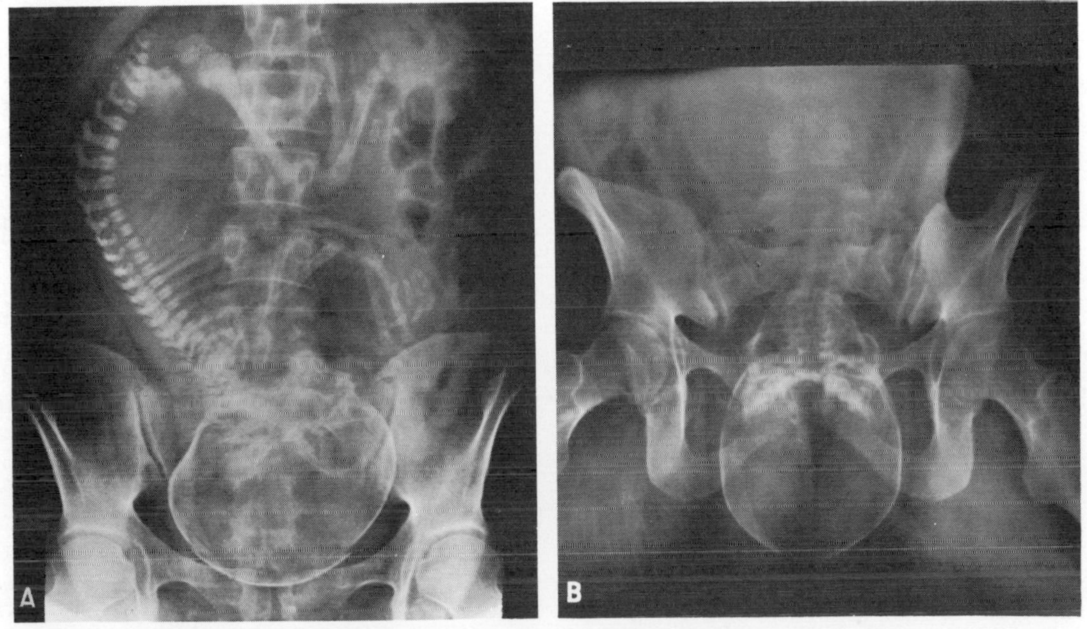

Figure 35–6 **A,** Fetus at term *in utero,* cephalic presentation. Note the fetal vertebrae and ribs, bones of the limbs, and skull. The parietal bones have overlapped the frontal bone at the coronal suture. **B,** Infant during birth. Cf. fig. 31–6, lowest position (**A,** Courtesy of Robert A. Arens, M.D., Chicago, Illinois. **B,** Courtesy of Robert P. Ball, M.D., Oak Ridge, Tennessee.)

come retroverted. **The cervix may be considered in two parts: (1) a supravaginal portion above the limits of the vagina and (2) a vaginal portion, which projects into the cavity of the vagina** (see fig. 35–4). The cavity of the body, which is somewhat triangular in coronal perspective (see fig. 35–2), is slit-like in sagittal section (see fig. 35–1). The canal of the cervix communicates with the vagina by the *ostium uteri* (formerly called the "external os"), which is bounded by *anterior* and *posterior lips*. The entire uterine cavity can be demonstrated radiographically by *hysterosalpingography* (see fig. 35–3). The uterus can be palpated bimanually *in vivo* (fig. 35–7). Dilatation (of the cervical canal) and curettage (scraping of the uterine lining) are performed for diagnostic or therapeutic purposes.

Peritoneal Relations. The uterus is supported by being anchored to the vagina and by its peritoneal and fascial attachments to nearby structures. The peritoneum is reflected from the bladder (uterovesical pouch) to the isthmus uteri and then over the fundus and onto the back of the cervix (recto-uterine pouch) and vagina (see figs. 35–1 and 35–4).

Ligaments. The peritoneum that covers the uterus continues laterally as a double fold known as the *broad ligament* (figs. 35–2 and 35–8). The ligament extends to the lateral wall of the pelvis and serves as a mesentery for the uterine tube, which lies between its two layers. This part is the *mesosalpinx,* whereas the part adjacent to the uterus is the *mesometrium.* The posterior layer of the broad ligament forms the *mesovarium.* In addition to the uterine tube, the broad ligament contains connective tissue (the *parametrium*), the uterine and ovarian vessels, the round and ovarian ligaments, and some embryonic remnants (e.g., the *epoophoron,* which consists largely of a duct parallel to and below the uterine tube) (fig. 35–8*B*). The *round ligament* is a fibrous band attached to the uterus immediately below the entrance of the uterine tube. It extends laterally and forward, hooks around the inferior epigastric artery, traverses the inguinal canal, and becomes lost in the labium majus. The round ligament is accompanied in the fetus, and occasionally in the adult, by a process of peritoneum, the *processus vaginalis.* The visceral pelvic fascia at the side of the cervix is thickened as the *lateral* (or *transverse*) *cervical* (or *cardinal*) *ligament* and as the *uterosacral ligament* (see fig. 35–5).

Blood Supply. The uterine arteries (fig. 35–

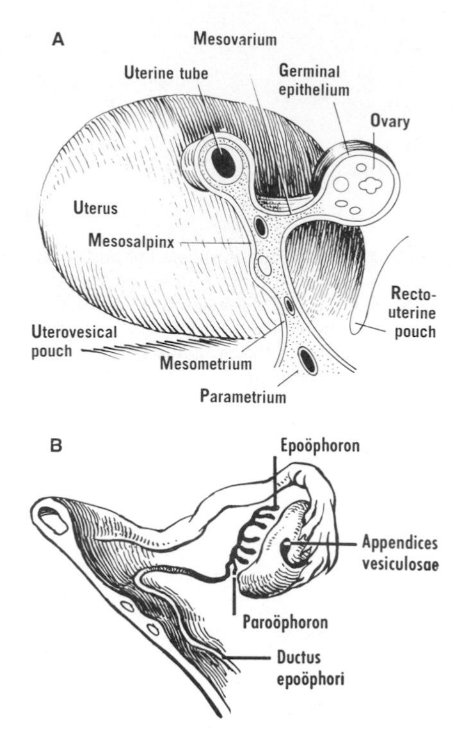

Figure 35–8 Sagittal sections of (**A**) left and (**B**) right broad ligaments, showing their relationships to the ovaries and uterine tubes. **B** shows embryonic remnants.

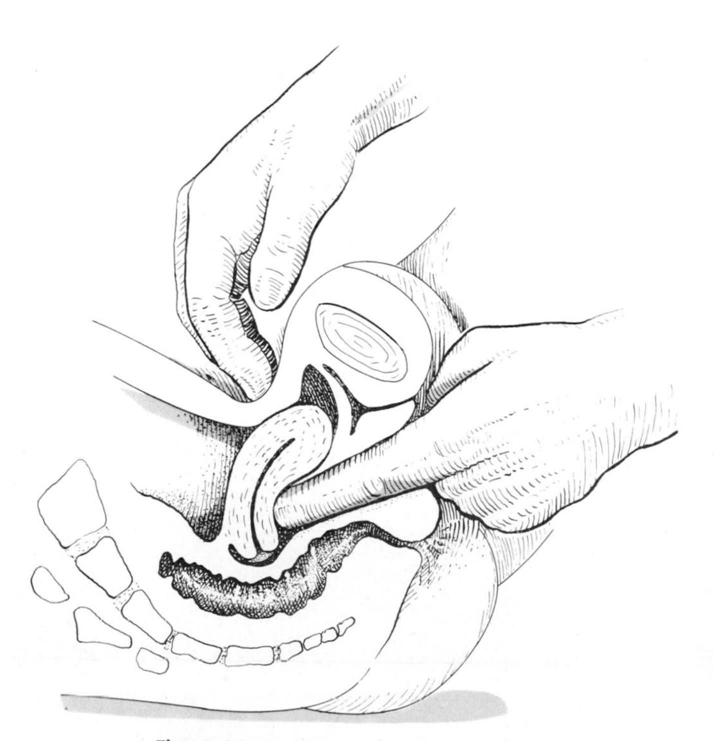

Figure 35–7 Bimanual palpation of the uterus. (After Kelly and Noble.)

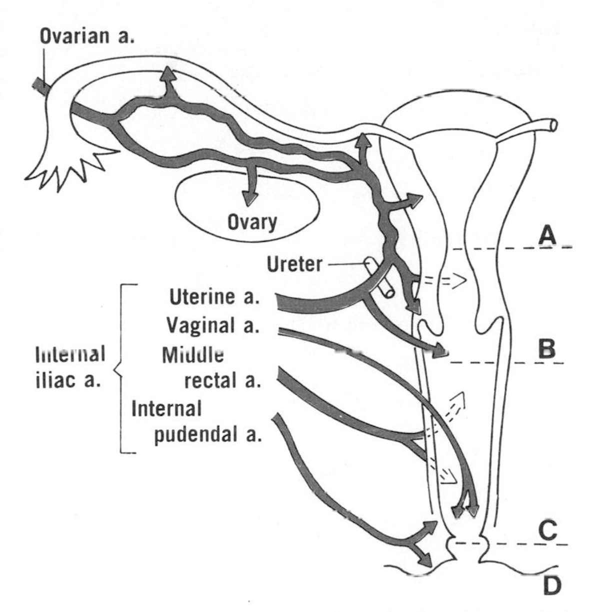

Figure 35–9 The blood supply to the female reproductive system. Extensive anastomoses occur between the ovarian and uterine arteries. Cervical branches of the uterine arteries anastomose across the median plane. The four-tiered concept of the reproductive system (*A,B,C,D*) is based on anatomical, physiological, and pathological data and may perhaps have embryological implications. For details see R. Contamin et al., Gynécol., *28*:235–252, 1977.

9) provide the main blood supply. Each artery ascends between the layers of the broad ligament, near the lateral margin of the body, and supplies branches to both front and back. The uterine venous plexus is connected with the superior rectal vein, thereby forming a portal-systemic anastomosis.

Lymphatic Drainage. The fundus and upper part of the body drain into the lumbar (aortic) nodes, the lower part of the body into the external iliac nodes, and the cervix into the external and internal iliac and the sacral nodes.

VAGINA (see figs. 35–1 and 35–2)

The vagina serves for copulation, as the lower end of the birth canal, and as the excretory duct for the products of menstruation. The cavity of the vagina communicates with that of the uterus above, and it opens into the vestibule below. The vagina extends downward and forward, parallel to the plane of the pelvic inlet.

The *anterior* and *posterior walls* of the vagina are about $7^1/_2$ and 9 cm long, respectively. They are highly distensible and are in contact below the cervix. The recess between the vagina and the vaginal part of the cervix consists of a continuous anterior, lateral, and posterior *fornix*. The posterior fornix, which is the deepest, is related to the recto-uterine pouch. The opening of the vagina into the vestibule may be partially closed by a fold called the *hymen* (see fig. 38–5).

Relations. The vagina is related anteriorly to the cervix, ureters, and bladder and is fused with the urethra. Posteriorly, the vagina is related to the recto-uterine pouch, the rectum, and the perineal body. The lateral fornix of the vagina is related to the ureter and uterine artery. The pubococcygeal muscles act as a sphincter for the vagina. The vagina is supplied by branches (including uterine and vaginal) of the internal iliac artery.

The vagina and cervix can be inspected through a speculum in the vagina. **Digital examination *per vaginam* may be combined with palpation through the anterior abdominal wall by the other hand (bimanual examination).** The following structures are palpable *per vaginam:*

Anteriorly—urethra, vaginal part of cervix, distended bladder, and body of uterus bimanually

Laterally—ureters and displaced or enlarged ovaries and uterine tubes bimanually

Posteriorly—rectum, any mass in the recto-uterine pouch, and sometimes sacral promontory

Vaginal (Papanicolaou) smears are used in histodiagnosis.

QUESTIONS

35–1 Where is the ovary situated?

35–2 Is the ovary covered by peritoneum?

35–3 What are the openings of the uterine tube?

35–4 What is the broad ligament?

35–5 Which Latin terms and Greek roots are associated with the ovary, uterine tube, and uterus?

35–6 What is the endometrium?

35–7 What are anteflexion and anteversion of the uterus?

35–8 What is the lower opening of the uterus termed?

35–9 What is hysterosalpingography?

35–10 What are the peritoneal relations of the uterus?

35–11 By what is the uterus supported?

35–12 What is the epoophoron?

35–13 What are the surgical approaches to the uterus?

36
THE RECTUM AND ANAL CANAL

RECTUM

At the level of the middle of the sacrum, **the sigmoid colon loses its mesentery and gradually becomes the rectum,** which, at the upper limit of the pelvic diaphragm, ends in the anal canal (fig. 36–1). The rectum, about 15 cm long, widens below as the *ampulla,* which is very distensible. Although variable in shape, the rectum follows the sacrococcygeal curve. At the anorectal junction, the gut curves backward and its concavity is held by the puborectal sling, which can be palpated *per anum.* The rectum presents three or more lateral curvatures, which correspond to *transverse rectal folds* in the interior of the gut. The rectum has neither mesentery nor haustra, and it has an almost complete outer longitudinal muscular coat rather than teniae.

Relations. In the upper third of the rectum, its front and sides are covered by peritoneum; in its middle third, the front only; its lower third is devoid of peritoneum. The rectovesical and recto-uterine pouches descend to within about 7 to 8 cm and 5 to 6 cm, respectively, of the anus (see figs. 31–9 and 35–4). Below the pouches, condensations of parietal pelvic fascia are found, and the rectum is surrounded by visceral pelvic fascia from the superior fascia of the pelvic diaphragm (see fig. 38–3).

Anteriorly, the rectum is related in the male to coils of small intestine in the rectovesical pouch above and to the back of the bladder, prostate, seminal vesicles, and ductus deferentes below; in the female, it is related to coils of small intestine in the recto-uterine pouch above and to the back of the vagina below. Laterally, the rectum is related to the ileum or sigmoid colon. Posteriorly, it is related to the sacrum, coccyx, and pelvic diaphragm.

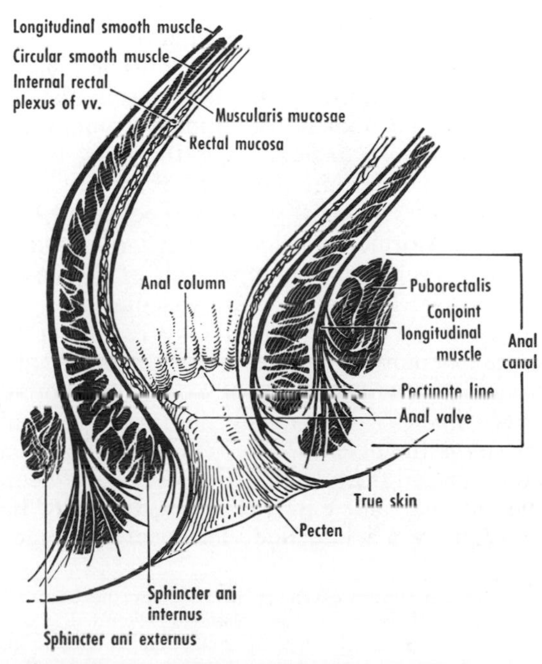

Figure 36–1 Median section of the rectum and anal canal. The anterior aspect is on the left side of the drawing. (Based on Morgan and Thompson and on Walls.)

ANAL CANAL

Anatomically, the anal canal extends from the level of the upper aspect of the pelvic diaphragm to the anus. In surgical usage, however, the anal canal is frequently limited to that part of the intestine which is below the pectinate line; this part differs from the part above the pectinate line in innervation, venous and lymphatic drainage, and possibly lining epithelium. The anal canal (fig. 36–1) is about 3 cm in length. The congenital anomaly in which the anal canal fails to communicate with the exterior is known as *imperforate anus*.

Relations. Surrounded by the levatores ani, the anal canal passes through the pelvic diaphragm (see fig. 38–3), and the anorectal junction is held forward by the puborectal sling (see fig. 37–1). Below the pelvic diaphragm, the anal canal is surrounded by the sphincter ani externus, and the ischiorectal fossae are situated laterally (see fig. 38–3).

Sphincter ani externus. The external sphincter (figs. 36–1 and 38–3) is usually described in three parts.* The *subcutaneous part* surrounds the lowermost portion of the canal. The *superficial part,* situated above the subcutaneous division, is attached to the perineal body and coccyx. The *deep part,* more or less continuous with the superficial division, surrounds the uppermost portion of the canal and is associated with the puborectalis posteriorly. The sphincter ani externus is supplied by the inferior rectal nerves and by a perineal branch of the fourth sacral nerve (S.N.4). The muscle is in variable tonic contraction during waking hours, and it can be contracted voluntarily. The *sphincter ani internus* is the thick lower end of the inner circular layer of the gut.

Interior. Several vertical mucosal folds, the *anal* (formerly called rectal) *columns,* are usually visible in the upper half of the canal (fig. 36–1). The columns are vascular, and enlargement of their venous plexus results in internal hemorrhoids. The anal columns are united below by *anal valves,* which bound *anal sinuses.* The lower limit of the anal valves is the *pectinate line,* below which is a zone termed the *pecten.* The interval between the internal and external sphincters may be marked by a white line. The pecten merges with the skin of the anus. An ischiorectal abscess may drain through a *fistula in ano* into the anal canal.

The interior of the lower portion of the intestine can be inspected by electrically lit instruments (*proctoscopy, sigmoidoscopy,* and *colonoscopy*). **Digital examination *per rectum* is an important clinical procedure, e.g., in assessment of the prostate.** The anal sphincters can be felt by an index finger inserted into the anal canal and rectum. The following structures are palpable *per rectum:*

Anteriorly in the male (see fig. 33–1)—rectovesical pouch, full bladder, seminal vesicles, displaced or enlarged ductus deferentes, membranous part of urethra when catheterized, and bulbo-urethral glands

Anteriorly in the female—vagina, cervix, ostium uteri, body of uterus when retroverted, recto-uterine fossa, and, pathologically, broad ligaments, uterine tubes, and ovaries

Laterally—ischial tuberosity and spine and sacrotuberous ligament

Posteriorly—pelvic surface of sacrum and coccyx

Blood Supply. The rectum and anal canal are supplied by the superior rectal artery (the

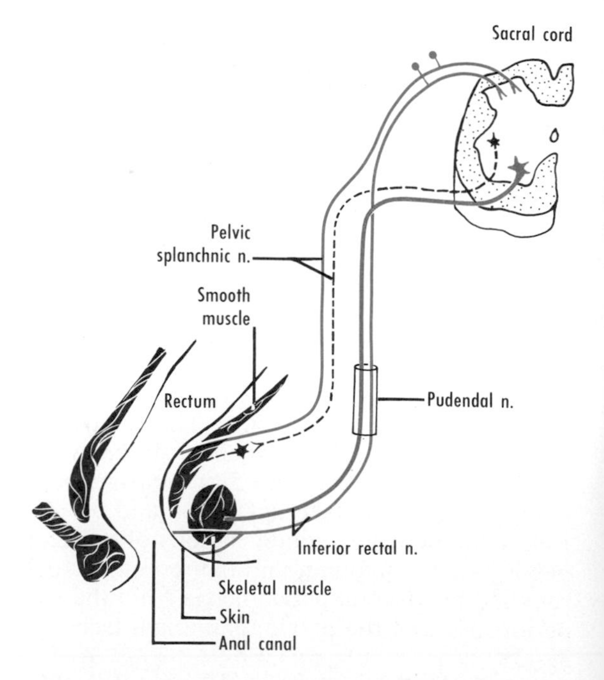

Figure 36–2 Nerve supply of the rectum and anal canal. Motor fibers are shown in red, parasympathetic fibers as interrupted lines, and sensory fibers in blue. The fibers in the pelvic splanchnic nerves reach the intestine by way of plexuses.

* Although it is customary to consider the external sphincter in three parts, it has also been regarded as one muscular mass (S. F. Ayoub. Acta Anat., *105*:25–36, 1979). Moreover, the proposal that the puborectalis is a part of the external sphincter (A. Shafik, Invest. Urol., *13*:175–182, 1975) rather than of the levator ani has been disputed.

continuation of the inferior mesenteric artery), with assistance from the middle and inferior rectal arteries, and by the median sacral artery. The submucosal venous plexus above the pectinate line drains into the superior rectal veins (portal system), which may become varicose, resulting in internal hemorrhoids or "piles." **The submucosal plexus below the pectinate line drains into the inferior rectal veins, which may become varicose, resulting in external hemorrhoids or piles. The unions of the superior with the middle and inferior rectal veins are important portal-systemic anastomoses.**

Innervation (fig. 36–2). Parasympathetic fibers supply the smooth muscle, including the internal sphincter. Sympathetic fibers are mainly vasomotor. Somatic motor fibers supply the external sphincter. Sensory fibers are concerned with the reflex control of the sphincters and with pain. The anal canal is very sensitive below the pectinate line, so that external hemorrhoids may be very painful.

Defecation. The rectosigmoid junction does not possess an anatomical sphincter but does contract; therefore the sigmoid colon usually contains feces and the rectum is empty. Fecal continence depends on colonic control and also on the reflex control of the sphincter ani internus and externus. When a sensation of fullness arrives, the abdominal muscles can be contracted, thereby increasing intra-abdominal pressure. The puborectal sling and the sphincters relax, and the rectal musculature contracts. The colon and rectum then descend, the rectum becomes elongated, feces are discharged, and the anal canal is closed by its sphincters.

QUESTIONS

36–1 Where does the rectum begin and end?

36–2 What is imperforate anus?

36–3 Where is the pectinate line?

36–4 What are hemorrhoids?

37 THE PELVIC DIAPHRAGM AND FASCIA

PELVIC DIAPHRAGM

The pelvic diaphragm is a muscular partition formed by the levatores ani and coccygei, with which may be included the parietal pelvic fascia on their upper and lower aspects. It separates the pelvic cavity above from the perineal region below.

The right and left *levatores ani* (figs. 37–1 and 37–2) lie almost horizontally in the floor of the pelvis, separated by a narrow gap that transmits the urethra, vagina, and anal canal. The levator ani is usually considered in three parts: pubococcygeus, puborectalis, and iliococcygeus. The *pubococcygeus,* the main part of the levator, runs backward from the body of the pubis toward the coccyx and may be damaged during parturition. Some fibers are inserted into the prostate, urethra, and vagina.

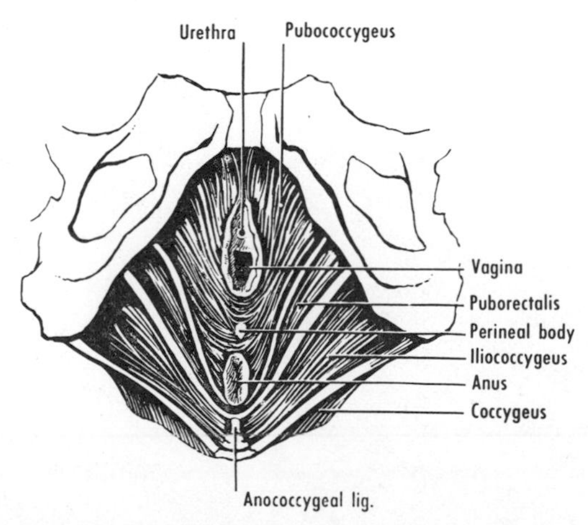

Figure 37–1 Muscles of the pelvic diaphragm in the female, from below. (After Milligan and Morgan.)

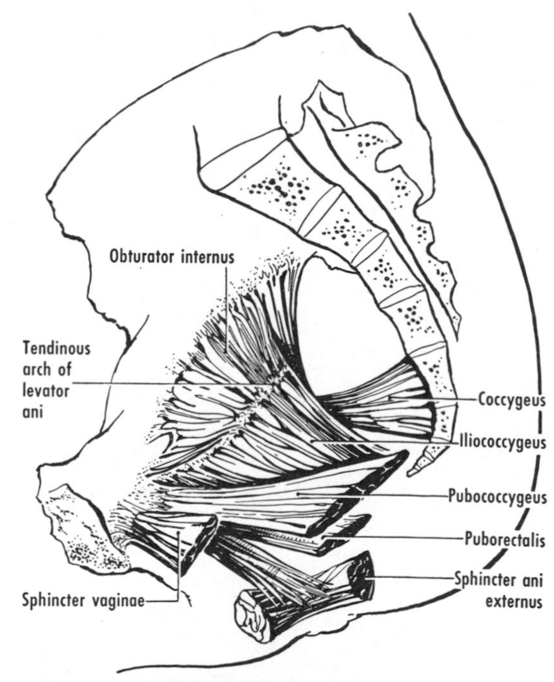

Figure 37–2 Muscles of the pelvic diaphragm, pelvic aspect, showing different parts of the levator ani. The pubococcygeus has several portions, sphincter vaginae, puborectalis, and pubococcygeus proper, depending upon the direction and insertion of the fibers. Some fibers of the puborectalis pass toward the sphincter ani externus and can elevate the anus.

The right and left puborectales unite behind the anorectal junction to form a muscular sling (see fig. 37–1). Some regard them as a part of the sphincter ani externus. The *iliococcygeus*, the most posterior part of the levator ani, is often poorly developed.

The *coccygeus*, situated behind the levator ani and frequently tendinous as much as muscular, extends from the ischial spine to the lateral margin of the sacrum and coccyx.

The pelvic diaphragm is supplied chiefly by the ventral rami of S.N.3,4. The diaphragm helps to support the pelvic viscera, resists increases in intra-abdominal pressure, and aids in micturition.

The fascia of the pelvic diaphragm, as part of the parietal pelvic fascia, is usually described in two layers (see figs. 38–2 and 38–3). The *superior fascia* covers the pelvic surface of the muscles and presents a tendinous arch that forms the medial puboprostatic ligament. The *inferior fascia* covers the lower surface of the muscles and forms the medial wall of the ischiorectal fossa (see fig. 38–3).

PELVIC FASCIA

The pelvic fascia (see figs. 38–2 and 38–3) has parietal and visceral divisions.

The parietal pelvic fascia is a part of the general lining of the abdominal and pelvic walls. It contributes to the floor of the pelvis as the superior and inferior fasciae of the pelvic diaphragm, and it lines the lateral pelvic wall as the obturator fascia. The obturator fascia lines the obturator internus, and, below the origin of the levator ani, it forms the lateral wall of the ischiorectal fossa (see fig. 38–3). In this wall, a fascial tunnel, the *pudendal canal*, houses the internal pudendal vessels and the pudendal nerve.

The visceral pelvic fascia is the extraperitoneal tissue that ensheathes the pelvic organs and vessels.

The membranous partition between the rectum and the bladder and prostate is termed the *rectovesical septum*. It provides a cleavage plane during surgery. The existence of a *rectovaginal septum* is disputed.

QUESTIONS

37–1 What is the pelvic diaphragm?

37–2 Which main structures pass between the right and left levatores ani?

37–3 List the diaphragms in the body.

38 THE PERINEAL REGION AND EXTERNAL GENITALIA

PERINEAL REGION (fig. 38–1)

The perineal region is that part of the trunk below the pelvic diaphragm. It is diamond shaped and has the same boundaries as the pelvic outlet. It is divided into an anterior, urogenital, region and a posterior, anal, region. The term "perineum" in obstetrics and gynecology is generally limited to the area between the anal and vaginal orifices.

The *tendinous center,* or *perineal body,* is a median, fibromuscular mass between the urogenital and anal regions. Several muscles and fasciae are anchored to it, including the levatores ani and the sphincter ani externus. **The perineal body may be damaged during parturition. To avoid injury, the opening for the passage of the fetal head may be enlarged by incising the posterior wall of the vagina and the nearby part of the perineum (episiotomy).**

Urogenital Region (fig. 38–1*A*, *C*, and *F*)

The male urogenital region is pierced by the urethra, whereas the female urogenital region is pierced by the vagina as well as the urethra and also contains the female external genitalia.

In the male, the region (fig. 38–2) comprises (1) skin, (2) superficial perineal fascia (which consists of fatty and membranous layers of subcutaneous tissue), (3) deep perineal fascia, (4) the *superficial perineal space* (which contains the root of the penis and the superficial perineal muscles), (5) inferior fascia of the urogenital diaphragm, (6) the *deep perineal space* (which contains the urogenital diaphragm, membranous urethra, and bulbourethral glands), and (7) superior fascia of the urogenital diaphragm (recently disputed).

In the female, the urogenital region com-

Figure 38–1 **A**, Boundaries and subdivisions of the perineal region, inferior view. **B**, Fasciae of the male perineal region. The superficial fascia has been removed at the right. **C**, Muscles of the superficial perineal space in the female after removal of the superficial and deep perineal fasciae. **D**, Similar view in the male. **E**, Muscles of the deep perineal space in the female. The inferior fascia of the urogenital diaphragm has been removed at the right. **F**, Similar view in the male.

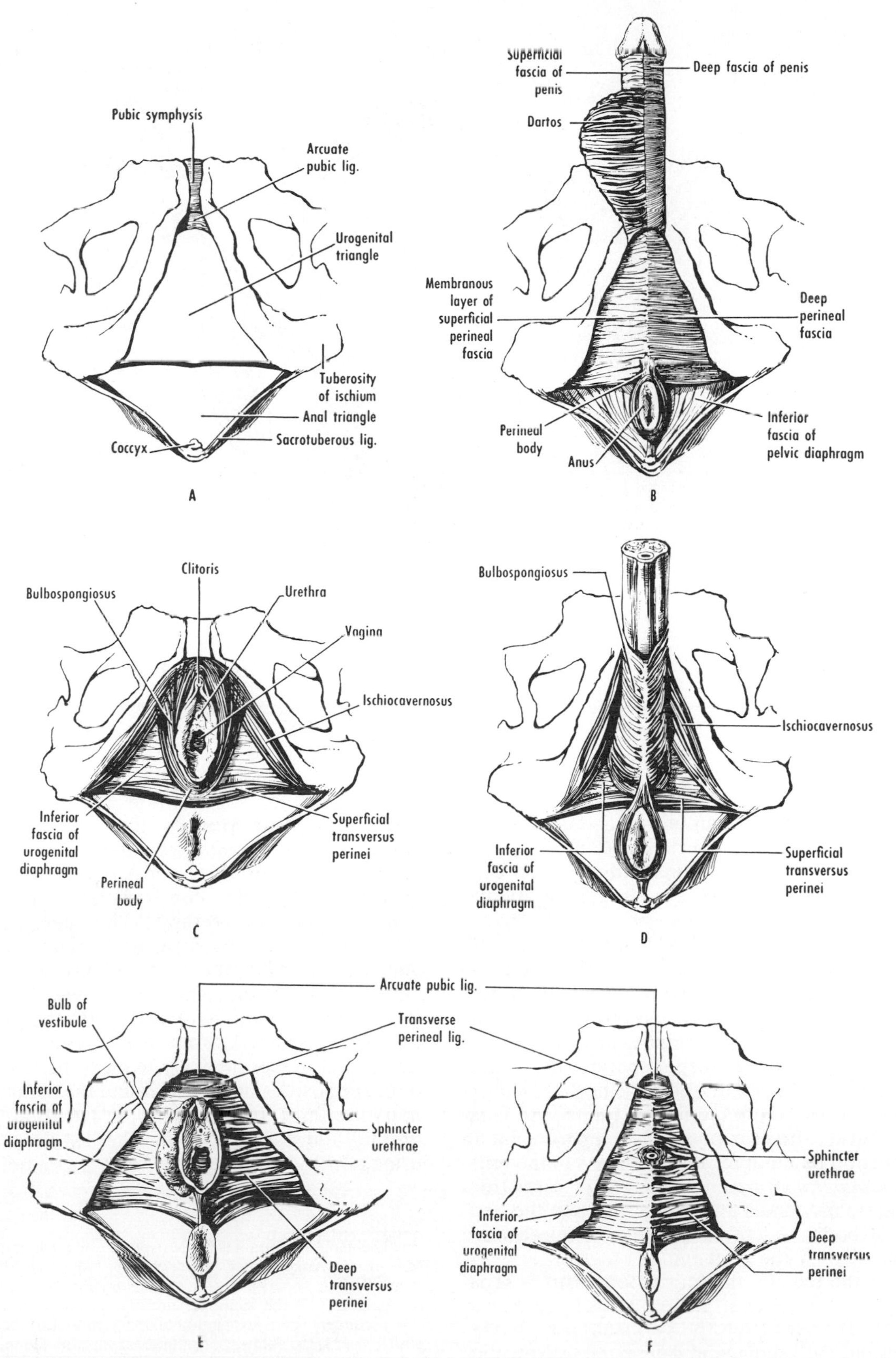

Figure 38–1 *See legend on opposite page.*

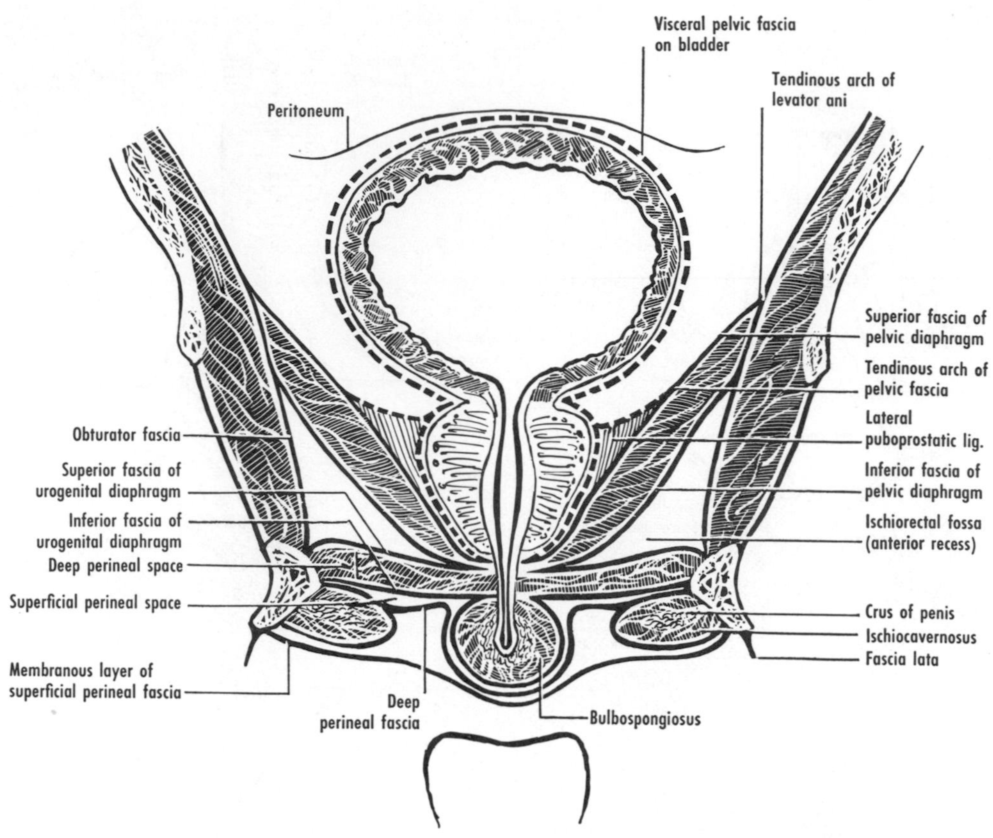

Figure 38–2 Fasciae of the male pelvis and urogenital region, as seen in a coronal section through the prostatic part of the urethra.

prises the seven components listed above for the male, but the superficial perineal space is interrupted by the vagina and the deep perineal space contains the greater vestibular glands.

Superficial Perineal Muscles (figs. 38–1*D* and 38–2). Three muscles are found bilaterally. In the male, the *bulbospongiosus* arises from the perineal body and the fibrous raphe on the bulb of the penis and is inserted into the upper aspect of the corpus spongiosum. It aids in expelling urine or semen. The *ischiocavernosus* arises from the ischial ramus and is inserted on the crus penis. It helps to maintain erection by compressing the veins in the crus. The *superficial transversus perinei* arises from the ischial ramus and is inserted into the perineal body. All three superficial muscles are supplied by the pudendal nerve.

In the female, the bulbospongiosus is separated from the contralateral muscle by the vagina. It arises from the perineal body, passes around the vagina, and is inserted into the clitoris. The ischiocavernosus is inserted on the crus clitoridis (see fig. 38–1*C*).

Deep Perineal Muscles (figs. 38–1*F* and 38–2). The *urogenital diaphragm,* which is pierced by the urethra, consists of two muscles bilaterally: the deep transversus perinei and the sphincter urethrae. The *deep transversus perinei* arises from the ischial ramus and is inserted into the perineal body, which it helps to fix. In the male, the *sphincter urethrae,* which may be fused with the deep transversus, arises from the inferior pubic ramus and passes medially to meet the muscle of the opposite side and surround the membranous urethra.* It constricts the membranous urethra and is said to expel the last drops of urine. Both of the deep muscles are supplied

* The usual account has been disputed. According to T. M. Oelrich (Am. J. Anat., *158*:229–246, 1980). (1) "there is no distinct superior fascia of the so-called urogenital diaphragm." (2) the sphincter urethrae is not horizontal ("the concept of a urogenital diaphragm is not borne out"), and (3) the sphincter urethrae (at least developmentally) is part of a continuous sheet that extends from the bladder to the perineal membrane and that lies within the pelvic cavity.

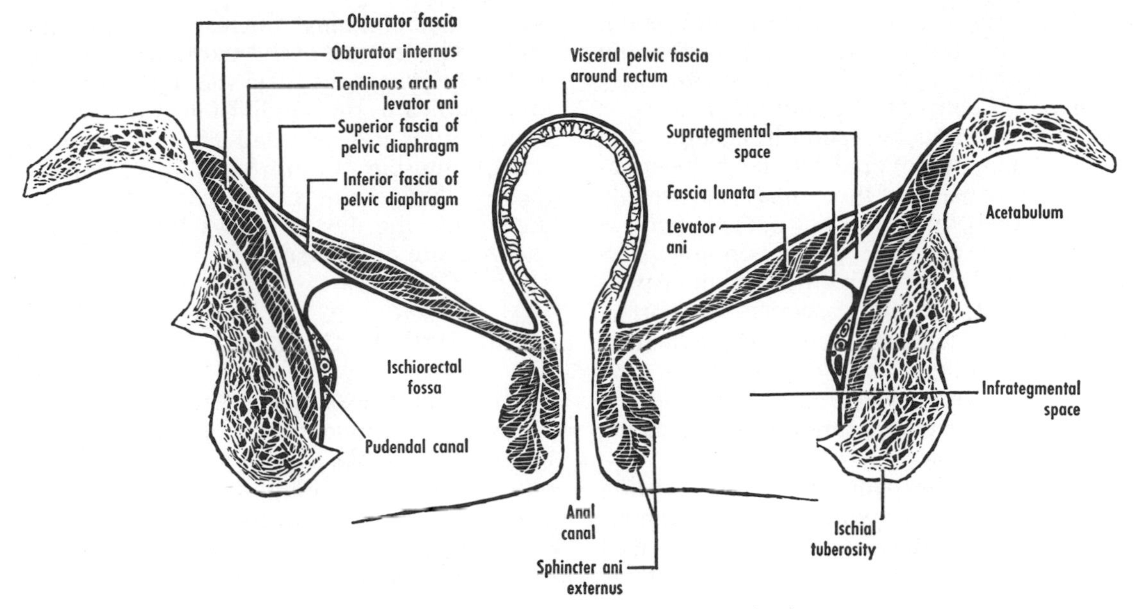

Figure 38–3 The ischiorectal fossae and pelvic diaphragm.

by the pudendal nerve (through the dorsal nerve of the penis).

In the female, the urogenital diaphragm is pierced by the vagina as well as the urethra. The sphincter urethrae is inserted mostly on the lateral wall of the vagina. The sphincters of the two sides do not surround the urethra and hence, despite their name, do not act as a sphincter urethrae (see fig. 38–1*E*).

ANAL REGION (see fig. 38–1*A*)

The anal region is the posterior part of the perineal region, and it contains the sphincter ani externus. The subcutaneous tissue ascends on each side of the anus as the *ischiorectal pad of fat,* which fills the ischiorectal fossa. **The ischiorectal fossa is the space between the skin of the anal region below and the pelvic diaphragm above. It is sometimes the site of an abscess, which may communicate with the rectum or anal canal.** The ischiorectal fossa, triangular in coronal sections (figs. 38–2 and 38–3), is situated between (1) the obturator internus and its fascia laterally and (2) the pelvic diaphragm and its inferior fascia, as well as the sphincter ani externus, medially. The *pudendal canal,* which contains the internal pudendal vessels and pudendal nerve, is a fascial compartment on the lateral wall of the ischiorectal fossa (fig. 38–3).

EXTERNAL GENITALIA

MALE EXTERNAL GENITALIA

The male external genitalia are the penis and scrotum.

The *penis* (see figs. 34–1 and 38–4), or male organ of copulation, becomes erect and en-

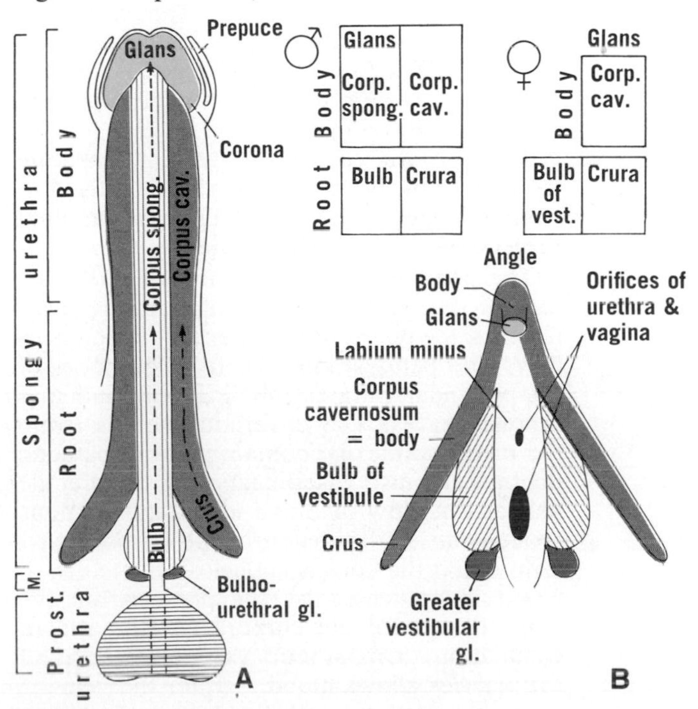

Figure 38–4 The male and female external genitalia. *M.*, membranous urethra.

larged as a result of engorgement with blood. The penis is attached to the linea alba and pubic symphysis by *fundiform* and *suspensory ligaments.* It consists of a root and a body.

The *root,* or attached portion, is situated in the perineum, is covered by the bulbospongiosus, and consists of three masses of erectile tissue: a bulb and two crura. The *bulb* is pierced by the urethra, which continues forward in it, and it is covered by the bulbospongiosi. In front, the bulb becomes the corpus spongiosum. Each *crus penis* is attached to the corresponding ischial ramus and is covered by the ischiocavernosus (see fig. 38–2). The two crura meet and turn downward as the corpora cavernosa.

The *body* of the penis is the free portion covered by skin. The surface that is continuous with that of the anterior abdominal wall is termed the *dorsum* of the penis. The other, or urethral, surface is characterized by a median *raphe* continuous with that of the scrotum. The body of the organ contains the corpus spongiosum and the paired corpora cavernosa.

The *corpus spongiosum* contains the urethra and ends in an expansion, the glans. The *glans penis* is limited circumferentially by a neck, adjacent to which is a rim known as the *corona. Preputial glands,* which secrete smegma, are present on the neck and corona. The *external urethral orifice* is a median slit near the tip of the glans. The *prepuce,* or foreskin, is a double layer of skin that extends from the neck to cover the glans. A median fold, the *frenulum* of the prepuce, is present near the urethral opening.

The *corpora cavernosa,* which constitute the bulk of the body of the penis, form the dorsum and sides of the organ. They end in blunt projections covered by the glans.

The penis is supplied by branches (the artery of the bulb and the dorsal and deep arteries of the penis) of the internal pudendal artery. The penis is innervated by branches of the pudendal, perineal, ilio-inguinal, and cavernous nerves. The cavernous nerves (from the prostatic plexus) contain parasympathetic fibers that cause vasodilatation in the erectile tissue. The flow of blood into the cavernous spaces causes distension of the corpora cavernosa and the corpus spongiosum. Return of blood is prevented by the pressure on the veins that drain the corpora. At the end of ejaculation, sympathetic vasoconstriction of the arteries allows blood to enter the veins.

The *scrotum* (see fig. 34–1) is a cutaneous pouch that contains the testes and epididymides. The left half of the scrotum is usually at a slightly lower level than the right. A median *raphe* indicates the subdivision of the scrotum by a *septum* into right and left compartments. Smooth muscle, the *dartos,* is firmly united to the overlying skin. Loose connective tissue underlying the dartos allows free movement and is a site of edema.

Female External Genitalia

The female external genitalia (fig. 38–5), known as the pudendum or vulva, are the mons pubis, labia majora and minora, vestibule of the vagina, bulb of the vestibule, vestibular glands, and clitoris.

The *mons pubis* is a fatty eminence in front of the pubic symphysis. Two elongated folds, the *labia majora* (*labium majus* in the singular), pass backward from the mons. They are usually united in front (by an *anterior commissure*) but not behind, although the forward projection of the perineal body may give the appearance of a posterior commissure. The round ligaments of the uterus end in the labia majora. Developmentally, the labia majora are comparable to the scrotum of the male.

The *labia minora (labium minus* in the singular) are two cutaneous folds situated between the labia majora and on each side of the vaginal opening. In front, they unite over the glans clitoridis to form a *prepuce,* and below it to form a *frenulum.* The labia minora join the

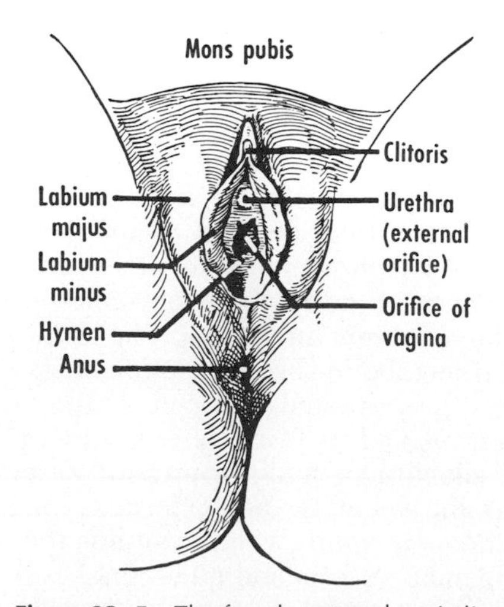

Figure 38–5 The female external genitalia.

labia majora behind and, in virgins, are usually united by a transverse fold known as the *frenulum of the labia,* or *fourchette.*

The *vestibule of the vagina* is the cleft between the labia minora. It presents the openings of the urethra, vagina, and ducts of the greater vestibular glands. The *external urethral orifice* is behind the clitoris. The *vaginal orifice,* immediately behind the urethral orifice, varies with the condition of the hymen.

The *bulb of the vestibule* consists of two elongated, erectile masses covered by the bulbospongiosi. The masses are united in front at the glans. The bulb of the vestibule is homologous with the bulb of the penis and the adjoining part of the corpus spongiosum (see fig. 38–4).

The two *greater vestibular glands* are situated at the back of the bulb of the vestibule. The duct of each gland opens between the vaginal orifice and the labium minus. The greater vestibular glands correspond to the bulbo-urethral glands of the male (see fig. 38–4).

The *clitoris,* which is homologous with the penis, consists of erectile tissue but is not traversed by the urethra. It is behind the anterior commissure and partly hidden by the labia minora. Each *crus clitoridis* (see fig. 38–4) is attached to the corresponding ischial ramus and is covered by the ischiocavernosus. The two crura meet and turn downward as the *corpora cavernosa.* The corpora constitute the body of the clitoris and end as the *glans clitoridis* (see fig. 38–4), which consists of erectile tissue and is highly sensitive. The clitoris is attached to the pubic symphysis by a *suspensory ligament.*

The blood supply and innervation of the female external genitalia are similar to those of the corresponding structures in the male.

QUESTIONS

38–1 What is the perineum?

38–2 What is the perineal body?

38–3 What is the urogenital diaphragm?

38–4 What is the ischiorectal fossa?

38–5 What is excision of the prepuce termed?

38–6 What is the pudendum?

Part 7

THE BACK

The term *back* is used for the posterior part of the trunk, including skin, muscles, vertebral column, spinal cord, and various nerves and vessels. The vertebral column extends from the base of the skull down to the tip of the coccyx. In addition to protecting the spinal cord, the column supports the weight and transmits it to the pelvis and lower limbs. Because most of the weight is anterior to the column, the latter is supported posteriorly by numerous and powerful muscles attached to strong levers (the spinous and transverse processes). The presence of the spinal cord within the framework provided by the vertebral column results in frequently combined neural and skeletal manifestations in abnormal conditions, such as spina bifida.

The vertebral levels summarized below are useful. The spinous processes of C.V.6 and 7 and T.V.1 are usually visible *in vivo* when the neck is flexed. The apex of the lung and the cupola of the pleura extend to C.V.7. The spine of the scapula is commonly in line with T.V.3, and the manubrium is generally opposite T.V.3 and 4. The trachea divides at the level of T.V.5 to 7 when a subject is in the erect position. The inferior angle of the scapula is frequently at the level of the spinous process of T.V.7. The xiphosternal junction is approximately opposite T.V.10. The transpyloric plane generally transects L.V.1, and the spinal cord usually ends at L.V.1 or 2. The supracristal plane (through the highest points of the iliac crests) at L.V.4 is used as a landmark for lumbar puncture, and the aorta divides at this level. The anterior and (frequently marked by a dimple) posterior superior iliac spines are at the level of S.V.2.

39
THE VERTEBRAL COLUMN

VERTEBRAL COLUMN IN GENERAL

The vertebral column generally consists of 33 vertebrae: 24 presacral vertebrae (7 cervical, 12 thoracic, and 5 lumbar) followed by the sacrum (5 fused sacral vertebrae) and the coccyx (4 frequently fused coccygeal vertebrae). The 24 presacral vertebrae allow movement and hence render the vertebral column flexible. Stability is provided by ligaments, muscles, and the form of the bones. The abbreviations C., T., L., S., and Co. are used for the regions, and these are frequently followed by V. for vertebra or N. for nerve.

Curvatures. **The adult vertebral column presents four anteroposterior curvatures: thoracic and sacral, both concave anteriorly, and cervical and lumbar, both concave posteriorly** (fig. 39–1). The thoracic and sacral curvatures, termed primary, appear during the embryonic period proper, whereas the cervical and lumbar curvatures, termed secondary, appear later (although before birth) and are accentuated in infancy by support of the head and adoption of an upright posture.

Parts of a Vertebra. **A typical vertebra consists of (1) a body and (2) a vertebral arch, which has several processes (articular, transverse, and spinous) for articular and muscular attachments. Between the body and the arch is the vertebral foramen: the sum of the vertebral foramina constitutes the vertebral canal, which houses the spinal cord** (fig. 39–2). In addition to the transverse and spinous processes,

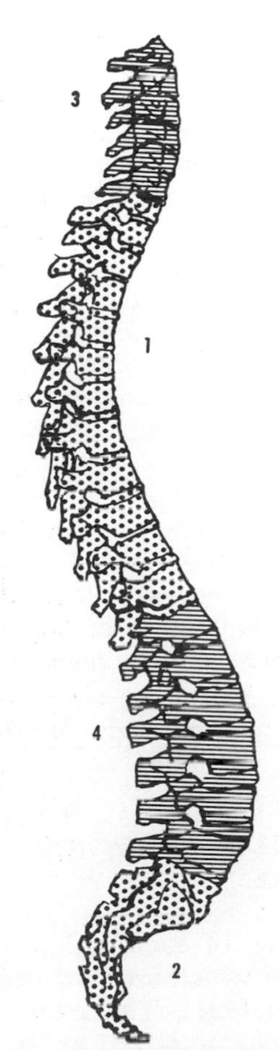

Figure 39–1 The primary (1, thoracic; 2, sacral) and secondary (3, cervical; 4, lumbar) curvatures of the vertebral column.

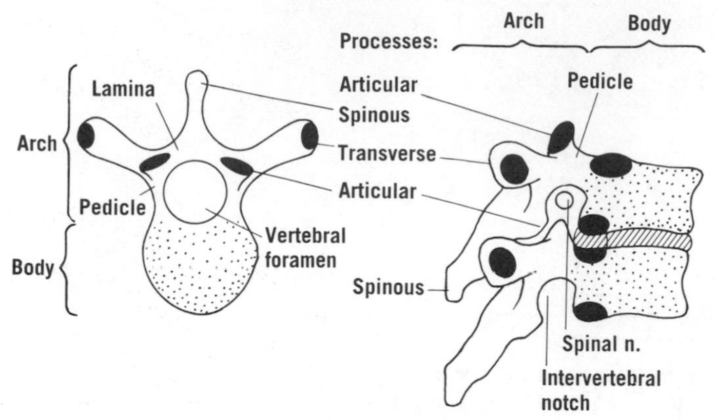

Figure 39–2 The parts of a vertebra (T.V.6) seen from above and from the right side. Adjacent intervertebral notches form intervertebral foramina for the transmission of nerves.

which serve as short levers, the 12 thoracic vertebrae are connected with paired, long levers, namely the ribs.

The bodies of the vertebrae are separated from each other by intervertebral discs. The *body* is mainly spongy bone and red marrow, but the margins of the upper and lower surfaces consist of a ring of compact bone. The body is marked by vascular foramina.

The *vertebral arch* consists of right and left *pedicles* (which connect it to the body) and right and left *laminae*. The *transverse processes* emerge at the junction of the pedicles and laminae, and the *spinous process* proceeds posteriorly from the union of the laminae. The *superior* and *inferior articular processes* on each side bear articular facets. Notches between adjacent pedicles form *intervertebral foramina,* each of which typically transmits a spinal ganglion and a ventral root of a spinal nerve.

Relationship of Spinal Nerves to Vertebrae. **C.N.1 emerges between the skull and the atlas, and C.N.2 to 7 continue to leave the vertebral canal above the correspondingly numbered vertebrae. C.N.8 emerges between C.V.7 and T.V.1, and the remaining spinal nerves leave below the correspondingly numbered vertebrae.**

REGIONAL CHARACTERISTICS OF VERTEBRAE

The vertebrae of each region have special characteristics, which are now described.

Cervical Vertebrae. The seven vertebrae of the neck are characterized by an opening in each transverse process known as a *foramen transversarium*. The upper six pairs of foramina transversaria transmit the vertebral artery. C.V.1, which supports the skull, is termed the *atlas,* and C.V.2, which serves as a pivot for the atlas, is termed the *axis*.

ATLAS (fig. 39–3). The atlas (C.V.1), which has neither body nor spinous process, consists of two lateral masses connected by a short anterior and a longer posterior arch. Each *lateral mass* presents upper and lower facets, for the occipital condyle of the skull and for the axis, respectively. The transverse processes are long and are vaguely palpable *in vivo* immediately below the auricle. The *anterior arch* presents an *anterior tubercle* in front (for the anterior longitudinal ligament) and a facet behind (for the dens of the axis). The *posterior arch* is grooved above for the vertebral artery and C.N.1 on each side, and it presents a *posterior tubercle* behind.

AXIS (fig. 39–4). The axis (C.V.2) is characterized by the *dens* (or *odontoid process*), which projects upward from the body and articulates with the anterior arch of the atlas. The dens is anchored to the occipital bone (by apical and alar ligaments) and is limited behind by the transverse ligament of the atlas (fig. 39–3). It is frequently claimed that the dens represents the body of the atlas, but this is doubtful.*

REMAINING CERVICAL VERTEBRAE (fig. 39–5). C.V.2 to 6 are typical and present short, bifid spinous processes. Each transverse process, pierced by a foramen transversarium, ends laterally in anterior and posterior tubercles, which are connected by an "intertubercular lamella" or bar.† The bars are grooved by the ventral rami of the spinal nerves, which pass behind the foramina transversaria. The anterior tubercles of C.V.6 are large and are termed the *carotid tubercles,* because the common carotid arteries can be compressed against them. C.V.7 has a long, non-bifid spinous process and is known as the *vertebra prominens*. (The spinous processes of C.V.6 and 7 and T.V.1 are usually visible *in vivo* when the neck is flexed.) The anterior tubercles (costal processes) of C.V.7 may develop separately as *cervical ribs*. (Lumbar ribs are less frequent.)

* See F. A. Jenkins, Anat. Rec., *164*:173–184, 1969.

† The part of a cervical vertebra that corresponds to a rib is probably the transverse process lateral to the foramen transversarium and including the anterior and posterior tubercles as well as the "intertubercular lamella" (or so-called costotransverse bar). See A. J. E. Cave, J. Zool., *177*:377–393, 1975. Only the anterior tubercle is shaded in figure 39–12.

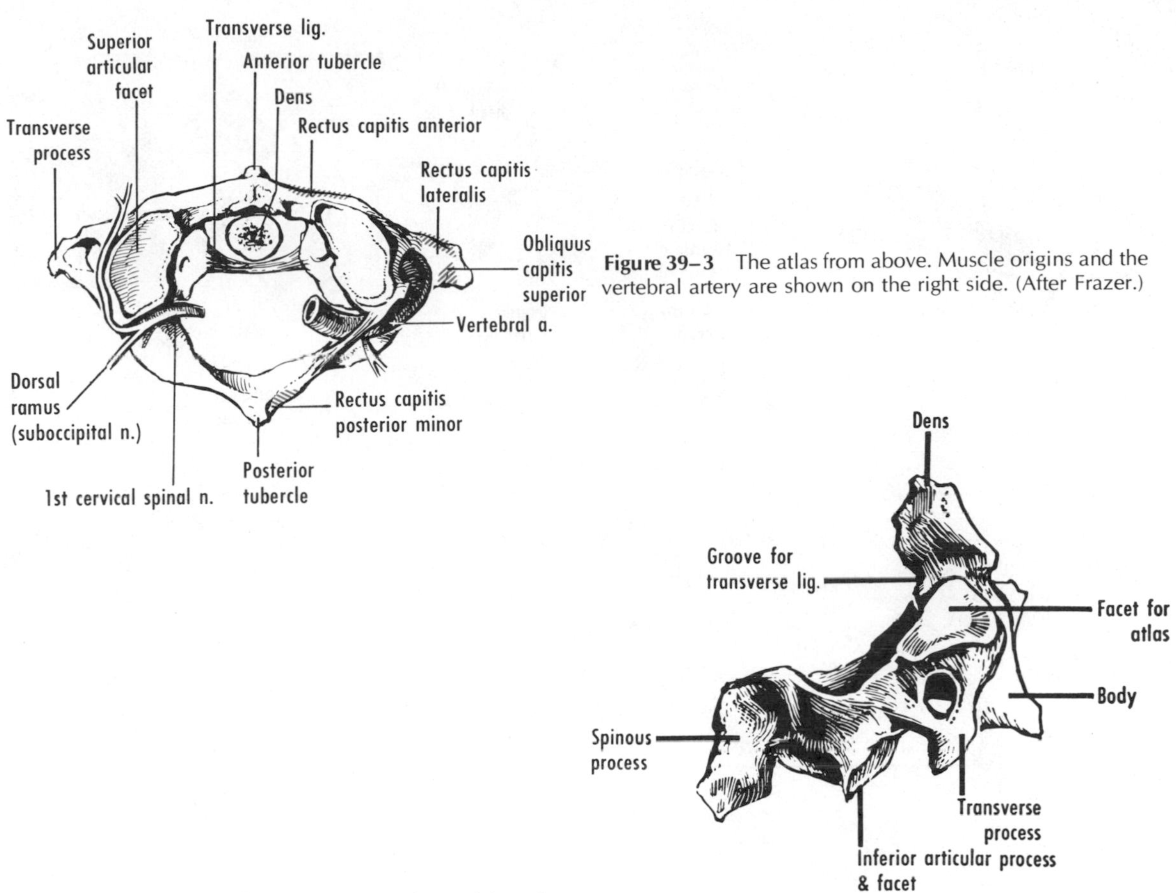

Figure 39–3 The atlas from above. Muscle origins and the vertebral artery are shown on the right side. (After Frazer.)

Figure 39–4 Lateral and posterosuperior views of the axis.

LATERAL

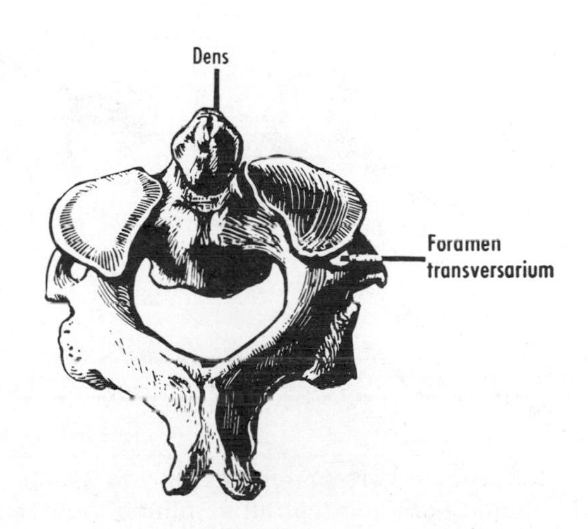

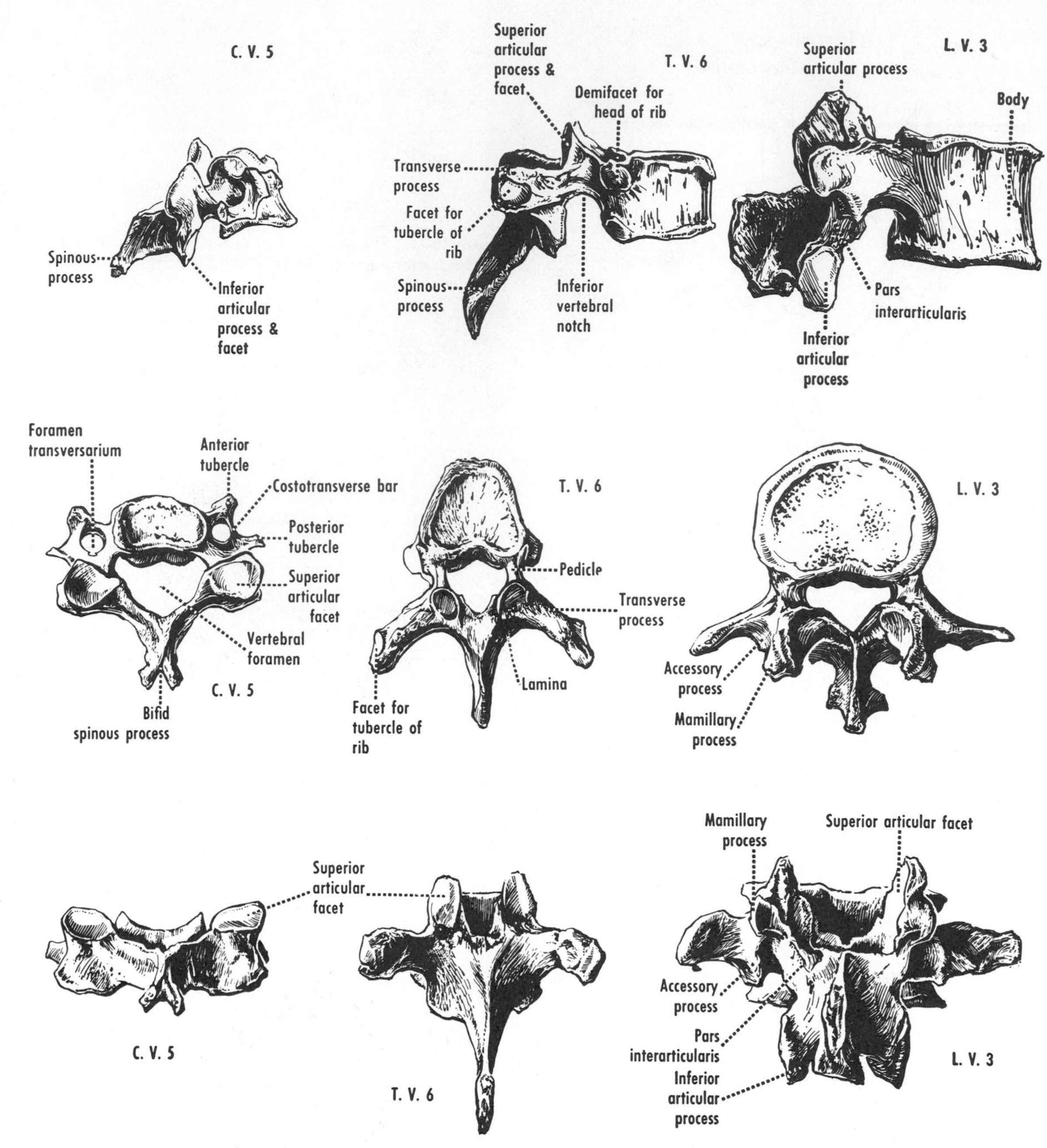

Figure 39–5 Various vertebrae from lateral, superior, and posterior aspects.

Thoracic Vertebrae (figs. 39–5 to 39–7). It should be noted that all vertebrae are dorsal, although only 12 are thoracic. The 12 vertebrae of the thorax bear the ribs. T.V.1 (like C.V.7) is transitional in appearance. T.V.2 to 8 are typical thoracic vertebrae with a reniform body (fig. 39–5). Demifacets for the heads of the ribs (see fig. 20–3) are found above and below at the junction of the body and pedicle. The transverse processes of T.V.1 to 10 have costal facets for the tubercles of the ribs. **The spinous processes are long, slender, and sloping: their tips lie opposite the subjacent body or even at the intervertebral disc below the subjacent body.** The various facets for the ribs on T.V.1 and on T.V.9 to 12 are arranged differently. T.V.11 and 12 are transitional in form, i.e., partly resembling lumbar vertebrae. A humped back is termed *kyphosis.*

Lumbar Vertebrae (figs. 39–8 and 39–9). The five vertebrae between the thorax and sacrum are large and present neither foramina transversaria nor costal facets (fig. 39–5). The body is reniform, and the pedicles and laminae are short and thick. The part of the lamina between the superior and inferior articular processes is known as the *pars interarticularis* and is liable to injury in some people (resulting in *spondylolisthesis,* i.e., a slipping forward of the body of one vertebra on the vertebra or sacrum below it). A *mamillary process* projects backward from the superior articular process. The transverse process, which corresponds to a rib, is long and thin, and an *accessory process* may project downward from its root. The spinous processes are quadrilateral and project horizontally backward. L.V.5, usually the largest vertebra, is mainly responsible for the *lumbosacral angle* (between the lumbar part of the column and the sacrum). Excessive "hollowing" of the back is termed *lordosis.*

Sacrum (figs. 39–10 and 39–11). Five (sometimes six) vertebrae are fused in the adult to form the sacrum, which can be felt below the "small of the back." The sacrum articulates above with L.V.5, laterally with the hip bones, and below with the coccyx. It has pelvic and dorsal surfaces, a lateral part on each side, and a base and apex. The *pelvic surface,* concave and facing antero-inferiorly, presents four paired sacral foramina for the ventral rami of S.N.1 to 4. The *dorsal surface,* convex and facing posterosuperiorly, presents a modified series of spinous processes termed the *median sacral crest* and four paired sacral foramina for the dorsal rami of S.N.1 to 4. Below, the sacrum shows right and left *sacral cornua,* which bound a variable gap termed the *sacral hiatus.* An anesthetic for the spinal nerves may be injected extradurally through the sacral hiatus (*caudal analgesia*). The cornua articulate with corresponding horns on the coccyx. The *lateral part* or *mass* of the sacrum, lateral to the sacral foramina, consists of the fused transverse processes (including their costal elements). Its upper surface is frequently termed the *ala.* The upper part of the lateral mass presents the *auricular surface* for articulation with the hip bone (sacro-iliac joint). The surface is limited behind by an area (*sacral tuberosity*) for interosseous ligaments. The *base,* formed by S.V.1, presents a prominent anterior margin termed the *promontory* (fig. 39–11). Superior articular processes articulate with L.V.5. The *sacral canal* (which contains the dura, cauda equina, and filum terminale) extends from the base to the sacral hiatus. The *apex* of the sacrum may be fused with the coccyx.

Coccyx (fig. 39–10). The vertebrae (usually four) below the sacrum are variably fused in the adult to form the coccyx, which resembles a miniature sacrum in shape.

DEVELOPMENT OF VERTEBRAE
(figs. 39–12 and 39–13)

Vertebrae develop in mesenchyme and cartilage during the embryonic period proper, and most begin to ossify during fetal life. Typically, a vertebra at birth shows three ossific areas, one for the centrum (defined in fig. 39–14) and one for each half of the neural arch. At about puberty, ossific centers appear in the margins of the upper and lower surfaces of the body (*ring epiphyses*) and at the tips of the various processes. Developmental failure of half a vertebra (*hemivertebra*) is one cause of lateral curvature (*scoliosis*).

Failure of fusion of the halves of one or more neural (future vertebral) arches is termed *spina bifida.* The spinal cord and meninges, or the meninges alone, may protrude through the defect (*spina bifida cystica*). When the defect is skeletal rather than neural, it may be concealed by the skin (although sometimes marked by a tuft of hair) and is termed *spina bifida occulta.* In the sacrum, this is quite common.

SURFACE ANATOMY

The spinous processes of the vertebrae are palpable in the median furrow of the back. The external occipital protuberance is palpable in adults. **The spines of C.V.6, C.V.7, and T.V.1 are usually prominent and palpable, and they are made more conspicuous by flexion of the neck and trunk.** In the thoracic region, the spinous process of each vertebra extends to the level of at least the body of the vertebra below. The inferior angle of the scapula is frequently at the level of the spinous process of T.V.7. **A horizontal plane between the highest points of the iliac crests (supracristal plane) is usually at the level of the spinous process of L.V.4, and this is used as a landmark for lumbar puncture.** A needle introduced here should enter the subarachnoid space after 4 to 6 cm. The posterior superior iliac spine is commonly marked by a skin dimple (fig. 39–15).

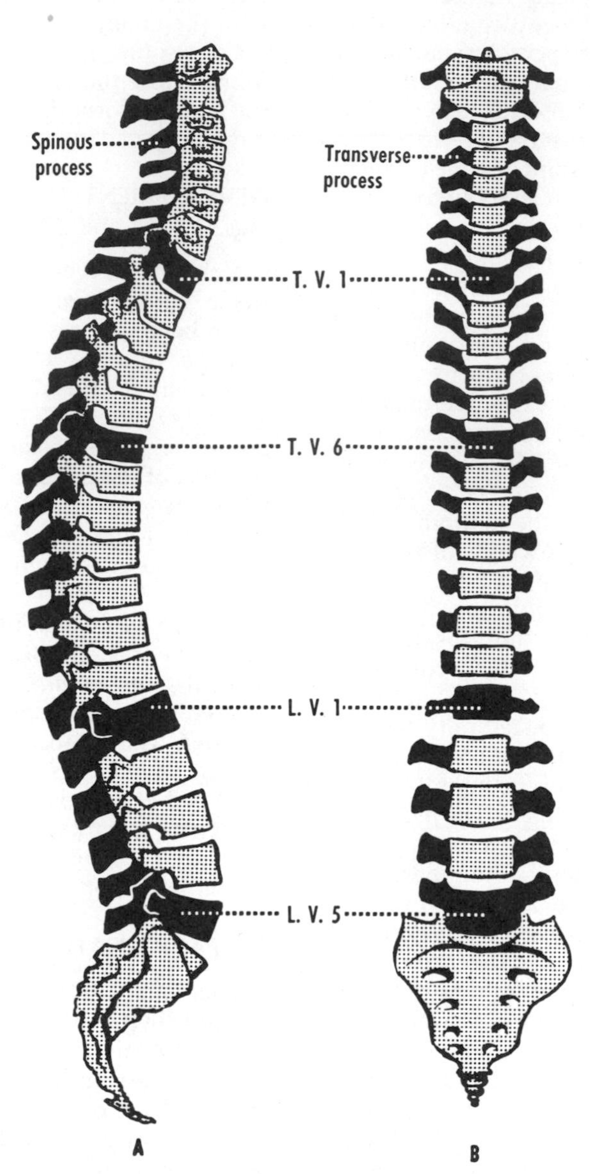

Figure 39–6 The positions, lengths, and directions of (**A**) the spinous processes and (**B**) the transverse processes. The vertebrae in black mark the levels at which a change in direction of curvature occurs.

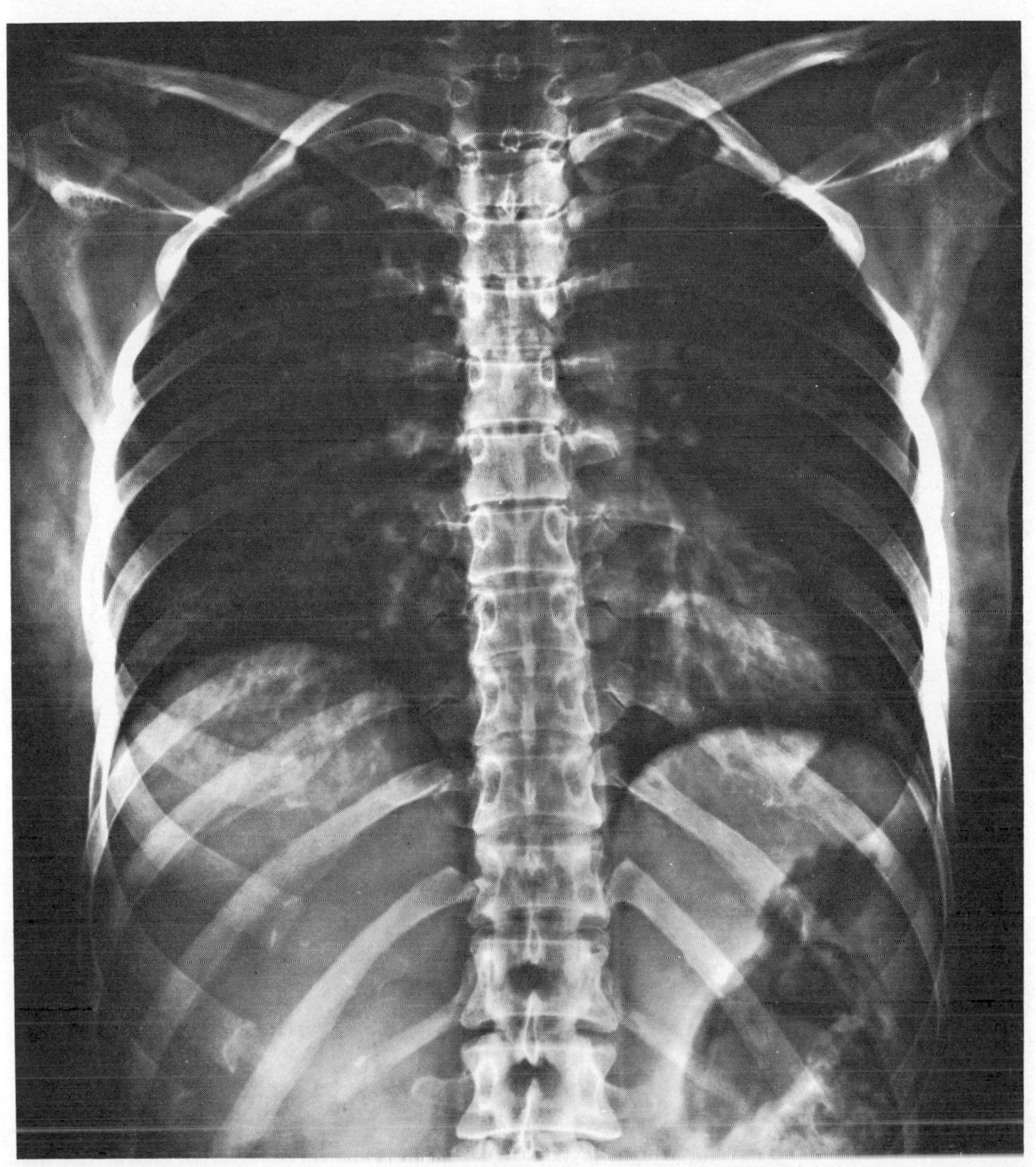

Figure 39–7 Thoracic vertebrae (and C.V.7 and L.V.1). Note the bodies, pedicles, transverse and spinous processes, and costrotransverse joints. (Courtesy of V.C. Johnson, M.D., Detroit, Michigan.)

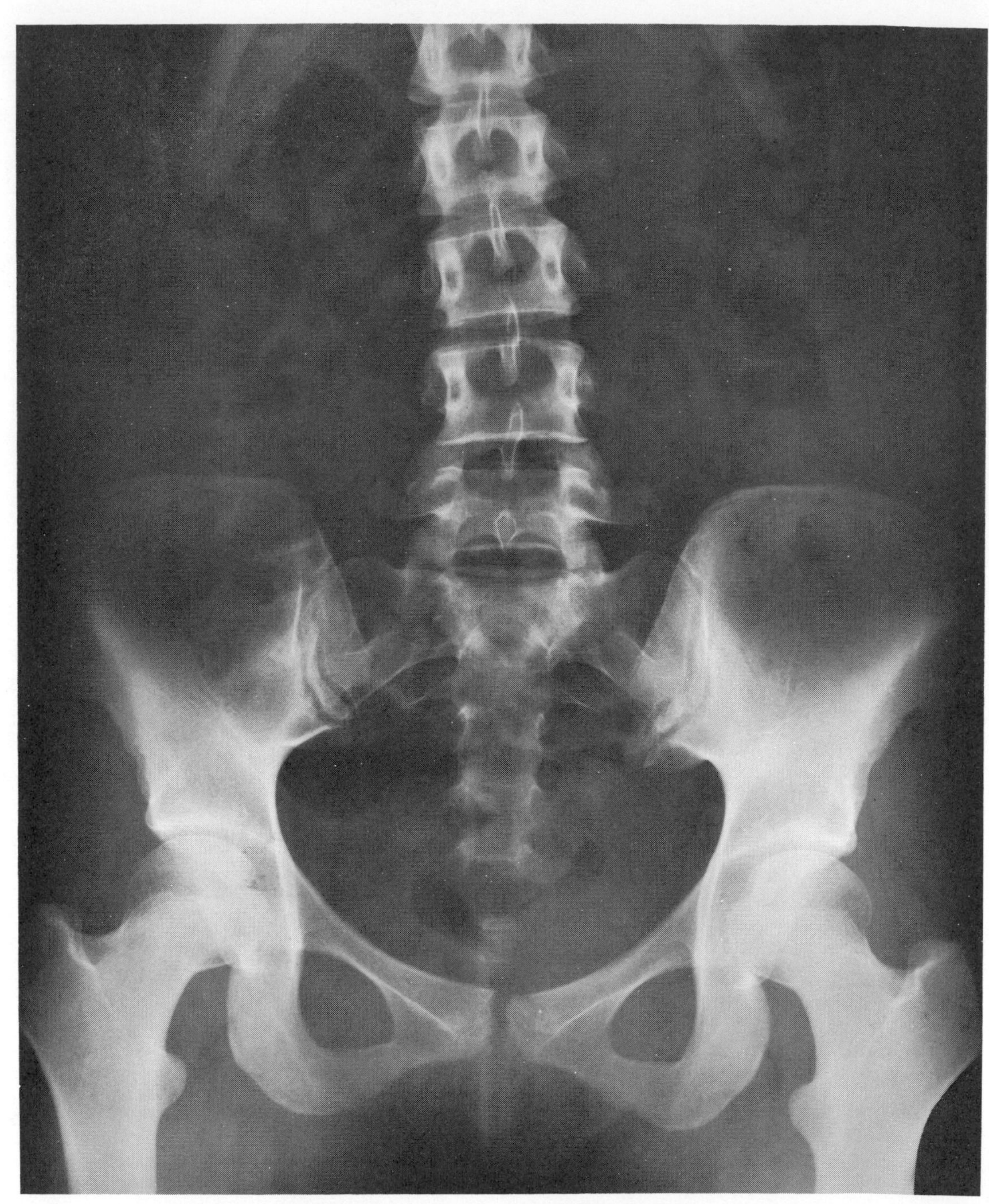

Figure 39–8 Lumbar vertebrae and female pelvis.

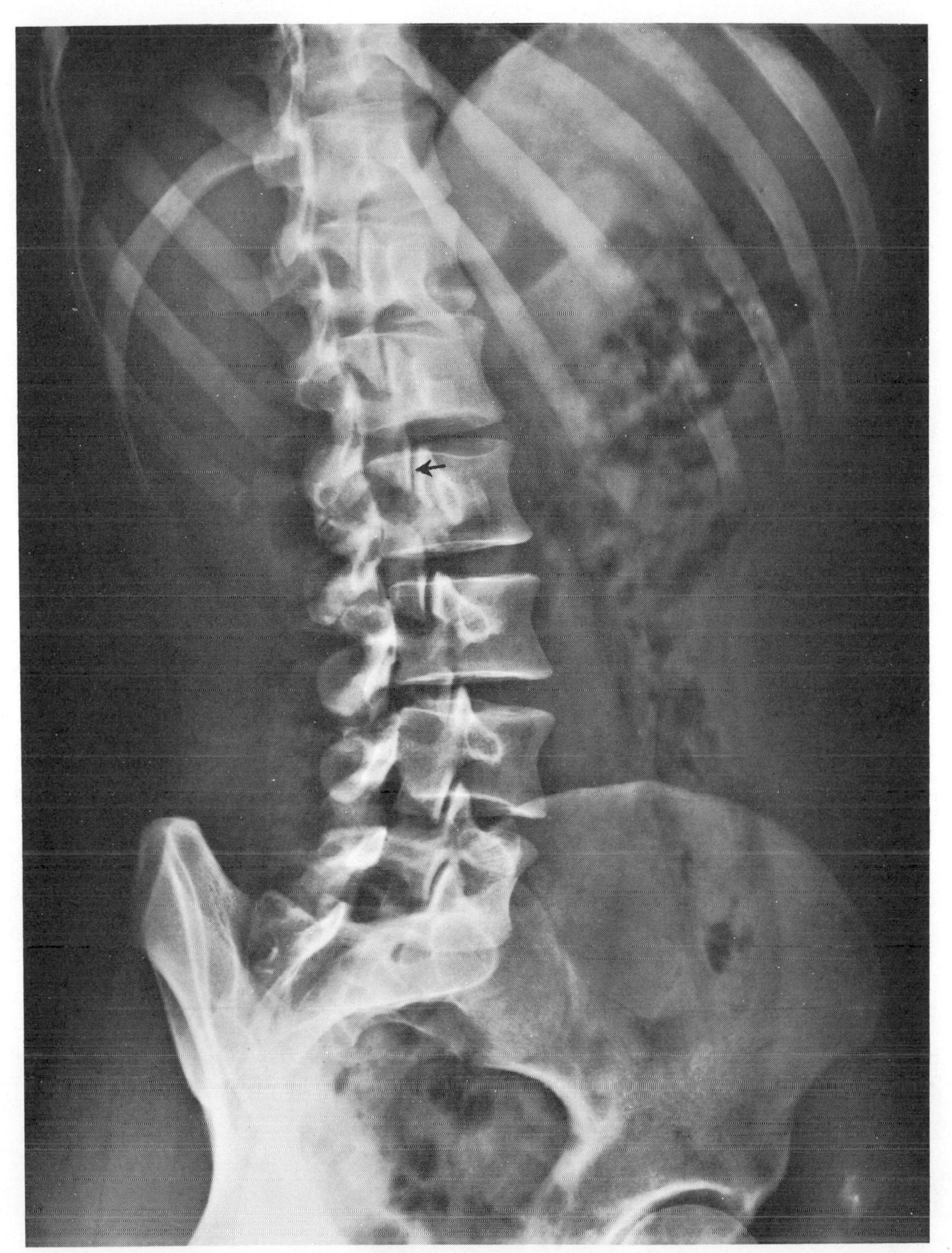

Figure 39–9 Oblique view of the lumbar vertebrae. Note the very small twelfth rib, the joints between the articular processes of the lumbar vertebrae (the arrow indicates the joint between L.V.1 and L.V.2), and the sacrum. In this view the outline of a Scotch terrier is formed by the transverse process (snout, overlapping the vertebral body), the superior articular process (ear), and the inferior articular process (forepaw). The neck of the dog corresponds to the important pars interarticularis, injury to which may result in spondylolisthesis.

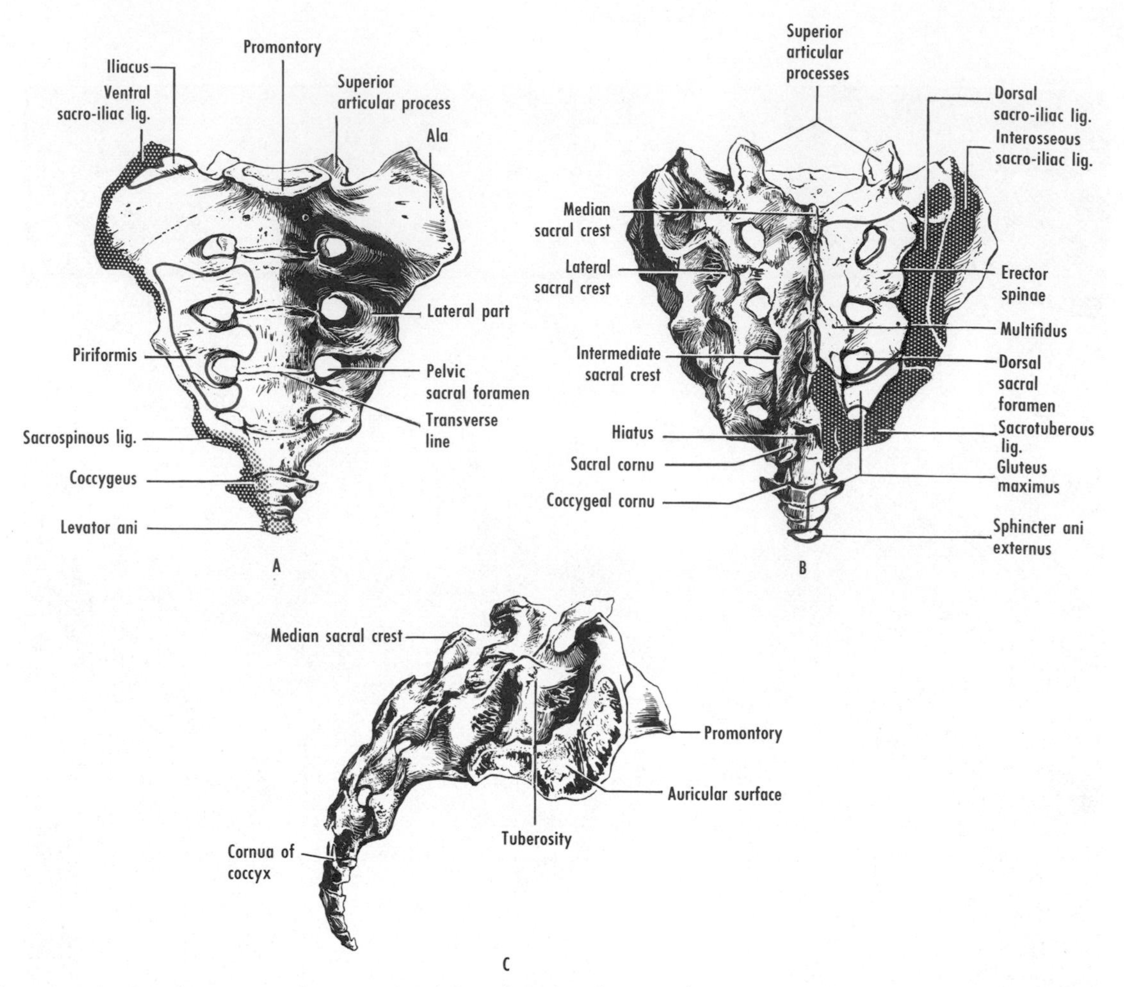

Figure 39–10 Female sacrum and coccyx. **A**, Pelvic and, **B**, dorsal aspects showing muscular and ligamentous attachments. **C**, Right lateral aspect in the anatomical position.

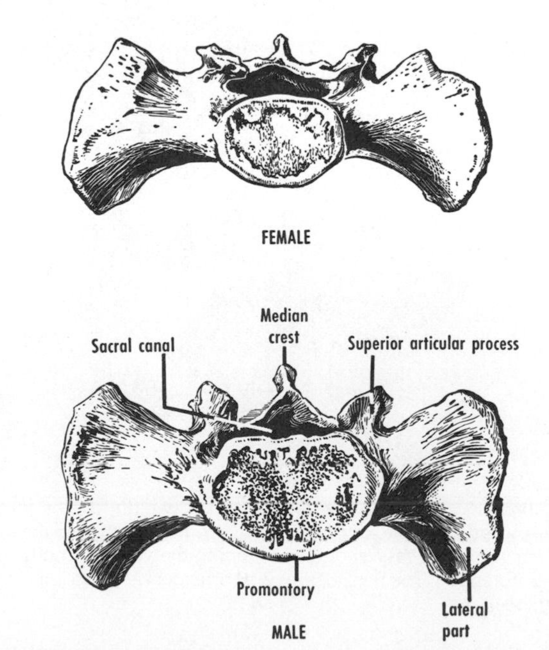

Figure 39–11 Female and male sacra from above. The superior aspect of the lateral part is the ala.

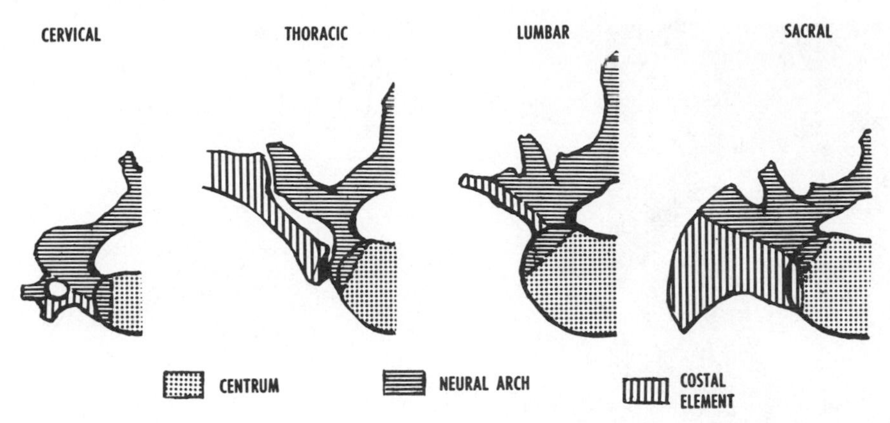

Figure 39–12 Scheme of horizontal sections of vertebrae, showing what are thought to be corresponding parts. Note that the costal element forms a part of the transverse process of a cervical vertebra. It forms the rib in the thoracic region, most of the transverse process in the lumbar region, and the greater portion of the lateral part of the sacrum. In the cervical vertebra, the posterior tubercle of the transverse process should probably also be shaded as part of the costal element.

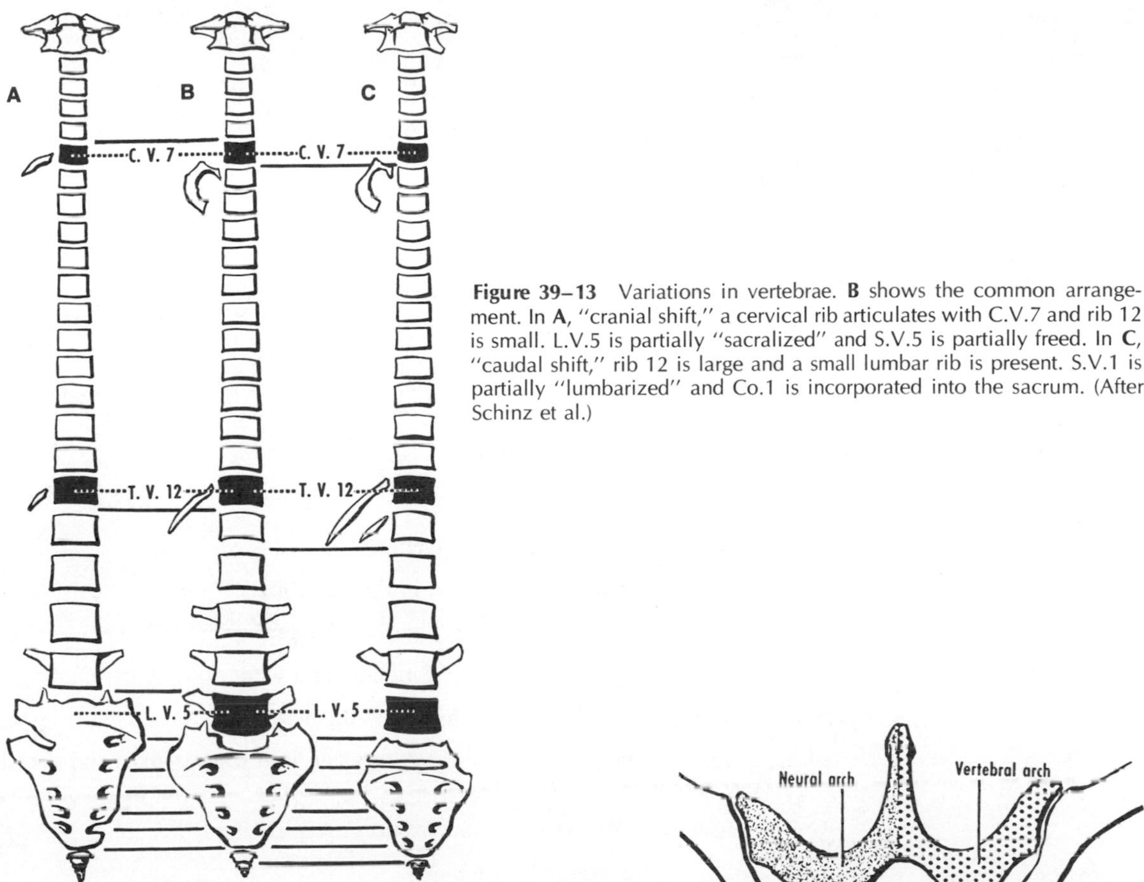

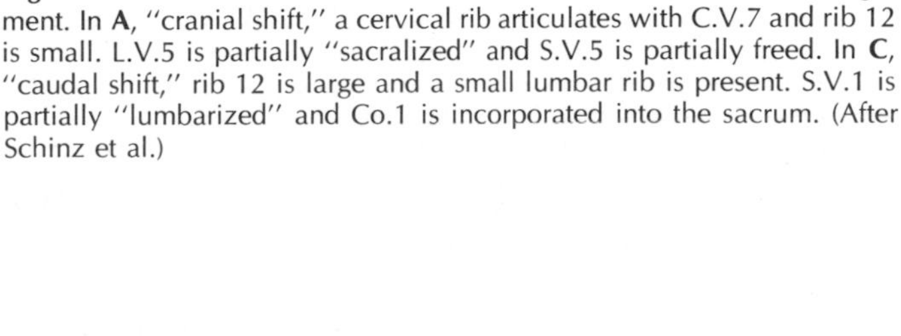

Figure 39–13 Variations in vertebrae. **B** shows the common arrangement. In **A**, "cranial shift," a cervical rib articulates with C.V.7 and rib 12 is small. L.V.5 is partially "sacralized" and S.V.5 is partially freed. In **C**, "caudal shift," rib 12 is large and a small lumbar rib is present. S.V.1 is partially "lumbarized" and Co.1 is incorporated into the sacrum. (After Schinz et al.)

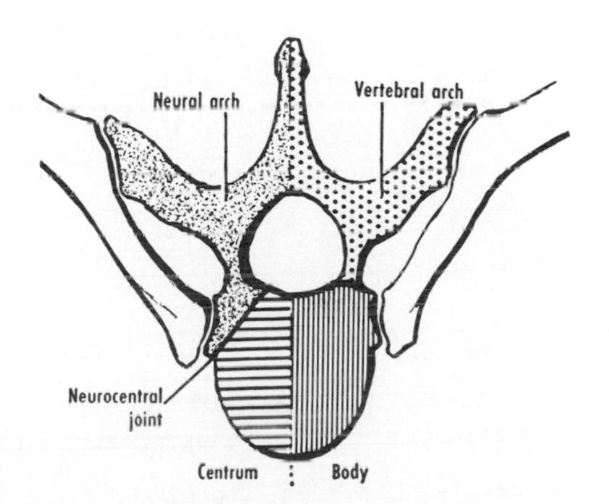

Figure 39–14 The neural arch and centrum (*left half of figure*), and the vertebral arch and body (*right half*). The terms *centrum* and *neural arch* refer to those parts of a vertebra ossified from primary centers. The terms *vertebral arch* and *body* are descriptive terms generally applied to adult vertebrae. The body of a vertebra includes the centrum and part of the neural arch. The vertebral arch, therefore, is less extensive than the neural arch. Note that the rib articulates with the neural arch and not with the centrum.

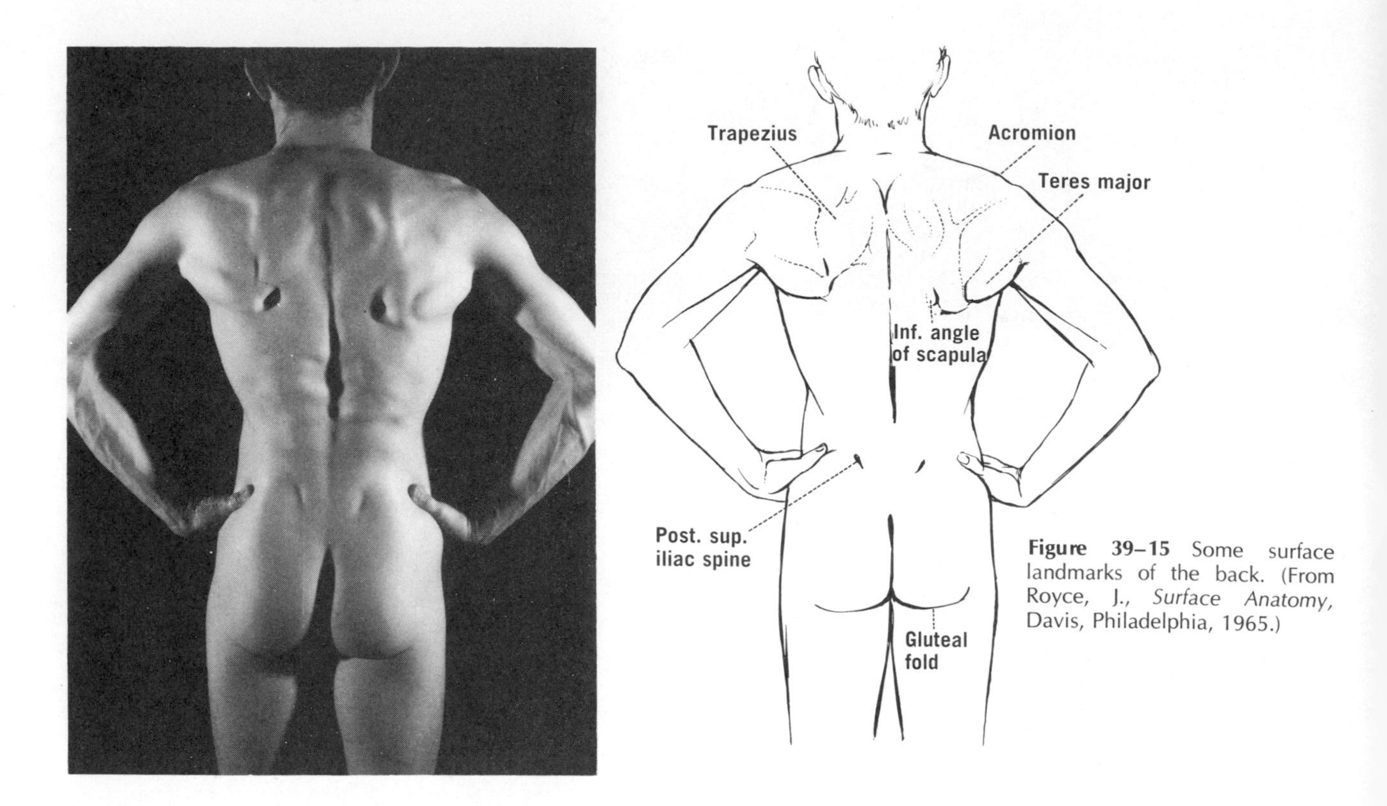

Figure 39–15 Some surface landmarks of the back. (From Royce, J., *Surface Anatomy,* Davis, Philadelphia, 1965.)

ADDITIONAL READING

Köhler, A., and Zimmer, E. A., *Borderlands of the Normal and Early Pathologic in Skeletal Roentgenology.* 3rd ed., trans. by S. P. Wilk, Grune & Stratton, New York, 1968. This classic study of the entire skeleton is available in a more recent edition in German.

Schmorl, G., and Junghanns, H., *The Human Spine in Health and Disease,* 2nd ed., trans. by E. F. Besemann, Grune & Stratton, New York, 1971. This well-known work is an important account of the normal and abnormal vertebral column.

QUESTIONS

39–1 How many vertebrae are movable?

39–2 Which curvatures first appear in the vertebral column?

39–3 How many processes characterize a vertebral arch?

39–4 Where are the intervertebral foramina and what do they contain?

39–5 Between which vertebrae does C.N.8 emerge?

39–6 What are the key features of the cervical, thoracic, lumbar, and sacral vertebrae?

39–7 How many dorsal vertebrae are present in the body?

39–8 What is the pars interarticularis?

39–9 What are the chief contents of the sacral canal?

39–10 What is the ossific status of a typical vertebra at birth?

39–11 How is surgical access to the spinal cord achieved?

39–12 What is spina bifida?

39–13 Which Latin and Greek roots are used with reference to the vertebral column?

40 MUSCLES, VESSELS, NERVES, AND JOINTS OF THE BACK

MUSCLES

The prevertebral muscles are those that are situated on the front of the column in the neck and abdomen and supplied by ventral rami. **The muscles of the back include (1) superficially the trapezius and latissimus dorsi; (2) then the levator scapulae, rhomboids, and serrati posteriores; and (3) the deep muscles** of the back, which are mostly supplied by dorsal rami. The fascia of the back is attached to the spines of the vertebrae, and it proceeds laterally, as the *thoracolumbar fascia,* to ensheathe such muscles as the latissimus dorsi. Two small muscles, the *serratus posterior superior* and *serratus posterior inferior,* extend from the thoracic spines to the ribs and are supplied by ventral rami. The serrati and the thoracolumbar fascia serve as retinacula that retain the underlying muscles.

The deep muscles of the back (1) occupy a "gutter" on each side of the vertebral column; (2) are practically all innervated by dorsal rami; (3) comprise longer and more vertical bundles superficially and shorter and more oblique components deeply; (4) produce extension and lateral flexion of the vertebral column by their longitudinal bundles and rotation by their oblique components; and (5) may be considered, in terms of attachments, as a spinotransverse system superficially and a transversospinal group deeply (fig. 40–1).

The "spinotransverse system" consists of the erector spinae and the splenius (fig. 40–2). **The erector spinae (or sacrospinalis) is the chief extensor of the back.** It ascends from the sacrum, ilium, and lumbar spinous processes and divides into three columns: (1) the *spinalis* medially; (2) the *longissimus,* which forms the

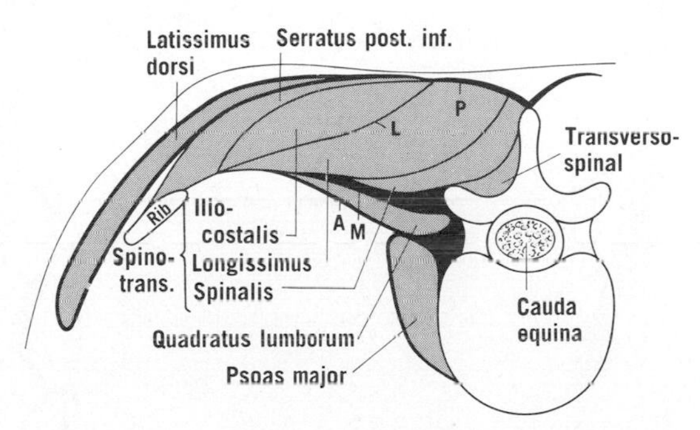

Figure 40–1 Horizontal section through the muscles of the back, showing the arrangement of the spinotransverse and transversospinal systems. The posterior (*P*) layer of the thoracolumbar fascia encloses the latissimus dorsi. The middle (*M*) and anterior (*A*) layers of the thoracolumbar fascia enclose the quadratus lumborum. See also fig. 29–5. *L,* "lumbar intermuscular aponeurosis" (N. Bogduk, J. Anat., *131*:525–540, 1980).

bulk of the erector and reaches the skull; and (3) the *iliocostalis* laterally. The three subdivisions interconnect the spinous and transverse processes and the ribs in a complicated manner and send extensions into the neck. The vertical muscles on the back of the neck are bandaged by the splenius, which ascends obliquely from the thoracic spinous processes and is inserted into (1) the mastoid part of the temporal bone and the superior nuchal line (*splenius capitis*) and (2) the cervical transverse processes (*splenius cervicis*). The splenius rotates the head. The splenii of the two sides, acting together, extend the head.

The "transversospinal system" consists of the semispinalis and a number of small underlying muscles (*multifidus* and *rotatores*), deep to which are *interspinous* and *intertransverse muscles*. The semispinalis, so called because it extends chiefly along the upper half of the vertebral column, is responsible for the longitudinal bulge on the back of the neck near the median plane. This part (*semispinalis capitis*) ascends from cervical and thoracic transverse processes to reach the occipital bone. It extends the head and neck and covers the suboccipital triangle and a deeper muscle, the *semispinalis cervicis* (fig. 40–2). The splenius, semispinalis capitis, and sternomastoid are the chief rotators of the head.

The *suboccipital triangle* (fig. 40–3) is delimited by three points: the spine of the axis,

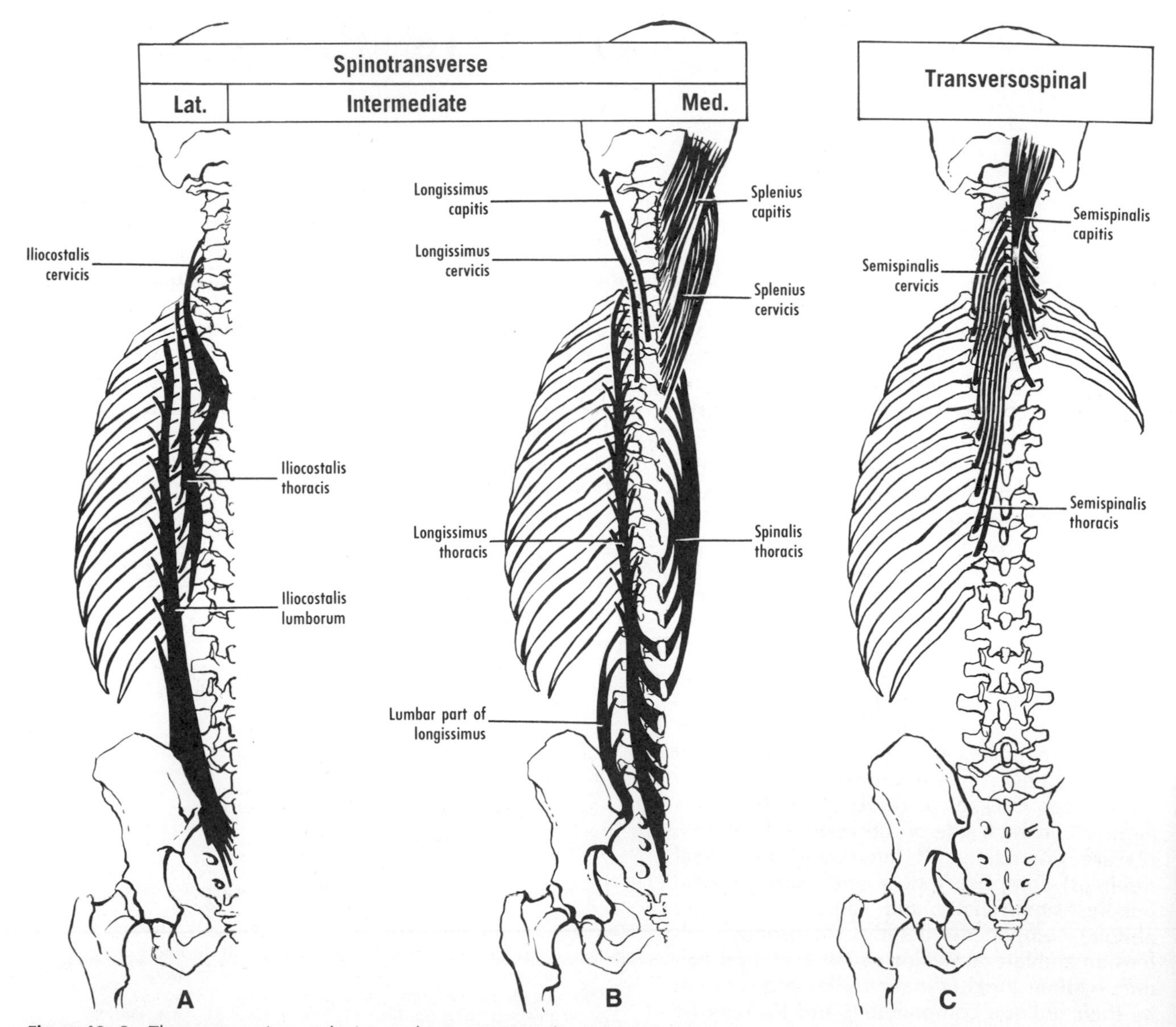

Figure 40–2 The erector spinae, splenius, and transversospinalis. (After Winckler.)

the transverse process of the atlas, and the lateral part of the occipital bone. The boundaries of the triangle are the muscles attached to these three points: the *obliquus capitis inferior, obliquus capitis superior,* and *rectus capitis posterior major*. The muscles, supplied chiefly by the suboccipital nerve, are mainly postural. The triangle is roofed by the semispinalis capitis, and its floor is the posterior atlanto-occipital membrane. **The suboccipital triangle contains the vertebral artery and the suboccipital nerve, both lying in a groove on the upper surface of the posterior arch of the atlas. The subarachnoid space can be tapped by inserting a needle at the back of the neck and piercing the posterior atlanto-occipital membrane. This procedure is termed cisternal puncture** (see fig. 41–1).

BLOOD VESSELS

The *vertebral artery* is one of the main vessels to the brain. It arises from the subclavian artery, and its course comprises four parts: cervical (p. 452), vertebral (p. 452), suboccipital, and intracranial (p. 385). The suboccipital part winds around the lateral mass of the atlas, lies in a groove on the posterior arch of the atlas (see fig. 39–3) within the suboccipital triangle, enters the vertebral canal, perforates the dura and arachnoid, and traverses the foramen magnum.

The vertebral venous system is a valveless network that extends between the cranial dural sinuses and the pelvic veins and is connected with the azygos and caval systems (see fig. 24–2*A*). It allows a flow of blood in either direction, depending on intrathoracic and intra-abdominal pressure, and it permits the spread of carcinoma, emboli, and infections. The divisions of the network (see fig. 41–3) are as follows: (1) the *internal vertebral plexus,* which surrounds the dura and is drained by veins in the intervertebral foramina; (2) the *basivertebral veins* on the back of the vertebral bodies, which drain a network in the marrow spaces of the vertebrae; and (3) the *external vertebral plexus,* which lies on the front of the vertebral bodies and on the vertebral arches. The *suboccipital plexus* is an extensive and complicated part of the external plexus. It lies on and in the suboccipital triangle and drains the scalp. The suboccipital plexus receives the occipital vein and sends tributaries to the vertebral vein, which descends through the foramina transversaria to end in the brachiocephalic vein.

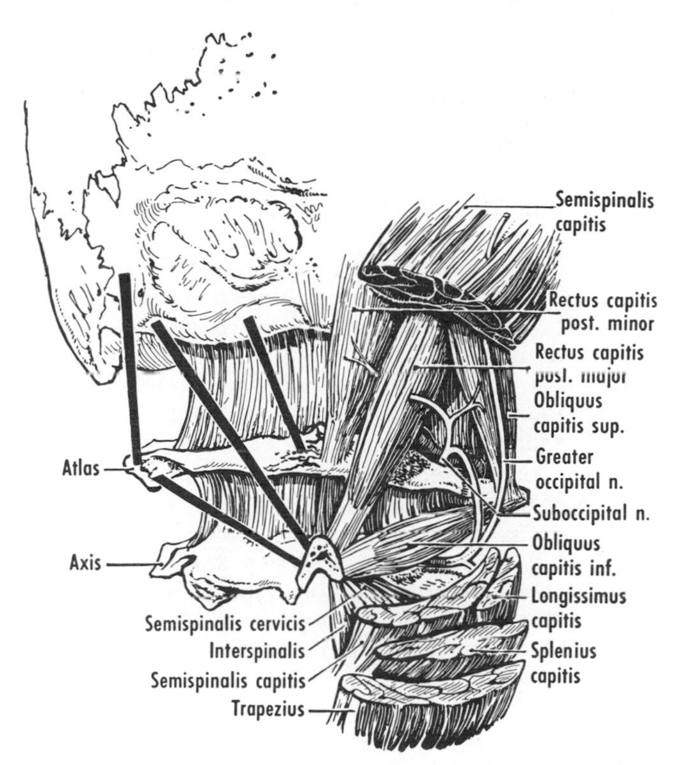

Figure 40–3 The suboccipital triangle. Most of the semispinalis capitis has been removed. Note the greater occipital nerve emerging at the lower border of the inferior oblique muscle. The vertebral artery and the suboccipital nerve are seen in the triangle. The massive suboccipital venous plexus has been omitted. On the left side, lines indicate the directions and attachments of the muscles that bound the triangle.

NERVES

The innervation of the back is from the meningeal branches and dorsal rami of the spinal nerves.

Each spinal nerve gives off a *meningeal branch* (or *sinuvertebral nerve*), which reenters the vertebral canal and supplies vasomotor and sensory fibers to dura, ligaments, periosteum, and blood vessels.

The dorsal rami of the spinal nerves contain motor, sensory, and sympathetic fibers. They supply the muscles, bones, joints, and skin of the back. Most dorsal rami divide into medial and lateral branches.

The dorsal ramus of C.N.1, known as the *suboccipital nerve,* usually has no cutaneous distribution. The medial branch of the dorsal ramus of C.N.2, termed the *greater occipital nerve,* supplies a considerable part of the scalp. The cutaneous division of the medial branch of the dorsal ramus of C.N.3 is known as the *third occipital nerve*. The dorsal rami of C.N.1 and C.N.6 to 8 usually give no cutaneous branches, so that the C5 and T1 dermatomes are adjacent. The lumbar and sacral dorsal rami supply the skin of the buttocks as a series of *clunial nerves*.

JOINTS

The joints of the vertebral column are (1) those between the bodies of adjacent vertebrae (intervertebral discs), (2) those of the vertebral arches (between articular processes), (3) the atlanto-occipital and atlanto-axial joints, (4) the costovertebral joints, and (5) the sacro-iliac joint.

Joints Between Bodies of Adjacent Vertebrae. The bodies of adjacent vertebrae are united by longitudinal ligaments and intervertebral discs. The *anterior longitudinal ligament* extends from the anterior tubercle of the atlas to the sacrum and is attached to the front of the vertebral bodies and discs. The *posterior longitudinal ligament* is a continuation of the membrana tectoria, extends from the occipital bone to the sacrum, lies within the vertebral canal, and is loosely attached to the back of the vertebral bodies (fig. 40–4).

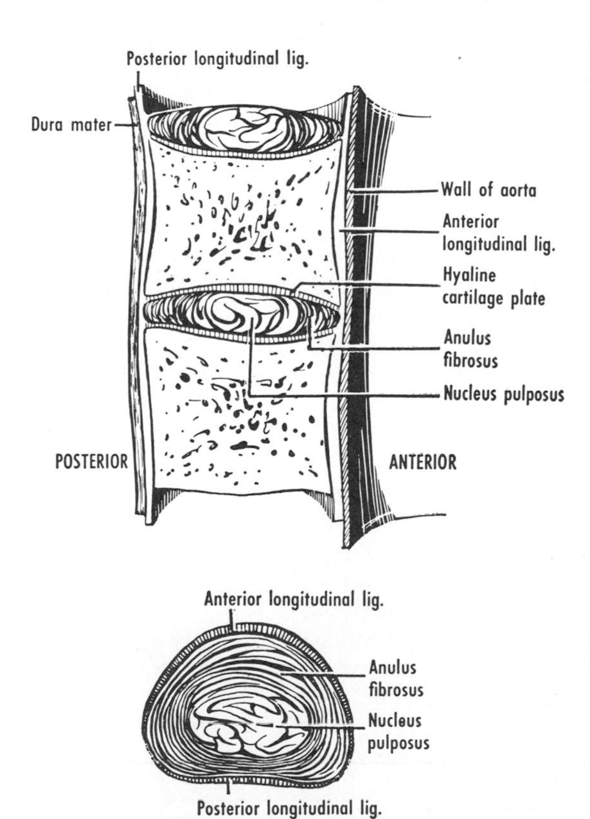

Figure 40–4 Intervertebral discs in median and horizontal section.

The intervertebral discs account for about a quarter of the length of the vertebral column. They are shock-absorbing, fibrocartilaginous joints between adjacent vertebrae. In young adults, each disc consists of a semigelatinous nucleus pulposus surrounded peripherally by an anulus fibrosus. Each disc is separated from the bone above and below by two growth plates of hyaline cartilage (fig. 40–4). The nucleus develops from the notochord in the embryo. The discs contain much water, diminution of which (temporarily during the day and permanently in advanced age) results in a slight decrease in stature. With advancing age, the entire disc tends to become fibrocartilaginous, and the distinction between nucleus and anulus becomes lost. **Pathological change in an intervertebral disc may be followed by herniation of the nucleus pulposus, which then compresses adjacent nerves.** For example, the L.V.4/5 disc may press on the roots of L.N.5. The cervical discs may develop fissures that have been dubiously called *uncovertebral joints*. They are situated laterally between lips on the adjacent upper and lower surfaces of the vertebral bodies.

Joints of Vertebral Arches. The articular processes of adjacent vertebrae are united by plane synovial joints. The vertebral arches are also connected by ligaments, particularly strong in the lumbar region, e.g., the *ligamenta flava* between the laminae. The spinous processes are united by the *supraspinous ligament,* which merges above with the *ligamentum nuchae*. This ligament is a median partition between the muscles of the two sides of the neck and is attached to the occipital bone.

Atlanto-occipital and Atlanto-axial Joints. The *atlanto-occipital joints* are between the lateral masses of the atlas and the occipital condyles. Synovial in type, the right and left joints allow nodding of the head around a transverse axis and sideways tilting of the head around an anteroposterior axis. *Anterior* and *posterior atlanto-occipital membranes* connect the respective arches of the atlas to the margins of the foramen magnum (figs. 40–5 and 40–6).

The *atlanto-axial joints,* synovial in type, unite the first two vertebrae. The two lateral joints are between the articular processes. The median atlanto-axial joint is between (1) the anterior arch and transverse ligament of the atlas and (2) the pivot formed by the dens of the axis (fig. 40–5).

The *transverse ligament of the atlas* unites the medial aspects of the lateral masses (fig.

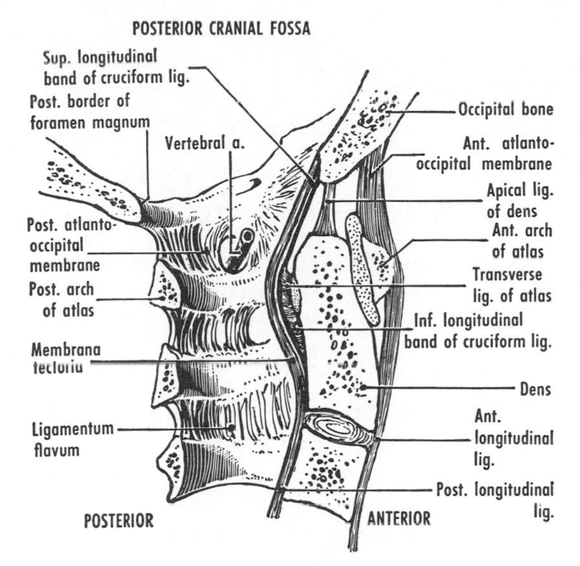

Figure 40–5 Median section of the atlas and axis. (After Poirier and Charpy.)

40–6) and, together with longitudinal bands to the foramen magnum and the body of the axis, constitutes the *cruciform ligament of the atlas* (fig. 40–6*B*). The apex of the dens is connected to the occipital bone by a median *apical ligament* and two lateral *alar ligaments*. Finally, the *membrana tectoria*, the upward continuation of the posterior longitudinal ligament, is anchored to the basilar part of the occipital bone (fig. 40–5).

Costovertebral Joints. The *costovertebral joints* (see fig. 20–6) are those between (1) the heads of the ribs and the vertebral bodies and (2) the tubercles of the ribs and the transverse processes (*costotransverse joints*).

Sacro-iliac Joint. The *sacro-iliac joint* (see fig. 31–7) is a plane synovial joint formed by the union of the auricular surfaces of the sacrum and ilium. The ilium is connected to the transverse process of L.V.5 by various strong bands collectively known as the *iliolumbar ligaments*. **The weight of the head, upper limbs, and trunk is transmitted through the sacrum and ilia to the femora when one is standing and to the ischial tuberosities when one is sitting.**

Movements. **The movements of the vertebral column are flexion (forward bending), extension (backward bending), lateral flexion to the right or left (bending to the side), and rotation (around a longitudinal axis). The axis of each type of movement runs through the nucleus pulposus. The cervical and lumbar regions are the most mobile and are frequent sites of aches.** Flexion and extension of the head

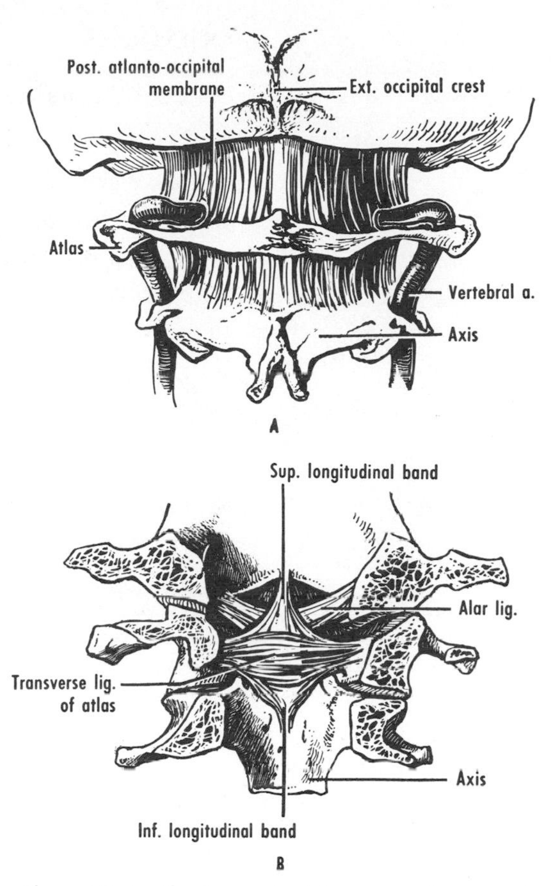

Figure 40–6 The ligaments of the atlas and axis, posterior view. **A** shows the vertebral arteries. **B** shows the interior of the vertebral canal after removal of portions of the skull and vertebrae.

occur mainly at the atlanto-occipital and atlanto-axial joints. The skull and the atlas rotate on the axis at the three atlanto-axial joints, pivoting on the dens like a ball-and-socket joint.

The chief flexors of the vertebral column are the prevertebral muscles, recti abdominis, iliopsoas, scaleni, and sternomastoids. Gravity may also be important, and the movement is controlled by the erectores spinae. The chief extensors are the erectores spinae. Lateral flexion is carried out mainly by the oblique muscles of one side of the abdominal wall. The muscles of the back are relatively inactive when one is standing at ease. Injury or inflammation may easily result in reflex spasm of the muscles of the back.

ADDITIONAL READING

De Palma, A. F., and Rothman, R. H., *The Intervertebral Disc*, W. B. Saunders Company, Philadelphia, 1970. An account of disc disease.

QUESTIONS

40–1 What are the prevertebral muscles?

40–2 Which muscles of the back can be most readily identified *in vivo?*

40–3 What are the main actions of the deep muscles of the back?

40–4 Which muscles are included in the "spinotransverse system"?

40–5 Which is the most prominent muscle of the "transversospinal system"?

40–6 What is the most important structure in the suboccipital triangle?

40–7 How is the vertebral venous system arranged?

40–8 How many intervertebral discs are present in the body?

40–9 What is an intervertebral disc?

40–10 What is a "slipped disc"?

40–11 On which nerve would herniation of the L.V.4/5 disc be likely to press?

40–12 Which joints are involved in (a) nodding in approval and (b) shaking the head in disapproval?

40–13 Which are the most mobile regions of the vertebral column?

41 THE SPINAL CORD AND MENINGES

SPINAL CORD

The spinal cord, about 45 cm in length, extends from the foramen magnum, where it is continuous with the medulla oblongata, to the level of L.V.1 or L.V.2. (The range is T.V.12 to L.V.3.) Below that level, the vertebral canal is occupied by spinal nerve roots and meninges. A fibrous strand, the *filum terminale,* continues from the spinal cord down to the coccyx (fig. 41–1).

The spinal cord presents a *cervical* and a *lumbar enlargement* at the levels of attachment of the nerves to the limbs. The lower end of the cord is conical and is termed the *conus medullaris.* The coccygeal nerves are attached to it. The cord presents a *posterior median sulcus* and an *anterior median fissure,* lateral to which the dorsal and ventral root filaments are attached (figs. 41–2 and 41–3). **The segment of spinal cord to which a given pair of dorsal and ventral roots is attached is a myelomere** (fig. 41–2). Because the adult spinal cord does not extend down as far as the vertebral column does, the lower myelomeres are not opposite their correspondingly numbered vertebrae. Thus myelomere S1 is opposite T.V.12 (see fig. 41–1).

Each dorsal root presents a swelling, the *spinal ganglion,* which lies near or within the intervertebral foramen. Distal to the ganglion, each dorsal root combines with the corresponding ventral root to form a spinal nerve (figs. 41–2 and 41–3). There are 31 pairs of spinal nerves: 8 cervical, 12 thoracic, 5 lumbar, 5 sacral, and 1 coccygeal. **The first pair of spinal nerves emerges between the atlas and the skull; hence C.N.1 to 7 leave the vertebral canal above the correspondingly numbered vertebrae. C.N.8 emerges below C.V.7, and all the remaining spinal nerves leave below the corresponding vertebrae.** The nerve roots below those of L.N.1, and those which occupy the vertebral canal below the cord, resemble a horse's tail and hence are termed collectively the *cauda equina* (fig. 41–4).

In section, the spinal cord is seen to consist of gray matter, which is shaped like the letter H, surrounded by white matter (see figs. 41–2 and 41–3). Regional differences occur; e.g., the contour of the gray matter varies, and the amount of white matter decreases as one descends the cord. In the median plane, the gray matter presents the *central canal,* which extends from the fourth ventricle of the brain to the upper part of the filum terminale.

Because of the discrepancy between the levels of the myelomeres and their corresponding vertebrae, the lower spinal roots become increasingly oblique and emerge from their intervertebral foramina at increasingly lower levels (figs. 41–1 and 41–4). Thus, because myelomere S1 is opposite T.V.12, the roots of S.N.1 must descend steeply in order that the rami can emerge through the first sacral foramina. The lumbosacral roots are the longest and the thickest. **The lumbar nerves increase in size from above downward, whereas the lumbar intervertebral foramina decrease in diameter. Thus L.N.5, the thickest, in travers-**

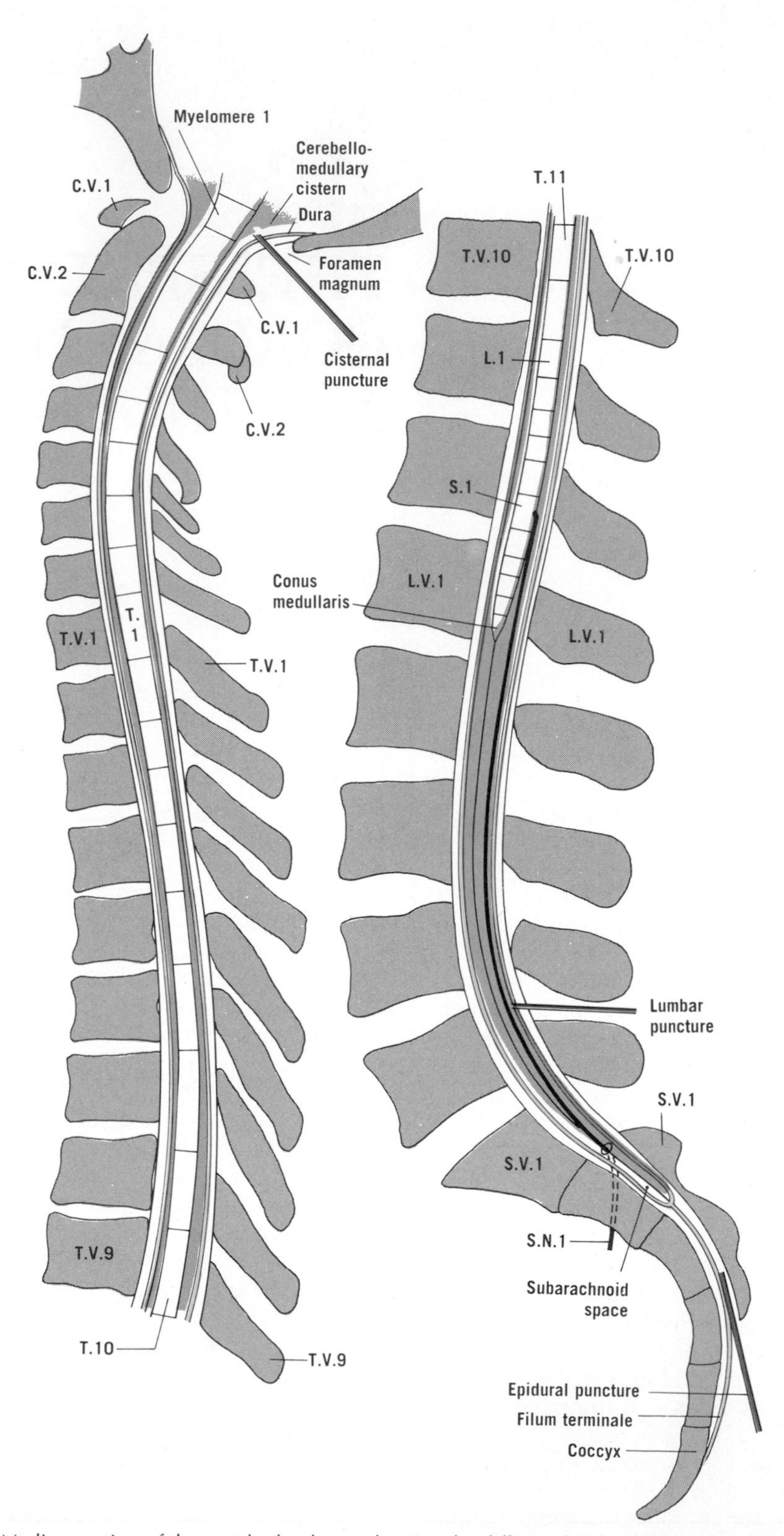

Figure 41–1 Median section of the vertebral column, showing the different levels of the vertebral bodies, myelomeres, and spinous processes. The spinal cord ends at L.V.1/2 and the subarachnoid space at S.V.1/2. Cisternal, lumbar, and epidural punctures are shown. As an example of a spinal nerve, S.N.1 can be seen arising from myelomere S.1 opposite T.V.12, descending (as part of the cauda equina), and emerging from intervertebral foramen S.V.1/2.

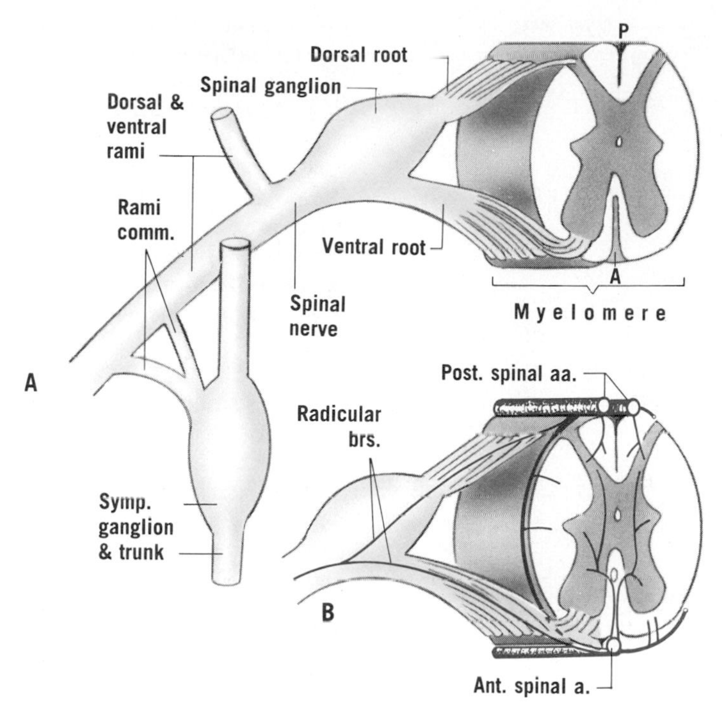

Figure 41–2 A myelomere of the spinal cord, and one of its two associated spinal nerves. In **A**: *A*, anterior median fissure; *P*, posterior median sulcus. **B** shows the arterial supply to the cord.

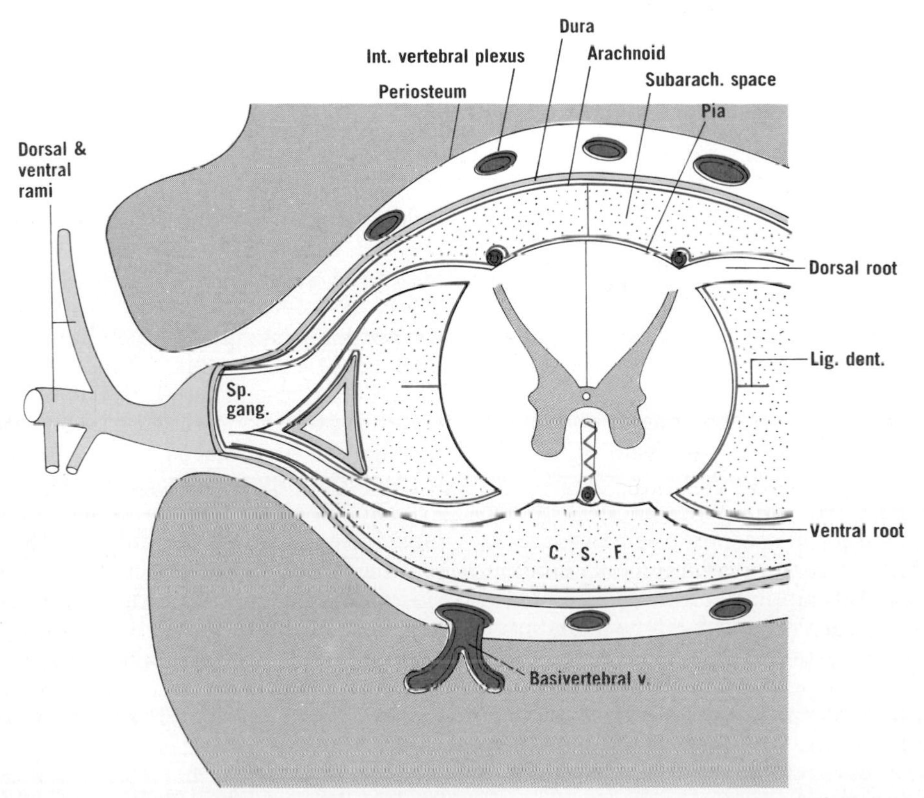

Figure 41–3 Horizontal section of the spinal cord showing the meninges. The dura is in yellow, the arachnoid in red, and the pia in blue. The anterior and posterior spinal arteries are shown. *C.S.F.*, cerebrospinal fluid in the subarachnoid space.

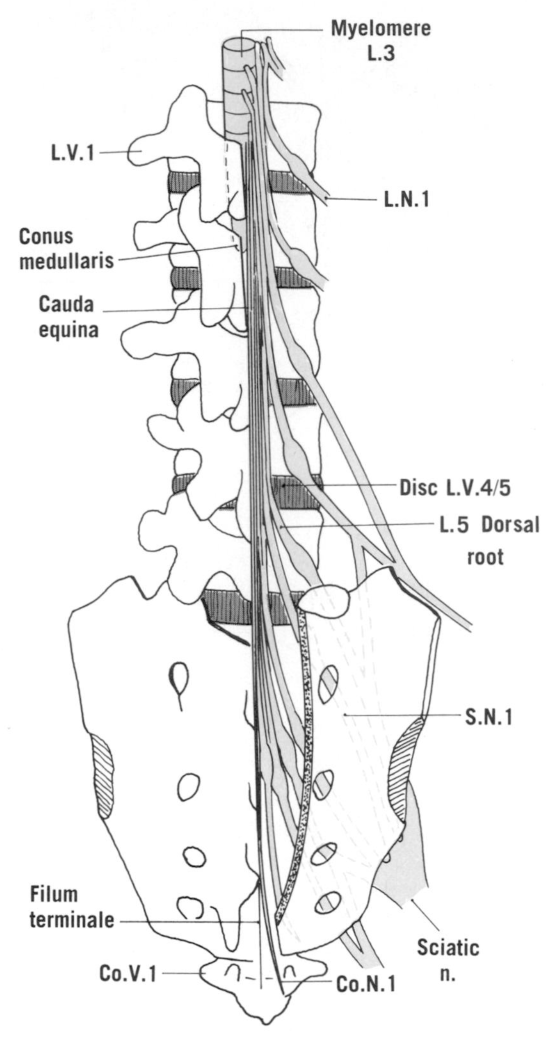

Figure 41–4 The spinal cord and cauda equina *in situ*, posterior aspect, made visible by a laminectomy on the right-hand side. The dorsal rami are omitted. The intervertebral discs are shown in blue. It can be seen that prolapse of disc L.V.4/5, for example, would be likely to damage L.5 roots. (Based partly on Pernkopf.)

ing the narrowest foramen, has as increased chance of compression in the event of herniation of the nucleus pulposus.

The spinal cord contains the descending motor tracts and the ascending sensory tracts. The cervical and lumbar enlargements contain the neurons that supply the limbs. The cervical part of the cord gives rise to the spinal part of the accessory nerve and contains the neurons that supply the diaphragm. The thoracic and upper lumbar parts of the cord provide the sympathetic outflow, and the sacral part contributes the parasympathetic outflow of the autonomic system.

Blood Supply (see figs. 41–2 and 41–3). **The spinal cord is supplied by three longitudinal arterial channels, which are reinforced by segmental (e.g. intercostal) arteries.** The anterior spinal artery (from the vertebral artery) lies in the anterior median fissure. Two posterior spinal arteries (also from the vertebral artery, directly or indirectly) descend lateral to the posterior median sulcus. **The segmental reinforcements are very important in feeding the longitudinal channels, and frequently a large vessel in the thoracic region descends to supply the lumbar enlargement.**

MENINGES (see fig. 41–3)

The spinal cord, like the brain, is surrounded by the three meninges. The *dura mater* extends from the foramen magnum to the sacrum and coccyx (see fig. 41–1). The dura is separated from the periosteum of the vertebral canal by the *epidural* (or *peridural* or *extradural*) *space*, which contains fat and the internal vertebral venous plexus. In caudal analgesia, an anesthetic solution injected into the sacral hiatus diffuses upward into the epidural space (see fig. 41–1) and may be used in surgical procedures relating to the vesical, prostatic, vaginal, external genital, and anorectal areas. Dural sheaths surround the roots and spinal ganglia, and continue into the epineurium of the spinal nerves.

The *arachnoid* invests the spinal cord loosely. Continuous with the cerebral arachnoid above, it traverses the foramen magnum and descends to about S.V.2. **The subarachnoid space, which contains cerebrospinal fluid (C.S.F.), is a wide interval between the arachnoid and pia. Because the spinal cord ends at about L.V.2, whereas the subarachnoid space continues to S.V.2, access can be gained to the C.S.F. by inserting a needle below the end of the cord, a procedure termed lumbar puncture** (see fig. 41–1). By this means, the pressure of C.S.F. can be measured, the fluid can be analyzed, a spinal anesthetic can be introduced, or fluid can be replaced by a contrast medium for radiography (*myelography*).

The *pia mater* invests the spinal cord closely, ensheathes the anterior spinal artery (as the *linea splendens*), and enters the anterior median fissure. Laterally, the pia forms a longitudinal septum, the *denticulate ligament* (see fig. 41–3), which sends about 21 tooth-like processes laterally to fuse with the arachnoid and dura on each side. The ligament is a surgical landmark in that it is attached to the spinal cord about midway between the attachments of dorsal and ventral roots.

Further details concerning the spinal cord should be sought in books on neuroanatomy.

QUESTIONS

41–1 At which level does the spinal cord end?

41–2 What is a myelomere?

41–3 How are myelomeres related to vertebrae in level?

41–4 Where are spinal ganglia found?

41–5 How many spinal nerves are present in the body?

41–6 Where do spinal nerves emerge in relation to their correspondingly numbered vertebrae?

41–7 What is the cauda equina?

41–8 At which levels is subarachnoid space found below the spinal cord?

41–9 Which Latin and Greek roots are used with reference to the spinal cord?

Part 8

THE HEAD AND NECK

The cervical vertebrae and the back of the neck have already been described. In this part the skull is examined, followed by the brain, ear, and eye, and ending with the mouth, nose, pharynx, and larynx. Included in the head and neck are a number of important structures, the diseases of which form the subject matter of various specialties: neurology, neurosurgery, and neuroradiology (brain, nerves, and skull), ophthalmology (eye), otology (ear), rhinolaryngology (nose and throat), and dentistry and oral surgery (teeth and jaws).

Acquaintance with the parts of the brain and with the names of the cranial nerves is a necessary preliminary to a study of the head and neck.

The main divisions of the brain (encephalon) are the forebrain (prosencephalon), midbrain (mesencephalon), and hindbrain (rhombencephalon). Important subdivisions are listed in table 43–1 and illustrated in figure 43–1.

Cranial nerves are encountered in each of the chapters that follow. Their names are:

1. Olfactory
2. Optic
3. Oculomotor
4. Trochlear
5. Trigeminal: ophthalmic, maxillary, and mandibular nerves
6. Abducent
7. Facial (including the nervus intermedius)
8. Vestibulocochlear: vestibular and cochlear parts
9. Glossopharyngeal
10. Vagus
11. Accessory: internal branch (cranial part) and external branch (spinal part)
12. Hypoglossal

Further details are provided in table 43–2.

ATLASES AND SPECIAL TEXTS

Aubaniac, R., and Porot, J., *Radio-anatomie générale de la tête,* Masson, Paris 1955. Key drawings and radiographs of coronal, sagittal, and horizontal sections through the head made 1 cm apart. Special atlases of the radiological anatomy of the skull are available.

Kampmeier, O. F., Cooper, A. R., and Jones, T. S., *A Frontal Section Anatomy of the Head and Neck,* University of Illinois Press, Urbana, 1957. Photographs of coronal sections made about 1 cm apart.

Lang, J., *Klinische Anatomie des Kopfes,* Springer, Berlin, 1981. A superbly illustrated, detailed account of the neurocranium, orbit, and craniocervical junction for specialists such as neurosurgeons.

Symington, J., *An Atlas Illustrating the Topographical Anatomy of the Head, Neck and Trunk,* Oliver & Boyd, Edinburgh, reprinted in 1956. Drawings of horizontal sections, including a dozen pertaining to the head and neck.

Truex, R. C., and Kellner, C. E., *Detailed Atlas of the Head and Neck,* Oxford University Press, New York, 1948. Colored regional drawings of dissections and coronal and horizontal sections.

42
THE SKULL AND HYOID BONE

The skeleton of the head and neck consists of the skull and hyoid bone and the cervical vertebrae (which have already been described).

SKULL

The skull, or *cranium,* protects the brain and the organs of special sensation, allows the passage of air and food, and supports the teeth. It consists of a series of bones, mostly united at immovable joints. The mandible, however, can move freely at a synovial articulation, the temporomandibular joint. Some bones of the skull are paired, whereas others are not. Each consists of *external* and *internal tables* of compact bone and a middle spongy layer, the *diploë*. The skull is covered by periosteum (*pericranium*) and lined by dura (*endocranium*). The top part (skull cap) is termed the *calvaria* (calvarium is incorrect) (fig. 42–1). The various bones are shown in figures 42–1 to 42–3 and 42–8.

The fibrous joints between the bones are termed *sutures,* and they allow growth at the calvaria. With increasing age, many sutures disappear by osseous fusion. Bony areas termed *sutural bones* are frequent in some sutures.

When a subject is in the anatomical position, the skull is oriented so that **the lower margins of the orbits and the upper margins of the external acoustic meatuses are horizontal, i.e., in the orbitomeatal plane.**

The most frequently used views in radiography of the head are the right and left lateral (fig. 42–4), postero-anterior (fig. 42–5), and

Text continues on page 351.

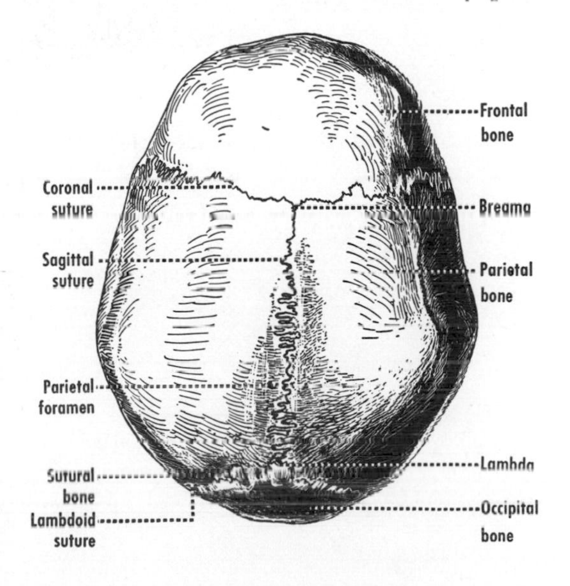

Figure 42–1 Superior aspect of the skull. Note that some portions of a suture are more serrated than others.

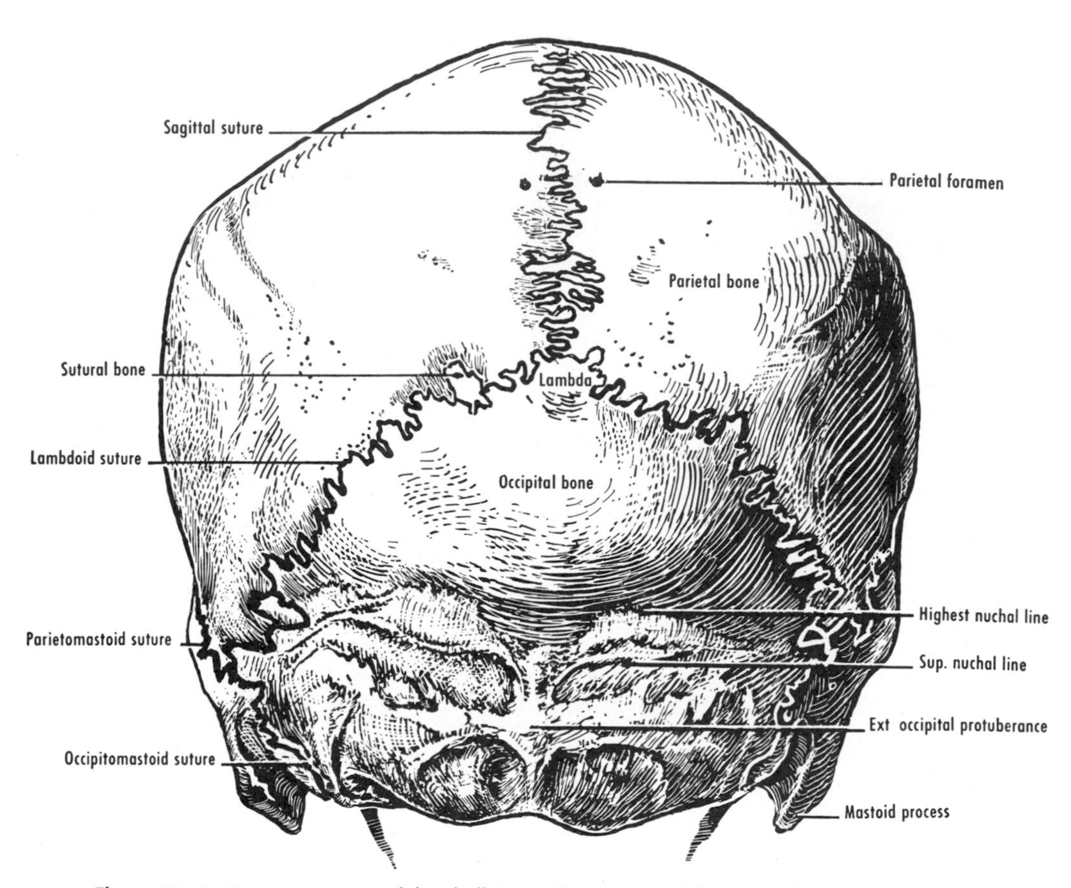

Figure 42–2 Posterior aspect of the skull. Note the two sutural bones on the lambdoid suture.

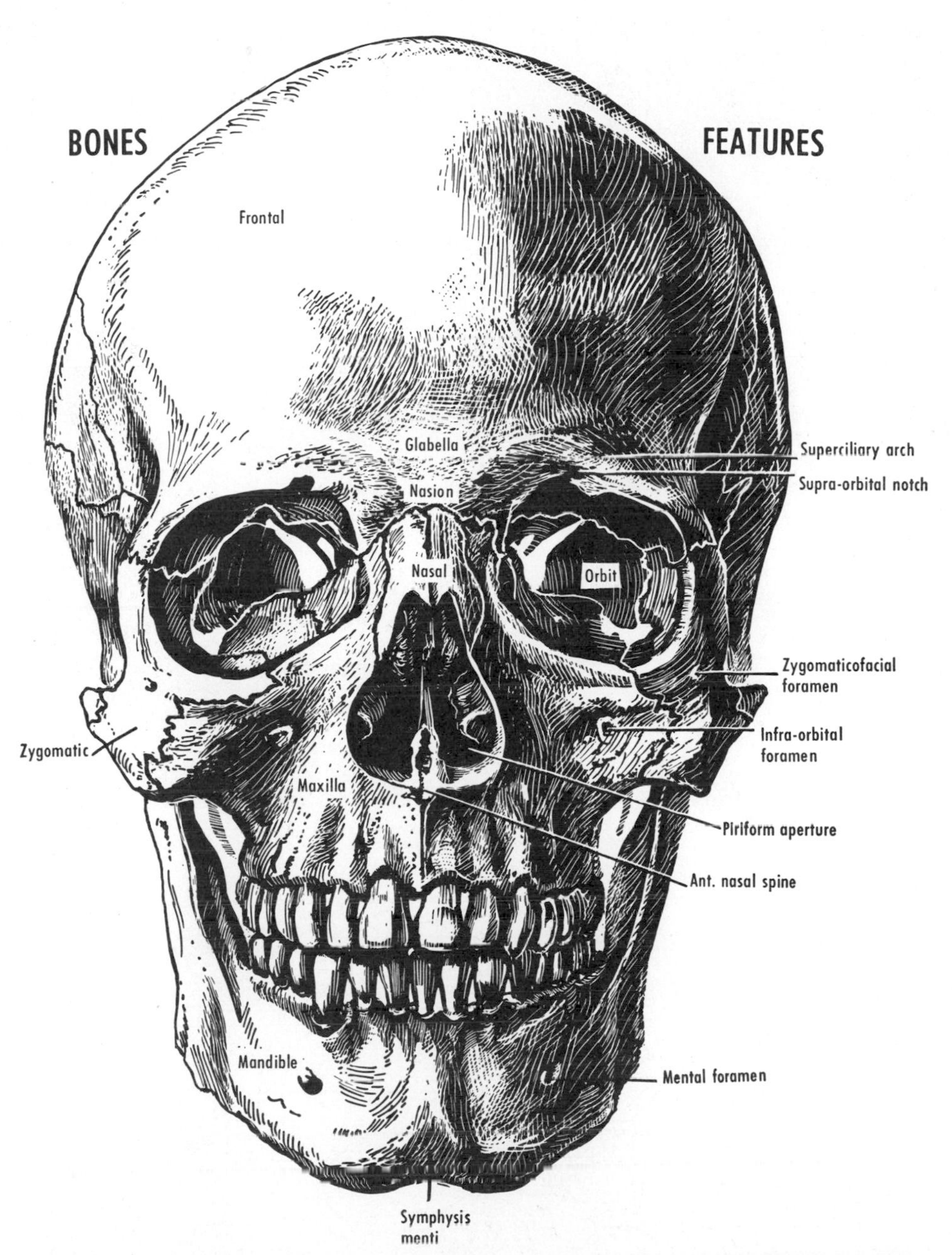

Figure 42–3 Anterior aspect of the skull. Observe that **the supra-orbital notch, infra-orbital foramen, and mental foramen are approximately in a vertical line. They transmit branches of divisions 1, 2, and 3 of the trigeminal nerve, respectively.**

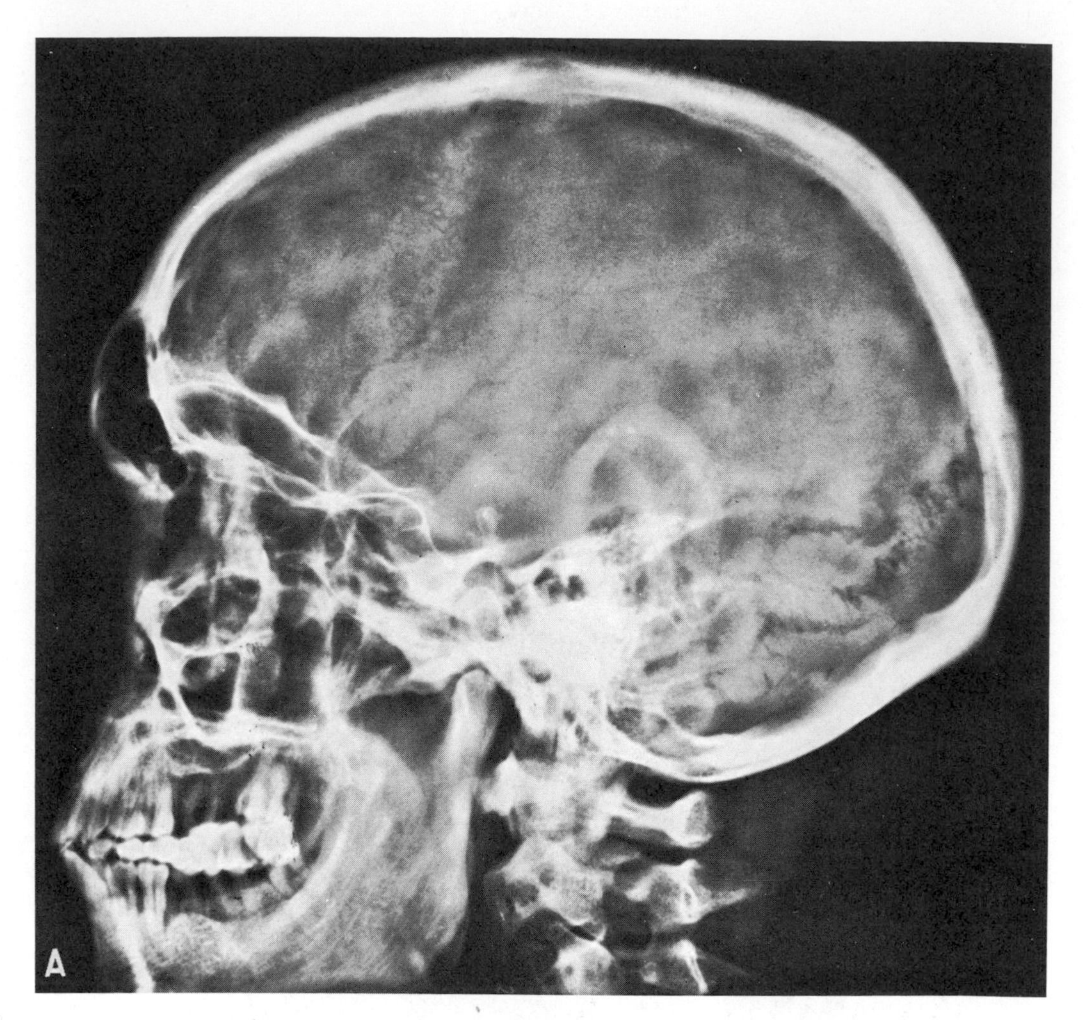

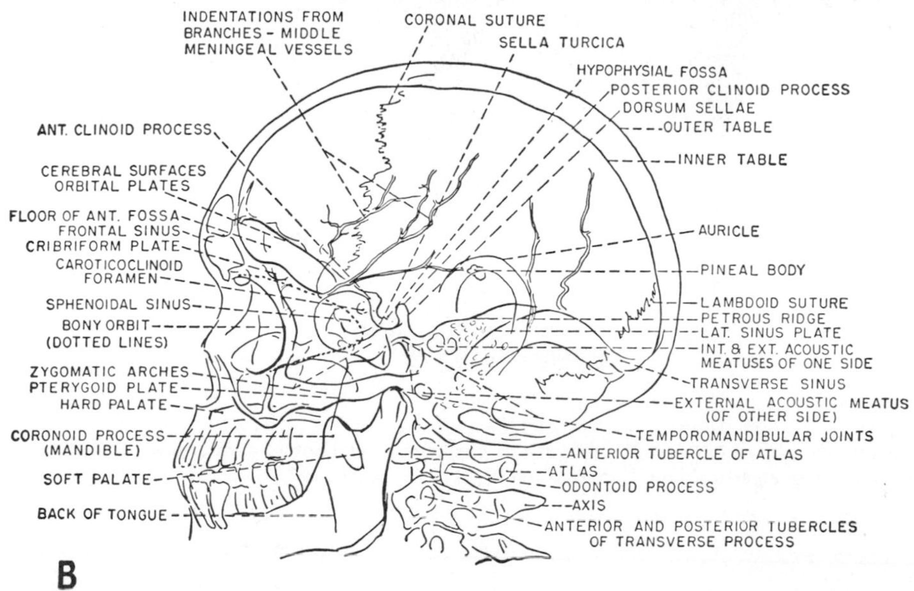

Figure 42–4 Lateral radiograph of the head. (From Meschan, I., *Normal Radiographic Anatomy,* 2nd ed., W. B. Saunders Company, Philadelphia, 1959; courtesy of the author.)

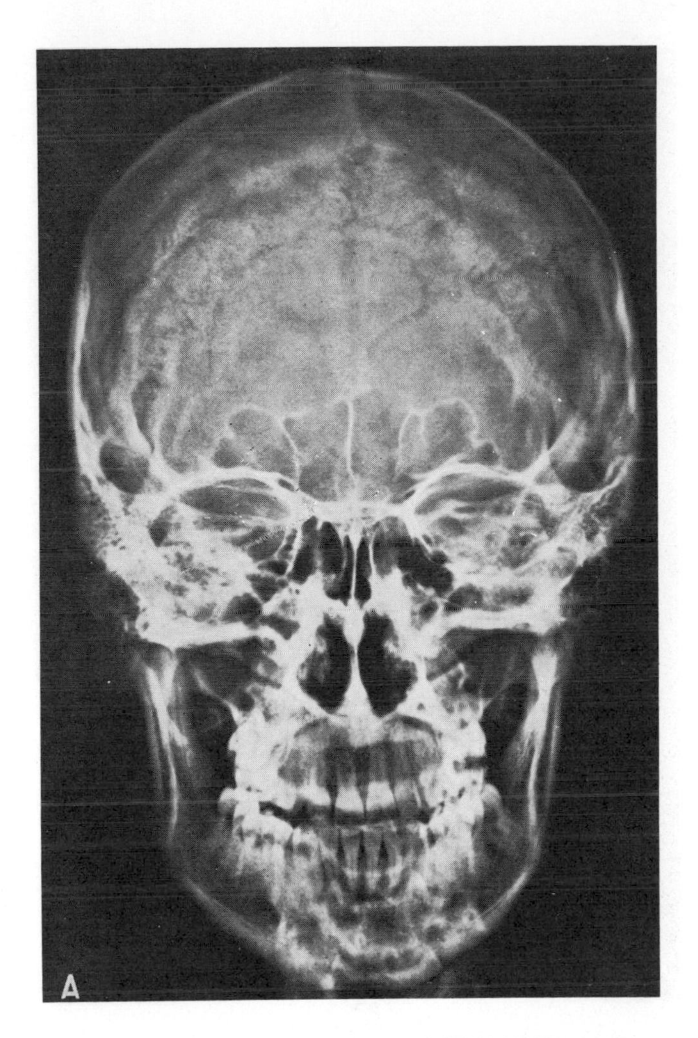

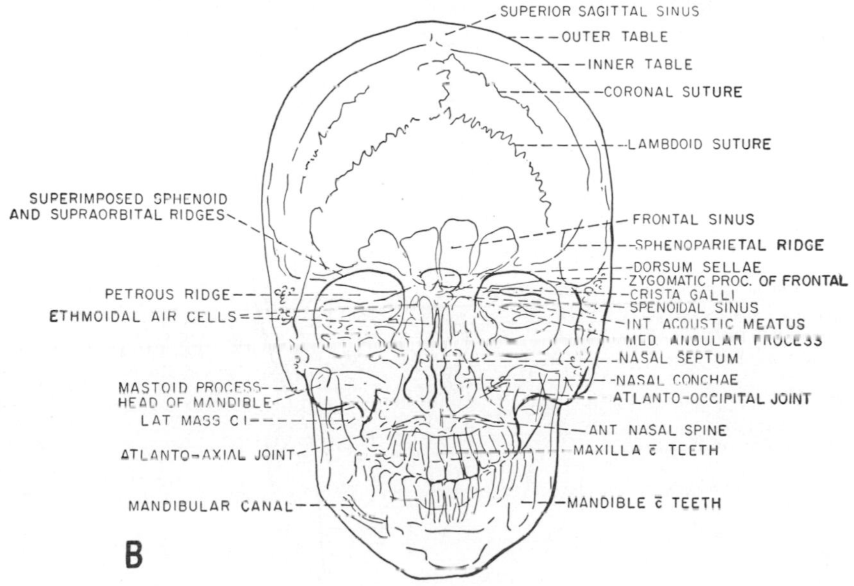

Figure 42–5 Postero-anterior radiograph. (From Meschan.)

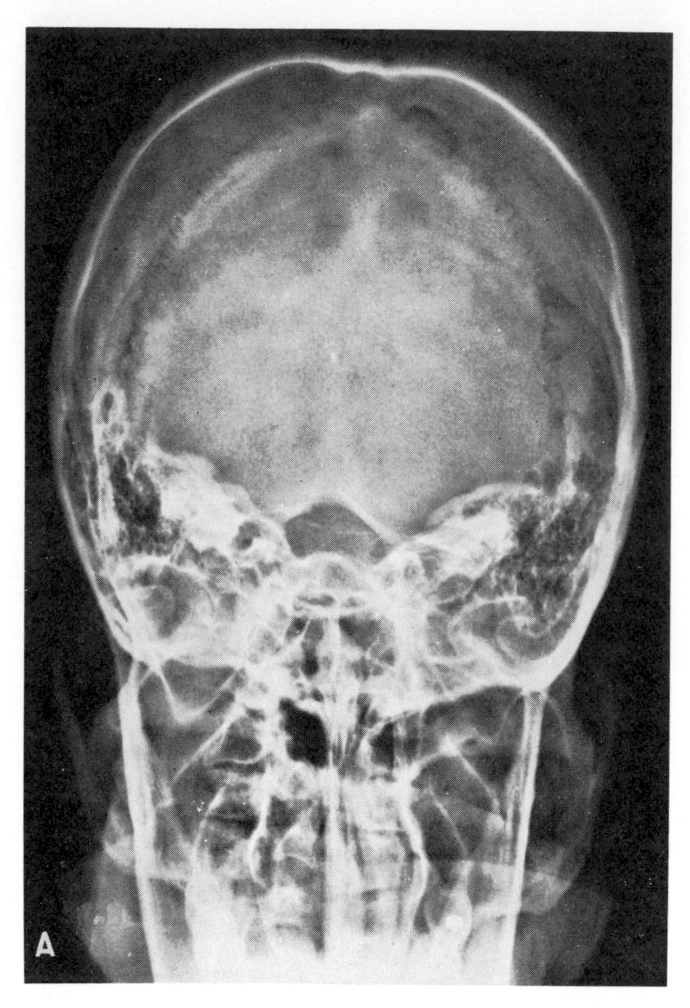

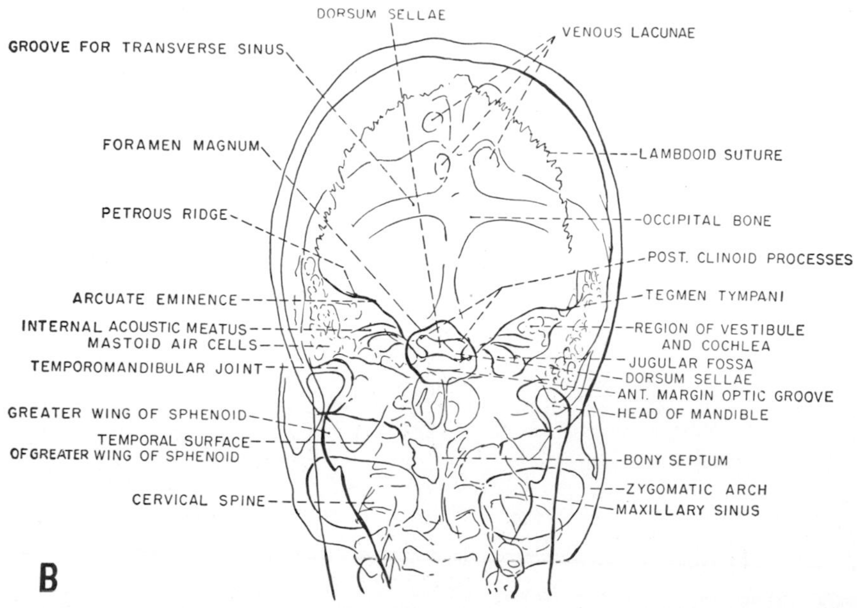

Figure 42–6 Anteroposterior radiograph (Towne's projection). The subject is supine, the back of the head rests on the film, and the central ray is directed slightly caudally from the forehead to the external occipital protuberance. (From Meschan.)

anteroposterior (fig. 42–6). Calcified areas may be found normally in the pineal body and in the choroid plexuses of the lateral ventricles.

Superior, Posterior, and Anterior Aspects of Skull

Superior Aspect (fig. 42–1). The two parietal bones are separated by the *sagittal suture*. The frontal and parietal bones are separated by the *coronal suture*. The occipital and parietal bones are separated by the *lambdoid suture*. An emissary opening, the *parietal foramen*, may be found on one or both sides of the sagittal suture posteriorly.

Posterior Aspect (fig. 42–2). The parietal and occipital bones meet the mastoid part of the temporal bone laterally. The *external occipital protuberance* is a median projection that is palpable *in vivo*. Its center is termed the *inion*. *Superior nuchal lines* extend laterally from the protuberance and mark the upper limit of the neck.

Anterior Aspect (see figs. 42–3 and 42–5). This aspect presents the forehead, the orbits,

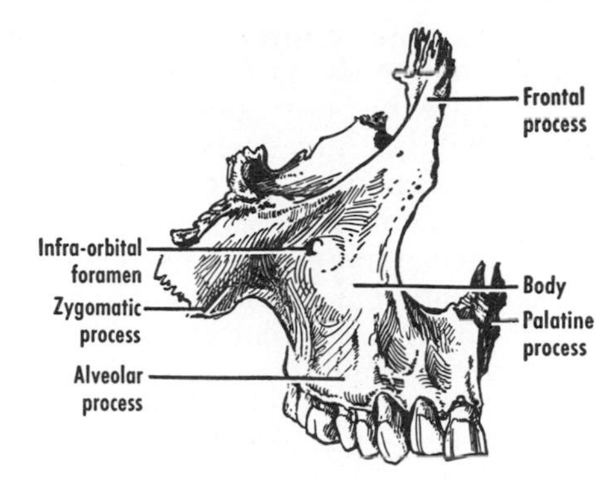

Figure 42–7 Anterior surface of the right maxilla. The maxilla consists of a body and four processes: zygomatic, frontal, palatine, and alveolar.

the prominence of the cheek, the bony external nose, and the upper and lower jaws.

The skeleton of the forehead is formed by the *frontal bone*, which articulates below with the nasal bones medially and with the zygomatic bones laterally. The bony area between the eyebrows is the *glabella*, from which an elevation, the *superciliary arch*, extends laterally on each side. The two halves of the frontal bone are separated until the age of about six

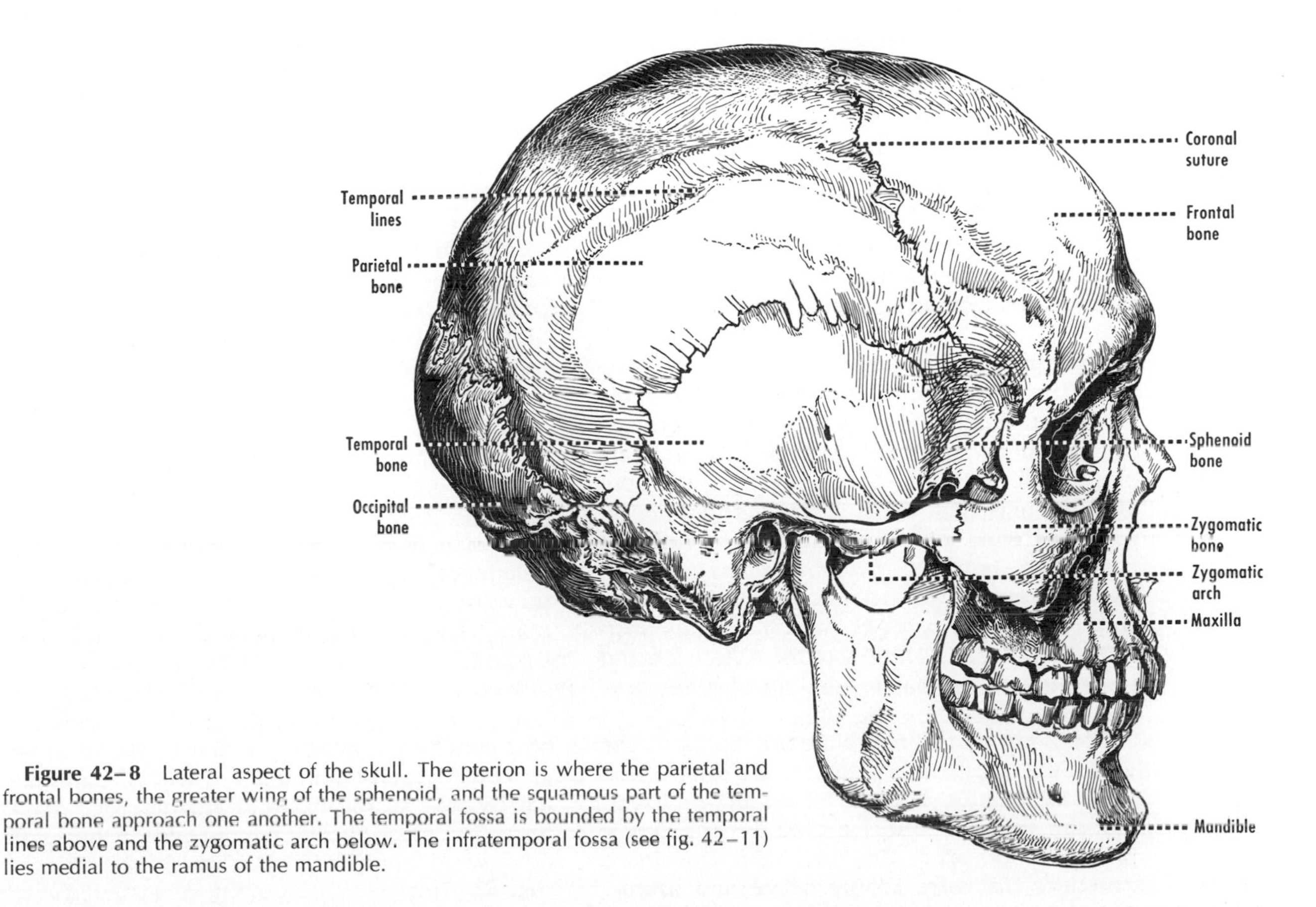

Figure 42–8 Lateral aspect of the skull. The pterion is where the parietal and frontal bones, the greater wing of the sphenoid, and the squamous part of the temporal bone approach one another. The temporal fossa is bounded by the temporal lines above and the zygomatic arch below. The infratemporal fossa (see fig. 42–11) lies medial to the ramus of the mandible.

years by the frontal suture, which sometimes persists into adulthood as the *metopic suture*.

The orbits are the bony cavities that contain the eyes, and they are described in a later chapter. At its junction with the face, the rim of the orbit is described as superior, lateral, inferior, and medial margins. The superior margin presents either a *supra-orbital notch* or *foramen* medially for the corresponding nerve and vessels. Laterally, the superior margin ends in the *zygomatic process of the frontal bone,* which is readily palpable *in vivo*. The lateral margin is formed by the frontal and zygomatic bones, and the inferior margin by the zygomatic bone and the maxilla. Below the inferior margin, the maxilla presents the *infraorbital* foramen, which transmits the corresponding nerve and artery (fig. 42–7). The medial margin, less clearly defined, is formed by the maxilla and the lacrimal and frontal bones.

The prominence of the cheek is formed by the *zygomatic* (formerly *malar) bone* (fig. 42–8), which contributes to the face, the orbit, and the temporal fossa. It presents two processes, which articulate with the zygomatic processes of the frontal and temporal bones, respectively.

The bony part of the external nose is formed by the nasal bones and the maxillae. Its opening, to which the cartilaginous part of the nose is anchored in the intact state, is the *piriform aperture*. Through this the *nasal cavity* can be seen, divided by the *nasal septum* into right and left portions. The septum is composed of cartilage in front and bone (ethmoid and vomer) behind (fig. 52–2*B*). Each lateral wall of the nasal cavity presents three curled bony plates termed *conchae* (formerly *turbinates*).

The upper jaw is composed of the two maxillae. **The growth of the maxillae is responsible for the vertical elongation of the face between the ages of six and twelve years.** Each maxilla (see fig. 42–7) consists of a *body,* which contains the maxillary sinus; a *zygomatic process,* which extends laterally to articulate with the zygomatic bone; a *frontal process,* which projects upward to articulate with the frontal bone; a *palatine process,* which extends horizontally to meet its fellow of the other side and contribute to the palate; and an *alveolar process,* which carries the upper teeth.

The body of the maxilla contributes to the lateral wall of the nasal cavity, the floor of the orbit, the front wall of the infratemporal fossa, and the face. About 1 cm below the lower margin of the orbit, the *infra-orbital foramen* transmits the infra-orbital nerve and artery. The two maxillae are united at the *intermaxillary suture*.

The lower jaw, or mandible, which carries the lower teeth in its *alveolar part,* is described later.

Lateral Aspect of Skull
(figs. 42–4 and 42–8)

The lateral aspect of the skull includes portions of the temporal bone and the temporal and infratemporal fossae.

Temporal Bone. **The temporal bone comprises squamous, tympanic, styloid, mastoid, and petrous parts** (fig. 42–9).

SQUAMOUS PART. The parietal bone articulates below with the squamous part of the temporal bone (see fig. 42–8). From the squamous portion, the *zygomatic process* (or *zygoma*) projects forward to meet the zygomatic bone and form the *zygomatic arch,* which is readily palpable *in vivo*. **The upper border of the zygomatic arch corresponds to the lower limit of the cerebral hemisphere** (see fig. 43–8). The lower border of the zygomatic arch, which gives origin to the masseter muscle, can be traced backward to the *tubercle of the root of the zygoma,* behind which the head of the mandible is lodged in the mandibular fossa. The *external acoustic meatus,* situated behind the head of the mandible, leads toward the middle ear, from which it is separated in the intact state by the tympanic membrane. The roof and adjacent part of the posterior wall of the meatus are formed by the squamous part of the temporal bone, whereas the other walls are formed by the tympanic part. A small depression, the *suprameatal triangle,* lies immediately above and behind the meatus. **The mastoid antrum, a cavity in the temporal bone, lies about 1 cm medial to the suprameatal triangle.**

TYMPANIC PART. The floor and anterior wall of the meatus are formed by a curved lamina, the *tympanic plate*. In children, it is merely a *tympanic ring*.

STYLOID PART. The *styloid process,* of variable length (fig. 42–9), extends downward and forward from under cover of the tympanic plate. The stylohyoid ligament on each side suspends the hyoid bone from the skull.

MASTOID PART. The mastoid is the back portion of the temporal. **In the adult, the mastoid part generally contains air spaces, the mastoid air cells, which communicate with the middle ear by way of the mastoid antrum.** The *mastoid process,* easily felt *in vivo,* projects downward and gives attachment to muscles (fig. 42–10).

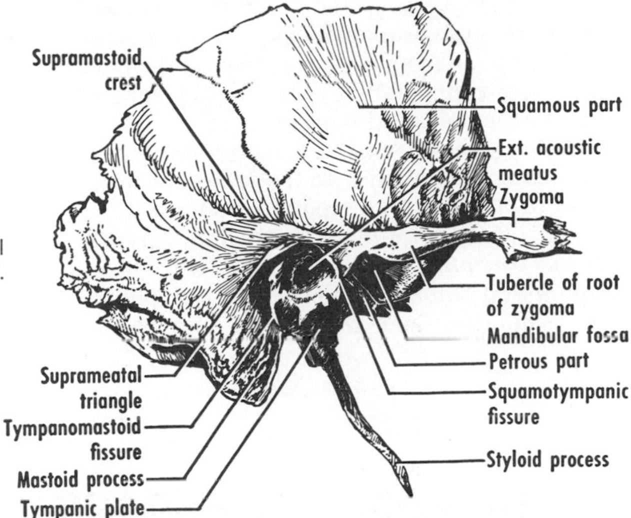

Figure 42–9 Lateral aspect of the right temporal bone. The styloid process is unusually long in this skull.

PETROUS PART. The petrous part is placed deeply and is described later.

Temporal Fossa. The *temporal line,* to which the temporal fascia is attached, arches across the frontal and parietal bones (see fig. 42–8) and joins a ridge (*supramastoid crest,* see fig. 42–9) on the temporal bone. The temporal fossa, in which the temporal muscle is located, is bounded by the temporal line above and the zygomatic arch below. The floor of the fossa includes portions of the parietal and frontal bones, the greater wing of the sphenoid bone, and the squamous part of the temporal bone. The area where these four bones approach each other is known as the *pterion* (see fig. 42–8). **The pterion overlies the anterior branch of the middle meningeal artery on the internal aspect of the skull, and it corresponds also to the stem of the lateral sulcus of the brain. The center of the pterion is about 4 cm above the midpoint of the zygomatic arch and nearly the same distance behind the zygomatic process of the frontal bone** (see fig. 43–8).

Infratemporal Fossa (fig. 42–11). The interval between the zygomatic arch and the rest of the skull is traversed by the temporal muscle,

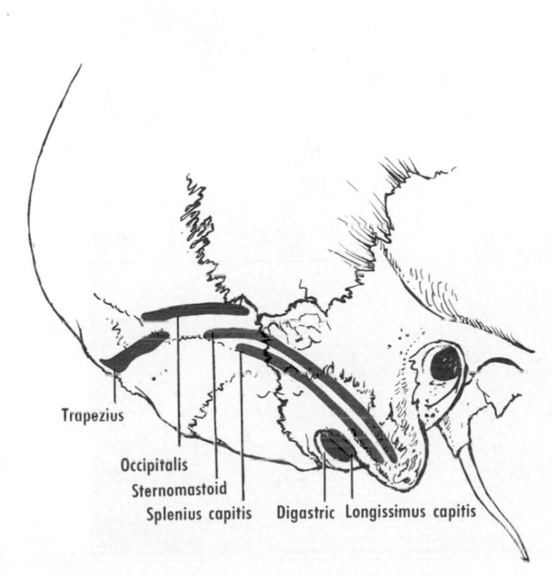

Figure 42–10 Lateral aspect of the occipitomastoid region of the skull, showing muscular attachments.

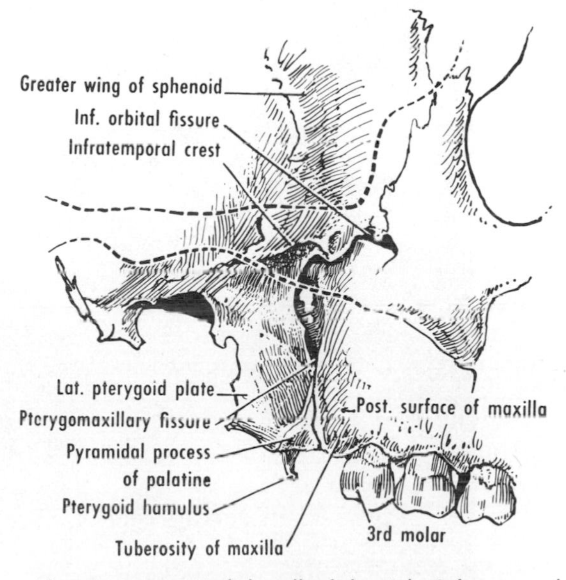

Figure 42–11 Medial wall of the right infratemporal fossa after removal of the mandible. The zygomatic arch is shown as if transparent; its borders are indicated by interrupted lines.

which thereby leaves the temporal fossa and enters the infratemporal fossa. The infratemporal fossa, an irregular space behind the maxilla, is limited medially by the lateral *pterygoid plate* of the sphenoid bone (see fig. 48–3) and laterally by the ramus of the mandible. Medial to its communication with the temporal fossa, the roof of the infratemporal fossa is formed by the *infratemporal surface of the greater wing of the sphenoid bone* (fig. 42–12). The infratemporal fossa contains the lower part of the temporalis and the lateral and medial pterygoid muscles; the maxillary artery and its branches and the pterygoid venous plexus; and the mandibular, maxillary, and chorda tympani nerves. The infratemporal fossa communicates with the orbit (table 42–1) through the inferior orbital fissure, which is continuous behind with the pterygomaxillary fissure. The infratemporal fossa communicates with the pterygopalatine fossa through the *pterygomaxillary fissure,* which is between the lateral pterygoid plate and the maxilla and which transmits the maxillary artery. The *pterygo-*

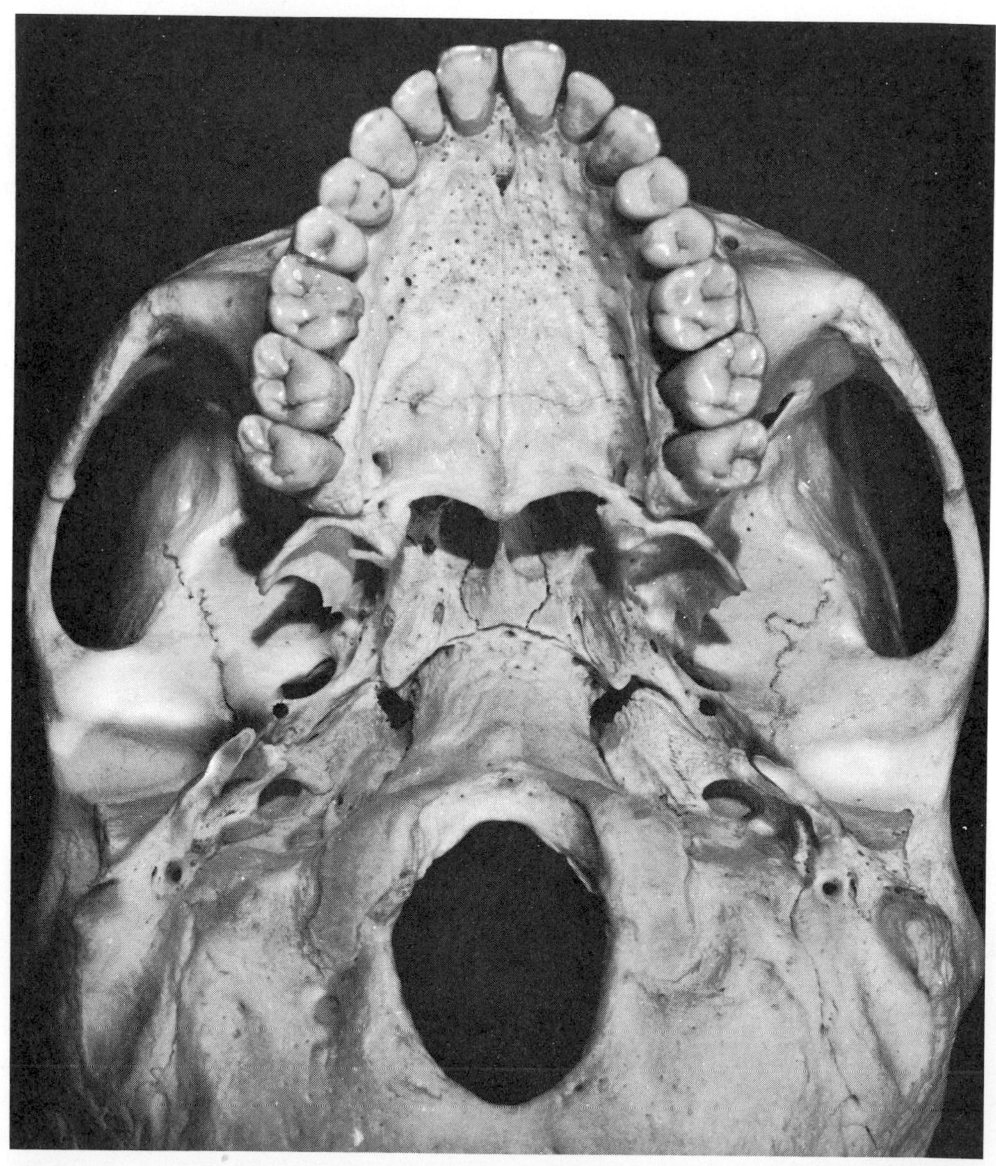

Figure 42–12 Photograph of the base of the skull. The features are identified in the following figure.

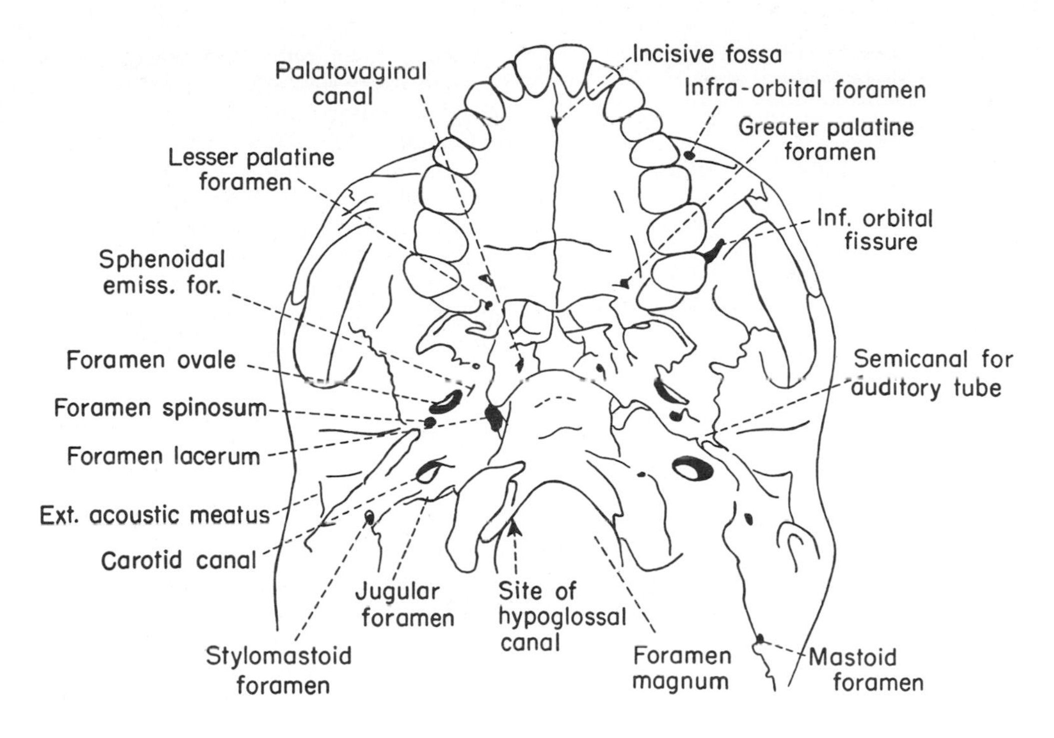

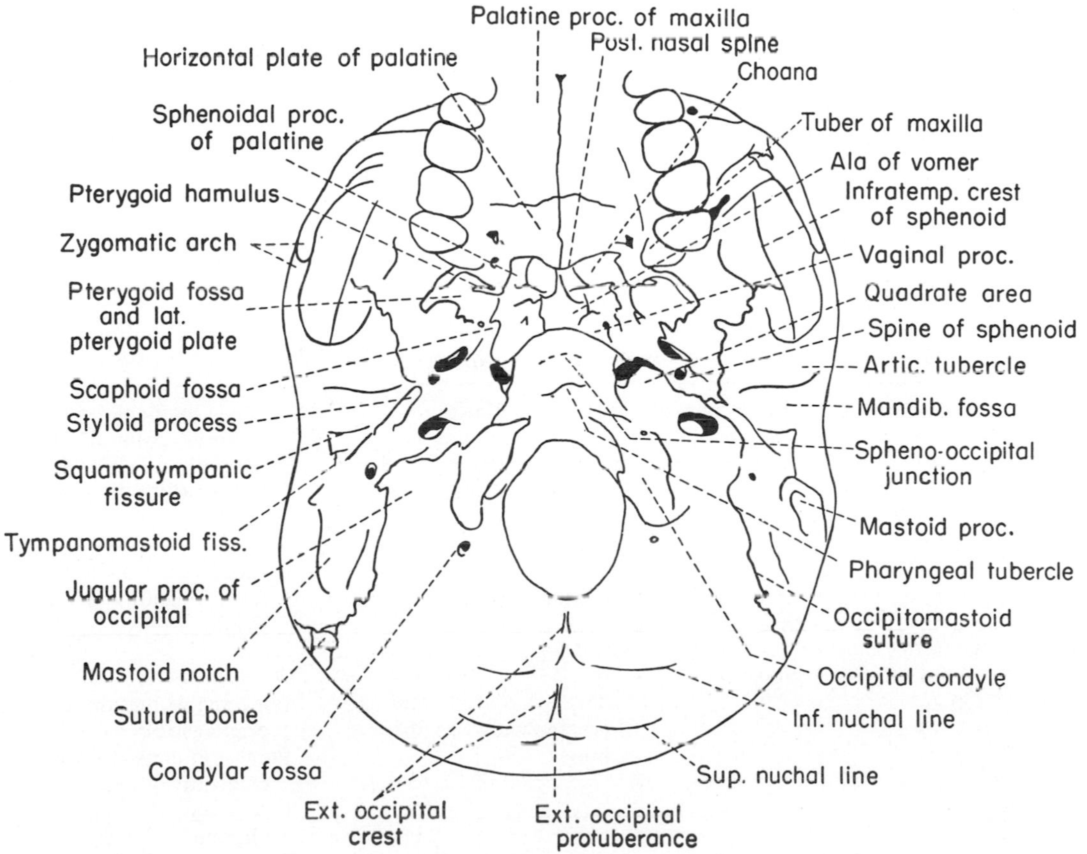

Figure 42–13 Keys to the previous figure. The upper drawing shows mainly foramina and canals.

TABLE 42–1 SUMMARY OF COMMUNICATIONS WITH INFRATEMPORAL FOSSA

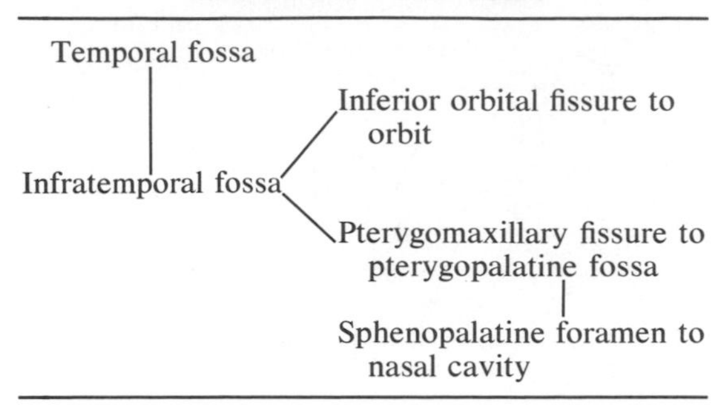

palatine fossa, between the pterygoid plates and the palatine bone, is below the apex of the orbit, and it contains the maxillary artery and nerve. It communicates with the nasal cavity through the sphenopalatine foramen (table 42–2).

INFERIOR ASPECT OF SKULL
(figs. 42–12 and 42–13)

Occipital Bone (fig. 42–14)

The lower surface of the base is formed behind by the occipital bone, which consists of **four parts arranged around the foramen magnum: a squamous part behind, a lateral part on each side, and a basilar part in front (see also p. 359). These four portions, separable at birth, become fused by the age of about six years. The foramen magnum is located midway between, and on a level with, the mastoid processes. Through it the posterior cranial fossa and contained brain become continuous with the vertebral canal and contained spinal cord.** The foramen magnum contains the medulla oblongata, tonsils of the cerebellum, meninges, and the subarachnoid space; the spinal roots of the accessory nerve, meningeal branches (C.N.1 to 3), and sympathetic plexuses; the vertebral and anterior and posterior spinal arteries; and several ligaments (apical ligament of the dens, cruciform ligament of the atlas, and membrana tectoria).

SQUAMOUS PART. The squamous part of the occipital bone continues from the base onto the back of the skull. The *external occipital crest,* to which the ligamentum nuchae is attached, extends from the foramen magnum to the *external occipital protuberance,* which is readily palpable *in vivo. Nuchal lines* extend laterally and delimit areas of muscular insertion.

LATERAL PARTS. The *lateral parts* of the occipital bone present the *occipital condyles* at the sides of the foramen magnum. The condyles articulate with the lateral masses of the atlas and, through them, the weight of the head is transmitted to the vertebral column. The *hypoglossal canal,* which transmits the hypoglossal nerve, lies hidden above the front of each condyle. The *jugular process* extends laterally from each condyle to the temporal bone, and its concave anterior border is the posterior boundary of the jugular foramen.

TABLE 42–2 BOUNDARIES OF AND OPENINGS FROM INFRATEMPORAL AND PTERYGOPALATINE FOSSAE

Aspect	Boundaries of and Openings from Infratemporal Fossa	Boundaries of Pterygopalatine Fossa	Openings from Pterygopalatine Fossa
Superior	Infratemporal surface of greater wing of sphenoid bone	Body of sphenoid bone and orbital process of palatine bone	Inferior orbital fissure to orbit
Inferior	Open	Meeting of anterior and posterior walls	Greater (and sometimes lesser) palatine canal to palate
Anterior	Posterior surface of maxilla and inferior orbital fissure	Posterior surface of maxilla	None
Posterior	Open	Lateral pterygoid plate and greater wing of sphenoid bone	Foramen rotundum to middle cranial fossa Pterygoid canal to foramen lacerum Palatovaginal canal to choana
Medial	Lateral pterygoid plate and pterygomaxillary fissure	Perpendicular plate of palatine bone	Sphenopalatine foramen to nasal cavity
Lateral	Ramus and coronoid process of mandible	Open	Pterygomaxillary fissure to infratemporal fossa

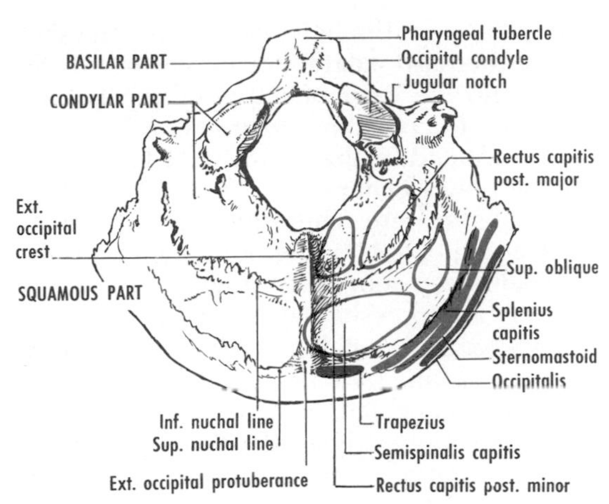

Figure 42–14 Inferior aspect of the occipital bone. The four chief parts—basilar, two lateral, and squamous—can be seen around the foramen magnum. The main muscular attachments are shown on the left side of the bone.

The transverse process of the atlas lies immediately below the jugular process and rarely may be fused with it.

BASILAR PART. The *basilar part* of the occipital bone joins the sphenoid bone (fig. 42–15) at the *spheno-occipital joint,* which is cartilaginous until about puberty, when bony fusion occurs.* In front of the foramen magnum, the ill-defined *pharyngeal tubercle* anchors the pharynx.

Goethe first pointed out clearly the resemblance of the occipital bone to a vertebra. The basilar part and foramen magnum represent the body and vertebral foramen, respectively; the squamous part, with its external occipital protuberance and crest, perhaps represents the laminae and spinous process; and the condyles and jugular processes are in series with the articular and transverse processes, respectively.

Temporal Bone (see figs. 42–9 and 42–19). All five divisions of the temporal bone are seen on the lower surface of the base (see fig. 42–12). **The temporal bone contains the middle and internal portions of the ear.**

SQUAMOUS PART. The *zygomatic process* extends forward from the squamous part of the temporal bone to form the *zygomatic arch.* (see also p. 352). **The mandibular fossa and articular tubercle are portions of the squamous part of the temporal bone.**

TYMPANIC PART. The tympanic plate is separated from the head of the mandible by a portion of parotid gland. In the mandibular fossa, the tympanic plate is separated from the squamous part of the temporal bone by the *squamotympanic fissure* (see fig. 42–13), the medial portion of which is generally occupied by a small portion of the petrous part (*tegmen tympani*). The fissure is thereby divided into *petrosquamous* and *petrotympanic fissures,* and the latter allows the exit of the chorda tympani.

STYLOID PART. The *styloid process* is separated from the mastoid process by the *stylomastoid foramen,* through which the facial nerve emerges.

MASTOID PART. The mastoid part is fused medially with the petrous part, from which it is developed. The *mastoid process* (see fig. 42–9) projects downward and is grooved medially by the *mastoid notch* for the digastric muscle (see fig. 42–10) and the *groove for the occipital artery.*

PETROUS PART. **The petrous part, shaped like a medially directed pyramid, contains the internal ear and contributes to the boundaries of the middle ear. Laterally, its base is fused with the rest of the temporal bone. The apex of the petrous part of the temporal bone is directed forward and medially between the sphenoid and occipital bones.** The petrous part of the temporal bone presents anterior and posterior surfaces, which are described with the cranial cavity (see pp. 359 and 362), and an inferior surface, which is considered here. The jugular foramen is formed by the petrous part of the temporal bone (*jugular fossa*) and by the occipital bone (*jugular notch*) (see fig. 42–13). **The jugular foramen is related to the carotid canal anteriorly, the styloid process laterally, the transverse process of the atlas posteriorly, and the hypoglossal canal medially. The foramen transmits the internal jugular vein (a continuation of the sigmoid sinus) and the glossopharyngeal, vagus, and accessory nerves,** as well as a tributary (inferior petrosal sinus). The internal jugular is dilated (superior bulb) at its commencement. The tympanic nerve (from cranial nerve 9) and the auricular branch of cranial nerve 10 pierce the skull in or near the jugular fossa. **The carotid canal, a tunnel in the petrous part of the temporal bone, transmits the internal carotid artery and its sympathetic plexus to the cranial cavity. The entrance to the canal lies immediately in front of the jugular foramen.** The canal is close to the internal ear, and the beating of the artery after exertion may be heard as a thundering sound. Further medially, the *qua-*

* This is earlier than was previously believed (B. Ingervall and B. Thilander, Acta Odontol. Scand., *30*:349–356, 1972; B. Melsen, Acta Anat., *83*:112–118, 1972).

drate area of the petrous part of the temporal bone gives origin to the levator veli palatini. The groove between the petrous part of the temporal bone and the greater wing of the sphenoid bone is occupied *in vivo* by the cartilaginous part of the auditory tube. The *foramen lacerum* is a jagged opening at the junction of the petrous part of the temporal bone and the sphenoid and occipital bones. It is closed by fibrous tissue *in vivo* and is a part of the prenatal cartilaginous skull rather than a foramen. The *pterygoid canal* passes from the anterior margin of the foramen lacerum to the pterygopalatine fossa.

Sphenoid Bone (fig. 42–15). **The sphenoid bone consists of a body and three pairs of processes or wings: greater wings, lesser wings, and pterygoid processes** (see also p. 359). The greater wing and pterygoid process are described here.

GREATER WING. *The infratemporal surface of the greater wing* (see figs. 42–12 and 42–13) forms the roof of the infratemporal fossa. It is continuous medially with the lateral pterygoid plate, and it presents two openings into the middle cranial fossa. The anterior and larger is the *foramen ovale,* which transmits the mandibular nerve. The posterior and smaller is the *foramen spinosum* (in front of a spur named the spine of the sphenoid), which transmits the middle meningeal vessels and the meningeal branch of the mandibular nerve. **Medial to the foramen ovale and foramen spinosum, the sphenoid bone is separated from the petrous part of the temporal bone by a groove for the cartilaginous part of the auditory tube.**

PTERYGOID PROCESS. The pterygoid process of each side projects downward from the greater wing, behind the maxilla. Each consists of a lateral and a medial plate separated by the *pterygoid fossa*. The *medial pterygoid plate,* to which the auditory tube is anchored, is prolonged as the *pterygoid hamulus*. Above, the medial plate presents the *scaphoid fossa,* which gives origin to the tensor veli palatini; the muscle hooks around the hamulus to gain the soft palate. The *lateral pterygoid plate* gives origin to the lateral and medial pterygoid muscles (see figs. 42–12 and 42–13).

Choanae. **The nasal cavities are continuous with the nasopharynx through the choanae.** They are two large openings above the posterior margin of the bony palate and are separated from each other by the back of the nasal septum (vomer). Laterally, each is bounded by the medial pterygoid plate. Above is found an overlapping complex of vomer, palatine, and sphenoid bones (see fig. 42–12).

Bony Palate. **The bony palate (i.e., the skeleton of the hard palate) lies in the roof of the mouth and the floor of the nasal cavity. It is formed by the palatine processes of the maxillae in front and by the horizontal plates of the palatine bones behind** (see fig. 42–12). These four processes are united by a cruciform suture (see fig. 42–13). The bony palate is covered below by mucoperiosteum. Behind the incisors, a depression (*incisive fossa*) allows the passage of nasopalatine nerves through *incisive canals* and *foramina*. The posterior border of the bony palate, which presents the *posterior nasal spine,* gives attachment to the soft palate (palatine aponeurosis). Posterolaterally, the bony palate allows the passage of palatine nerves and vessels through *greater* and *lesser palatine foramina* and *canals*. **A cleft palate is one in which the right and left halves have failed to meet in the median plane during development.**

CRANIAL CAVITY

The cranial cavity lodges the brain and its meninges, cranial nerves, and blood vessels. It is roofed by the skull cap, and its floor is the upper surface of the base of the skull. **The floor of the cranial cavity is divisible into three "steps," known as the anterior, middle, and posterior cranial fossae, separated by two prominent bony ledges on each side: the "sphenoidal ridge" (the posterior border of the lesser wing of the sphenoid bone) in front and the "petrous ridge" (the superior border of the petrous part of the temporal bone) behind** (fig. 42–17).

Calvaria (fig. 42–16). In young persons, portions of the coronal, sagittal, and lambdoid sutures, as well as the emissary parietal foramina, can be seen. *Digital impressions* corresponding to gyri of the brain may be visible. A median *sagittal groove* for the superior sagittal sinus runs backward on the internal surface of the vault. Depressions termed *granular pits* are found on each side of the groove, and they lodge lateral lacunae and arachnoid granulations. Numerous vascular grooves for the meningeal vessels are present on the internal surface of the vault.

Anterior Cranial Fossa (fig. 42–17). The frontal lobes of the brain rest on the ethmoid, frontal, and sphenoid bones. The *crista galli* projects up from the ethmoid bone and gives attachment to the falx cerebri. Behind and at each side of the crista galli, the *cribriform plate* of the ethmoid bone transmits the filaments of the olfactory nerves from the nasal mucosa to the olfactory bulbs, which lie on the plate. The ethmoid bone articulates behind with the *jugum sphenoidale,* a part of the sphenoid bone that unites the right and left lesser wings. Laterally, the *orbital plate of the frontal bone* roofs the orbit and ethmoidal air-sinuses and articulates behind with the lesser wing of the sphenoid bones. The sphenoidal ridge of the lesser wing projects into the lateral sulcus of the brain and ends medially in the *anterior clinoid process.*

Middle Cranial Fossa. The floor resembles a butterfly in that a smaller median part is expanded on each side. The *body of the sphenoid bone,* which is united to the occipital bone posteriorly, supports the hypophysis above. **The features of the cranial fossae seen in a median section are the jugum sphenoidale, limbus sphenoidalis, chiasmatic groove, tuberculum sellae, hypophysial fossa, dorsum sellae, and clivus** (fig. 42–18). The shallow *chiasmatic groove* is close to but does not lodge the optic chiasma. The *optic canal* for the optic nerve and ophthalmic artery leads forward and laterally into the orbit. It is bounded by the body of the sphenoid bone and the two roots of the lesser wing. The *sella turcica* is the upper surface of the body of the sphenoid bone between (and including) the *tuberculum sellae* (the posterior limit of the chiasmatic groove) and the *dorsum sellae,* which juts upward and presents a *posterior clinoid process* on each side. The seat of the "Turkish saddle" is the hypophysial fossa, which lodges the hypophysis (pituitary gland) and roofs the sphenoidal air-sinuses. Laterally, the *carotid groove* ascends from the foramen lacerum, then runs forward along the side of the body of the sphenoid bone, and finally ascends medial to the anterior clinoid process. **The carotid groove contains the internal carotid artery, embedded in the cavernous sinus.** Near each end of the tuberculum sellae a *middle clinoid process* may be detectable; it is united by a ligament (in some cases by bone) to the anterior clinoid process (*caroticoclinoid foramen*).

The lateral part of the middle cranial fossa is formed by the greater wing of the sphenoid bone and by the squamous and petrous parts of the temporal bone. It lodges the temporal lobe of the brain. It is limited by the sphenoidal ridge in front and by the petrous ridge behind. These ridges are closely related to venous sinuses (sphenoparietal and superior petrosal, respectively). The *superior orbital fissure* is a slit between the greater and lesser wings of the sphenoid bone. It transmits the oculomotor, trochlear, and abducent nerves, and the branches of the ophthalmic nerve (see fig. 45–3). The *foramen rotundum,* immediately below the medial end of the superior orbital fissure, transmits the maxillary nerve to the pterygopalatine fossa. Behind the foramen rotundum, the *foramen ovale* transmits the mandibular nerve to the infratemporal fossa. Nearby, the *foramen spinosum* transmits the middle meningeal vessels, a groove for which can be traced laterally and forward. The **superior orbital fissure, foramen rotundum, foramen ovale, and foramen spinosum are arranged in a crescent on the greater wing of the sphenoid bone** (see fig. 42–15). Only the last two of these four openings can be seen on the lower surface of the base.

The anterior surface of the petrous part of the temporal bone (fig. 42–19) lodges the trigeminal ganglion on the *trigeminal impression* medially. The abducent nerve bends sharply forward across the apex of the petrous part, medial to the trigeminal ganglion. Lateral to the ganglion, a rounded elevation (*arcuate eminence*) indicates the underlying anterior semicircular canal. Nearby, the *hiatus for the greater petrosal nerve* proceeds toward the foramen lacerum. The tympanic cavity and mastoid antrum are roofed by the petrous part of the temporal bone, this part being called the *tegmen tympani.* The *foramen lacerum,* occupied by fibrous tissue *in vivo,* is at the junction of the petrous part and the sphenoid bone. It marks the end of the carotid canal and the beginning of the carotid groove.

Posterior Cranial Fossa. The hindbrain (cerebellum, pons, and medulla oblongata) occupies the posterior cranial fossa, which is formed by the sphenoid, temporal, parietal, and occipital bones (see fig. 42–17). The fossa is limited above by a dural fold, the tentorium cerebelli, which is attached to the petrous ridge and which separates the cerebellum from the occipital lobes of the brain. Below, the *foramen magnum* is evident, and, above the margin of its front part, the *hypoglossal canal* on each side transmits cranial nerve 12. In front of the foramen magnum, the basilar part of the occipital bone ascends to meet the body of the sphenoid bone. This sloping surface, termed the *clivus* (see fig. 42–18), is related to the pons and medulla. Behind the foramen magnum, the *internal occipital crest* leads up to the *internal occipital protuberance,* near which the superior sagittal, straight, and transverse sinuses form a confluence (see fig. 43–21). A *groove for the transverse sinus* runs laterally on each side and then turns downward

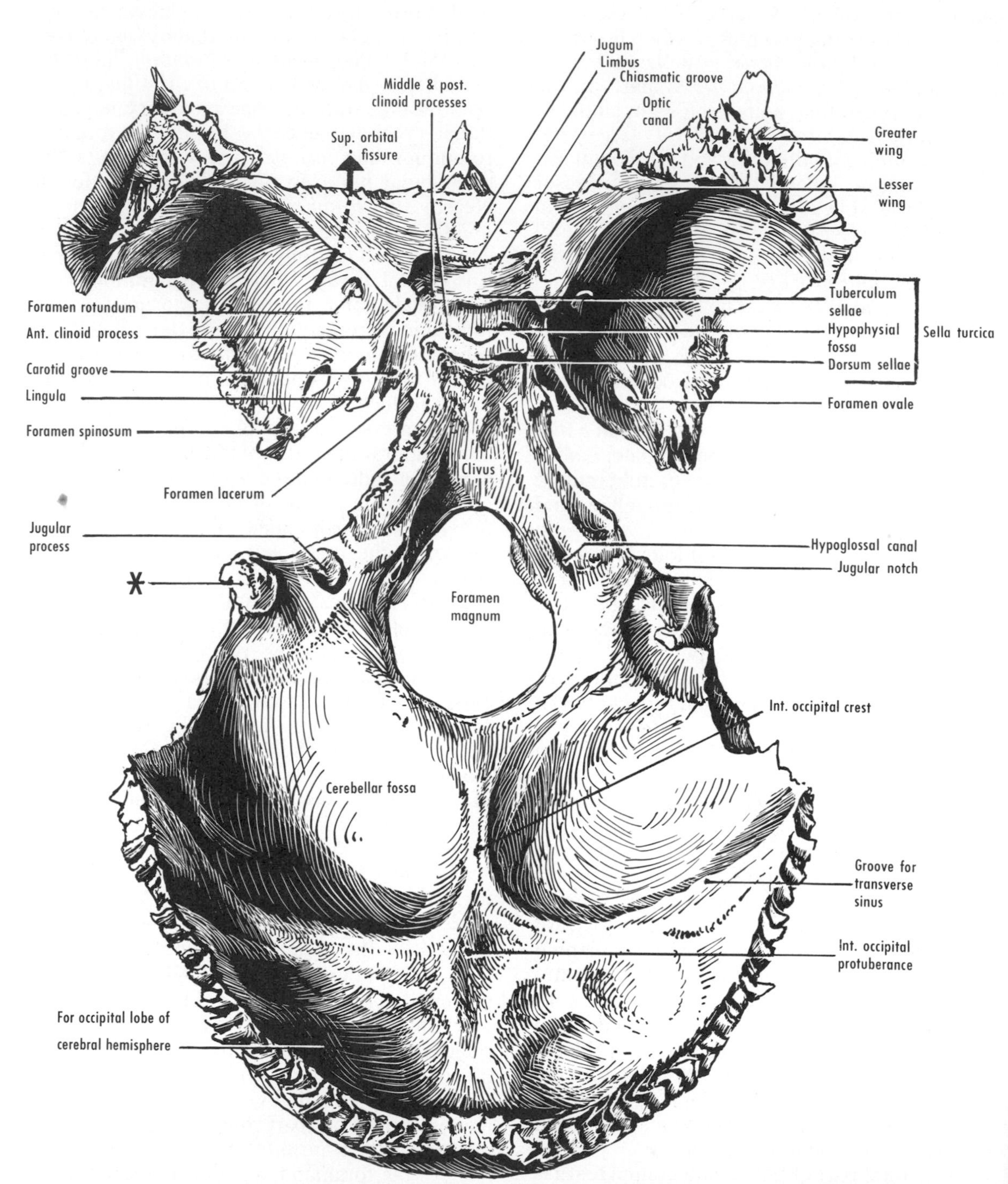

Figure 42–15 Superior aspect of the sphenoid and occipital bones. These two elements, separated by cartilage in the child, become united by bone at about puberty. The body, paired greater wings, and paired lesser wings of the sphenoid can be identified, but the pterygoid processes are not seen in this view. Note the crescentic row of openings in the greater wing: superior orbital fissure (*arrow*), foramen rotundum, foramen ovale, and foramen spinosum. Of these openings, only the last two would be visible from below (see fig. 42–13). The four chief parts of the occipital bone (basilar, two lateral, and squamous) can be seen around the foramen magnum. The area marked with an asterisk articulates with a corresponding area on the temporal bone. The wide gap between the greater wing of the sphenoid and the basilar part of the occipital bone would be occupied by the petrous part of the temporal bone (see fig. 42–19). The apical area at the junction of the three bones, however, remains cartilaginous and, in the dried skull, is known as the foramen lacerum.

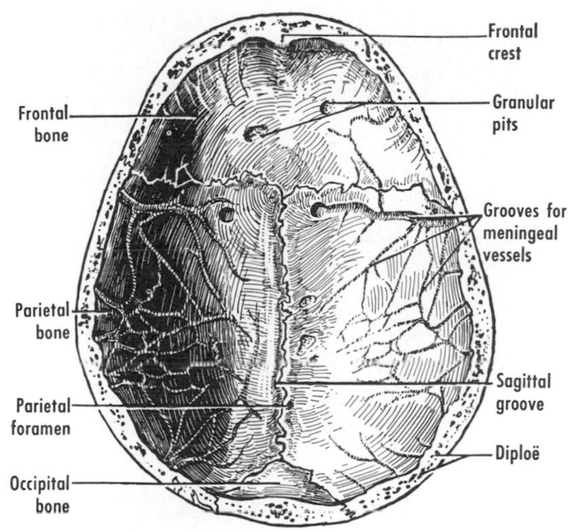

Figure 42–16 Internal aspect of the calvaria.

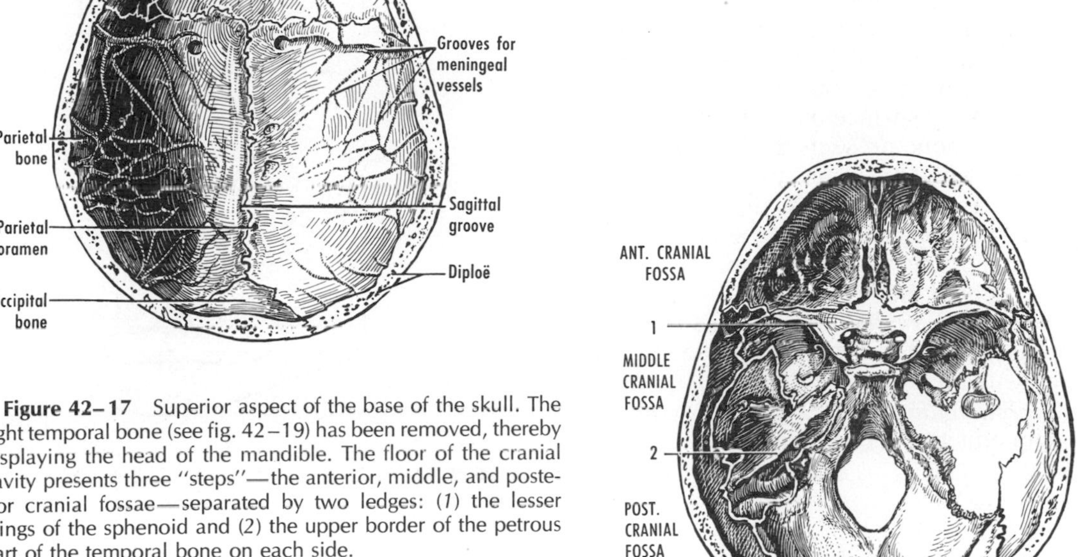

Figure 42–17 Superior aspect of the base of the skull. The right temporal bone (see fig. 42–19) has been removed, thereby displaying the head of the mandible. The floor of the cranial cavity presents three "steps"—the anterior, middle, and posterior cranial fossae—separated by two ledges: (*1*) the lesser wings of the sphenoid and (*2*) the upper border of the petrous part of the temporal bone on each side.

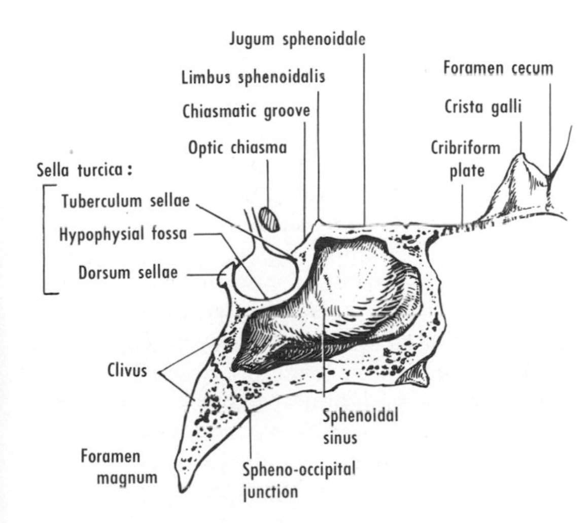

Figure 42–18 Median section through the base of the skull, showing important features of the cranial fossae.

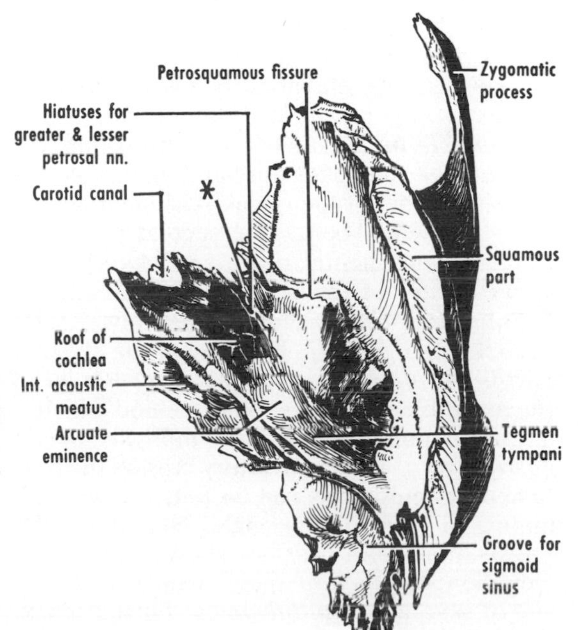

Figure 42–19 Superior aspect of the right temporal bone. (Cf. fig. 42–17.) The asterisk marks the portion of the petrous part (tegmen tympani) that turns downward into the squamotympanic fissure. The trigeminal impression, which lodges the trigeminal ganglion, is immediately behind the apex of the bone and the termination of the carotid canal.

as the *groove for the sigmoid sinus,* which leads medially and forward into the *jugular foramen* (where the sigmoid sinus becomes the internal jugular vein), accompanied by cranial nerves 9, 10, and 11. The cerebellar fossa on each side lies between the transverse and sigmoid grooves and the foramen magnum. The occipital lobes of the brain lie on the occipital bone above the transverse sinuses.

The posterior surface of the petrous part of the temporal bone presents the *internal acoustic meatus,* which transmits the facial and vestibulocochlear nerves and the labyrinthine vessels. The internal meatus is almost directly medial to (in line with) the external acoustic meatus. Behind the internal meatus, a slit (the *aqueduct of the vestibule*) transmits the endolymphatic duct of the internal ear (see fig. 44–8). Below the internal meatus, a notch (the *cochlear canaliculus*) lodges the perilymphatic duct (aqueduct of the cochlea).

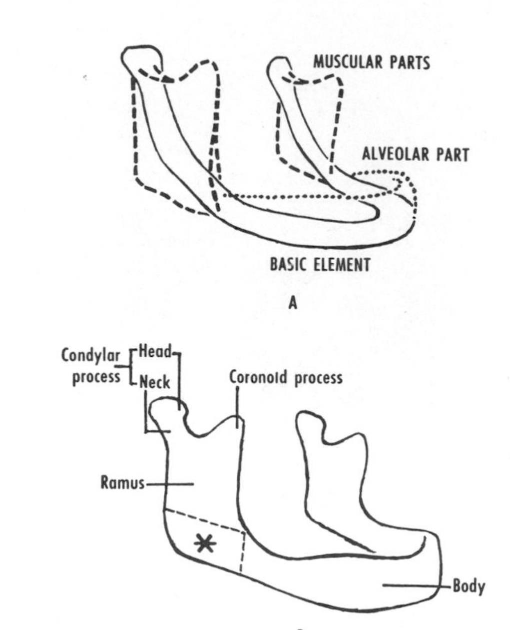

Figure 42–20 Schemes of the mandible. **A** shows the muscular and alveolar parts added to a basic element. **B** shows the main structural portions. The area marked by the asterisk may be classified as a part of either the ramus or the body of the bone. (**A** is based on Symons.)

Mandible

The mandible (lower jaw) presents a body and a pair of rami (fig. 42–20). The obtuse *angle of the mandible* can readily be felt *in vivo:* its most prominent point is called the *gonion.* The mandible develops bilaterally, but bony union between the two halves occurs during the first postnatal year.

Body. The U-shaped body (fig. 42–21) presents external and internal surfaces and superior and inferior borders. The line of fusion of the two halves of the mandible at the *symphysis menti* is generally visible on the *external surface.* An elevation below, termed the *mental protuberance,* leads laterally to a *mental tubercle* on each side, and, from this, the *oblique line* runs backward and upward to the anterior border of the ramus. The *mental foramen,* frequently below the second premolar, transmits the mental nerve and vessels. The *upper border* of the body of the mandible is the *alveolar part* and contains the lower teeth in sockets, or *alveoli.* The edge of the alveolar part is the *alveolar arch.* The *lower border* of the mandible is its *base,* on or behind which is a rough depression near the symphysis, the *digastric fossa.* **The facial artery crosses the base (where its pulsations can be felt) a few centimeters in front of the angle.** Near the symphysis, the *internal surface* shows an irregular elevation, the *mental spine,* which may consist of several *genial tubercles.* Further back, the *mylohyoid line* runs backward and upward to a point behind the third molar. Below the line, the *submandibular fovea* lodges the submandibular gland, whereas, further forward, the *sublingual fovea* (for the sublingual gland) lies above the mylohyoid line.

Ramus. The ramus (fig. 42–21) is a quadrilateral plate with lateral and medial surfaces. The *lateral surface* gives insertion to the masseter. The *medial surface* shows the *mandibular foramen,* which leads downward and forward into the mandibular canal and transmits the inferior alveolar nerve and vessels. The foramen is limited medially by a projection, the *lingula,* to which the sphenomandibular ligament is attached. The mandibular canal gives off a side canal that opens at the mental foramen and then runs as far as the median plane. The *mylohyoid groove* (for the mylohyoid nerves and vessels) begins behind the lingula and descends to the submandibular fovea. The medial surface near the angle gives insertion to the medial pterygoid muscle. The sharp anterior border of the ramus can be felt inside the mouth. The rounded posterior border is related to the parotid gland. The concave upper border of the ramus, the *mandibular notch,* is bounded in front by the *coronoid process,* into which the temporalis is inserted.

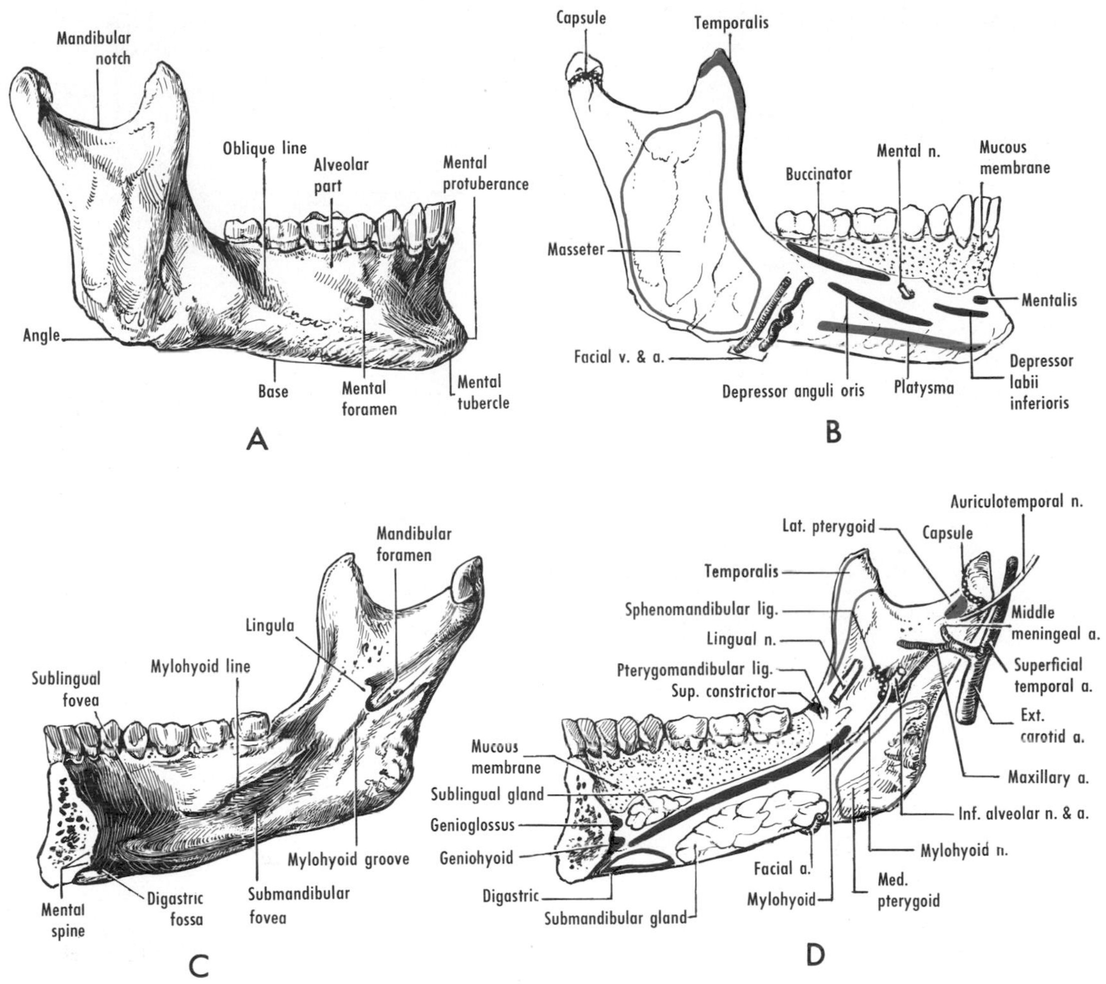

Figure 42–21 **A** and **B**, Right lateral aspect of the mandible. **C** and **D**, Medial aspect of the right half of the mandible. **A** and **C** show the main structural features. **B** and **D** show attachments and relations and are based on *Frazer's Anatomy of the Human Skeleton*.

The notch is limited behind by the *condylar process,* that is, the head and neck of the mandible. The *head,* or *condyle,* covered with fibrocartilage, articulates indirectly with the squamous part of the temporal bone at the temporomandibular joint. The lateral end of the condyle can be felt *in vivo*. The *neck* gives insertion to the lateral pterygoid muscle.

Development of Skull (fig. 42–22)

The vault and portions of the base ossify intramembranously, whereas most of the base ossifies endochondrally. Postnatally, growth of the skull takes place at sutures and at the spheno-occipital junction, nasal septum, and condylar processes of the mandible.

Fonticuli, or *fontanelles,* are temporary membranous areas that bridge the gaps between the angles or margins of some of the ossifying bones of the skull. Usually six are present at birth, and these are situated at the angles of the parietal bones. **The anterior fontanelle, the largest, is commonly seen to pulsate (because of the cerebral arteries) and is readily palpable in an infant. It is usually obliterated by the age of two years.** It may be used to palpate the position of the fetal head, to estimate abnormal intracranial pressure in infancy, to assess cranial development, and to obtain blood from the superior sagittal sinus.

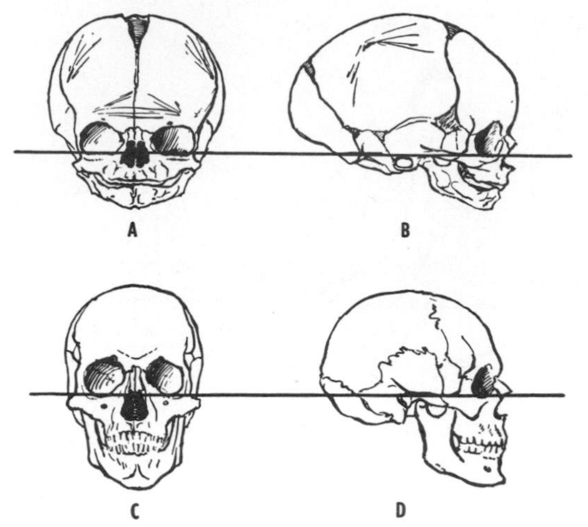

Figure 42–22 The growth of the skull. A and B, Neonatal skull. C and D, Adult skull. The scale used for the adult is half that of the neonatal skull. The horizontal lines indicate the orbitomeatal plane. In the infant's skull, although the vault appears large, the facial region (chiefly below the horizontal line) is relatively small. The jaws, nasal cavities and paranasal sinuses are all small, and the orbits and teeth are close together. In the adult, however, the horizontal line approximately bisects the vertical height of the skull. Note the fontanelles in A and B. (Based chiefly on the work of J. C. Brash.)

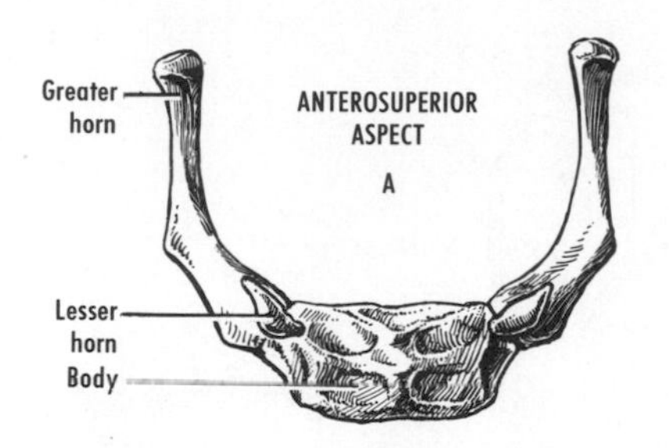

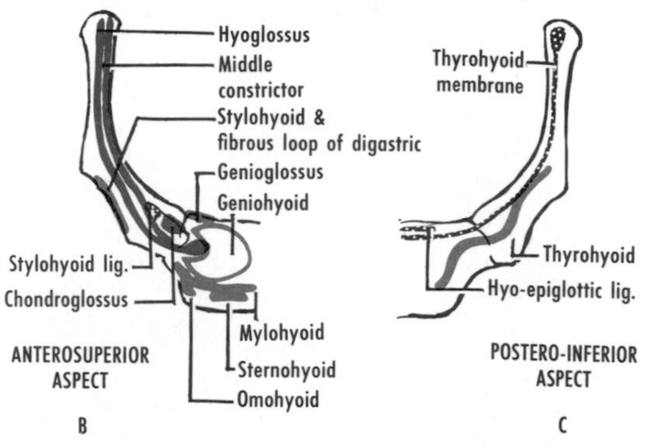

Figure 42–23 The hyoid bone. A, Anterosuperior aspect, showing the parts of the bone. B and C, Anterosuperior and postero-inferior aspects of right half, showing attachments. (After Frazer.)

HYOID BONE (fig. 42–23)

The hyoid bone lies in the front of the neck at the upper limit of the larynx. It is suspended from the skull by the stylohyoid ligaments. It presents a body and paired greater and lesser horns. Each *greater horn* or *cornu* projects backward and upward from the body, to which it is united by cartilage or bone. **When the neck is relaxed, the two greater horns can be gripped between index finger and thumb, and the hyoid bone can be moved from side to side.** Each *lesser horn* or *cornu* projects upward from the junction of the body and the greater horn and gives attachment to the stylohyoid ligament.

ADDITIONAL READING

Enlow, D. H., *Handbook of Facial Growth,* W. B. Saunders Company, Philadelphia, 1975. A nicely illustrated account of the developing skull.

QUESTIONS

42–1 What is the cranium?

42–2 What is the orbitomeatal plane?

42–3 Where may calcified areas be found normally in the interior of the head?

42–4 What forms the skeleton of the nasal septum?

42–5 What are the main parts of the temporal bone?

42–6 How far inferiorly do the cerebral hemispheres extend in terms of craniocerebral topography?

42–7 What are the limits of the temporal fossa?

42–8 Where is the pterion and what is its importance?

42–9 Where is the infratemporal fossa?

42–10 What are the main parts of the occipital bone?

42–11 In which ways does the occipital bone resemble a vertebra?

42–12 To which part of the temporal bone do the mandibular fossa and articular tubercle belong?

42–13 What is the petrous part of the temporal bone?

42–14 What are the main parts of the sphenoid bone?

42–15 What lies below and between the petrous part of the temporal bone and the greater wing of the sphenoid bone?

42–16 What are the choanae?

42–17 What forms the bony palate?

42–18 What is cleft palate?

42–19 What forms the "sphenoidal ridge" and the "petrous ridge"?

42–20 What is the calvaria?

42–21 What is the meaning of *sella turcica* and *clinoid processes?*

42–22 What are the main parts of the mandible, and how does the ramus terminate above?

42–23 List some important sites of growth of the skull.

42–24 Where are the main fontanelles situated?

42–25 What are the main parts of the hyoid bone?

42–26 Review the cranial exits of the cranial nerves.

43 THE BRAIN, CRANIAL NERVES, AND MENINGES

BRAIN

The nervous system in general is described in chapter 3. The divisions of the brain are summarized here in table 43–1. This chapter is limited to a brief description of the gross structure of the brain, an account of the ventricles, and some general remarks on the cranial nerves, the meninges, and the blood supply. Further details, including information about the internal structure of the brain, should be sought in books and atlases on neuroanatomy.

TABLE 43–1 DIVISIONS OF THE BRAIN (ENCEPHALON)

	Divisions	Cavities
Prosencephalon	Telencephalon	Two lateral ventricles & rostral part of 3rd ventricle
	Diencephalon†	Most of 3rd ventricle
Mesencephalon*		Aqueduct
Rhombencephalon	Metencephalon: Cerebellum, Pons*	Fourth ventricle & central canal
	Myelencephalon: Medulla*	

* Brain stem
† Frequently included in brain stem

GROSS STRUCTURE OF BRAIN

The brain comprises, from below upward, the hindbrain, midbrain, and forebrain.

In their attachment to the brain, the first two cranial nerves are associated with the forebrain, nerves 3 and 4 with the midbrain, and nerves 5 to 12 with the hindbrain.

Hindbrain

The hindbrain, or rhombencephalon, consists of the medulla oblongata, the pons, and the cerebellum.

Medulla Oblongata (fig. 43–1). The uppermost part of the spinal cord expands, passes through the foramen magnum, and becomes the medulla oblongata. The medulla rests anteriorly on the basilar part of the occipital bone. Posteriorly, the medulla is largely covered by the cerebellum (see fig. 43–3).

The lower half of the medulla contains the continuation of the central canal of the spinal cord, which, in the upper half of the medulla, widens to become the fourth ventricle.

The medulla presents an *anterior median fissure,* the lower part of which is interrupted

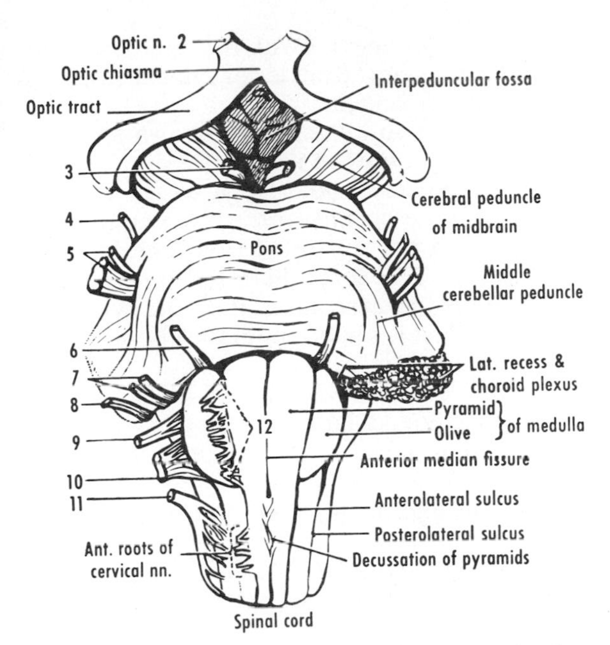

Figure 43–1 The brain stem, anterior aspect, showing the cranial nerves.

by the *decussation of the pyramids*, where about 75 to 90 per cent of the descending pyramidal fibers cross the median plane. The portion of the medulla adjacent to the upper part of the anterior median fissure on each side is termed the *pyramid*. It contains the fibers of the *pyramidal* (*corticospinal*) *tract*. Lateral to each pyramid, an elevation termed the *olive* is found.

The *hypoglossal* (*twelfth cranial*) *nerve* emerges from the medulla between the pyramid and the olive, whereas the *accessory* (*eleventh cranial*), *vagus* (*tenth cranial*), and *glossopharyngeal* (*ninth cranial*) *nerves* emerge posterolateral to the olive.

The dorsal aspect of the medulla presents a *posterior median sulcus*. On each side, two tracts from the spinal cord (the *fasciculus gracilis* medially and *fasciculus cuneatus* laterally) terminate in eminences known as the *gracile* and *cuneate tubercles*, respectively. Higher up, the lower part of the fourth ventricle is bounded laterally by the *inferior cerebellar peduncles*, which comprise fibers connecting the medulla and spinal cord with the cerebellum (fig. 43–2).

The medulla contains very important nerve centers associated with functions such as respiration and circulation.

Pons (fig. 43–1). The pons lies between, and is sharply demarcated from, the medulla and midbrain. It is situated in front of the cerebellum and appears superficially to bridge (hence its name) the two cerebellar hemispheres. As seen from the front, the transverse fibers of the pons form the *middle cerebellar peduncle* on each side and enter the cerebellum. The middle peduncle actually comprises fibers that connect one cerebellar hemisphere with the contralateral half of the pons.

The front of the pons rests on the basilar part of the occipital bone and on the dorsum sellae. The pons is grooved longitudinally in front, and this groove is frequently occupied by the basilar artery.

The *vestibulocochlear* (*eighth cranial*), *facial* (*seventh cranial*), and *abducent* (*sixth cranial*) *nerves* emerge in the groove between the pons and the medulla (see fig. 43–1). Higher up, the *trigeminal* (*fifth cranial*) *nerve* emerges from the side of the pons by a large sensory and a smaller motor root (fig. 43–3).

The back of the pons forms the floor of the upper part of the fourth ventricle (fig. 43–2), which is bounded laterally by the *superior cerebellar peduncles*. Each superior peduncle comprises fibers that connect the cerebellum with the midbrain and thalamus.

The *trigeminal* (*fifth cranial*) *nerve* is large and complicated. It is sensory from the face,

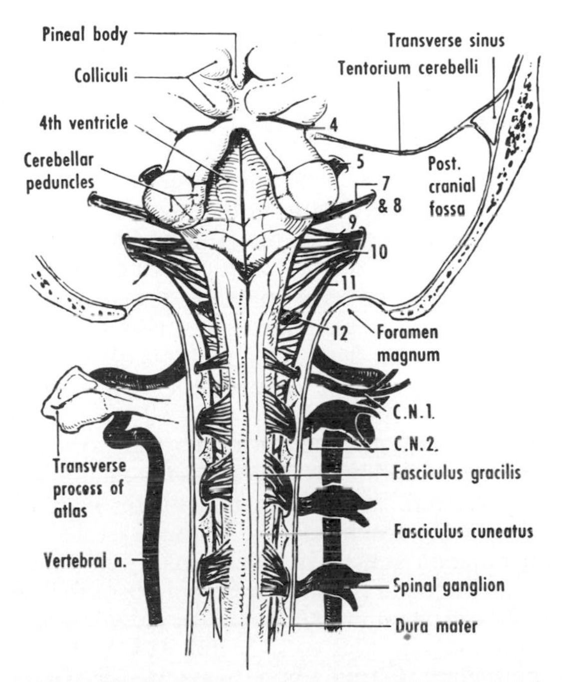

Figure 43–2 Posterior aspect of the brain stem and the upper part of the spinal cord after removal of the cerebellum, which displays the floor of the fourth ventricle. The vertebral artery is visible on each side, together with certain cranial and spinal nerves.

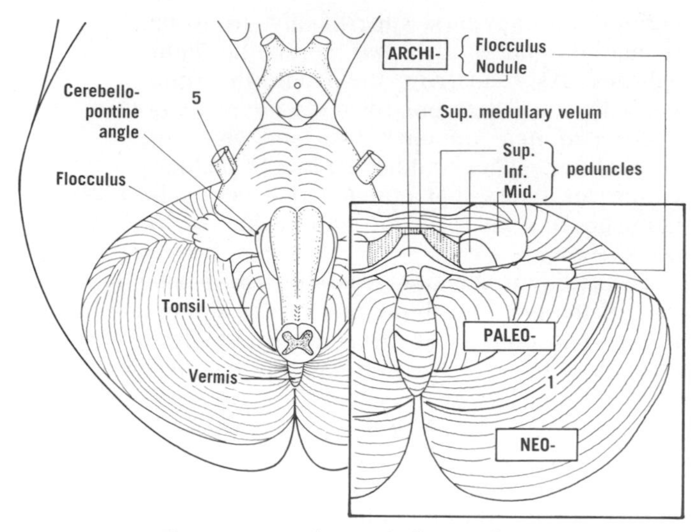

Figure 43–3 The cerebellum, inferior aspect. In the inset, the brain stem has been removed by sectioning of the cerebellar peduncles. *1*, fissura prima; *5*, trigeminal nerve.

teeth, mouth, and nasal cavity, and motor to the muscles of mastication. It emerges from the side of the pons as a *sensory* and a *motor root,* generally with some intermediate fibers. The two portions proceed from the posterior to the middle cranial fossa by passing beneath the attachment of the tentorium cerebelli to the petrous part of the temporal bone and also by passing usually beneath the superior petrosal sinus. The sensory root expands into a large, flat *trigeminal* (*semilunar*) *ganglion,* which contains the cells of origin of most of the sensory fibers. The ganglion overlies the foramen lacerum, and the roots of the nerve occupy an impression on the anterior surface of the petrous part of the temporal bone near its apex. Most of the ganglion is enclosed in the cavity of the dura known as the *cavum trigeminale.* The ganglion gives rise to three large divisions: the ophthalmic, maxillary, and mandibular nerves. The motor root, which contains proprioceptive as well as motor fibers, continues beneath the ganglion and joins the mandibular nerve. The ganglion can be "blocked" by passing a needle through the mandibular notch and the foramen ovale and injecting an anesthetic. The sensory root may be sectioned in the middle cranial fossa for the relief of trigeminal neuralgia (*tic douloureux*).

Cerebellum (figs. 43–3 and 43–16). The cerebellum is situated on the back of the brain stem, to which it is attached by the three cerebellar peduncles on each side. The inferior peduncles connect the cerebellum with the medulla; the middle connect it with the pons; and the superior connect it with the midbrain. The cerebellum is in the posterior cranial fossa. It comprises a median portion, termed the *vermis,* and two lateral parts known as *cerebellar hemispheres.* Like the cerebral hemispheres, the cerebellum has a cortex of gray matter. The *cerebellar cortex* is folded to form *folia,* which are separated from one another by *fissures.* The cerebellum is connected by tracts with the cerebral cortex and with the spinal cord. It is important in the coordination of muscular activities.

The *archicerebellum* (figs. 43–3 and 43–16) is the flocculonodular lobe, which is vestibular in function (i.e., concerned with equilibration). The *paleocerebellum* comprises most of the vermis and the adjacent parts of the hemispheres and is spinal in function (i.e., concerned with posture and tonus). The *neocerebellum* consists of the lateral parts of the hemispheres and is corticopontine in function (i.e., concerned with the control of voluntary movements).

Midbrain (see figs. 43–1 and 43–16)

The midbrain, or mesencephalon, connects the hindbrain with the forebrain. It is located in the tentorial notch of the dura mater (see fig. 43–15). It consists of a ventral part, the cerebral peduncles, and a dorsal part, the tectum.

The *cerebral peduncles* are two large bundles that converge as they descend from the cerebral hemispheres, in which each is continuous with a band of white matter termed the *internal capsule.* The front portion of each peduncle is termed the *crus cerebri,* whereas the back portion is the *tegmentum.* The upper part of each peduncle is crossed by the corresponding optic tract. The right and left optic tracts emerge from the optic chiasma, which is formed by the junction of the two optic nerves. The depression behind the chiasma and bounded by the optic tracts and the cerebral peduncles is termed the *interpeduncular fossa* (see fig. 43–7). The interpeduncular fossa contains, from before backward: (1) the tuber cinereum and the infundibular stem of the hypophysis, (2) the mamillary bodies, and (3) the posterior perforated substance.

The *oculomotor* (*third cranial*) *nerve* emerges at the upper border of the pons and at the medial border of the corresponding cerebral peduncle.

The *tectum,* or posterior part of the midbrain, consists of four hillocks known as the *superior* and *inferior colliculi* (see fig. 43–2). The superior colliculi are concerned with visual functions, the inferior with auditory functions. The pineal body is attached to the forebrain above the superior colliculi.

The *trochlear* (*fourth cranial*) *nerve* decussates and emerges from the dorsal aspect of the midbrain below the corresponding inferior colliculus (see fig. 43–2).

The midbrain is traversed by the aqueduct, which connects the fourth with the third ventricle (see fig. 43–9).

The medulla, pons, midbrain, and (frequently) diencephalon (that part of the forebrain that is adjacent to the midbrain) are collectively known as the brain stem (see fig. 43–1).

Forebrain (figs. 43–4 to 43–6)

The forebrain, or prosencephalon, comprises a smaller part, the diencephalon, and a massive portion, the telencephalon.

Diencephalon. The diencephalon largely bounds the third ventricle. A small portion of the third ventricle, however, is telencephalic.

The diencephalon includes the (1) thalami, (2) medial and lateral geniculate bodies, (3) pineal body and habenulae, (4) hypothalamus, and (5) subthalamus.

The *thalami* are two large masses of gray matter situated one on each side of the third ventricle. Each thalamus includes many nuclei and acts as an important sensory correlation center, but each has motor functions also.

The *medial* and *lateral geniculate bodies* are two elevations on each side of the colliculi, to which they are connected. They lie under cover of the posterior portion of the thalamus.

The *pineal body,* or *epiphysis,* is located below the splenium of the corpus callosum (see fig. 43–16*A*) and is frequently visible radiographically.

The term *hypothalamus* is restricted functionally to the anterior part of the floor and the lower part of the lateral walls of the third ventricle. This region is concerned with autonomic and neuro-endocrinological functions. Anatomically, certain adjacent areas are generally included: the optic chiasma, *tuber cinereum* (to which the infundibular stem of the hypophysis is attached), the hypophysis, and the *mamillary bodies* (see fig. 43–7).

Telencephalon. The term *telencephalon* is virtually synonymous with *cerebral hemispheres* (figs. 43–4 to 43–6). The term *cerebrum,* however, refers either to the brain as a whole or to merely the forebrain or the forebrain and midbrain together. Each hemisphere contains a cavity known as the lateral ventricle.

As seen from above, the cerebral hemispheres conceal the other parts of the brain from view. Each hemisphere presents a superolateral, a medial, and an inferior surface. The right and left cerebral hemispheres are partly separated from each other by the *longitudinal fissure,* which is occupied by a fold of dura mater, the falx cerebri. The *corpus callosum* (figs. 43–6 and 43–16*A*), found in the depths of the longitudinal fissure, is a bundle of fibers connecting the hemispheres. It forms the roof of the central part and of the anterior horn of the lateral ventricle of each side. It is curved sagittally and consists, from before backward, of the *rostrum, genu, trunk,* and *splenium.*

Each hemisphere presents *frontal, occipital,* and *temporal poles.* These poles are located, respectively, in the anterior, posterior, and middle cranial fossae, and they are related to the frontal and occipital bones and the greater wing of the sphenoid bone.

The gray matter of the surface of each hemisphere is termed the *cerebral cortex.* It is folded or convoluted into *gyri,* which are separated from each other by *sulci.* The pattern is variable, and it is necessary to remove the pia-arachnoid to identify individual gyri and sulci.

The *lateral sulcus* begins on the inferior surface of the brain. It extends laterally and, on gaining the superolateral surface of the hemisphere, proceeds backward between (1) the frontal and parietal lobes and (2) the temporal lobe. (The *posterior ramus* is being referred to here; small *anterior* and *ascending rami* arise from the stem of the lateral sulcus where the sulcus reaches the superolateral surface of the hemisphere.) A portion of the cerebral cortex termed the *insula* lies buried in the depths of the lateral sulcus.

In disorders of speech (*aphasia*), either a portion of the frontal lobe or a larger portion of the temporal lobe or both, adjacent to the lateral sulcus, are frequently involved, usually on the left side of brain (fig. 43–6*A*).

The *central sulcus* begins on the medial surface of the hemisphere and, on gaining the superolateral surface, descends between the frontal and parietal lobes. The area of cortex immediately in front of the central sulcus is known as the motor area and is concerned with muscular activity, mostly in the opposite half of the body. The contralateral control

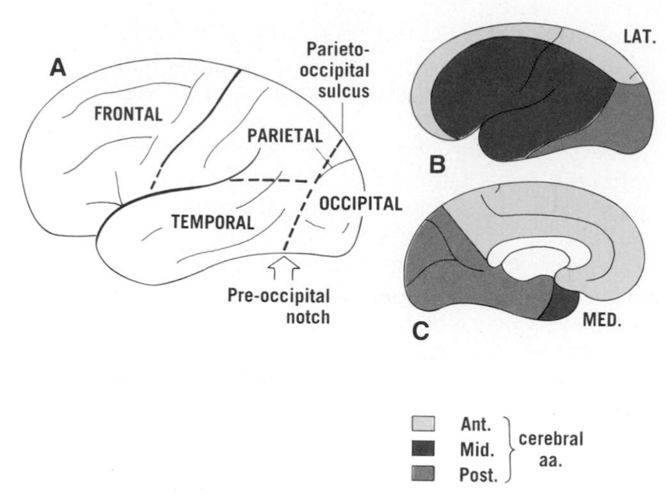

Figure 43–4 **A**, The lobes of the brain, left lateral aspect. **B** and **C**, The territories supplied by the cerebral arteries, lateral and medial aspects of the left cerebral hemisphere.

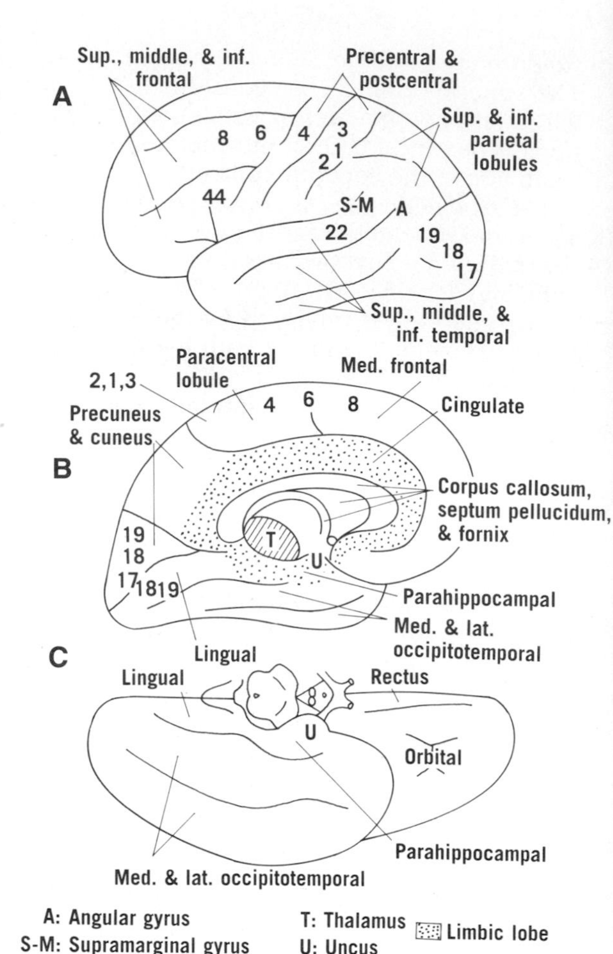

Figure 43–6 Gyri. **A**, **B**, and **C**: lateral, medial, and inferior aspects, respectively, of the left cerebral hemisphere. The chief speech area (of Broca) is mainly area 44 and is usually on the left side of the brain. A receptive speech area (of Wernicke) is found mainly in area 22.

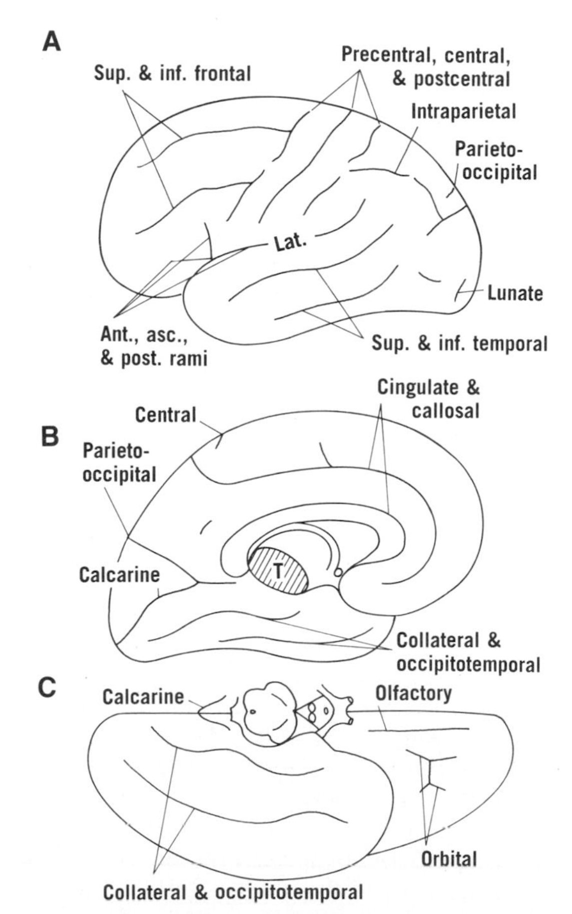

Figure 43–5 Sulci. **A**, **B**, and **C**: lateral, medial, and inferior aspects, respectively, of the left cerebral hemisphere. In **B**, *T* represents the cut surface of the thalamus.

may be demonstrated by artificial stimulation of this area (particularly of that part known as the *precentral gyrus*, or area 4), as a result of which movements of the opposite half of the body take place. Furthermore, the body is represented in an inverted position in the motor area. That is, stimulation of the upper part of the motor area gives rise predominantly to movements of the opposite lower limb; stimulation of the middle part, to movements of the upper limb; and stimulation of the lower part, to movements of the head and neck. The area of cortex immediately behind the central sulcus (the *postcentral gyrus*) is an important primary receptive area, to which afferent pathways project by means of relays in the thalamus.

The cortex of each cerebral hemisphere is arbitrarily divided into frontal, parietal, occipital, and temporal lobes. The *frontal lobe* is bounded by the central and lateral sulci. It lies in the anterior cranial fossa. The *parietal lobe* extends from the central sulcus in front to an arbitrary line behind (between a groove above, the *parieto-occipital sulcus,* and a vague indentation below, the *pre-occipital notch*). The *occipital lobe* lies behind this line. The *temporal lobe* is situated in front of this line and below the lateral sulcus. It lies in the middle cranial fossa. The gyri and sulci characteristic of these lobes are shown in figures 43–5 and 43–6. The *calcarine sulcus* is on the medial surface of the occipital lobe but may extend onto the superolateral surface of the hemisphere. When the cerebellum is displaced, portions of each of the four lobes can be seen also on the medial and inferior surfaces of the hemisphere. The occipital lobe is concerned especially with vision.

The *olfactory* (*first cranial*) *nerves* are groups of nerve filaments that, on leaving the nose and passing through the base of the skull (cribriform plate of the ethmoid bone), end in the olfactory bulbs. Each *olfactory bulb* lies on the inferior aspect of the corresponding frontal lobe and gives rise to an *olfactory tract* (fig. 43–7) that runs backward and is attached to the brain.

The *optic* (*second cranial*) *nerves* leave the orbits through the optic canals and unite to form the *optic chiasma* (fig. 43–7). The chiasma gives rise to the right and left *optic tracts,* which proceed backward and around the cerebral peduncles. The optic chiasma and

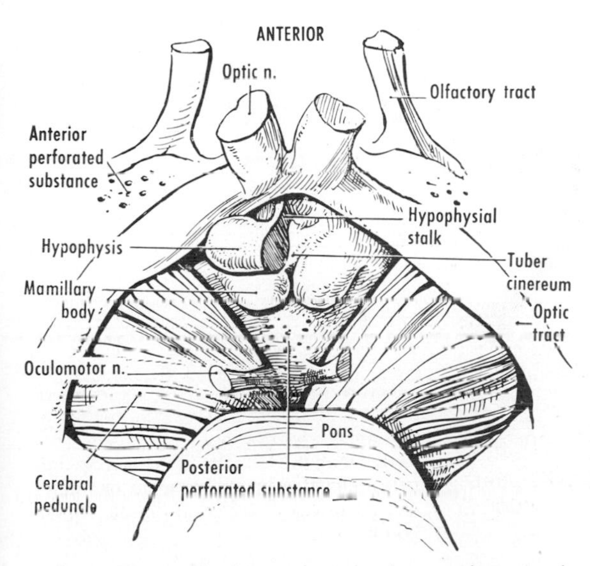

Figure 43–7 The interpeduncular fossa and surroundings, antero-inferior aspect. The left half of the hypophysis has been removed. (From a photograph by David L. Bassett, M.D.).

the interpeduncular fossa are contained within a very important arterial anastomosis known as the circulus arteriosus. The infundibular stem of the neurohypophysis emerges from the tuber cinereum in the interpeduncular fossa, in front of the mamillary bodies. The area immediately anterolateral to each optic tract is pierced by branches of the anterior and middle cerebral arteries and is known as the *anterior perforated substance.*

The *basal nuclei* (or "ganglia") are certain masses of gray matter within the white substance of the cerebral hemispheres, especially the corpus striatum, amygdaloid body, and claustrum.

The *corpus striatum* comprises the caudate and lentiform nuclei. The *caudate nucleus* bulges into the lateral ventricle and presents a head, a body, and a tail. It has an arched form, for which reason it is often seen twice in a section. The *head* lies anteriorly, behind the genu of the corpus callosum; the *body* extends backward, above and lateral to the thalamus; and the tail of the nucleus curves downward and forward into the temporal lobe to end in the *amygdaloid body.* The *lentiform* (or *lenticular*) *nucleus* lies lateral to the head of the caudate nucleus and to the thalamus. In front, it is connected with the head of the caudate nucleus by bars of gray matter, hence the name *corpus striatum* for the two nuclei. The lateral part of the lentiform nucleus, known as the *putamen,* is related laterally to the *claustrum* and the *insula.* The two medial parts of the lentiform nucleus are called the *globus pallidus.*

The *internal capsule* is a broad band of white matter situated between (1) the lentiform nucleus laterally and (2) the head of the caudate nucleus and the thalamus medially. The internal capsule consists of an *anterior limb* (between the lentiform and caudate nuclei), a *genu,* a *posterior limb* (between the lentiform nucleus and the thalamus), and *retrolentiform* and *sublentiform parts* (behind and below the lentiform nucleus, respectively).

The fibers of the internal capsule, on being traced upward, spread out in the hemisphere to form a fan-shaped arrangement termed the *corona radiata.* The fibers of the corona are intersected by those of the corpus callosum.

Craniocerebral Topography (fig. 43–8)

Considerable variation occurs in the precise relations of the brain to the skull, so only an

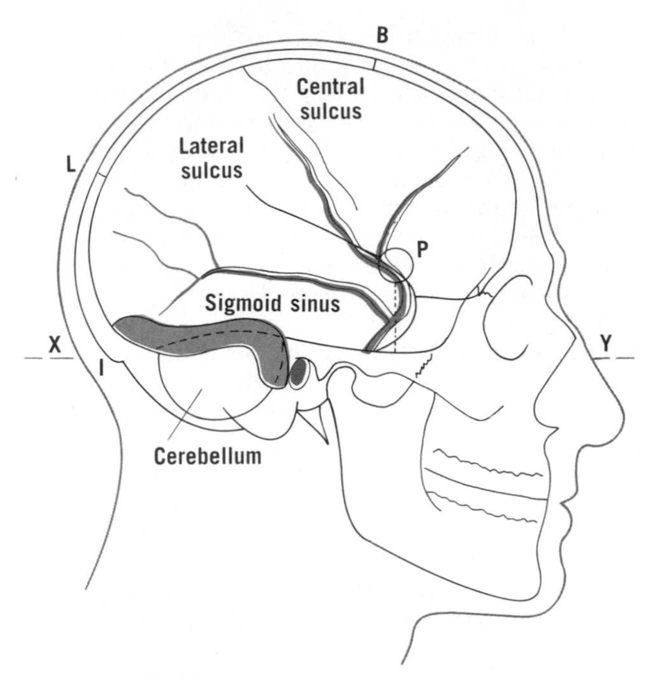

Figure 43–8 Craniocerebral topography. The middle meningeal artery proceeds from the middle of the zygomatic arch upward (*dotted line*) about 4 to 5 cm to the pterion (*P*). The external acoustic meatus is shown in black. *B*, bregma; *I*, inion; *L*, lambda; *XY*, orbitomeatal plane. (After von Lanz and Wachsmuth.)

approximate localization of the parts of the brain is possible on the surface of the body.

The inferior limit of the cerebral hemisphere lies above the eyebrow, zygomatic arch, external acoustic meatus, and external occipital protuberance. Hence **the hemispheres lie above the orbitomeatal plane. A considerable portion of the cerebellum, however, lies below the level of that plane.**

The central sulcus begins at 1 cm behind the vertex, that is, behind the midpoint of a line on the head between the nasion and the inion. The sulcus runs downward, forward, and laterally for about 10 cm toward the midpoint of the zygomatic arch. The sulcus makes about three quarters of a right angle with the median plane.

The lateral sulcus, on the superolateral surface of the hemisphere, extends from the pterion backward and slightly upward and ends below the parietal eminence.

Ventricles (figs. 43–9 to 43–11 and 43–16)

The two lateral ventricles (fig. 43–9) communicate with the third ventricle by an inter-

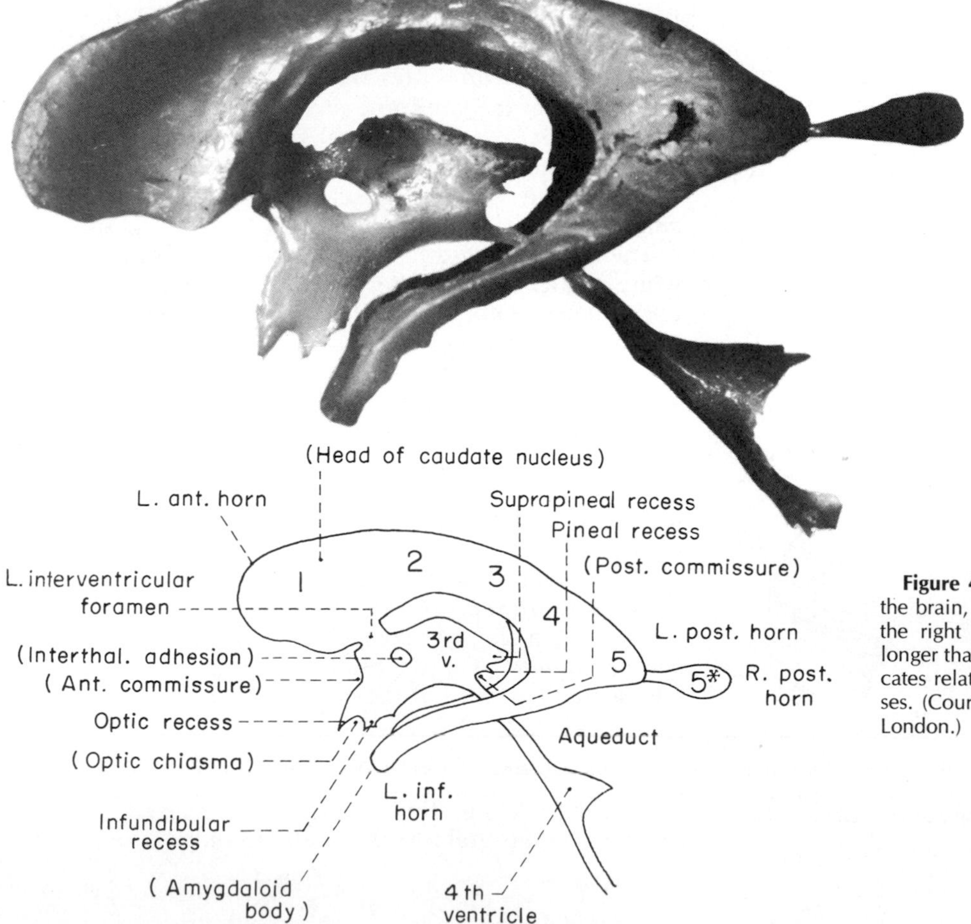

Figure 43–9 Cast of the ventricles of the brain, left lateral aspect. In this brain, the right posterior horn is considerably longer than the left. The key drawing indicates related solid structures in parentheses. (Courtesy of David Tompsett, Ph.D., London.)

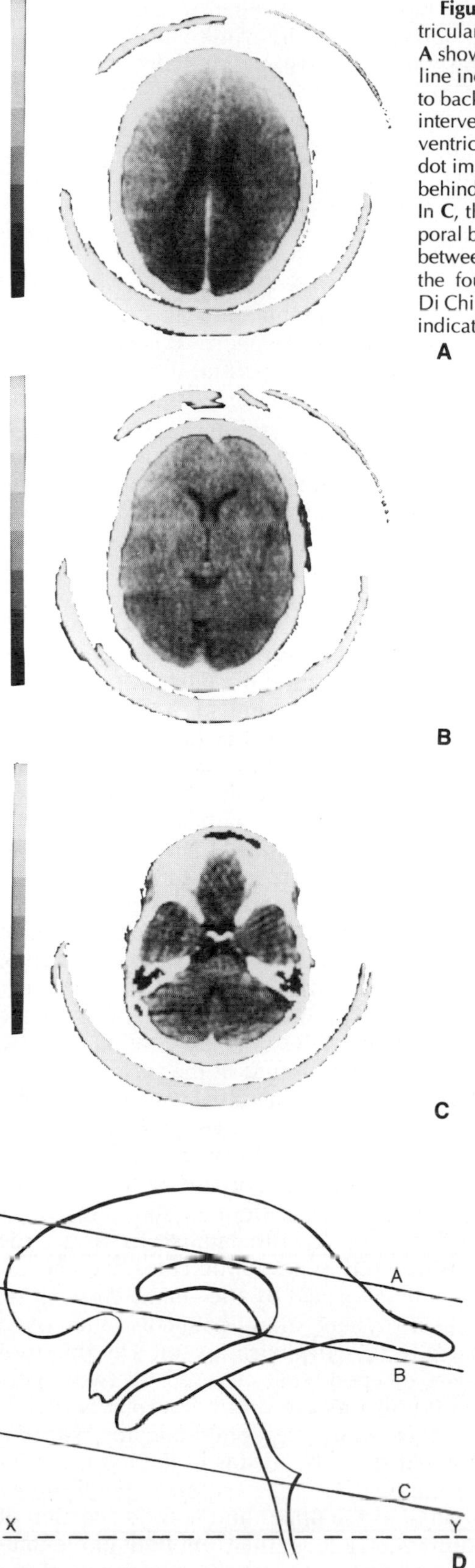

Figure 43–10 Computerized axial tomograms of the head, showing the ventricular system as seen in horizontal sections. **A** and **B** are from the same subject. **A** shows the body and posterior horn of each lateral ventricle. The median white line indicates the falx cerebri. The following structures are visible in **B**, from front to back: the anterior horns curved around the head of each caudate nucleus; the interventricular foramina, barely visible between the anterior horns and the third ventricle, which appears as a median slit; a calcified pineal body, seen as a white dot immediately posterior to the third ventricle; and a transverse, crescentic area behind this, which is the subarachnoid space behind the tectum of the midbrain. In **C**, the lesser wings of the sphenoid bone and the petrous portions of the temporal bones delimit the cranial fossae. The dorsum sellae appears as a white band between the shadows of the temporal lobes. Between the cerebellar hemispheres, the fourth ventricle is visible as a dark, inverted U. (Courtesy of Giovanni Di Chiro, M.D., Bethesda, Maryland.) In **D**, the approximate planes of section are indicated. *XY*, the orbitomeatal plane.

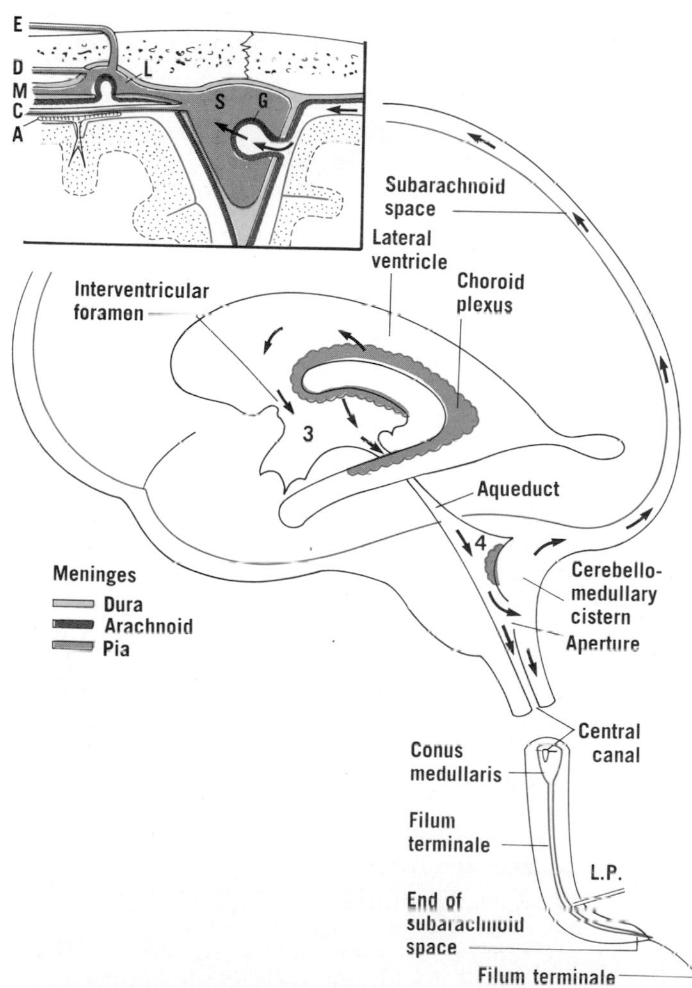

Figure 43–11 The course of the cerebrospinal fluid (C.S.F.). Arrows lead from the choroid plexuses of the lateral and third ventricles toward the aqueduct. The fluid thereby formed is joined by that produced in the fourth ventricle and passes through the median aperture to the cerebello medullary cistern of the subarachnoid space. The fluid then extends (1) upward around the brain and (2) downward around the spinal cord. The inset (a coronal section at the sagittal suture) shows the drainage of the C.S.F. into the venous system (superior sagittal sinus, *S*, and lateral lacunae, *L*, by way of arachnoid granulations, *G*). Various adjacent vessels are also included. *A*, cerebral artery; *C*, cerebral vein; *D*, diploic vein; *E*, emissary vein; M, meningeal vein; *3* and *4*, third and fourth ventricles.

The lowermost part of the figure shows the caudal end of the spinal cord. Lumbar puncture (*L.P.*) is performed in the part of the subarachnoid space that lies below the termination of the spinal cord. Cf. fig. 41–1.

ventricular foramen on each side. The third ventricle communicates with the fourth ventricle through the aqueduct. The fourth ventricle becomes continuous with the central canal of the medulla and spinal cord and opens by means of apertures into the subarachnoid space.

The neuroglia that lines the ventricles of the brain and the central canal of the spinal cord is termed *ependyma*. In the ventricles, vascular fringes of pia mater, known as the *tela choroidea,* invaginate their covering of modified ependyma and project into the ventricular cavities. This combination of vascular tela and cuboidal ependyma is termed the *choroid plexus* (see fig. 43–11). The plexuses are invaginated into the cavities of the lateral, third, and fourth ventricles, and they are concerned with the formation of cerebrospinal fluid.

The term *blood–cerebrospinal fluid barrier* refers to the tissues that intervene between the blood and the cerebrospinal fluid.

Lateral Ventricles. Each lateral ventricle is a cavity in the interior of a cerebral hemisphere, and each communicates with the third ventricle by means of an interventricular foramen. The portion of the lateral ventricle in front of the foramen is termed its anterior horn. Behind this is the central part of the ventricle, which divides into the posterior and inferior horns. The anterior, posterior, and inferior horns are found in the frontal, occipital, and temporal lobes of the cerebral hemisphere, respectively (fig. 43–9).

The *anterior horn* of the lateral ventricle is bounded below by the rostrum, in front by the genu, and above by the trunk, of the corpus callosum. Laterally, it is limited by the bulging head of the caudate nucleus. Medially, it is separated from the lateral ventricle of the opposite side by a thin vertical partition, the *septum pellucidum*.

The *central part* of the lateral ventricle lies beneath the trunk of the corpus callosum and upon the thalamus and the body of the caudate nucleus. Medially, the two lateral ventricles are separated from each other by the posterior portion of the septum pellucidum and the *fornix* (which is an arched band of fibers). In the angle between the diverging posterior and inferior horns, the floor of the cavity presents a triangular elevation, the *collateral trigone,* associated with an underlying groove (generally the *collateral sulcus*).

The variable *posterior horn* tapers backward into the occipital lobe of the hemisphere. Above and on the lateral side, each posterior horn is bounded by a sheet of fibers (the *tapetum*) derived from the trunk and the splenium of the corpus callosum. Medially, two elevations may project laterally into the posterior horn. The upper elevation (*bulb of the posterior horn*) is produced by fibers (*forceps major*) derived from the splenium. The lower (*calcar avis*) is associated with a groove (*calcarine sulcus*) on the exterior of the hemisphere.

The *inferior horn* extends downward and forward behind the thalamus and into the temporal lobe of the hemisphere. It is bounded laterally by fibers (the tapetum) derived from the corpus callosum. Inferiorly, the most noticeable feature is an elevation known as the *hippocampus,* which is partly covered by the choroid plexus. Superiorly, the tail of the caudate nucleus runs forward to end in the amygdaloid body.

The choroid plexus of each lateral ventricle is invaginated along a curved line known as the *choroid fissure*. The fissure extends from the interventricular foramen in front, and in an arched manner around the posterior end of the thalamus, as far as the end of the inferior horn. The choroid plexus of the lateral ventricle is practically confined to the central part and the inferior horn. It is best developed at the junction of the central part with the inferior horn, and it is there known as the *glomus choroideum*. Calcified areas (*corpora amylacea*) are frequent in the glomera. The vessels of the plexus are derived from the internal carotid (anterior choroid) and the posterior cerebral (posterior choroid) arteries. At the interventricular foramina the choroid plexuses of the two lateral ventricles become continuous with each other and with that of the third ventricle.

Third Ventricle. The third ventricle is a narrow cleft between the two thalami. Over a variable area the thalami are frequently adherent to each other, giving rise to the *interthalamic adhesion*. The floor of the ventricle is formed by the hypothalamus. In front, the floor is crossed by the optic chiasma. The anterior wall is formed by the *lamina terminalis,* a delicate sheet that connects the optic chiasma to the corpus callosum. The thin roof consists of ependyma covered by two layers of pia (known as the *velum interpositum*).

The third ventricle communicates with the lateral ventricles by means of the interventricular foramina. Each *interventricular foramen* is situated at the upper and anterior portion of the third ventricle, at the front limit of the thalamus, and at the site of the outgrowth of the cerebral hemisphere in the embryo. From it a shallow groove, the *hypothalamic sulcus*, may be traced backward to the aqueduct. The sulcus marks the boundary between the thalamus above and the hypothalamus below.

The third ventricle presents several recesses (fig. 43–9): *optic, infundibular, pineal,* and *suprapineal.*

The choroid plexuses of the third ventricle invaginate the roof of the ventricle on each side of the median plane (see fig. 43–16*B*). At the interventricular foramina they become continuous with those of the lateral ventricles. Their vessels (posterior choroid arteries) are derived from the posterior cerebral.

The *aqueduct* is the narrow channel in the midbrain that connects the third and fourth ventricles.

Fourth Ventricle. The fourth ventricle is a rhomboidal cavity (see fig. 43–2) in the posterior portions of the pons and medulla. Above, it narrows to become continuous with the aqueduct of the midbrain. Below, it narrows and leads into the central canal of the medulla, which, in turn, is continuous with the central canal of the spinal cord. Laterally, the widest portion of the ventricle is prolonged on each side as the *lateral recess* (see fig. 43–1). The superior and inferior cerebellar peduncles form the lateral boundaries of the ventricle.

The anterior boundary or floor (the *rhomboid fossa*) of the fourth ventricle is formed by the pons above and by the medulla below (see fig. 43–2). It is related directly or indirectly to the nuclei of origin of the last eight cranial nerves. A *median groove* divides the floor into right and left halves. Each half is divided by a longitudinal groove (the *sulcus limitans*) into medial (basal) and lateral (alar) portions. The medial portion, known as the *medial eminence,* overlies certain motor nuclei, e.g., those of the abducent and hypoglossal nerves. The area lateral to the sulcus limitans overlies certain afferent nuclei, e.g., that of the vestibular part of the vestibulocochlear nerve.

The lowermost portion of the floor of the fourth ventricle is shaped like the point of a pen (*calamus scriptorius*) and contains the important respiratory, cardiac, vasomotor, and deglutition centers.

The posterior boundary or roof of the fourth ventricle is extremely thin and concealed by the cerebellum (see fig. 43–16*A*). It consists of sheets of white matter (*superior* and *inferior medullary vela*), which are lined by ependyma and stretch between the two superior and the two inferior cerebellar peduncles. The lower portion of the roof presents a deficiency, the *median aperture* of the fourth ventricle, through which the ventricular cavity communicates with the subarachnoid space. The ends of the lateral recesses have similar openings, the *lateral apertures.* The median and lateral apertures are the only means by which cerebrospinal fluid formed in the ventricles enters the subarachnoid space. In the event of occlusion of the apertures, the ventricles become distended (*hydrocephalus*).

The choroid plexuses of the fourth ventricle invaginate the roof on each side of the median plane. A prolongation of each plexus protrudes through the corresponding lateral aperture (see fig. 43–1). The vessels to the plexus are derived from cerebellar branches of the vertebral and basilar arteries.

The ventricular system can be examined radiographically *in vivo* (fig. 43–10).

Cerebrospinal Fluid

It is generally held that the cerebrospinal fluid (C.S.F.) is formed chiefly by the choroid plexuses. The course of the C.S.F. is shown in figure 43–11. The arachnoid villi and arachnoid granulations seem to be responsible for the drainage of C.S.F. into the venous sinuses of the cranial dura and the spinal veins.

The functions of C.S.F. are not entirely clear. The liquid acts as a fluid buffer for the protection of the nervous tissue. It also compensates for changes in blood volume within the cranium.

The C.S.F. may be examined by means of lumbar puncture (see fig. 41–1).

HYPOPHYSIS CEREBRI

The hypophysis cerebri, or *pituitary gland* (fig. 43–12), is an important endocrine organ, the main portion of which is situated in the hypophysial fossa of the sphenoid bone, where it generally remains after removal of the brain. This main portion is connected to the brain by the infundibulum (figs. 43–7 and 43–12*B*). The diaphragma sellae (see fig. 43–14*A*) forms a dural roof for the greater part of the gland and is pierced by the infundibulum. The organ is surrounded by a fibrous capsule fused with the endosteum.

The hypophysis is related above to the optic chiasma, below to an intercavernous venous sinus and the sphenoidal air-sinus (through which it can be approached endonasally), and

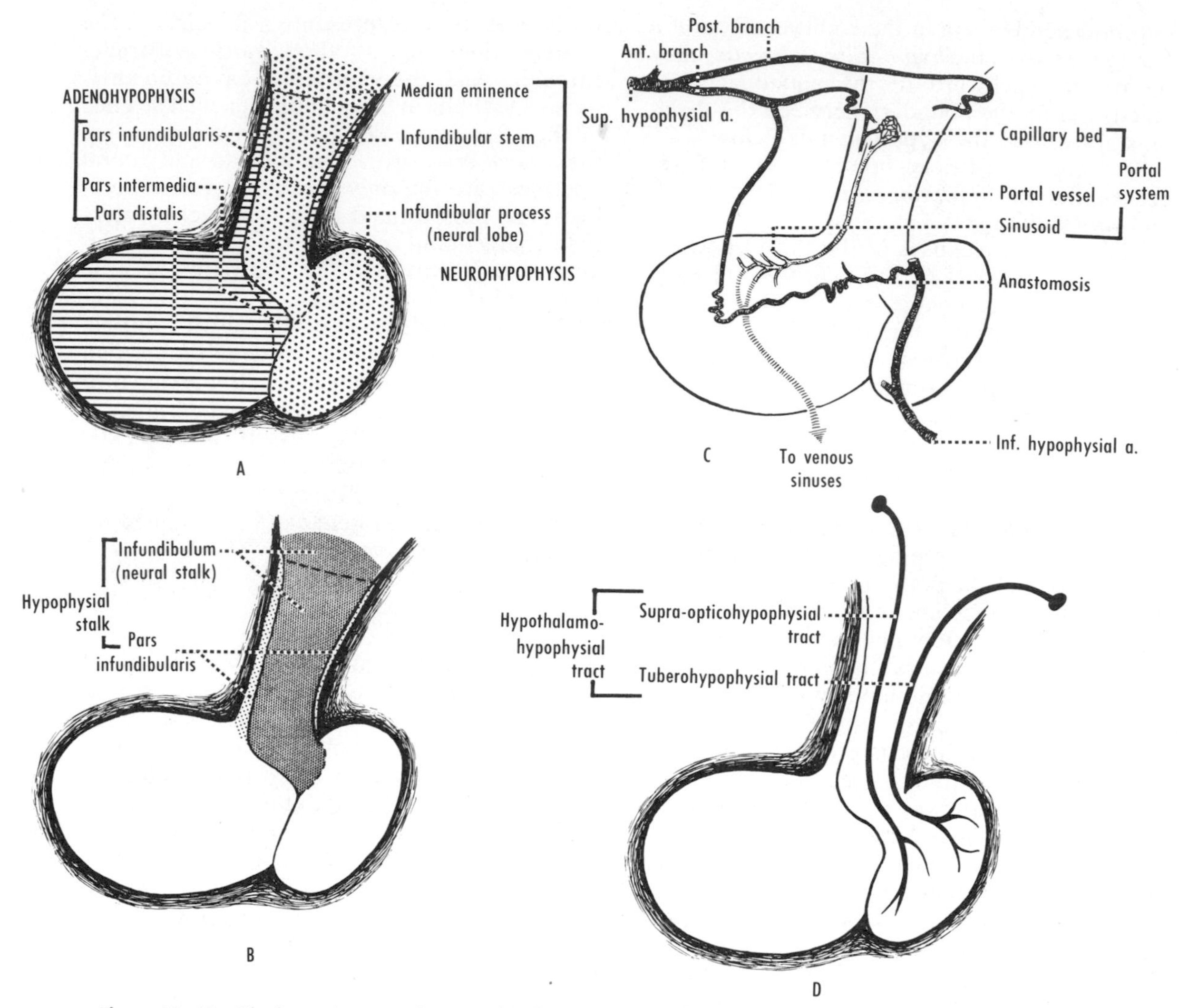

Figure 43–12 The hypophysis cerebri. **A** and **B** illustrate the terminology. **C** shows the blood supply. Arteries in the hypophysial stalk break up into capillary loops, which drain into hypophysial portal vessels. These, on reaching the pars distalis, drain into sinusoids, which enter the venous sinuses around the gland. **D** illustrates schematically the hypothalamohypophysial tract.

laterally to the cavernous sinus and the structures contained therein (see fig. 43–22). Hypophysial tumors, by causing pressure on the chiasma, commonly result in visual defects.

Terminology. The hypophysis* is best divided on embryological grounds into two main portions: the adenohypophysis and the neurohypophysis (fig. 43–12*A*). The *adenohypophysis* comprises the *pars infundibularis* (*pars tuberalis*), *the pars intermedia,* and the *pars distalis*.

The *neurohypophysis* comprises the *median eminence,* the *infundibular stem,* and the *infundibular process* (*neural lobe*). The median eminence is frequently classified also as a part of the tuber cinereum. The term *infundibulum* (*neural stalk*) is used for the median eminence and the infundibular stem. The term *hypophysial stalk* refers to the pars infundibularis and the infundibulum (fig. 43–12*B*).

Development. The adenohypophysis which develops as a diverticulum of the buccopharyngeal region, is an endocrine gland, the pars distalis of which secretes a number of hormones.

The neurohypophysis develops as a diverticulum of the floor of the third ventricle. It is

* The terms *anterior lobe* and *posterior lobe* are best avoided because they are defined variously. The pars intermedia is included in the anterior lobe by most anatomists but in the posterior lobe by many physiologists.

a storehouse for neurosecretions produced by the hypothalamus and carried down axons, e.g., the supra-opticohypophysial tract.

Blood Supply and Innervation. The hypophysis is supplied by a series of hypophysial arteries from the internal carotid arteries (fig. 43–12*C*).

The maintenance and regulation of the activity of the adenohypophysis are dependent on the blood supply by way of the hypophysial portal system. It is likely that nerve fibers from the hypothalamus liberate releasing factors into the capillary bed in the infundibulum and that these substances are then carried by the portal vessels to the pars distalis of the gland, which they affect.

The neurohypophysis receives its main nerve supply from the hypothalamus by way of fibers known collectively as the *hypothalamohypophysial tract* (fig. 43–12*D*). This contains two sets of fibers, the *supra-opticohypophysial tract* and the *tuberohypophysial tract*.

CRANIAL NERVES

The cranial nerves, that is, the nerves attached to the brain (see fig. 43–1 and tables 43–2 and 43–3) are twelve on each side. They are numbered and named as follows:

1. Olfactory nerve (see fig. 52–5)
2. Optic nerve (see figs. 43–7 and 45–5*B*)
3. Oculomotor nerve (see fig. 43–15)
4. Trochlear nerve (see fig. 45–5*A*)
5. Trigeminal nerve (see figs. 43–22 and 48–10)
   (a) Ophthalmic nerve (see figs. 45–4 and 45–5*A*)
   (b) Maxillary nerve (see fig. 48–6)
   (c) Mandibular nerve (see fig. 48–8)
6. Abducent nerve (see figs. 43–1, 43–22, and 45–3)
7. Facial nerve (see figs. 44–7 and 47–5*A*)
8. Vestibulocochlear nerve (see fig. 44–9)
9. Glossopharyngeal nerve (see figs. 50–13 to 50–16)
10. Vagus nerve (see figs. 50–15 and 50–17 to 50–19)
11. Accessory nerve (see figs. 50–15 and 50–19)
12. Hypoglossal nerve (see figs. 50–14 and 50–20)

Functional Components. Some of the cranial nerves are exclusively or largely afferent (1, 2, and 8), others are largely efferent (3, 4, 6, 11, and 12), and still others are mixed, that is, contain both afferent and efferent fibers (5, 7, 9, and 10). The efferent fibers of the cranial nerves arise within the brain from groups of nerve cells termed *motor nuclei*. The afferent fibers arise outside the brain from groups of nerve cells, generally in a ganglion along the course of the nerve. The central processes of these nerve cells then enter the brain, where they end in groups of nerve cells termed *sensory nuclei*.

The four functional types of fibers found in spinal nerves are present also in some of the cranial nerves: somatic afferent, visceral afferent, visceral efferent, and somatic efferent. These four types are termed "general." In certain cranial nerves, however, components that are "special" to the cranial nerves are present. The special afferent fibers comprise visual, auditory, equilibratory, olfactory, taste, and visceral reflex fibers. (The first three are usually classified as somatic, and the last three as visceral.) The special efferent fibers (which are classified as visceral) are those to skeletal muscles either known or thought to be derived from the pharyngeal arches (muscles of mastication, facial muscles, muscles of pharynx and larynx, sternomastoid, and trapezius).

The cranial nerves may be grouped as follows:

- *Olfactory, optic, and vestibulocochlear nerves* (1, 2, and 8) pertain to organs of special sense (special afferent).
- *Oculomotor, trochlear, abducent, and hypoglossal nerves* (3, 4, 6, and 12) supply skeletal muscle of specific regions of the head (eyeballs in the case of 3, 4, and 6; tongue in the case of 12). Nerve 3 also contains parasympathetic fibers to the smooth muscle of the sphincter pupillae and the ciliary muscle (general visceral efferent).
- *Trigeminal nerve* (5) contains motor fibers to the muscles of mastication (special visceral efferent) and sensory fibers from various parts of the head, e.g., face, nasal cavity, tongue, and teeth (general somatic afferent).
- *Facial, glossopharyngeal, vagus, and accessory nerves* (7, 9, 10, and 11) contain several components:
  (a) Motor fibers to the muscles of facial expression (7) and the muscles of the pharynx and larynx (9 and 10) (special visceral efferent). Many of the fibers to the pharynx and larynx are derived from nerve 11 (internal branch) and travel by

TABLE 43–2 SUMMARY OF CRANIAL NERVES

Nerve	Attachment to Brain	Cranial Exit	Cells of Origin	Chief Components	Chief Functions
1. Olfactory	Olfactory bulb	Cribriform plate	Nasal mucosa	Special visceral or somatic afferent	Smell
2. Optic	Optic chiasma	Optic canal	Retina (ganglion cells)	Special somatic afferent	Vision
3. Oculomotor	Midbrain, at medial border of cerebral peduncle	Superior orbital fissure	Midbrain	Somatic efferent	Movements of eyes
			Midbrain	General visceral efferent (parasympathetic)	Miosis and accommodation
4. Trochlear	Midbrain, below inferior colliculus	Superior orbital fissure	Midbrain	Somatic efferent	Movements of eyes
5. Trigeminal	Side of pons	Superior orbital fissure, foramen rotundum, and foramen ovale	Pons	Special visceral efferent	Chiefly movements of mandible
			Trigeminal ganglion	General somatic afferent	Sensation in head
6. Abducent	Lower border of pons	Superior orbital fissure	Pons	Somatic efferent	Movements of eyes
7. Facial	Lower border of pons	Stylomastoid foramen	Pons	Special visceral efferent	Facial expression
			Pons	General visceral efferent (parasympathetic)	Secretion of tears and saliva*
			Geniculate ganglion	Special visceral afferent	Taste*
8. Vestibulo-cochlear	Lower border of pons	Does not leave skull	Vestibular ganglion	Special somatic afferent	Equilibration
			Spiral ganglion	Special somatic afferent	Hearing
9. Glosso-pharyngeal	Medulla, lateral to olive	Jugular foramen	Medulla (nucleus ambiguus)	Special visceral efferent	Elevation of pharynx
			Medulla (dorsal nucleus)	General visceral efferent (parasympathetic)	Secretion of saliva
			Inferior ganglion	General visceral afferent	Sensation in tongue and pharynx; visceral reflexes
			Inferior ganglion	Special visceral afferent	Taste
			Inferior ganglion	General somatic afferent	Sensation in external and middle ear

TABLE 43–2 Summary of Cranial Nerves *(Continued)*

Nerve	Attachment to Brain	Cranial Exit	Cells of Origin	Chief Components	Chief Functions
10. Vagus	Medulla, lateral to olive	Jugular foramen	Medulla (nucleus ambiguus)	Special visceral efferent	Movements of larynx
			Medulla (dorsal nucleus)	General visceral efferent (parasympathetic)	Movements and secretion of thoracic and abdominal viscera
			Inferior ganglion	General visceral afferent	Sensation in pharynx, larynx, and thoracic and abdominal viscera. Also visceral reflexes
			Inferior ganglion	Special visceral afferent	Taste
			Superior ganglion	General somatic afferent	Sensation in external ear
11. Accessory	Medulla, lateral to olive	Jugular foramen	Medulla (dorsal nucleus)	Special visceral (?) efferent	Movements of pharynx and larynx
			Spinal cord (cervical)	Special visceral (?) efferent	Movements of head and shoulder
			Medulla (dorsal nucleus)	General visceral efferent	Movements and secretion of thoracic and abdominal viscera
12. Hypoglossal	Medulla, between pyramid and olive	Hypoglossal canal	Medulla	Somatic efferent	Movements of tongue

* These fibers of the facial nerve (in the chorda tympani) are important components of the nervus intermedius.

TABLE 43–3 Parasympathetic Ganglia Associated with Cranial Nerves

Ganglion	Figure	Location	Parasympathetic Root	Sympathetic Root	Chief Distribution
Ciliary	45–6	Lateral to optic n.	Oculomotor n.	Internal carotid plexus	Ciliary m. and sphincter pupillae Dilator pupillae and tarsal muscles
Pterygopalatine	48–7	In pterygopalatine fossa	Greater petrosal (7) and n. of pterygoid canal	Internal carotid plexus	Lacrimal gland
Otic	48–9	Below foramen ovale	Lesser petrosal (9)	Plexus on middle meningeal a.	Parotid gland
Submandibular	49–3	On hyoglossus	Chorda tympani (7) by way of lingual n.	Plexus on facial a.	Submandibular and sublingual glands

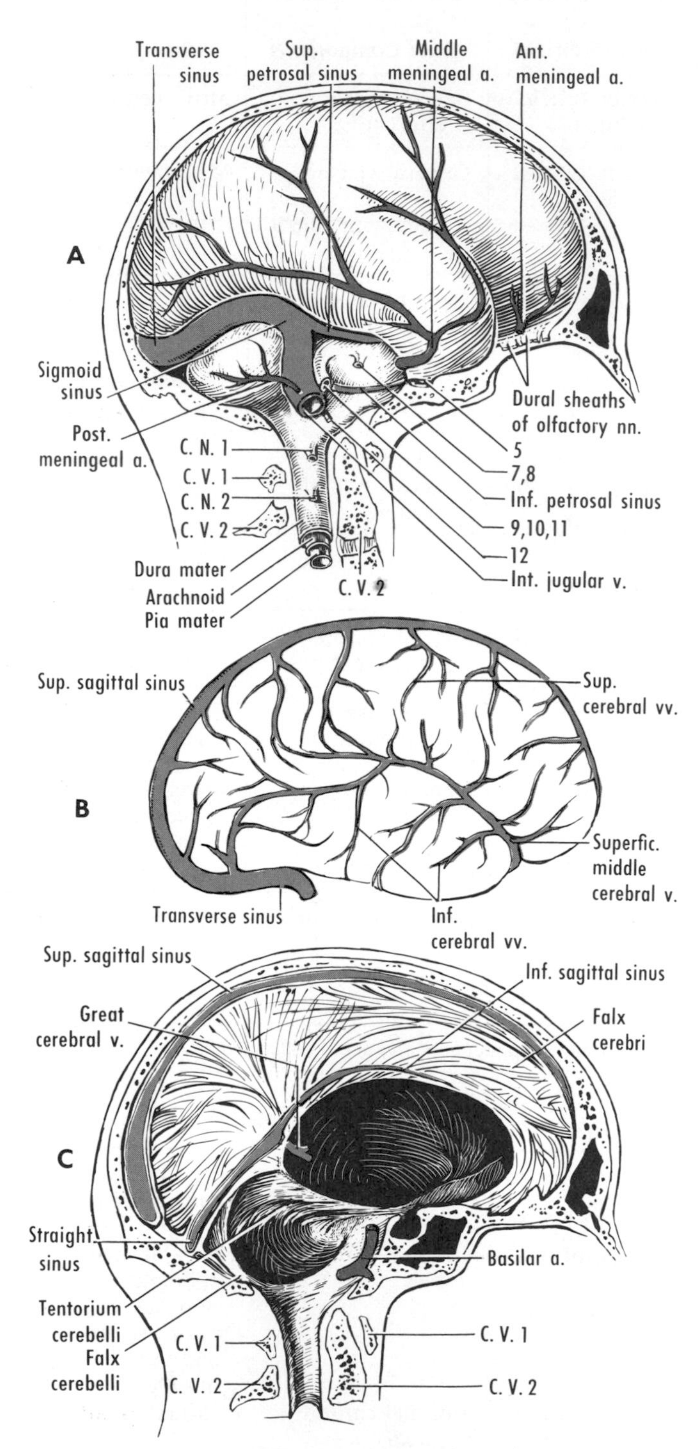

Figure 43–13 The meninges and associated vessels. **A**, lateral aspect of the intact dural sac. **B**, cerebral veins as seen through the arachnoid after removal of the dura. **C**, processes of the dura after removal of the brain and spinal cord. (**A** is based on Strong and Elwyn.)

way of nerve 10 (hence 11 is "accessory" to the vagus).

(b) Parasympathetic secretory fibers to the lacrimal and salivary glands (nervus intermedius of 7), the salivary glands (9), and certain glands associated with the respiratory and digestive systems (10) (general visceral efferent). Nerve 10 also supplies most of the smooth muscle of the respiratory and digestive systems, as well as cardiac muscle.

(c) Taste fibers (nervus intermedius of 7; also 9 and 10) (special visceral afferent).

(d) Fibers from the mucous membrane of the tongue and pharynx (hence the name *glossopharyngeal*) and of much of the respiratory and digestive systems (general visceral afferent) are contained in nerves 9 and 10.

(e) The spinal part of nerve 11 supplies the sternomastoid and trapezius, two muscles of disputed development.

Parasympathetic Ganglia Associated with Cranial Nerves (see fig. 48–11). The ciliary, pterygopalatine, otic, and submandibular ganglia are associated with certain of the cranial nerves. In these ganglia, parasympathetic fibers synapse, whereas sympathetic and other fibers merely pass through. The chief features of the ganglia are summarized in table 43–3.

MENINGES (figs. 43–11 and 43–13 to 43–16)

PACHYMENINX OR DURA MATER

The *dura mater* that surrounds the brain is frequently described as consisting of two layers: an external, or endosteal, and an internal, or meningeal, layer. Because the two layers are indistinguishable except in a few areas, however, it is simpler to consider the dura as one layer, which serves as both endocranium and meninx.* Instead of being considered as separating two layers, the venous sinuses are here described as being situated within the (single) dura.

The dura is particularly adherent at the base and also at the sutures and foramina, where it becomes continuous with the pericranium. Four processes are sent internally from the

* See L. C. Rogers and E. E. Payne (J. Anat., *95*:586–588, 1961), whose views have been disputed by A. B. Beasley and H. Kuhlenbeck (Anat. Rec., *154*:315, 1966).

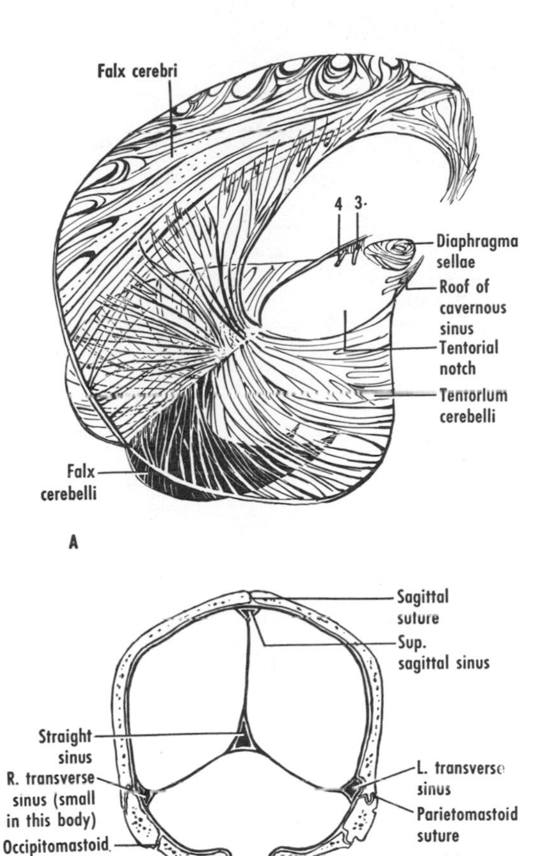

Figure 43–14 Processes of the dura. **A**, Right lateral aspect seen also from above and behind. **B**, Coronal section through the foramen magnum.

dura: the falx cerebri, tentorium cerebelli, falx cerebelli, and diaphragma sellae (fig. 43–14).

The *falx cerebri,* median and sickle shaped, occupies the longitudinal fissure between the two cerebral hemispheres. Anchored in front to the crista galli, its upper and lower borders enclose the superior and inferior sagittal sinuses, respectively.

The *tentorium cerebelli* separates the occipital lobes of the cerebral hemispheres from the cerebellum. Its internal, concave border contributes to the tentorial notch. (See below.) The external, convex border encloses the transverse sinus. Beyond the "petrous ridge," the tentorium is anchored to the anterior and posterior clinoid processes.

The *tentorial notch* (fig. 43–15), which contains the midbrain, a part of the cerebellum, and the subarachnoid space, is bounded by the tentorium and the dorsum sellae. **Space-occupying intracranial lesions may cause herniation of the brain upward or downward through the notch, and distortion of the midbrain may ensue.**

Near the apex of the petrous part of the temporal bone, the dura of the posterior cranial fossa bulges forward beneath that of the middle cranial fossa to form a recess, the *cavum trigeminale,* which contains the trigeminal ganglion.

The *falx cerebelli,* median and sickle shaped, lies below the tentorium and projects between the cerebellar hemispheres.

The *diaphragma sellae* is the small, circular, horizontal roof of the sella turcica.

Meningeal Innervation and Vessels. The dura, like the scalp, is supplied by both cranial (chiefly the trigeminal) and cervical nerves. **The brain itself is normally insensitive, and headaches are commonly either of vascular (intracranial or extracranial) or dural origin.**

The meningeal vessels are nutrient to the bones of the skull. Small anterior and posterior branches are provided by the internal carotid and vertebral arteries, but the middle artery is of much greater significance. **The middle meningeal artery is clinically the most important branch of the maxillary artery, be-**

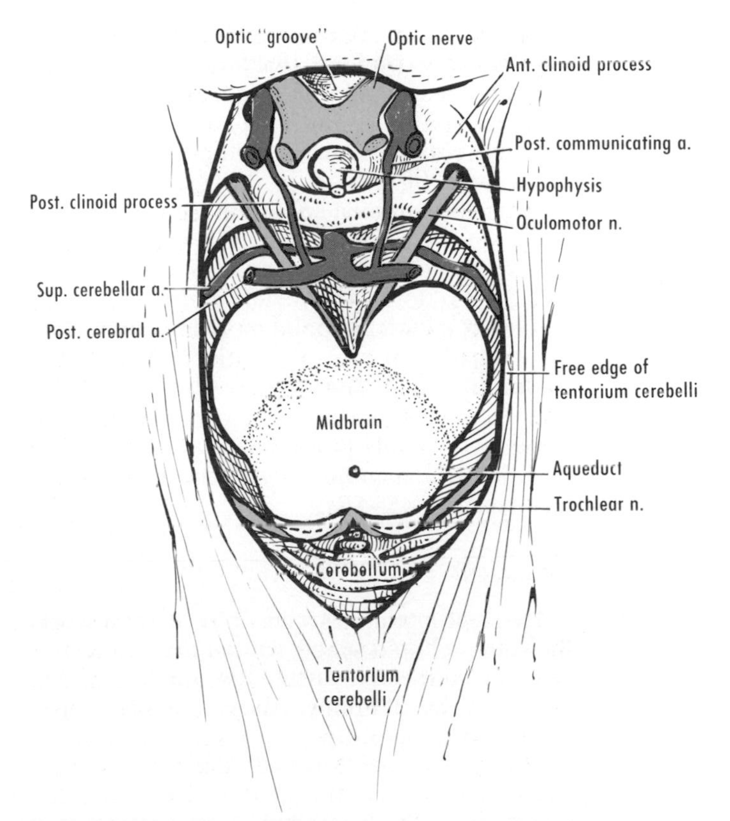

Figure 43–15 The tentorial notch, superior aspect. The notch is bounded in front by the dorsum sellae, between the two posterior clinoid processes.

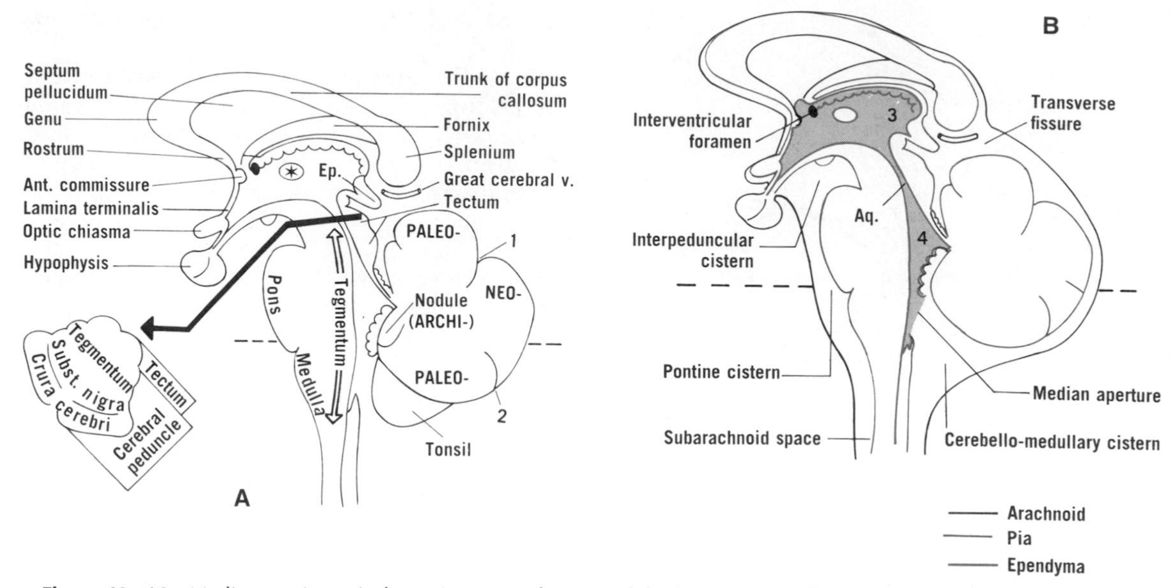

Figure 43–16 Median sections. **A** shows important features of the brain stem and a cross section of the midbrain. The interthalamic adhesion is indicated by an asterisk. **B** emphasizes the meninges, subarachnoid space, and cisterns. *Aq.*, aqueduct; *Ep.*, epiphysis (pineal body); *1*, fissura prima; *2*, fissura secunda; *3* and *4*, third and fourth ventricles. The interrupted line indicates the orbitomeatal plane, which has been used for orientation, thereby rendering the brain stem almost vertical.

cause, in head injuries, tearing of this vessel may cause extradural (epidural) hemorrhage. This may result in brain compression and contralateral paralysis and may necessitate trephining. From its origin in the infratemporal fossa (see fig. 48–3*C*), the artery ascends through the foramen spinosum, runs anterolaterally on the wall of the middle cranial fossa, and divides into a frontal and a parietal branch (see fig. 43–13*A*). **The middle meningeal artery divides at a variable point on a line connecting the midpoint of the zygomatic arch with the posterior end of the pterion** (see fig. 43–8). The meningeal vessels occupy grooves and sometimes canals in the bones. The branches include an anastomosis with the lacrimal artery (see fig. 45–5*B*).

LEPTOMENINGES (fig. 43–16)

The leptomeninges constitute a meshwork, the subarachnoid space, limited externally by a layer of connective tissue, the arachnoid, and internally by a thinner layer, the pia mater. C.S.F. circulates in the subarachnoid space.

The *arachnoid* surrounds the brain loosely and is separable from the dura by a potential space into which subdural hemorrhage may occur. The arachnoid dips into the longitudinal fissure but not into the sulci. Near the dural venous sinuses, the arachnoid presents microscopic projections, termed *arachnoid villi,* which are believed to be concerned with the absorption of C.S.F. Enlargements of the villi, known as *arachnoid granulations,* enter some of the sinuses (especially the superior sagittal) and their associated lateral lacunae and are visible to the naked eye. Both the granulations and the lacunae lie in granular pits on the internal aspect of the calvaria.

The *pia* covers the brain and dips between the gyri of the cerebral hemispheres and the folia of the cerebellum.

The *subarachnoid space* contains the C.S.F. and the cerebral vessels. **The subarachnoid space communicates with the fourth ventricle by means of apertures: a median one and two lateral ones.** At certain areas on the base of the brain, the subarachnoid space is expanded into *cisternae* (fig. 43–16*B*). The most important of these is the *cerebellomedullary cisterna* (or *cisterna magna*), which can be "tapped" by a needle inserted through the posterior atlanto-occipital membrane, a procedure known as cisternal puncture (see fig. 41–1). Cisternae that include important vessels are found on the front of the pons (basilar artery), between the cerebral peduncles (circulus arteriosus), and above the cerebellum (great cerebral vein).

BLOOD SUPPLY OF BRAIN

ARTERIES (figs. 43–17 to 43–20, 43–22, and table 43–4)

The brain is supplied by the two internal carotid and the two vertebral arteries. The former supply chiefly the frontal, parietal, and temporal lobes, the latter the temporal and occipital lobes, together with the midbrain and the hindbrain. On the inferior surface of the brain the four arteries form an anastomosis, the circulus arteriosus.

The tissues that intervene between the blood and the neurons include capillary endothelial cells (and their basement membranes), which form the "blood-brain barrier."

Internal Carotid Artery (Petrous, Cavernous, and Cerebral Parts)

The cervical part of the internal carotid artery enters the carotid canal in the petrous part of the temporal bone. The petrous part of the artery first ascends and then curves forward and medially. It is closely related to the cochlea, the middle ear, the auditory tube, and the trigeminal ganglion. The subsequent directions of the petrous, cavernous, and cerebral parts of the vessel may be numbered from 5 to 1, as follows (fig. 43–18, *inset*):

5. At the foramen lacerum, the petrous part of the internal carotid artery ascends to a point medial to the lingula of the sphenoid bone.

4. The artery then enters the cavernous sinus but is covered by the endothelial lining (see fig. 43–22). This is the cavernous part of the artery. In the sinus the vessel passes forward along the side of the sella turcica.

3. It next ascends and pierces the dural roof of the sinus between the anterior and middle clinoid processes.

2. The cerebral part of the internal carotid artery turns backward in the subarachnoid space below the optic nerve. The U-shaped bend, convex forward, formed by parts 2, 3, and 4 is termed the "carotid siphon" (fig. 43–19).

1. The artery finally ascends and, at the medial end of the lateral sulcus, divides into the anterior and middle cerebral arteries.

The internal carotid artery and its branches, including the cerebral arteries, are surrounded and supplied by a sympathetic plexus derived from the superior cervical ganglion.

Branches (fig 43–18). The internal carotid artery gives no named branches in the neck. Within the cranial cavity, it supplies the hypophysis, the orbit, and much of the brain. The carotid siphon gives off three branches: the ophthalmic, posterior communicating, and anterior choroid arteries.

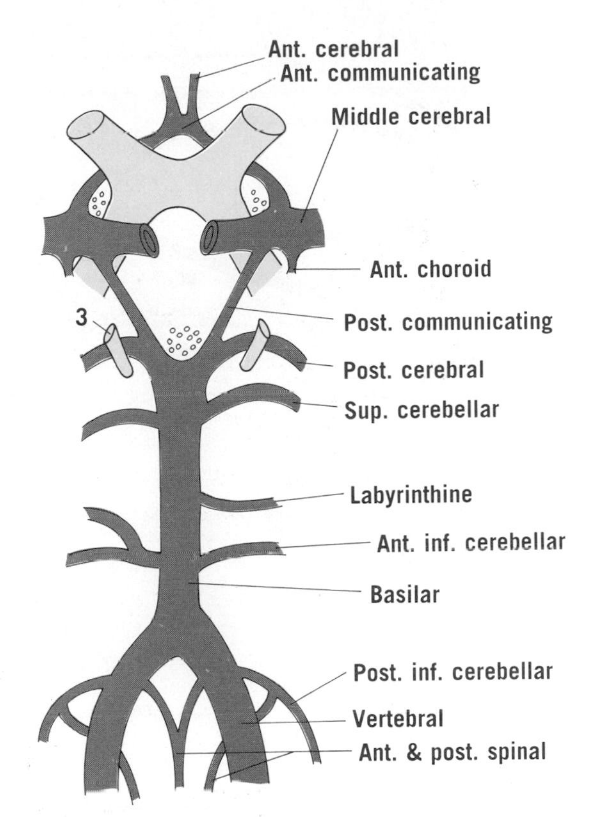

Figure 43–17 The carotid-vertebral circulation and the circulus arteriosus. The optic chiasma and the anterior and posterior perforated substances have been included.

The *ophthalmic artery* is described with the orbit (see fig. 45–5*B*).

The *posterior communicating artery* connects the internal carotid artery with the posterior cerebral artery and thereby forms a part of the circulus arteriosus.

The *anterior choroid artery* passes backward along the optic tract and enters the choroid fissure. It gives numerous small branches to the interior of the brain, including the choroid plexus of the lateral ventricle. The anterior choroid artery is frequently the site of thrombosis.

The terminal branches of the internal carotid artery are the anterior and middle cerebral arteries.

The *anterior cerebral artery* passes medially above the optic chiasma and enters the longitudinal fissure of the brain. Here it is connected with its fellow of the opposite side by the anterior communicating artery (which is

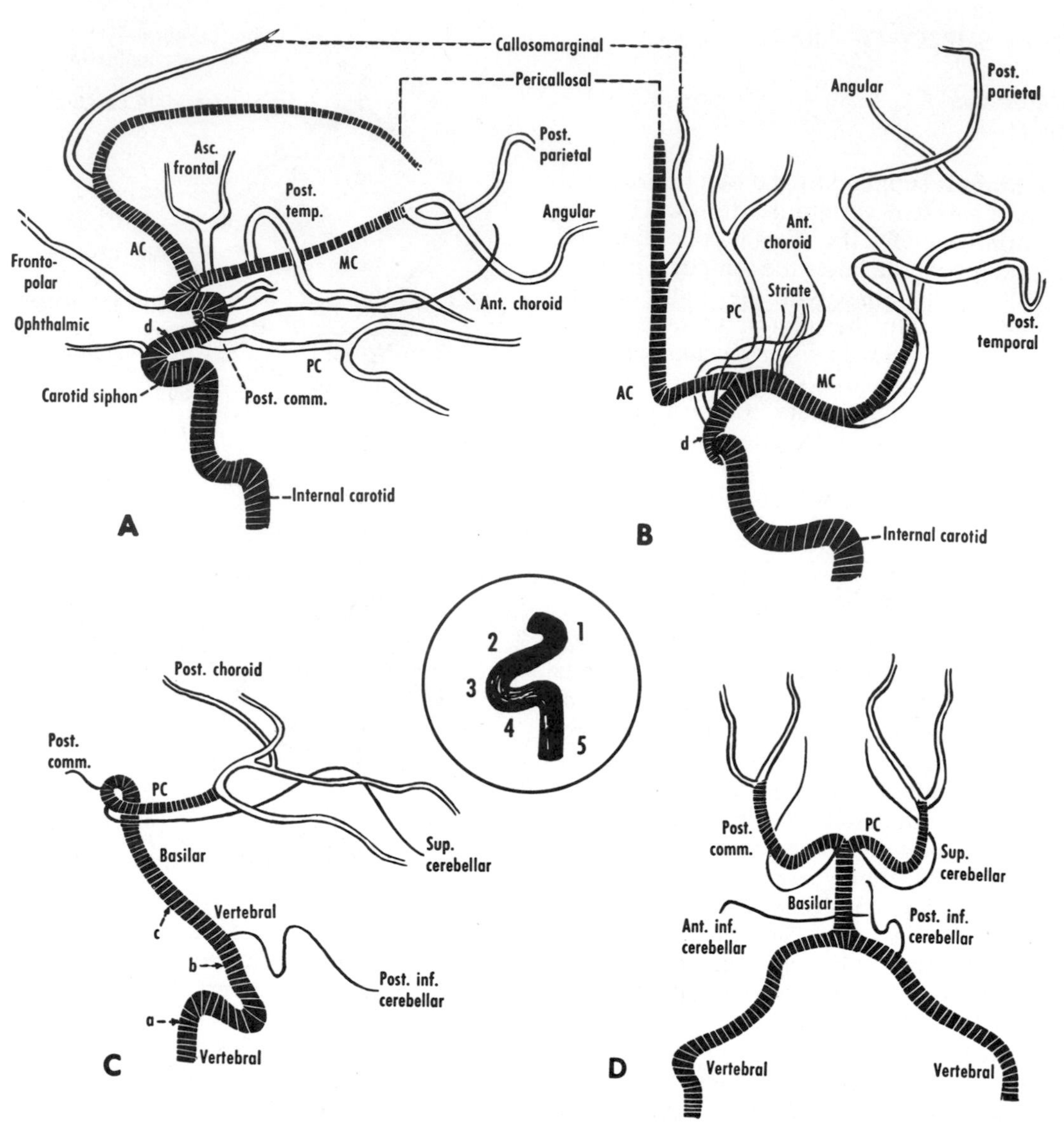

Figure 43–18 The chief arteries to the brain, as seen on angiography. The main vessels are shaded. **A** and **B**, Carotid arteriograms in lateral and anteroposterior projections. The internal carotid pierces the dura at *d*. The inset shows, in left lateral view, the numbered portions of the internal carotid artery. **C** and **D**, Vertebral arteriograms in lateral and half-axial (fronto-occipital) projections. In **C**, *a* is the site of the foramen transversarium of the atlas, *b* is the foramen magnum, and *c* is at the junction of the right and left vertebral arteries, i.e., the beginning of the basilar artery. *AC*, anterior cerebral artery; *MC*, middle cerebral artery; *PC*, posterior cerebral artery. (Based largely on Greitz and Lindgren.)

frequently double and which sometimes gives off a median anterior cerebral artery). It then runs successively forward, upward, and backward, usually lying on the corpus callosum, and ends by turning upward on the medial surface of the hemisphere in front of the parieto-occipital sulcus.

The *middle cerebral artery,* the larger terminal branch of the internal carotid artery, is frequently regarded as the continuation of that vessel. It passes laterally in the lateral sulcus and gives rise to numerous branches on the surface of the insula. It supplies the motor and premotor areas and the sensory and auditory areas. **Central branches (lenticulostriate arteries) enter the anterior perforated substance and are liable to rupture (Charcot's "artery of cerebral hemorrhage").** Occlusion of the middle cerebral artery causes a contralateral paralysis (*hemiplegia*) and a sensory defect. The paralysis is least marked in the lower limb (territory of anterior cerebral artery). When the

TABLE 43–4 BRANCHES OF INTERNAL CAROTID ARTERY

Part		Chief Branches
Cervical		No named branches
Petrous		Caroticotympanic branches
Cavernous		Meningohypophysial trunk
		Tentorial branch
		Meningeal branch
		Inferior hypophysial artery
		Cavernous branch
Cerebral		Superior hypophysial arteries
	From siphon	Ophthalmic artery
	From siphon	Posterior communicating artery
	From siphon	Anterior choroid artery
		Anterior cerebral artery
		Middle cerebral artery

dominant (usually left) side is involved, there are also disturbances of speech (*aphasia*).

The general distribution of the cerebral arteries is shown in figure 43–4*B* and *C*.

The chief branches of the internal carotid artery are summarized in table 43–4.

Vertebral Artery (Intracranial Part) and Basilar Artery (see figs. 43–2, 43–15, 43–17, and 43–18)

The vertebral and basilar arteries and their branches supply the upper part of the spinal cord, the brain stem, the cerebellum, and much of the postero-inferior portion of the cerebral cortex. The branches to the brain stem are functionally end-arteries.

Vertebral Arteries. The vertebral artery, a branch of the subclavian artery, may be considered in four parts: cervical, vertebral, suboccipital, and intracranial. The suboccipital part of the vertebral artery perforates the dura and arachnoid and passes through the foramen magnum (see fig. 43–2). The intracranial part of each vertebral artery ascends medially in front of the medulla and, at approximately the lower border of the pons, the two vertebral arteries unite to form the basilar artery (see fig. 50–23).

BRANCHES OF INTRACRANIAL PART. The vertebral artery, which gives off muscular and spinal branches in the neck, supplies chiefly the posterior part of the brain, both directly and, of greater importance, by way of the basilar artery.

The *anterior spinal artery* descends in front of the medulla and unites with the vessel of the opposite side to form a median trunk that contributes to the supply of the medulla and spinal cord.

The *posterior inferior cerebellar artery* winds backward around the olive and gives branches to the medulla, the choroid plexus of the fourth ventricle, and the cerebellum. The *posterior spinal artery* is usually a branch of the posterior inferior cerebellar artery, but it may come directly from the vertebral artery.

Basilar Artery. The basilar artery is formed by the union of the right and left vertebral arteries. It begins at approximately the lower border of the pons and ends near the upper border by dividing into the two posterior cerebral arteries (fig. 43–17). It passes through the cisterna pontis and lies frequently in a longitudinal groove on the front of the pons.

BRANCHES (fig. 43–18). Branches of the basilar artery are distributed to the pons, cerebellum, internal ear, midbrain, and cerebral hemispheres.

The paired *anterior inferior cerebellar arteries* pass backward on the lower surface of the cerebellum and supply the cerebellum and pons.

The paired *labyrinthine* (*internal auditory*) *arteries* may arise from either the basilar or the anterior inferior cerebellar artery, more commonly the latter. Each enters the corresponding internal acoustic meatus and is distributed to the internal ear.

The paired *superior cerebellar arteries* pass laterally below the oculomotor and trochlear nerves and are distributed to the cerebellum.

The two *posterior cerebral arteries* are the terminal branches of the basilar artery. They supply much of the temporal and most of the occipital lobes (see fig. 43–4). Each is connected with the corresponding internal carotid artery by a posterior communicating artery; occasionally the posterior cerebral arises as a branch of the internal carotid artery (an arrangement referred to as trifurcation of the internal carotid artery). The posterior cerebral artery runs backward, above and parallel to the superior cerebellar artery, from which it is separated by the oculomotor and trochlear nerves. Among the branches are the *posterior choroid branches,* which supply the choroid plexuses of the third and lateral ventricles.

Circulus Arteriosus (fig. 43–17)

Branches of the three cerebral arteries to the cerebral cortex have important anastomoses with one another on the surface of the brain. The arterial system in the brain, therefore, is not strictly terminal. However, in the event of occlusion, these microscopic anasto

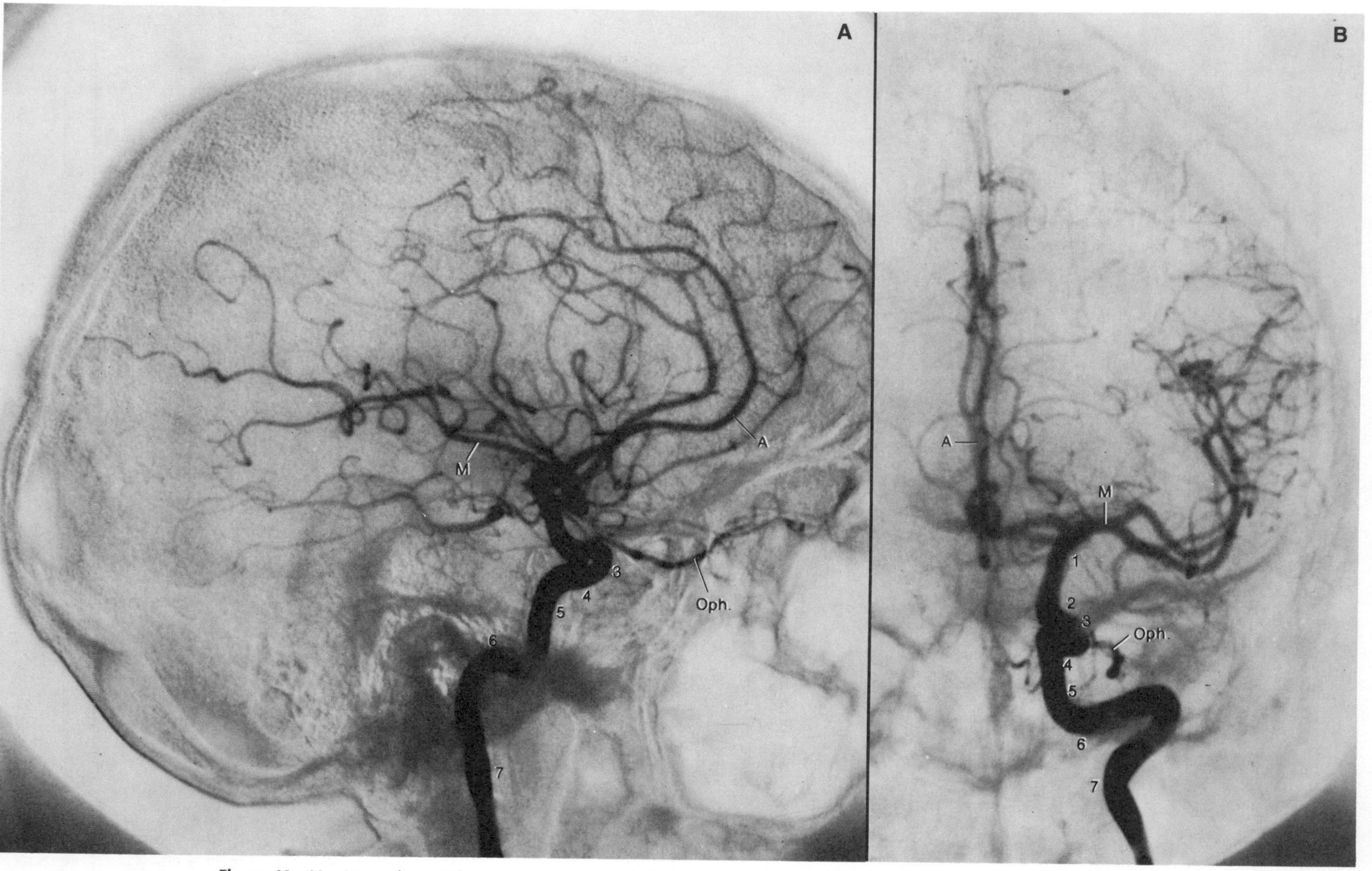

Figure 43–19 Internal carotid angiograms *in vivo*. **A**, lateral view. **B**, Postero-anterior view. The numerals from 1 to 7 refer to successive parts of the internal carotid artery. Number 7 is in the neck, number 6 lies within the carotid canal, number 5 is medial to the trigeminal ganglion, numbers 4 to 2 constitute the carotid siphon (which gives off the ophthalmic artery, *Oph.*), and number 1 ascends to the division into anterior (*A*) and middle (*M*) cerebral arteries. (Courtesy of Arthur B. Dublin, M.D., Sacramento, California.)

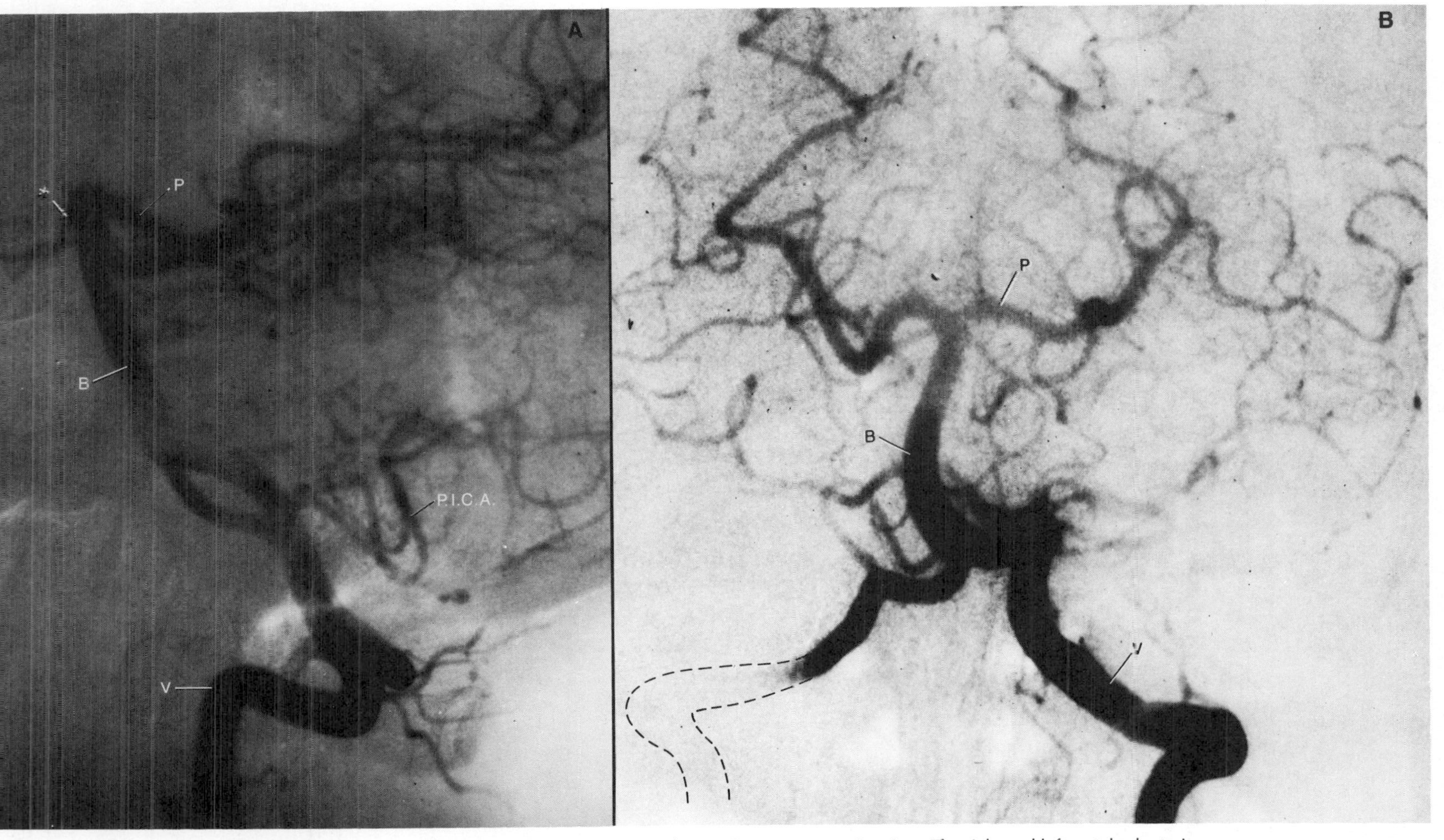

Figure 43–20 Vertebral angiograms *in vivo*. **A**, Lateral view. **B**, Postero-anterior view. The right and left vertebral arteries (*V*) unite to form the basilar artery (*B*), which divides into the two posterior cerebral arteries (*P*). The posterior communicating arteries (*asterisk in* **A**) run forward to take part in the circulus arteriosus (see fig. 43–17). An important branch of the vertebral, the posterior inferior cerebellar artery (*P.I.C.A.*), forms a characteristic loop. (Courtesy of Arthur B. Dublin, M.D., Sacramento, California.)

moses are not capable of providing an alternate circulation for the ischemic brain tissue.

The circulus arteriosus, described by Thomas Willis in 1664, is an important polygonal anastomosis between the four arteries that supply the brain: the two vertebral and the two internal carotid arteries. It is formed by the posterior cerebral, posterior communicating, internal carotid, anterior cerebral, and anterior communicating arteries. **The circulus forms an important means of collateral circulation in the event of obstruction.** Variations in the size of the vessels that constitute the circulus are very common.

The blood vessels of the brain may be demonstrated radiographically *in vivo* (*cerebral angiography,* figs. 43–18 to 43–20).

Venous Drainage

Veins of Brain (see fig. 43–13*B* and *C*)

The veins of the brain pierce the arachnoid and dura and open into the venous sinuses of the dura.

The *superior cerebral veins* drain into the superior sagittal sinus. The *superficial middle cerebral vein* follows the lateral sulcus, sends *superior* and *inferior anastomotic veins* to the superior sagittal and transverse sinuses, respectively, and ends in the cavernous sinus. The *inferior cerebral veins* drain the inferior aspect of the hemispheres and join nearby sinuses.

The *basal vein* is formed by the union of several veins, including those that accompany the anterior and middle cerebral arteries. It winds around the cerebral peduncle and ends in the great cerebral vein.

The single *great cerebral vein* (see fig. 43–13*C*) is formed by the union of two internal cerebral veins. It receives, directly or indirectly, a number of vessels from the interior of the cerebral hemispheres and also the basal veins. It ends in the straight sinus.

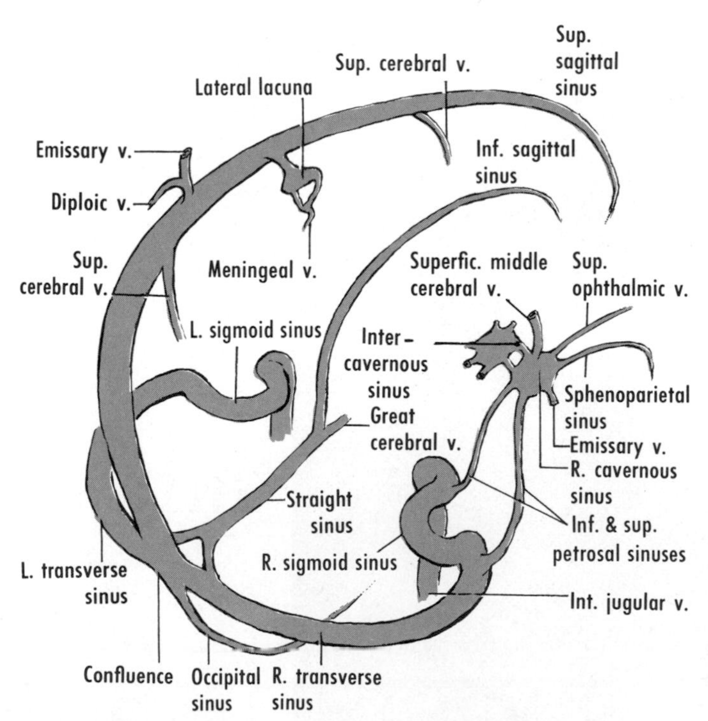

Figure 43–21 The venous sinuses of the dura. The orientation is similar to that of fig. 43–14*A*.

Venous Sinuses of Dura Mater (fig. 43–21)

The blood from the brain drains into sinuses that are situated within the dura mater and that empty ultimately into the internal jugular veins.

The *superior sagittal sinus* (see fig. 43–13*B* and *C*) lies in the convex border of the falx cerebri. From its commencement in front of the crista galli, the sinus runs backward and, near the internal occipital protuberance, enters in a variable manner one or both transverse sinuses. It receives the superior cerebral veins and communicates with lateral lacunae.

The *confluence of the sinuses* (or *torcular*) is the junction of the superior sagittal, straight, and right and left transverse sinuses (fig. 43–21). It is situated near the internal occipital protuberance. The pattern of the constituent sinuses varies, and dominance of one side in drainage (e.g., the right) is usual.

The *inferior sagittal sinus* lies in the concave border of the falx cerebri (the blade of the sickle) and ends in the straight sinus, which receives the great cerebral vein, runs backward between the falx and tentorium, and joins the confluence.

The *transverse sinuses* begin in the confluence, and each curves laterally in the convex border of the tentorium. At the petrous part of the temporal bone, the transverse becomes the *sigmoid sinus* (see fig. 43–13*A*), which grooves the mastoid part of the temporal bone and traverses the jugular foramen to become the internal jugular vein. Smaller channels (*petrosal sinuses*) connect the cavernous sinus with the transverse sinus and jugular vein.

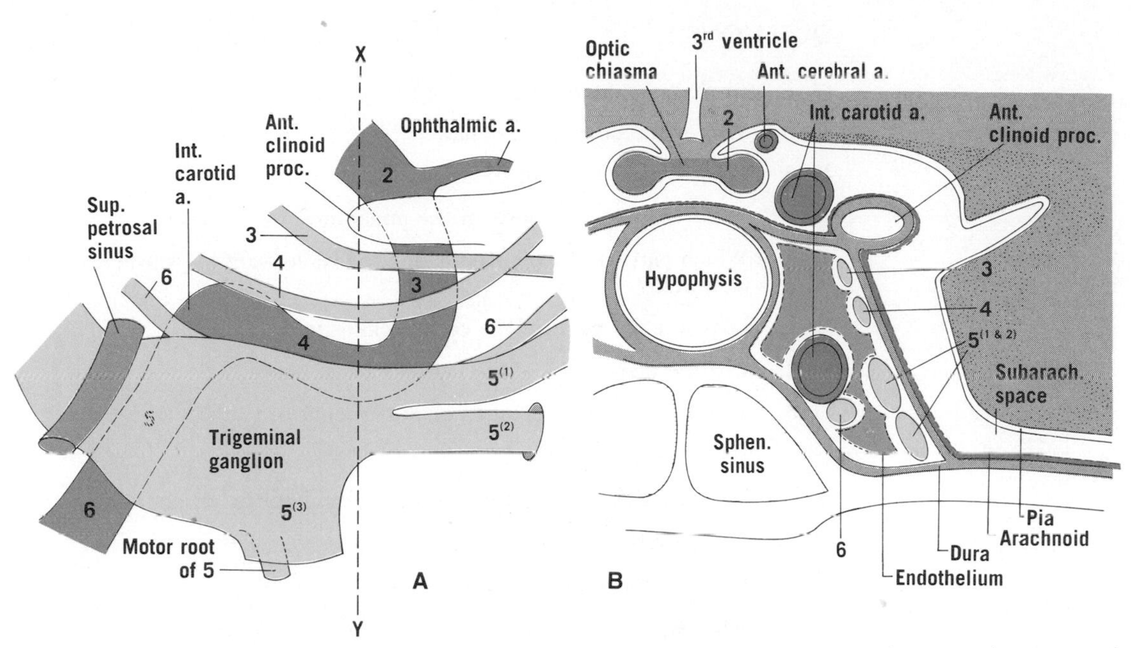

Figure 43–22 A, Right lateral aspect showing the structures related to the cavernous sinus. The numerals 2 to 6 on the internal carotid artery indicate the successive parts of the artery; parts 2 to 4 constitute the anteriorly directed carotid siphon. (After Cunningham.) **B**, Coronal section through the cavernous sinus along the plane *XY* in **A**. Although the maxillary nerve is shown here in rather close relation to the sinus (as it is usually depicted), according to W. R. Henderson (J. Anat., *100:*905, 1966) the nerve is embedded in the dura of the middle cranial fossa, lateral to the sinus.

The cavernous sinus comprises one or more venous channels (sometimes a plexus*) in a dural compartment bounded by the body of the sphenoid bone and the front portion of the tentorium. In addition to the venous channels, the dural compartment contains (outside the endothelium) the internal carotid artery, sympathetic plexus, abducent nerve, and, further laterally, the oculomotor, trochlear, and ophthalmic nerves (fig. 43–22). The cavernous sinus extends from the superior orbital fissure in front to the apex of the petrous part of the temporal bone behind. It receives several veins (superior ophthalmic, superficial middle cerebral, and sphenoparietal sinus) and communicates (by the petrosal sinuses) with the transverse sinus and internal jugular vein, as well as with the opposite cavernous sinus. The facial vein (via the superior ophthalmic vein) communicates with the cavernous sinus and hence allows infection around the nose and upper lip to spread to intracranial structures.

Lateral lacunae (see fig. 43–11) are venous meshworks within the dura near the superior sagittal sinus, and both the lacunae and sinus occupy the granular pits of the calvaria. The lacunae receive (1) emissary veins, (2) diploic veins, (3) meningeal veins, and (4) sometimes some cerebral veins. It should be noted that the emissary veins, which pass through foramina in the skull, connect the deeper vessels with the veins of the scalp and hence also allow infection to spread from the scalp to intracranial structures.

* The cavernous sinus may be a venous plexus in the newborn (P.-E. Duroux, A. Bouchet, J. Bossy, and F. Calas, C. R. Assoc. Anat., *42*:486–490, 1956), but it is usually a non-cavernous orbitotemporal sinus in the adult, with few or no trabeculae (M. A. Bedford, Br. J. Ophthalmol., *50*:41–46, 1966).

ADDITIONAL READING

Blinkov, S. M., and Glezer, I. I., *The Human Brain in Figures and Tables,* trans. by B. Haigh, Basic Books, New York, 1968.

Bossy, J., *Atlas du système nerveux. Aspects macroscopiques de l'encéphale,* Éditions Offidoc, Paris, 1972. Beautiful photographs of the brain.

Gardner, E., *Fundamentals of Neurology* 6th ed., W. B. Saunders Company, Philadelphia, 1975. A good introduction to the nervous system.

Ludwig, E., and Klingler, J., *Atlas cerebri humani,* Karger, Basel, 1956. Excellent photographs of superb dissections.

Stephens, R. B., and Stilwell, D. L., *Arteries and Veins of The Human Brain,* Thomas, Springfield, Illinois, 1969.

QUESTIONS

43–1 Review the major divisions of the brain.

43–2 Why is the pons so designated?

43–3 What are the roots of the trigeminal nerve?

43–4 What are the main parts of the midbrain?

43–5 Which parts of the diencephalon are visible from the surface of an intact brain?

43–6 What are the parts of the corpus callosum?

43–7 List the main lobes of the cerebral hemisphere.

43–8 Which are the two most frequently mentioned sulci?

43–9 Where is the insula?

43–10 How are the various areas of the cerebral cortex numbered?

43–11 What are the first pair of cranial nerves?

43–12 How are the ventricles numbered?

43–13 What are the main parts of the lateral ventricles?

43–14 List the chief recesses of the third ventricle.

43–15 Where are the lateral recesses of the fourth ventricle?

43–16 What are the subdivisions of the hypophysis cerebri?

43–17 Review the cranial nerves and their components.

43–18 List the chief parasympathetic ganglia associated with cranial nerves.

43–19 What are the chief processes of the cerebral dura mater?

43–20 What is the tentorial notch?

43–21 Which is the most important meningeal vessel clinically?

43–22 What is the subarachnoid space?

43–23 Which is the most important subarachnoid cisterna?

43–24 What is the carotid siphon?

43–25 What is the continuation of the internal carotid artery?

43–26 Which vessel supplies the medulla behind the olive?

43–27 What is the circulus arteriosus?

43–28 How is the great cerebral vein formed?

43–29 What is the confluence of the sinuses?

43–30 Where is the cavernous sinus?

43–31 Which veins drain into the lateral lacunae?

44
THE EAR

The ears are vestibulocochlear, that is, concerned with equilibration as well as hearing. Adjectives relating to the ear are *aural* and *auditory* (L.) and *otic* and *acoustic* (Gk). Each ear comprises three parts: external, middle, and internal (fig. 44–1).

EXTERNAL EAR

The external ear, which conducts sound and protects the deeper parts, consists of the auricle (described with the face) and the external acoustic meatus.

External Acoustic Meatus. The external meatus, about 25 mm long, extends from the concha to the tympanic membrane. Its lateral part is cartilaginous and continuous with the auricle; the longer medial part is bony. **The cartilaginous part of the meatus is slightly concave anteriorly, and a speculum may be inserted more readily when the auricle is pulled backward and upward.** The meatus is lined by the skin of the auricle, which presents hairs, and sebaceous and ceruminous glands.

Sensory Innervation and Blood Supply. The external ear is supplied by the auriculotemporal (fifth cranial) nerve (and probably by contributions from cranial nerves 7, 9, and 10) and the great auricular nerve. The posterior auricular and superficial temporal arteries of the external carotid artery provide the blood supply.

Tympanic Membrane. The ear drum (fig. 44–2), about 1 cm in diameter, separates the

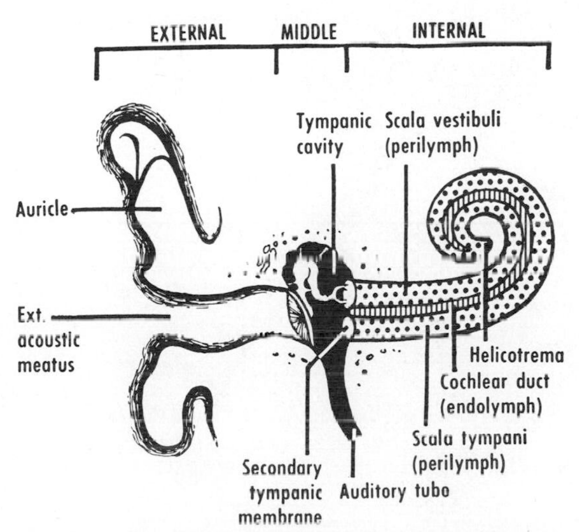

Figure 44–1 Basic structure of the ear. The base of the stapes closes the fenestra vestibuli, and the secondary tympanic membrane closes the fenestra cochleae.

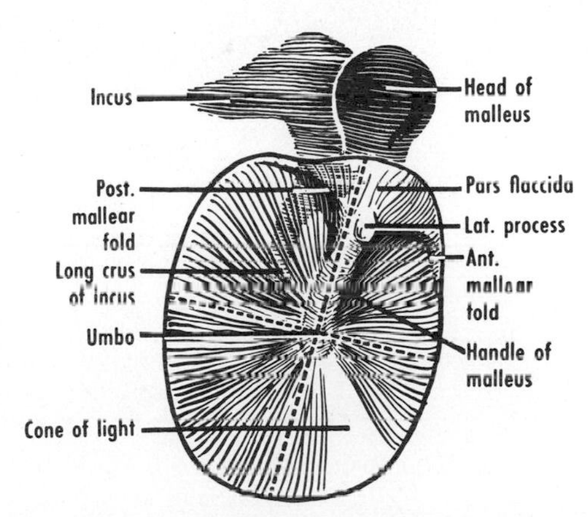

Figure 44–2 Right tympanic membrane, lateral aspect. The interrupted lines delimit the quadrants. The umbo is approximately opposite the promontory of the middle ear. The head of the malleus and the body and short crus of the incus are in the epitympanic recess, above the level of the tympanic membrane. Cf. fig. 53–11A.

external meatus from the tympanic cavity. Its fibrous basis is attached to the tympanic plate of the temporal bone and is covered laterally by epidermis and medially by the mucous membrane of the middle ear. The larger portion of the membrane is its *tense part;* the anterosuperior corner, or *flaccid part,* is bounded by *anterior* and *posterior mallear folds*. The tympanic membrane is set very obliquely (fig. 44–1). Its lateral surface is concave, and its deepest point is the *umbo*. **The handle and lateral process of the malleus are attached to the medial surface of the tympanic membrane.** Incisions through the membrane are made in its postero-inferior quadrant to avoid the ossicles and chorda tympani. The membrane can be examined *in vivo* by a speculum (fig. 53–11*A*). The tympanic membrane is highly sensitive (cranial nerves 5 and 10 laterally and 9 medially).

MIDDLE EAR

The middle ear consists largely of an air space in the temporal bone. This tympanic cavity contains the auditory ossicles and communicates with (1) the mastoid air cells and mastoid antrum by means of the aditus and (2) the nasopharynx by means of the auditory tube (figs. 44–3 and 44–4). The tympanic cavity and auditory tube develop as a recess of the embryonic pharynx. Mucous membrane covers the structures in the tympanic cavity.

Boundaries (fig. 44–5). The *lateral wall* is the tympanic membrane. However, **a portion of the tympanic cavity, the epitympanic recess, forms an attic above the level of the tympanic membrane and communicates with the aditus.**

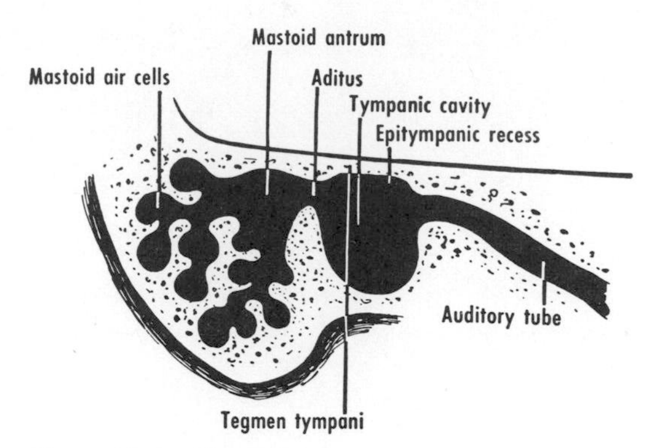

Figure 44–3 The tympanic cavity and communicating air spaces. Right lateral aspect.

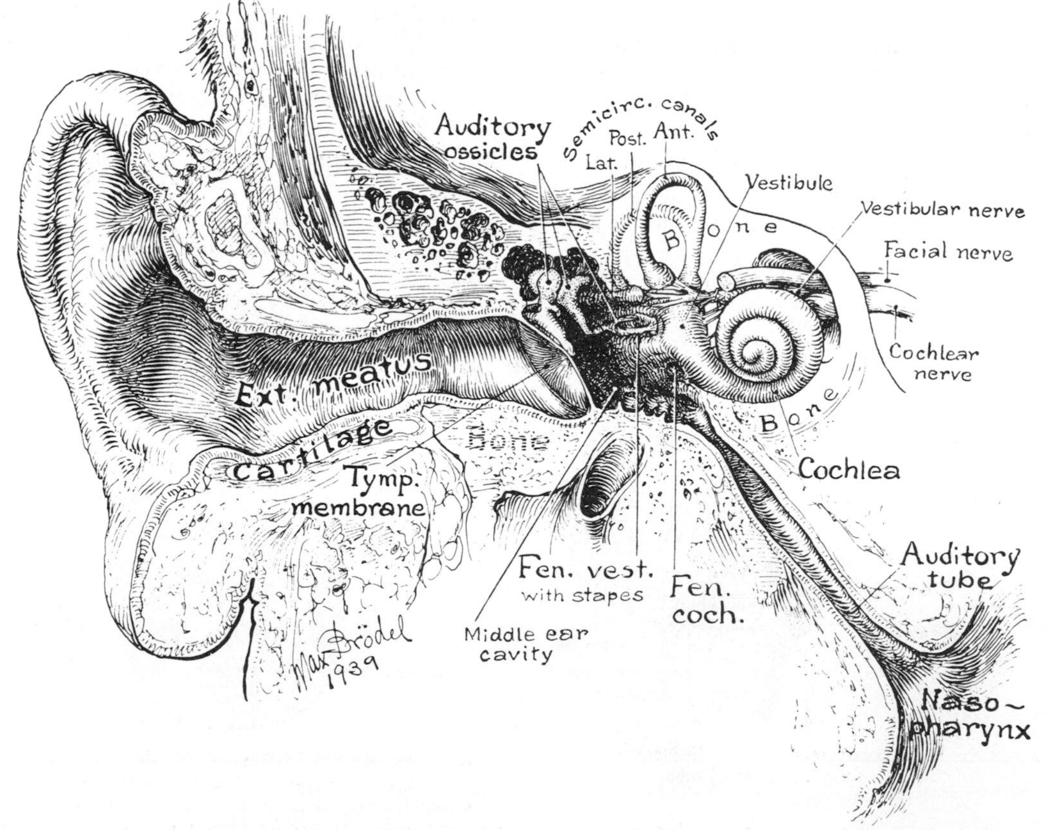

Figure 44–4 Drawing of the ear by Max Brödel (with lettering modified), showing the osseous labyrinth, cranial nerves 7 and 8, and the auditory tube.

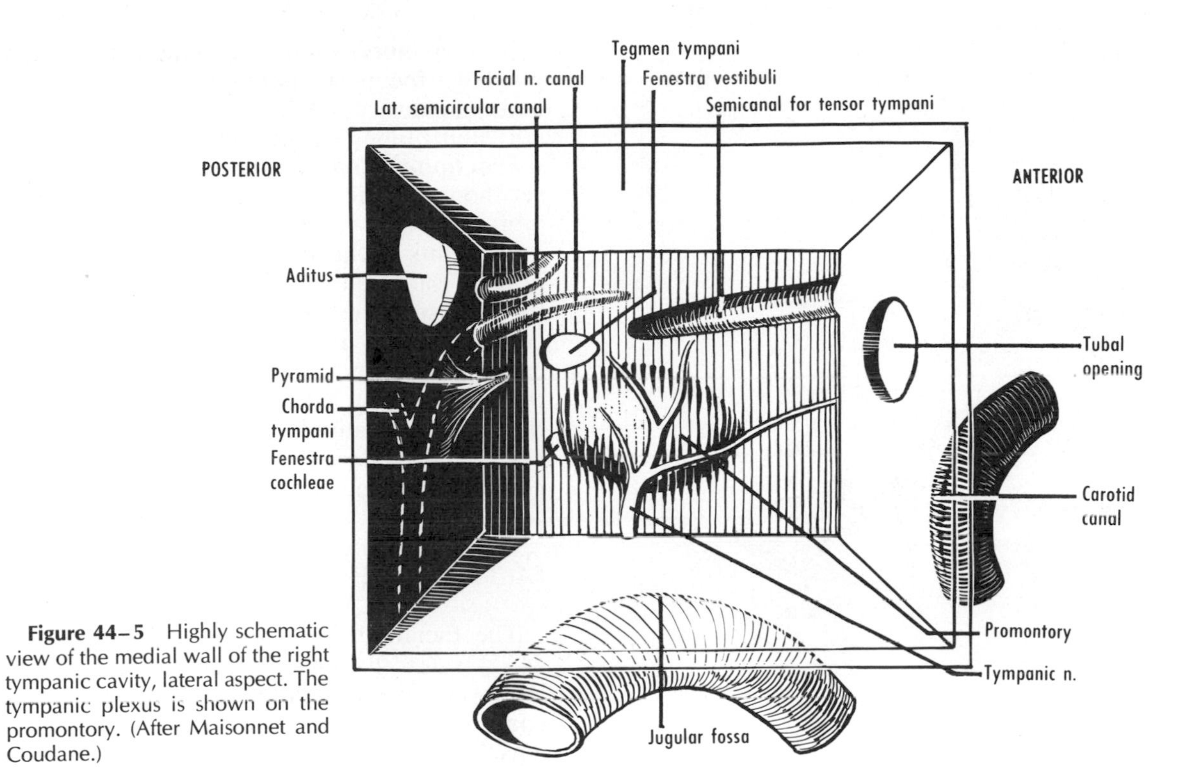

Figure 44–5 Highly schematic view of the medial wall of the right tympanic cavity, lateral aspect. The tympanic plexus is shown on the promontory. (After Maisonnet and Coudane.)

The epitympanic recess contains the head of the malleus and the body and short crus of the incus. The *roof* is formed by a portion of the petrous part of the temporal bone known as the *tegmen tympani,* which separates the middle ear from the middle cranial fossa. The *floor* is the jugular fossa, in which is lodged the superior bulb of the internal jugular vein. The *anterior wall* presents the *semicanal for the tensor tympani muscle,* the opening of the auditory tube, and the *carotid canal,* in which is lodged the internal carotid artery. The *posterior wall* presents the *aditus,* which leads to the *mastoid antrum,* and the *pyramidal eminence,* which contains the stapedius muscle. The mastoid portion of the temporal bone, especially the mastoid process, is usually hollowed out by air cells, which communicate with each other and with the mastoid antrum and are lined by mucoperiosteum. The *medial wall* (fig. 44–5) presents (1) the *prominence of the lateral semicircular canal* and the *prominence of the facial nerve canal;* (2) the *fenestra vestibuli* (*oval window*), closed by the base of the stapes, and the *processus cochleariformis;* (3) the *promontory,* formed by the basal turn of the cochlea; and (4) the *fenestra cochleae* (*round window*), closed by mucous membrane. The *tympanic plexus,* situated on the promontory, is formed chiefly by the tympanic nerve (from cranial nerve 9), which gives sensory fibers to the middle ear and secretomotor fibers to the parotid gland (see fig. 48–9).

Auditory Ossicles, Joints, and Muscles. The ossicles (fig. 44–6) are the malleus (hammer), incus (anvil), and stapes (stirrup). The *malleus* presents a *head* and *neck;* a *handle* (*manubrium*) and *lateral process,* embedded in the tympanic membrane; and an *anterior process,* attached to the petrotympanic fissure. The *incus* presents a *body, short crus,* and *long crus.* The *stapes* presents a *head, anterior* and *posterior crura,* and a *base* or footplate, which is attached by an annular ligament to the margin of the fenestra vestibuli. The *incudomallear* and *incudostapedial* joints are saddle and ball-and-socket synovial joints, respectively. The *tensor tympani* arises from the cartilaginous part of the auditory tube, enters a semicanal, turns laterally around the processus cochleariformis, and is inserted on the handle of the malleus. Supplied by the mandibular nerve and tympanic plexus, the tensor draws the handle of the malleus medially, thereby tensing the tympanic membrane. The *stapedius* arises within the pyramidal eminence and is inserted on the neck of the stapes. Supplied by the facial nerve, the stapedius draws the stapes laterally. Both the tensor and the stapedius attenuate sound transmission through the middle ear.

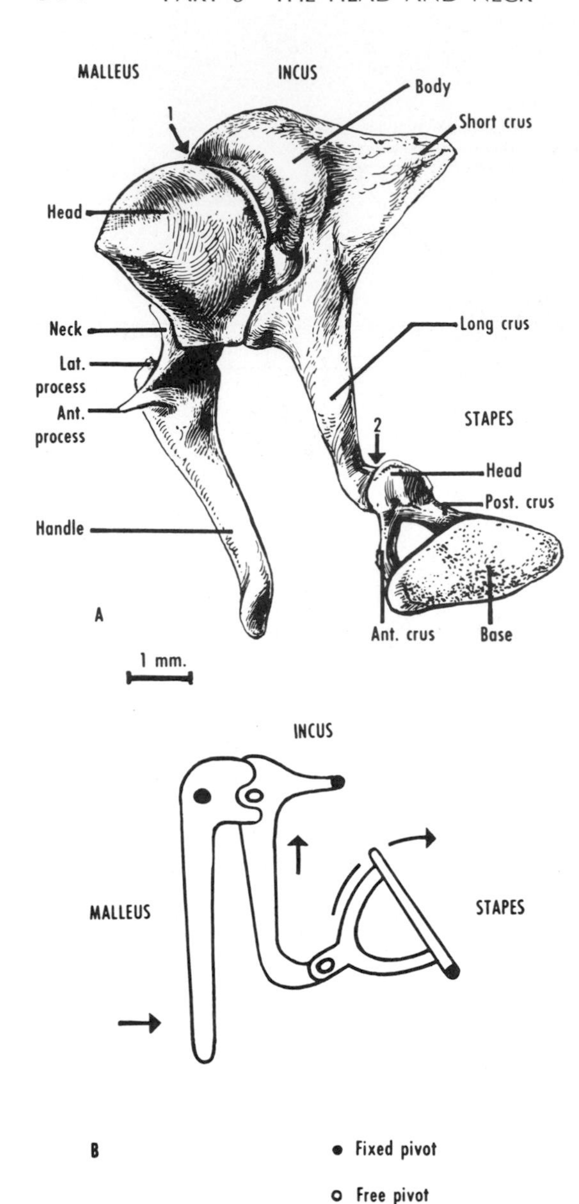

Figure 44–6 Ossicles of the right ear, medial aspect. **A**, Processes and articulations: *1*, incudomallear; *2*, incudostapedial. **B**, Diagram of the lever-like action of the ossicles. For simplification, movements (*arrows*) are shown as though they take place in one plane. The fixed pivots (*in black*) are anchored to the bony walls of the tympanic cavity. (**A** is based on Bassett, **B** on Guelke and Keen.)

Sensory Innervation. The middle ear is supplied by the auriculotemporal (fifth cranial) and tympanic (ninth cranial) nerves and by the auricular branch of the vagus.

Functional Considerations. Sound waves set the tympanic membrane in motion. These vibrations are converted into intensified movements of the stapes by the lever-like action of the auditory ossicles. Movements at the fenestra vestibuli are accompanied by compensatory movements at the fenestra cochleae. **Sound vibrations are transmitted to the inner ear by (1) the auditory ossicles and the fenestra vestibuli, (2) air in the tympanic cavity and the fenestra cochleae, and (3) bone conduction through the skull.**

FACIAL NERVE

The facial, or seventh cranial, nerve (fig. 44–7) is described here because of its close relationship to the middle ear. **The facial nerve comprises a larger part, which supplies the muscles of facial expression, and a smaller part termed the nervus intermedius, which contains taste fibers for the anterior two thirds of the tongue, secretomotor fibers for the lacrimal and salivary glands, and some pain fibers.** The

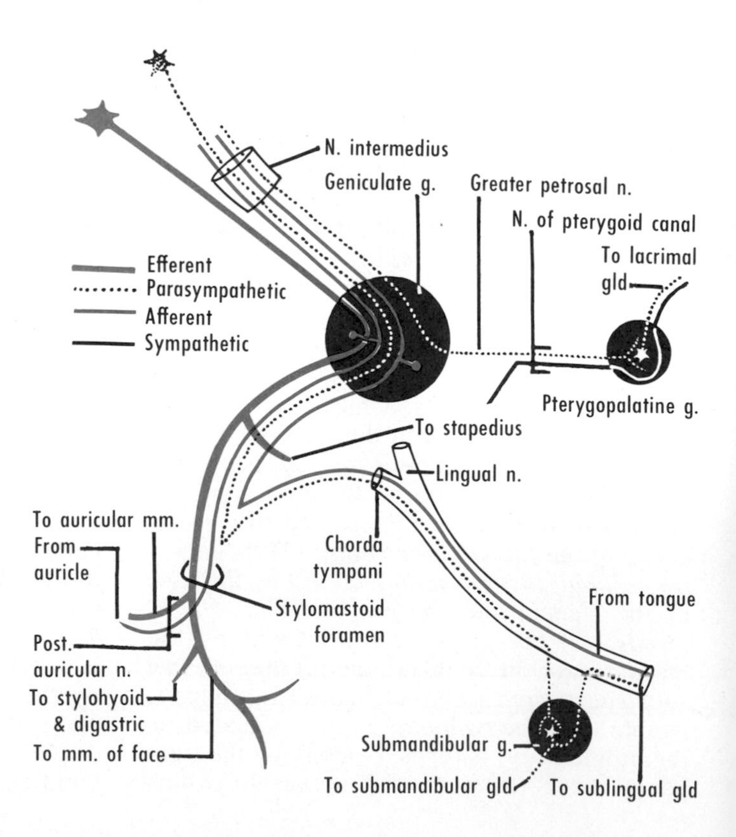

Figure 44–7 The facial nerve and its components.

two parts leave the brain at the lower border of the pons (cerebellopontine angle) and enter (with the eighth cranial nerve) the internal acoustic meatus. The facial nerve traverses the *facial canal* in the temporal bone. Above the promontory on the medial wall of the middle ear, the nerve expands to form the *geniculate ganglion* (see fig. 44–9), which contains the cells of origin of its taste fibers. The nerve turns sharply backward and then sweeps downward behind the middle ear (see fig. 44–5) and emerges at the stylomastoid foramen. Finally, the facial nerve enters the parotid gland, forms the parotid plexus, and gives terminal branches to the facial muscles (see fig. 47–5*A*). **The facial nerve traverses in succession the (1) posterior cranial fossa, (2) internal acoustic meatus, (3) facial canal, and (4) parotid gland and face.**

Branches (Table 44–1). The facial nerve gives a number of communicating branches. The chief named branches arise in the facial canal, below the base of the skull, and in the parotid gland.

In the facial canal, the geniculate ganglion gives rise to the *greater petrosal nerve*. This branch passes forward to join the deep petrosal nerve (from the carotid sympathetic plexus) and form the *nerve of the pterygoid canal,* which reaches the pterygopalatine ganglion. **The greater petrosal nerve contains secretomotor fibers for the lacrimal and nasal glands** and a number of afferent fibers of uncertain distribution and function.

In the facial canal, the facial nerve as it descends gives off the *nerve to the stapedius* and the *chorda tympani*. The chorda enters the tympanic cavity, passes medial to the tympanic membrane and the handle of the malleus, and again enters the temporal bone. It leaves the skull through the petrotympanic fissure and descends in the infratemporal fossa. **The chorda tympani joins the lingual nerve, with which it is distributed to the anterior two thirds of the side and dorsum of the tongue.** The chorda contains (1) taste fibers from the anterior two thirds of the tongue and from the soft palate and (2) preganglionic secretory fibers, which synapse in the submandibular ganglion, the postganglionic fibers supplying the submandibular, sublingual, and lingual glands.

Immediately below the base of the skull, the facial nerve gives off *muscular branches* to the stylohyoid muscle and the posterior belly of the digastric muscle and the posterior auricular nerve, which supplies motor fibers to the auricular muscles and occipitalis and sensory fibers to the auricle.

Within the parotid gland, the facial nerve forms the *parotid plexus,* from which terminal branches radiate forward in the face to supply the muscles of facial expression, including the platysma. The branches are usually classified as *temporal, zygomatic, buccal, marginal,* and *cervical*. They contain afferent (probably proprioceptive or pain or both) as well as motor fibers.

Examination. For examination of the facial muscles, the subject is asked to show the teeth, puff out the cheeks, whistle, frown, and close the eyes tightly. **The level of a lesion of**

TABLE 44–1 Branches of Facial Nerve

Region	Branches	Abbreviation in Fig. 48–10
In internal acoustic meatus	Communicates with 8th cranial nerve	
In facial canal, from geniculate ganglion	Greater petrosal nerve	GP
	Communicating branch to tympanic plexus	
	External petrosal nerve	
In facial canal, beyond geniculate ganglion	Nerve to stapedius	St
	Chorda tympani	Ch Ty
Below base of skull	Stylohyoid and digastric branches	SH & D
	Communicating branches, e.g., with 9th and 10th cranial nerves	
	Posterior auricular nerve	PA
On face	Temporal branches } Zygomatic branches }	T & Z
	Buccal branches	B
	Marginal branch of mandible	M
	Cervical branch	C

the facial nerve is inferred from effects that depend on whether specific branches are intact. Involvement of the facial nerve in the facial canal (as in *Bell's palsy*) results in ipsilateral drooping of the mouth, which is produced by the unopposed contraction of the muscles of the sound side (see fig. 47–4*D*).

INTERNAL EAR

The internal ear, situated within the petrous part of the temporal bone, consists of a complex series of fluid-filled spaces, the membranous labyrinth, lodged within a similarly arranged cavity, the bony (osseous) labyrinth (fig. 44–8). The cochlea is the essential organ of hearing. Other portions (the utricle and semicircular ducts) of the internal ear constitute the vestibular apparatus, although equilibrium is also maintained by vision and proprioceptive impulses. The chief components of the internal ear are summarized in table 44–2.

Osseous Labyrinth. The osseous labyrinth comprises a layer of dense bone (*otic capsule*) in the petrous part of the temporal bone and the enclosed perilymphatic space, which contains perilymph. The perilymphatic space consists of a series of continuous cavities: semicircular canals, vestibule, and cochlea (fig. 44–9).

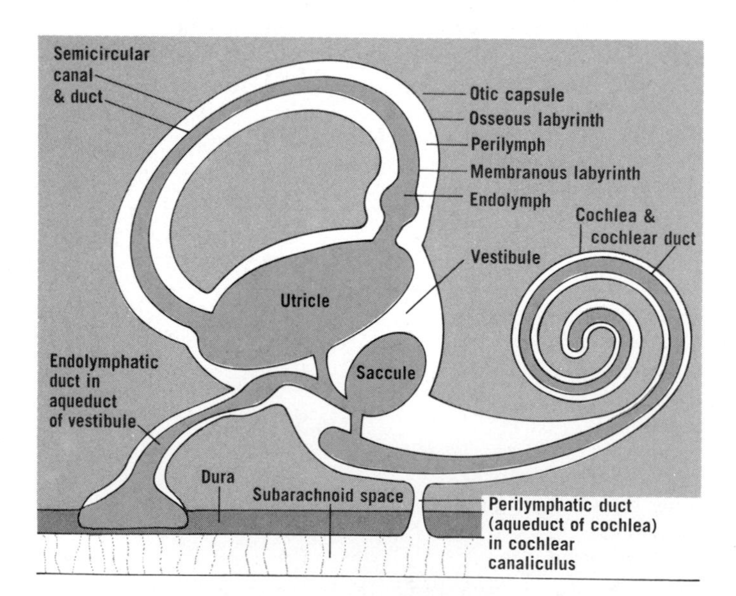

Figure 44–8 Osseous and membranous labyrinths, showing the basic arrangement and terminology of the internal ear. The membranous labyrinth and endolymph are shown in blue. That the aqueduct of the cochlea communicates with the subarachnoid space is questionable.

TABLE 44–2 COMPONENTS OF INTERNAL EAR AND THEIR OPENINGS

Osseous Labyrinth	Membranous Labyrinth
3 SEMICIRCULAR CANALS	3 SEMICIRCULAR DUCTS
VESTIBULE	UTRICLE
Five openings of semicircular canals	Five openings of semicircular ducts
Aqueduct of vestibule containing endolymphatic duct	SACCULE
	Endolymphatic duct and sac arise via utricular and saccular ducts
Fenestra vestibuli closed by base of stapes	
COCHLEA	COCHLEAR DUCT
Cochlear canaliculus containing perilymphatic duct (aqueduct of cochlea)	Continuous with saccule via ductus reuniens
Fenestra cochleae closed by secondary tympanic membrane	

SEMICIRCULAR CANALS. The *anterior, posterior,* and *lateral semicircular canals* are at right angles one to another (see fig. 44–4). The lateral canal is not quite horizontal.

VESTIBULE. The term *vestibule* is restricted here to the middle part of the bony labyrinth, immediately medial to the tympanic cavity (see fig. 44–4). It contains the utricle and saccule of the membranous labyrinth. The *fenestra vestibuli,* situated laterally, is closed by the base of the stapes. The *aqueduct of the vestibule* transmits the endolymphatic duct (see fig. 44–8).

COCHLEA. The cochlea (fig. 44–10), named for its resemblance to the shell of a snail, is a helical tube of about 2½ turns. Its base lies against the lateral end of the internal acoustic meatus, its basal coil forms the promontory of the middle ear, and its apex is directed anterolaterally (see fig. 44–9). A bony core, the *modiolus,* transmits the cochlear nerve and contains the spiral ganglion (fig. 44–10). A winding shelf, the *osseous spiral lamina,* projects from the modiolus like the flange of a screw. The cochlear duct extends from this lamina to the wall of the cochlea so that the space in the cochlea is divided into two: a scala vestibuli anteriorly and a scala tympani

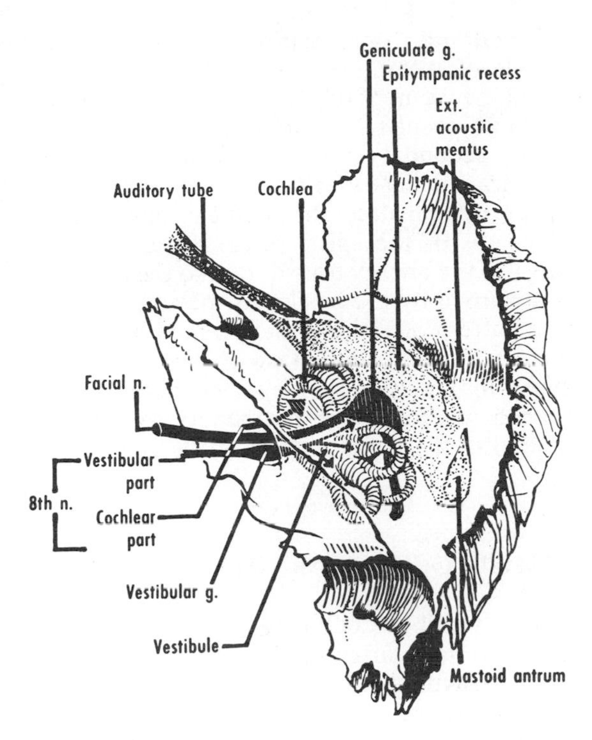

Figure 44–9 Right temporal bone viewed from above. The osseous labyrinth, tympanic cavity, and air spaces are shown as if the surrounding bone were transparent.

posteriorly. The *scala vestibuli* begins in the vestibule and passes to the apex of the cochlea, where the two scalae communicate with each other (at the *helicotrema;* see fig. 44–1); the *scala tympani* returns to end blindly near the fenestra cochleae, which is closed by the secondary tympanic membrane.

The *perilymphatic duct,* or *aqueduct of the cochlea,* is situated in a bony channel, the *cochlear canaliculus,* and is frequently said to connect the scala tympani with the subarachnoid space (see fig. 44–8).

Membranous Labyrinth. The membranous labyrinth (fig. 44–11) lies within the bony labyrinth and contains a different fluid, endolymph. The membranous labyrinth consists of a series of continuous cavities: semicircular ducts, utricle and saccule, and cochlear duct.

SEMICIRCULAR DUCTS. The *anterior, posterior,* and *lateral semicircular ducts* are situated eccentrically in the corresponding canals. One end of each duct is enlarged as an ampulla (fig. 44–11), each of which possesses a neuroepithelial *ampullary crest,* which is stimulated by movement of the endolymph.

UTRICLE AND SACCULE. The utricle and saccule lie in the vestibule and communicate with

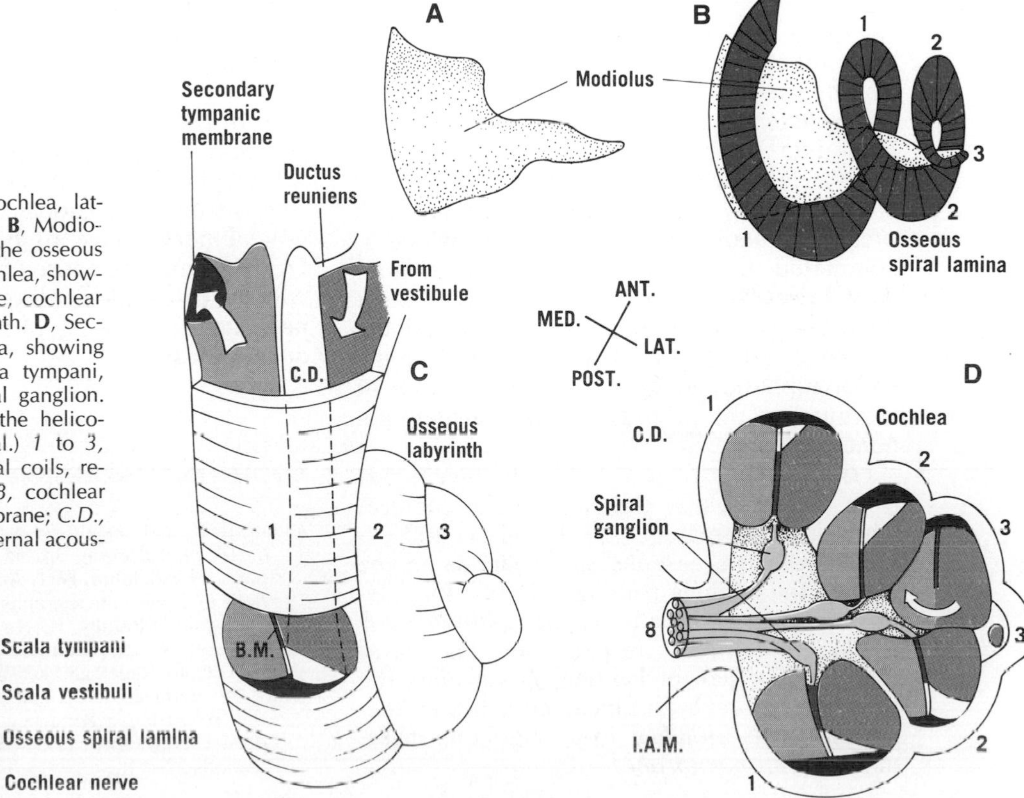

Figure 44–10 The cochlea, lateral aspect. **A**, Modiolus. **B**, Modiolus with the addition of the osseous spiral lamina. **C**, The cochlea, showing the basilar membrane, cochlear duct, and osseous labyrinth. **D**, Section through the cochlea, showing the scala vestibuli, scala tympani, cochlear duct, and spiral ganglion. An arrow is placed in the helicotrema. (After Wolff et al.) *1* to *3*, basal, middle, and apical coils, respectively, of cochlea; *8*, cochlear nerve; *B.M.*, basilar membrane; *C.D.*, cochlear duct; *I.A.M.*, internal acoustic meatus.

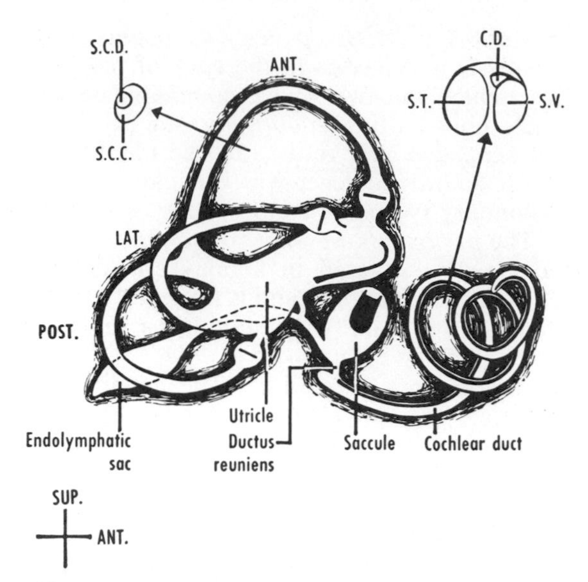

Figure 44–11 Membranous labyrinth. Note the ampullae of the semicircular ducts. The black lines and areas in the ampullae, utricle, saccule, and cochlear duct represent the neuro-epithelium: the ampullary crests, maculae, and spiral organ, respectively. Two cross sections are included. *C.D.*, cochlear duct; *S.C.C.*, semicircular canal; *S.C.D.*, semicircular duct; *S.T.*, scala tympani; *S.V.*, scala vestibuli.

each other by utricular and saccular ducts. The utricle has five openings for the semicircular ducts: the anterior and posterior semicircular ducts have one opening in common. The saccule is joined to the cochlear duct by the *ductus reuniens*. The utricle and saccule each present a neuro-epithelial *macula*, which is stimulated by gravity. The *endolymphatic duct* arises from the utricular and saccular ducts and is transmitted by the aqueduct of the vestibule (see fig. 44–8). The duct ends in the endolymphatic sac (fig. 44–11), under cover of the dura on the petrous part of the temporal bone.

COCHLEAR DUCT. The cochlear duct winds from the saccule to the apex of the cochlea, where it ends blindly (see fig. 44–1) and extends from the osseous spiral lamina to the wall of the cochlea (see fig. 44–10). Its anterior and posterior walls are the *vestibular* and *basilar membranes*, respectively. The *spiral organ*, the organ of hearing, lies against the basilar membrane and includes neuro-epithelial hair cells attached to a gelatinous mass, the *tectorial membrane*.

Functional Considerations. The functional details of the internal ear are poorly understood. Movement of the stapes in the fenestra vestibuli results in compensatory movement of the secondary tympanic membrane in the fenestra cochleae (see fig. 44–1). Vibrations in the perilymph affect the cochlear duct, and the hair cells of the spiral organ are stimulated. Most of the energy transmitted to the inner ear is absorbed by the basilar membrane, the motion pattern of which is determined by the frequency of the sound. Low-frequency sounds cause maximum activity in the basilar membrane, whereas high-frequency sounds are limited to the basal portion of the cochlea.

VESTIBULOCOCHLEAR NERVE

The vestibulocochlear, or eighth cranial, nerve contains afferent fibers from the internal ear. It leaves the brain at the lower border of the pons (cerebellopontine angle; see fig. 43–3) and enters (with cranial nerve 7) the internal acoustic meatus. The *vestibular part*, concerned with equilibration, is distributed to the maculae of the utricle and saccule and to the ampullary crests of the semicircular ducts. The vestibular fibers arise in bipolar cells in the *vestibular ganglion* in the internal acoustic meatus. The *cochlear part* of the eighth cranial nerve, concerned with hearing, is distributed to the hair cells of the spiral organ. The cochlear fibers arise in bipolar cells in the *spiral ganglion* in the modiolus (see fig. 44–10). The vestibular and cochlear parts of the eighth cranial nerve have to be tested separately. **Deafness may be caused by (1) a lesion of the cochlea or cochlear nerve fibers (neural deafness) or (2) a disease of the middle ear (conductive deafness).**

ADDITIONAL READING

Anson, B. J., and Donaldson, J. A., *Surgical Anatomy of the Temporal Bone,* 3rd ed., W. B. Saunders Company, Philadelphia, 1981. An atlas of detailed illustrations and photomicrographs.

Vidić, B., and O'Rahilly, R., *An Atlas of the Anatomy of the Ear,* W. B. Saunders Company, Philadelphia, 1971. Color slides and key drawings.

Wolff, D., Bellucci, R. J., and Eggston, A. A., *Surgical and Microscopic Anatomy of the Temporal Bone,* Hafner, New York, 1971. Photomicrographs of serial sections.

QUESTIONS

44–1 What is the structure of the tympanic membrane and where is the membrane incised?

44–2 With what does the tympanic cavity communicate?

44–3 What does the epitympanic recess contain?

44–4 What is the tegmen tympani?

44–5 What closes the fenestra vestibuli, and what closes the fenestra cochleae?

44–6 Of which type are the joints of the middle ear?

44–7 What is the nervus intermedius?

44–8 Which fibers arise in the geniculate ganglion?

44–9 Which are the main directions taken by the facial nerve (a) in approaching the geniculate ganglion, (b) after it leaves the ganglion, and (c) in approaching the stylomastoid foramen?

44–10 What is the chorda tympani?

44–11 How is the level of a lesion of the facial nerve determined?

44–12 What are the chief signs of facial palsy?

44–13 What is the otic capsule?

44–14 What are the main parts of (a) the osseous and (b) the membranous labyrinth?

44–15 Is the lateral semicircular canal horizontal?

44–16 In which direction does the apex of the cochlea point?

44–17 Which structures separate the scalae from each other and from the cochlear duct?

44–18 Where are the neuro-epithelial areas in the membranous labyrinth?

45
THE ORBIT

BONY ORBIT

The orbits (figs. 45–1 and 45–2) are two bony cavities occupied by the eyes and associated muscles, nerves, blood vessels, fat, and much of the lacrimal apparatus. Each orbit is shaped like a pear or a four-sided pyramid, with its apex situated posteriorly and its base anteriorly. **The orbit is related (1) above to the anterior cranial fossa and usually to the frontal sinus, (2) laterally to the temporal fossa in front and to the middle cranial fossa behind, (3) below to the maxillary sinus, and (4) medially to the ethmoidal and usually the sphenoidal sinuses.**

The margin of the orbit, readily palpable *in*

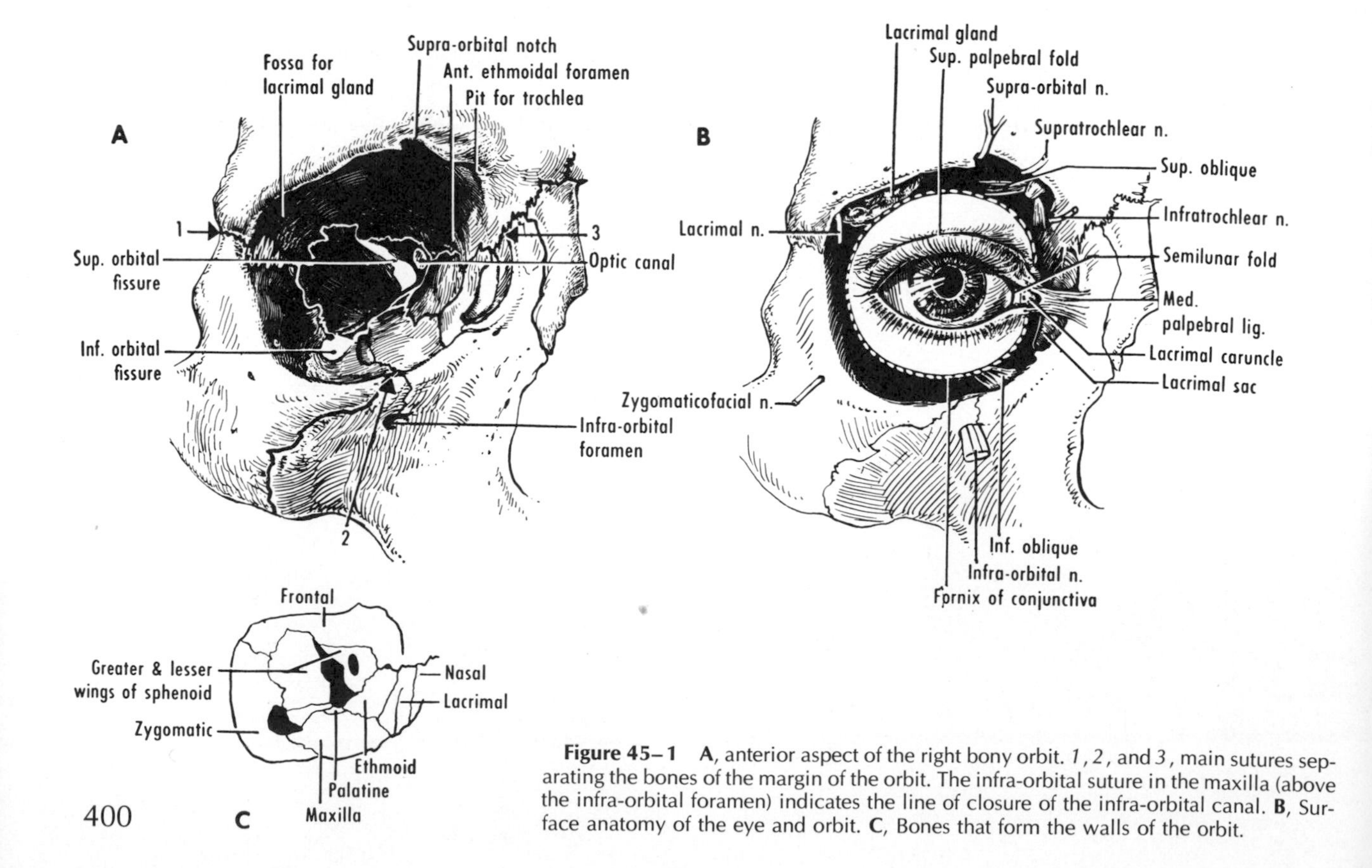

Figure 45–1 **A**, anterior aspect of the right bony orbit. *1*, *2*, and *3*, main sutures separating the bones of the margin of the orbit. The infra-orbital suture in the maxilla (above the infra-orbital foramen) indicates the line of closure of the infra-orbital canal. **B**, Surface anatomy of the eye and orbit. **C**, Bones that form the walls of the orbit.

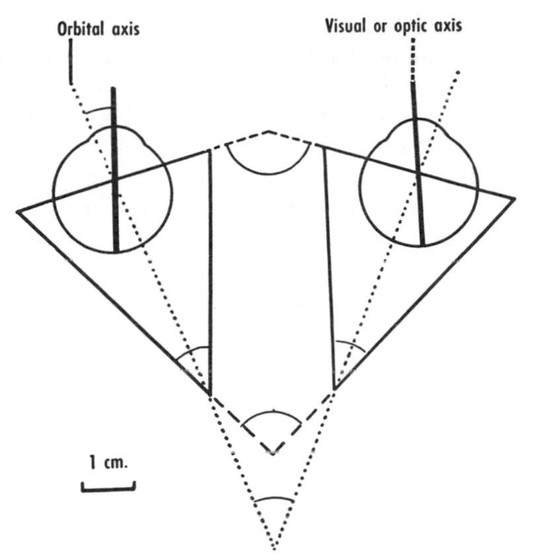

Figure 45–2 Horizontal scheme of the orbits, showing the angles formed by the walls. The dotted lines indicate the longitudinal axes of the orbits. The visual axes, here shown directed toward a distant object, are nearly parallel. (After Whitnall.)

vivo, is formed by the frontal, zygomatic, and maxillary bones (fig. 45–1*A*). It may be considered in four parts: superior, lateral, inferior, and medial.

The *superior margin,* formed by the frontal bone, presents near its medial end either a *supra-orbital notch* or a *supra-orbital foramen,* which transmits the nerve and vessels of the same name.

The *lateral margin* is formed by the zygomatic process of the frontal bone and the frontal process of the zygomatic bone.

The *inferior margin* is formed by the zygomatic and maxillary bones. The *infra-orbital foramen,* for the nerve and artery of the same name, is less than 1 cm below the inferior margin.

The *medial margin,* formed by the maxilla and the lacrimal and frontal bones, is expanded as the *fossa for the lacrimal sac*. The fossa descends through the floor of the orbit as the *nasolacrimal canal,* which transmits the nasolacrimal duct from the lacrimal sac to the inferior meatus of the nose.

The orbit possesses four walls (fig. 45–1*A* and *C*): a roof, lateral wall, floor, and medial wall.

The *roof* (frontal and sphenoid bones) presents the *fossa for the lacrimal gland* anterolaterally and the *trochlear pit* for the pulley of the superior oblique muscle anteromedially. **The optic canal lies in the back of the roof, between the roots of the lesser wing of the sphenoid bone. It transmits the optic nerve and ophthalmic artery from the middle cranial fossa.**

The *lateral wall* (zygomatic and sphenoid bones) is demarcated behind by the superior and inferior orbital fissures. **The superior orbital fissure lies between the greater and lesser wings of the sphenoid bones. It communicates with the middle cranial fossa and transmits cranial nerves 3, 4, and 6, the three branches of the ophthalmic nerve, and the ophthalmic veins** (fig. 45–3). *The inferior orbital fissure* communicates with the infratemporal and pterygopalatine fossae and transmits the infraorbital (maxillary) nerve and artery and the zygomatic nerve. **The lateral walls of the two orbits are set at approximately a right angle, whereas the medial walls are nearly parallel to each other** (see fig. 45–2).

The *floor* (maxilla and zygomatic and palatine bones) presents the *infra-orbital groove* and *canal* for the nerve and artery of the same name. The inferior oblique muscle arises anteromedially just lateral to the nasolacrimal canal.

The medial wall (ethmoid, lacrimal, and frontal bones) is very thin. At the junction of the medial wall with the roof, the *anterior* and *posterior ethmoidal foramina* transmit the nerves and arteries of the same name to the anterior cranial fossa.

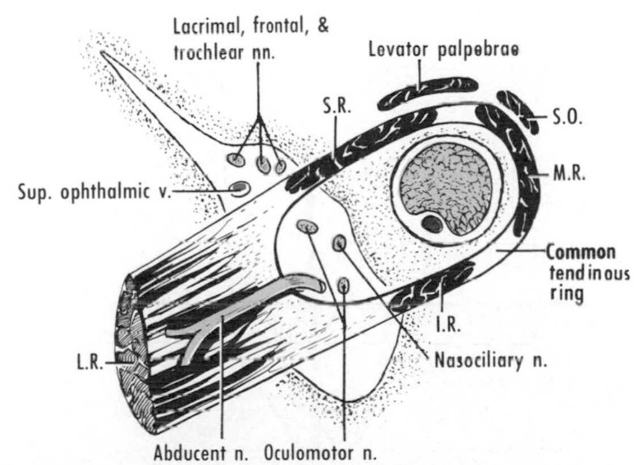

Figure 45–3 Superior orbital fissure and optic canal, anterior aspect. The optic canal and adjacent part of the fissure are surrounded by the common tendinous ring, from which the four recti arise. The structures that pass through the ring all lie at first within the cone formed by the muscles. The lacrimal, frontal, and trochlear nerves enter the orbit above the ring and therefore above the muscular cone. The optic canal contains the optic nerve and ophthalmic artery. Cf. fig. 45–5. (Based on Wolff and on Whitnall.)

OPHTHALMIC NERVE

The ophthalmic nerve (first division of the trigeminal nerve) is an afferent nerve that supplies the globe and conjunctiva, lacrimal gland and sac, nasal mucosa and frontal sinus, external nose, upper eyelid, forehead, and scalp. It arises from the trigeminal ganglion and divides near the superior orbital fissure into the lacrimal, frontal, and nasociliary nerves. These pass through the superior orbital fissure and traverse the orbit (fig. 45–4). The lacrimal and frontal nerves lie above the muscles of the globe, whereas the nasociliary nerve enters the orbit inside the cone of muscles (see fig. 45–3).

The *lacrimal nerve* proceeds along the upper border of the lateral rectus and supplies the lacrimal gland, conjunctiva, and upper eyelid. A communication with the zygomatic nerve carries some secretory fibers to the lacrimal gland.

The *frontal nerve* passes forward on the levator palpebrae superioris and divides into the supra-orbital and supratrochlear nerves. The *supra-orbital nerve* leaves the orbit through the supra-orbital notch or foramen and supplies the forehead, scalp, upper eyelid, and frontal sinus. The *supratrochlear nerve,* much smaller, supplies the forehead and upper eyelid (fig. 45–5*A*).

The *nasociliary nerve* is within the cone of muscles and is therefore on a lower plane than the lacrimal and frontal nerves. **The nasociliary nerve is the sensory nerve to the eye and is accompanied by the ophthalmic artery.** It courses forward below the superior rectus, crosses the optic nerve, and is continued medially as the anterior ethmoidal nerve. The nasociliary nerve gives off a *communicating branch* to the ciliary ganglion, *long ciliary nerves* (which convey sympathetic fibers to the dilator pupillae), and the *infratrochlear* and *anterior ethmoidal nerves*. The last-named contributes branches to the nasal cavity and external nose.

The area of skin supplied by the ophthalmic nerve (see fig. 47–6) is tested for sensation by cotton wool and a pin. Blowing on or touching lightly the cornea causes "closure of the eyes" (corneal reflex) as a result of bilateral contraction of the orbicularis oculi muscles. The afferent limb includes the nasociliary nerve, and the efferent limb is the facial nerve.

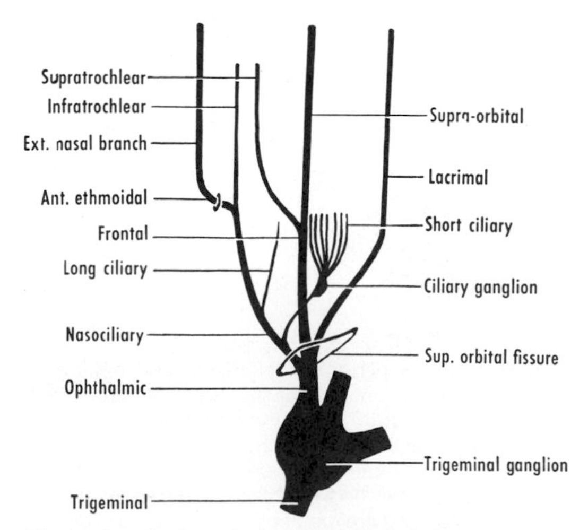

Figure 45–4 Branches of the ophthalmic nerve, superior aspect.

OPHTHALMIC VESSELS

The *ophthalmic artery* (fig. 45–5*B*) arises from the internal carotid artery medial to the anterior clinoid process and passes through the optic canal, below the optic nerve. Accompanied by the nasociliary nerve, it turns medially, below the superior rectus and usually above the optic nerve. It then turns forward, above the medial rectus. It gives numerous branches, only some of which are significant.

The important *central artery of the retina* pierces the optic nerve and supplies the retina (see figs. 46–5 and 46–6). **Its terminal branches are virtually end arteries.** *Long* and *short posterior ciliary arteries* pierce the sclera and supply the uvea. The *lacrimal artery,* which supplies the lacrimal gland and eyelids, gives off a *recurrent meningeal branch,* which anastomoses with the middle meningeal artery and hence provides an **anastomosis between the internal and external carotid arteries.** Muscular branches are important in that they give off the *anterior ciliary arteries,* which pierce the sclera and supply the iris and ciliary body (see fig. 46–6).

Additional branches include the *supra-orbital, anterior ethmoidal,* and *palpebral arteries*. The terminal branches of the ophthalmic artery are the *supratrochlear* and *dorsal nasal arteries*. The latter anastomoses with branches of the facial artery and provides another example of an **anastomosis between the internal and external carotid arteries.**

The superior and inferior ophthalmic veins drain the orbit and have important communications with the facial vein, pterygoid plexus,

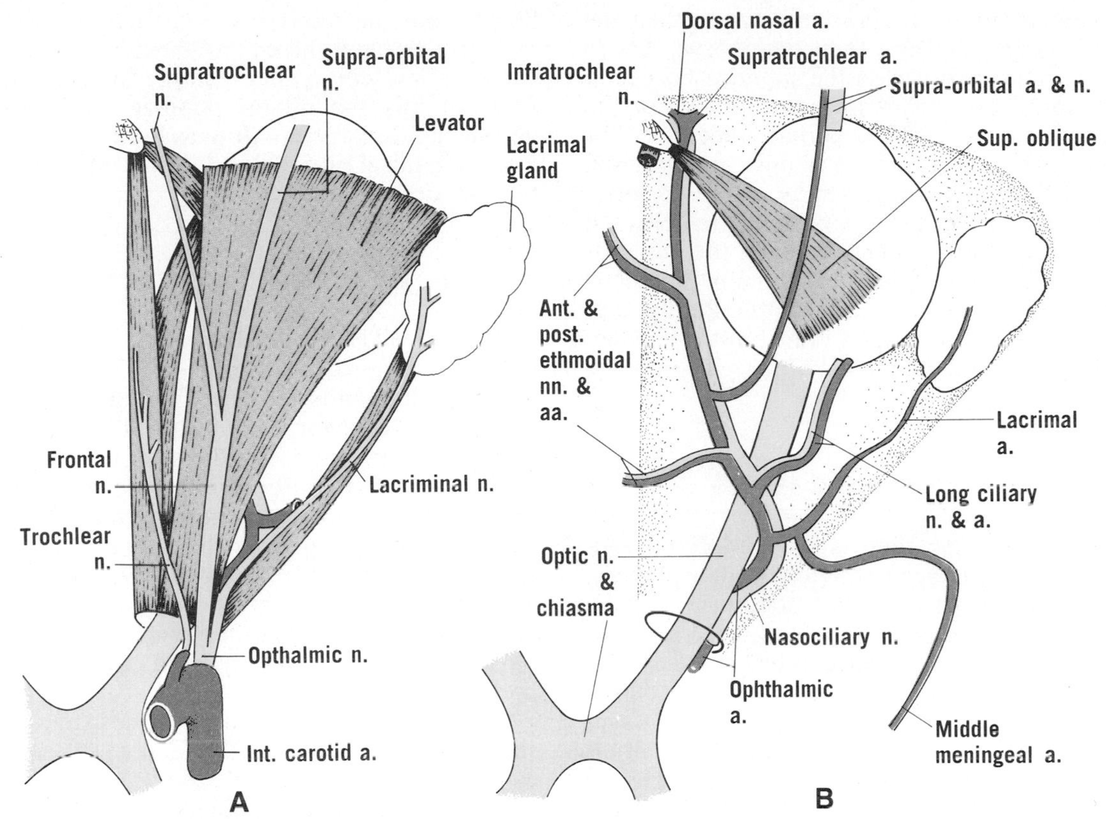

Figure 45–5 Right orbit from above. **A** shows the three nerves that enter above the muscular cone. In **B**, the levator and superior rectus have been removed to show the artery and nerves that enter within the muscular cone. Cf. fig. 45–3.

and cavernous sinus, in which they end directly or indirectly. The superior ophthalmic vein is formed near the root of the nose and allows the **spread of superficial infections of the face to the cavernous sinus.** The *venae vorticosae* (see fig. 46–6) are four veins that drain the uvea and end in the ophthalmic veins.

OCULOMOTOR, TROCHLEAR, AND ABDUCENT NERVES

The oculomotor (third cranial) nerve supplies all the muscles of the eyeball except the superior oblique muscle and the lateral rectus. It emerges from the brain stem, passes lateral to the posterior clinoid process, traverses the cavernous sinus, and divides into superior and inferior divisions, which pass through the superior orbital fissure within the common tendinous ring. The oculomotor nerve supplies the levator palpebrae superioris and superior rectus (by its superior division) and the medial rectus, inferior rectus, and inferior oblique muscle (by its inferior division). From the branch to the inferior oblique muscle, **a parasympathetic communication joins the ciliary ganglion and conveys motor fibers to the sphincter pupillae and ciliary muscle. In the act of focusing the eyes on a near object, the oculomotor nerves are involved in adduction (medial recti), accommodation (ciliary muscle), and miosis (sphincter pupillae).** Paralysis of the oculomotor nerve results in ptosis (paralysis of the levator), abduction (unopposed lateral rectus), and other signs.

The trochlear (fourth cranial) nerve supplies only the superior oblique muscle of the eyeball. It emerges from the back of the brain stem, winds around the cerebral peduncle, traverses the cavernous sinus, and passes through the superior orbital fissure. It then lies above the levator and enters the superior oblique muscle. The trochlear nerve is tested by asking the subject to look downward when the eye is in adduction.

The abducent (sixth cranial) nerve supplies only the lateral rectus muscle of the eyeball. It emerges from the brain stem between the pons and medulla. The abducent nerve bends sharply forward across the superior border of the apical portion of the petrous part of the temporal bone, which perhaps accounts for **abducent involvement in almost any cerebral lesion that is accompanied by increased intracranial pressure.** The nerve next traverses the cavernous sinus, passes through the superior orbital fissure within the common tendinous ring, and enters the lateral rectus. Paralysis of the lateral rectus results in inability to abduct the eye beyond the middle of the palpebral fissure.

CILIARY GANGLION (fig. 45–6)

The ciliary ganglion is the peripheral ganglion of the parasympathetic system of the eye. It is situated between the optic nerve and the lateral rectus. Communications with the nasociliary nerve convey afferent fibers from the eye. A parasympathetic root from the oculomotor nerve contains the only fibers that synapse in the ganglion. The **postganglionic fibers pass to the short ciliary nerves (which are branches of the ganglion) and supply the ciliary muscle and sphincter pupillae.** Sympathetic fibers from the internal carotid plexus reach and pass through the ciliary ganglion. By way of the short ciliary nerves, the **sympathetic fibers supply the dilator pupillae and blood vessels, as well as smooth muscle in the eyelid (superior tarsal muscle) and in the inferior orbital fissure (orbitalis).**

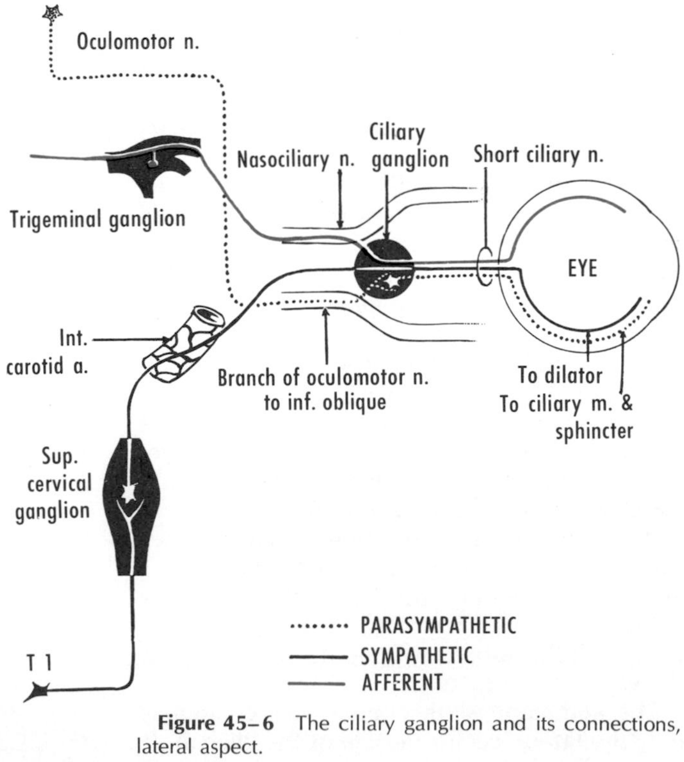

Figure 45–6 The ciliary ganglion and its connections, lateral aspect.

MUSCLES OF EYEBALL

The eyeball is moved chiefly by six muscles: four recti and two oblique muscles (fig. 45–7). These skeletal muscles arise from the back of the orbit (except for the inferior oblique muscle) and are inserted into the sclera.

The four recti arise from a common tendinous ring, which surrounds the optic canal and a part of the superior orbital fissure. All the structures that enter the orbit through the optic canal and adjacent part of the fissure lie at first within the cone of the recti (see fig. 45–3). The four muscles are inserted into the *front* portion of the sclera.

The *superior oblique muscle* arises from the sphenoid bone above and medial to the optic canal. It passes forward, above the medial rectus, and through a cartilaginous pulley (the *trochlea*) attached to the frontal bone. The tendon is thereby turned backward and is inserted into the *back* of the sclera. The *inferior oblique muscle* arises from the maxilla at the front of the orbit, passes backward, and is inserted into the *back* of the sclera.

The superior oblique muscle is supplied by the trochlear nerve, the lateral rectus by the abducent nerve, and the others by the oculomotor nerve: SO_4, LR_6, $remainder_3$.

Actions of Muscles of Eyeball. The eye is poised in the fascia and fat of the orbit, and equilibrium is maintained by all the muscles, none of which ever acts alone. Movements may be considered to be around a vertical axis (abduction and adduction), a lateromedial axis (elevation and depression) and even an anteroposterior axis (extorsion and intorsion).

The recti extend from the back of the orbit to the front of the sclera. **The lateral and medial recti are purely an abductor and adductor, respectively. The superior and inferior recti elevate and depress, respectively, and, because of their lateral course, are the only muscles that can do so when the eye is abducted.** The trochlea of the superior oblique muscle serves as its functional origin, and hence the two oblique muscles may be said to extend from the front of the orbit to the back of the sclera.

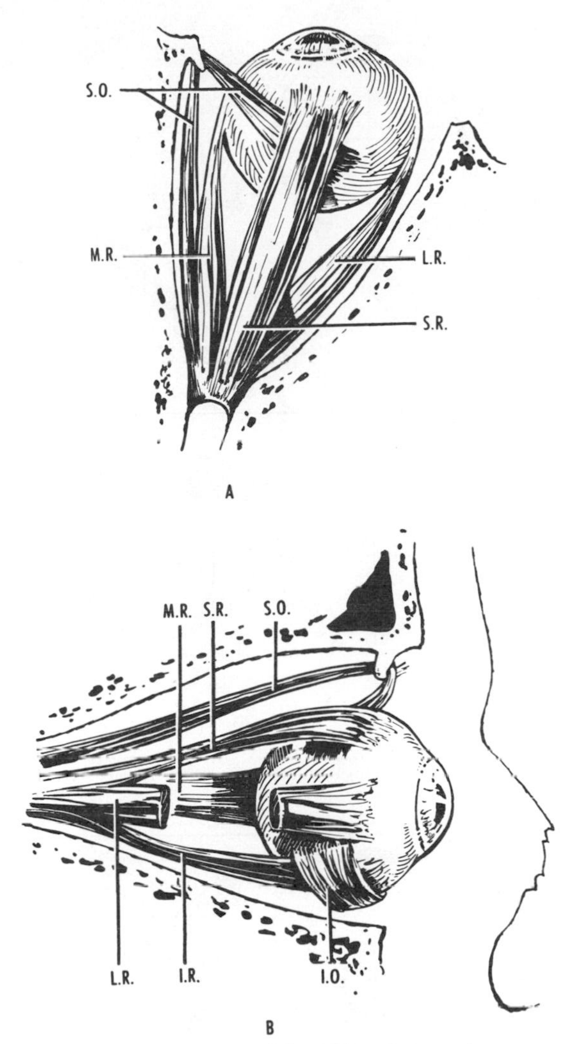

Figure 45–7 Muscles of the globe. **A**, Superior aspect. **B**, Lateral aspect. Note that the two oblique muscles pass below the corresponding recti. (**A** is based on Krimsky, **B** on Cogan.)

The superior and inferior oblique muscles depress and elevate, respectively, and, because of their lateral course, are the only muscles that can do so when the eye is adducted (fig. 45–8).

Paralysis of a muscle is noted by (1) limitation of movement in the field of action of the paralyzed muscle and (2) the presence of two images (*diplopia*) that are separated maximally when an attempt is made to use the paralyzed muscle.

OPTIC NERVE

The optic (second cranial) nerve is the nerve of sight, and it extends from the eye to the optic chiasma. Developmentally, it may be considered as a tract from the retina (a derivative of the brain) to the brain. The nerve fibers, which arise in the retina, converge on the optic disc, pierce the layers of the eye, and receive myelin sheaths. The nerve is surrounded by sheaths continuous with the meninges. The optic nerve lies within the cone of the recti, is crossed by the ophthalmic artery and nasociliary nerve, is pierced by the central vessels of the retina, and passes through the optic canal. **The optic nerve is related below to the internal carotid and ophthalmic arteries and to the hypophysis. The nerve ends in the optic chiasma, where the medial fibers decussate. (The decussating fibers are those that come from the nasal side of the retina and represent the temporal side of the visual field.)** Examination of the optic nerve includes ophthalmoscopy, testing of visual acuity, and plotting of the visual field.

EYELIDS

The eyelids (*palpebrae*) (fig. 45–9) are musculofibrous folds in front of each orbit. The upper eyelid, more extensive and mobile, meets the lower at the medial and lateral *angles* (*canthi*). In some people, chiefly Oriental, the medial canthus is covered by a fold of skin (*epicanthus*). The *palpebral fissure*, between the lids, is the mouth of the conjuctival sac. The free margin of each lid possesses hairs termed *eyelashes* (*cilia*), near which are *ciliary glands*. Infection of a ciliary gland may result in a sty(e). Medially, the lid margin presents the lacrimal punctum and, between the lids, an area termed the *lacrimal lake*. The floor of the lake presents a "fleshy" mass, the *lacrimal caruncle*. The caruncle lies on a conjunctival fold, the *plica semilunaris*.

The upper eyelid (fig. 45–9) is composed of skin and subcutaneous tissue, muscle (the palpebral part of the orbicularis oculi and levator palpebrae superioris), fibrous tissue (including the tarsal plate), and mucous membrane (the palpebral part of the conjunctiva).

The subcutaneous tissue usually contains no fat, and fluid can readily accumulate there. The *levator palpebrae superioris* arises from the sphenoid bone above the optic canal and is inserted into the skin of the upper lid and the upper border of the tarsal plate by means of the superior tarsal muscle (fig. 45–10). The levator is supplied by the oculomotor nerve; the innervation of the tarsal muscle is sympathetic.

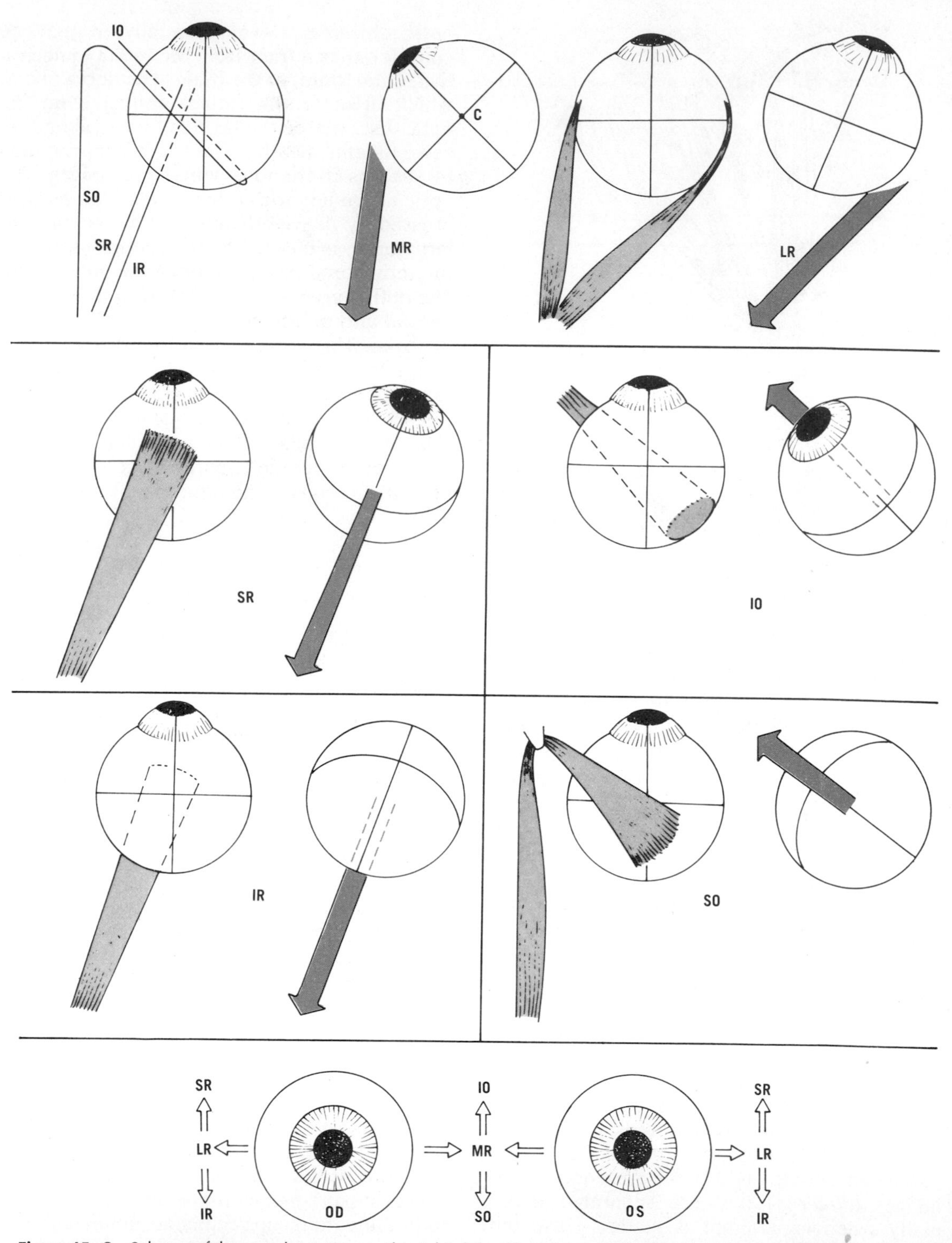

Figure 45–8 Scheme of the muscles acting on the right globe. Note the actions of the superior (*SR*) and inferior (*IR*) recti on the abducted eye, and those of the inferior (*IO*) and superior (*SO*) obliqui on the adducted eye. The lowest scheme summarizes the muscles responsible for vertical movements of the abducted and adducted eyes, anterior aspect. The arrows must be followed in strict sequence, i.e., beginning with the horizontal arrows. *MR*, medial rectus; *OD*, oculus dexter (right eye); *OS*, oculus sinister (left eye).

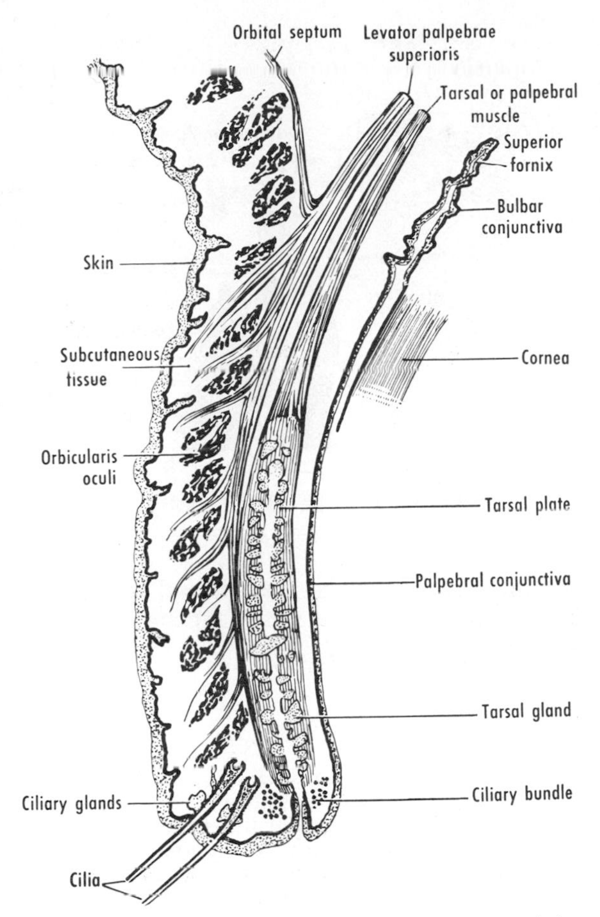

Figure 45–9 Sagittal section through the upper eyelid.

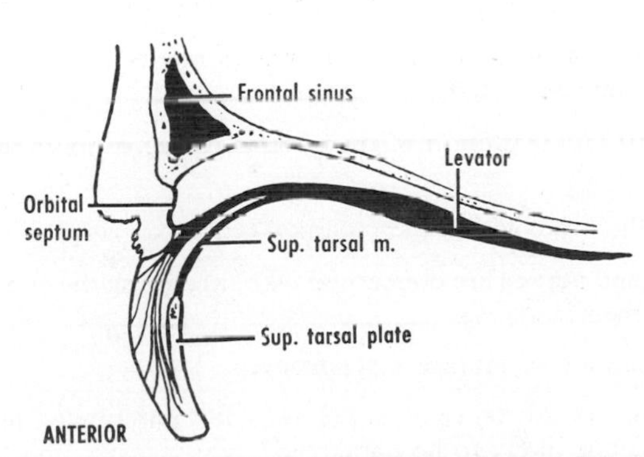

Figure 45–10 Sagittal section through the eyelid and roof of the orbit to show the levator palpebrae superioris. (After Whitnall.)

Paralysis of the levator results in drooping (ptosis) of the upper lid. The *tarsal plate* is a fibrous support related posteriorly to *tarsal glands*. The ends of the plates are anchored to the orbital margin by *lateral* and *medial palpebral ligaments*. **The medial palpebral ligament, identifiable *in vivo* on drawing the lids laterally, is in front of the lacrimal sac, to which it serves as a guide** (see fig. 45–1*B*). The *superior tarsal muscle* connects the levator with the tarsal plate, consists of smooth muscle, and is supplied by sympathetic fibers. **A lesion of the cervical sympathetic trunk may result in ptosis of the upper lid (Horner's syndrome, which also includes seeming recession of the eye and redness and increased temperature of the skin).**

CONJUNCTIVA

The conjunctiva is the mucous membrane that lines the back of the eyelids (palpebral conjunctiva) and the front of the globe (bulbar conjunctiva) (see figs. 45–9 and 46–3). The capillary interval, lined by conjunctiva, between the lids and the globe is termed the *conjunctival sac* (see fig. 46–3). The mouth of the sac is the *palpebral fissure,* which varies in size according to the degree to which the "eye is open." The reflections of the conjunctiva from the lids to the globe are known as *fornices*. The superior fornix receives the openings of the lacrimal glands.

The *palpebral conjunctiva* contains the openings of the lacrimal canaliculi, thereby allowing the conjunctival sac to communicate with the nasal cavity. **The palpebral conjunctiva is red and vascular and is examined when anemia is suspected.**

The *bulbar conjunctiva* is transparent, thereby allowing the sclera to show through as the "white of the eye." It is colorless, except when its vessels are dilated as a result of inflammation. Centrally, it is continuous at the limbus with the anterior epithelium of the cornea. The *plica semilunaris,* a conjunctival fold at the medial angle of the eye, helps to intercept foreign bodies.

Innervation and Blood Supply. The conjunctiva is supplied by branches of the ophthalmic nerve. **The vessels of the bulbar conjunctiva are visible *in vivo*.** They arise from (1) a peripheral palpebral arcade and (2) the anterior ciliary arteries (see fig. 46–6). **In conjunctivitis (e.g., from the wind) the bulbar conjunctiva becomes brick-red. In deeper conditions (e.g., diseases of the iris or ciliary**

body), in which branches of the anterior ciliary arteries are dilated, a rose-pink band of "ciliary injection" is produced.

LACRIMAL APPARATUS

The lacrimal apparatus comprises (1) the lacrimal gland and its ducts and (2) associated passages: the lacrimal canaliculi and sac and the nasolacrimal duct (fig. 45–11).

The *lacrimal gland,* lodged in a fossa anterolaterally at the roof of the orbit, rests on the lateral rectus and the levator. The main portion is the *orbital part,* but a process called the *palpebral part* projects into the upper lid. A dozen *lacrimal ducts* leave the palpebral part to enter the superior conjunctival fornix. Tears are secreted by the gland, and they keep the eye moist. The half that does not evaporate drains into the lacrimal sac.

The secretory fibers to the lacrimal gland are derived from the greater petrosal nerve (of the facial nerve) and from the nerve of the pterygoid canal (fig. 48–7). The fibers synapse in the pterygopalatine ganglion and reach the gland both directly and by a connection between the zygomatic and lacrimal nerves.

The *lacrimal canaliculi,* one in each lid, each begin at a *lacrimal punctum* on a *papilla.* They open into the lacrimal sac, which is continuous with the nasolacrimal duct. The sac, lodged in a fossa at the medial margin of the orbit, is partly covered by the medial palpebral ligament (see fig. 45–1*B*). The *nasolacrimal duct* extends from the lacrimal sac to the inferior meatus of the nose, and its lumen is marked by valve-like folds.

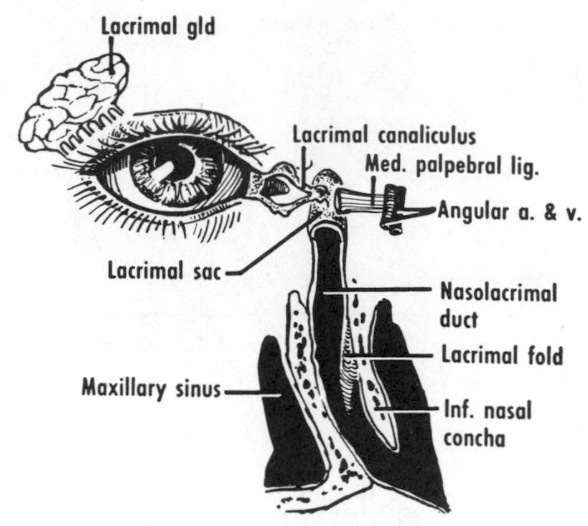

Figure 45–11 The lacrimal apparatus.

ADDITIONAL READING

Duke-Elder, S., and Wybar, K. C., *The Anatomy of the Visual System,* vol. 2 of *System of Ophthalmology,* ed. by S. Duke-Elder, Kimpton, London, 1961. An excellent work of reference for the orbit and the eye.

Whitnall, S. E., *The Anatomy of the Human Orbit and Accessory Organs of Vision,* 2nd ed., Oxford University Press, London, 1932. The classic study of the orbit.

Wolff, E., *The Anatomy of the Eye and Orbit,* 7th ed., rev. by R. Warwick, Lewis, London, 1976. An attractive, well-illustrated text.

QUESTIONS

45–1 Force applied to the rim of the orbit may be transmitted toward the side of the nose. Which thin bones are likely to be splintered?

45–2 Which nerve accompanies the ophthalmic artery?

45–3 Which is the most important branch of the ophthalmic artery?

45–4 Which cranial nerve is "the weakling of the cranial contents" because of its likely damage from increased intracranial pressure?

45–5 Where is the peripheral relay station of the parasympathetic fibers to the eye?

45–6 Which nerves enter the orbit within the common tendinous ring?

45–7 Which chief muscles and nerves are concerned with (a) closing the eyelids and (b) opening them?

45–8 What are the main features of Horner's syndrome?

45–9 On looking downward and to the right, a patient's left pupil failed to descend. Which muscle is likely to be paralyzed?

45–10 On looking upward and to the left, a patient's right pupil failed to ascend. Which muscle is most likely to be involved?

46
THE EYE

The eye (L., *oculus;* Gk, *ophthalmos*) (fig. 46–1) lies in the cavity of the orbit and measures 24 mm in diameter. The anteroposterior diameter may be greater (as in *myopia,* or shortsightedness) or less (as in *hypermetropia,* or longsightedness) than the normal. The midpoints of the two pupils lie about 60 mm apart. The central points of the corneal and scleral curvatures are known as the *anterior* and *posterior poles,* respectively.

TUNICS OF EYE (fig. 46–2)

The *eyeball* (*globe, bulb*) has three concentric coverings (fig. 46–3): (1) an external, fibrous tunic comprising the cornea and sclera; (2) a middle, vascular tunic comprising the iris, ciliary body, and choroid; and (3) an internal, nervous tunic, or retina.

EXTERNAL FIBROUS TUNIC

The *cornea* is the anterior, transparent part of the eye. **Most of the refraction by the eye takes place not in the lens but at the surface of the cornea.** When the cornea is not a part of a sphere but is more curved in one axis than in another, the condition is termed *astigmatism.* The cornea is continuous with the conjunctiva and the sclera, and the junctional region is

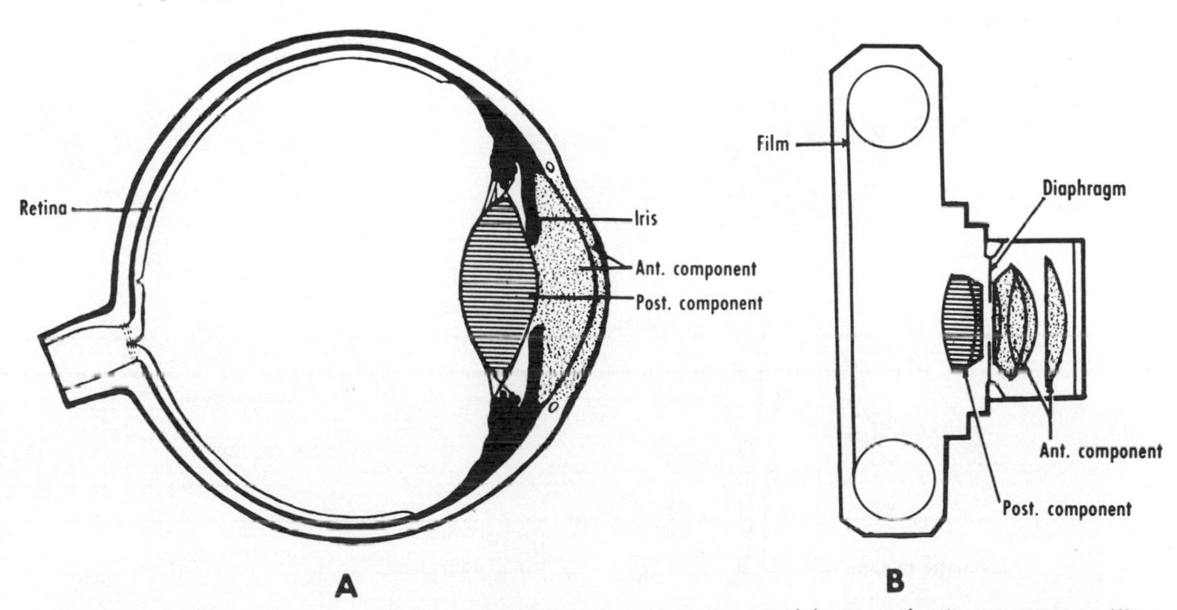

Figure 46–1 Optical comparison of (**A**) the eye and (**B**) a miniature camera. Each has two refractive components: (1) an anterior (cornea and aqueous humor) and (2) a posterior (lens in the eye), separated by an iris diaphragm. The lens systems have a focal length of 2 cm and 5 cm and an aperture range of f2.5 to f11 and f2 to f22, respectively. The visual image, however, is not imprinted as on a film but is coded and transmitted more as in television.

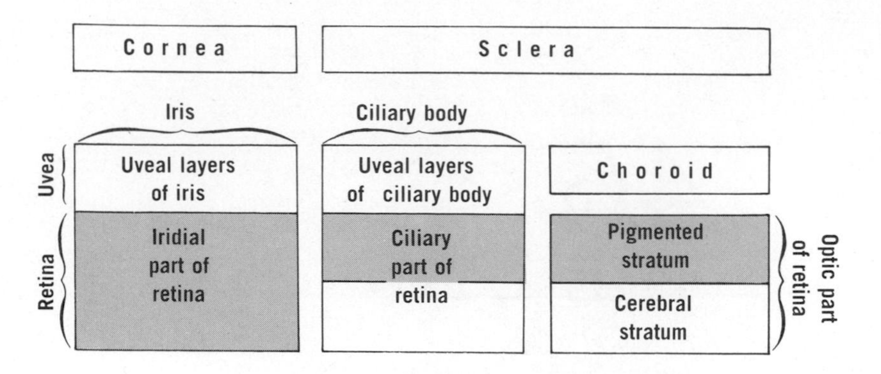

Figure 46–2 Scheme of the tunics of the eye: cornea and sclera, uvea, and retina. From left to right in the drawing represents a progression from anterior to posterior in the eye.

known as the *limbus*. **The cornea is supplied by the ophthalmic nerve (from the fifth cranial nerve) by means of its ciliary branches.** The eyelids close on stimulation of the cornea (corneal reflex). The cornea consists of five layers histologically, and it is avascular.

In inflammation of the conjunctiva, the posterior conjunctival vessels (from the palpebral) become dilated, whereas in inflammation of the cornea, iris, or ciliary body, the anterior ciliary vessels become dilated. These last-named vessels do not move when the conjunctiva is moved.

The *sclera* is the posterior, opaque part of

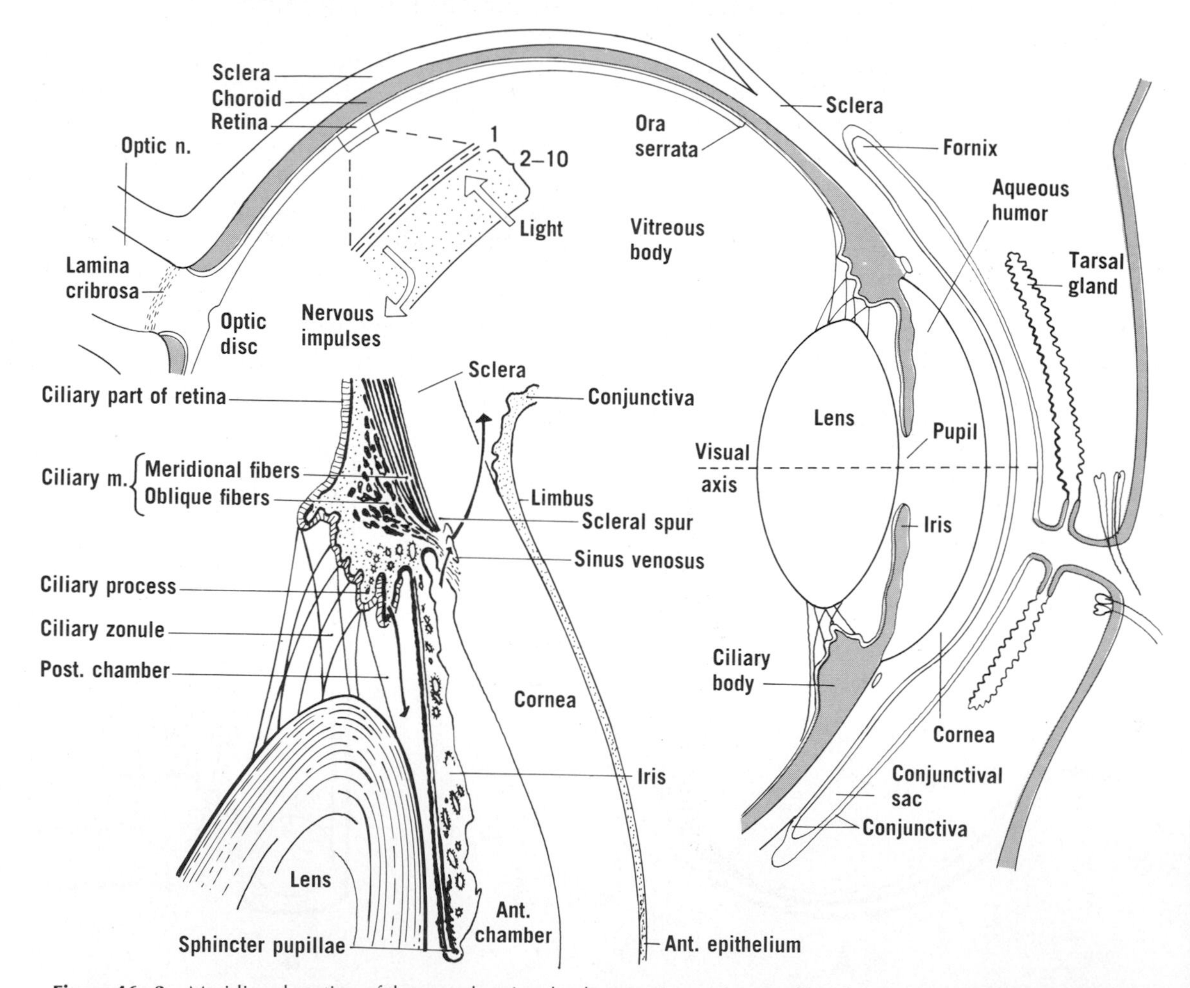

Figure 46–3 Meridional section of the eye, showing the three tunics, and a meridional section of the ciliary region, showing the iridocorneal angle. Arrows indicate the formation and drainage of the aqueous humor. The trabecular meshwork lies between the sinus venosus and the iridocorneal angle. A narrow line in front of the pigmented epithelium of the iris represents the dilator pupillae.

the external tunic. Its front part can be seen through the conjunctiva as "the white of the eye." The sclera consists of fibrous tissue, and it receives the tendons of the muscles of the eyeball. Posteriorly, the fibers of the optic nerve pierce the sclera at the *lamina cribrosa*. External to the sclera, the eye lies in a socket of fascia bulbi within the orbital fat.

An important, circular canal termed the *sinus venosus sclerae* is situated at the sclero-corneal junction (fig. 46–3). **The aqueous humor, formed by the ciliary processes, filters through channels leading from the anterior chamber to the sinus venosus and drains by means of aqueous veins into scleral plexuses. The iridocorneal angle (between the iris and the cornea), also known as the angle of the anterior chamber or as the filtration angle, is very important physiologically and pathologically.**

Middle Vascular Tunic

The middle tunic, frequently termed the uvea, comprises the choroid, ciliary body, and iris, from behind forward.

(a) The *choroid* is a vascular, pigmented (brown) coat that lines most of the sclera.

(b) The *ciliary body* connects the choroid with the iris, and it contains the ciliary muscle and the ciliary processes. It is lined by the ciliary part of the retina.

The ciliary muscle comprises two main sets of smooth-muscle fibers (fig. 46–3): (1) longitudinal fibers connect the sclera in front to the choroid behind, and (2) oblique fibers enter the base of the ciliary processes. **The ciliary muscle is supplied chiefly by parasympathetic fibers by way of the ciliary nerves. On contraction, the ciliary body moves forward. This presumably decreases the tension on the fibers of the ciliary zonule so that the central part of the lens becomes more curved and the eye can be focused on near objects, a process known as accommodation.**

The ciliary processes, about 70 in number, are arranged in a circle behind the iris (fig. 46–4). They form the aqueous humor.

(c) The *iris* is a circular, pigmented diaphragm that lies in front of the lens in a more or less coronal plane. It is anchored peripherally to the ciliary body, whereas its central border is free and bounds an aperture known as the *pupil*. The iris divides the space between the cornea and the lens into two chambers (see fig. 46–3). **The anterior chamber is bounded**

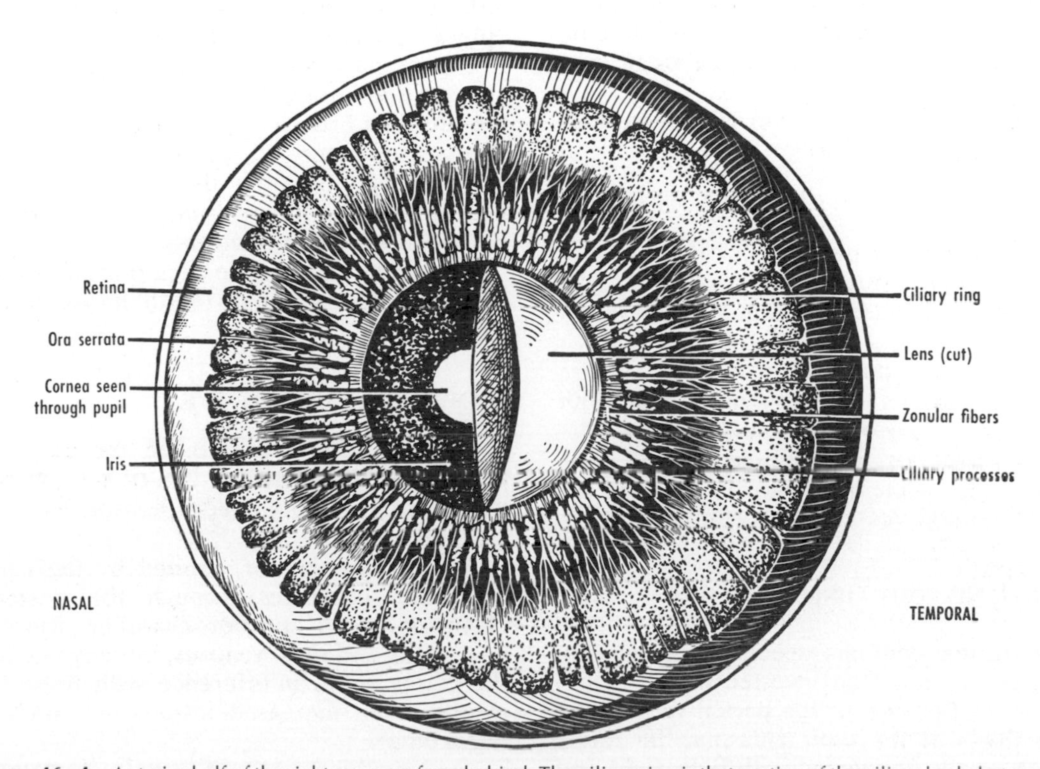

Figure 46–4 Anterior half of the right eye seen from behind. The ciliary ring is that portion of the ciliary body between the ora serrata and the ciliary processes. (After Wolff.)

largely by the cornea and iris. It communicates through the pupil with the posterior chamber, which is bounded by the iris, ciliary processes and zonule, and lens. Both chambers are filled with aqueous humor.

The anterior surface of the iris presents a fringe known as the *collarette*. The stroma of the iris contains pigment, especially in brown irides. A defect of the iris is termed a *coloboma*.

The *sphincter pupillae* is situated in the posterior part of the iris, near the pupil, and consists of smooth muscle. The **sphincter pupillae is supplied by parasympathetic fibers by way of the ciliary nerves, and its contraction results in constriction of the pupil (miosis). The iris contracts reflexly when light reaches the retina (the light reflex) and during focusing on a near object (the accommodation reaction). A drop of an atropine-like drug placed on the eye annuls the action of the ciliary muscle and the sphincter pupillae, both of which are under parasympathetic control. The resultant dilatation of the pupil (caused by overaction of the dilator) is of use in the examination of the eye.**

The dilator pupillae consists of smooth muscle in front of the pigmented epithelium on the back of the iris, which constitutes the iridial part of the retina. **The dilator pupillae is supplied by sympathetic fibers (roots from C8 to T4 by way of the ciliary nerves), and its contraction results in dilatation of the pupil (mydriasis).**

The autonomic innervation of the eye may be summarized in the following manner:

1. Parasympathetic (synapses in ciliary ganglion)
   - Sphincter pupillae
   - Ciliary muscle
2. Sympathetic (synapses in superior cervical ganglion)
   - Dilator pupillae
   - Orbitalis (smooth muscle of inferior orbital fissure)
   - Superior tarsal muscle (smooth muscle in eyelid)
   - Blood vessels of choroid and retina

Internal Nervous Tunic, or Retina

The retina contains special receptors on which is projected an inverted image of objects seen. **Because of the partial crossing of nerve fibers at the optic chiasma, the retina of each eye is connected with both right and left visual areas of the forebrain.** The retina is shaped like a sphere that has had its anterior segment removed, leaving an irregular margin termed the *ora serrata*. Many of the layers of the retina end at the ora, but a pigmented continuation lines the ciliary body and iris as the *ciliary* and *iridial parts of the retina*.

Basically, the retina comprises two main strata: (1) an external, pigmented stratum derived from the external lamina of the optic cup and (2) an internal, transparent, cerebral stratum derived from the inverted lamina of the optic cup. **The cerebral stratum of the retina may become detached pathologically (or in the course of preparing histological sections) from the pigmented layer along a plane that represents the cavity of the embryonic optic vesicle.**

The *macula* is a pigmented area of the retina on the temporal side of the optic disc. It contains a pit, the *fovea centralis*, which in turn presents a depression, the foveola. **The foveola functions in detailed vision,** when an object is looked at specifically.

The entering optic nerve fibers form the *optic disc*, or "blind spot." **The optic disc, insensitive to light, is situated nasal to the posterior pole of the eye and to the fovea centralis. Normally it is flat and does not form a papilla, but, near its center, a variable depression, the "physiological cup," is present.**

The retina is nourished externally by the choroid and internally by the central artery of the retina, a branch of the ophthalmic artery. The central artery travels in the optic nerve and divides at the optic disc. **The branches of the central artery do not anastomose, and their occlusion results in blindness.**

The *fundus oculi* is the back part of the interior of the eye as seen on ophthalmoscopy (fig. 46–5).

DIOPTRIC MEDIA OF EYE

The refractive apparatus of the eye comprises the cornea (which contributes most of the optical power), aqueous humor, lens, and vitreous body.

The aqueous humor, formed by the ciliary processes, circulates through the posterior chamber, pupil, anterior chamber, iridocorneal angle, and sinus venosus, thereby reaching the ciliary veins. Interference with resorption results in an increased intra-ocular pressure (glaucoma).

The *lens*, biconvex and 1 cm in diameter, is covered by a capsule and consists largely of

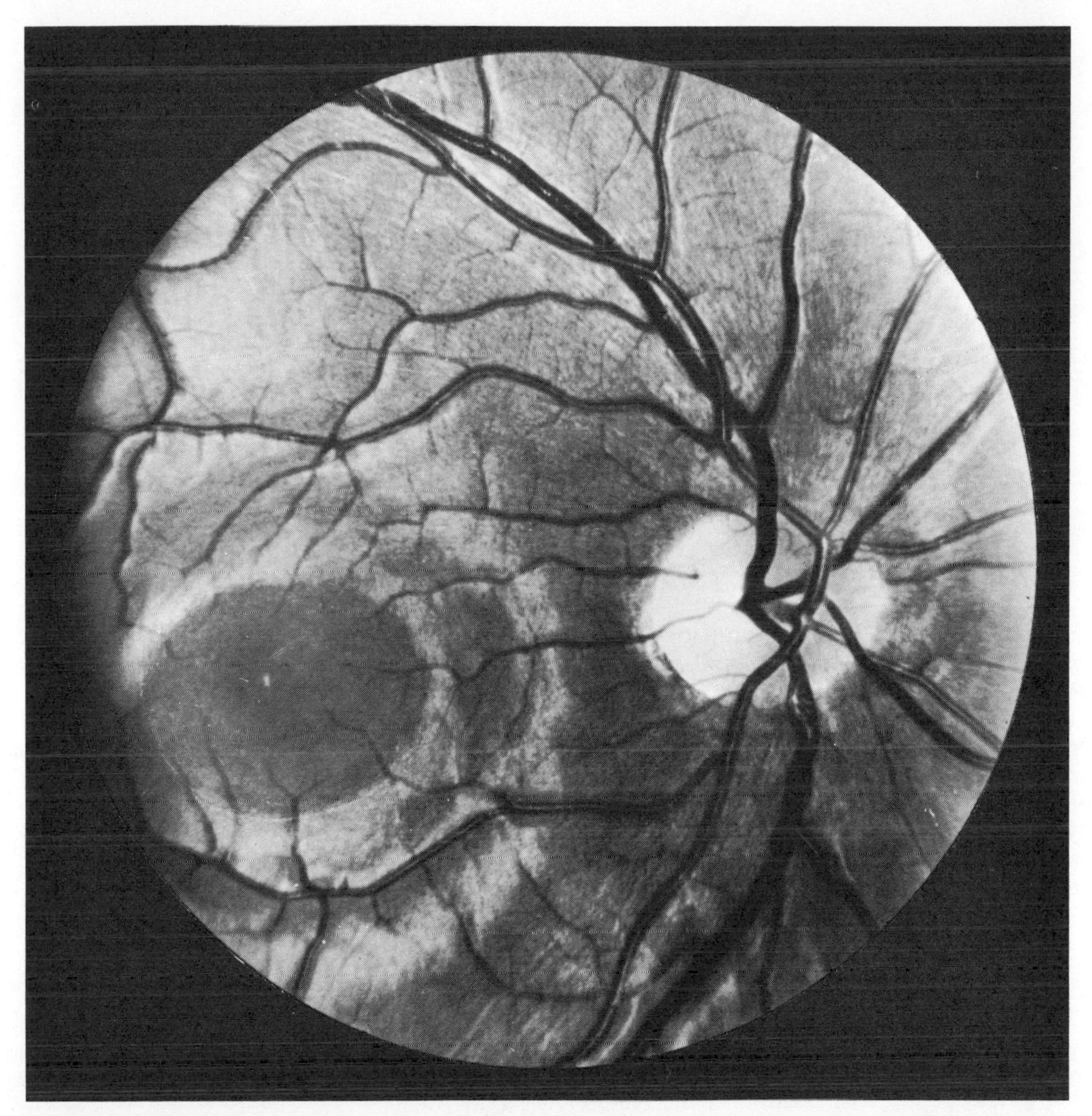

Figure 46–5 Right fundus oculi *in vivo*. The optic disc is on the right side of the photograph (i.e., medially). The whitish component of its (pink) color is produced by the lamina cribrosa. The lateral border of the disc is sharper than the medial. The retinal vessels radiate from the disc. The arteries show a light streak along their middle; the veins are darker and wider. The central vein is lateral to the central artery at the disc. The macula situated lateral (*on the left side of the photograph*) to the optic disc, appears as a dark (red) oval area. The foveola appears here as a whitish spot in the macula. Striations caused by the nerve fibers in the retina proceed downward and medially and upward and medially, toward the disc. (Courtesy of Dr. Hans Littmann and Carl Zeiss Inc.)

lens fibers. **The lens becomes harder with age, so that the power of accommodation is lessened (presbyopia) and convex spectacles may be required. An opacity of the lens is termed a cataract.** The lens capsule is anchored to the ciliary body by its *suspensory ligament*, or *ciliary zonule*. When distant objects are being looked at, elastic fibers in the choroid pull on the ciliary body, which, in turn, keeps the zonular fibers and also the lens capsule under tension, thereby keeping the curvatures of the lens minimal.

The *vitreous body* is a transparent, gelatinous mass that fills the eyeball behind the lens. The movement of specks in the vitreous body is sometimes seen as *muscae volitantes* (L., "flitting flies").

GENERAL SENSORY INNERVATION AND BLOOD SUPPLY OF EYE

Sensory fibers from the cornea and uvea reach the nasociliary nerve (of the ophthalmic

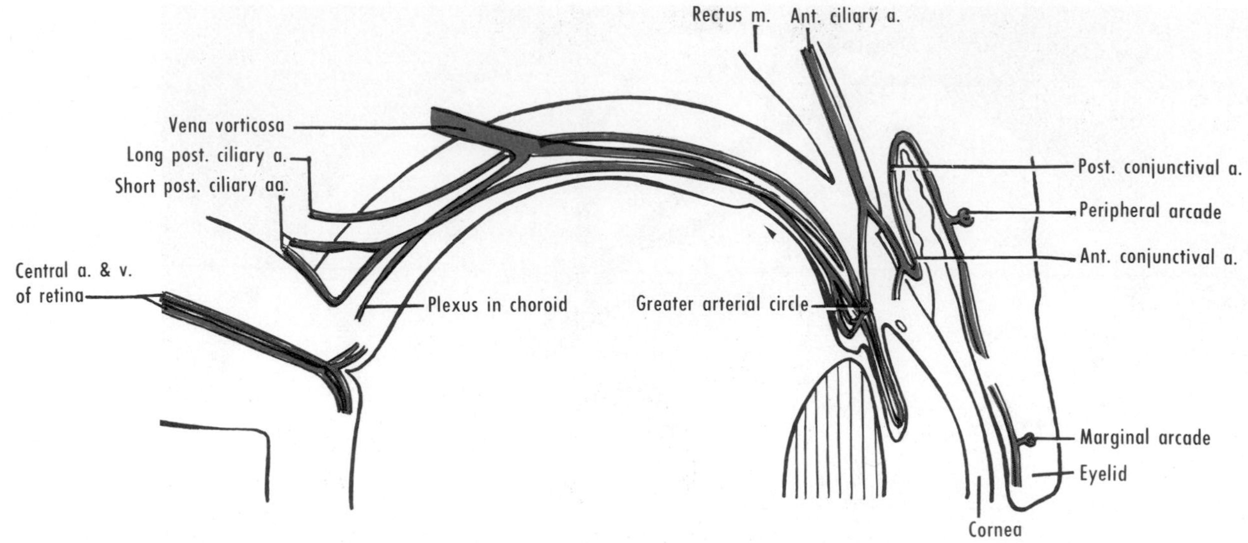

Figure 46–6 Blood supply of the eye. (Based on Wolff.)

nerve) by way of the short and long ciliary nerves. The eye receives its blood supply (fig. 46–6) from the ophthalmic artery by way of the central artery of the retina, short and long posterior ciliary arteries, and the anterior ciliary arteries (from muscular branches of the ophthalmic artery). Most of the veins from the eye accompany the arteries and drain into the cavernous sinus by way of the ophthalmic veins.

ADDITIONAL READING

Adler's Physiology of the Eye. Clinical Applications, 7th ed., ed. by R. A. Moses, Mosby, St. Louis, 1981. A good text on functional aspects.

Duke-Elder, S., and Wybar, K. C., *The Anatomy of the Visual System,* vol. 2 of *System of Ophthalmology,* ed. by S. Duke-Elder, Kimpton, London, 1961. An excellent work of reference for the orbit and the eye.

Wolff, E., *The Anatomy of the Eye and Orbit,* 7th ed., rev. by R. Warwick, Lewis, London, 1976. An attractive, well-illustrated text.

QUESTIONS

46–1 Where is most of the optical power of the eye concentrated?

46–2 Is the cornea covered by conjunctiva?

46–3 Into which channel does the aqueous humor in the anterior chamber drain?

46–4 When "drops" are used in the examination of the eye, which muscles are removed from action?

46–5 Where do the posterior and anterior chambers communicate?

46–6 Which disease is characterized by increased intra-ocular pressure?

46–7 In obstruction of the aqueous pathway, what effect would excision of a portion of the iris have?

46–8 Which muscle and nerve are involved in accommodation?

46–9 What is the autonomic innervation of the muscles of the eye?

46–10 Where does detachment of the retina occur?

46–11 On which side of the optic disc is the macula?

46–12 What is the name for an opacity of the lens?

47
THE SCALP, AURICLE, AND FACE

SCALP

Layers. The scalp (fig. 47–1) comprises five layers:

1. Skin, usually hairy.
2. Close subcutaneous tissue containing vessels and nerves. **The scalp gapes when cut, and the blood vessels do not contract, resulting in bleeding that should be arrested by pressure.**
3. Aponeurosis and occipitofrontalis muscle. The *galea aponeurotica* (*epicranial aponeurosis*) is a fibrous helmet between the occipitalis and frontalis muscles and is anchored to the occipital bone. The *occipitofrontalis* comprises two occipital bellies (attached to the occipital and temporal bones) and two frontal bellies (which end in the skin of the forehead), united by the galea and supplied by the facial nerve. The muscle moves the scalp and elevates the eyebrows.
4. Loose subaponeurotic tissue containing the emissary veins and allowing free movement of layers 1 to 3 as a unit. **Layer 4 is a "dangerous area" because it allows spread of infection even, by way of the emissary veins, to intracranial structures.**
5. Pericranium, the periosteum on the outside of the skull.

It should be noticed that the initial letters of the names of the layers form the word *scalp*.

Innervation and Blood Supply. The nerves and vessels of the scalp (fig. 47–2) ascend in layer 2. Surgical flaps of the scalp are cut so as to remain attached inferiorly. The scalp is supplied successively by the branches of divisions 1 to 3 of the trigeminal nerve and by the cervical plexus and dorsal rami of the cervical nerves (see fig. 47–6). **The trigeminal and cervical territories are commonly equal in area, but overlapping does occur. The arterial supply, although partly by the internal carotid artery, is mainly by branches of the external**

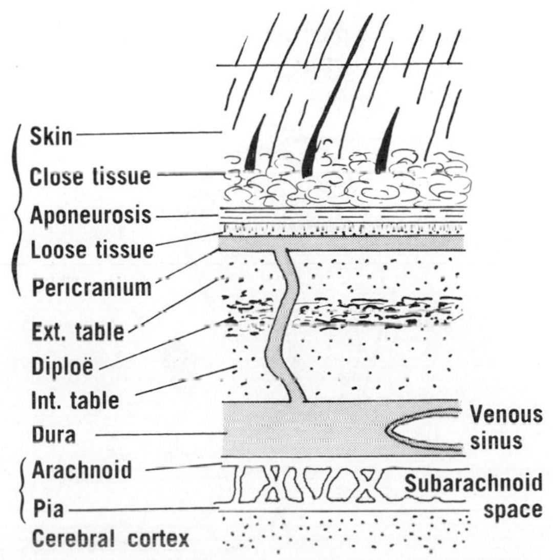

Figure 47–1 Section through the scalp, skull, and meninges.

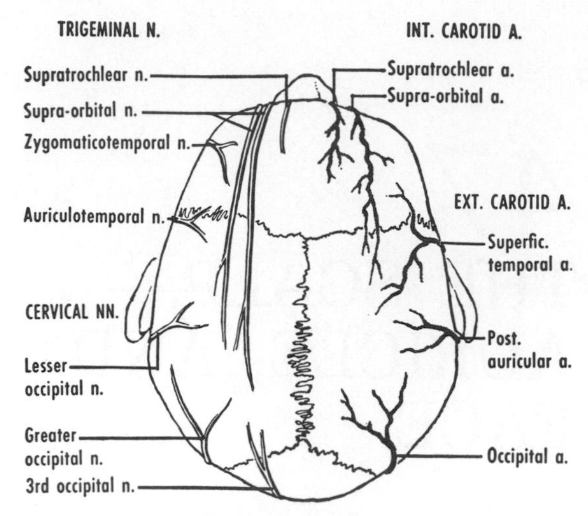

Figure 47–2 Innervation and blood supply of the scalp, superior aspect. In this instance the trigeminal territory extends behind the vertex of the head. The abundant arterial anastomoses are omitted. (After Grant.)

carotid artery. Anastomoses are abundant, so partially detached pieces of scalp may be replaced successfully. The *superficial temporal artery,* a terminal branch of the external carotid artery, arises in the parotid gland, crosses the zygomatic arch, and divides into frontal and parietal branches. **The pulsations of the superficial temporal artery can be felt over the zygomatic arch.** The *posterior auricular* and *occipital arteries,* also branches of the external carotid artery, run upward and backward to the scalp.

AURICLE

The auricle, a part of the external ear, consists of elastic cartilage covered by skin. The chief depressions and elevations of the auricle are shown in figure 47–3. The lobule consists merely of fibrous tissue and fat. The auricle is connected with the fascia on the side of the skull by unimportant anterior, superior, and posterior auricular muscles, which are supplied by the facial nerve.

The auricle is supplied by both cranial (auriculotemporal nerve from cranial nerve 5; probably also twigs from cranial nerves 7, 9, and 10) and spinal (lesser occipital and great auricular) nerves (see fig. 47–6).

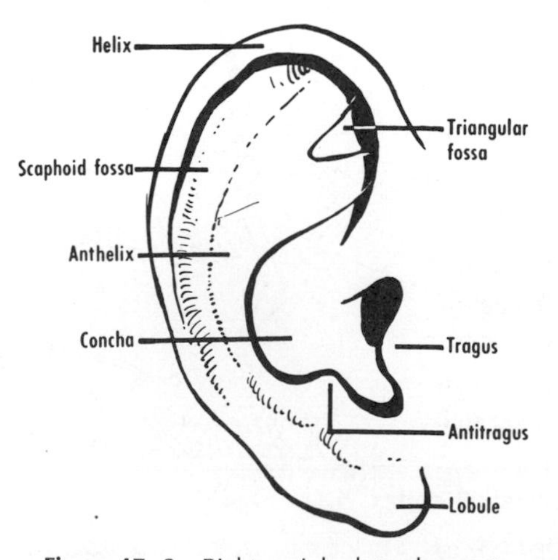

Figure 47–3 Right auricle, lateral aspect.

FACE

The eyelids, nose, and lips are discussed elsewhere.

Muscles of Facial Expression (fig. 47–4). The muscles of facial expression are very superficial (being attached to or influencing the skin) and are all supplied by the facial nerve. In addition to those of the scalp and auricle, already mentioned, muscles are arranged around the openings of the orbits, nose, and mouth.

The *orbicularis oculi* (fig. 47–4*A*), a sphincter around the rim of the orbit, comprises (1) an *orbital part,* attached mainly to the medial margin of the orbit; (2) a *palpebral part,* contained in the eyelids and anchored medially and laterally; and (3) a *lacrimal part,* behind the lacrimal sac. The orbicularis protects the eye and brings the eyelids together in blinking and sleep. Its antagonists are the levator palpebrae superioris and the frontalis. **Paralysis of the orbicularis causes drooping of the lower eyelid (ectropion) and spilling of tears (epiphora), as occurs in facial palsy.**

The muscles around the nose and mouth are shown in figure 47–4*B* and *C*. The *orbicularis oris* is a complicated sphincter that closes the lips and can also protrude them. The *buccinator* arises from the maxilla and mandible and is inserted in a complicated manner into the orbicularis oris and lips. It is overlain by the *buccal pad of fat,* and both are pierced by the parotid duct. The buccinator keeps the cheek taut, preventing injury from the teeth.

Facial Nerve. **All the muscles of facial expression develop from pharyngeal arch 2 and hence are supplied by the facial nerve. The facial nerve traverses successively (1) the posterior cranial fossa, (2) the internal acoustic meatus, (3) the facial canal in the temporal bone, and (4) the parotid gland and face.** After it gives off the posterior auricular nerve, the fa-

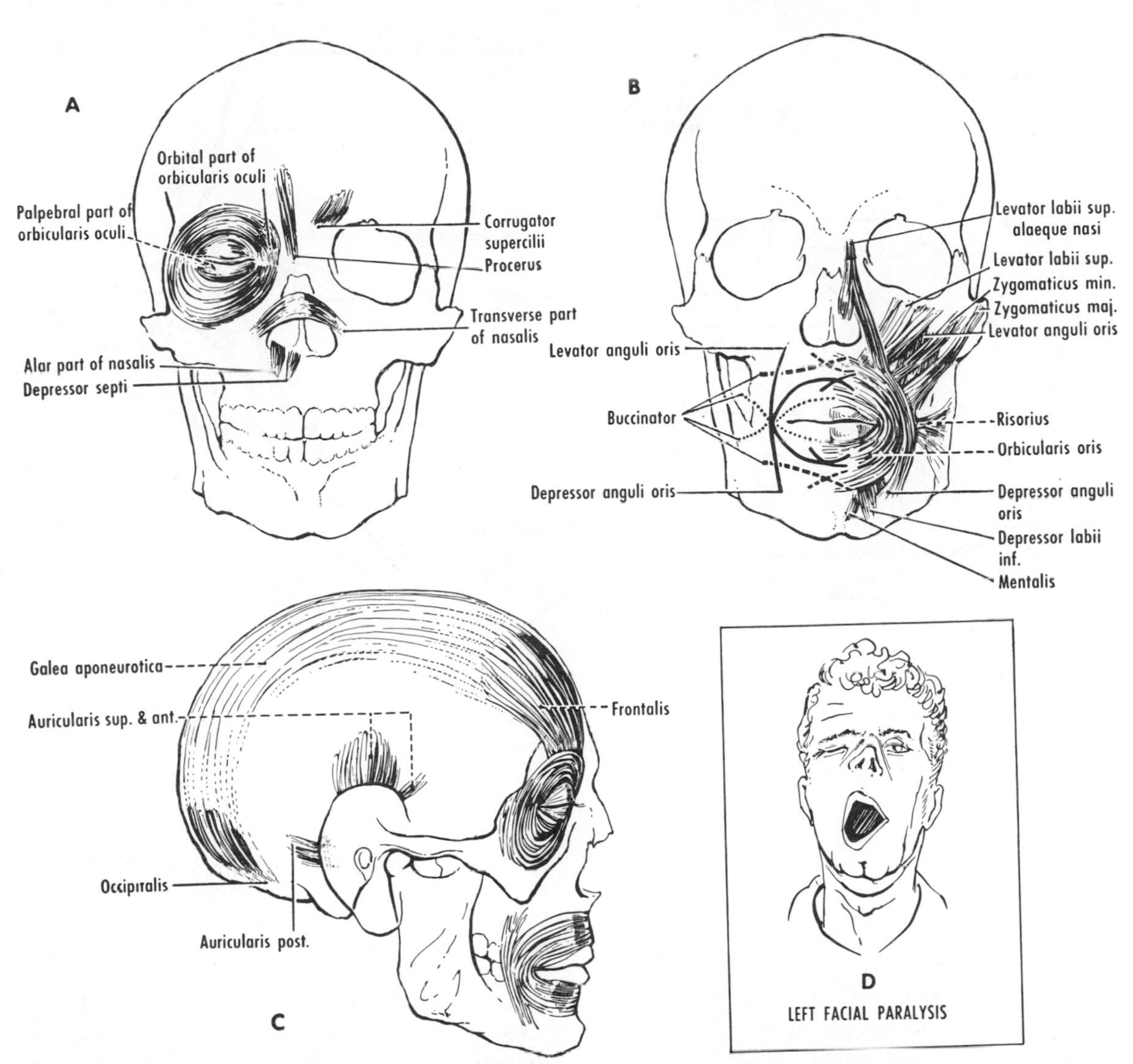

Figure 47–4 Muscles of facial expression. **A**, Anterior aspect, showing the muscles around the openings of the orbit and nose. **B**, Muscles of the mouth. The course of the fibers that constitute the orbicularis oris is shown schematically on the right half of the face. **C**, Lateral aspect, showing the muscles of the scalp and auricle. The orbicularis oculi and orbicularis oris are shown also. In **A**, **B**, and **C** the unbroken leaders indicate the bony attachments of the muscles. **D**, Left-sided facial paralysis caused by a lesion of the facial nerve at its exit from the skull. The patient has been asked to "shut his eyes" tightly and to open his mouth. Note the deviation of the lips, characteristic triangular shape of the mouth, and failure to close the affected eye. (From a photograph by Pitres and Testut.)

cial nerve divides within the parotid gland into its terminal branches, which form the parotid plexus. The branches emerge under cover of the parotid gland, radiate forward in the face, communicate with the branches of the trigeminal nerve, and supply the muscles of facial expression, including the platysma. The terminal branches are usually classified as *temporal, zygomatic, buccal, marginal* (along the mandible), and *cervical* (fig. 47–5*A*).

Facial Artery. The facial artery, a branch of the external carotid artery in the neck, winds around the lower border of the mandible and proceeds upward and forward on the face (fig. 47–5*B*). It is very tortuous and takes part in many anastomoses. In the face, the facial artery supplies the lips (by *inferior* and *superior labial arteries*) and nose and ends as the *angular artery* at the medial angle of the eye by anastomosing with branches of the ophthalmic

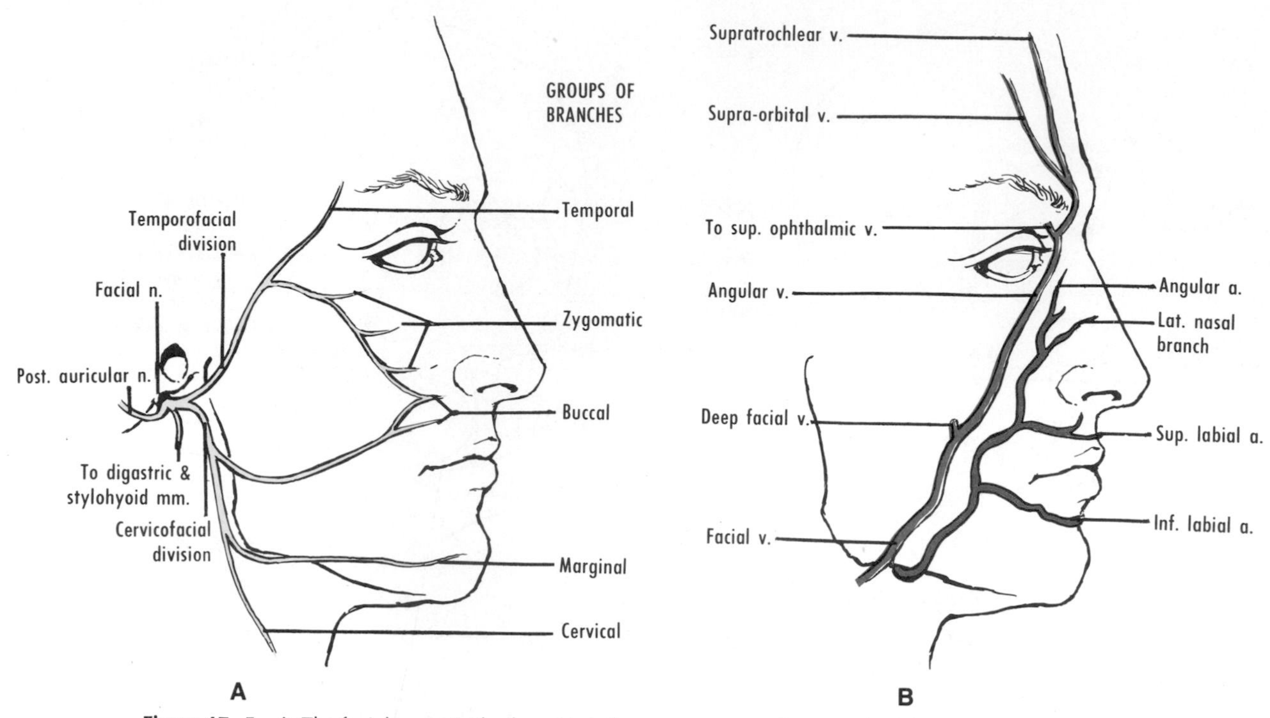

Figure 47–5 **A**, The facial nerve in the face. Variations are common, but two chief divisions (temporofacial and cervicofacial) are generally found. The intricacies of the parotid plexus have been omitted. **B**, The facial vessels in the face. The vein is posterior, more superficial, and less tortuous.

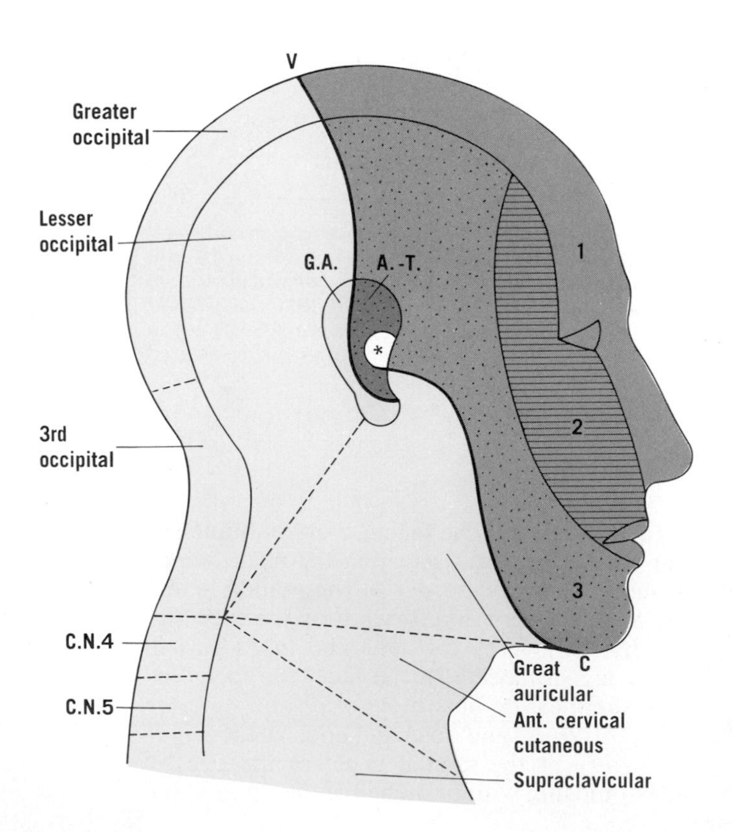

artery, thereby establishing a **communication between the external and internal carotid arteries. The labial arteries of the two sides anastomose across the median plane: hemorrhage is controlled by compressing both parts of a cut lip between the index fingers and thumbs.**

Facial Vein. The facial vein (fig. 47–5*B*) is behind the artery and is not as tortuous. Its commencement, the *angular vein,* communicates with the cavernous sinus by way of the ophthalmic veins. In the cheek, the facial vein receives the *deep facial vein* from the pterygoid plexus, and it usually ends directly or indirectly in the internal jugular vein. **Because of**

Figure 47–6 Cutaneous innervation of the head and neck. The vertex-ear-chin line separates the trigeminal territory from that of the cervical nerves. Although the central portion of the face is supplied by the trigeminal nerve only, it is believed that most of the skin of the face and neck "is supplied from a combination of trigeminal and upper cervical nerve roots" (L. Kruger and R. F. Young, in *The Cranial Nerves,* ed. by M. Samii and P. J. Jannetta, Springer, Berlin, 1981). *A.-T.,* auriculotemporal n. (from the mandibular); *G.A.,* great auricular n. (C.N. 2, 3); *asterisk,* probably branches from cranial nerves and 10; *1, 2, 3,* ophthalmic, maxillary, and mandibular nn.; V-C, vertex-ear-chin line.

its connections with the cavernous sinus and the pterygoid plexus and the consequent possibility of spread of infection, the territory of the facial vein around the nose and upper lip is termed the "danger area" of the face.

CUTANEOUS INNERVATION OF HEAD AND NECK

The "vertex-ear-chin line" indicates the boundary between the cranial (trigeminal) and spinal innervations (fig. 47–6). The ophthalmic, maxillary, and mandibular nerves separate before emerging from the base of the skull, and their cutaneous distributions must be tested separately; this is usually done over the forehead, the prominence of the cheek, and the chin.

The spinal innervation of the skin (see fig. 3–3) may be considered as (1) successive areas of distribution for each spinal nerve, both ventral and dorsal rami, or (2) areas of distribution of named nerves, because ventral rami combine to form plexuses (e.g., the cervical plexus) in which the individual rami become regrouped to form named nerves (e.g., great auricular) (fig. 47–6).

QUESTIONS

47–1 List the layers of the scalp.

47–2 What are the antagonists of the orbicularis oculi?

47–3 Which is the main muscle of the cheek?

47–4 What is the orbicularis oris?

47–5 How do the facial muscles develop?

47–6 How is hemorrhage from a cut lip controlled?

47–7 Why is the area around the nose and upper lip termed the "danger area" of the face?

47–8 What is the boundary line between the cranial and spinal areas of cutaneous innervation of the head and neck?

47–9 What are the two main modes of spinal innervation of the skin?

47–10 Where is the trigeminal ganglion and what is its significance?

48

THE PAROTID, TEMPORAL, AND INFRATEMPORAL REGIONS

The parotid region (see fig. 50–2*C*) comprises the parotid gland and its bed, which includes muscles and part of the skull. The temporal region is on the side of the head (temple). The infratemporal region is medial to the ramus of the mandible.

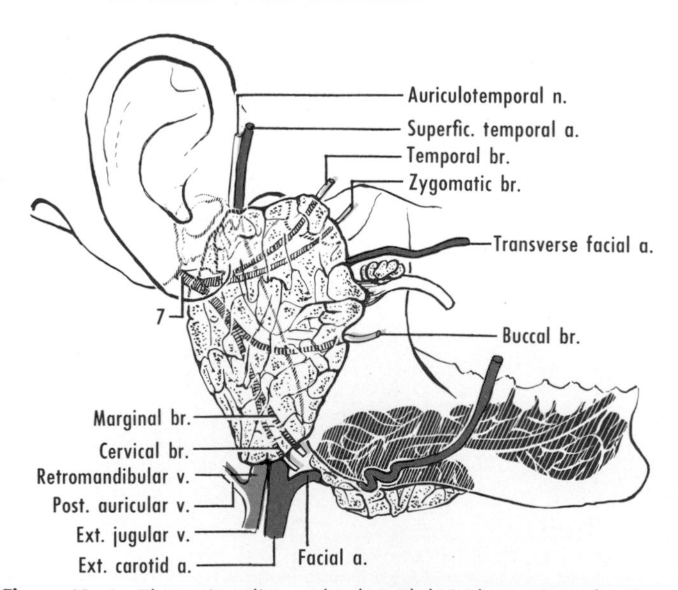

Figure 48–1 The main salivary glands and their ducts: parotid, submandibular, and sublingual. The accessory parotid gland lies above the parotid duct. The chief branches of the facial nerve are shown, but the details of the parotid plexus have been omitted. For submandibular and sublingual glands, see also fig. 49–1.

PAROTID GLAND

The parotid gland is the largest of the salivary glands, which include the also-paired submandibular and sublingual glands and numerous small glands in the tongue, lips, cheeks, and palate (fig. 48–1). Their combined secretion is termed *saliva*. The parotid, a serous compound tubulo-alveolar gland, is yellowish, lobulated, and irregular in shape. It occupies the interval between the sternomastoid muscle and the mandible.

Surface Anatomy. **The parotid gland lies below the zygomatic arch, below and in front of the external acoustic meatus, in front of the mastoid process, and behind the ramus of the mandible.** Its close relationship to the mandible is indicated by pain on mastication in viral parotitis (mumps).

Relations. The parotid gland is enclosed in a sheath (parotid fascia) and is shaped roughly like an inverted pyramid, with three (or four) sides (fig. 48–2*A*). It presents a *base* (from which the superficial temporal vessels and auriculotemporal nerve emerge), *apex* (which descends below and behind the angle of the mandible), and lateral, anterior, and posterior (or posterior and medial) surfaces. The *lateral surface* is superficial and contains lymph nodes. The *anterior surface* is grooved by the ramus of the mandible and masseter (fig. 48–2*B*), producing a medial lip (from which the maxillary artery emerges) and a lateral lip, under cover of which the parotid duct, branches of the facial nerve, and the transverse facial artery emerge (see fig. 48–1). The *posterior surface* is grooved by (1) the mastoid process and the sternomastoid and digastric muscles and (2) more medially by the styloid process and its attached muscles. Medially, the gland is pierced by the facial nerve above and by the external carotid artery below.

The following structures lie partly within the parotid gland, from superficial to deep.

1. The facial nerve forms the parotid plexus (*pes anserinus*) within the gland and separates the glandular tissue partially into superficial and deep layers ("lobes"). **In surgical excision of the parotid gland (e.g., for a tumor), damage to the facial nerve ("the hostage of parotid surgery") is a possibility.**
2. The superficial temporal and maxillary veins unite in the gland to form the retromandibular vein, which contributes in a variable manner to the formation of the external jugular vein (see fig. 50–5).
3. **The external carotid artery divides within the parotid gland into the superficial temporal and maxillary arteries.**

Parotid Duct. The parotid duct, emerging under cover of the lateral surface, runs forward on the masseter and turns medially to pierce the buccinator. The branching of the duct can be examined radiographically after injection of a radio-opaque medium (sialography, see fig. 48–4*A*). **The parotid duct, which is palpable, opens into the oral cavity on the parotid papilla opposite the upper second molar** (see fig. 51–1*A*).

Innervation of Parotid Gland (see fig. 48–9). Parasympathetic secretomotor fibers (from the glossopharyngeal, tympanic, and lesser petrosal nerves) synapse in the otic ganglion. Postganglionic fibers enter the auriculotemporal nerve and so reach the gland. Cranial nerves 7 and 9 communicate, so that **secretory fibers to each of the three major salivary glands may travel in both the facial and glossopharyngeal nerves.** The sympathetic supply to the salivary glands includes vasomotor fibers.

SUPERFICIAL TEMPORAL ARTERY (see fig. 48–1)

The superficial temporal artery, the smaller terminal branch of the external carotid artery, arises in the parotid gland, behind the neck of the mandible. Accompanied by the auriculotemporal nerve, it crosses the zygomatic arch and divides into frontal and parietal branches. Pulsations of the artery can be felt against the zygomatic arch. A lateral, surgical flap of the scalp is made like an inverted horseshoe, so that it contains the intact superficial temporal artery. The *transverse facial artery* arises from the superficial temporal artery within the parotid gland and runs forward below the zygomatic arch.

TEMPORAL AND INFRATEMPORAL FOSSAE (fig. 48–3)

The temporal fossa is bounded by the temporal line, the frontal process of the zygomatic bone, and the zygomatic arch. Deep to the arch, it communicates with the infratemporal fossa. The infratemporal fossa is bounded in front by the back of the maxilla, above by the

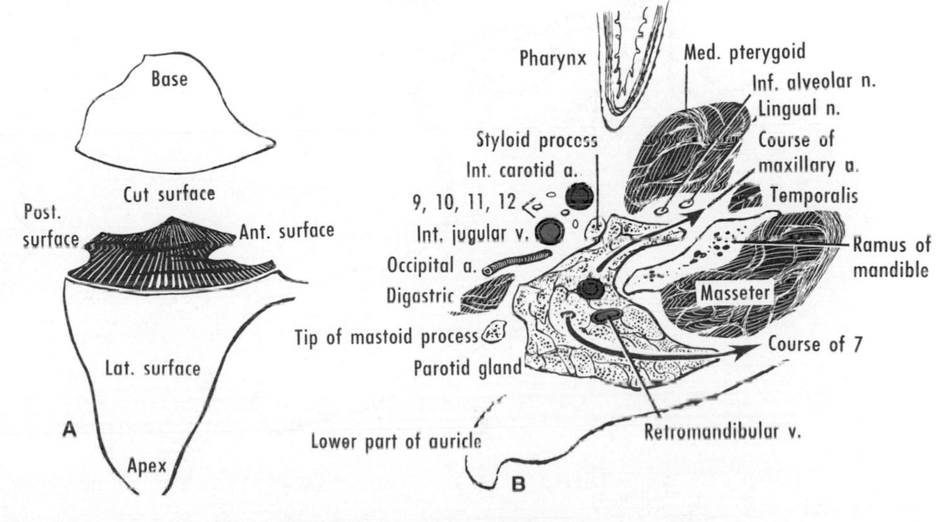

Figure 48–2 **A**, The parotid gland, lateral aspect, sectioned horizontally to show the surfaces. **B**, Horizontal section at the level of the atlas. (**B** is based on Truex and Kellner and on Parsons.)

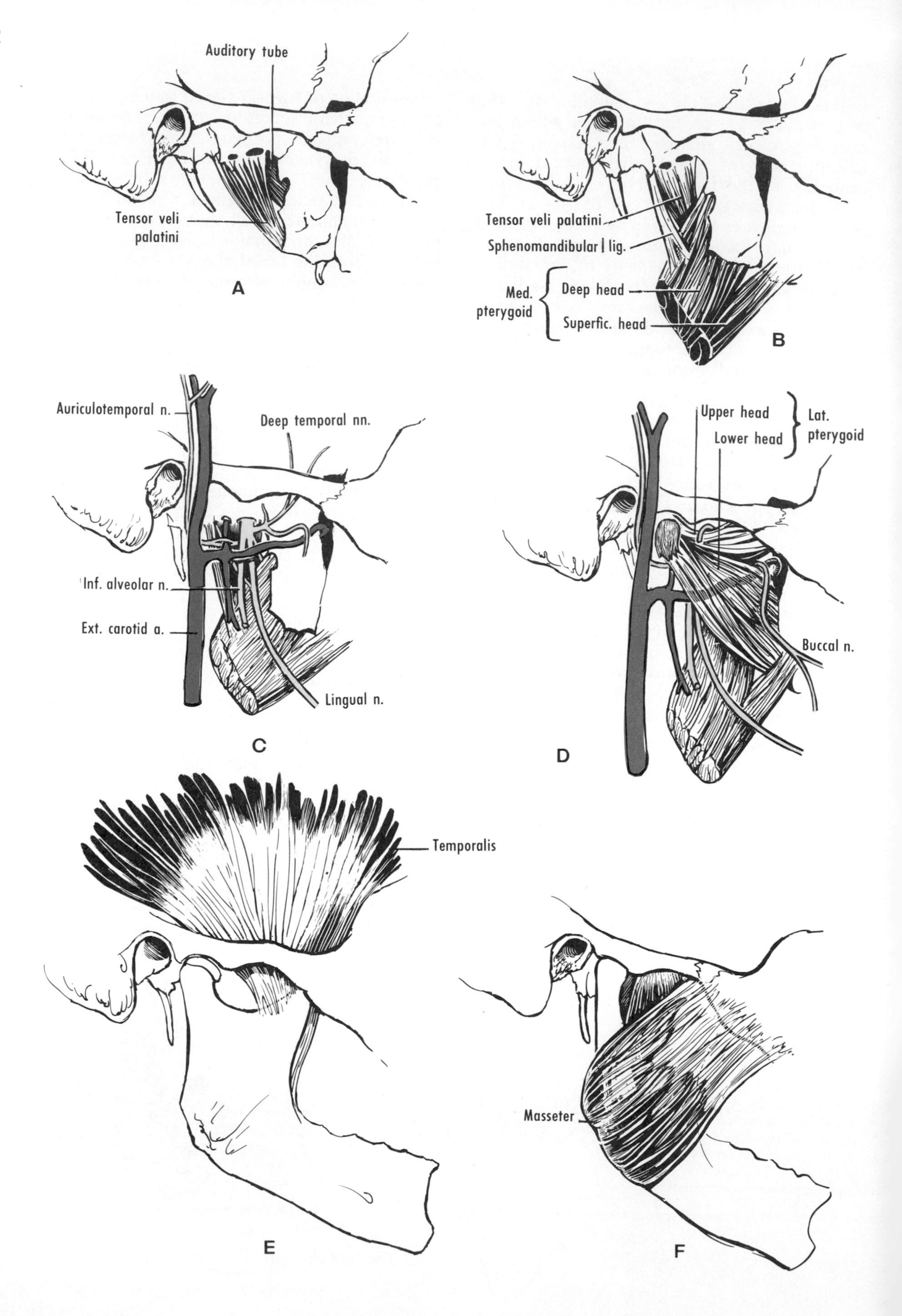
Auditory tube
Tensor veli palatini
A
Tensor veli palatini
Sphenomandibular lig.
Med. pterygoid
Deep head
Superfic. head
B
Auriculotemporal n.
Deep temporal nn.
Inf. alveolar n.
Ext. carotid a.
Lingual n.
C
Upper head
Lower head
Lat. pterygoid
Buccal n.
D
Temporalis
E
Masseter
F

TABLE 48–1 MUSCLES OF MASTICATION

Muscle	Origin	Insertion	Innervation	Action
Masseter	Zygomatic arch (inferior border and medial surface)	Lateral aspect of ramus of mandible	Mandibular nerve	Elevates mandible
Temporalis	Floor of temporal fossa; temporal fascia	Coronoid process and anterior border of ramus	Mandibular nerve	Chiefly maintains mandibular posture; posterior fibers retract mandible
Medial pterygoid	Deep head from medial surface of lateral pterygoid plate Superfical head from adjacent palatine bone and maxilla	Medial surface of mandible	Mandibular nerve	Acts as synergist of masseter
Lateral pterygoid	Lower head from lateral surface of lateral pterygoid plate Upper head from infratemporal surface of greater wing of sphenoid bone	Capsule, articular disc, and front of neck of mandible	Mandibular nerve	Protracts mandible

infratemporal surface of the greater wing of the sphenoid bone (see fig. 42–12), medially by the lateral pterygoid plate (see fig. 42–11), and laterally by the ramus of the mandible. The temporalis is largely in the temporal fossa, and **the infratemporal fossa contains chiefly the two pterygoid muscles, the maxillary artery and the pterygoid venous plexus, and the mandibular nerve and chorda tympani.**

MUSCLES OF MASTICATION (fig. 48–3)

The muscles of mastication are the masseter, temporalis, medial pterygoid and lateral ptery-

Figure 48–3 Six successively more superficial planes in the infratemporal region. **A**, Medial wall of infratemporal fossa (cf. fig. 42–11). **B**, Medial pterygoid muscle. **C**, Mandibular nerve and maxillary artery. Note the chorda tympani and the middle meningeal artery. **D**, Lateral pterygoid muscle. The second part of the maxillary artery may be deep (as shown here) or superficial to the lower head of the lateral pterygoid muscle. The buccal nerve emerges between the heads of the lateral pterygoid. The masseteric nerve emerges above the upper head of the lateral pterygoid. **E**, Temporalis. The anterior fibers are attached to the ramus of the mandible. **F**, Masseter. The deep fibers (visible above) proceed directly downward.

goid. They are all supplied by the mandibular nerve, a division of the trigeminal nerve. The muscles are summarized in table 48–1.

The *masseter* (fig. 48–3*F*) is conveniently described with the infratemporal region, although it is lateral to the ramus of the mandible. It is a quadrate muscle that can be divided partially into superficial, middle, and deep portions. The masseteric nerve from the anterior trunk of the mandibular nerve reaches the muscle by traversing the mandibular notch. The masseter is a powerful elevator of the mandible, and it can be palpated during clenching of the teeth.

The *temporalis* (fig. 48–3*E*) is fan-shaped and arises from the deep surface of the temporal fascia as well as from the floor of the temporal fossa. Its tendon passes deep to the zygomatic arch to reach the coronoid process. It is supplied by deep temporal branches of the anterior trunk of the mandibular nerve. The temporalis maintains mandibular posture and elevates the mandible in molar occlusion. Its posterior fibers pull the head of the mandible back into the mandibular fossa during closure of the mouth.

The *medial pterygoid muscle* (fig. 48–3*B*) lies on the medial aspect of the ramus of the mandible. The *lateral pterygoid muscle* (fig. 48–3*D*) occupies the infratemporal fossa. The pterygoid muscles have two heads of origin each, and the two heads of the medial pterygoid embrace the lower head of the lateral pterygoid. The lateral and medial pterygoid muscles, acting together, protrude the mandible. The lateral pterygoid controls the articular disc, to which it is attached.

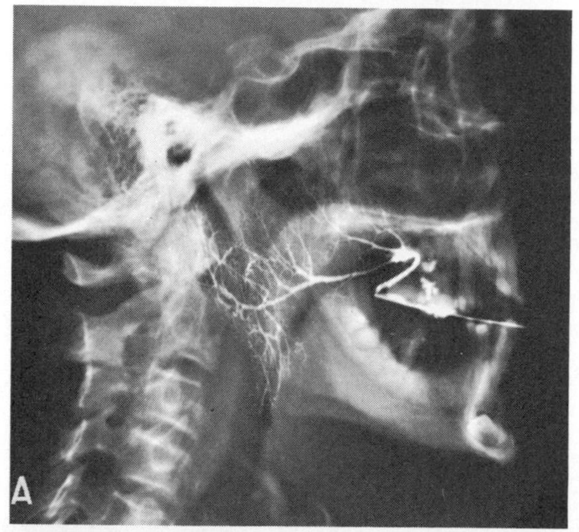

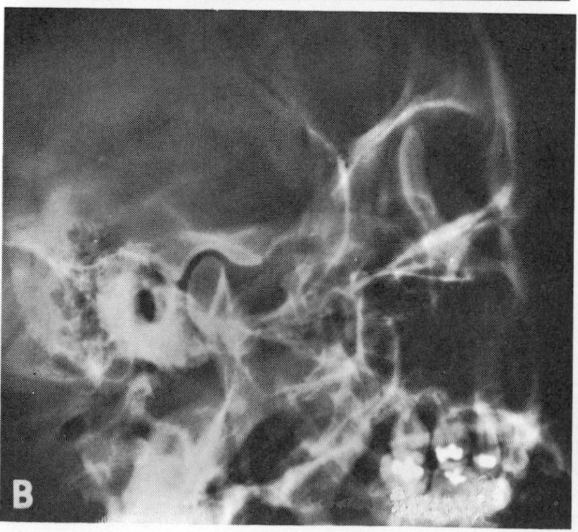

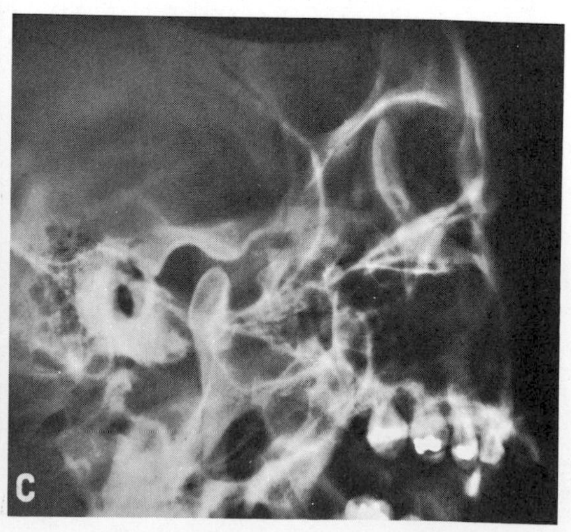

Figure 48–4 **A**, Parotid sialogram *in vivo*. Iodized oil has been injected through the parotid duct. **B** and **C**, The temporomandibular joint with the mouth closed (head in the mandibular fossa) and with the mouth open (head on the articular tubercle). The black oval area behind the head of the mandible is the external acoustic meatus. (Courtesy of Mr. John A. Hill, Birkenhead, England.)

TEMPOROMANDIBULAR JOINT (figs. 48–4 and 48–5)

The temporomandibular joint is a synovial joint between (1) the articular tubercle, mandibular fossa, and postglenoid tubercle of the temporal bone above and (2) the head of the mandible below. The articular surfaces are covered with fibrous tissue. The joint is related behind to the parotid gland, auriculotemporal nerve, and superficial temporal vessels.

The articular capsule is loose but is strengthened laterally by a *lateral ligament*. The *articular disc*, which is mostly fibrous, divides the joint into two separate compartments. Both the capsule and the disc receive in front a part of the insertion of the lateral pterygoid muscle. A separate synovial membrane is present peripherally in each of the two compartments of the joint.

The *sphenomandibular ligament*, from the spine of the sphenoid bone to the lingula of the mandible, lies medial to the joint. It is believed to develop from the sheath of the cartilage of the first pharyngeal arch. A fascial thickening between the styloid process and the angle of

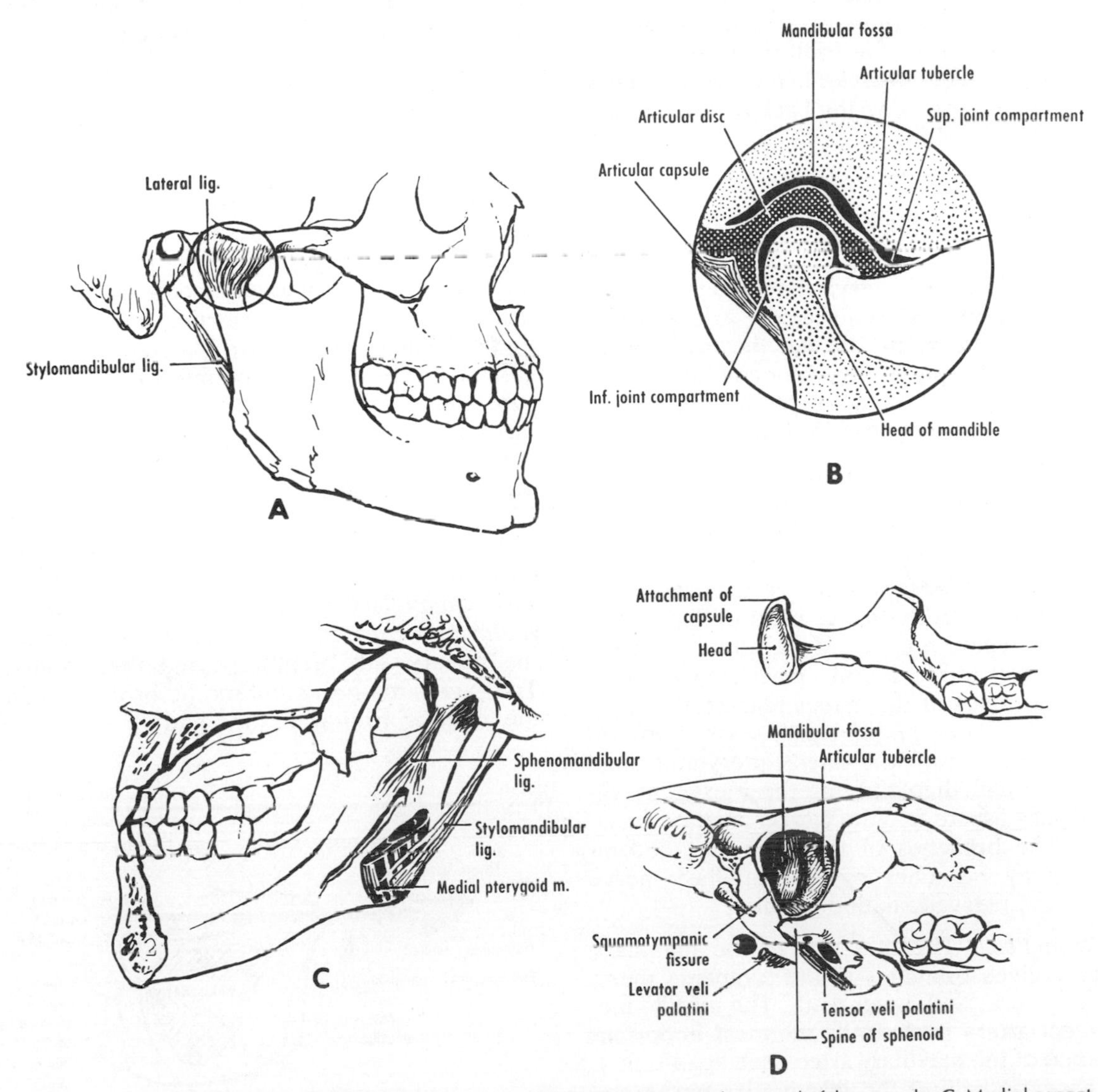

Figure 40–5 The temporomandibular joint. **A**, Lateral aspect. **B**, After partial removal of the capsule. **C**, Medial aspect. **D**, Superior aspect of the condylar process of the mandible and inferior aspect of the mandibular fossa and articular tubercle. Note the attachment of the capsule in both drawings. (After Sicher and Tandler.)

the mandible constitutes the *stylomandibular ligament*.

The temporomandibular joint is supplied by branches of the mandibular nerve (such as the auriculotemporal nerve) and by the terminal divisions of the external carotid artery.

Movements of the Mandible (fig. 48–4). The movements of the mandible are controlled largely by the play of muscles. Depression is produced by the lateral pterygoid and digastric muscles and by gravity. Elevation is due to the temporalis, masseter, and medial pterygoid muscle. Protrusion is caused by the pterygoid muscles and the masseter. Retraction results from the posterior fibers of the temporalis. Lateral movements depend on an interaction between all the muscles of mastication.

Anterior dislocation may result when the mouth is open, so the head may slip forward off the articular tubercle. Reduction is accomplished by depressing the back of the jaw and elevating the chin.

MAXILLARY ARTERY

The maxillary artery, the larger terminal branch of the external carotid artery, arises in the parotid gland, behind the neck of the mandible. It supplies the upper and lower jaws, the muscles of mastication, the palate, and the nose. Its course is described in three parts.

1. The *mandibular part* runs forward medial to the neck of the mandible.
2. The *pterygoid part* runs forward and upward under cover of the temporalis and either superficial or deep (see fig. 48–3*D*) to the lower head of the lateral pterygoid muscle. Most branches of the first and second parts accompany branches of the mandibular nerve.
3. The *pterygopalatine part* passes between the heads of the lateral pterygoid muscle and then through the pterygomaxillary fissure into the pterygopalatine fossa. The branches of the third part accompany branches of the maxillary nerve and pterygopalatine ganglion.

Branches. The first part of the maxillary artery gives branches to the tympanic membrane, dura, and lower teeth. **The middle meningeal artery is clinically the most important branch of the maxillary artery** (see fig. 48–3*C*). It ascends behind the mandibular nerve and enters the cranial cavity by traversing the foramen spinosum. The *inferior alveolar artery* accompanies the corresponding nerve, enters the mandibular foramen and canal, and supplies mucosa and teeth.

The second part of the maxillary artery supplies the muscles of mastication by *deep temporal, pterygoid, masseteric,* and *buccal arteries.*

The third part of the maxillary artery supplies the upper teeth, face, orbit, palate, and nasal cavity. The chief branches are several *superior alveolar arteries,* the *infra-orbital* and *descending palatine arteries,* the *artery of the pterygoid canal,* and the sphenopalatine artery. **The sphenopalatine artery is the termination of the maxillary artery. It enters the nasal cavity through the sphenopalatine foramen, supplies the nose and paranasal sinuses, and is important in nasal bleeding (epistaxis).**

MAXILLARY NERVE (fig. 48–6)

The maxillary nerve (second division of the trigeminal nerve) arises from the trigeminal ganglion, traverses the foramen rotundum, and enters the pterygopalatine fossa (where it can be "blocked" by passing a needle through the mandibular notch and injecting a local anesthetic). Then, as the infra-orbital nerve, it gains the orbit through the inferior orbital fissure and ends on the face by emerging from the infra-orbital foramen. The area of skin supplied by the maxillary nerve is tested by the use of cotton wool and a pin (see fig. 47–6).

Branches. Communications with the pterygopalatine ganglion enable the ganglion to distribute maxillary fibers. *Posterior superior alveolar branches* enter the maxilla and supply the molars and premolars and their gums. The *zygomatic nerve* enters the orbit through the inferior orbital fissure and divides into

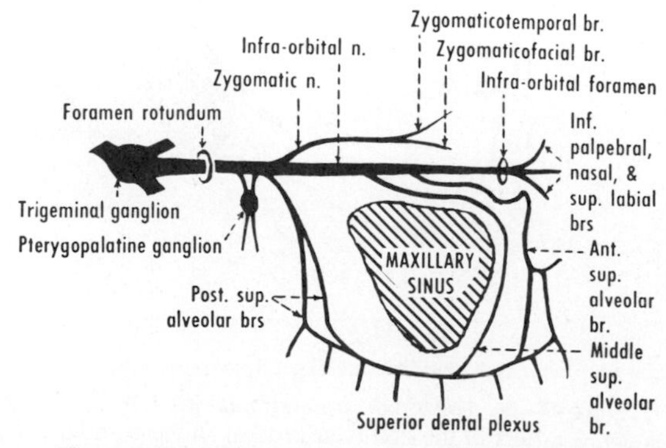

Figure 48–6 The maxillary nerve, lateral aspect.

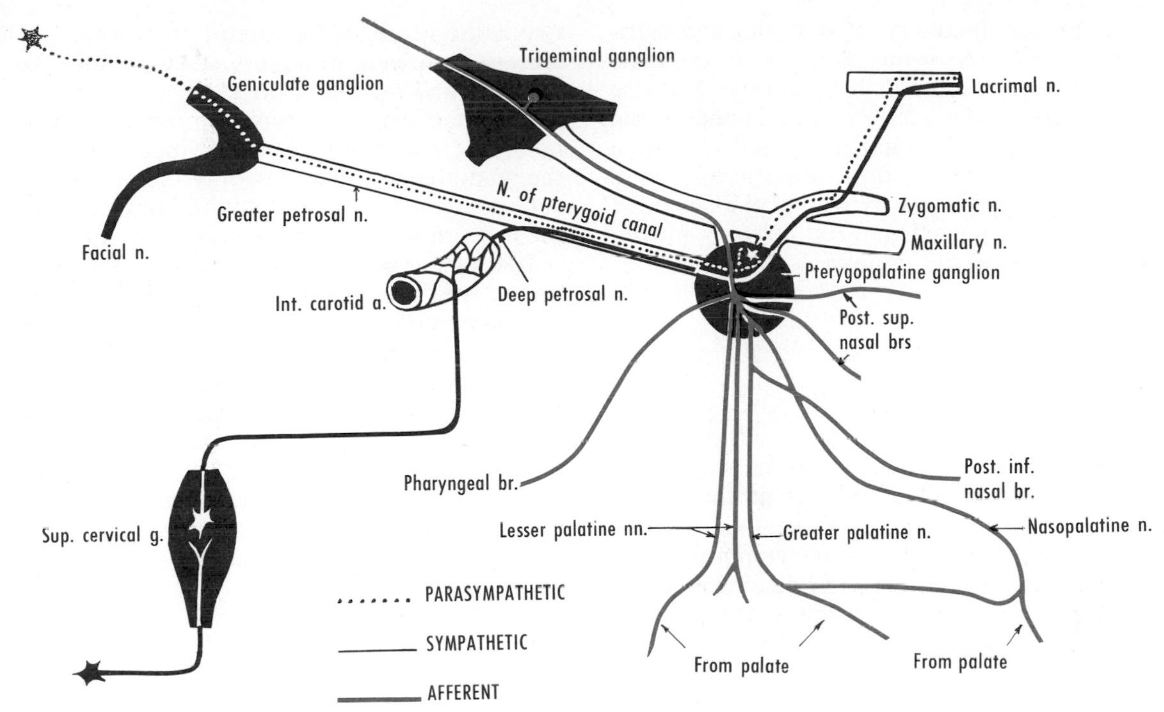

Figure 48–7 The pterygopalatine ganglion and its connections, lateral aspect.

zygomaticotemporal and *zygomaticofacial branches* that end on the temple and face, respectively. A communication with the lacrimal nerve probably enables secretory fibers to reach the lacrimal gland (fig. 48–7). **The infra-orbital nerve is the continuation of the maxillary nerve.** It traverses the inferior orbital fissure and occupies the infra-orbital groove, canal, and foramen. It ends on the face as *inferior palpebral* (to the lower eyelid), *nasal* (to the skin of the nose), and *superior labial branches* (to the mucosa). The infra-orbital nerve gives off a *middle superior alveolar branch,* which runs in the wall of the maxillary sinus, and an *anterior superior alveolar branch,* which descends sinuously in the front wall of the maxillary sinus and supplies the canine and incisors. The various superior alveolar nerves form a *superior dental plexus* within the maxilla. The branches of the maxillary nerve are listed in table 48–2.

TABLE 48–2 Summary of Branches of Maxillary and Infra-orbital Nerves

Location	Branches
In middle cranial fossa	Meningeal branch
In pterygopalatine fossa	Communicating branches (pterygopalatine nerves) to pterygopalatine ganglion
	Posterior superior alveolar branches
	Zygomatic nerve
	Zygomaticotemporal branch
	Zygomaticofacial branch
In infra-orbital canal	Middle superior alveolar branch
	Anterior superior alveolar branch
On face	Inferior palpebral branches
	Nasal branches
	Superior labial branches

PTERYGOPALATINE GANGLION (fig. 48–7)

The pterygopalatine ganglion is in the pterygopalatine fossa, below the maxillary nerve and lateral to the sphenopalatine foramen. It can be injected through the mandibular notch. A parasympathetic root from the greater petrosal nerve and the nerve of the pterygoid canal conveys fibers from the facial nerve. These synapse in the ganglion, and the postganglionic fibers pass to the lacrimal gland (e.g., by the maxillary, zygomatic, and lacrimal nerves). A sympathetic root contains postganglionic fibers that merely traverse the ganglion. An afferent root connects the ganglion with the maxillary nerve. These fibers come from the orbit, nasal cavity, palate, and nasopharynx by way of the so-called branches of the ganglion, which are mostly fibers of the maxillary nerve.

MANDIBULAR NERVE (figs. 48–3C and 48–8)

The mandibular nerve (third division of the trigeminal nerve) arises from the trigeminal ganglion and, together with the motor root of the trigeminal nerve, traverses the foramen ovale and enters the infratemporal fossa (where it can be "blocked" by passing a needle through the mandibular notch and injecting a local anesthetic). At the base of the skull, the mandibular nerve is joined by the motor root and then divides into branches classified as anterior and posterior divisions. The area of skin supplied by the mandibular nerve is tested by the use of cotton wool and a pin (see fig. 47–6). The muscles of mastication are tested by palpating the temporalis and masseter on clenching of the teeth.

Branches. The trunk of the mandibular nerve gives off (1) a *meningeal branch,* the *nervus spinosus,* which accompanies the middle meningeal artery, and (2) the *nerve to the medial pterygoid,* which may also supply the tensor tympani and tensor veli palatini by way of the otic ganglion.

The anterior division of the mandibular nerve comprises several small branches, namely the *buccal nerve* (which emerges between the heads of the lateral pterygoid muscle and descends to supply sensory fibers to the skin and mucosa of the cheek, the gums, and perhaps the first two molars and premolars), the *masseteric nerve* (which traverses the mandibular notch to supply the masseter), *deep temporal nerves* (to the temporalis), and the *nerve to the lateral pterygoid muscle.*

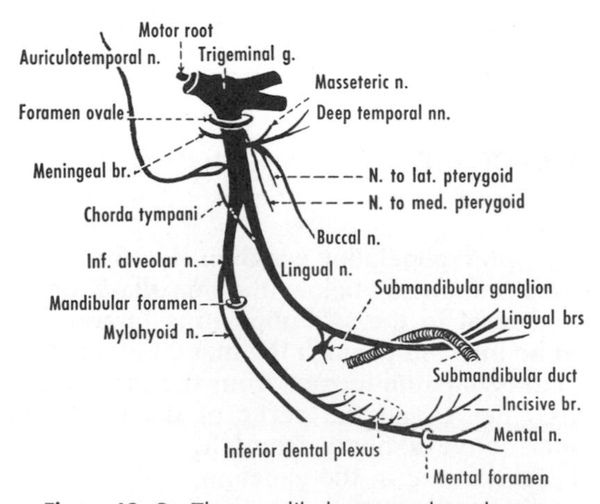

Figure 48–8 The mandibular nerve, lateral aspect.

The posterior division of the mandibular nerve, which is chiefly sensory, gives off the auriculotemporal nerve and divides into the lingual and inferior alveolar nerves.

The *auriculotemporal nerve,* which usually arises by two roots that encircle the middle meningeal artery, proceeds backward deep to the neck of the mandible and closely related to the parotid gland (which it supplies).* It ascends behind the temporomandibular joint (which it supplies). It is accompanied by the superficial temporal artery, and both supply the scalp. The auriculotemporal nerve receives communications from the otic ganglion, conveying secretory fibers from the glossopharyngeal nerve to the parotid gland. **Pain from disease of a tooth or the tongue is sometimes referred to the distribution of the auriculotemporal nerve to the ear.**

The *lingual nerve* descends medial to the lateral pterygoid muscle and is joined by the chorda tympani (a branch of the facial nerve that contains taste fibers). The lingual nerve (which lies in front of the inferior alveolar nerve) passes between the medial pterygoid muscle and the ramus of the mandible. It then lies under cover of the oral mucosa, where it is palpable below and behind the third molar. It crosses the lateral surface of the hyoglossus, passes deep to the mylohyoid muscle, crosses the submandibular duct, and curves upward on the genioglossus. It gives several small branches (e.g., to the submandibular gland) and supplies sensory fibers to the tongue, gums, and first molar and premolar.

The *inferior alveolar nerve,* with its companion artery, descends deep to the lateral pterygoid muscle and then traverses the mandibular foramen and canal. **Above the mandibular foramen, the inferior alveolar nerve can be "blocked" intra-orally with a local anes-**

* The two roots of the auriculotemporal nerve usually enclose a V-shaped interval for the middle meningeal artery and then form a short trunk which breaks up into a spray of branches: two of these are communications with the facial nerve (J. J. Baumel, J. P. Vanderheiden, and J. E. McElenney, Am. J. Anat., *130*:431–440, 1971).

thetic. The inferior alveolar nerve gives off the *mylohyoid nerve* (which descends in a groove on the ramus of the mandible and supplies the mylohyoid muscle and the anterior belly of the digastric muscle), *inferior dental branches* (which form the *inferior dental plexus* and supply the lower teeth), *gingival branches* (to the gums), the *mental nerve* (which emerges through the mental foramen to supply skin), and the *incisive branch,* which is the termination and supplies the canines (sometimes) and incisors, including frequently the incisors of the opposite side.

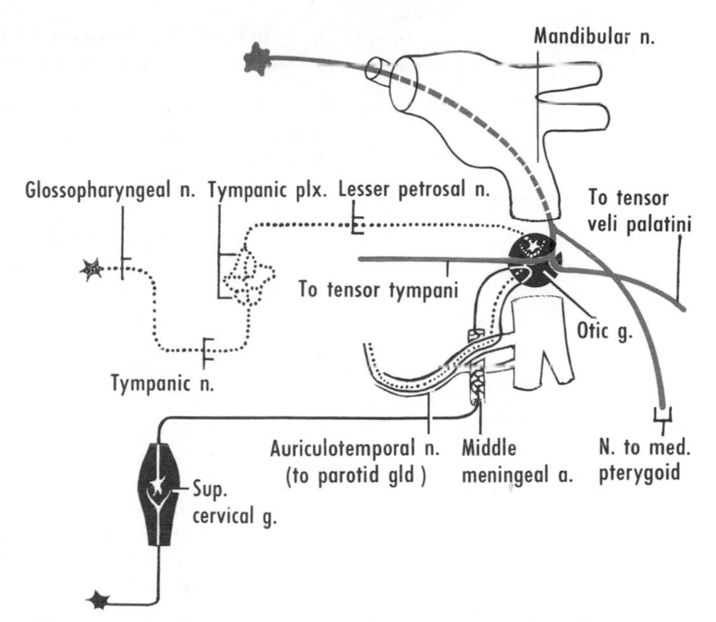

Figure 48–9 The otic ganglion and its connections, lateral aspect. Interrupted lines, parasympathetic fibers. Continuous lines, sympathetic fibers. Red lines, motor fibers.

OTIC GANGLION (fig. 48–9)

The otic ganglion is in the infratemporal fossa, immediately below the foramen ovale and medial to the mandibular nerve. The parasympathetic root is the lesser petrosal nerve. These preganglionic fibers (from the glossopharyngeal nerve) synapse in the ganglion, and the postganglionic fibers pass to the auriculotemporal nerve, through which they supply the parotid gland. A sympathetic root (from the plexus on the middle meningeal artery) contains postganglionic fibers that merely traverse the ganglion. An efferent root from the nerve to the medial pterygoid muscle is believed to supply the tensor tympani and tensor veli palatini. Some taste fibers from the tongue may also pass through the ganglion.

TRIGEMINAL NERVE (fig. 48–10 and table 48–3)

The trigeminal nerve is sensory from the face, anterior half of the scalp, teeth, mouth, nasal cavity, and paranasal sinuses, and motor to the muscles of mastication. It is attached to the side of the pons by a sensory and a motor root. The sensory root expands into the trigeminal ganglion, which gives rise to three large divisions: the ophthalmic, maxillary, and mandibular. The motor root, which also contains afferent fibers from the muscles of mastication, joins the mandibular division. The attachment of the trigeminal roots to the pons is in an area termed the *cerebellopontine angle*. In this vicinity, space-occupying lesions (e.g., tumors) generally involve several or all of the local nerves, namely, the trigeminal, facial, and vestibulocochlear, and sometimes the glossopharyngeal and vagus also.

The branches of the trigeminal nerve are

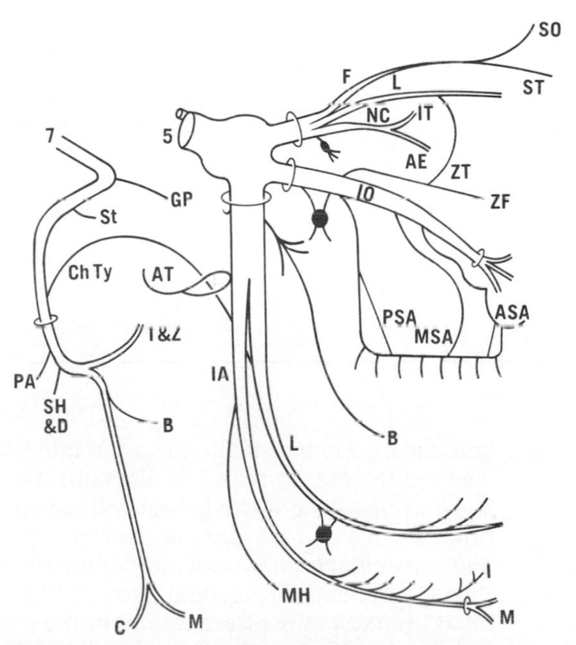

Figure 48–10 Summary of the facial and trigeminal nerves and their branches. For abbreviations, see tables 44–1 and 48–3.

TABLE 48–3 SUMMARY OF BRANCHES OF TRIGEMINAL NERVE

Nerves and Branches	Abbreviation in figure 48–10
1. *Ophthalmic nerve*	
Lacrimal nerve	L
Frontal nerve	F
Supra-orbital nerve	SO
Supratrochlear nerve	ST
Nasociliary nerve	NC
Communicating branch	
Long ciliary nerves	
Infratrochlear nerve	IT
Posterior ethmoidal nerve	
Anterior ethmoidal nerve	AE
Nasal branches	
2. *Maxillary & infra-orbital nerves*	IO
Meningeal branch	
Communicating branches	
Posterior superior alveolar branches	PSA
Zygomatic nerve	Z
Zygomaticotemporal branch	ZT
Zygomaticofacial branch	ZF
Middle superior alveolar branch	MSA
Anterior superior alveolar branch	ASA
Inferior palpebral branches	
Nasal branches	
Superior labial branches	
3. *Mandibular nerve*	
Meningeal branch	
Nerve to medial pterygoid muscle	
Anterior division	
Buccal nerve	B
Masseteric nerve	
Deep temporal nerves	
Nerve to lateral pterygoid muscle	
Posterior division	
Auriculotemporal nerve	AT
Lingual nerve	L
Communicating branches	
Inferior alveolar nerve	IA
Mylohyoid nerve	MH
Inferior dental branches	
Gingival branches	
Mental nerve	M
Incisive branch	I

summarized in figure 48–10 and in table 48–3. The ophthalmic nerve (first division) runs forward in the dura of the lateral wall of the cavernous sinus and divides into lacrimal, frontal, and nasociliary nerves, which enter the orbit through the superior orbital fissure. The nasociliary nerve is the afferent limb of the corneal reflex; the efferent limb is the facial nerve. The maxillary nerve (second division) lies in the dura lateral to the cavernous sinus. It passes through the foramen rotundum and enters the pterygopalatine fossa. Then, as the infra-orbital nerve, it gains the orbit through the inferior orbital fissure and ends on the face by emerging through the infra-orbital foramen. The mandibular nerve (third division), together with the motor root, passes through the foramen ovale to the infratemporal fossa. As it

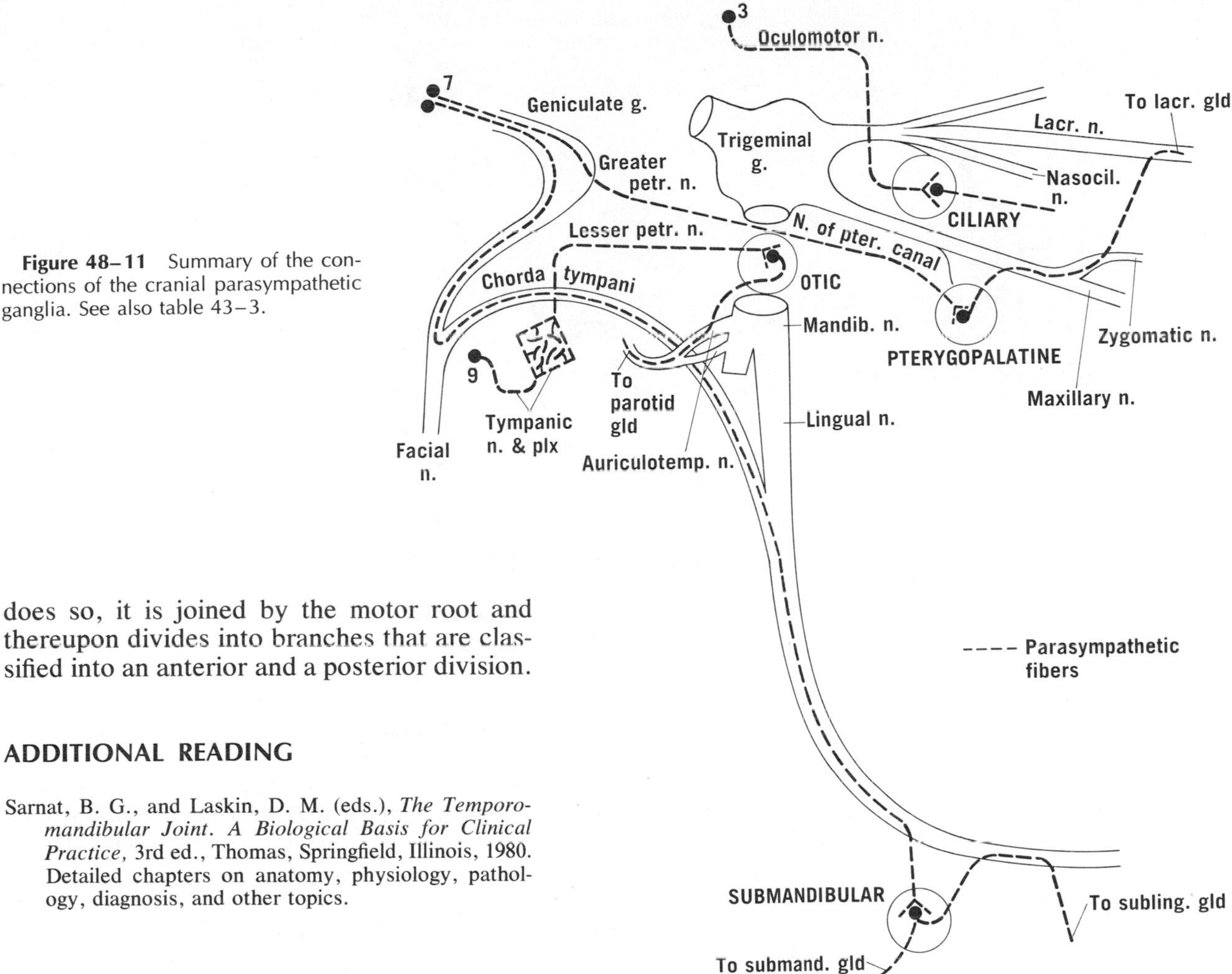

Figure 48–11 Summary of the connections of the cranial parasympathetic ganglia. See also table 43–3.

does so, it is joined by the motor root and thereupon divides into branches that are classified into an anterior and a posterior division.

ADDITIONAL READING

Sarnat, B. G., and Laskin, D. M. (eds.), *The Temporomandibular Joint. A Biological Basis for Clinical Practice,* 3rd ed., Thomas, Springfield, Illinois, 1980. Detailed chapters on anatomy, physiology, pathology, diagnosis, and other topics.

QUESTIONS

48–1 What is "the hostage of parotid surgery?"

48–2 Where do the temporal and infratemporal fossae communicate?

48–3 Which muscles of mastication have at least two heads?

48–4 Which nerve supplies the muscles of mastication?

48–5 Which major part of the temporal bone takes part in the temporomandibular joint?

48–6 What is the significance of the sphenomandibular ligament?

48–7 Which muscles produce retraction of the mandible?

48–8 How is dislocation of the temporomandibular joint reduced?

48–9 Which is clinically the most important branch of the maxillary artery?

48–10 From which structures may pain be referred to the distribution of the auriculotemporal nerve?

48–11 What is the source of the afferent fibers that converge on the pterygopalatine ganglion?

48–12 What is the fundamental nature of the ganglia associated with the trigeminal nerve?

49
THE SUBMANDIBULAR REGION

The region between the mandible and the hyoid bone contains the submandibular and sublingual glands, suprahyoid muscles, submandibular ganglion, and lingual artery. The lingual and hypoglossal nerves and the facial artery are discussed elsewhere.

Submandibular Gland. The large, paired salivary glands are the parotid, submandibular, and sublingual glands. The submandibular gland is usually scarcely palpable. It has a larger superficial part (body) and a smaller deep process (fig. 49–1). **The two parts are continuous with each other around the posterior border of the mylohyoid muscle. The body of the gland is in and below the digastric triangle and also partly under cover of the mandible.** It has three surfaces: *inferior* (covered by skin and platysma), *lateral* (related to the medial surface of the mandible), and *medial* (related to the mylohyoid, hyoglossus, and digastric muscles). The deep process lies between the mylohyoid and hyoglossus muscles and gives exit to the submandibular duct (fig. 49–1), which is crossed by the lingual nerve. **The submandibular duct opens by one to three orifices into the oral cavity on the sublingual papilla, at the side of the frenulum linguae.** The branches of the duct can be examined radiographically after injection of a radio-opaque medium (*sialography*). The submandibular gland is supplied by parasympathetic, secretomotor fibers derived from the submandibular ganglion (see fig. 49–3). The preganglionic fibers are derived from the chorda tympani (from the facial nerve) by way of the lingual nerve.

Sublingual Gland. The sublingual gland (fig. 49–1) is below the mucosa of the floor of the mouth and between the deep process of the submandibular gland behind and the contralateral sublingual gland in front. The 10 to 30 sublingual ducts open mostly separately into the oral cavity (fig. 49–1). The innervation of the gland from the submandibular ganglion is similar to that of the submandibular gland.

Suprahyoid Muscles (table 49–1). The suprahyoid muscles (see fig. 50–6), which connect the hyoid bone to the skull, are the digastric, stylohyoid, mylohyoid, and geniohyoid

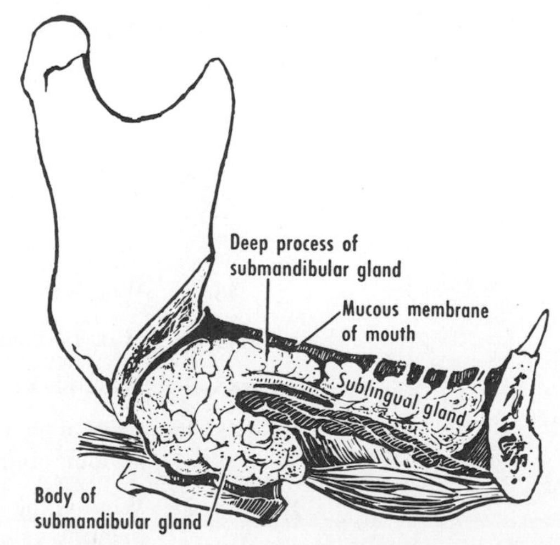

Figure 49–1 The submandibular and sublingual glands, right lateral aspect, after resection of a portion of the mandible.

TABLE 49–1 SUPRAHYOID MUSCLES

Muscle	Origin	Insertion	Innervation	Action
Digastric	Posterior belly from mastoid notch of temporal bone Anterior belly from digastric fossa of mandible	Middle tendon to body & greater horn of hyoid bone	Facial Mylohyoid branch of inferior alveolar	Pulls chin backward & downward in opening mouth
Stylohyoid	Styloid process	Body of hyoid bone	Facial	Retracts hyoid bone & elongates floor of mouth
Mylohyoid	Mylohyoid line of mandible	Raphe and body of hyoid bone	Mylohyoid branch of inferior alveolar	Elevates floor of mouth & tongue
Geniohyoid	Inferior genial tubercle of mandible	Body of hyoid bone	C.N.1 in branch of hypoglossal	Protrudes hyoid bone & shortens floor of mouth

muscles. The genioglossus and hyoglossus are described with the tongue.

The *digastric muscle* (see fig. 50–2, *C* and *D*) consists of two bellies united by an intervening tendon. The *anterior belly,* from the mandible, and the *posterior belly,* from the mastoid region, develop from pharyngeal arches 1 and 2, respectively, and hence are supplied by cranial nerves 5 and 7. The middle tendon, which is anchored to the hyoid bone, commonly passes through the stylohyoid muscle. **The posterior belly of the digastric muscle and the stylohyoid muscle are crossed superficially by the facial vein, the great auricular nerve, and the cervical branch of the facial nerve. The external and internal carotid arteries, the internal jugular vein, cranial nerves 10 to 12, and the sympathetic trunk lie deeply.**

The *stylohyoid muscle* (see fig. 50–6) lies along the upper border of the posterior belly of the digastric muscle and has a similar innervation.

The *mylohyoid* muscle (fig. 49–2) lies above the anterior belly of the digastric muscle and has a similar innervation. The right and left mylohyoid muscles extend from the mandible to a median raphe and form a muscular floor (*diaphragma oris*) beneath the front of the mouth. This muscular sling supports the tongue and is important in forcing both solids and liquids from the oropharynx to the laryngopharynx.

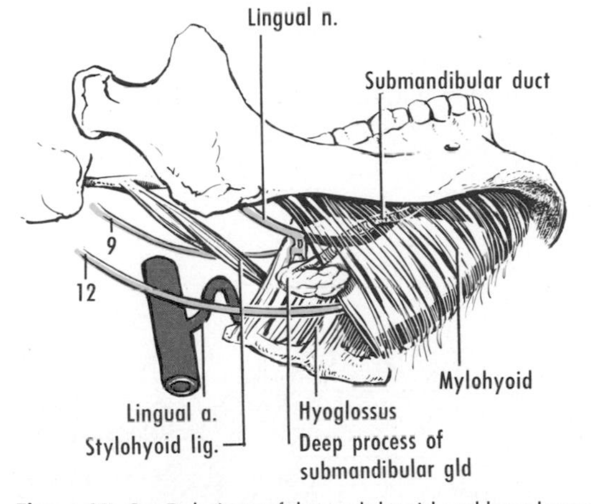

Figure 49–2 Relations of the mylohyoid and hyoglossus muscles. The lingual nerve, the deep process of the submandibular gland and the submandibular duct, and the hypoglossal nerve pass deep to the posterior border of the mylohyoid. The glossopharyngeal nerve, the stylohyoid ligament, and the lingual artery pass deep to the posterior border of the hyoglossus. The submandibular ganglion is shown suspended from the lingual nerve.

The *geniohyoid muscle* (see figs. 49–4 and 53–4) lies above the mylohyoid muscle and is in contact or fused with the muscle of the opposite side.

Submandibular Ganglion (fig. 49–3). **The submandibular ganglion lies on the lateral surface of the hyoglossus muscle, medial to the**

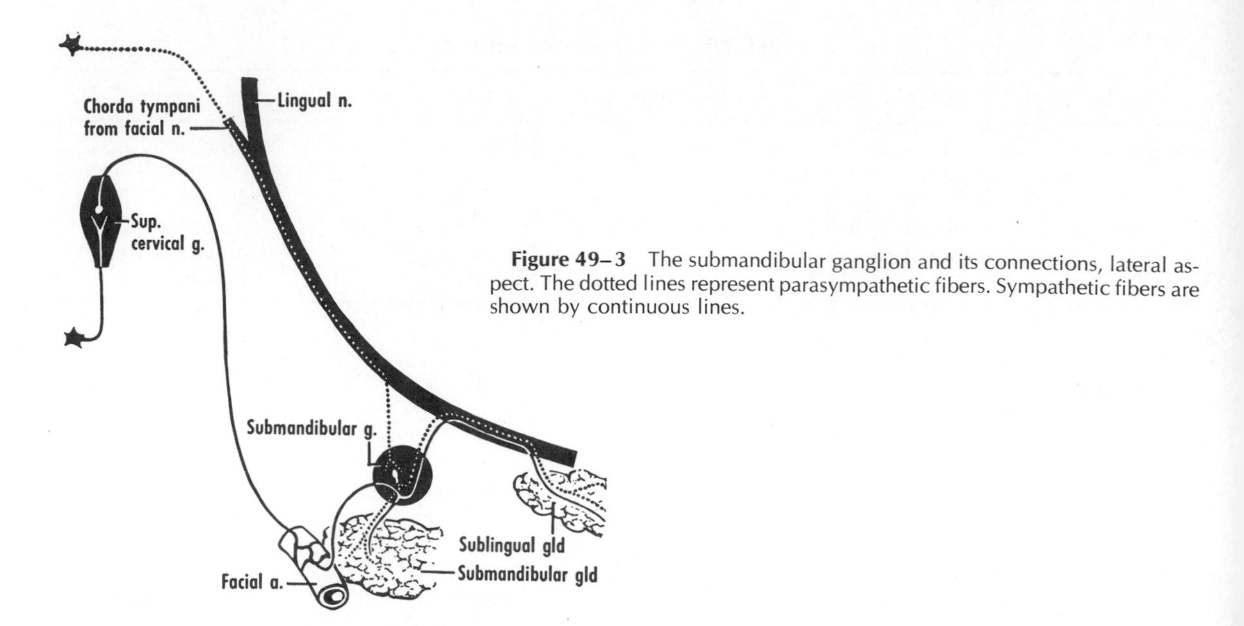

Figure 49–3 The submandibular ganglion and its connections, lateral aspect. The dotted lines represent parasympathetic fibers. Sympathetic fibers are shown by continuous lines.

mylohyoid muscle, above the submandibular duct and hypoglossal nerve, and below the lingual nerve, from which it is suspended by several branches. Preganglionic parasympathetic fibers derived from the chorda tympani travel in the lingual nerve and synapse in the submandibular ganglion. Some of the postganglionic secretory fibers enter the submandibular gland; others, by entering the lingual nerve, reach the sublingual gland. Postganglionic sympathetic fibers (from the superior cervical ganglion) pass through the submandibular ganglion and are distributed with the parasympathetic fibers.

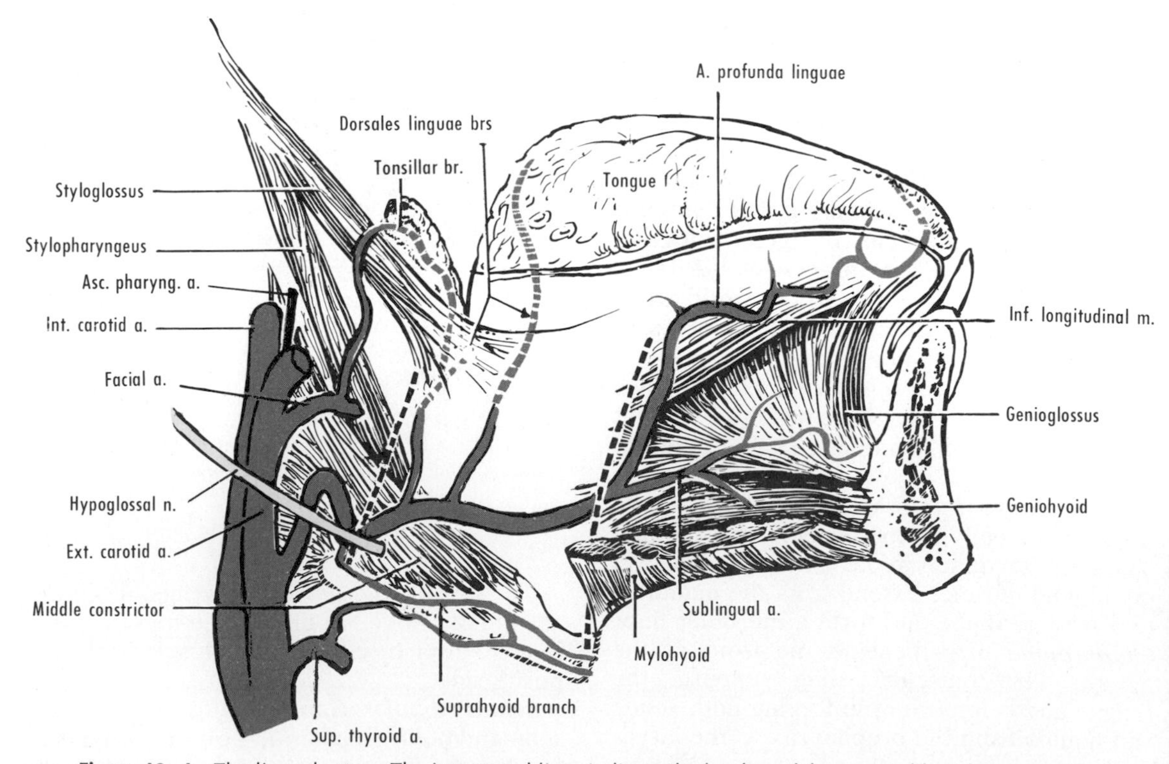

Figure 49–4 The lingual artery. The interrupted lines indicate the borders of the resected hyoglossus muscle.

Lingual Artery (fig. 49–4). The lingual artery arises from the external carotid artery at or above the level of the hyoid bone. It passes successively (1) posterior, (2) deep, and (3) anterior to the hyoglossus muscle. The *first part* of the artery lies mainly in the carotid triangle. It forms a loop on the middle constrictor and is crossed by the hypoglossal nerve. The *second part* runs deeply above the hyoid bone and gives branches to the dorsum of the tongue. The *third part* (*arteria profunda linguae*) ascends between the muscles of the tongue and anastomoses with its fellow of the opposite side.

QUESTIONS

49–1 Around which muscle is the submandibular gland wrapped?

49–2 What is the significance of the innervation of the digastric muscle?

49–3 How are the mylohyoid muscles inserted?

49–4 From what is the submandibular ganglion suspended?

49–5 Which muscle covers the middle (second) part of the lingual artery directly?

50
THE NECK

SUPERFICIAL STRUCTURES

STERNOMASTOID AND TRAPEZIUS MUSCLES

The *sternocleidomastoid,* or more simply *sternomastoid, muscle* arises by two heads from the front of the manubrium sterni and the upper surface of the medial third of the clavicle (see figs. 6–2, 19–2, and 50–8). It ascends obliquely in the neck (figs. 50–1 and 50–2) and is inserted into the lateral surface of the mastoid process (see fig. 42–10) and the lateral part of the superior nuchal line on the occipital bone (see fig. 42–14). A variable interval between the two heads of origin (fig. 50–1) lies over the termination of the internal jugular vein.

The sternomastoid muscle is crossed by the platysma and the external jugular vein, and it covers the great vessels of the neck, the cervical plexus, and the cupola of the pleura.

The *trapezius muscle* (see fig. 8–4) arises

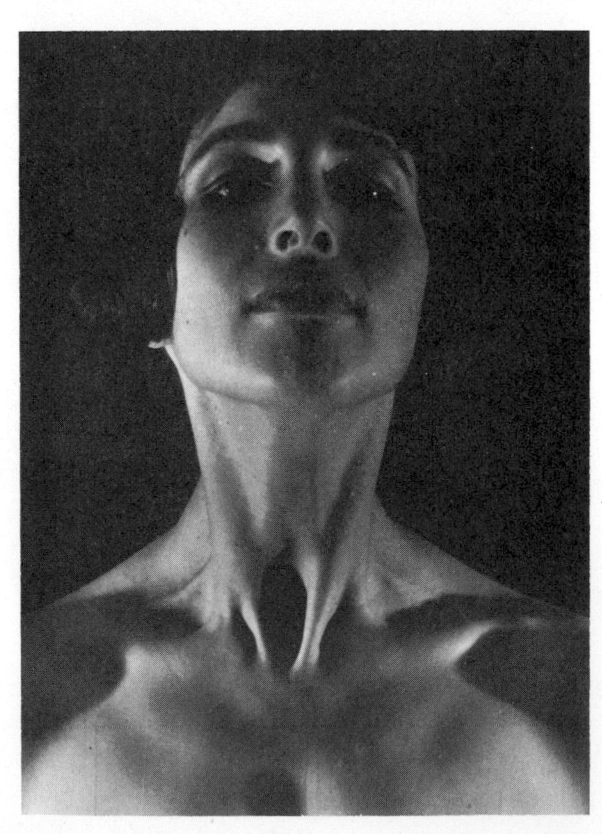

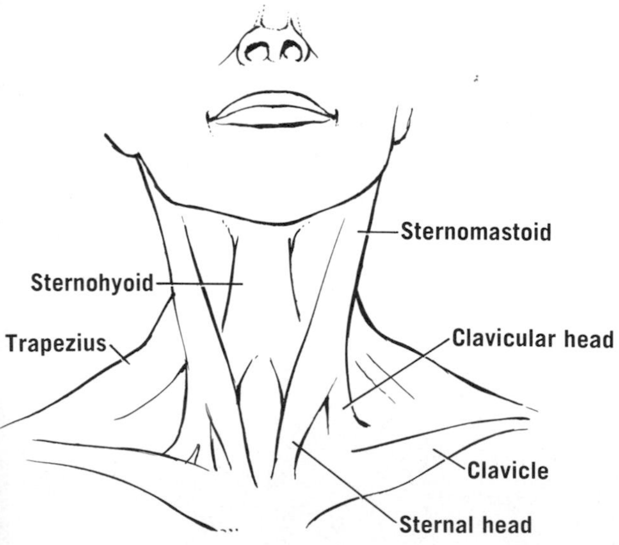

Figure 50–1 Surface anatomy of the neck. The sternal and clavicular heads of the sternomastoid muscles are clearly visible. On each side, the anterior triangle of the neck is bounded by the anterior border of the sternomastoid, the anterior median line of the neck, and the lower border of the mandible. (From Royce, J., *Surface Anatomy,* Davis, Philadelphia, 1965, courtesy of the author and publisher.)

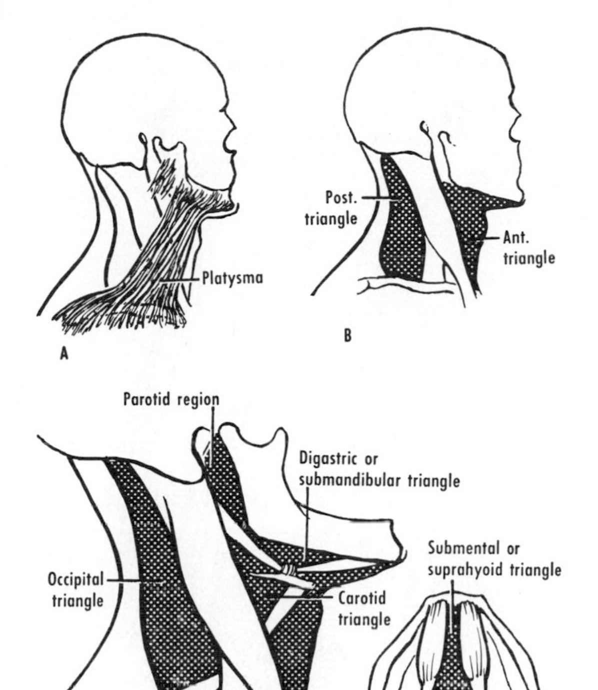

Figure 50–2 Triangles of the neck. **A** shows the platysma, which roofs parts of both the anterior and posterior triangles. **B** shows the division of the neck by the sternomastoid into anterior and posterior triangles. **C** and **D** show the subdivisions of the triangles.

from the superior nuchal line, the external occipital protuberance, the ligamentum nuchae, and the spinous processes of the last cervical and all the thoracic vertebrae. The upper fibers are inserted into the upper surface of the lateral third of the clavicle, and the remaining fibers go to the scapula, as described with the upper limb.

The sternomastoid and trapezius muscles are supplied mainly by the accessory (eleventh cranial) nerve. Cervical nerves 2 to 4 also contribute.

The sternomastoid muscles, acting together, bend the head forward against resistance. Flexion of the head is usually performed by gravity, however. One sternomastoid muscle inclines the head ipsilaterally while the head is rotated contralaterally. Spasm of a sternomastoid produces wry neck (*torticollis*). The trapezius elevates the scapula. **The accessory nerve (external branch) is tested by asking the subject to shrug the shoulders (trapezius) and then to rotate the head (sternomastoid).**

TRIANGLES

The sternomastoid divides the quadrilateral area of the side of the neck into anterior and posterior triangles (fig. 50–2). The posterior triangle is bounded by the sternomastoid and trapezius and by the clavicle. The anterior triangle is bounded by the sternomastoid, anterior median line of the neck, and inferior border of the mandible.

Posterior Triangle (fig. 50–3)

The posterior triangle is crossed by the inferior belly of the omohyoid muscle, which separates an occipital triangle above from a supraclavicular triangle below (fig. 50–2*C*).

The *roof* of the posterior triangle consists of fascia and the platysma. The *floor* is formed by a series of longitudinal muscles—the splenius capitis, levator scapulae, and scaleni medius and posterior—all covered by the prevertebral fascia.

The most important contents of the posterior triangle are the accessory nerve, brachial plexus, third part of the subclavian artery, and lymph nodes. The accessory nerve (external branch), which crosses the transverse process of the atlas, either pierces or runs deep to the sternomastoid, which it supplies. **Above the middle of the posterior border of the sternomastoid, the accessory nerve crosses the posterior triangle obliquely** (fig. 50–3*B*). It then passes deep to the anterior border of the trapezius and supplies that muscle.

The brachial plexus is formed by C.N.5 to T.N.1, which are sandwiched between the scaleni anterior and medius. In the posterior triangle, the brachial plexus is found in the angle between the posterior border of the sternomastoid and the clavicle, where it can be "blocked" by injection of a local anesthetic, thereby rendering insensitive all the deep structures of the upper limb and the skin distal to the middle of the arm.

The *cervical plexus* (see table 50–5) is situated deeply in the upper part of the neck, under cover of the internal jugular vein and the sternomastoid. Formed by C.N.1 to 4, it gives rise to superficial branches, which emerge near the middle of the posterior border of the sternomastoid (fig. 50–4). The branches are (1) the lesser occipital nerve, which hooks around the accessory nerve and ascends to the auricle; (2) the great auricular nerve which crosses the sternomastoid and ascends to the

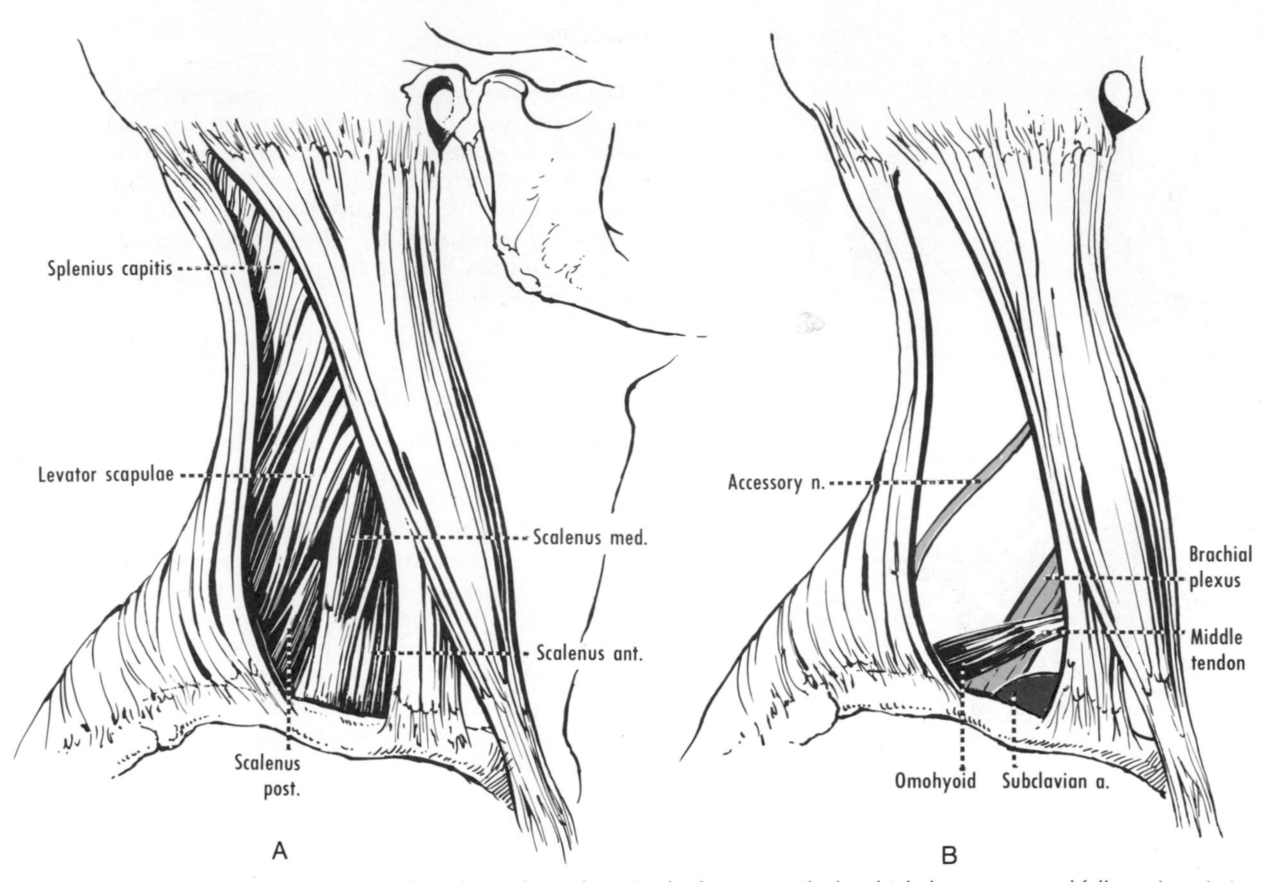

Figure 50–3 Posterior triangle of the neck. **A,** Floor. **B,** Chief contents. The brachial plexus meets and follows the subclavian artery. The third part of the subclavian artery is the site for compression.

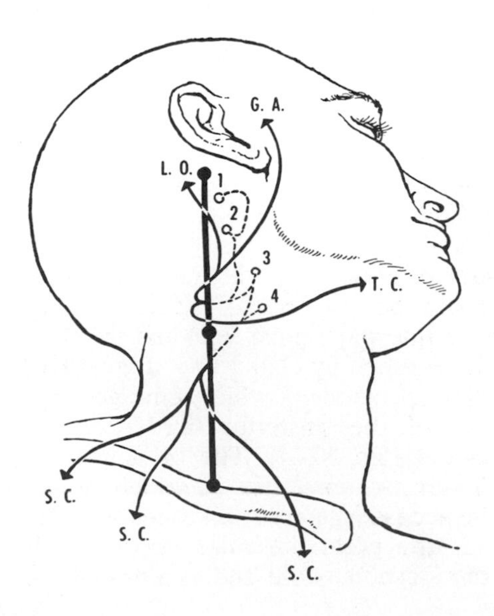

Figure 50–4 The cutaneous branches of the cervical plexus. The vertical line represents the posterior border of the sternomastoid: the branches of the plexus emerge from the middle of this border. *G.A.*, great auricular nerve; *L.O.*, lesser occipital nerve; *S.C.*, supraclavicular nerves. *T.C.*, transverse cervical nerve; *1* to *4*, cervical nerves. (After von Lanz and Wachsmuth.)

parotid region; (3) the transverse cervical nerve, which crosses the sternomastoid and gains the front of the neck; and (4) the supraclavicular nerves, which descend as three nerves that cross the clavicle and supply the skin over the shoulder.

The external jugular vein drains most of the scalp and face and also contains a significant amount of cerebral blood. It begins below or in the parotid gland by a highly variable union of smaller veins (fig. 50–5). It descends under cover of the platysma and obliquely crosses the sternomastoid muscle, where it is frequently visible. It ends in the subclavian or internal jugular vein. The tributaries of the external jugular vein, which are very variable, include communications with the internal jugular and an inconstant anterior jugular vein that descends on the front of the neck.

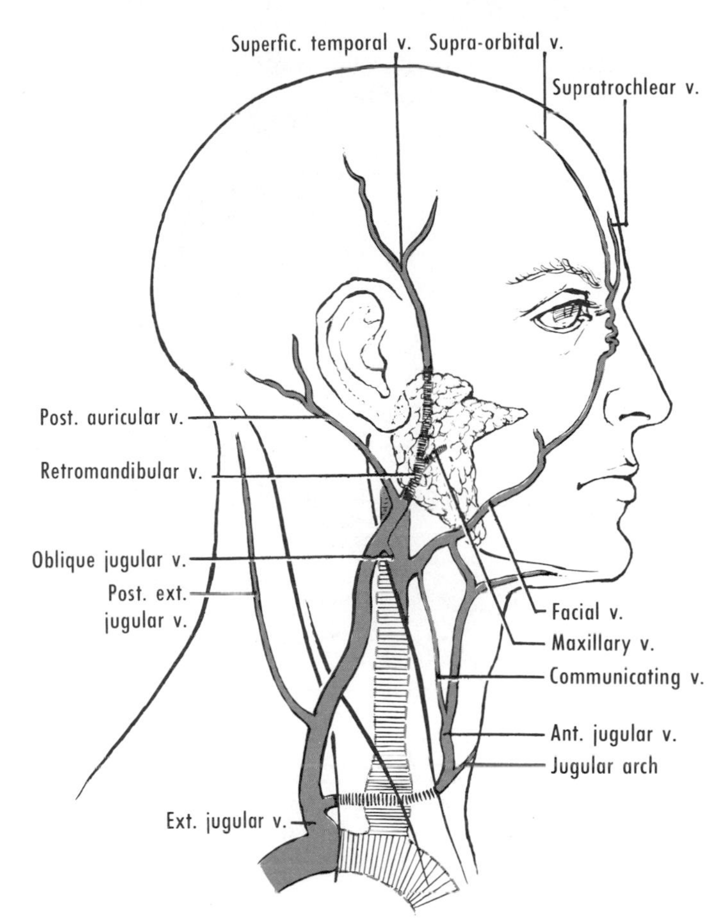

Figure 50–5 Superficial veins of the head and neck. Variations are very common. The internal jugular vein can be seen deep to the sternomastoid.

Anterior Triangle

The anterior triangle is bounded by the sternomastoid, the anterior median line of the neck, and the inferior border of the mandible. The anterior triangle is crossed by the digastric and stylohyoid muscles and by the superior belly of the omohyoid. These muscles allow further subdivisions of the triangle to be made, such as the carotid triangle, which is bounded by the sternomastoid, the posterior belly of the digastric, and the superior belly of the omohyoid (see fig. 50–2*B*).

The *roof* of the anterior triangle consists of fascia and the platysma. The *platysma* (see fig. 50–2*A*) is a subcutaneous, quadrilateral muscular sheet. It arises from the skin over the deltoid muscle and the pectoralis major and is inserted into the lower border of the mandible and the skin around the mouth. It is supplied by the cervical branch of the facial nerve. It raises the skin, thereby probably relieving pressure on the underlying veins.

The *floor* of the anterior triangle is formed by a series of muscles, including the mylohyoid and hyoglossus, infrahyoid muscles, and the constrictors of the pharynx. **The carotid triangle contains a portion of the external carotid artery and its branches. The common and internal carotid arteries and the internal jugular vein tend to be overlapped by the anterior border of the sternomastoid** (see fig. 50–12).

The *infrahyoid muscles* (fig. 50–6 and table 50–1) are four strap-like muscles that anchor the hyoid bone. They are arranged in (1) a superficial plane comprising the sternohyoid and omohyoid muscles and (2) a deep plane comprising the sternothyroid and thyrohyoid muscles. The first three muscles are innervated by the ansa cervicalis and its superior root, whereas the thyrohyoid muscle is supplied by cervical fibers carried in the hypoglossal nerve. All the fibers to the infrahyoid muscles are derived ultimately from C.N.1 to 3. The muscles act, according to circumstances, either to depress the larynx, hyoid bone, and floor of the mouth or to resist their elevation.

The *omohyoid muscle* consists of two bellies, an inferior belly from the scapula, which ends in a middle tendon, and a superior belly, which continues from the tendon to the hyoid bone. The *middle tendon,* situated deep to the sternomastoid, is attached by fascia to the manubrium, first costal cartilage, and clavicle.

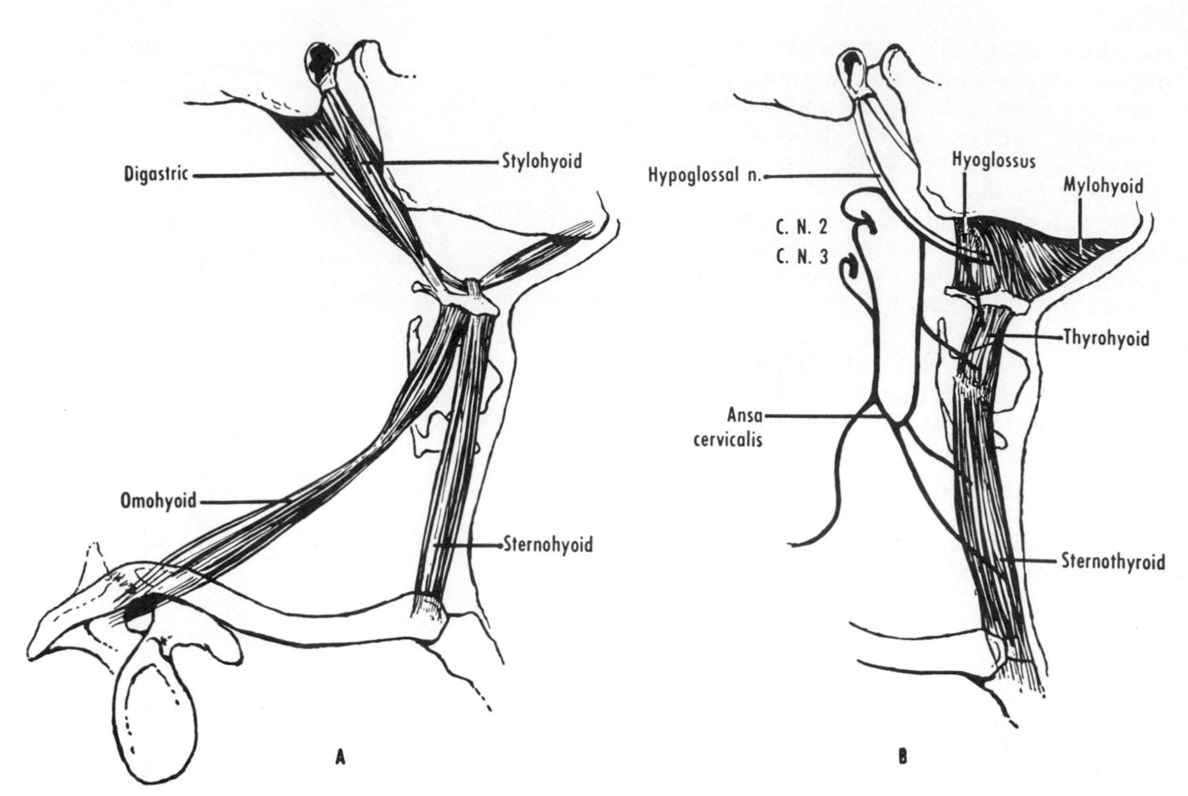

Figure 50–6 Suprahyoid and infrahyoid muscles. **A,** Superficial plane. For the geniohyoid, see figs. 49–4 and 53–4. **B,** Deeper plane. The infrahyoid muscles are innervated mainly by the ansa cervicalis.

TABLE 50–1 Infrahyoid Muscles

Muscle	Origin	Insertion	Innervation	Action
Sternohyoid	Manubrium, clavicle, or both	Body of hyoid bone	Ansa cervicalis & its superior root	Depress larynx, hyoid bone, & floor of mouth
Omohyoid	Scapula near suprascapular notch	Body of hyoid bone	Ansa cervicalis & its superior root	Depress larynx, hyoid bone, & floor of mouth
Sternothyroid	Manubrium	Oblique line of lamina of thyroid cartilage	Ansa cervicalis & its superior root	Depress larynx, hyoid bone, & floor of mouth
Thyrohyoid	Oblique line of lamina of thyroid cartilage	Greater horn of hyoid bone	Hypoglossal	Depress larynx, hyoid bone, & floor of mouth

DEEP STRUCTURES

The cervical vertebrae (fig. 50–7) are described with the back. The lower part of the neck is a junctional region between the thorax and the upper limbs (fig. 50–8). **The inlet (superior aperture) of the thorax is a reniform interval bounded by T.V.1, the first ribs and costal cartilages, and the manubrium sterni. The chief structures that pass through the inlet are vessels (brachiocephalic trunk and left common carotid, left subclavian, and internal thoracic arteries), nerves (phrenic, vagus, recurrent laryngeal, and sympathetic trunk), the**

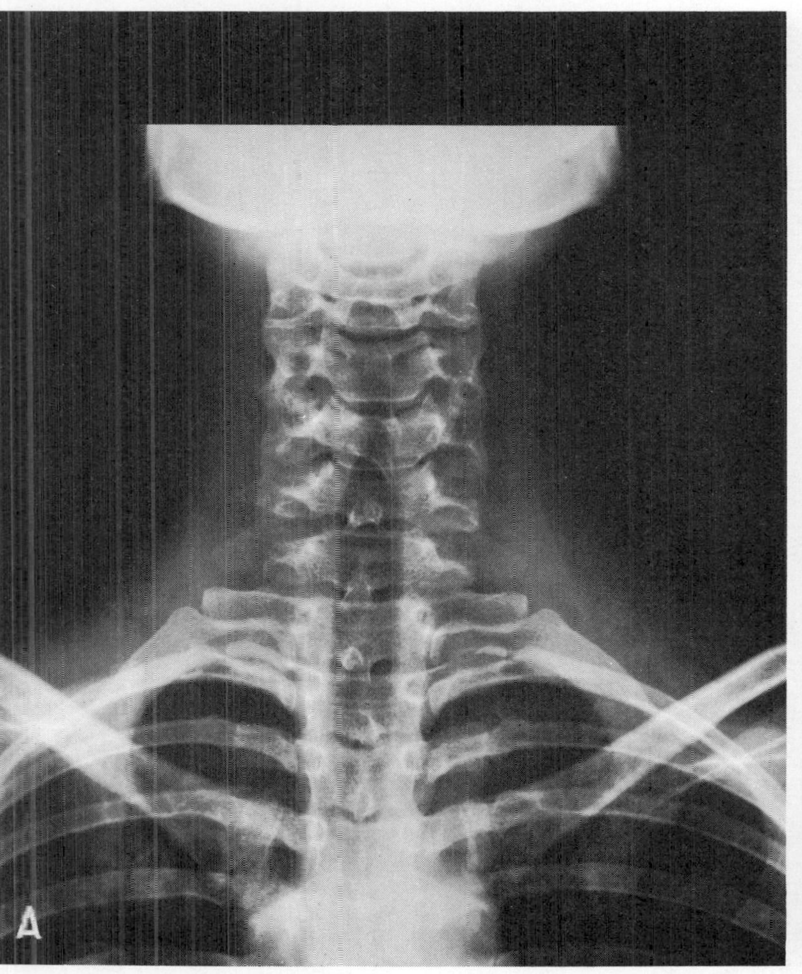

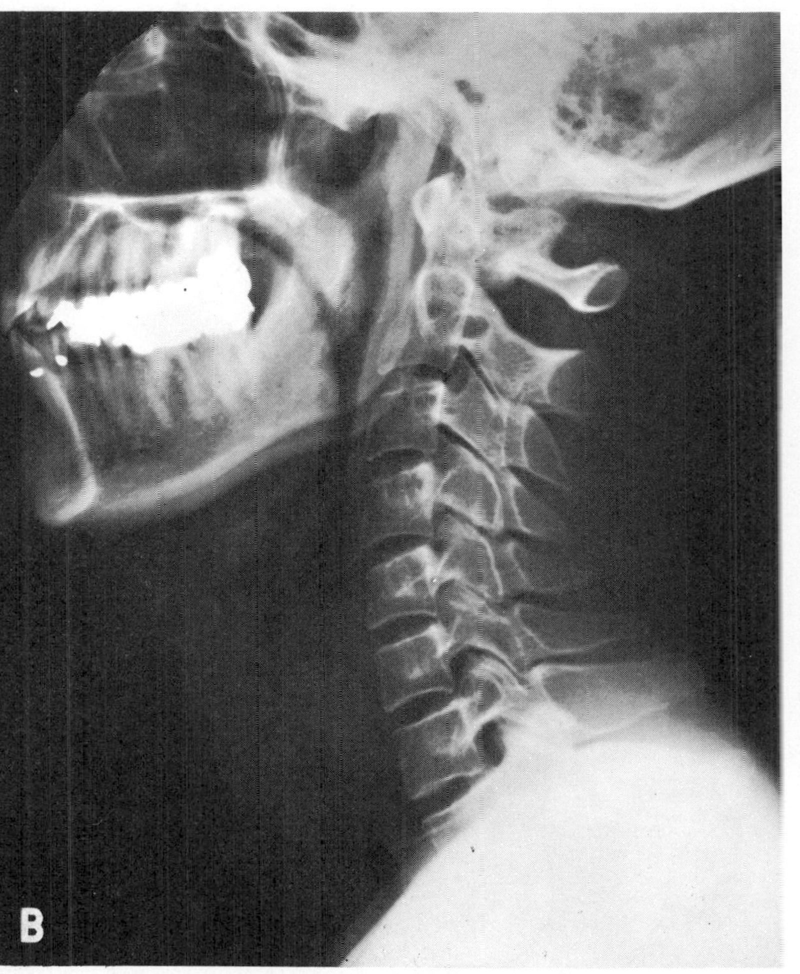

Figure 50–7 Cervical vertebrae. **A,** Anteroposterior view. Note the translucency of the larynx and trachea. **B,** Lateral view. Note the anterior and posterior arches of the atlas, the curve of the cervical column, and the slopes of the articular facets. The teeth display metallic fillings.

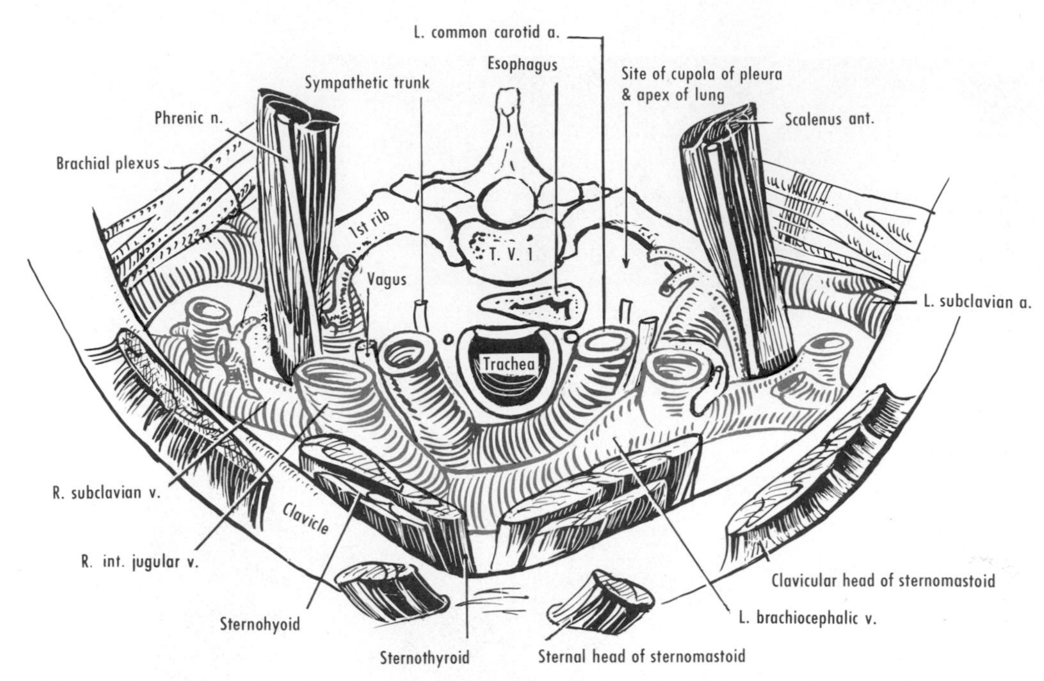

Figure 50–8 The main structures that cross the thoracic inlet. See text. In addition to various vessels, note the recurrent laryngeal nerves and (on the left side) the thoracic duct. (After von Lanz and Wachsmuth.)

trachea and esophagus, cupola of the pleura, apex of the lung, and thymus. The apex of the axilla is bounded by the upper border of the scapula, the external border of the first rib, and the posterior surface of the clavicle. The chief structures that pass through this interval are the brachial plexus and the axillary artery and vein.

The thymus, described with the thorax, has a cervical part on the front and sides of the trachea behind the sternohyoid and sternothyroid muscles.

THYROID GLAND

The thyroid gland is an endocrine organ in the neck; it may become enlarged to form a goiter. The gland is covered by (1) an adherent, fibrous *capsule* and (2) a *sheath* ("false capsule") derived from the deep cervical fascia.

The thyroid gland, seen from the front, is H or U shaped, consisting of right and left lobes connected by an isthmus (figs. 50–9 to 50–11). **The lobes can be palpated *in vivo*.** Each lobe has an *apex,* which ascends between the sternothyroid muscle and the inferior constrictor of the pharynx; a *base* directed downward; and three surfaces. The *lateral surface* is covered by infrahyoid muscles; the *medial surface* is related to the larynx, pharynx, trachea, and esophagus; the *posterior surface* (fig. 50–9*B*) is related to the carotid sheath and its contents and to the prevertebral muscles. **The isthmus connects the right and left lobes and generally covers rings 2 to 4 of the trachea.** The inconstant *pyramidal lobule* (or "lobe") ascends from the isthmus (fig. 50–9*A*) and may be anchored to the hyoid bone by fibrous or muscular tissue (*levator glandulae thyroideae*).

Blood Supply (fig. 50–11). The thyroid gland is highly vascular and is supplied mainly by the superior thyroid artery (from the external carotid) and the inferior thyroid artery (from the thyrocervical trunk of the subclavian). The arteria thyroidea ima is an inconstant branch of variable origin (e.g., from the brachiocephalic trunk) that ascends to the isthmus. The thyroid plexus on the surface of the gland and on the front of the trachea is drained by superior and middle thyroid veins into the internal jugular vein and by inferior thyroid veins into the brachiocephalic vein.

Lymphatic Drainage. The lymph vessels drain (1) upward to the deep cervical nodes and (2) downward to the paratracheal nodes. The isthmus drains into the prelaryngeal and pretracheal nodes.

Development. **The thyroid gland develops largely as a median diverticulum from the floor**

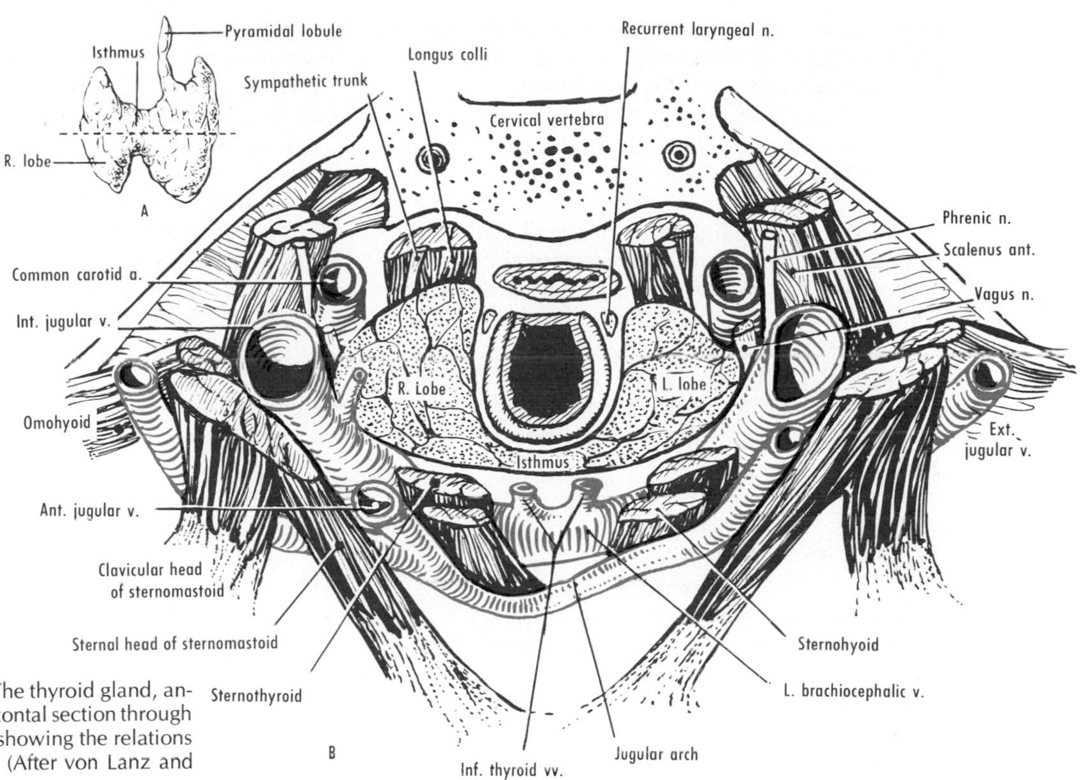

Figure 50–9 **A,** The thyroid gland, anterior aspect. **B,** Horizontal section through the line shown in **A,** showing the relations of the thyroid gland. (After von Lanz and Wachsmuth.)

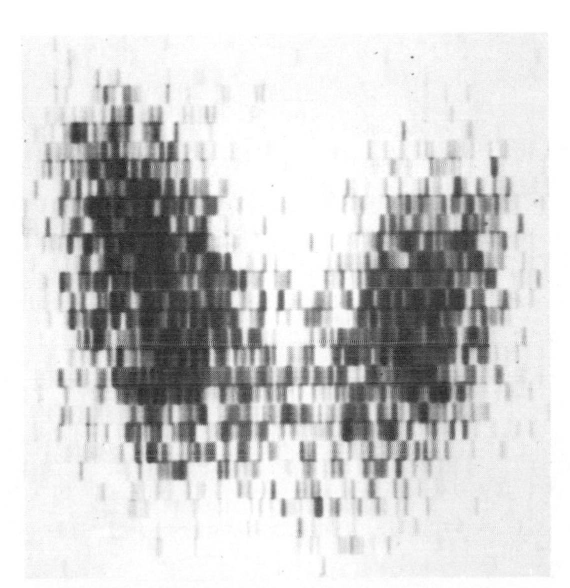

Figure 50–10 Scintigram of the thyroid gland produced by the uptake of a radio-isotope. The right and left lobes are united by the isthmus. (From DeLand, F. H., and Wagner, H. N., *Atlas of Nuclear Medicine,* vol. 3, W. B. Saunders Company, Philadelphia, 1972, courtesy of the authors.)

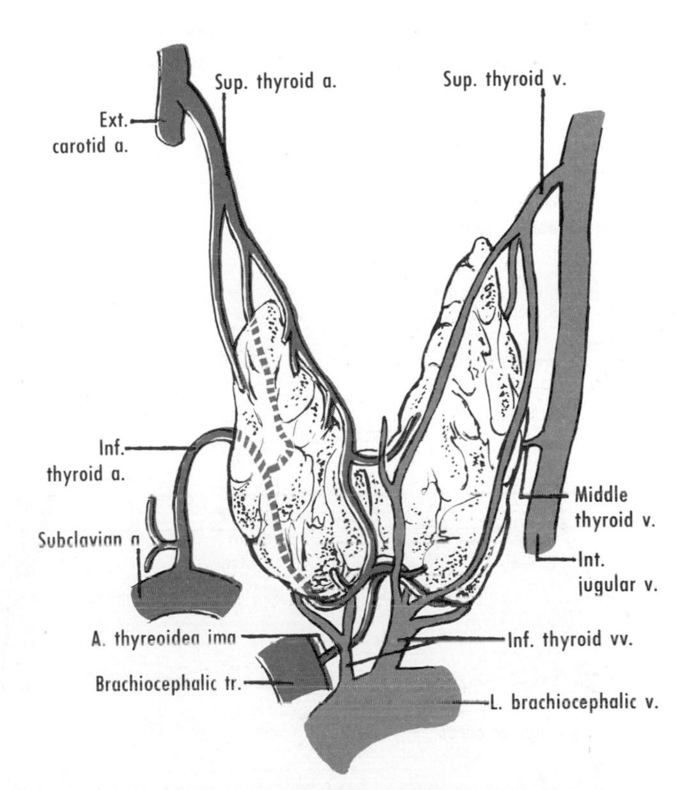

Figure 50–11 The blood supply of the thyroid gland. Only the arteries are shown on one side and only the veins on the other. Most anastomotic vessels are omitted.

of the pharynx. Parts of the embryonic thyroglossal duct may remain as cysts, the pyramidal lobule, and accessory thyroid tissue.

Parathyroid Glands

The parathyroid glands are small endocrine organs that are essential to life and that therefore must not be removed during thyroidectomy. They usually lie outside the thyroid capsule and on the medial half of the posterior surface of each lobe of the thyroid gland. Although usually four (*superior* and *inferior* on each side), they vary from two to six in number.

Trachea and Esophagus

The *trachea,* with its incomplete rings of hyaline cartilage (see fig. 53–9), lies partly in the neck and partly in the thorax (q.v.). **The trachea extends from C.V.6 to about T.V.6 or 7 *in vivo*.** The cervical part (see fig. 50–9) is related in front to infrahyoid muscles, the thyroid isthmus (generally over rings 2 to 4), and vessels (e.g., inferior thyroid veins). **Tracheotomy, the making of an artificial opening in the trachea, is sometimes necessary for treating respiratory obstruction. For non-surgeons, however, cricothyrotomy is preferable.**

The *esophagus* lies in the neck, thorax, and abdomen (q.v.). **The esophagus extends from C.V.6 to about T.V.11. Its narrowest point is adjacent to the inferior constrictor of the pharynx, about 15 cm from the upper incisors.** The cricopharyngeal fibers of the inferior constrictor aid a more distally placed sphincter for the esophagus (''pharyngo-esophageal sphincter''). The upper part of the esophagus consists of skeletal muscle that is attached to the back of the lamina of the cricoid cartilage. The esophagus is behind the trachea and in front of the longus colli and vertebral column.

Carotid Arteries

The main vessels of the head and neck (fig. 50–12) are the right and left common carotid arteries, each of which divides in the carotid triangle into (1) an external carotid artery, which supplies the structures external to the skull as well as the face and most of the neck, and (2) an internal carotid artery, which is distributed within the cranial cavity and the orbit. **The common and internal carotid arteries lie in a cleft bounded by (1) the cervical vertebrae and their attached muscles, (2) the pharynx**

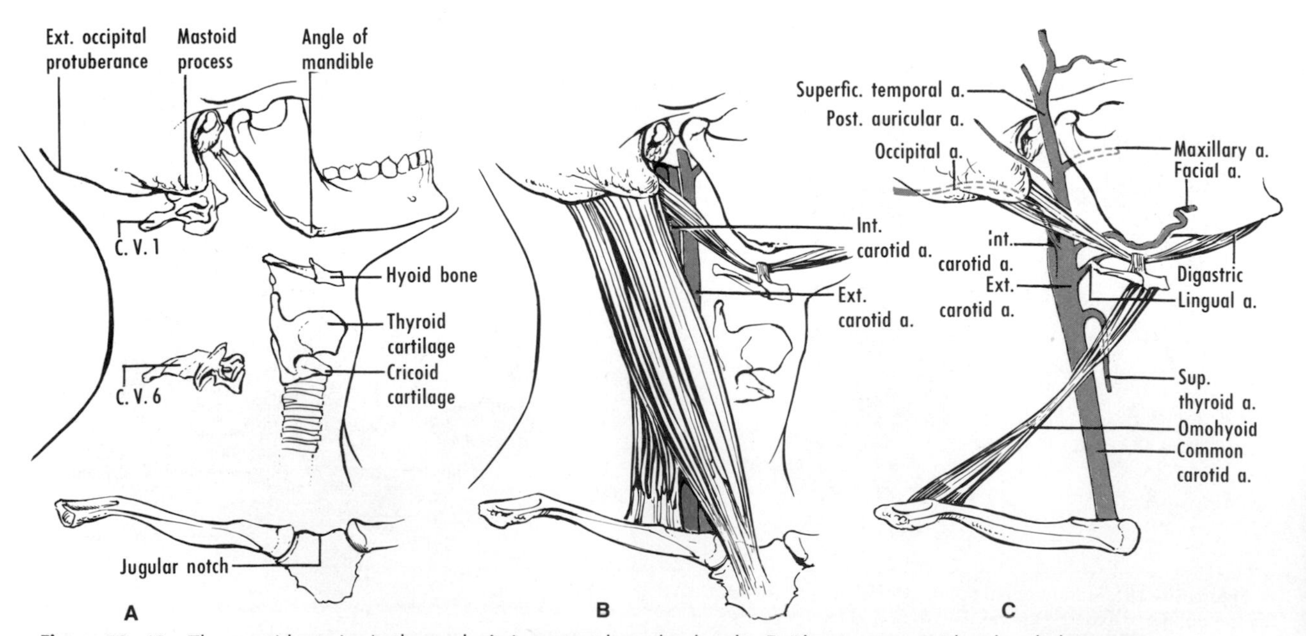

Figure 50–12 The carotid arteries in the neck. **A,** Important bony landmarks. **B,** The sternomastoid and underlying great vessels. Note the internal jugular vein in the interval between the heads of the sternomastoid below. **C,** The carotid arteries. Note the branches of the external carotid artery.

and esophagus, larynx and trachea, and thyroid gland, and (3) the sternomastoid (see fig. 50–29*C*).

Surface Anatomy. **The line of the carotid arteries ascends from (1) the sternoclavicular joint, along the anterior border of the sternomastoid muscle, to (2) a point medial to the lobule of the auricle. The left common carotid has also a thoracic part. Each common carotid artery is crossed by the corresponding omohyoid muscle opposite the cricoid cartilage (C.V.6), and this is the site for compression. The common carotid artery divides usually at the level of the upper border of the lamina of the thyroid cartilage. The pulsation of the common and external carotid arteries can be felt along the anterior border of the sternomastoid muscle.**

Common Carotid Artery

The right common carotid artery arises from the brachiocephalic trunk (behind the right sternoclavicular joint), whereas the left common carotid is a branch of the arch of the aorta and hence has a thoracic portion before it reaches the neck (behind the left sternoclavicular joint). The common carotid artery is crossed by the omohyoid at the level of the cricoid cartilage (C.V.6), and then lies under cover of the anterior border of the sternomastoid. The artery is related behind to the cervical vertebrae. **The common carotid artery can be compressed against the transverse processes of the cervical vertebrae by pressing medially and posteriorly with the thumb.** The common carotid usually gives off no named branches in the neck. **Each common carotid artery divides usually at the level of the upper border of the lamina of the thyroid cartilage (C.V.4).**

External Carotid Artery (see fig. 50–14)

The external carotid artery extends from the division of the common carotid artery in the carotid triangle to behind the neck of the mandible. At first generally anteromedial to the internal carotid, it inclines backward to adopt a lateral position. The external carotid artery is crossed by the hypoglossal nerve and passes deep to the posterior belly of the digastric and the stylohoid muscle. The constrictors of the pharynx lie medially. The external carotid divides in the substance of the parotid gland into the superficial temporal and maxillary arteries.

TABLE 50–2 Branches of External Carotid Artery

Aspect	Branches
Anterior	1. Superior thyroid 2. Lingual 3. Facial
Posterior	4. Occipital 5. Posterior auricular
Medial	6. Ascending pharyngeal
Terminal	7. Superficial temporal 8. Maxillary

Branches. The eight branches of the external carotid artery are described in the following paragraphs and summarized in table 50–2. The first four generally arise in the carotid triangle.

1. The *superior thyroid artery* (see fig. 50–11) arises in the carotid triangle, descends deep to the infrahyoid muscles, and gains the apex of the corresponding lobe of the thyroid gland, where it divides into glandular branches. It generally gives off the superior laryngeal artery, which accompanies the internal laryngeal nerve.

2. The *lingual artery* (see fig. 49–4) may arise in common with the facial artery. Its description in three parts is based on its relationship to the hyoglossus muscle. The first part forms a loop in the carotid triangle and is crossed by the hypoglossal nerve. The second part passes deep to the hyoglossus and lies on the middle constrictor. The third part (*arteria profunda linguae*) runs along the lower surface of the tongue and anastomoses with its fellow of the opposite side.

3. The *facial artery* (see fig. 47–5*B*), in its cervical part, ascends in the carotid triangle and enters a groove on the back of the submandibular gland. It then descends to wind around the lower border of the mandible at the anterior margin of the masseter. The facial artery, in its facial part, proceeds tortuously upward and forward on the face. It supplies the muscles of facial expression, to which it is variably related. The facial artery ends at the medial angle of the eye by anastomosing with branches of the ophthalmic artery. The facial vein is posterior and straighter.

The cervical part of the facial artery gives off *palatine, tonsillar, glandular* (submandibular), and *muscular branches*. The facial part gives off the *inferior* and *superior labial arteries* and a *nasal branch* and ends as the *angular artery*.

4. The *occipital artery* (fig. 50–12*C*) arises posteriorly. Its description in three parts is based on its relationship to the sternomastoid muscle. In the carotid triangle, the hypoglossal nerve winds around the artery, which ascends deep to the lower border of the posterior belly of the digastric muscle. Deep to the sternomastoid, the artery occupies a groove medial to the mastoid process. Behind the sternomastoid, the occipital artery divides into branches on the scalp. The most important branch descends (as superficial and deep divisions) to anastomose with transverse and deep cervical branches of the subclavian artery (see fig. 50–21). **The descending branch of the occipital artery provides the chief collateral circulation after ligation of the external carotid or the subclavian artery.**

5. The *posterior auricular artery* arises posteriorly above the digastric muscle. It ends by dividing into auricular and occipital branches.

6. The *ascending pharyngeal artery* (see fig. 49–4) arises medially and ascends on the wall of the pharynx.

7 and 8. The *superficial temporal* and *maxillary arteries* have already been described.

Internal Carotid Artery (Cervical Part) (figs. 50–12 to 50–14)

The internal carotid artery begins at the level of the upper border of the lamina of the thyroid cartilage. It enters the skull through the carotid canal of the temporal bone, and it ends in the middle cranial fossa by dividing into the anterior and middle cerebral arteries.

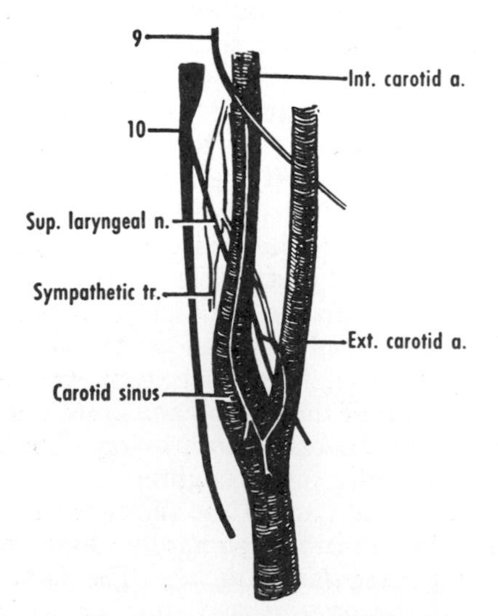

Figure 50–13 The carotid sinus and its innnervation from the glossopharyngeal and vagus nerves and from the sympathetic trunk.

The *carotid sinus* (fig. 50–13) is a slight dilatation of the internal carotid artery or of the common and internal carotid arteries near the point of division. Its wall contains receptors sensitive to changes in blood pressure. The *carotid body* lies in the angle of bifurcation of the common carotid artery and is sensitive to anoxemia.

The internal carotid artery gives no named branches in the neck. Its course may be considered in four parts: cervical, petrous, cavernous, and cerebral. The last three have been described already.

The internal carotid artery begins in the carotid triangle, which it leaves by passing deep to the digastric and stylohyoid muscles. The vessel is closely related to cranial nerves 9, 10, and 12 and to the sympathetic trunk. The internal carotid artery lies on the transverse processes of the cervical vertebrae and against the wall of the pharynx. The external carotid artery is at first generally anteromedial or anterior but becomes lateral higher up.

Glossopharyngeal Nerve (figs. 50–14 to 50–16)

The glossopharyngeal (ninth cranial) nerve is afferent from the tongue and pharynx (hence its name) and efferent to the stylopharyngeus and parotid gland. It emerges from the medulla and passes through the jugular foramen, where it presents *superior* (*jugular*) and *inferior* (*petrous*) ganglia. These ganglia contain the cell bodies of the afferent fibers. The nerve then passes between the internal jugular vein and the internal carotid artery and descends deep to the styloid process and muscles. **The glossopharyngeal nerve curves forward around the stylopharyngeus,** runs deep to the posterior border of the hyoglossus (see Fig. 49–2), and passes between the superior and middle constrictors of the pharynx.

Branches. The branches of the glossopharyngeal nerve are described in the following paragraphs.

1. The *tympanic nerve* is secretomotor and vasodilator to the parotid gland. It enters the bone (tympanic canaliculus) to reach the tympanic cavity, where it divides to form the *tympanic plexus* on the promontory on the medial wall of the middle ear (see fig. 44–5). Having supplied the adjacent mucosa, the plexus forms the *lesser petrosal nerve,* which traverses the temporal bone, passes through or near the foramen ovale, and joins the otic gan-

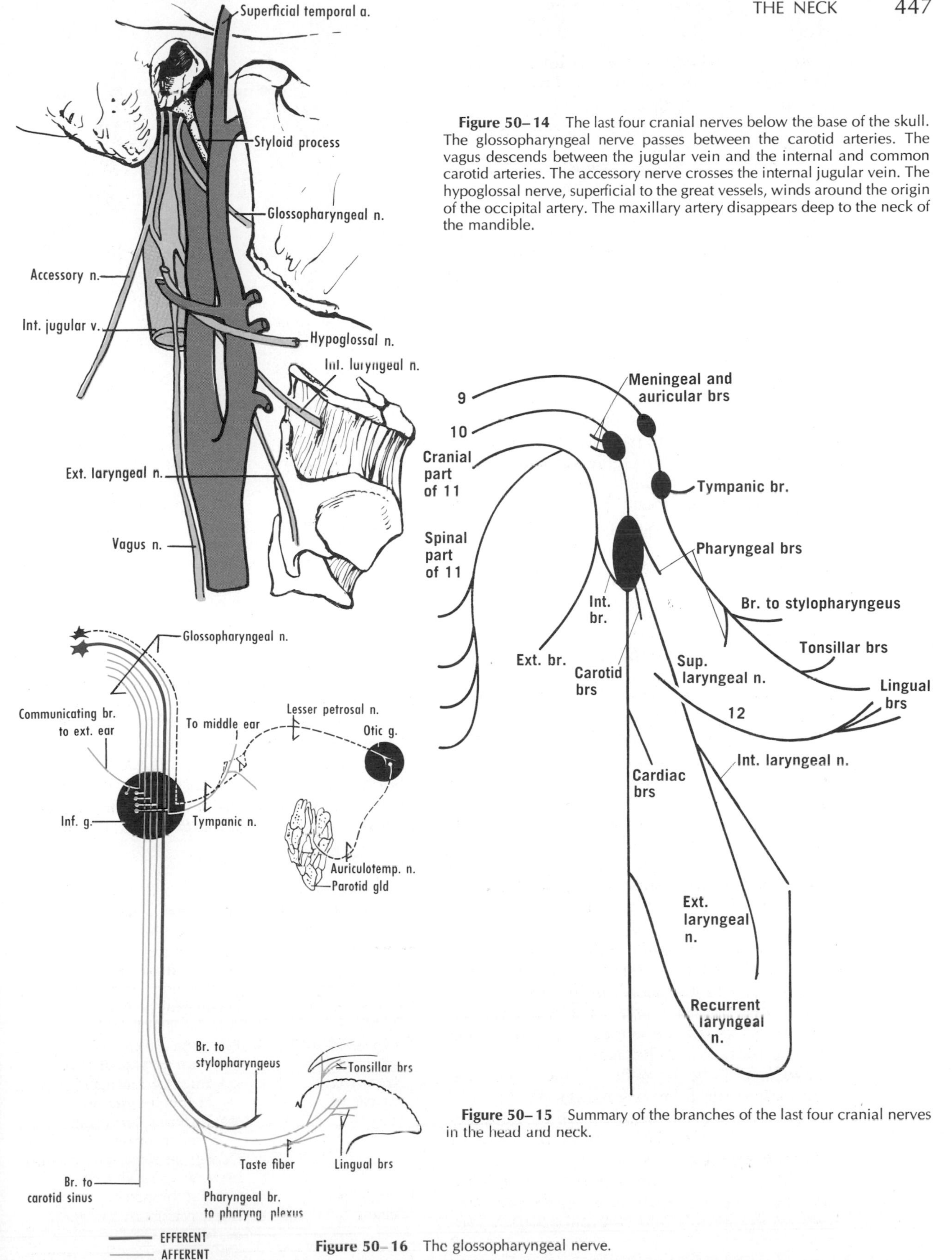

Figure 50–14 The last four cranial nerves below the base of the skull. The glossopharyngeal nerve passes between the carotid arteries. The vagus descends between the jugular vein and the internal and common carotid arteries. The accessory nerve crosses the internal jugular vein. The hypoglossal nerve, superficial to the great vessels, winds around the origin of the occipital artery. The maxillary artery disappears deep to the neck of the mandible.

Figure 50–15 Summary of the branches of the last four cranial nerves in the head and neck.

Figure 50–16 The glossopharyngeal nerve.

glion. Postganglionic secretomotor fibers arise there and reach the parotid gland by the auriculotemporal nerve (see fig. 48–9).

2. A *communicating branch* joins the auricular branch of the vagus.

3. The *branch to the carotid sinus* (see fig. 50–13) supplies the carotid sinus and body with afferent fibers.

4. The *pharyngeal branches* unite on the middle constrictor with vagal and sympathetic branches. They are sensory to the pharyngeal mucosa.

5. The motor *branch to the stylopharyngeus* arises at that muscle.

6. *Tonsillar branches* are sensory to the mucosa.

7. *Lingual branches* supply taste and general sensory fibers to the posterior third of the tongue and the vallate papillae. The ninth cranial nerve is tested by tactile sensation in the vault of the pharynx and sometimes by taste on the back of the tongue. Touching the posterior wall of the oropharynx normally induces elevation of the palate and contraction of the pharyngeal constrictors (*gag reflex*).

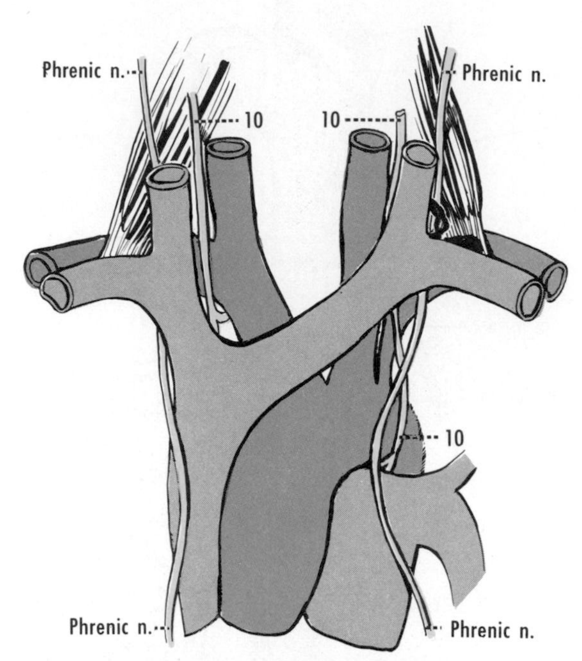

Figure 50–17 The vagus and phrenic nerves, anterior aspect. Note the different levels of origin of the right and left recurrent laryngeal nerves. The scalenus anterior muscle is depicted on each side. The thoracic duct can be seen terminating on the left side of the body.

VAGUS (figs. 50–14 and 50–15, 50–17 and 50–18, and table 50–3)

The vagus (tenth cranial nerve), which is mostly afferent, is "vagrant" in the head, neck, thorax, and abdomen. It is both afferent and efferent to the pharynx and larynx. It emerges from the medulla and passes through the jugular foramen, where it presents *superior* (*jugular*) and *inferior* (*nodose*) *ganglia*. These ganglia contain the cell bodies of the afferent fibers. The vagus is then joined by the internal branch of the accessory nerve, the fibers of which are distributed with branches of the vagus.

The vagus descends within the carotid sheath, between the internal jugular vein and the internal and common carotid arteries. The vagus then crosses the first part of the subclavian artery; its subsequent course is described with the thorax and abdomen.

Branches. The branches of the vagus are described in the following paragraphs.

1. The *meningeal branch,* which contains spinal fibers (C.N.1,2), supplies the dura of the posterior cranial fossa.

2. The *auricular branch* passes through the temporal bone (tympanomastoid fissure) and supplies the auricle and external acoustic meatus.

3. **The pharyngeal branches are the chief motor nerves to the pharynx and soft palate. Most of the fibers are derived from the internal branch of the accessory nerve.** Along with glossopharyngeal and sympathetic nerve fibers, these branches form the *pharyngeal plexus* on the constrictors of the pharynx. The plexus supplies most of the muscles of the

TABLE 50–3 SUMMARY OF BRANCHES OF VAGUS NERVE IN HEAD AND NECK

Part	Branches
From superior ganglion	1. Meningeal branch 2. Auricular branch
From inferior ganglion or from trunk of vagus	3. Pharyngeal branches 4. Superior laryngeal nerve (*a*) Internal laryngeal branch or nerve (*b*) External laryngeal branch or nerve 5. Depressor nerves or carotid branches (inconstant)
From trunk of vagus	6. Cardiac branches 7. Right recurrent laryngeal nerve

pharynx and soft palate. Accessory fibers in the vagus are tested by asking the subject to say "ah": the uvula should proceed backward in the median plane.

4. The *superior laryngeal nerve* descends along the side of the pharynx and divides into internal and external branches. **The internal laryngeal branch (or nerve) is afferent from the mucosa of the larynx as far down as the vocal folds.** The nerve pierces the thyrohyoid membrane, divides into terminal branches and communicates with the recurrent laryngeal nerve. **The external laryngeal branch (or nerve) pierces the inferior constrictor of the pharynx and enters the cricothyroid, supplying both muscles.**

5. *Depressor nerves* (*carotid branches*) assist the glossopharyngeal nerve in supplying the carotid sinus and body.

6. *Cardiac branches* arise in the neck and thorax and are often considered in superior, middle, and inferior groups. They end in the cardiac plexus.

7. **The recurrent laryngeal nerve supplies the mucosa of the larynx below the vocal folds and all the muscles of the larynx except the cricothyroid. Most of its fibers are derived from the cranial part of the accessory nerve. Correlated with the development of the aortic arches in the embryo, the right and left recurrent nerves arise at different levels, the right nerve being related to the first part of the subclavian artery and the left nerve to the arch of the aorta and the ligamentum arteriosum** (fig. 50–17). **Both recurrent nerves ascend in or near the groove between the trachea and the esophagus, and both are closely related to the thyroid gland and inferior thyroid artery; therefore they are in danger in thyroid operations. Damage to one recurrent nerve renders the ipsilateral vocal fold motionless and results in alteration of the voice.** The recurrent nerve gives off cardiac, tracheal, and esophageal branches and a sensory branch to the laryngopharynx. It divides into two or more branches before passing deep to the lower border of the inferior constrictor of the pharynx and behind the cricothyroid joint to gain the larynx.

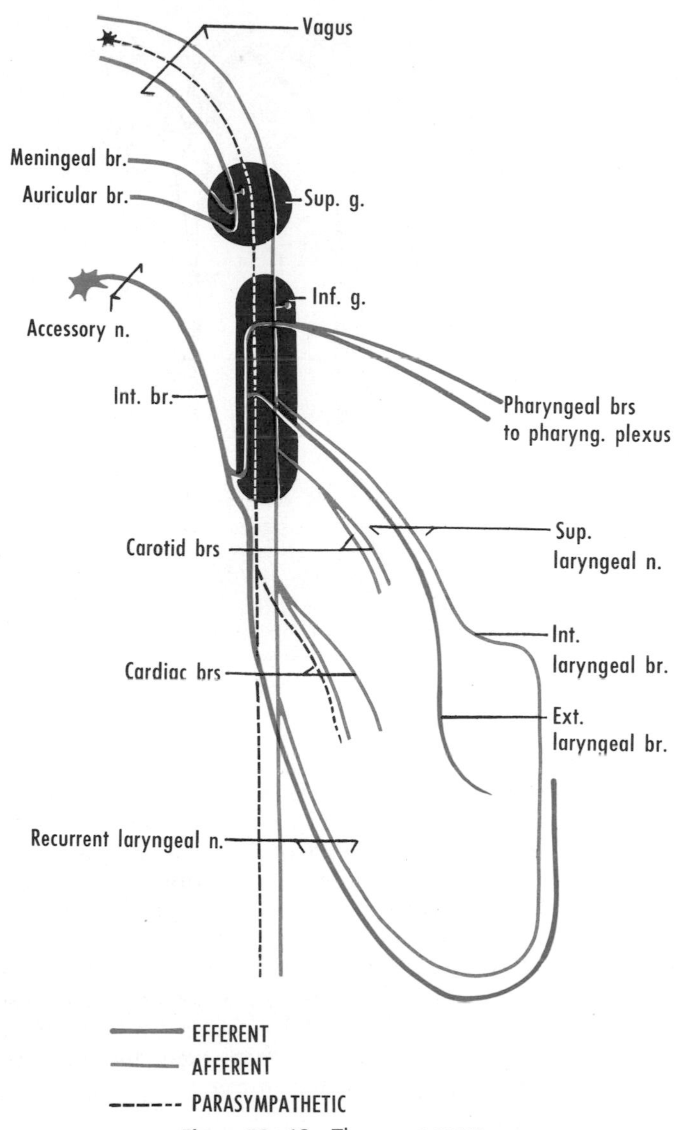

Figure 50–18 The vagus nerve.

ACCESSORY NERVE (fig. 50–19)

The accessory (eleventh cranial) nerve arises by the union of a cranial and a spinal part. The cranial roots emerge from the medulla. The spinal roots emerge from the spinal cord (sometimes as far down as C.N.7) and form a trunk that ascends in the vertebral canal and passes through the foramen magnum. Both parts, cranial and spinal, traverse the jugular foramen. **Below the jugular foramen the cranial part, or internal branch, joins the vagus. It contains motor fibers to skeletal muscles and is best regarded as a portion of the vagus.** By means of the pharyngeal and laryngeal branches of the vagus, it is distributed to the soft palate, constrictors of the pharynx, and larynx. **The spinal part, or external branch, of the accessory nerve is distributed to the sternomastoid and trapezius muscles.** It crosses the transverse process of the atlas, descends deep to the styloid process and the posterior belly of the digastric, and usually pierces the

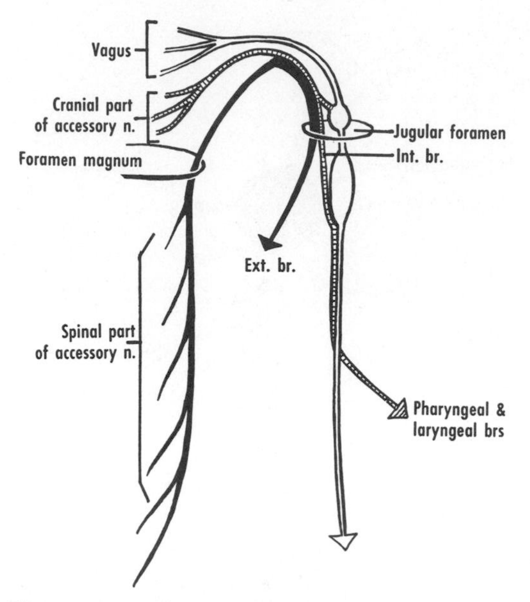

Figure 50–19 The vagus and accessory nerves. The superior and inferior ganglia of the vagus are shown above and below the jugular foramen, respectively.

sternomastoid, which it supplies. **Above the middle of the posterior border of the sternomastoid muscle, the accessory nerve crosses the posterior triangle of the neck obliquely** (see fig. 50–3*B*), **lying on the levator scapulae and in relation to lymph nodes.** It passes deep to the anterior border of the trapezius and supplies that muscle. The nerve communicates with C.N.2 to 4. **The spinal part of the accessory nerve is tested by asking the subject to shrug the shoulders (trapezius) and then to rotate the head (sternomastoid).**

Hypoglossal Nerve (fig. 50–20)

The hypoglossal (twelfth cranial) nerve is chiefly the motor nerve to the tongue. It emerges from the medulla and traverses the hypoglossal canal of the occipital bone. It then descends between the internal carotid artery and the internal jugular vein, deep to the posterior belly of the digastric. **The hypoglossal nerve loops forward around the occipital artery and crosses the internal carotid, external carotid, and lingual arteries** (see figs 49–4 and 50–14). It lies on the hyoglossus and passes deep to the digastric and mylohyoid muscles (see fig. 49–2).

Branches. Some of the branches of the hypoglossal nerve are hypoglossal in composition, whereas others are spinal and are merely carried in the hypoglossal nerve. The meningeal branches, the superior root of the ansa cervicalis, the nerve to the thyrohyoid muscle, and the branch to the geniohyoid muscle consist of cervical fibers. The branches of the hypoglossal nerve are described in the following paragraphs.

1. *Meningeal branches* supply the dura of the posterior cranial fossa.
2. The *superior root of the ansa cervicalis* descends from the hypoglossal nerve to the ansa cervicalis and supplies infrahyoid muscles (see fig. 50–6*B*).
3. The *thyrohyoid branch* arises in the carotid triangle and supplies the thyrohyoid muscle.
4. The terminal *lingual branches* supply the extrinsic and intrinsic muscles of the tongue and communicate with the lingual nerve. The hypoglossal nerve is tested by asking the subject to protrude the tongue (genioglossus and intrinsic muscles). A lesion of one hypoglossal nerve would result in deviation of the protruded tongue toward the affected side.

Subclavian Artery (fig. 50–21 and table 50–4)

The main artery to the upper limb is called by various names (subclavian, axillary, and brachial) during its course. The territory supplied by the subclavian artery extends as far as

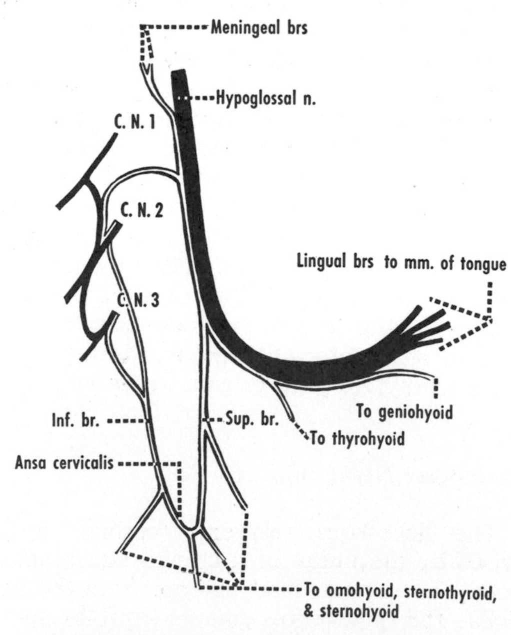

Figure 50–20 The hypoglossal nerve.

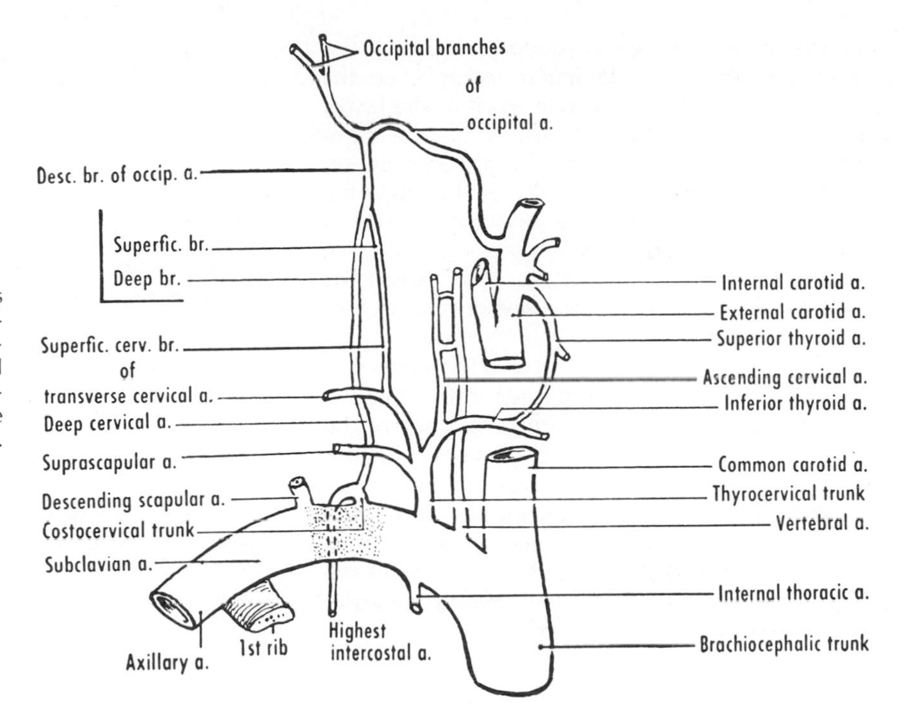

Figure 50–21 The branches of the subclavian artery: vertebral, internal thoracic, thyrocervical trunk, costocervical trunk, and descending scapular. The second part of the subclavian artery is shaded.

the forebrain, abdominal wall, and fingers. The left subclavian artery arises directly from the arch of the aorta, whereas the right subclavian is from the brachiocephalic trunk. The course of each subclavian artery may be considered in three parts: (1) from the origin of the vessel to the medial border of the scalenus anterior, (2) behind that muscle, and (3) to the external border of the first rib, where the subclavian is renamed the axillary artery.

The *first part* arches upward and laterally from behind the sternoclavicular joint and is deeply placed under cover of the sternomastoid, sternohyoid, and sternothyroid muscles. **The second part of the subclavian artery, which is normally narrowed between the scaleni, extends a few centimeters above the clavicle.** It lies in front of the apex of the lung and cupola of the pleura. **The third part of the subclavian artery is the most superficial, and its**

TABLE 50–4 BRANCHES OF SUBCLAVIAN ARTERY

Branch of Subclavian Artery	Branches
Vertebral artery	
Cervical part	Muscular branches
Vertebral part	Spinal branches
Suboccipital part	Muscular & meningeal branches
Intracranial part	Spinal & posterior inferior cerebellar arteries
Internal thoracic artery	Superior epigastric & musculophrenic arteries & others
Thyrocervical trunk	
Inferior thyroid artery	Ascending cervical artery
	Inferior laryngeal artery
	Tracheal, pharyngeal, & esophageal branches
	Glandular branches
Suprascapular artery	Suprasternal branch
	Acromial branch
	Articular branches
Transverse cervical artery	Superficial cervical artery
	A deep branch may replace descending scapular artery
Costocervical trunk	Deep cervical artery
	Highest intercostal artery
	Posterior intercostal arteries
Descending scapular (dorsal scapular) artery	Muscular branches

pulsations can be felt on deep pressure. It lies mainly in the supraclavicular triangle, on the first rib (see fig. 8–7), **and in front of the lower trunk of the brachial plexus. It can be compressed against the first rib by pressing downward, backward, and medially in the angle between the clavicle and the posterior border of the sternomastoid muscle** (see fig. 50–3*B*). **This is also the site of ligation, after which the collateral circulation to the upper limb is generally adequate.**

The subclavian vein, which is the continuation of the axillary vein, passes in front of the scalenus anterior (and seldom rises above the clavicle) and unites with the internal jugular vein to form the brachiocephalic vein.

Neurovascular Compression. **Abnormal compression of the subclavian or axillary vessels, the brachial plexus, or both produces the signs and symptoms of the "neurovascular compression syndromes" of the upper limb.** The features, which may include pain, paresthesia (prickling), numbness, weakness, discoloration, swelling, ulceration, and gangrene, may be produced also by other causes. The neurovascular bundle to the upper limb is liable to be compressed: (1) between the scalenus anterior and the scalenus medius (see fig. 50–28*B*), where the compression may be produced or accentuated by a cervical rib (fig. 50–22); (2) between the first rib and the clavicle (see fig. 8–5); or (3) near the coracoid process, where the neurovascular bundle is crossed by the pectoralis minor (see fig. 8–7).

Branches (see fig. 50–21 and table 50–4). Most of the branches arise from the first part

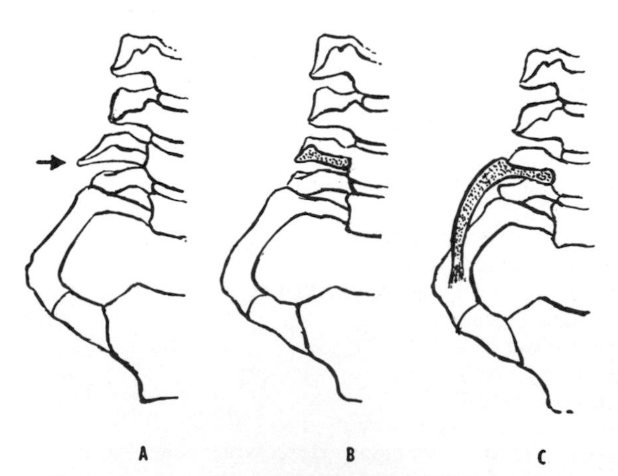

Figure 50–22 Cervical ribs. **A,** An unusually long transverse process of C.V.7 (*arrow*). **B,** A minute cervical rib with head, neck, and tubercle. **C,** A cervical rib bound to the first rib: in other instances it may end freely. (After von Lanz and Wachsmuth.)

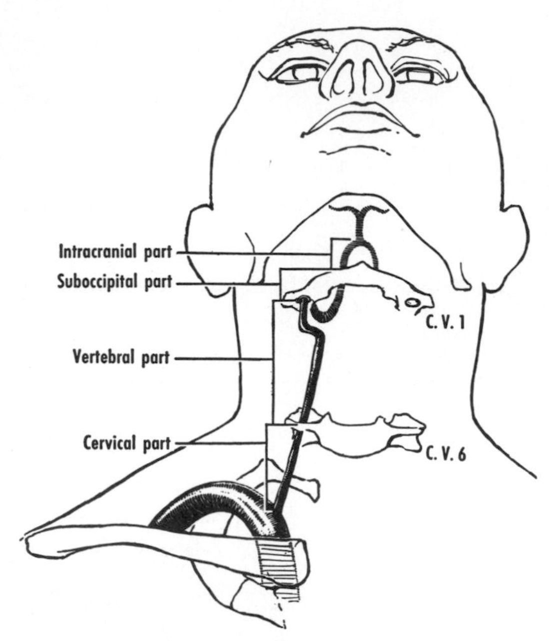

Figure 50–23 The vertebral artery, a branch of the subclavian, presents cervical, vertebral, suboccipital, and intracranial parts. It unites with its fellow of the opposite side to form the basilar artery, which divides into the right and left posterior cerebral arteries.

of the subclavian artery. They are described in the following paragraphs.

1. **The vertebral artery** (fig. 50–23), **despite its name, supplies chiefly the posterior part of the brain. Arising medial to the scalenus anterior, it ascends through the foramina transversaria of C.V.6 to C.V.1, winds behind the lateral mass of the atlas** (see fig. 39–3). **and enters the cranial cavity through the foramen magnum. At the lower border of the pons, it unites with the vessel of the opposite side to form the basilar artery, which divides into the two posterior cerebral arteries. The course of the vertebral artery may be considered in four parts: cervical, vertebral, suboccipital, and intracranial.** The cervical part ascends behind the common carotid artery in the pyramidal space between the longus colli and scalenus anterior (see Fig. 50–28*B*). The vertebral part of the artery, accompanied by a venous plexus and sympathetic filaments, gives branches to the spinal cord and vertebrae. The suboccipital and intracranial parts of the vertebral artery have been described already.

2. The *internal thoracic artery* is described with the thorax.

3. The *thyrocervical trunk* divides almost at once into three branches: the inferior thyroid, suprascapular, and transverse cervical arteries.

(a) The *inferior thyroid artery* ascends in front of the scalenus anterior and arches medially in front of the vertebral vessels and behind the carotid sheath. It is closely related to the middle cervical ganglion and the recurrent laryngeal nerve. It enters the posterior surface of the thyroid gland. The inferior thyroid artery gives branches to the vertebrae (*ascending cervical artery*), larynx (*inferior laryngeal artery*), trachea, pharynx, esophagus, and thyroid gland.

(b) The *suprascapular artery* passes laterally across the scalenus anterior and phrenic nerve and then across the subclavian artery and cords of the brachial plexus, finally taking part in the anastomosis around the scapula. It gives off *suprasternal, acromial,* and *articular branches*.

(c) The *transverse cervical artery* passes laterally (at a higher level than the suprascapular artery) across the scalenus anterior and phrenic nerve and then across the trunks of the brachial plexus in the posterior triangle. It supplies the trapezius (as the *superficial cervical artery*).

4. The *costocervical trunk* arches over the cupola of the pleura to reach the neck of the first rib, where it divides into two branches: (a) the *deep cervical artery,* which ascends to anastomose with the descending branch of the occipital artery, and (b) the *highest intercostal artery,* which descends behind the pleura and usually gives off the first two posterior intercostal arteries.

5. The *descending scapular* (*dorsal scapular*) *artery* generally passes between the trunks of the brachial plexus and accompanies the dorsal scapular nerve to the rhomboid muscles. It may, however, be replaced by a deep branch of the transverse cervical artery.

CUPOLA OF PLEURA (see fig. 50–28A)

The *cupola* is the *cervical pleura,* i.e., the continuation of the costal and mediastinal pleura over the apex of the lung. It begins at the inlet of the thorax along the sloping internal border of the first rib (see fig. 8–5). **The cupola and apex of the lung project into the root of the neck up to about 3 cm above the medial third of the clavicle.** The cupola is covered by fascia, the *suprapleural membrane* (attached to the first rib and to C.V.7 and T.V.1); muscle fibers (*scalenus minimus*) are sometimes present. The cupola and apex of the lung occupy the pyramidal interval between the scaleni and the longus colli and are related in front to the subclavian vessels and scalenus anterior. They are entirely under cover of the sternomastoid muscle.

SYMPATHETIC TRUNK (see figs. 32–5 and 50–24)

The sympathetic supply to the head and neck arises in segments T1 and 2 (and sometimes C8) of the spinal cord. The preganglionic fibers leave in the ventral roots and pass through

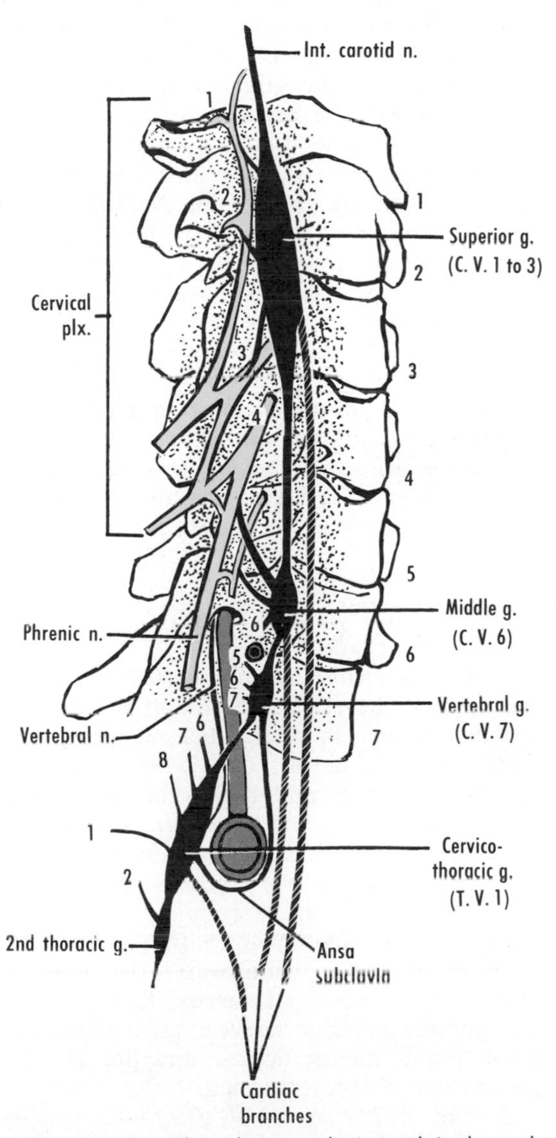

Figure 50–24 The right sympathetic trunk in the neck, lateral aspect. Only cervical nerves 1 to 5 are shown. The numerals on the left side of the drawing refer to those cervical and thoracic nerves to which rami communicantes (postganglionic fibers) are given. The numerals on the right side refer to the cervical vertebrae. The subclavian (in transverse section) and vertebral arteries are shown.

rami communicantes to the thoracic part of the sympathetic trunk. They then ascend to the cervical part of the sympathetic trunk, where they synapse and from which postganglionic fibers are distributed to the blood vessels, smooth muscle, and glands of the head and neck.

The cervical part of the sympathetic trunk consists of three or four ganglia connected by an intervening cord or cords. Postganglionic fibers leave the trunk by variable rami communicantes and also in branches that go directly to blood vessels or viscera. **Interruption of the cervical part of the sympathetic trunk cuts off impulses to the superior cervical ganglion and is followed by Horner's syndrome:** constriction of the pupil (meiosis, caused by unopposed parasympathetic action), drooping of the upper eyelid (ptosis, caused by paralysis of smooth muscle), an illusion that the eye has receded (enophthalmos), redness and increased temperature of the skin (vasodilatation), and absence of sweating (anhidrosis). The preganglionic fibers for the eye and orbit are probably from T1 (range: C8 to T4), and they probably enter the cervicothoracic ganglion. **A local anesthetic injected near the cervicothoracic ganglion will "block" the cervical and upper thoracic ganglia (stellate ganglion block), thereby relieving vascular spasm involving the brain or an upper limb.**

Cervical Ganglia
(see figs. 24–8, 32–5, and 50–24)

1. The *superior cervical ganglion* lies below the base of the skull and behind the internal carotid artery. It distributes postganglionic fibers to cranial nerves 9 to 12 and C.N.1 to 4, the carotid sinus and body, the pharyngeal plexus and larynx, and the heart. A plexus on the external carotid artery is continued on its branches to the salivary glands. A large ascending branch from the ganglion, the *internal carotid nerve,* accompanies the internal carotid artery and forms a plexus that contributes to several cranial nerves, the tympanic and greater petrosal nerves, the ciliary ganglion (pupillodilator fibers), and the anterior and middle cerebral arteries.

2. The *middle cervical ganglion,* usually above the arch of the inferior thyroid artery, is very variable.

3. The *vertebral ganglion* generally lies in front of the vertebral artery and below the arch of the inferior thyroid artery. Cords connect this ganglion with those above and below, and another cord, the *ansa subclavia,* loops around and forms a plexus on the first part of the subclavian artery.

4. The *cervicothoracic* (*stellate*) *ganglion* comprises two variably fused components: inferior cervical and first thoracic ganglia. It lies behind the vertebral artery and in front of the C.V.7 transverse process and the neck of the first rib. Preganglionic rami come from T.N.1, and postganglionic rami go to C.N.6 to 8 and T.N.1. These fibers enter the brachial plexus and are distributed to the upper limb. Other branches of the ganglion go to the heart and to the subclavian and vertebral arteries. The vertebral plexus is ultimately distributed along the basilar artery. The plexuses on the posterior cerebral arteries may be derived from the vertebral or internal carotid plexuses.

INTERNAL JUGULAR VEIN (fig. 50–25)

The internal jugular vein drains the brain, neck, and face. It commences in the jugular foramen as a continuation of the sigmoid sinus. At the base of the skull, the internal carotid artery (in the carotid canal) lies in front of the internal jugular vein (in the jugular foramen), and the two vessels are there separated by cranial nerves 9 to 12 (see fig. 48–2*B*). The internal jugular vein descends in the carotid sheath and is hidden by the sternomastoid muscle. The internal and common carotid arteries ac-

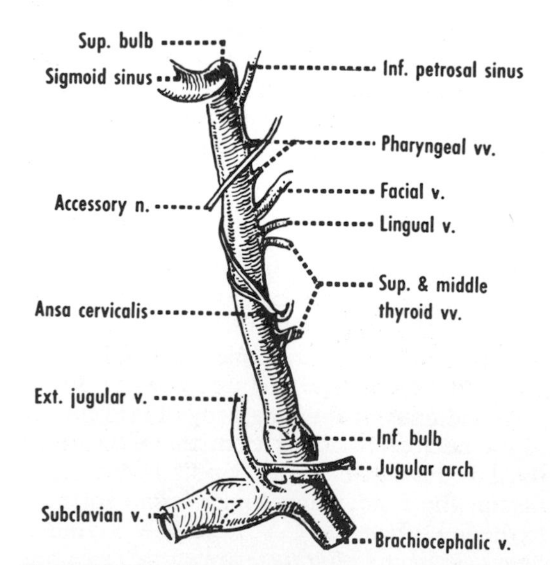

Figure 50–25 The internal jugular vein and its tributaries. Note the valves at the terminations of the subclavian and internal jugular veins: these are the last valves before the blood reaches the heart. (After Grant.)

company the vein medially, and the vagus lies behind and between the vein and the arteries. **The deep cervical lymph nodes lie along the course of the internal jugular vein.** The vein passes deep to the interval between the two heads of the sternomastoid muscle (see fig. 50–12*B*) and ends behind the medial end of the clavicle by uniting with the subclavian vein to form the brachiocephalic vein. Dilatations are found at its beginning and near its end (*superior* and *inferior bulbs*).

The tributaries, which are variable, include the inferior petrosal sinus and the pharyngeal, lingual, and superior and middle thyroid veins. The right lymphatic duct or (on the left) the thoracic duct opens usually into the internal jugular vein at or near its junction with the subclavian vein.

THORACIC DUCT

The thoracic duct receives the lymph from most of the body, including the left side of the head and neck. The duct (fig. 50–26), on leaving the thorax (q.v.), arches laterally in front of the left vertebral artery, phrenic nerve, and scalenus anterior and behind the carotid sheath. It receives the left jugular trunk and ends variably in front of the first part of the left subclavian artery in or near the angle between the left internal jugular and subclavian veins. **The right lymphatic duct receives the lymph from the right side of the head and neck, right upper limb, and right side of the thorax.** The duct, which is seldom present as a single structure (fig. 50–27*B*), has components (the right jugular, subclavian, and bronchomediastinal trunks) that usually open separately into the right internal jugular vein, the subclavian vein, or both.

LYMPHATIC DRAINAGE OF HEAD AND NECK

All the lymphatic vessels from the head and neck drain into the deep cervical nodes, either (1) directly from the tissues or (2) indirectly after traversing an outlying group of nodes. Several outlying groups of lymph nodes form a "pericervical collar" at the junction of the head and neck (fig. 50–27*A*): these are the occipital, retro-auricular (mastoid), parotid, submandibular, buccal (facial), and submental nodes. The superficial tissues drain into these groups and also into the superficial cervical nodes.

The *superficial cervical nodes* are in (1) the posterior triangle along the external jugular vein and (2) the anterior triangle along the anterior jugular vein.

The deep cervical nodes include several groups, the most important of which forms a chain along the internal jugular vein, mostly under cover of the sternomastoid muscle. The *jugulodigastric node* lies on the internal jugular vein immediately below the posterior belly of the digastric. It receives important afferents

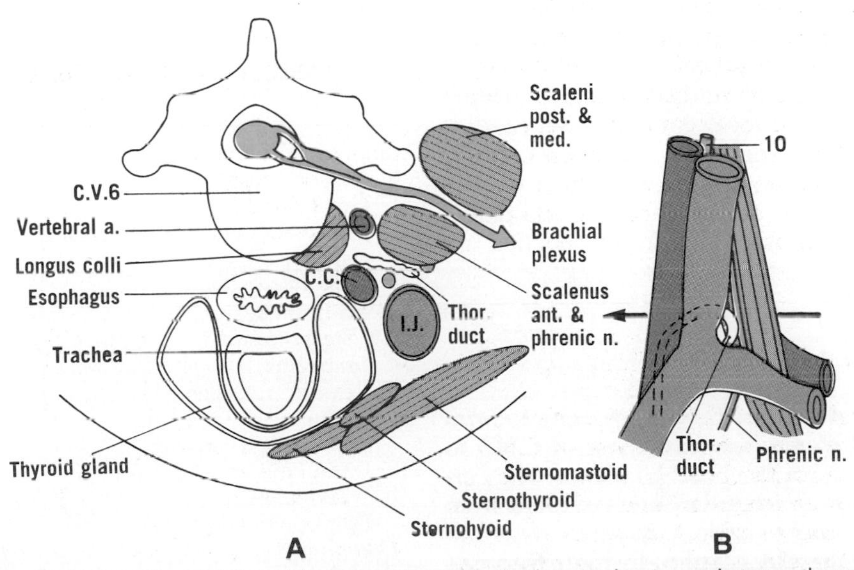

Figure 50–26 **A,** A horizontal section in which the arch formed by the thoracic duct is seen between the scalenus anterior muscle behind and the internal jugular vein and common carotid artery in front. The phrenic nerve is visible on the scalenus anterior. **B,** The termination of the thoracic duct, anterior aspect.

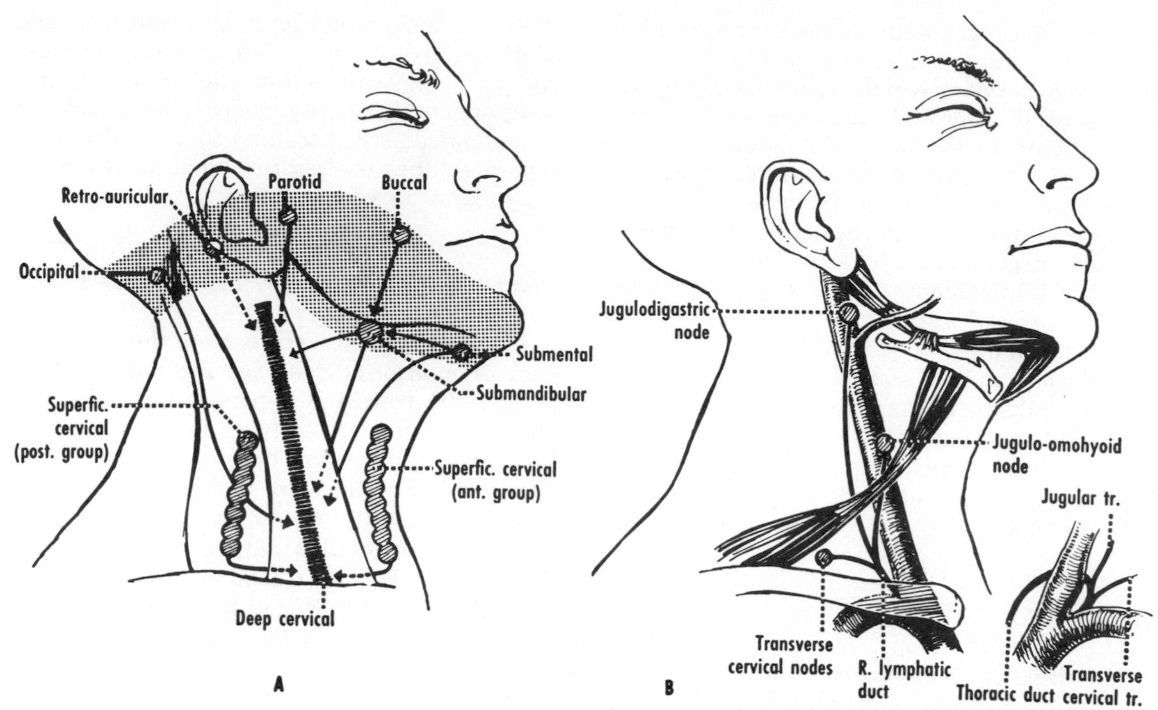

Figure 50–27 The lymphatic drainage of the head and neck. **A,** The superficial groups of cervical lymph nodes. The wide, shaded band indicates the "pericervical collar" of nodes. Each circle represents a group of nodes. The arrows show the direction of drainage. **B,** The deep cervical lymph nodes. The drawing at the lower right shows one of many patterns that may be found on the left side of the body.

from the back of the tongue and from the tonsils. The *jugulo-omohyoid node* lies on the vein immediately above the middle tendon of the omohyoid. It receives afferents from the tongue. One group of deep nodes is found in the posterior triangle and is related to the accessory nerve. Other groups are prelaryngeal, pretracheal, paratracheal, and retropharyngeal. These take part in the drainage of deeper structures (e.g., middle ear and nasal cavity). The efferent vessels from the deep cervical nodes form the *jugular trunk,* which usually joins the thoracic duct on the left and enters the internal jugular–subclavian junction on the right.

Cervical Plexus

The ventral rami of C.N.1 to 4 unite to form the cervical plexus, whereas those of C.N.5 to 8 and T.N.1 form the brachial plexus. The cervical plexus is an irregular series of loops from which the branches arise. Cutaneous areas and muscles are thereby supplied by more than one spinal nerve (table 50–5). **The cutaneous branches all emerge near the middle of the posterior border of the sternomastoid muscle** (see fig. 50–4). The cervical plexus lies in front of the levator scapulae and scalenus medius, under cover of the internal jugular vein and the sternomastoid. The ventral rami

TABLE 50–5 Summary of Branches of Cervical Plexus

Branch	Nerves
Superficial branches:	
Lesser occipital	C.N.2, 3
Great auricular	C.N.2, 3
Transverse cervical (anterior cervical cutaneous)	C.N.2, 3
Supraclavicular	C.N.3, 4
Deep branches:	
To Sternomastoid	C.N.2, 3
Trapezius	C.N.3, 4
Levator scapulae	C.N.3, 4
Scaleni	C.N.3, 4
Prevertebral muscles	C.N.1 to 4
Infrahyoid muscles by ansa cervicalis	C.N.1 to 3
Diaphragm by phrenic nerve	C.N.3 to 5
Communicating to cranial nerves (10, 11, and 12) and to sympathetic trunk and superior cervical ganglion	

receive postganglionic rami communicantes from the cervical sympathetic ganglia.

The *ansa cervicalis* is a loop on or in the carotid sheath. It is formed by fibers of C.N.1 to 3 (see figs. 50–6*B* and 50–20). It has a *superior root,* which descends from the hypoglossal nerve (but consists of spinal fibers), and an *inferior root,* which connects the ansa with C.N.2 and 3. The ansa and its superior root supply the infrahyoid muscles (but the thyrohyoid receives its cervical fibers directly from the hypoglossal nerve).

The phrenic nerve arises chiefly from C4 and supplies the diaphragm and the serosa of the thorax and abdomen. It sometimes has a root from C.N.3 and usually an accessory root from C.N.5 (see fig. 50–24) that may reach the phrenic through the nerve to the subclavius. The phrenic nerve, formed at the lateral border of the scalenus anterior, descends on the front of that muscle (see fig. 50–8) under cover of the internal jugular vein and the sternomastoid. It lies behind the prevertebral fascia, is crossed by the transverse cervical and suprascapular arteries (fig. 50–28*B*), and is accompanied by the ascending cervical artery from the inferior thyroid. It passes between the subclavian artery and vein (see fig. 50–17), crosses the internal thoracic artery, and proceeds through the thorax (q.v.). **Surgical interruption of the phrenic nerve on the scalenus anterior collapses a lung by paralyzing and thereby elevating the hemidiaphragm.**

SCALENE MUSCLES (fig. 50–28 and table 50–6)

The scalenus anterior (except at its insertion) lies entirely under cover of the sternomastoid.

TABLE 50–6 SCALENI AND PREVERTEBRAL MUSCLES

Muscle	Origin	Insertion	Innervation	Action
Scalenus anterior	Anterior tubercles of transverse processes of C.V.3 to 6	Internal border of first rib	Ventral rami of cervical nerves	Flex cervical column laterally & act in respiration
Scalenus medius	Posterior tubercles of transverse processes of C.V.1 to 7	Superior surface of first rib		
Scalenus posterior	Posterior tubercles of transverse processes of C.V.4 to 6	External surface of second rib		
Scalenus minimus	Transverse process of C.V.7	Internal border of first rib & cupola of pleura		Tenses cupola
Longus capitis	Transverse processes of C.V.3 to 6	Inferior surface of occipital bone		Flexes head
Longus colli	Bodies of upper T.V. & lower C.V.	Bodies of upper C.V.		Flexes cervical column
	Bodies of upper T.V.	Transverse processes of lower C.V.		
	Transverse processes of upper C.V.	Anterior arch of atlas		
Rectus capitis anterior	Transverse process of atlas	Basilar part of occipital bone		Flexes head
Rectus capitis lateralis	Transverse process of atlas	Jugular process of occipital bone		Flexes head laterally

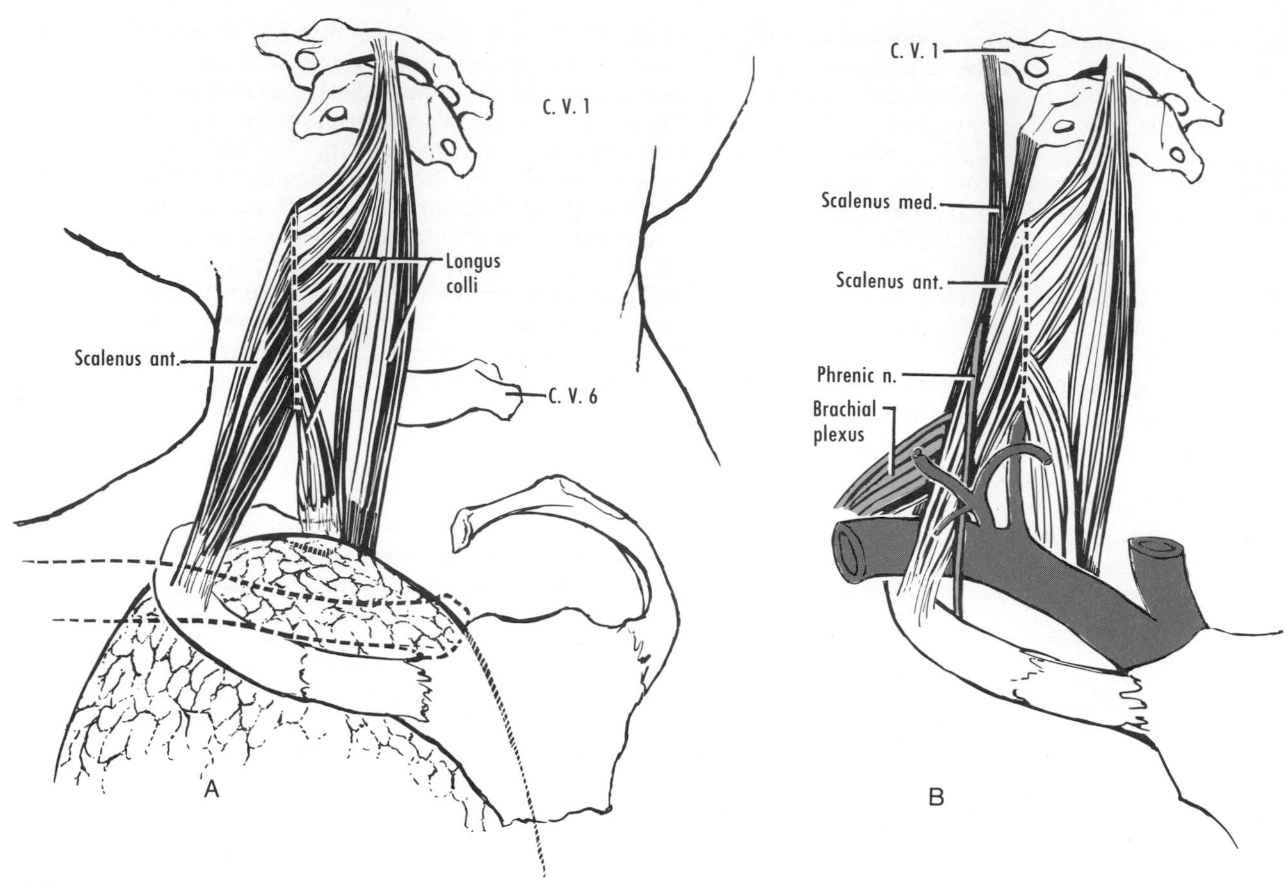

Figure 50–28 **A,** The cupola of the pleura, scalenus anterior, and longus colli; anterior aspect and slightly from the right side. The cupola projects upward between the scaleni and the longus colli, about 3 cm above the medial third of the clavicle (*interrupted line*). **B,** The third part of the subclavian artery and the brachial plexus between the scalenus anterior and the scalenus medius. The lower trunk of the plexus lies behind the artery. The phrenic nerve, which descends almost vertically on the obliquely set scalenus anterior, is bound to the front of the muscle by the transverse cervical and suprascapular arteries. The scalenus anterior, longus colli, and first part of the subclavian artery form a triangle. The carotid tubercle of C.V.6 lies at the apex, and the vertebral artery ascends through the triangle to reach the foramen transversarium of C.V.6. (**A** is based on von Lanz and Wachsmuth.)

The subclavian artery passes behind the scalenus anterior, whereas the phrenic nerve lies on the muscle. The scalenus anterior arises from the anterior tubercles; the scaleni medius and posterior (often absent or blended with the medius) arise from the posterior tubercles of the cervical transverse processes. The ventral rami of the cervical nerves emerge between the anterior and posterior tubercles; hence **the brachial plexus emerges between the scalenus anterior and the scalenus medius.** The scaleni may act as muscles of inspiration even during quiet breathing; they become active during strong expiratory effects, and they may be important in coughing and straining. **A pyramidal interval occurs between the scaleni laterally and the longus colli medially, and into this the pleura and apex of the lung project upward** (fig. 50–28*A*).

Cervical Fascia (fig. 50–29)

The cervical fascia "affords that slipperiness which enables structures to move and pass over one another without difficulty, as in swallowing, and allowing twisting of the neck without it creaking like a manilla rope—a looseness, moreover, that provides the easiest pathways for vessels and nerves to reach their destinations" (Whitnall).

The fascia of the neck comprises three layers: the investing, pretracheal, and prevertebral layers.

The *investing layer* is attached to the major bony prominences: the external occipital protuberance, superior nuchal line, ligamentum nuchae, cervical spinous processes, mastoid and styloid processes, lower border of the mandible, zygomatic arch, hyoid bone, acromion, clavicle, and manubrium. The layer surrounds the trapezius, roofs the posterior triangle, surrounds the sternomastoid, and roofs the anterior triangle. It forms the sheaths of the parotid and submandibular glands. At the manubrium, it bounds the *suprasternal space,* which encloses the sternal heads of the sternomastoid and the jugular venous arch.

The *pretracheal layer,* limited to the front of the neck, is more extensive than its name suggests. It lies below the hyoid bone and is attached to the oblique lines of the thyroid cartilage and to the cricoid cartilage. It surrounds the thyroid gland, forming its sheath, and it invests the infrahyoid muscles and the air and food passages. **Infections from the head and**

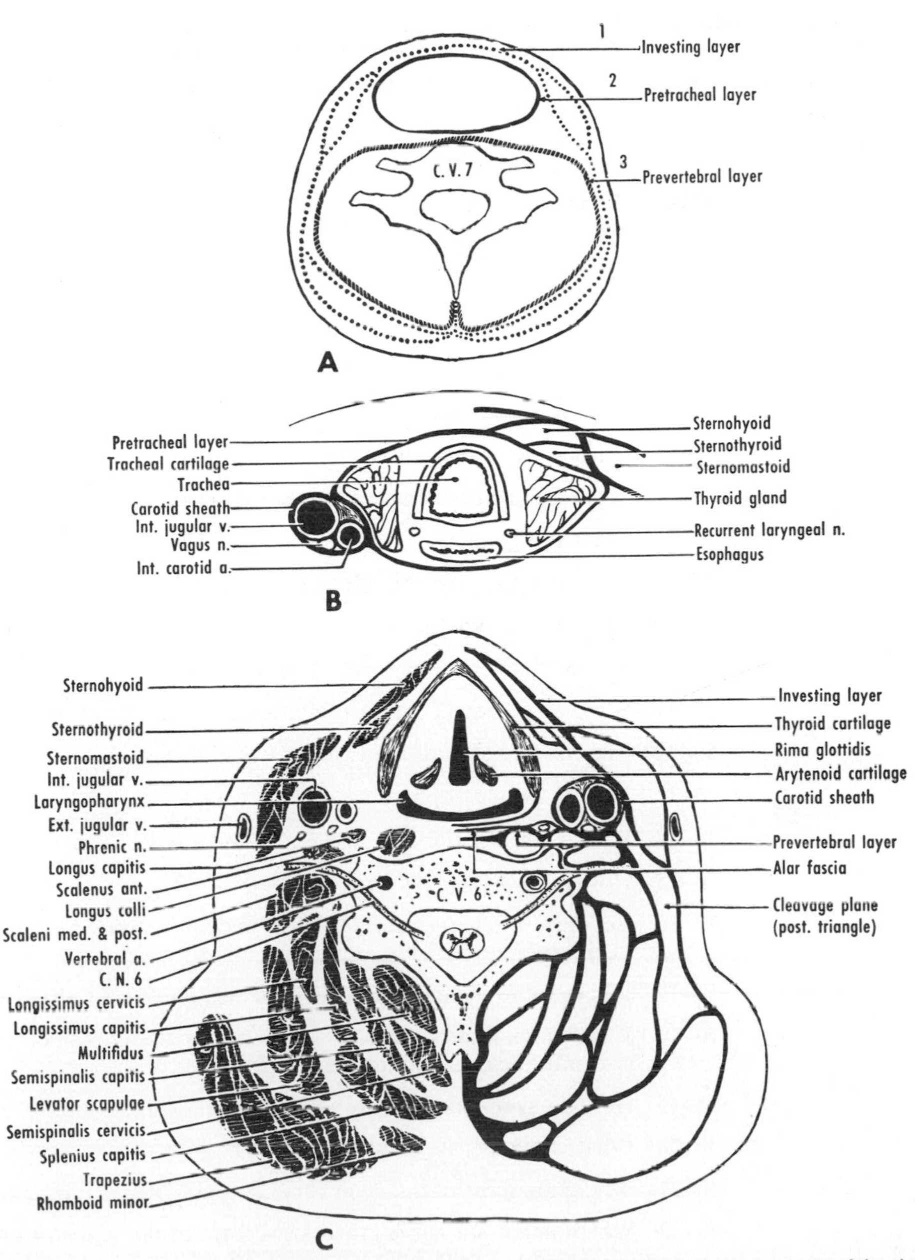

Figure 50–29 Horizontal sections illustrating the cervical fascia. **A** shows the general arrangement of the three layers. **B** presents the pretracheal layer at the level of C.V.7. **C** shows the three layers at the level of C.V.6, on the right. (**C** is based on Truex and Kellner.)

neck can spread in front of the trachea or behind the esophagus and reach the superior mediastinum in the thorax.

The *prevertebral layer* is attached to the base of the skull and to the transverse processes of the cervical vertebrae. It covers the prevertebral muscles, scaleni, phrenic nerve, and deep muscles of the back, and therefore the floor of the posterior triangle. In front of the subclavian artery, it is prolonged laterally as the axillary sheath, which also invests the brachial plexus.

The *carotid sheath,* which is fused with all three layers of the cervical fascia, is a condensation around the common and internal carotid arteries, internal jugular vein, and vagus.

PREVERTEBRAL MUSCLES (see table 50–6)

The *longus capitis,* which covers the uppermost part of the longus colli, extends from the lower cervical vertebrae to the occipital bone. The *longus colli* (see fig. 50–28) consists of a vertical bundle spanning vertebral bodies and of oblique fasciculi that connect cervical transverse processes to vertebral bodies. The *recti capitis anterior* and *lateralis* connect C.V.1 to the occipital bone. The longus colli is active during talking, coughing, and swallowing. The prevertebral muscles and sternomastoid muscles act with, and as antagonists to, the upper deep muscles of the back.

QUESTIONS

50–1 Why is the sternomastoid known officially as the sternocleidomastoid muscle?

50–2 How is the accessory nerve tested?

50–3 What are the most important contents of the posterior triangle?

50–4 Why is the carotid triangle so called?

50–5 What is the significance of *-oid* in *thyroid?*

50–6 What are the coverings of the thyroid gland?

50–7 What is a goiter?

50–8 Why may the voice be affected by thyroid surgery?

50–9 List some anomalies of thyroid development.

50–10 At which level is tracheotomy performed?

50–11 What is the line of the carotid arteries in surface anatomy?

50–12 List the branches of the external carotid artery.

50–13 How is the glossopharyngeal nerve tested?

50–14 What is the relationship between cranial nerves 10 and 11?

50–15 How is the hypoglossal nerve tested?

50–16 Between which two muscles is the middle (second) part of the subclavian artery lodged?

50–17 Where is the site of compression and ligation of the subclavian artery?

50–18 Which important structures are found in the pyramidal interval between the scaleni and longus colli?

50–19 How do sympathetic fibers reach the cervical ganglia?

50–20 What is the stellate ganglion?

50–21 Where are most of the deep cervical lymph nodes concentrated?

50–22 Which important nerve crosses the front of the scalenus anterior?

50–23 Which are the three main layers of the cervical fascia?

50–24 What are the contents of the carotid sheath?

51
THE MOUTH, TONGUE, AND TEETH

MOUTH AND PALATE

Oral Cavity. The mouth is lined by stratified squamous epithelium, from which oral smears may be taken for chromosomal studies. The temperature in the mouth is about 37° C. (98.6° F.). Mouth-to-mouth and mouth-to-nose are important methods of artificial respiration. The oral cavity comprises the vestibule and the oral cavity proper.

The *vestibule* is the cleft between the lips and cheeks externally and the gums and teeth internally. **The parotid duct opens opposite the upper second molar. When the teeth are in contact, the vestibule and oral cavity proper communicate only by a gap between the last molars and the ramus of the mandible.**

The *oral cavity proper* (fig. 51–1) is bounded by the alveolar arches, teeth and gums, and palate and tongue. It communicates

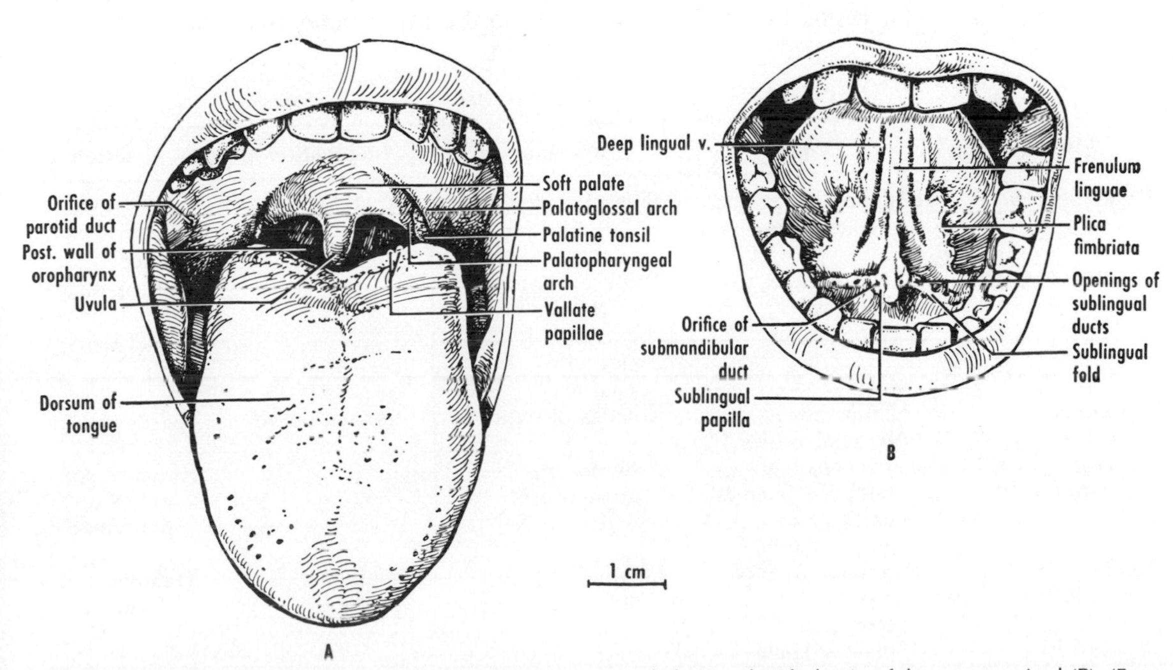

Figure 51–1 Views of the open mouth with the tongue protruded (**A**) and with the tip of the tongue raised (**B**). (From photographs by Bassett.)

behind with the oropharynx by an opening termed the *faucial isthmus,* between the palatoglossal arches. The lower surface of the tongue is connected to the floor of the mouth by a median fold of mucous membrane, the *frenulum* (fig. 51–1*B*). The submandibular duct opens on an elevation, the *sublingual papilla,* on the side of the frenulum. Laterally, the *sublingual fold,* which is produced by the sublingual gland, contains the openings of the sublingual ducts.

Lips and Cheeks. The lips are musculofibrous folds that are connected to the gums by superior and inferior *frenula.* The median part of the upper lip shows a shallow external groove, the *philtrum.* The lips consist chiefly of skin, the orbicularis oris muscle, labial glands, and mucosa. **Harelip is most frequent in the upper lip in a paramedian position, and it is often associated with cleft palate.** The cheeks, which contain the buccinator and buccal glands, resemble the lips in structure.

Palate. **The palate is the roof of the mouth and the floor of the nasal cavity.** It extends backward into the pharynx (see fig. 53–4). The palate has "an extravagant arterial supply" (from branches of the maxillary artery) and many sensory nerves (branches of the pterygopalatine ganglion). **The palate comprises the hard palate, or anterior two thirds, and the soft palate, or posterior third.**

The hard palate contains the *bony palate,* formed by the palatine processes of the maxillae and the horizontal plates of the palatine bones (see fig. 52–2). The mucoperiosteum of the hard palate contains many *palatine glands,* a median *raphe,* and *transverse palatine folds* or *rugae.*

The soft palate (velum palatinum) is a mobile, fibromuscular fold suspended from the hard palate posteriorly and ending in the uvula. It separates partially the nasopharynx and oropharynx and aids in closing the pharyngeal isthmus in swallowing and speech. The soft palate is continuous laterally with two folds, the palatoglossal and palatopharyngeal arches.

The muscles of the soft palate are listed in table 51–1. The *palatine aponeurosis* is an expansion of the front part of the soft palate to which the muscles of the palate are attached. **With the exception of the tensor, all the muscles of the palate are thought to be supplied through the pharyngeal plexus by the internal branch of the accessory nerve.** Other possible contributions are from cranial nerves 7, 9, and 12. The muscles aid in closing the oral cavity from the pharyngeal cavity and the oropharynx from the nasopharynx. They take part in phonation and swallowing. The tensor is perhaps responsible for opening the auditory tube.

TONGUE (figs. 51–1 and 51–2)

The tongue (L., lingua; Gk, glossa), situated in the floor of the mouth, is attached by muscles to the hyoid bone, mandible, styloid pro-

TABLE 51–1 MUSCLES OF SOFT PALATE

Muscle	Origin	Insertion	Innervation	Action
Palatoglossus	Palatine aponeurosis	Side of tongue	Pharyngeal plexus (11th cranial nerve)	Approximates palatoglossal folds
Palatopharyngeus	Bony palate & palatine aponeurosis	Posterior border of thyroid cartilage & side of pharynx	Pharyngeal plexus (11th cranial nerve)	Approximates palatopharyngeal folds
Musculus uvulae	Palatine bones & aponeurosis	Mucosa of uvula	Pharyngeal plexus (11th cranial nerve)	Raises uvula
Levator veli palatini	Petrous part of temporal bone & cartilage of auditory tube	Palatine aponeurosis	Pharyngeal plexus (11th cranial nerve)	Elevates soft palate & pharynx
Tensor veli palatini	Scaphoid fossa of medial pterygoid plate & spine of sphenoid bone	Palatine aponeurosis	Mandibular	Tightens soft palate

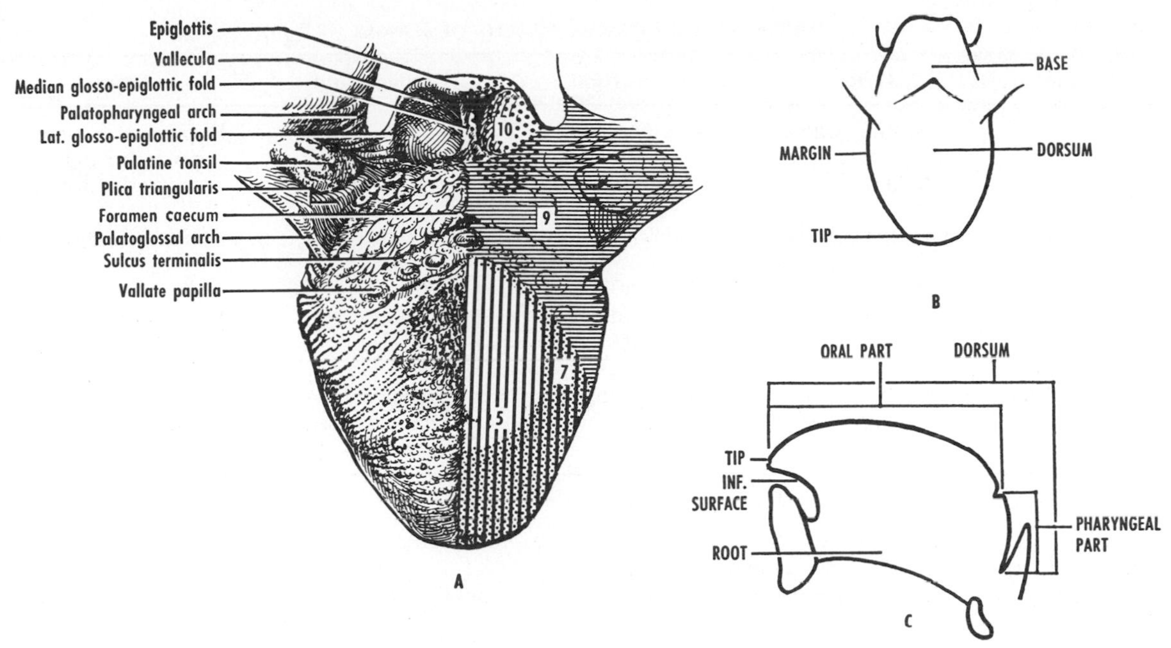

Figure 51–2 **A,** Dorsum of the tongue, showing the sensory innervation on one side. The numbers refer to cranial nerves. **B** and **C,** Diagrams showing the parts of the tongue.

cesses, and pharynx. The tongue is important in taste, mastication, swallowing, and speech. It is composed chiefly of skeletal muscle, is partly covered by mucous membrane, and presents a tip and margin, dorsum, inferior surface, and root (fig. 51–2*B* and *C*).

The *tip,* or *apex,* usually rests against the incisors and continues on each side into the *margin.*

The *dorsum* (fig. 51–2*A*) extends from the oral cavity into the oropharynx. A V-shaped groove, the *sulcus terminalis,* runs laterally and forward from a small pit, the *foramen cecum.* **The sulcus terminalis is the boundary between (1) the oral part, or anterior two thirds, and (2) the pharyngeal part, or posterior third, of the tongue. The foramen cecum, when present, indicates the site of origin of the embryonic thyroglossal duct.**

The oral part of the dorsum may show a shallow *median groove.* The mucosa has numerous minute *lingual papillae:* (1) the *filiform papillae,* the narrowest and most numerous; (2) the *fungiform papillae,* with rounded heads and containing taste buds; (3) the *vallate papillae,* about a dozen large projections arranged in a V-shaped row in front of the sulcus terminalis and containing numerous taste buds; and (4) the *folia,* inconstant grooves and ridges at the margin posteriorly.

The pharyngeal part of the dorsum faces posteriorly. **The base of the tongue forms the anterior wall of the oropharynx and can be inspected by downward pressure on the tongue with a spatula or by a mirror.** Lymphatic follicles in the submucosa are collectively known as the *lingual tonsil.* The mucosa is reflected onto the front of the epiglottis (*median glosso-epiglottic fold*) and onto the lateral wall of the pharynx (*lateral glosso-epiglottic fold*). **The space on each side of the median glosso-epiglottic fold is termed the vallecula.**

The *inferior surface* of the tongue (see fig. 51–1*B*) is connected to the floor of the mouth by the *frenulum,* lateral to which the profunda linguae vein can be seen through the mucosa. Lateral to the vein is a fringed fold, the *plica fimbriata.* The tongue contains a number of *lingual glands.*

The *root* of the tongue rests on the floor of the mouth and is attached to the mandible and hyoid bone. **The nerves, vessels, and extrinsic muscles enter or leave the tongue through its root.**

Muscles of Tongue. All the muscles are bilateral, being partially separated by a median

TABLE 51–2 EXTRINSIC MUSCLES OF TONGUE

Muscle	Origin	Insertion	Innervation	Action
Genioglossus	Superior genial tubercle of mandible	Inferior aspect of tongue & body of hyoid bone	Hypoglossal nerve	Depresses tongue; posterior part protrudes tongue
Hyoglossus	Greater horn & body of hyoid bone	Side & inferior aspect of tongue	Hypoglossal nerve	Retracts tongue
Styloglossus	Styloid process	Side & inferior aspect of tongue	Hypoglossal nerve	Retracts tongue
Palatoglossus	Palatine aponeurosis	Side of tongue	Pharyngeal plexus (11th cranial nerve)	Approximates palatoglossal folds

septum. Intrinsic muscles are arranged in several planes. The chief extrinsic muscles (fig. 51–3) are listed in table 51–2. The genioglossus is a vertically placed fan in contact with its fellow medially. **The attachment of the genioglossi to the mandible prevents the tongue from falling backward and obstructing respiration. Anesthetists keep the tongue forward by pulling the mandible forward. The hyoglossus, flat and quadrilateral, is largely concealed by the mylohyoid muscle. The glossopharyngeal nerve, stylohyoid ligament, and lingual artery pass deep to the posterior border of the hyoglossus** (see fig. 49–2). **With the exception of the palatoglossus, all the muscles of the tongue are supplied by the hypoglossal nerve.**

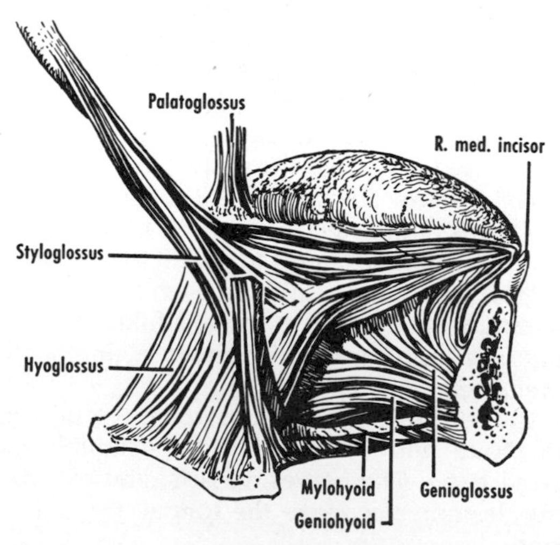

Figure 51–3 Extrinsic muscles of the tongue, right lateral aspect. Most of the right half of the mandible and of the mylohyoid muscle has been removed.

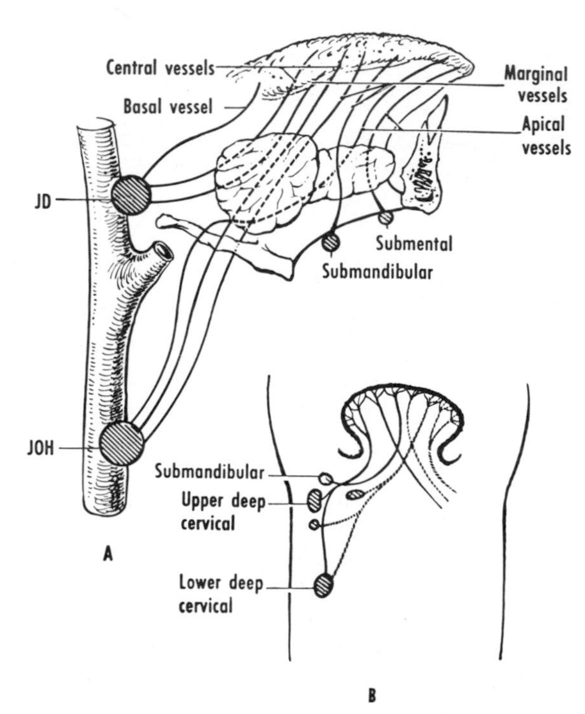

Figure 51–4 Lymphatic drainage of the tongue. **A,** Right lateral aspect. Note the submandibular and sublingual salivary glands. Each cross-hatched circle represents a group of nodes. *JD,* jugulodigastric nodes; *JOH,* jugulo-omohyoid nodes. **B,** Schematic coronal section. (**A** is based on Rouvière, **B** on Jamieson and Dobson.)

Blood Supply. The main artery is the lingual artery (fig. 49–4), a branch of the external carotid. It is accompanied by lingual veins. The *vena profunda linguae* (or *ranine vein*) can be seen *in vivo* at the side of the frenulum. The various veins of the tongue drain ultimately into the internal jugular.

Lymphatic Drainage. **The lymphatic drainage is important in the early spread of carcinoma of the tongue.** The drainage is to the submental, submandibular, and deep cervical nodes (fig. 51–4), and extensive communications occur across the median plane.

Sensory Innervation (see fig. 51–2*A*). **The anterior two thirds of the tongue is supplied by (1) the lingual nerve (of the mandibular nerve) for general sensation and by (2) the chorda tympani (a branch of the facial nerve that runs in the lingual nerve) for taste. The posterior third of the tongue and the vallate papillae are supplied by the glossopharyngeal nerve for both general sensation and taste. The nerves for taste are cranial nerves 7, 9, and 10.** The internal branch of the vagus is responsible for general sensation and taste near the epiglottis.

TEETH (table 51–3)

Each tooth (Gk, *odous, odontos;* L., *dens, dentis*) is composed of connective tissue, the *pulp,* covered by three calcified tissues: *dentin(e), enamel,* and *cement(um).* The pulp occupies the *pulp cavity,* which comprises a *pulp chamber* in the crown and one or more *root canals* in the root(s). The root canals open by *apical foramina,* which transmit nerves and vessels to the pulp. Enamel is highly radio-opaque (see fig. 51–10). The cement is connected to the alveolar bone by *peridontium* to

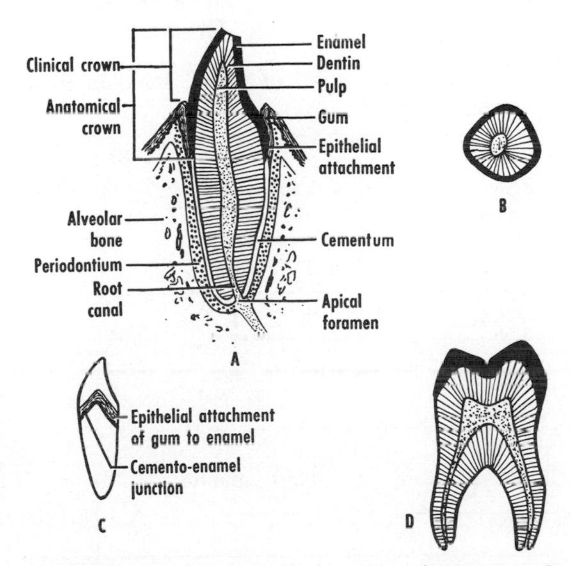

Figure 51–5 **A,** Longitudinal section of an incisor. **B,** Cross-section of the crown of an incisor, showing enamel, dentin, and pulp. **C,** Side view of an incisor, showing the area of epithelial attachment and the line of the cemento-enamel junction. **D,** Longitudinal section of a molar, showing the bifurcation of the pulp cavity.

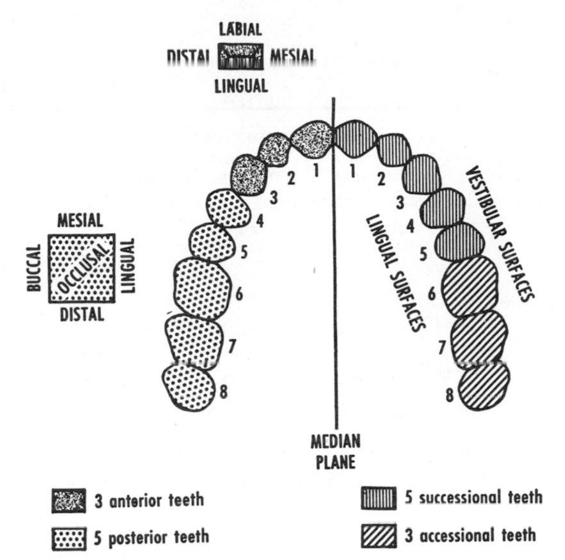

Figure 51–6 Dental terminology. Upper permanent teeth viewed from below.

form a fibrous joint between a tooth and its socket (*alveolus*). The *gums* (*gingivae*) are composed of dense fibrous tissue covered by oral mucosa.

The anatomical crown of a tooth is the part covered by enamel, whereas the *clinical crown* is the part that projects into the oral cavity (fig. 51–5). The *root* is covered by cement and includes the *neck,* adjacent to the crown. Some teeth have two or three roots.

The teeth are classified as incisors, canines, premolars, and molars. The eight *incisors* cut food by their edges. The four *canines* ("cuspids" or "eye-teeth") assist in cutting. The eight *premolars* ("bicuspids") assist in crushing food. They replace the deciduous molars. The twelve *molars* crush and grind food. **The roots of the upper molars are closely related to the floor of the maxillary sinus. Hence pulpal infection may cause sinusitis, or sinusitis may cause toothache. The third molar ("wisdom tooth") is highly variable.**

Most of the teeth in an adult are successional, i.e., they succeed a corresponding number of milk teeth. The permanent molars, however, are accessional, i.e., they are added behind the milk teeth during development.

Primary, or Deciduous, Dentition (figs. 51–7 and 51–8). **No functioning teeth have penetrated the oral cavity (i.e., erupted) at birth. The "milk teeth" appear in the oral cavity between the ages of 6 months and 2½ years. The first teeth to erupt are the lower medial incisors, at about 6 months.** All of the deciduous teeth have been shed by about 12 years. The 20 deciduous teeth are arranged in quadrants

TABLE 51–3 Dental Data

A. Deciduous Teeth

Calcification begins during fourth month of intrauterine life (in the sequence A, D, B, C, E).

Extent of calcification at birth:

cusps	occlusal surf.	⅓ crown	⅔ crown	⅔ crown
Similar to upper teeth				

Enamel of crowns is completed during first year.

Teeth erupt into mouth cavity (median age in years):

2½	1½	1¾	1	½
Similar to upper teeth				

Roots are completed about 1 to 1½ years after eruption.

Resorption of roots begins about 5 years after eruption.

Resorption of roots ends, and crowns are shed, between 5 and 15 years.

Usual number of cusps of deciduous teeth:

4-5	2-4	1	—	—
5	4	1	—	—

Usual number of roots of deciduous teeth:

3	3	1	1	1
2	2	1	1	1

B. Permanent Teeth

Calcification begins (in years; B = birth):

7-9	3	B	2	2	½	1	⅓
8-10	3	B	2	2	½	⅓	⅓

Enamel of crowns completed (in years):

12-16	7-8	3	6-7	5-6	6-7	4-5	4-5
Similar to upper teeth							

Teeth erupt into mouth cavity (median age in years):

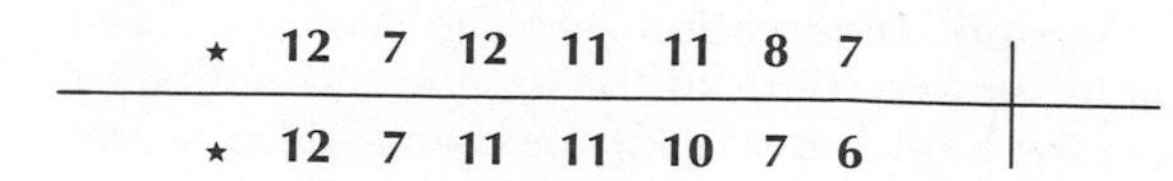

⋆	12	7	12	11	11	8	7
⋆	12	7	11	11	10	7	6

⋆ The highly variable third molar may erupt from 17 years onward, or not at all.

Roots are completed about 2 to 3 years after eruption.

Usual number of cusps:

3	4	4-5	2	2	1	—	—
4-5	4-5	5	2-3	2	1	—	—

Usual number of roots:

1-3	3	3	1-2	1-2	1	1	1
1-2	2	2	1	1	1	1	1

Frequent pattern of innervation of teeth:

Post. sup. alveolar: 8 7 6 5 4	Ant. sup. alveolar: 3 2 1	
Inf. alveolar: 8 7 6 5 4	Incisive br. of inf. alveolar: 3 2 1	1

Additional innervation of gums, alveolar bone, and periodontium:

(*a*) Vestibular aspect—

Post. sup. alveolar: 8 7 6 5 4	Labial br. of infra-orbital: 3 2 1
Buccal: 8 7 6 5	Mental: 4 3 2 1

(*b*) Lingual aspect—

Greater palatine: 8 7 6 5 4	Nasopalatine: 3 2 1
Lingual: 8 7 6 5 4 3 2 1	

Figure 51–7 Right lateral aspect of the maxilla and mandible of a 5-year-old child, showing the position of the deciduous and permanent teeth.

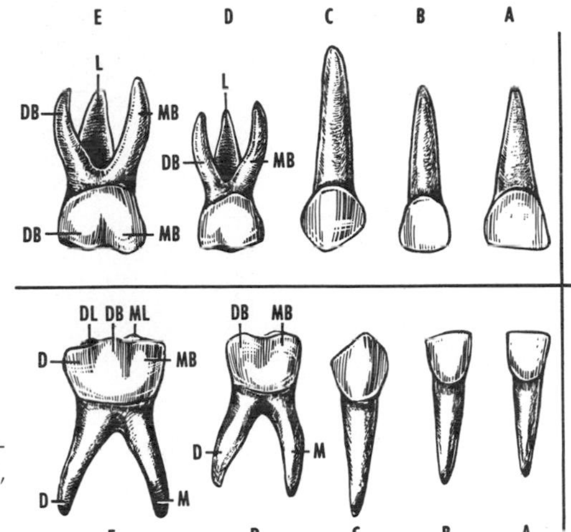

Figure 51–8 Vestibular surfaces of the right deciduous teeth. Abbreviations for cusps and roots: *B*, buccal; *D*, distal; *L*, lingual; *M*, mesial. (After Wheeler.)

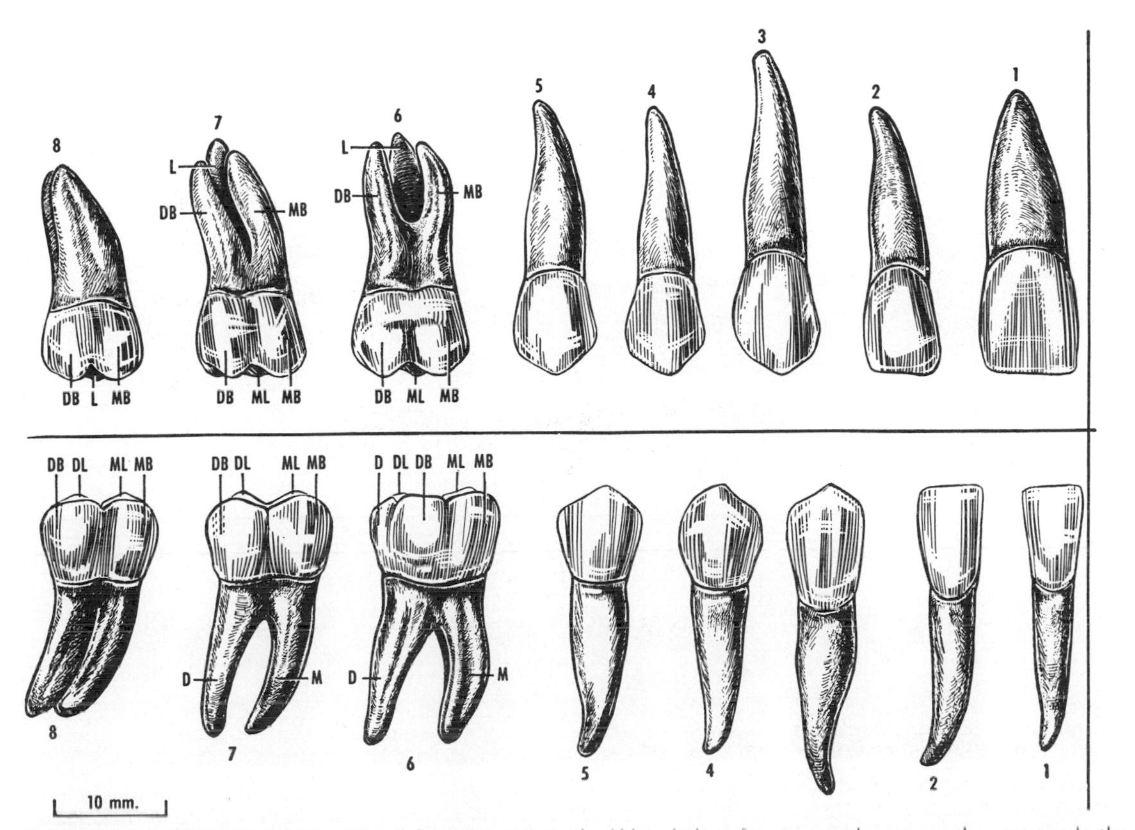

Figure 51–9 Vestibular surfaces of the right permanent teeth. Abbreviations for cusps and roots are the same as in the preceding figure. (After Wheeler.)

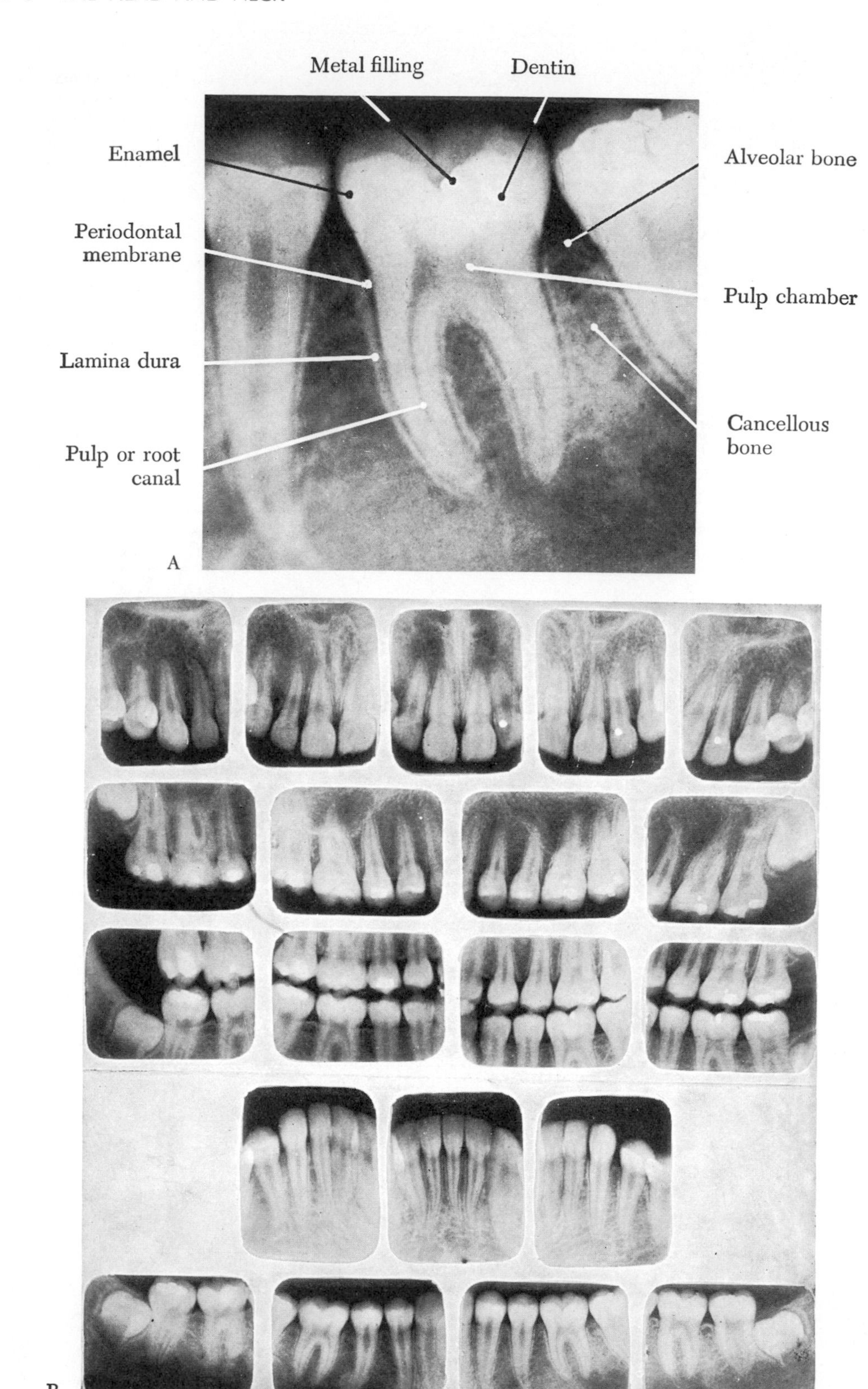

Figure 51–10 **A,** Dental and periodontal tissues from the right side of the mandible of an adolescent. **B,** Twenty intra-oral films of the permanent teeth. The teeth are being viewed as they would appear from within the oral cavity. (From McCall, J. O., and Wald, S. S., *Clinical Dental Roentgenology,* 4th ed., W. B. Saunders Company, Philadelphia, 1957, courtesy of the authors.)

of 5 each: 2 incisors, 1 canine, and 2 molars. They can conveniently be given letters A to E:

E D C B A	A B C D E
E D C B A	A B C D E

Thus, B| represents the right lower lateral incisor and |D indicates the first left upper molar.

Permanent Dentition (figs. 51–9 and 51–10). **The so-called permanent teeth begin to appear in the oral cavity at the age of about 6 years, and they have replaced the deciduous teeth by about 12 years. The first to erupt is the first molar, at about 6 to 7 years (the 6-year molar); the second molar erupts at about age 12 (the 12-year molar); the third molar may erupt from 17 years onward, or not at all.** The 32 permanent teeth are arranged in quadrants of 8 each: 2 incisors, 1 canine, 2 premolars, and 3 molars. The permanent molars have no deciduous predecessors. The teeth can conveniently be numbered from 1 to 8:

8 7 6 5 4 3 2 1	1 2 3 4 5 6 7 8
8 7 6 5 4 3 2 1	1 2 3 4 5 6 7 8

Thus, 6| represents the right lower first molar and |3 indicates the left upper canine.

Alignment and Occlusion. The teeth are arranged, or "aligned," in two *arches,* or *arcades,* one in each jaw. The term *occlusion* is used for any functional relation established when the upper and lower teeth come into contact with each other. Abnormal occlusion is termed malocclusion, the prevention and correction of which is orthodontics.

ADDITIONAL READING

Berkovitz, B. K. B., Holland, G. R., and Moxham, B. J., *A Colour Atlas and Textbook of Oral Anatomy,* Wolfe, London, 1978. Nice illustrations of gross anatomy, histology, and embryology.

Scott, J. H., and Symons, N. B. B., *Introduction to Dental Anatomy,* 8th ed., Churchill Livingstone, Edinburgh, 1977. A readable text on teeth, including an account of the development and growth of the face, jaws, and teeth.

Wheeler, R. C., *An Atlas of Tooth Form* 4th ed., W. B. Saunders Company Philadelphia, 1969.

Wheeler, R. C., *Dental Anatomy, Physiology and Occlusion,* 5th ed., W. B. Saunders Company, 1974. These two books are valuable for dental students.

QUESTIONS

51–1 Where does the parotid duct open?

51–2 What is the opening between the oral cavity and the oropharynx termed?

51–3 Which nerve supplies the muscles of the palate?

51–4 What is the anatomical significance of the sulcus terminalis on the back of the tongue?

51–5 Where are the vallate papillae?

51–6 Where are the valleculae?

51–7 Why is the attachment of the tongue to the mandible important clinically?

51–8 Into which chief lymphatic nodes does the tongue drain?

51–9 What is the difference between the anatomical crown and the clinical crown of a tooth?

51–10 What is the clinical importance of the relationship between the upper molar roots and the floor of the maxillary sinus?

51–11 Why are some teeth called successional and others accessional?

51–12 When do milk teeth appear?

51–13 Which are the first permanent teeth to erupt?

51–14 How are teeth numbered?

51–15 What is occlusion?

52 THE NOSE AND PARANASAL SINUSES

The nose (L., *nasus;* Gk, *rhis, rhinos*) includes the external nose on the face and the nasal cavity, which extends considerably further back. The nose functions in smell and provides filtered, warm, moist air for inspiration.

EXTERNAL NOSE

The external nose presents a *root* (or *bridge*), a *dorsum,* and a free *tip* or *apex.* The two inferior openings are the *nostrils* (or *nares*), bounded laterally by the *ala* and medially by the *nasal septum.* The upper part of the nose is supported by the nasal, frontal, and maxillary bones; the lower part includes several cartilages. The continuous free margin of the nasal bones and maxillae in a dried skull is termed the *piriform aperture.*

NASAL CAVITY

The nasal cavity extends from the nostrils, or nares, in front, to the choanae behind. **The choanae are the posterior apertures of the nose.** Each choana is bounded medially by the vomer, inferiorly by the horizontal plate of the palatine bone, laterally by the medial pterygoid plate, and superiorly by the body of the sphenoid bone (see figs. 42–12 and 42–13). **Posteriorly, the nasal cavity communicates with the nasopharynx, which in many respects may be regarded as the back portion of the cavity.** The nasal cavity is related to the anterior and middle cranial fossae, orbit, and paranasal sinuses and is separated from the oral cavity by the hard palate. In addition to the nostrils and choanae, the nasal cavity presents openings for the paranasal sinuses and the nasolacrimal duct. Further openings, covered by mucosa *in vivo,* are found in a dried skull, e.g., the sphenopalatine foramen. The nasal cavity is divided into right and left halves (each of which may be termed a nasal cavity) by the nasal septum. Each half has a roof, floor, and medial and lateral walls.

The *roof* is formed by nasal cartilages and several bones, chiefly the nasal and frontal bones, the cribriform plate of the ethmoid (fig. 52–1), and the body of the sphenoid. The *floor,* wider than the roof, is formed by the palatine process of the maxilla and the horizontal plate of the palatine bone, i.e., by the palate. The *medial wall,* or nasal septum, is formed from before backward by (1) the septal cartilage (destroyed in a dried skull), (2) the perpendicular plate of the ethmoid bone, and (3) the vomer (fig. 52–2*B*). It is usually deviated to one side. The lowest part (*the columella*) is membranous and mobile.

The *lateral wall,* uneven and complicated, is formed by several bones: nasal, maxilla, lacrimal and ethmoid, inferior nasal concha, perpendicular plate of palatine, and medial pterygoid plate of sphenoid (fig. 52–2*A*). **The lateral wall presents three or four medial projections**

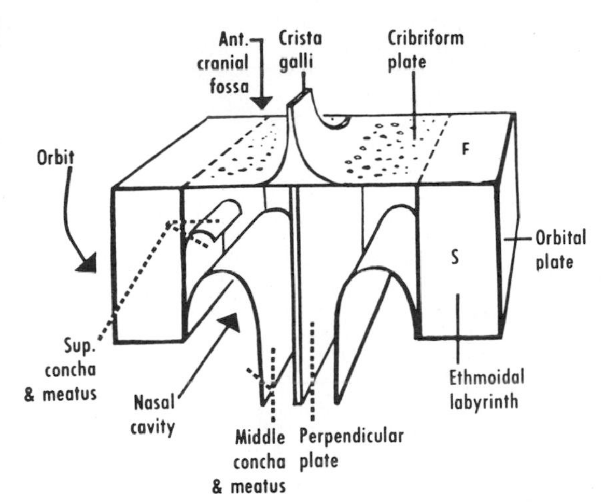

Figure 52–1 Scheme of the ethmoid bone viewed from behind. The two ethmoidal labyrinths are united by the cribriform plate. The perpendicular plate, which forms the upper part of the nasal septum, is set at a right angle to the horizontally placed cribriform plate. The lateral surface of each labyrinth forms a part of the medial wall of the orbit and is termed the orbital plate of the ethmoid bone. *F* indicates the portion of the labyrinth completed by the frontal bone; *S*, the portion completed by the sphenoid bone. The ethmoidal labyrinth contains ethmoidal air cells known collectively as the ethmoidal sinus. (After Grant.)

termed nasal conchae, which overlie meatuses. The inferior concha is a separate bone; the others are portions of the ethmoid bone. The conchae were formerly known as turbinates. It should be noted that the plural of meatus is meatus in Latin and meatuses in English (cf fetus).

The *spheno-ethmoidal recess,* above and behind the superior concha, receives the opening of the sphenoidal sinus. The *superior meatus,* under cover of the *superior concha,* receives the openings of the posterior ethmoidal cells and (in a dried skull) the sphenopalatine foramen. **The middle meatus, under cover of the middle concha, receives the openings of the maxillary and frontal sinuses.** Most anterior ethmoidal cells open on an elevation (*ethmoidal bulla,* fig. 52–3*B*). A curved slit (*hiatus semilunaris*) below the bulla receives the opening of the maxillary sinus. The frontal sinus and some anterior ethmoidal cells open either into an extension (*ethmoidal infundibulum*) of the hiatus or directly into the front part (*frontal recess*) of the middle meatus. **The inferior meatus, which lies between the inferior concha and the palate, receives the termination of the nasolacrimal duct.**

The nasal cavity can be examined *in vivo* either through a nostril or through the pharynx. A nasal speculum in a nostril is used in *anterior rhinoscopy*. A postnasal mirror inserted through the mouth and pharynx enables the choanae to be inspected in *posterior rhinoscopy* (figs. 52–4 and 53–11*B*).

Subdivisions. It is convenient to divide the nasal cavity into a vestibule, a respiratory region, and an olfactory region. The *vestibule* is a slight dilatation inside the nostril (see fig. 52–3*A*), limited by a ridge (*limen nasi*) over which skin becomes continuous with mucosa. The *respiratory region* is covered by mucoperichondrium and mucoperiosteum. The posterior two thirds show active ciliary motion for rapid drainage backward and downward into the nasopharynx. The nasal mucosa is highly vascular, and it warms and moistens the incoming air. The mucosa contains large venous-like spaces ("swell bodies"), which may become congested during a "cold in the nose." The middle meatus is continuous in front with a depression (*atrium;* see fig. 52–3*A*) that is limited above by a ridge (*agger nasi*). The *olfactory region* is bounded by the superior concha and the upper third of the nasal septum. The olfactory mucosa contains bipolar neurons (olfactory cells), the dendrites of which reach the surface. **The axons are collected into about 20 bundles known collectively as the olfactory (first cranial) nerve. These pass through the cribriform plate of the ethmoid bone and synapse in the olfactory bulb.** The olfactory nerve is tested by closing one nostril and presenting a test substance to the other.

Innervation. **The nerves of ordinary sensation are derived from the first two divisions of the trigeminal nerve** (fig. 52–5*A*, *B*, and *C*). The nerves for the posterior and larger portion of the nasal cavity come from branches of the pterygopalatine ganglion that are derived from the maxillary nerve. **The chief sympathetic (vasoconstrictor) and parasympathetic (vasodilator and secretory) innervation of the nasal cavity is from branches of the pterygopalatine ganglion, but some sympathetic fibers are carried along the walls of arteries.**

Blood Supply and Lymphatic Drainage. **The most important arteries to the nasal cavity are the sphenopalatine (from the maxillary) artery and the anterior ethmoidal (from the ophthalmic) artery. Bleeding from the nose (epistaxis) occurs usually from the junction between septal branches of the superior labial and sphenopalatine arteries.** The lymph vessels drain into deep cervical nodes. Communications probably occur between the nasal lymphatics and the subarachnoid space, probably through the sheath of the olfactory nerve.

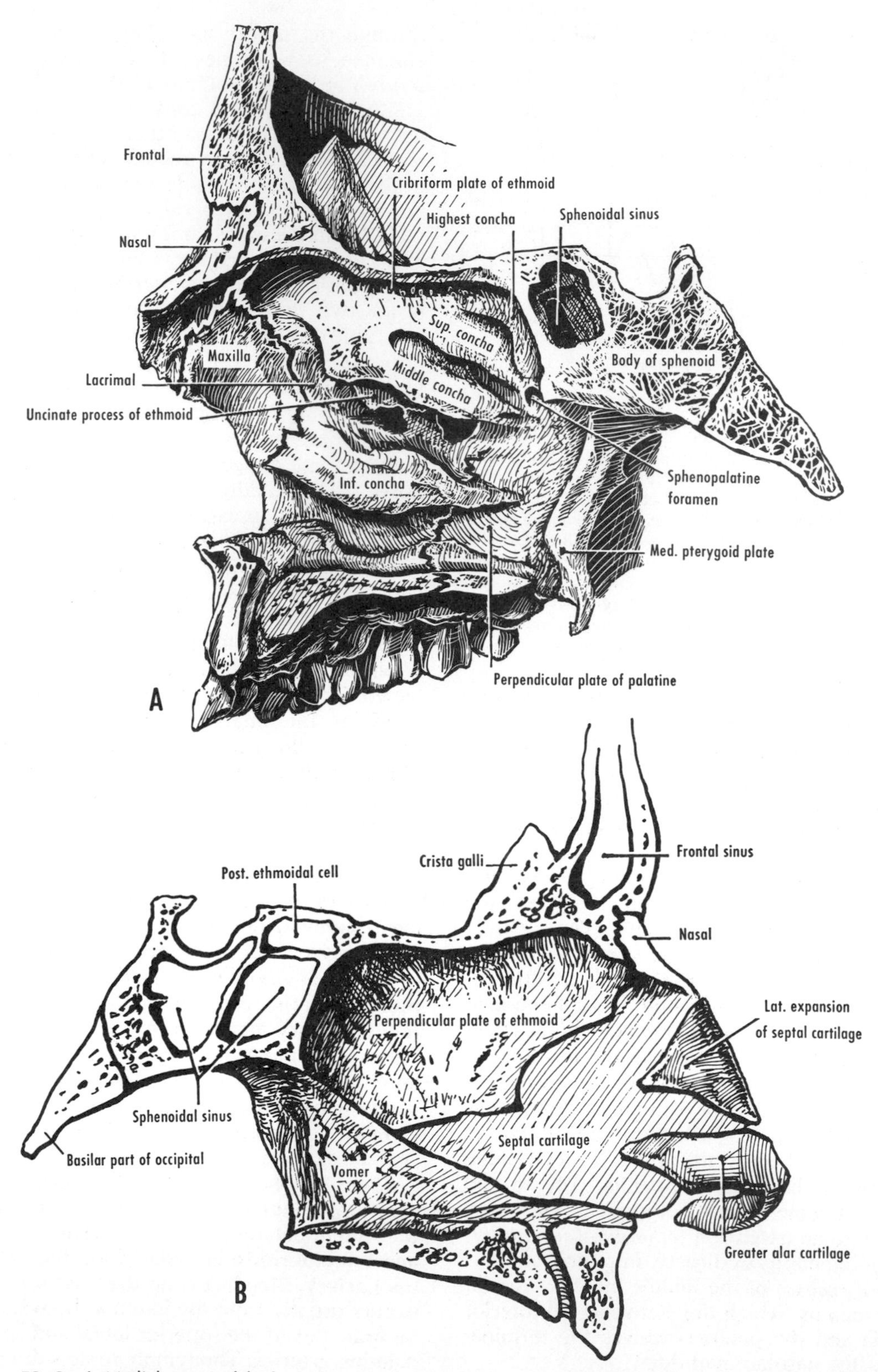

Figure 52–2 **A,** Medial aspect of the bony framework of the lateral wall of the right nasal cavity. The lateral boundary of the piriform aperture is formed by the nasal bone and the maxilla; that of the choana is formed by the medial pterygoid plate of the sphenoid bone. Note the line of the spheno-occipital junction. **B,** Lateral aspect of the medial wall (nasal septum) of the right nasal cavity. The lower limit of the ethmoidal contribution to the septum varies widely. The attachment of the septal cartilage to the vomer and the maxilla allows considerable movement without dislocation.

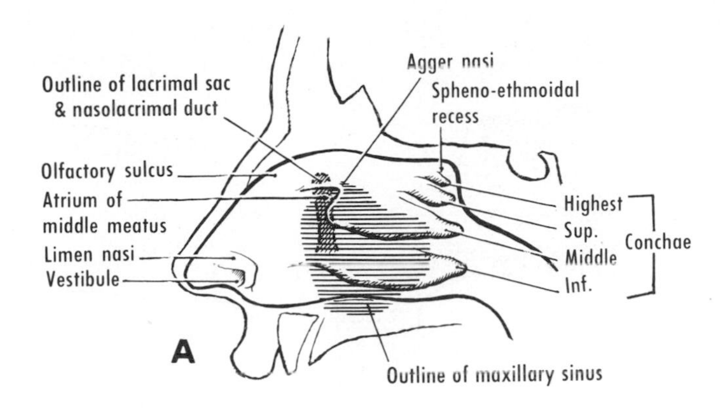

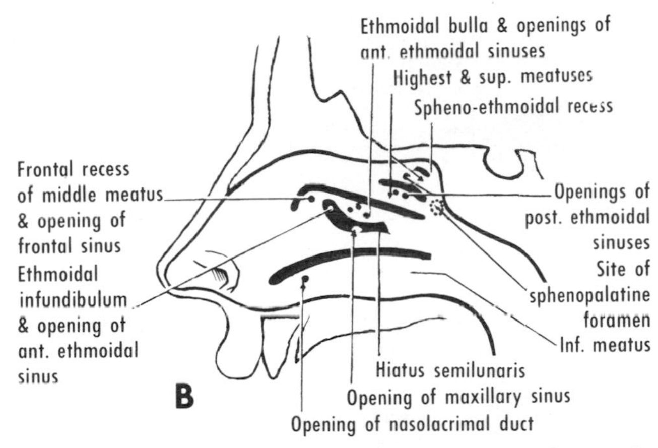

Figure 52–3 Medial aspect of the lateral wall of the right nasal cavity. **A** shows the four conchae. Note that each meatus is named after the concha that forms its roof. In **B,** the conchae have been largely removed. The frontal sinus may open into (1) the frontal recess (as shown here) or (2) the ethmoidal infundibulum.

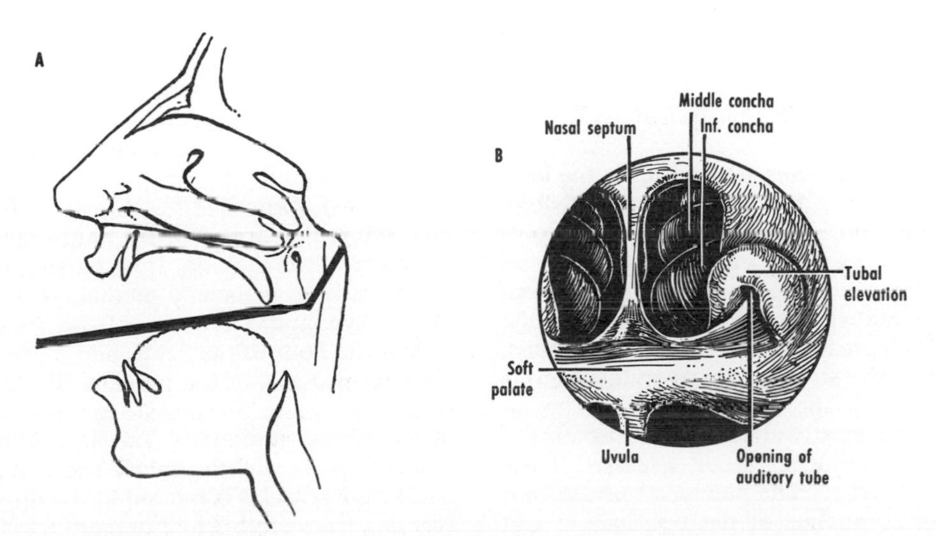

Figure 52–4 Posterior rhinoscopy. **A** shows the placement of the mirror, and **B** the structures seen. Cf. fig. 53–11*B*.

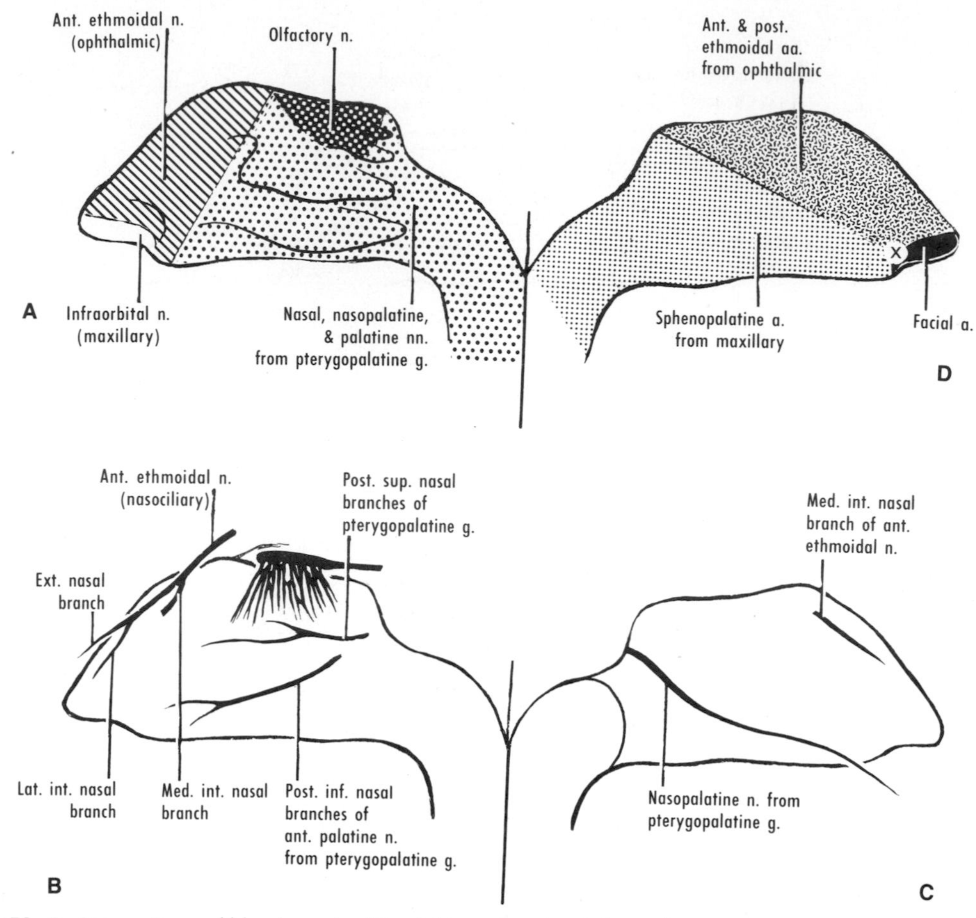

Figure 52–5 Innervation and blood supply of the right nasal cavity. **A** shows the main nerve territories on the lateral wall: those on the medial wall are similar. Apart from the olfactory region, the chief innervation of the nasal cavities is from the ophthalmic and maxillary nerves, the latter by way of the pterygopalatine ganglion. **B** and **C** show the nerves in the lateral and medial walls, respectively. **D** shows the main arterial territories on the medial wall: those on the lateral wall are similar. On the nasal septum, a septal branch of the superior labial artery (from the facial) and septal branches of the sphenopalatine artery anastomose near the point *X*, which is the chief site from which bleeding from the nose occurs.

PARANASAL SINUSES (fig. 52–6)

The paranasal sinuses are cavities in the interior of the maxilla and the frontal, sphenoid, and ethmoid bones. The sinuses develop as outgrowths from the nasal cavity; hence they all drain directly or indirectly into the nose. Nasal infection (rhinitis), e.g., during a "cold in the head," may spread to the sinuses (sinusitis). The lining of the sinuses (muco-endosteum) is continuous with the nasal mucosa. The sinuses develop mostly after birth, and their degree of development varies greatly. Their function is obscure. The paranasal sinuses are supplied by branches of the ophthalmic and maxillary nerves. The sinuses can be examined radiographically, and a light placed inside the mouth enables the maxillary sinus to be transilluminated.

Maxillary Sinus (see fig. 42–6). The maxillary sinus, the largest of the sinuses, is within the body of the maxilla. It is shaped like a pyramid; its base is usually medial, with its apex in the zygomatic process of the maxilla. Its roof is the floor of the orbit, and its floor is the alveolar process of the maxilla. **The maxillary sinus drains into the middle meatus by means of the hiatus semilunaris. The floor of the maxillary sinus is slightly below the level of the nasal cavity, and it is related to the upper teeth (varying from teeth 3 to 8 to teeth 6 to 8). Maxillary sinusitis is frequently accompanied by**

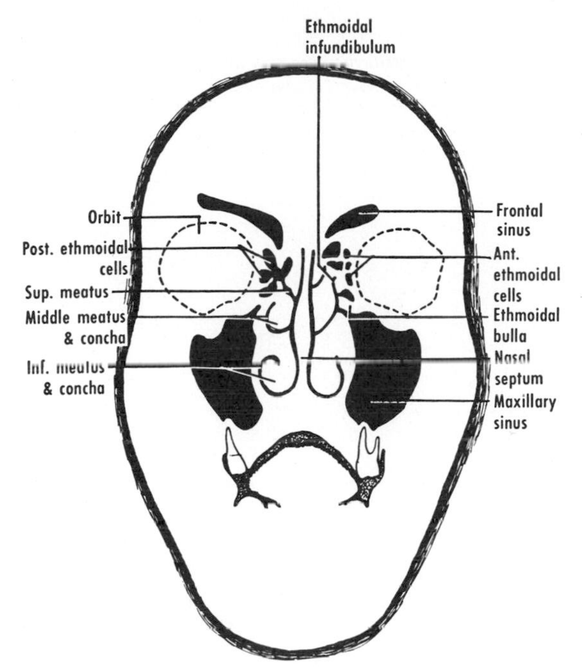

Figure 52–6 Coronal scheme of the nasal cavity to show the conchae, meatuses, and certain paranasal sinuses. The size of the nasal passages is based on an instructive laminogram by Proetz.

toothache. Infection may spread among the various sinuses, the nasal cavity, and the teeth. The opening of the maxillary sinus can be cannulated *in vivo* through the nostril.

Ethmoidal Sinus. **The ethmoidal sinus comprises numerous small cavities (ethmoidal cells) in the ethmoidal labyrinth.** The walls of these cavities are completed by the surrounding bones, which tend to be invaded by "the struggle of the ethmoids" (Seydel). Anterior and posterior groups drain into the middle and superior meatuses, respectively (see fig. 52–3).

Frontal Sinus (see figs. 42–4, 42–5, and 53–4). **The frontal sinus may be regarded as an anterior ethmoidal cell that has invaded the frontal bone postnatally. The right and left frontal sinuses, frequently of different sizes, are separated by a bony septum that is usually deviated to one side. The frontal sinus drains into the middle meatus in a variable manner directly or by a frontonasal duct,** which opens into the frontal recess or the ethmoidal infundibulum. The frontal sinus commonly extends backward in the roof of the orbit (see fig. 45–10).

Sphenoidal Sinus (see figs. 42–18, 52–2, and 53–4). **The sphenoidal sinus is in the body of the sphenoid bone, and it varies greatly in size. It is related above to the hypophysis (pituitary) and the optic nerves and chiasma and laterally to the cavernous sinus and internal carotid artery** (fig. 43–22). The sphenoidal sinus drains into the spheno-ethmoidal recess above the superior concha. The sinus is divided into right and left parts by a bony septum.

ADDITIONAL READING

Proetz, A. W., *Essays on the Applied Physiology of the Nose,* 2nd ed., Annals Publishing Co., St. Louis, 1953. An interesting functional account.

Ritter, F. N., *The Paranasal Sinuses,* Mosby, St. Louis, 1973. Includes a photographic atlas of coronal sections and radiograms.

Schaeffer, J. P., *The Nose, Paranasal Sinuses, Nasolacrimal Passageways, and Olfactory Organ in Man,* Blakiston, Philadelphia, 1920. A classic text on the anatomy and development of these structures.

Terracol, J., and Ardouin, P., *Anatomie des fosses nasales et des cavités annexes,* Maloine, Paris, 1965. A detailed account.

QUESTIONS

52–1 What are the anterior and posterior apertures of the nose termed?

52–2 What are the nasal conchae?

52–3 List the main openings into the nasal meatuses.

52–4 Where is the olfactory region of the nasal cavity?

52–5 Which nerves are responsible for general sensation in the nasal cavity?

52–6 Which are the most important arteries to the nasal cavity?

52–7 How do the paranasal sinuses develop?

52–8 List the chief paranasal sinuses.

52–9 What is the ethmoidal labyrinth?

52–10 What can be seen on posterior rhinoscopy?

53
THE PHARYNX AND LARYNX

PHARYNX

The word *throat* is used for the parts of the neck in front of the vertebral column, especially the pharynx and the larynx. **The pharynx is the part of the digestive system situated behind the nasal and oral cavities and behind the larynx. It is therefore divisible into nasal, oral, and laryngeal parts: the (1) nasopharynx, (2) oropharynx, and (3) laryngopharynx. The pharynx extends from the base of the skull down to the lower border of the cricoid cartilage**

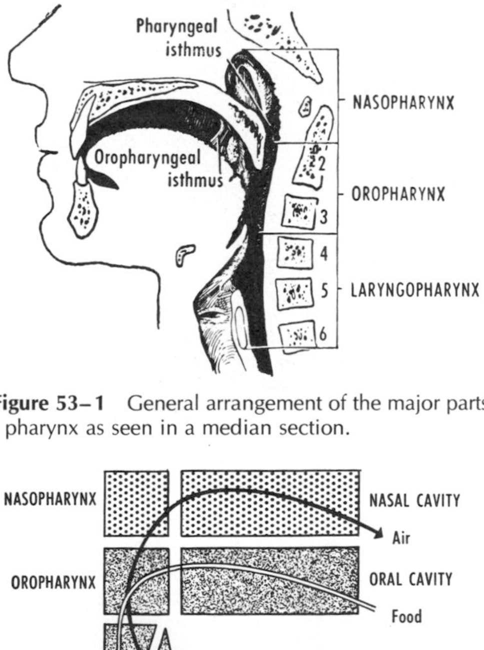

Figure 53–1 General arrangement of the major parts of the pharynx as seen in a median section.

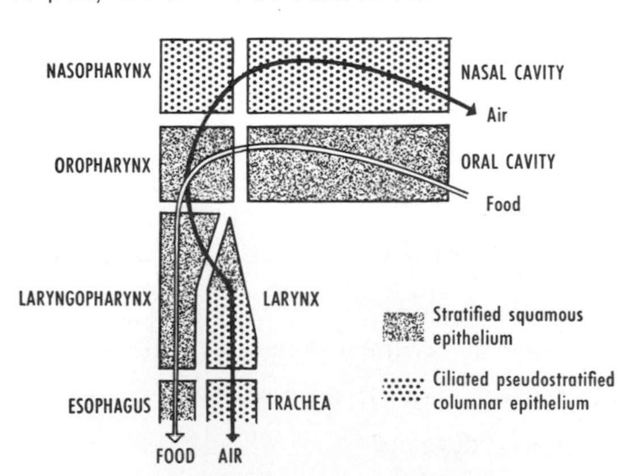

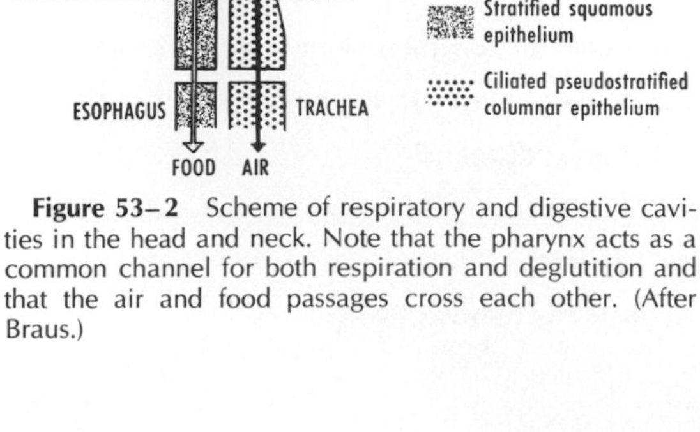

Figure 53–2 Scheme of respiratory and digestive cavities in the head and neck. Note that the pharynx acts as a common channel for both respiration and deglutition and that the air and food passages cross each other. (After Braus.)

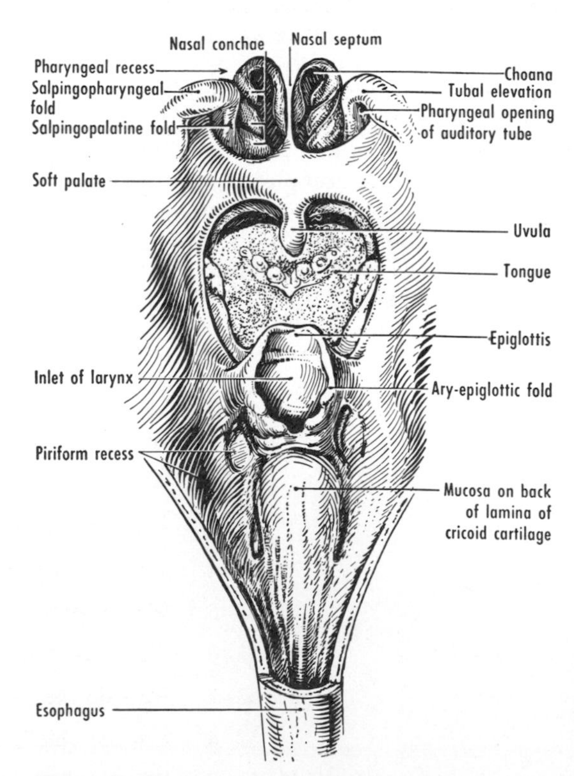

Figure 53–3 Anterior wall of the pharynx viewed from behind. The pharynx communicates with the nasal cavity, auditory tubes, oral cavity, larynx, and esophagus.

(C.V.6), where it becomes continuous with the esophagus. It is related above to the sphenoid and occipital bones and behind to the prevertebral fascia and muscles and the upper six cervical vertebrae. The pharynx (figs. 53–1 to 53–4) is a fibromuscular tube lined by mucous membrane.

The pharynx is the common channel for deglutition (swallowing) and respiration, and the food and air pathways cross each other in the pharynx. In the anesthetized patient, the passage of air through the pharynx is facilitated by extension of the head.

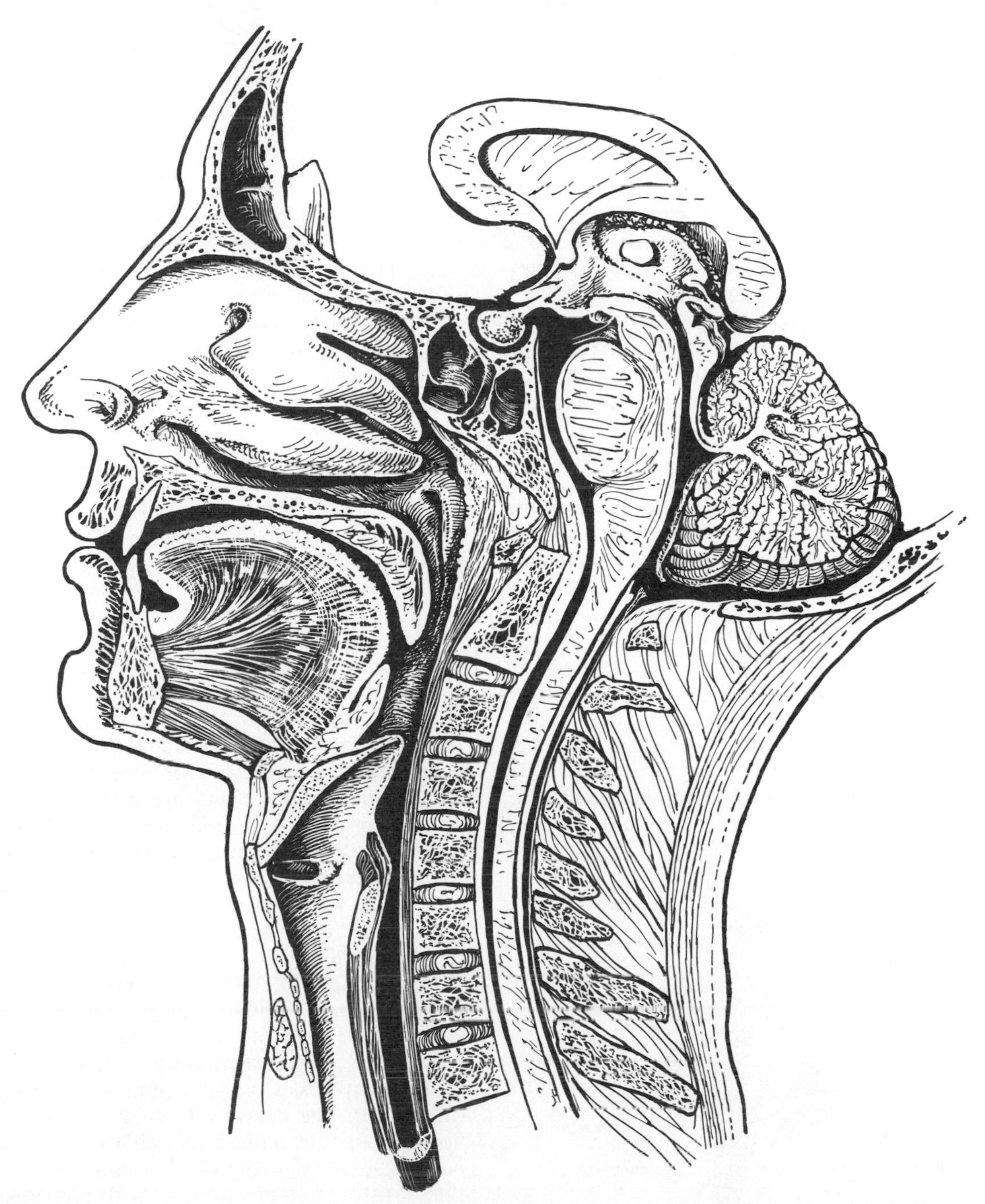

Figure 53–4 Sagittal (almost median) section of the head and neck, with a portion of the brain included. The various structures shown in this illustration have been given labels in other figures. Note the hypophysis, corpus callosum, septum pellucidum, pineal body, third ventricle, aqueduct, fourth ventricle, pons, cerebellum, medulla, and spinal cord; C.V.1 to 7 and T.V.1; the frontal and sphenoidal sinuses; the nasal conchae, palate, and opening of the auditory tube; the genioglossus and geniohyoid muscles; the larynx and trachea and pharynx and esophagus.

SUBDIVISIONS

Nasopharynx. **The nasopharynx, at least in its anterior part, may be regarded as the back portion of the nasal cavity, with which it is a component of the respiratory system. The nasopharynx communicates with the oropharynx through the pharyngeal isthmus, which is bounded by the soft palate, the palatopharyngeal arches, and the posterior wall of the pharynx.** The isthmus is closed by muscular action during swallowing. The choanae are the junction between nasopharynx and nasal cavity proper.

A mass of lymphoid tissue, the *(naso)pharyngeal tonsil* is embedded in the mucous membrane of the posterior wall of the nasopharynx. **Enlarged (naso)pharyngeal tonsils are termed "adenoids" and may cause respiratory obstruction.** Higher up, a minute *pharyngeal hypophysis* (resembling the adenohypophysis) may be found (see fig. 53–5).

Each lateral wall of the nasopharynx presents the pharyngeal opening of the auditory tube, located about 1 to 1½ cm (1) below the roof of the pharynx, (2) in front of the posterior wall of the pharynx, (3) above the level of the palate, and (4) behind the inferior nasal concha and the nasal septum (fig. 53–5). The auditory tube can be catheterized through a nostril. The opening is limited above by a *tubal elevation,* from which mucosal folds descend to the palate and side wall of the pharynx. **The part of the pharyngeal cavity behind the tubal elevation is termed the pharyngeal recess.** Nearby lymphoid tissue is referred to as the *tubal tonsil.*

The auditory tube is pharyngotympanic; i.e., it connects the nasopharynx to the tympanic cavity. Hence, infections may spread along this route. The tube equalizes the pressure of the external air and that in the tympanic cavity. The auditory tube, about 3 to 4 cm in length, extends backward, laterally, and upward. It consists of (1) a cartilaginous part, the anteromedial two thirds, which is a diverticulum of the pharynx, and (2) an osseous part, the posterolateral third, which is a forward prolongation of the tympanic cavity. The cartilaginous part lies on the inferior aspect of the skull, in a groove between the greater wing of the sphenoid bone and the petrous part of the temporal (see fig. 42–12). **The cartilaginous part of the auditory tube remains closed except on swallowing or yawning, when its opening prevents excessive pressure in the middle ear.** The osseous part of the tube is within the petrous part of the temporal bone.

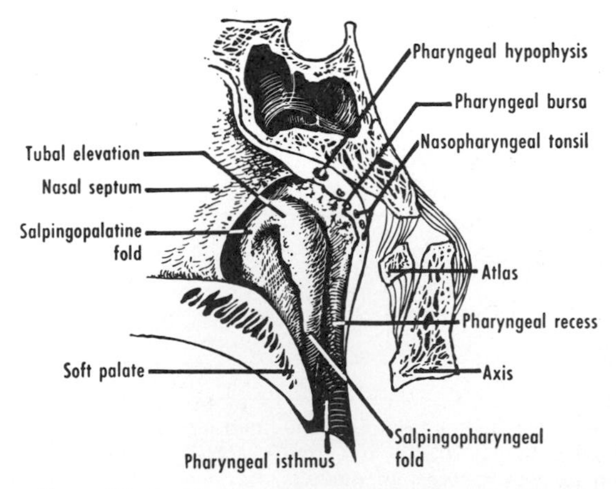

Figure 53–5 Medial view of the right lateral wall of the nasopharynx. See the preceding figure for orientation.

Oropharynx. **The oropharynx extends from the soft palate above to the superior border of the epiglottis below. It communicates in front with the oral cavity by the faucial (oropharyngeal) isthmus, which is bounded above by the soft palate, laterally by the palatoglossal arches, and below by the tongue** (see fig. 53–1). This area is characterized by a *lymphatic ring* composed of the nasopharyngeal, tubal, palatine, and lingual tonsils.

The mucous membrane of the epiglottis is reflected onto the base of the tongue and onto the lateral wall of the pharynx. **The space on each side of the median glosso-epiglottic fold is termed the epiglottic vallecula.**

Each lateral wall of the oropharynx presents the diverging palatoglossal and palatopharyngeal arches, which are produced by the similarly named muscles and are often called the anterior and posterior pillars of the fauces, respectively. The triangular recess (tonsillar fossa) between the two arches lodges the palatine tonsil, which is often referred to as merely "the tonsil" (see fig. 53–1). (A tonsil is a mass of lympoid tissue containing reaction or germinal centers and related to an epithelial surface in the pharynx.) The medial surface of the tonsil usually presents the *intratonsillar cleft* (commonly but inaccurately called the "supratonsillar fossa") and a number of *crypts* (fig. 53–6). The lateral surface is covered by a fibrous capsule and is related to fascia, the paratonsillar vein (the chief source of hemorrhage after *tonsillectomy*), and pharyngeal musculature. The tonsil is supplied by the tonsillar branch of the facial artery, and it drains into the facial vein. Involution of the tonsil begins at puberty.

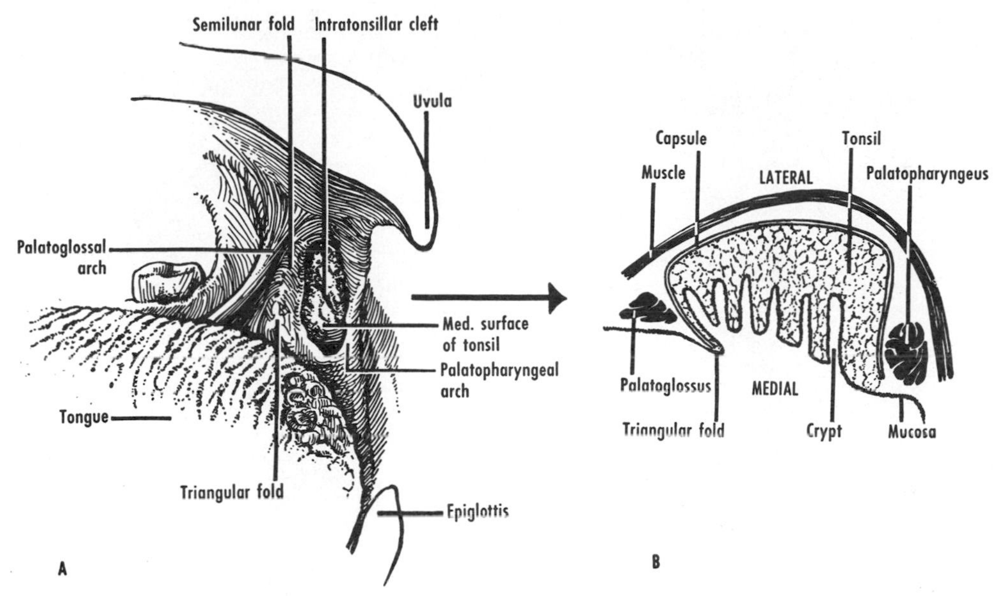

Figure 53–6 **A,** Right palatine tonsil and its surroundings, medial aspect. **B,** Horizontal section through the tonsil, at a greater magnification. (After Fetterolf.)

Laryngopharynx. **The laryngopharynx extends from the upper border of the epiglottis to the lower border of the cricoid cartilage, where it becomes continuous with the esophagus. Anteriorly, it presents the inlet of the larynx and the backs of the arytenoid and cricoid cartilages. The piriform recess, in which foreign bodies may become lodged, is the part of the cavity of the laryngopharynx situated on each side of the inlet of the larynx** (see fig. 53–3).

MUSCLES

The pharynx consists of four coats, from within outward: (1) a mucous membrane continuous with that of the auditory tubes and the nasal, oral, and laryngeal cavities; (2) a fibrous coat, thick above (*pharyngobasilar fascia*) and forming a median raphe posteriorly; (3) a muscular coat, described below, and (4) a fascial coat (*buccopharyngeal fascia*) covering the muscles.

The wall of the pharynx is composed mainly of two layers of skeletal muscles. The external, circular layer comprises three constrictors (fig. 53–7 and table 53–1). The internal, chiefly longitudinal layer consists of two levators: the stylopharyngeus and the palatopharyngeus.

The constrictors of the pharynx have their fixed points in front, where they are attached to bones or cartilages, whereas they expand behind, overlap one another from below upward, and end in a median tendinous raphe posteriorly. Their overlapping has been compared with that of three flower pots placed one inside another. The inferior constrictor arises from the cricoid and thyroid cartilages. The cricopharyngeal fibers are horizontal and continuous with the circular fibers of the esophagus. They act as a sphincter and prevent air from entering the esophagus. **A pharyngeal diverticulum may form posteriorly through the fibers of the inferior constrictor.** The middle constrictor arises from the hyoid bone, whereas the superior constrictor arises from the mandible and sphenoid bone.* The constrictors are inserted into the median raphe posteriorly.

The *palatopharyngeus* arises from the palate, forms the palatopharyngeal fold, and is inserted into the thyroid cartilage and the side of the pharynx. The *stylopharyngeus* arises from the styloid process, passes between the superior and middle constrictors, and is inserted with the palatopharyngeus. The stylopharyngeus is supplied by the glossopharyngeal nerve, whereas the palatopharyngeus and the constrictors are innervated by the pharyngeal branch of the vagus (probably fibers from the accessory nerve) through the pharyngeal plexus on the middle constrictor.

* The existence of a pterygomandibular raphe between the superior constrictor and buccinator has been denied by G. R. L. Gaughran (Anat. Rec., *184*:410, 1976).

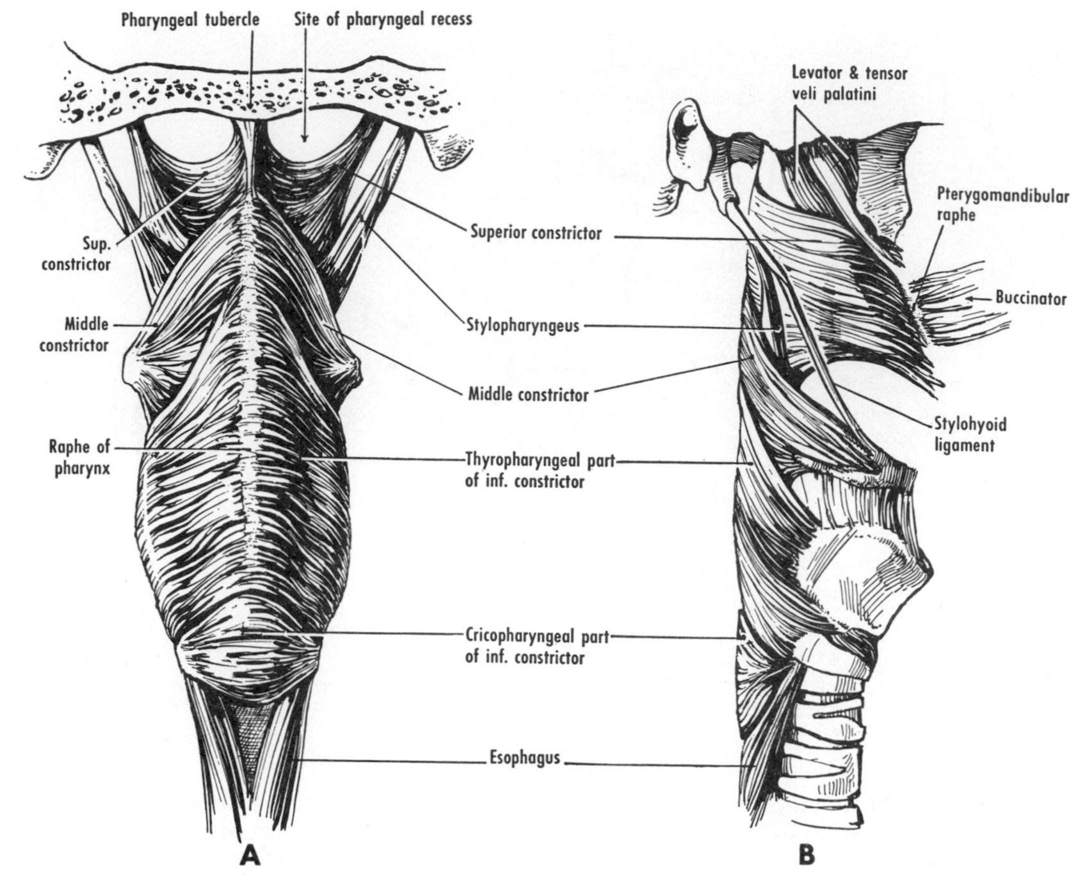

Figure 53–7 Muscles of the pharynx. **A,** Posterior aspect. **B,** Right lateral aspect.

TABLE 53–1 Muscles of Pharynx

Muscle	Origin	Insertion	Innervation	Action
Superior constrictor	Mylohyoid line of mandible, buccinator muscle, & pterygoid hamulus			
Middle constrictor	Angle between greater & lesser horns of hyoid bone	Median raphe posteriorly	Pharyngeal branch of vagus through pharyngeal plexus	Constrict wall of pharynx
Inferior constrictor	Oblique line of thyroid cartilage & arch of cricoid cartilage			
Palatopharyngeus	Bony palate & palatine aponeurosis	Posterior border of thyroid cartilage & side of pharynx & esophagus		Elevate pharynx & larynx
Stylopharyngeus	Styloid process		Glossopharyngeal	

The chief action in which the muscles of the pharynx combine is deglutition (or swallowing), a complicated, neuromuscular act whereby food is transferred from (1) the mouth through (2) the pharynx and (3) the esophagus to the stomach. The pharyngeal stage is the most rapid and most complex phase of deglutition. During swallowing, the nasopharynx and vestibule of the larynx are sealed but the epiglottis adopts a variable position. Food is usually deviated laterally by the epiglottis and ary-epiglottic folds into the piriform recesses of the laryngopharynx. The *pharyngeal ridge* is an elevation or bar on the posterior wall of the pharynx below the level of the soft palate; it is produced during swallowing by transverse muscle fibers.

Innervation and Blood Supply

The motor and most of the sensory supply to the pharynx is by way of the pharyngeal plexus, which, situated chiefly on the middle constrictor, is formed by the pharyngeal branches of the vagus and glossopharyngeal nerves and also by sympathetic fibers. The motor fibers in the plexus are from the accessory nerve but are carried by the vagus and supply all the muscles of the pharynx and soft palate except the stylopharyngeus (supplied by cranial nerve 9) and tensor veli palatini (supplied by cranial nerve 5). The sensory fibers in the plexus are from the glossopharyngeal nerve, and they supply the greater portion of all three parts of the pharynx. The pharynx is supplied by branches of the external carotid (ascending pharyngeal) and subclavian (inferior thyroid) arteries.

LARYNX

The larynx is the organ that connects the lower part of the pharynx with the trachea. It serves (1) as a valve to guard the air passages, especially during swallowing, (2) for the maintenance of a patent airway, and (3) for vocalization.

Superficial anteriorly (fig. 53–8), the larynx is related posteriorly to the laryngopharynx, the prevertebral fascia and muscles, and the bodies of cervical vertebrae 3 to 6. Laterally, the larynx is related to the carotid sheath, in-

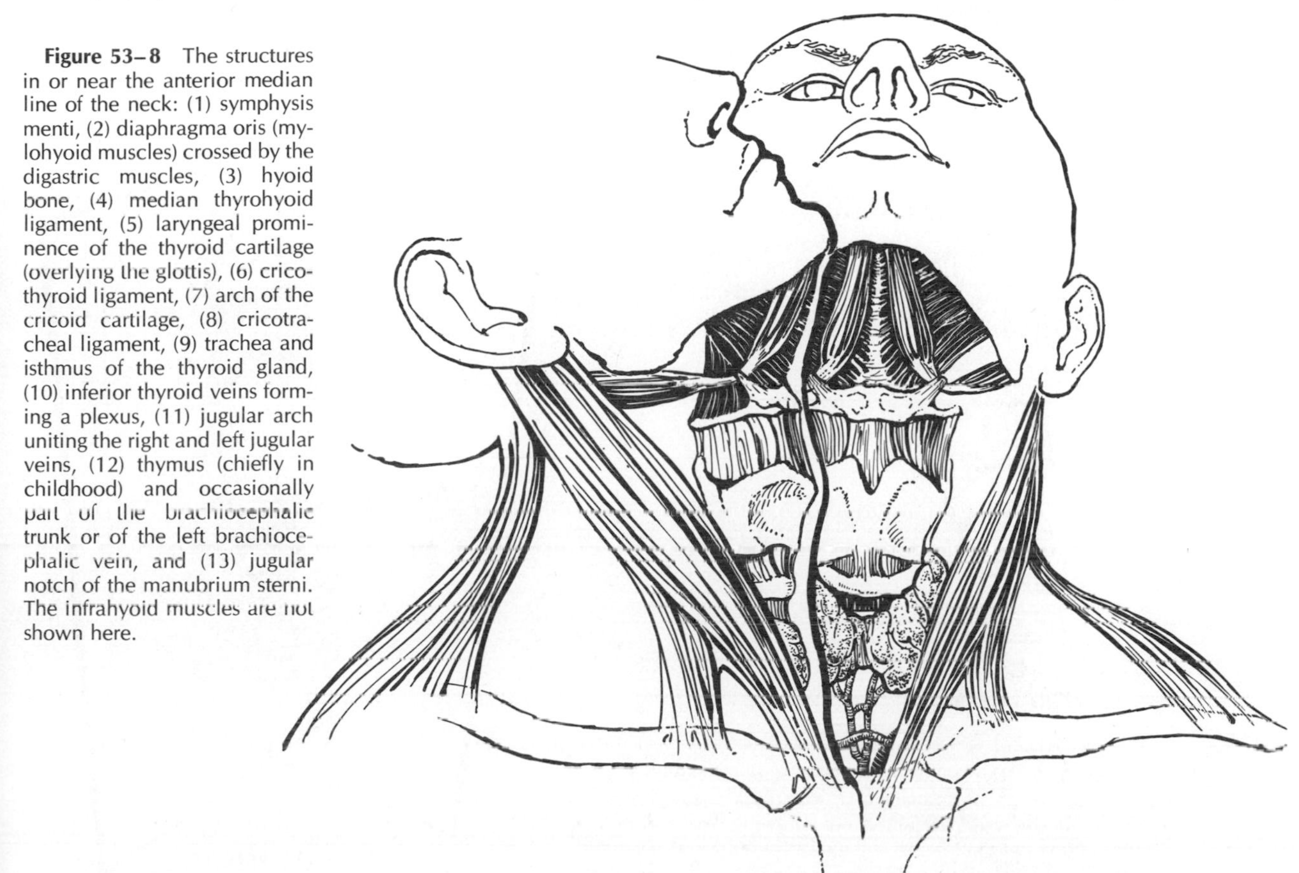

Figure 53–8 The structures in or near the anterior median line of the neck: (1) symphysis menti, (2) diaphragma oris (mylohyoid muscles) crossed by the digastric muscles, (3) hyoid bone, (4) median thyrohyoid ligament, (5) laryngeal prominence of the thyroid cartilage (overlying the glottis), (6) cricothyroid ligament, (7) arch of the cricoid cartilage, (8) cricotracheal ligament, (9) trachea and isthmus of the thyroid gland, (10) inferior thyroid veins forming a plexus, (11) jugular arch uniting the right and left jugular veins, (12) thymus (chiefly in childhood) and occasionally part of the brachiocephalic trunk or of the left brachiocephalic vein, and (13) jugular notch of the manubrium sterni. The infrahyoid muscles are not shown here.

frahyoid muscles, sternomastoid, and thyroid gland. The larynx is elevated (particularly by the palatopharyngeus) during extension of the head and during deglutition.

The larynx can be examined *in vivo* by means of a mirror (indirect laryngoscopy) or a tubular instrument (direct laryngoscopy) (see figs. 53–11*C* and *D* and 53–12).

CARTILAGES (figs. 53–8 to 53–10 and 53–14)

The larynx possesses three single cartilages (thyroid, cricoid, and epiglottic) and three paired cartilages (arytenoid, corniculate, and cuneiform). The thyroid, cricoid, and arytenoid cartilages are composed of hyaline carti-

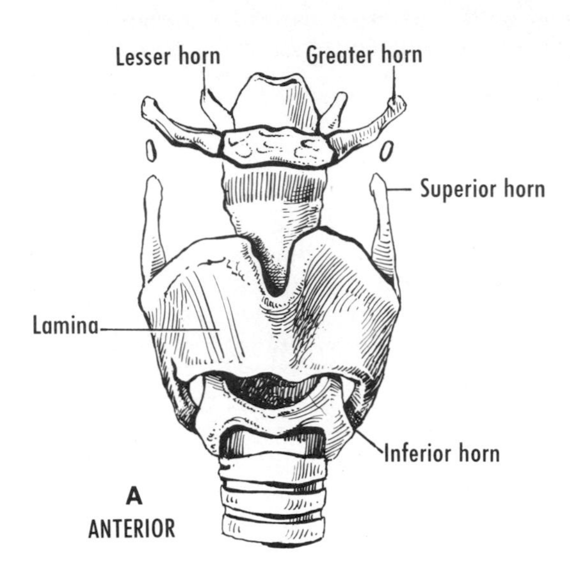

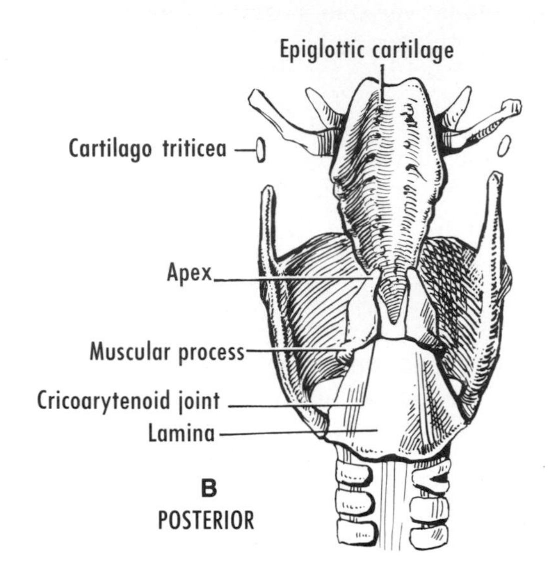

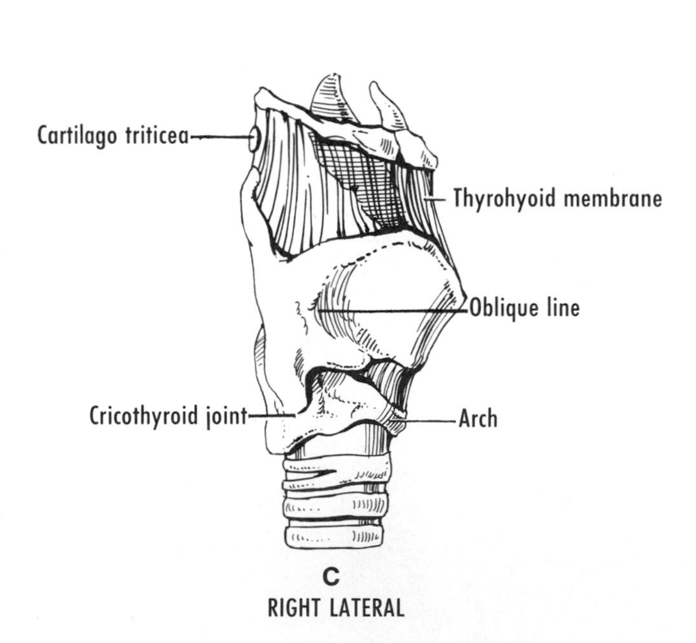

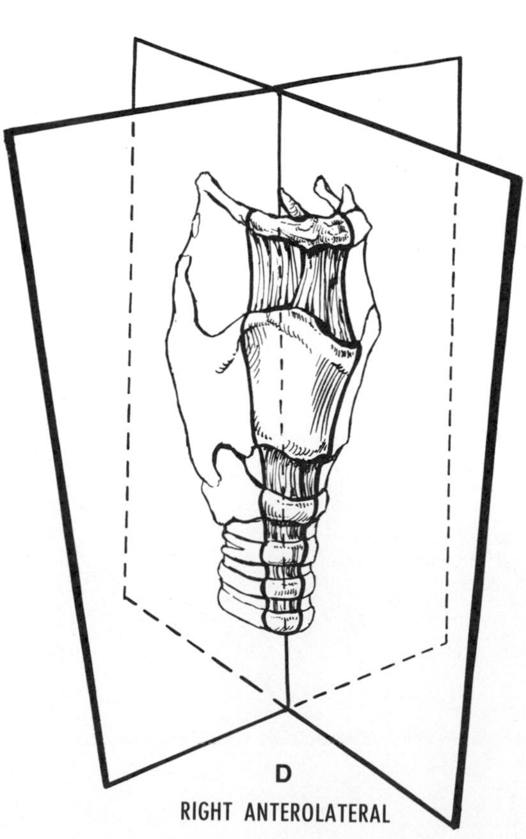

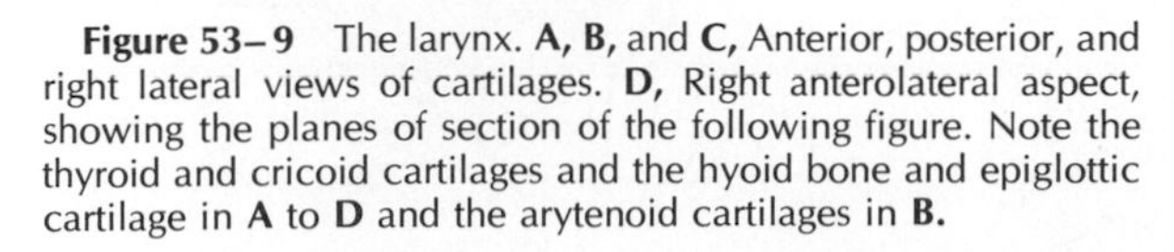

Figure 53–9 The larynx. **A, B,** and **C,** Anterior, posterior, and right lateral views of cartilages. **D,** Right anterolateral aspect, showing the planes of section of the following figure. Note the thyroid and cricoid cartilages and the hyoid bone and epiglottic cartilage in **A** to **D** and the arytenoid cartilages in **B.**

lage and may undergo calcification, endochondral ossification, or both, thereby becoming visible radiographically. The other cartilages are elastic in type.

The *thyroid cartilage* (fig. 53–9) comprises two spring-like plates termed laminae, which are fused in front but divergent behind. **The laminae produce a median elevation termed the laryngeal prominence ("Adam's apple"), which is palpable and frequently visible *in vivo.*** The posterior border of each *lamina* is prolonged upward and downward as cornua, or horns. The *superior horn* is anchored to the tip of the greater horn of the hyoid bone. The *inferior horn* articulates medially with the cricoid cartilage. The lateral surface of each lamina is crossed by an *oblique line* for the attachment of muscles.

The *cricoid cartilage* (fig. 53–9) is shaped like a signet ring. It comprises a posterior plate, the *lamina,* and a narrow, anterior part, the *arch.* The lamina articulates superolaterally with the arytenoid cartilages. **The cricoid cartilage is at the level of C.V.6, and its arch is palpable *in vivo.* The lower border of the cricoid cartilage marks the end of the pharynx and larynx and hence the commencement of the esophagus and trachea.**

The *arytenoid cartilages* (fig. 53–9*B*) articulate with the upper border of the lamina of the cricoid cartilage. Each presents an *apex* above (which supports the corniculate cartilage) and a base below. The *base* sends a *vocal process* forward (for attachment to the vocal ligament) and a *muscular process* laterally (for muscular attachments). The *corniculate* and (inconstant) *cuneiform cartilages* are nodules in the ary-epiglottic folds (figs. 53–10*B* and 53–12).

The *epiglottic cartilage* (see fig. 53–9) is covered by mucous membrane to form the epiglottis. **The epiglottis is situated behind the root of the tongue and the body of the hyoid bone and in front of the inlet of the larynx.** The lower end, or *stalk,* of the leaf-shaped cartilage is anchored to the back of the thyroid cartilage. Taste buds are present in the posterior surface of the epiglottis.

JOINTS (figs. 53–9)

Two synovial joints are present on each side. The *cricothyroid joint,* between the side of the cricoid cartilage and the inferior horn of the thyroid cartilage, allows mainly rotation of the thyroid cartilage around a horizontal axis through the joints of the two sides. The *crico-arytenoid joint,* between the upper border of the lamina of the cricoid cartilage and the base of the arytenoid cartilages, allows gliding and rotation of the arytenoid cartilages.

LIGAMENTS

The *thyrohyoid membrane* connects the thyroid cartilage with the upper border of the hyoid bone (see fig. 53–9*C*). The median part is thickened to form a ligament. The membrane is pierced on each side by the internal

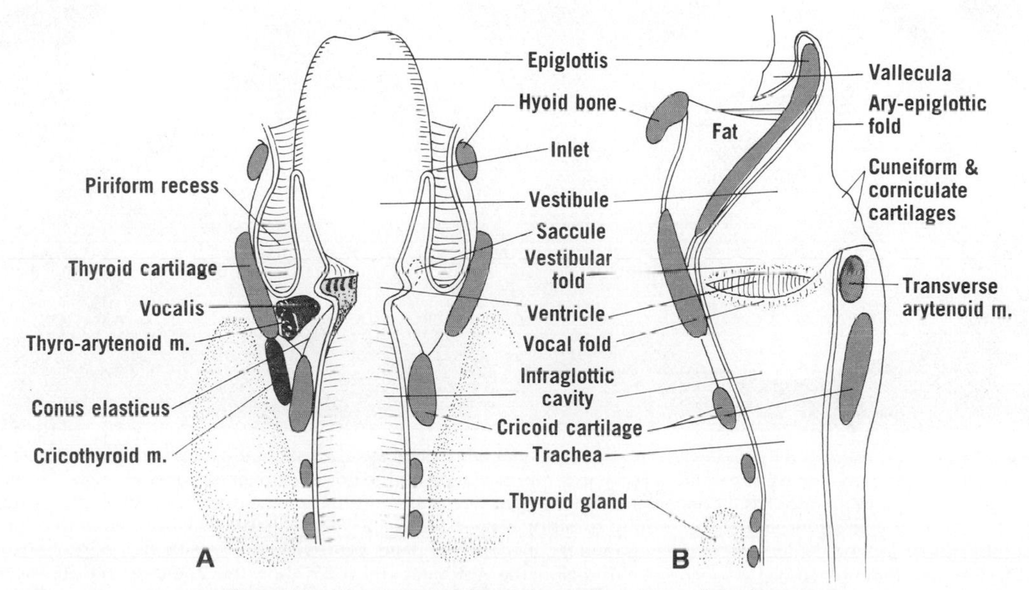

Figure 53–10 **A,** Coronal and, **B,** median views of the larynx.

laryngeal nerve and the superior laryngeal vessels.

The *cricothyroid ligament* (see fig. 53–8) connects the arch of the cricoid cartilage with the thyroid cartilage. The term *conus elasticus* (fig. 53–10*A*) is used for elastic fibers that extend upward from the cricoid cartilage to the vocal ligaments (*cricovocal membrane*). **In acute respiratory obstruction, cricothyrotomy is preferable to tracheotomy for the non-surgeon.**

The *vocal ligament* on each side extends from the thyroid cartilage in front to the vocal process of the arytenoid cartilage behind. It may be regarded as the upper border of the conus elasticus. Composed of elastic fibers, it is covered tightly by the vocal fold of mucous membrane (fig. 53–10*B*). The *vestibular ligament* on each side is an indefinite band situated above the vocal ligament and covered loosely by the vestibular fold.

The epiglottis is attached by ligaments to

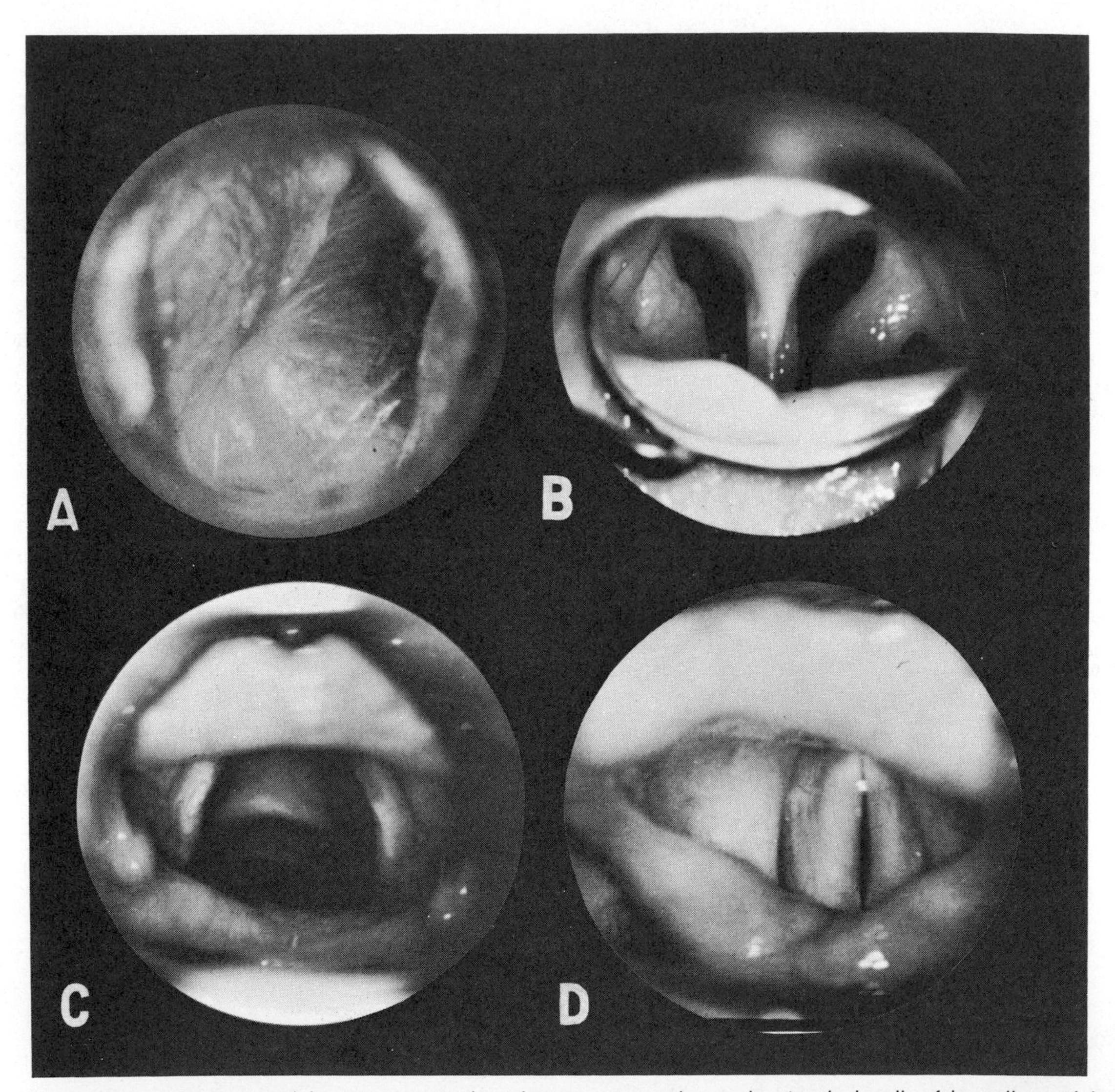

Figure 53–11 Ear, nose, and throat *in vivo*. **A,** The right tympanic membrane, showing the handle of the malleus. Cf. fig. 44–2. **B,** The nasopharynx and nasal cavities as seen in a mirror placed on the posterior pharyngeal wall. Note the posterior edge of the nasal septum, inferior nasal concha, and (on the right side of the illustration) the opening of the auditory tube. Cf. fig. 52–4. **C,** The larynx on inspiration, as seen in a mirror placed on the posterior pharyngeal wall. Note the epiglottis, ary-epiglottic folds, and (on the left side of the illustration) cuneiform cartilage, vestibular and vocal folds, and trachea. Cf. fig. 53–2. **D,** The larynx on phonation, as seen in a mirror. Note the vestibular and vocal folds: the latter are now approximated. (All photographs courtesy of Paul H. Holinger, M.D., Chicago, Illinois.)

the hyoid bone, the back of the tongue, the side of the pharynx, and the thyroid cartilage.

Inlet

The inlet, or aditus, of the larynx leads from the laryngopharynx into the cavity of the larynx. It is set obliquely, facing largely backward. It is bounded anteriorly by the upper border of the epiglottis, on each side by the ary-epiglottic folds, and below and behind by an interarytenoid fold (fig. 53–11*C*). The inlet is related laterally to the piriform recess of the laryngopharynx (see fig. 53–3). The ary-epiglottic folds provide lateral food channels that lead down the sides of the epiglottis, through the piriform recesses, and to the esophagus (fig. 53–12). Closure of the inlet protects the respiratory passages against the invasion of food and foreign bodies.

Cavity

The cavity of the larynx is divided into three portions—the vestibule, the ventricles and the area between them, and the infraglottic cavity—by two pairs of horizontal folds—the vestibular and the vocal folds (see fig. 53–10).

(1) The *vestibule* extends from the inlet to the vestibular folds.

(2) The *ventricle* extends laterally in the interval between the vestibular and vocal folds. Each ventricle resembles a canoe laid on its side, and the two communicate with each other through the median portion of the laryngeal cavity. A small diverticulum, the *saccule,* which extends upward from the front of each ventricle, possesses mixed glands and has been termed the "oil can" of the vocal folds.

The *vestibular folds* (see fig. 53–10*A* and *B*), or "false vocal cords," contain the vestibular ligaments and are protective rather than vocal in function. **The vocal folds, or "true vocal cords," which contain the vocal ligaments, are musculomembranous shelves below and medial to the vestibular folds. They extend from the angle of the thyroid cartilage in front to the vocal processes of the arytenoid cartilages behind. The bulk of each vocal fold is formed by the vocalis, which is a part of the thyro-arytenoid muscle. The vocal folds and processes, together with the interval (rima glottidis) between them, are collectively termed the glottis. The rima glottidis is the narrowest part of the laryngeal cavity** and can be seen between the more separated vestibular folds on laryngoscopy (see fig. 53–11*D*). The mucous membrane over each vocal ligament presents nonkeratinizing, stratified squamous epithelium, is firmly bound down, and appears white. The vocal folds control the stream of air passing through the rima and hence are important in

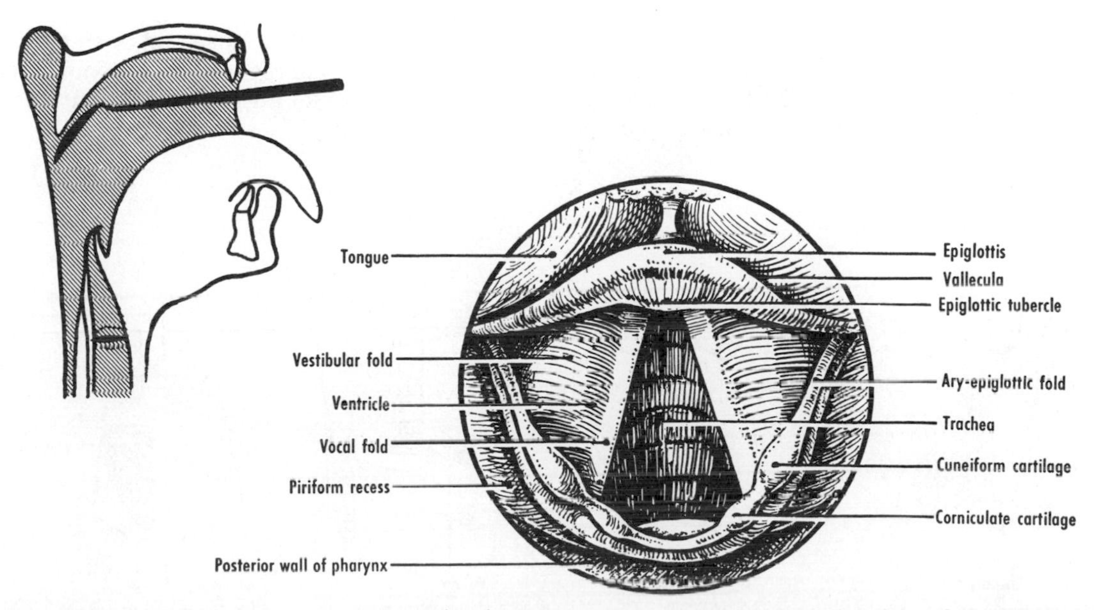

Figure 53–2 Indirect laryngoscopy. **A** shows the placement of the mirror in the pharynx. **B** shows the structures seen during respiration. The upper part of the trachea can be seen through the open glottis. Cf. fig. 53–11C.

voice production. The anterior, intermembranous part of the rima lies between the vocal folds, whereas the posterior, intercartilaginous part is situated between the arytenoid cartilages (see fig. 53–14). The shape and size of the rima are altered by movements of the arytenoid cartilages. The rima is wider during inspiration and quiet breathing and narrower during expiration and phonation. **In surface anatomy, the rima glottidis is approximately on the level of the midpoint of the anterior margin of the thyroid cartilage.**

(3) The *infraglottic cavity* extends from the rima glottidis to the trachea.

Closure

Three levels or tiers in the larynx can be closed by sphincteric muscles: (1) the inlet, which is closed during deglutition and protects the respiratory passages against the invasion of food; (2) the vestibular folds, closure of which traps the air below and makes possible an increase of intrathoracic pressure (as in coughing) or intra-abdominal pressure (as in micturition and defecation); and (3) the vocal folds, which are approximated in phonation. The presence of a foreign body is the commonest cause of laryngeal spasm, which usually involves not only the glottis but all of the sphincteric musculature of the larynx.

Mucous Membrane

The mucosa of the larynx, which is continuous with that of the laryngopharynx and trachea, is loose except over the back of the epiglottis and over the vocal ligaments. Hence it may become raised abnormally by submucous fluid, as in edema of the larynx. **The edema does not spread below the level of the vocal folds, being limited by the tight attachment of the mucosa to the vocal ligaments.**

Sensory Innervation and Blood Supply

The mucosa of the larynx is supplied on each side chiefly by the internal laryngeal branch of the superior laryngeal nerve, which supplies the larynx as far down as the vocal folds. The lower part of the larynx receives sensory fibers from the recurrent laryngeal nerve.

The larynx is supplied by (1) the superior laryngeal artery (from the superior thyroid), which accompanies the internal laryngeal nerve, and (2) the inferior laryngeal artery (from the inferior thyroid), which accompanies the recurrent laryngeal nerve.

Muscles of Larynx

The larynx as a whole can be elevated and depressed by extrinsic muscles (e.g., the stylopharyngeus and palatopharyngeus and the infrahyoid muscles).

The intrinsic laryngeal muscles are complicated, but they may be classified as follows:

1. **The sphincters of the inlet: transverse arytenoid; oblique arytenoid and ary-epiglottic.**
2. **The muscles that close and open the rima glottidis: lateral crico-arytenoid (adductor) and posterior crico-arytenoid (abductor).**

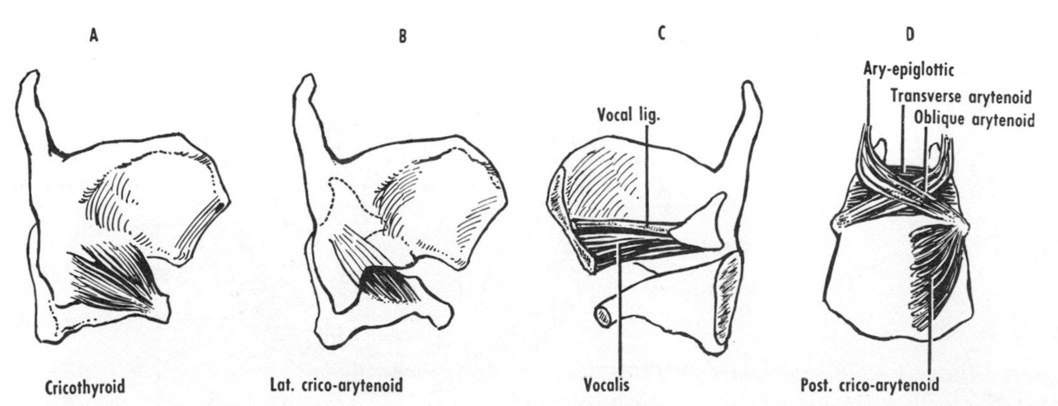

Figure 53–13 Intrinsic muscles of the larynx. **A** and **B,** Right lateral aspect of the thyroid and cricoid cartilages. **C,** Medial aspect of the right half of the thyroid and cricoid cartilages. **D,** Posterior aspect of the arytenoid and cricoid cartilages.

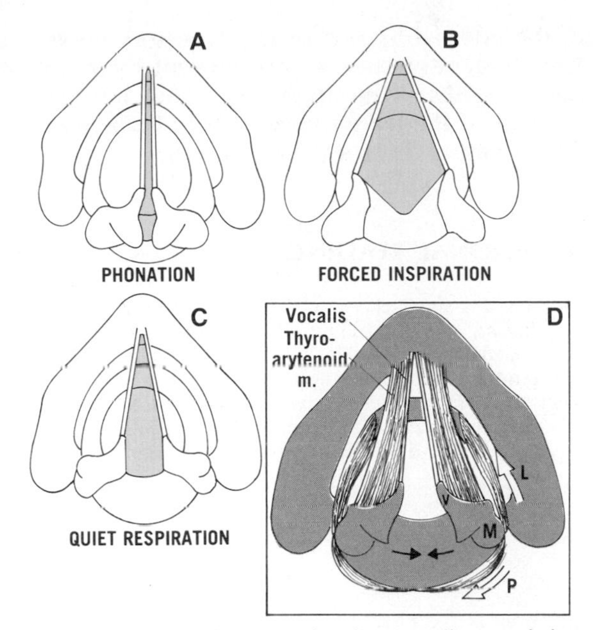

Figure 53–14 The rima glottidis (*in yellow*) and the vocal ligaments in **(A)** phonation, **(B)** forced inspiration, and **(C)** quiet respiration. Note the rotation and lateral sliding of the arytenoid cartilages and the different shapes of the glottis. **D,** Muscles of the larynx seen from above. The white arrows *L* and *P* show the direction of action of the lateral and posterior crico-arytenoid muscles, respectively. The black arrows show the direction of action of the transverse arytenoid muscle. *M*, muscular process of the arytenoid cartilage; *V*, vocal process of the arytenoid cartilage. It should be noticed that the apex of the A formed by the vocal ligaments is located anteriorly.

3. The muscles that regulate the vocal ligaments: thyro-artenoid and vocalis; cricothyroid.

The muscles of the larynx are illustrated in figures 53–13 and 53–14*D* and summarized in table 53–2.

Three muscles arise from the cricoid cartilage: the cricothyroid, passing backward to the lamina and inferior horn of the thyroid cartilage; the lateral crico-arytenoid, extending backward to the muscular process; and the posterior crico-arytenoid, extending laterally to the muscular process (fig. 53–13). Two muscles, closely related to each other, connect the thyroid and arytenoid cartilages: the thyro-arytenoid and the vocalis (fig. 52–14*D*). Two muscles unite the arytenoid cartilages: the transverse and oblique arytenoids (fig. 53–13*D*).

The muscles are concerned with widening the chink of the glottis (abduction), as in respiration, or closing it (adduction), as in phonation (fig. 53–14*A*, *B* and *C*). After closure of the glottis, the vocal folds can be tightened and lengthened by the cricothyroid muscles. Abduction is carried out solely by the posterior crico-arytenoid muscles, which, extending laterally from the back of the cricoid cartilage to the muscular processes, rotate the arytenoid cartilages laterally (fig. 53–14*B* and

TABLE 53–2 Muscles of Larynx

Muscle	Origin	Insertion	Innervation	Action
Cricothyroid	Arch of cricoid cartilage laterally	Lamina & anterior horn of thyroid cartilage	External laryngeal	Lengthens, tenses, & adducts vocal fold
Posterior crico-arytenoid	Lamina of cricoid cartilage posteriorly	Muscular process of arytenoid cartilage	Recurrent laryngeal	Abducts vocal fold
Lateral crico-arytenoid	Arch of cricoid cartilage	Muscular process of arytenoid cartilage	Recurrent laryngeal	Adduct vocal fold
Transverse & oblique arytenoid	One arytenoid cartilage	Opposite arytenoid cartilage	Recurrent laryngeal	Adduct vocal fold
Ary-epiglottic	Continuation of oblique arytenoid muscle into ary-epiglottic fold		Recurrent laryngeal	Adducts ary-epiglottic fold
Thyro-arytenoid	Lamina of thyroid cartilage medially	Muscular process of arytenoid cartilage	Recurrent laryngeal	Probably relaxes vocal fold
Vocalis	Angle between laminae of thyroid cartilage	Vocal process of arytenoid cartilage	Recurrent laryngeal	Alters vocal fold in phonation

C). Adduction is carried out by the lateral crico-arytenoid muscles, which, extending backward from the arch of the cricoid cartilage to the muscular processes, rotate the arytenoid cartilages medially (fig. 53–14*D*).

MOTOR INNERVATION

All the intrinsic muscles, with the exception of the cricothyroid, are supplied by the recurrent laryngeal nerve from the vagus. The cricothyroid is supplied by the external laryngeal branch of the superior laryngeal nerve from the vagus. The fibers of the various laryngeal muscles are believed to reach the vagus by way of the internal branch of the accessory nerve. Unilateral severence of a recurrent laryngeal nerve results in paralysis of all the intrinsic muscles of the larynx except the cricothyroid, which moves the vocal fold mediad.

ADDITIONAL READING

Jackson, C., and Jackson, C. L., *Diseases of the Nose, Throat, and Ear,* 2nd ed., W. B. Saunders Company, Philadelphia, 1959. This classic text contains an interesting chapter on laryngeal paralyses.

Tucker, G. F., *Human Larynx. Coronal Section Atlas,* Armed Forces Institute of Pathology, Washington, D.C., 1971. Black-and-white photomicrographs with labels.

QUESTIONS

53–1 In front of which vertebrae is the pharynx situated?

53–2 What is the nasopharynx?

53–3 What are the boundaries of the pharyngeal isthmus?

53–4 What are adenoids?

53–5 What is the pharyngeal hypophysis?

53–6 Where are the openings of the auditory tube?

53–7 What are the boundaries of the faucial isthmus?

53–8 List the components of the pharyngeal lymphatic ring.

53–9 Where is the tonsil?

53–10 What is the piriform recess?

53–11 Where does a pharyngeal diverticulum usually form?

53–12 What is the motor innervation of the pharynx?

53–13 How may the interior of the larynx be viewed *in vivo?*

53–14 At which vertebral level does the larynx end?

53–15 Is the hyoid bone a part of the larynx?

53–16 Are the arytenoid cartilages fixed or mobile?

53–17 Where are the corniculate and cuneiform cartilages?

53–18 How may the larynx be entered in acute respiratory obstruction?

53–19 Why have the vestibular folds been termed "false vocal cords"?

53–20 Which is the narrowest part of the laryngeal cavity?

53–21 What is the commonest cause of laryngeal spasm?

53–22 Why does laryngeal edema not extend below the glottis?

53–23 What are the afferent fibers involved in the cough reflex?

53–24 What are the results of injury (e.g., during thyroid surgery) to a recurrent laryngeal nerve?

ANSWERS

The text provides a relatively brief description of the structure of the body *a capite ad calcem,* "from head (L., *caput*) to foot (L., *calx, calcis,* heel)." Answers to the questions posed at the end of each chapter are provided below. Although no attempt at completeness has been made, these answers can be used independently for review and for seeking some further details.

PART 1 GENERAL ANATOMY

1 INTRODUCTION

1–1 A sagittal plane, especially the median plane, could include the entire length of the vertebral column. The column is too curved to be seen in its entirety in a coronal section.

1–2 Horizontal and coronal planes could pass through both shoulder joints.

1–3 Planes transverse to the little finger or neck would be horizontal. A plane transverse to the big toe would be coronal.

1–4 In flexion, abduction, and lateral rotation, the thigh is moved forward and laterally and is rotated around a longitudinal axis so that its front surface moves laterally. To refer to this as "external" rotation is incorrect usage. It should be noted that, in the absence of fractures and dislocations, the position of the foot should indicate whether the extended limb is rotated (laterally or medially) at the hip.

1–5 The *Fabrica* of Vesalius, published in Basel in 1543, is one of the most important scientific treatises ever published.

1–6 The circulation of the blood was not appreciated until nearly a century after Vesalius, when it was discovered by Harvey (1628).

2 THE LOCOMOTOR SYSTEM

2–1 Cartilage and membrane bones are similar in the adult.

2–2 Red marrow in the adult is found chiefly in the ribs, vertebrae, sternum, hip bones, and upper ends of the humeri and femora. The usual sites for bone marrow biopsy are the sternum and ilium.

2–3 The hand, for which special radiographic atlases are available for reference as to norms, is commonly used in assessing skeletal maturation.

2–4 The epiphysial center for the distal end of the femur is almost always visible at birth; that for the proximal end is frequently to be seen at full term. This can be important medicolegally when an infant's body is found.

2–5 The articular surfaces of the limb bones remain cartilaginous.

2–6 Shortening of the bones of the limbs, as seen, for example, in "circus" dwarfs (*achondroplasia*), would be expected from premature closure of epiphysial plates.

2–7 (a) Most carpal and tarsal joints are plane. (b) The elbow is a hinge. (c) The proximal radio-ulnar joint is a pivot. (d) The wrist is ellipsoidal. (e) The carpometacarpal joint of the thumb is a saddle joint. (f) The knee is condylar. (g) The shoulder and hip are ball-and-socket joints.

2–8 Synovial fluid arises in synovial membrane and functions chiefly in the lubrication and nourishment of articular cartilage.

2–9 Where an epiphysial line is intra-articular, as in the hip joint, infection can spread from the epiphysis (e.g., head of the femur) into the joint, and disease in the diaphysis can reach the joint before being stopped by the epiphysial plate. See figure 12–18.

2–10 Pennate muscles have relatively more fibers and hence are more powerful, because the power of a muscle is directly proportional to the number of its constituent fibers.

2–11 The total number of bones and muscles in the body is a rather meaningless question. The number of bones, sometimes given as 206 or 208, depends on what is included (e.g., auditory ossicles may be included but sesamoid bones may be excluded) and on age (the number of skeletal pieces is much greater in children and becomes reduced by fusion in later life). Similarly, the number of skeletal muscles, given variously as 346 to 501, depends on what is included (e.g., muscular strips that are not fully separated). Moreover, variations (absent and supernumerary structures) are common in the musculoskeletal system.

3 THE NERVOUS SYSTEM

3–1 *Encephalon* is from the Greek *en* (in) and *kephale* (head); hence, it means the main structure (brain) within the head. Prefixes indicate subdivisions. Thus, the addition of *tele* (far; cf. telephone, sound at a distance) gives *telencephalon,* or endbrain.

3–2 The diencephalon may conveniently be included in the brain stem, although some prefer to restrict the stem to the midbrain, pons, and medulla. See figure 3–1.

3–3 The spinal nerves are subdivided as follows: 8 cervical, 12 thoracic, 5 lumbar, 5 sacral, and 1 coccygeal. The arrangement of the vertebrae is similar except above (7 cervical) and below (4 coccygeal).

3–4 Ventral roots of spinal nerves are efferent (Bell, 1811); dorsal roots are afferent (Magendie, 1822). This is known as the Bell-Magendie law.

3–5 Sensory fibers in spinal nerves are arranged (1) as dermatomes, i.e., according to the (numbered) spinal nerves, e.g., Tl; or (2) according to the named peripheral nerves, e.g., ulnar nerve. See figure 3–3.

3–6 A dermatome is the area of skin supplied by the sensory fibers of a single dorsal root.

3–7 In *herpes zoster* (shingles), the blisters that follow the inflammation of roots and spinal ganglia may outline a dermatome.

3–8 The cranial nerves are so called because all 12 enter and (except for No. 8) leave the skull. The synonym *cerebral nerves* was popular at one time. The present numbering of the nerves (von Soemmering, 1778) is rather arbitrary but firmly established.

3–9 The cranial nerves (as compared with spinal nerves) are associated with the brain, are not formed of dorsal and ventral roots, and differ considerably among themselves: e.g., some have more than one ganglion, whereas others have none; and some are mixed (5, 7, 9, 10), whereas others are (almost) purely afferent (1, 2, 8) or (almost) purely efferent (3, 4, 6, 11, 12).

3–10 Classically, the autonomic nervous system (a term introduced by Langley in 1898) comprises the motor fibers that supply cardiac muscle, smooth muscle, and glands. Now known to be more complex and to involve cerebral as well as spinal levels, the autonomic nervous system is usually considered to include those visceral afferents and centers that are connected with the efferent pathways.

4 THE SKIN, HAIR, AND NAILS

4–1 The two layers of the skin differ structurally and developmentally. The epidermis is a stratified epithelium of ectodermal origin. The dermis is a connective tissue largely of mesodermal origin.

4–2 A hypodermic injection is given into the subcutaneous tissue, formerly called "superficial fascia."

4–3 The deepest layer of the epidermis, stratum basale, is normally its germinative layer.

4–4 In Dupuytren's classification (1832) a second-degree burn affects part of the epidermis, resulting in vesiculation (blisters) in addition to redness (erythema). See figure 4–1.

4–5 Water leaves the skin not only through the glands as perspiration but also by diffusion through the epidermis (insensible perspiration).

4–6 The arrectores pilorum are bundles of smooth muscle associated with hair follicles. On exposure to cold or under emotion (sympathetic nerves), they elevate the hair in a lever-like manner, and the nearby skin is depressed, whereas the surrounding skin is elevated ("goose skin").

4–7 Sebum (from sebaceous glands) keeps the stratum corneum pliable and conserves body heat by hindering evaporation. Seborrhea involves an excessive secretion of sebum, which may collect on the surface as scales (dandruff).

4–8 Different types of nerve endings of the skin can be related to basic kinds of sensation in only a general way. Correlations between type of sensation and a specific type of nerve ending are not justified.

5 RADIOLOGICAL ANATOMY

5–1 X-rays were discovered by Röntgen in 1895. Many accounts of this fascinating discovery are available, e.g., W. R. Nitske, *The Life of Wilhelm Conrad Röntgen,* University of Arizona Press, Tucson, 1971.

5–2 For a left anterior oblique (L.A.O.) view of the chest, the film would be placed obliquely on the left and front side of the chest (see fig. 23–21).

5–3 In radiology, an epiphysial line is the uncalcified portion of an epiphysial disc. The radiological joint space is the interval between the radio-opaque epiphysial regions of two bones, and it consists almost entirely of the adjacent layers of articular cartilage. It is important to appreciate that when a bone is said to be not yet present in radiology, it means that its cartilaginous or membranous precursor is not yet radio-opaque.

PART 2 THE UPPER LIMB

6 THE BONES OF THE UPPER LIMB

6–1 The clavicle is the first bone to ossify, specifically during the seventh embryonic week. It is followed closely by the mandible and maxilla.

6–2 The clavicle is likely to fracture at the junction of its medial two thirds and lateral third, i.e., where its two curves meet.

6–3 The greater tubercle of the humerus is the most lateral bony point of the shoulder.

6–4 Depending on the level, the axillary, radial, and ulnar nerves, all of which make direct contact with the humerus, are liable to injury in fractures.

6–5 As seen posteriorly, the epicondyles and the olecranon are in a straight line when the forearm is extended, but they form an equilateral triangle when the forearm is flexed.

6–6 To diagnose fractures and verify their correct reduction, it is important to appreciate that the styloid process of the radius ends more distally than that of the ulna.

6–7 Fracture of the lower end of the radius was described by Abraham Colles in 1814. It is caused by a fall on the palm of the outstretched hand. The lower end of the radius is displaced dorsally (''dinner-fork deformity'') and driven into the shaft so that the styloid processes are approximately at the same level.

6–8 The scaphoid bone may be fractured across its ''waist'' following a fall on the outstretched hand. Tenderness is marked over the scaphoid in the anatomical snuff-box.

6–9 The lunate may be dislocated anteriorly.

6–10 Ossification is sometimes found in the carpus at birth. More usually, however, the capitate and hamate do not begin to ossify until during the first postnatal year.

7 VESSELS, LYMPHATIC DRAINAGE, AND THE BREAST

7–1 Bloodletting was performed from the median cubital (or a nearby) vein, which is also the most frequent site for an intravenous injection. *Phlebotomy,* or *venesection,* is mentioned in Egyptian papyri and in the Hippocratic writings. The bleeding and bandaging of the barber-surgeons were represented by a red and white pole. See O. H. Wangensteen and S. D. Wangensteen, *The Rise of Surgery,* University of Minnesota Press, Minneapolis, 1978, Chapter 14.

7–2 A cardiac catheter can be passed through the median cubital (or a nearby) vein and then through the basilic, axillary, subclavian, and brachiocephalic veins and the superior vena cava to the right atrium.

7–3 On opening and closing the fist (contraction of the muscles of the forearm), the superficial veins are filled with blood. The muscular bellies push the blood proximally and, when a tourniquet is applied to the (upper) arm, the distended veins and their valves become prominent.

7–4 The mammary gland is situated in the subcutaneous tissue, into which the gland extends developmentally from the overlying ectoderm.

7–5 The suspensory ligaments are fibrous processes that subdivide the fat and anchor the glandular tissue to the skin.

7–6 The ''axillary tail'' is an extension of the mammary gland into the axilla, where it may be palpable (and even visible) and mistaken for enlarged lymph nodes or an abnormal swelling.

7–7 Most of the lymphatics of the breast drain into the axillary nodes. Hence the importance of examining the axilla.

7–8 Lymphatics from the breast penetrate the pectoral muscles; hence these muscles, as well as the axillary nodes, are removed in radical mastectomy.

8 THE SHOULDER AND AXILLA

8–1 The pectoralis major is involved in throwing and pushing, the latissimus dorsi in climbing and hammering, and the deltoid muscle in drawing a line across a blackboard.

8–2 Normally, the serratus anterior keeps the scapula close to the thoracic wall in pushing. When the muscle is paralyzed, e.g., by injury to the relatively exposed long thoracic nerve, the medial border and inferior angle of the scapula move laterally and dorsally (winging).

8–3 Paralysis of the trapezius results in inability to shrug the shoulder and to raise the arm above the horizontal.

8–4 The deltoid gives roundness to the shoulder. When the muscle is atrophied, the shoulder appears "squared."

8–5 The tendons of the supraspinatus, infraspinatus, teres minor, and subscapularis form the musculotendinous cuff. Degeneration or calcified deposits may cause pain on abduction (*painful arc syndrome*).

8–6 The lower trunk of the brachial plexus lies on the first rib, behind the subclavian artery (see figs. 8–7 and 20–5). In rare instances, a variable cervical rib (i.e., a rib associated with C.V.7) is found either unilaterally or bilaterally. Angulation of the subclavian artery, brachial plexus, or both over a cervical rib is one example of the "neurovascular compression syndrome" of the upper limb. Features may include pain on movement of the arm and numbness of the fingers. It is of interest to note that a mesenchymal or cartilaginous cervical rib is normally present bilaterally *in fetu*.

8–7 In "upper type" injuries to the brachial plexus, the upper limb is in medial rotation because the subscapularis (medial rotator) is unopposed by the paralyzed lateral rotators (infraspinatus and teres minor, supplied by C.N.5,6).

8–8 "Claw hand" may follow a "lower type" injury to the brachial plexus because the flexors and extensor are unopposed by the paralyzed interossei and lumbricals (supplied by the ulnar nerve). See figure 11–8.

8–9 In deltoid paralysis, abduction of the arm by the supraspinatus alone is usually incomplete, although defective abduction at the shoulder joint may be masked by lateral rotation of the scapula caused by the trapezius. Complete division of the axillary nerve also results in some loss of sensation over the deltoid muscle. See figure 8–13.

8–10 The subcoracoid and subscapular bursae usually communicate with the cavity of the shoulder joint.

8–11 The muscles of the musculotendinous cuff, which hold the head of the humerus in place, resist any tendency of the shoulder to become dislocated. When dislocation occurs, the head of the humerus escapes through the lower and front part of the capsule, where it is thinnest and weakest. The dislocation is primarily subglenoid and then usually becomes subcoracoid.

9 THE ARM AND ELBOW

9–1 The compartments of the arm are the anterior, for the flexors (musculocutaneous nerve), and the posterior, for the extensor (radial nerve). See figure 9–1.

9–2 The biceps is a flexor and powerful supinator of the forearm. It "puts a corkscrew in and pulls the cork out."

9–3 C.N.5 and 6 (biceps and brachialis) are involved in flexion, and chiefly C.N.6 and 7 (triceps) are concerned with extension at the elbow. See table 8–4.

9–4 The ulnar nerve is in contact with the back of the medial epicondyle ("funny bone"). When the nerve is under tension in ulnar neuritis, it is possible surgically to transpose the nerve to the front of the elbow.

9–5 Near the elbow, the radial nerve lies between the brachioradialis and the brachialis. The median nerve enters the forearm between the heads of the pronator teres; the ulnar nerve enters between the heads of the flexor carpi ulnaris.

9–6 The brachial artery and the median nerve lie medial to the biceps tendon in the cubital fossa. Occasionally the artery and nerve pass under cover of a hook-shaped spur of bone, the *supracondylar process,* which may project from the humerus above the medial epicondyle and is visible radiographically.

9–7 In children, a sudden jerk on the limb may cause the head of the radius to be subluxated through the annular ligament.

9–8 The word *cubital* comes from the Latin *cubitus,* elbow or forearm, on which the Romans rested in a position called "decubitus." A cubit, or ell, is the distance from the elbow to the fingertips. An L-shaped bow is an el-bow. Cf. ankle, the part at an angle.

10 THE FOREARM

10–1 Injury to the brachial artery, e.g., from a supracondylar fracture of the humerus, may cause damage to the deep flexors (pollicis longus and digitorum profundus) and flexion deformity of the wrist and fingers (*Volkmann's ischemic contracture*).

10–2 The compartments of the forearm are the anterior, for the flexors (median and ulnar nerves), and the posterior, for the extensors (radial and posterior interosseous nerves). See figure 10–1.

10–3 The superficial extensors arise from the lateral epicondyle of the humerus, which may become affected (*epicondylitis*) at the attachment (*tendinitis*) in "tennis elbow."

10–4 Clinically, a ganglion is a cystic swelling near, and often communicating with, a synovial sheath or joint. The dorsum of the wrist is a common site. Anatomically, the term *ganglion* (Gk, knot) is used for a collection of neurons located outside the central nervous system.

10–5 The scaphoid bone and trapezium form the floor of the snuff-box. Tenderness here is found in fractures of the scaphoid.

10–6 Wrist drop, i.e., flexion at the wrist when the pronated forearm is flexed at the elbow, is seen in radial nerve palsy. See figure 8–14.

10–7 At the wrist, the median nerve is between the flexor digitorum superficialis medially and the flexor carpi radialis laterally. It may be partly covered by the inconstant palmaris longus, which, however, "is a mere decoy and has no value as a landmark" (Henry). The nerve is liable to injury in lacerations at the wrist.

10–8 The flexor pollicis longus, flexor digitorum profundus (partly), and the pronator quadratus are supplied by the anterior interosseous nerve. Injury to the median nerve may result in weakness of the two long flexors and a characteristic pinch attitude on attempted flexion of the distal phalanges of the thumb and index ("anterior interosseous nerve syndrome"). See K. K. Nakano, C. Lundergan, and M. M. Okihiro, Arch. Neurol., *34*:477, 1977; E. Wiens and S. C. K. Lau, Can. J. Surg., *21*:354, 1978.

10–9 Fibers to the small muscles of the hand may, in some people, be transferred from the median to the ulnar nerve, or vice versa, resulting in unusual signs in the event of injury. It is not uncommon for the ulnar nerve to encroach upon the territory of the median nerve.

10–10 The flexor carpi radialis tendon is the guide to the radial artery. The vessel is lateral to the tendon and is used to feel the pulse.

11 THE HAND

11–1 When the hand has to be immobilized, it should be placed in the "position of rest" (see fig. 11–1). When the hand is at rest, the wrist is extended slightly and the fingers are flexed, as when the inactive hand is resting on a horizontal surface. In the "functional position" that precedes activity, the wrist is extended much further, and the metacarpophalangeal joints are extended also.

11–2 The carpal canal or tunnel (see fig. 11–2*B*) contains the flexor tendons, their synovial sheaths, and the median nerve. Compression of the median nerve behind the flexor retinaculum (*carpal tunnel syndrome*) may occur, e.g., from inflammation of the tendon sheaths (*tenovaginitis*) but generally from unknown causes. Numbness may be noticed in the lateral 3 and a half fingers (see fig. 8–11), and there may be weakness of the thenar muscles. The condition is treated by dividing the flexor retinaculum.

11–3 *Dupuytren's contracture,* described by the famous French surgeon in 1832, is commonly attributed to thickening and shortening of the palmar aponeurosis, although the causation is obscure. The little and ring fingers are most affected. The aponeurosis can be excised.

11–4 The "midpalmar space" and the "thenar space" are clinical terms for the medial and lateral parts, respectively, of the central compartment of the palm; they are situated behind the flexor tendons and are regarded as important sites for the accumulation of pus. See figure 11–2*A*.

11–5 The "ulnar bursa" and the "radial bursa" are clinical terms for, respectively, the synovial sheath common to all the superficialis and profundus tendons and the sheath of the flexor pollicis longus. See figure 11–3.

11–6 The small muscles of the hand are supplied chiefly by T1. See Table 8–4.

11–7 Usually, the thenar triad (abductor pollicis brevis, flexor pollicis brevis, and opponens pollicis) is supplied by the median nerve (recurrent branch; see fig. 10–2). The adductor pollicis, together with the deep head of the flexor brevis and the first palmar interosseus, is supplied by the ulnar nerve. See figure 11–9.

11–8 Section of the median nerve at the wrist causes anesthesia (see fig. 8–11) and impaired movements of the thumb.

11–9 The interossei and lumbricals (see fig. 11–8*B*) produce the "Z- position" of the hand. In ulnar nerve palsy, paralysis of these muscles results in "claw hand" (*main en griffe;* see fig. 11–8*C*).

11–10 In ulnar nerve lesions, in addition to the interossei and lumbricals 3 and 4, the adductor pollicis is also paralyzed. In attempting to grasp a sheet of paper between thumb and index finger, the thumb is incapable of adequate adduction. This is masked by flexion at the interphalangeal joint of the thumb, which is accentuated when an attempt is made to pull the paper away ("Froment's sign").

11–11 Sensation distally along the lateral side of the index finger and the medial side of the little finger is mediated by the median (see fig. 8–11) and ulnar (see fig. 8–12) nerves, respectively. These areas are reliable because they are unlikely to be altered by variation or by overlapping of adjacent nerves.

11–12 A fracture of the base of the first metacarpal is serious because it generally involves the carpometacarpal joint of the thumb. This fracture-dislocation, produced by a blow to the tip of the thumb, was described in 1881 by Edward Hallaran Bennett on the basis of his anatomical studies some 14 years before the discovery of x-rays.

PART 3 THE LOWER LIMB

12 THE BONES OF THE LOWER LIMB

12–1 When a subject is in the anatomical position, the pubic tubercle and the anterior superior iliac spine are in a coronal plane and the articular surface of the pubis is sagittal.

12–2 When a subject is sitting, the body weight rests on the ischial tuberosites and not on the glutei.

12–3 The pubic tubercle is a guide to the superficial inguinal ring, femoral ring, and saphenous opening.

12–4 The acetabulum is formed developmentally by three bones. As compared with the glenoid cavity of the scapula, it is directed downward as well as laterally and forward, is deeper, and is not covered completely by articular cartilage (but only on a lunate surface). The hip joint is stabilized chiefly by its capsule and ligaments, the shoulder joint by the musculotendinous cuff.

12–5 The blood supply to the head of the femur (see fig. 15–15) may be interrupted after fracture of the neck of the bone, resulting in avascular necrosis and collapse of the head. The neck is partly intracapsular and partly extracapsular (see fig. 12–18), and fractures may be classified accordingly.

12–6 One kneels chiefly on the tuberosity of the tibia (separated from the skin by the subcutaneous infrapatellar bursa) and the ligamentum patellae.

12–7 By tracing downward the tendon of the adductor magnus, one arrives at the adductor tubercle of the femur. By tracing downward the tendon of the biceps femoris, one reaches the head of the fibula.

12–8 The lateral malleolus is more prominent and more posterior and extends more distally than does the medial.

12–9 Fracture of the lower end of the fibula is frequently combined with tearing of the deltoid ligament, medial malleolus, or both, with consequent instability of the tibiofibular mortice of the ankle joint. Fracture-dislocation of the ankle was described by Percivall Pott in 1765: he himself suffered from a tibial (but not a Pott's!) fracture on falling from his horse.

12–10 The calcaneus, talus, and frequently the cuboid (i.e., the three largest tarsals) begin to ossify before birth. It is rare to find ossification before birth in the lateral cuneiform or in the head of the femur, whereas an ossific center may be seen in the head of the humerus around the time of birth. The center for the distal end of the femur is almost always present at full term, and that for the proximal end of the tibia is frequently to be seen then also.

13 VESSELS AND LYMPHATIC DRAINAGE OF THE LOWER LIMB

13–1 The return of venous blood from the lower limbs to the heart requires a pump (the muscles) and non-return valves. Loss of valvular competence (varicose veins) is not uncommon in the great and small saphenous veins. A similar phenomenon may occur at other sites, e.g., the spermatic cord (*varicocele*), esophagus (*varices*), and anal canal (*hemorrhoids*).

13–2 Superficial veins at the sides of the abdomen become noticeably enlarged in obstruction of the inferior vena cava. Blood from the lower limbs (femoral veins) by-passes the inferior vena cava and travels by way of the abdominal and thoracic walls to tributaries of the axillary veins and thence to the superior vena cava.

13–3 When inguinal lymph nodes are found to be enlarged, their entire territory of drainage should be examined. This comprises not only the lower limb but also the trunk below the level of the umbilicus, including the perineum, external genitalia (note venereal disease), and anus.

13–4 Collateral circulation is that carried on through secondary vessels, which enlarge after obstruction of the main vessel that supplies a part. For example, if the external iliac artery is ligated, many vessels become enlarged, e.g., branches of the internal iliac artery and vessels of the anterior abdominal wall anastomosing with branches of the external iliac and femoral arteries. Such an operation was performed in 1808 by Sir Astley Cooper, who provided excellent drawings of the collateral circulation. See R. C. Brock, *The Life and Work of Astley Cooper,* Livingstone, Edinburgh, 1952, plate 3.

13–5 A catheter inserted in the thigh can be passed through the femoral artery, the external and common iliac arteries, and the aorta, then through the left subclavian, left common carotid, or brachiocephalic artery, and finally through the vertebral or internal carotid artery for cerebral angiography.

14 THE GLUTEAL REGION

14–1 An intramuscular injection is generally given into the (upper) arm (deltoid muscle) or buttock (gluteal muscles). See figure 14–1*B*.

14–2 The glutei medius and minimus, on the side of the grounded limb, abduct the pelvis, i.e., tilt it so that the swinging limb can clear the ground (see fig. 18–3). When the abducting mechanism of the hip is affected, the pelvis drops on the unsupported (contralateral) side (Trendelenburg's sign, 1895), resulting in a waddling gait.

14–3 The sciatic nerve is derived from L4 to S3 (see fig. 30–6). The sciatic nerve, which is related to the region of the hip (Gk, *ischion;* hence the word *sciatic*), is subject to neuralgia (*sciatica*). In a supine patient, passive raising of the lower limb (hip flexion) and passive dorsiflexion of the foot by the examiner, while the knee is kept fully extended, places the sciatic nerve (lumbosacral roots) under tension and hence will cause pain in such conditions as herniation of a lumbar intervertebral disc ("straight-leg-raising test").

14–4 The piriformis, below which the sciatic nerve emerges (see fig. 14–1*A*), may be pierced by the peroneal component of the sciatic nerve. The piriformis is the key to the relationships in the gluteal region.

15 THE THIGH AND KNEE

15–1 The compartments of the thigh are the anterior, for the quadriceps (femoral nerve); the medial, for the adductors (obturator nerve); and the posterior, for the hamstrings (sciatic nerve). See figure 15–2.

15–2 The iliotibial tract is an important thickening of the fascia lata laterally. It extends from the iliac crest to the lateral condyle of the tibia, it receives the insertions of the gluteus maximus and tensor fasciae latae, and it stabilizes the knee. The combination of tensor, tract, and gluteus maximus may be thought of as the deltoid of the thigh, i.e., comparable to the anterior, middle, and posterior fibers, respectively, in the upper limb.

15–3 The femoral ring and canal are the site of femoral hernia. The neck of a femoral hernia is found immediately inferolateral to the pubic tubercle.

15–4 The femoral triangle contains the femoral nerve and vessels; the adductor canal contains the femoral vessels and saphenous nerve. John Hunter, who proposed ligature of an artery well above an aneurysm, ligated the femoral artery in the adductor (Hunter's) canal in 1785. See J. Dobson, *John Hunter,* Livingstone, Edinburgh, 1969, p. 260.

15–5 All four parts of the quadriceps extend the knee. The muscle is involved in many movements, e.g., rising from a chair and walking. In quadriceps paralysis, a patient can stand erect, because the body weight tends to hyperextend the knee. Walking, however, is then performed in short steps that avoid extending the hip and flexing the knee, and such a patient may press the knee backward with the hand.

15–6 The center for the knee jerk is chiefly in the L3 segment of the spinal cord. The quadriceps is innervated by L.N.2 to 4 (see table 15–4).

15–7 The saphenous nerve, which may be regarded as the termination of the femoral nerve (L.N.2 to 4), generally supplies skin as far as the medial side of the foot.

15–8 The profunda femoris is the most important branch of the femoral artery. Sudden occlusion of the femoral artery above the origin of the profunda almost always causes gangrene; blockage below the profunda origin seldom does so.

15–9 The tendon of the biceps is the guide to the common peroneal nerve. Injury to the nerve, e.g., from a fracture or from a tight plaster cast, results characteristically in footdrop (loss of both dorsiflexion and eversion of the foot). In walking, the foot tends to slap the ground ("steppage" gait). The superficial peroneal nerve supplies skin on the lateral side of the leg and a variable amount of the dorsum of the foot. The deep peroneal nerve supplies the skin between the first and second toes dorsally.

15–10 The stability of the hip joint depends on the shape of the articular surfaces, the capsule, ligaments, and muscles. In dislocation of the hip, the head of the femur is usually displaced posteriorly. Congenital dislocation of the hip is an important but obscure condition.

15–11 Both hip and knee joints are supplied ultimately by branches of the femoral, sciatic, and obturator nerves. Hip disease is an important cause of pain referred to the knee.

15–12 In the knee joint (as compared with the elbow), the fibula is excluded (cf. the ulna), the soft tissues (rather than the bony structure) limit extension, and rotation occurs.

15–13 The knee joint is condylar rather than a hinge because the axis of flexion and extension shifts forward during extension, and extension is associated with medial rotation of the femur.

15–14 In full extension of the knee, the articular surfaces are in maximal contact, the ligaments are taut, and the joint is "locked" or "screwed home."

15–15 Usually the suprapatellar pouch and bursae related to the popliteus, gastrocnemius, and semimembranosus communicate with the cavity of the knee joint.

16 THE LEG

16–1 The compartments of the leg are the anterior, for the extensors (deep peroneal nerve); the lateral, for the peronei (superficial peroneal nerve); and the posterior, for the flexors (tibial nerve). See figure 16–1. Increased pressure within a compartment may require surgical incision (*fasciotomy*) for ischemia of the muscles and nerves of the leg.

16–2 The deep peroneal nerve accompanies the anterior tibial artery, and the tibial nerve accompanies the posterior tibial artery (see Figs. 15–10 and 16–5).

16–3 The tibialis anterior and tibialis posterior are the chief invertors; the peronei longus and brevis and the extensor digitorum longus are the chief evertors (see fig. 16–5).

16–4 The triceps surae comprises the gastrocnemius (two heads) and the underlying soleus. The triceps is inserted by the tendo calcaneus and is the chief plantar-flexor of the foot.

16–5 The center for the ankle jerk is generally in the S1 segment of the spinal cord. The triceps surae is supplied by S.N.1,2.

16–6 The popliteus is usually regarded as the muscle that "unlocks" the knee joint at the beginning of flexion of the fully extended knee. It is believed to pull the lateral meniscus backward at the beginning of flexion and so protect it from being pinched between the femur and tibia. This may perhaps help to explain the much higher frequency of tears of the medial meniscus.

16–7 Section of the tibial nerve causes sensory loss in the sole of the foot, which interferes with posture and locomotion. Motor loss varies with the level, but in time the small muscles of the foot atrophy.

17 THE ANKLE AND FOOT

17–1 The plantar reflex consists of plantar flexion of the toes (S1) when the skin of the sole is stroked slowly along its lateral border.

17–2 The Babinski sign, named after a French neurologist (1896), consists of dorsiflexion of the great toe on stroking the sole slowly along its lateral border. "There is perhaps no more important single physical sign in clinical neurology" (Walshe). It is found in lesions of the upper motor neuron. It may be accompanied by dorsiflexion of the foot and flexion at the knee and hip; i.e., it is part of a more extensive flexion response, or of withdrawal of the limb. Hence it is incorrect to refer to Babinski's sign as the "extensor plantar response."

17–3 The territory of the femoral nerve extends (through the saphenous nerve) to the medial side of the foot; the obturator may reach almost as far as the knee; the sciatic distribution (through the plantar nerves) proceeds to the toes. The sole is supplied by the tibial and medial and lateral plantar nerves (L4,5; S1,2).

17–4 The following is a mnemonic for the dermatomes of the hand and foot: L4,5 (foot); C6,7,8 (hand); S1,2 (foot). See figures 8–10 and 15–11.

17–5 A pulse at the ankle and foot is sought chiefly in the posterior tibial artery (see fig. 17–4*A*) and the dorsalis pedis (see fig. 17–4*C*).

17–6 Between the medial malleolus and the heel are the tibialis posterior and flexor digitorum longus (which occupy a groove on the medial malleolus and lie on the medial, or deltoid, ligament of the ankle); the posterior tibial artery and tibial nerve; and the flexor hallucis longus (see figs. 16–5 and 17–4*A*). These structures lie under cover of the flexor retinaculum.

17–7 The following cross the ankle joint from lateral to medial: the extensor digitorum longus, anterior tibial artery, deep peroneal nerve, and extensor hallucis longus (see fig. 17–4*C*). These structures are anchored by the extensor retinacula.

17–8 The peroneus longus and, in front of it, the peroneus brevis are situated behind the lateral malleolus (see fig. 17–4*B*). Both tendons wind around the lateral malleolus and lie on the lateral ligament (calcaneofibular part) under cover of the peroneal retinacula.

17–9 The deltoid ligament is the medial ligament of the ankle joint. It extends between the medial malleolus and the navicular, calcaneus (sustentaculum tali), and talus (see fig. 17–5).

17–10 The lateral ligament connects the lateral malleolus with the talus in front and behind and with the calcaneus in between (see fig. 17–5). Although a sprain is generally a minor ligamentous injury, forcible inversion may cause rupture of the lateral ligament without a fracture. The ankle may seem to be normal unless the foot is inverted by the examiner, when the talus becomes tilted out of the tibiofibular mortice. Immobilization is then indicated to avoid recurrent dislocation of the ankle joint.

17–11 The subtalar and transverse tarsal (talocalcaneonavicular and calcaneocuboid) are the most important intertarsal joints (see fig. 17–7). The chief movements are (a) inversion and (b) eversion (see fig. 17–6), carried out by (a) the tibialis anterior and posterior and (b) the peronei longus and brevis and the extensor digitorum longus (see fig. 16–5).

17–12 The most frequent type of clubfoot is talipes equinovarus, a deformity in which the foot is plantar flexed (*equinus,* resembling the stance of a horse), supinated, and adducted (*varus,* bent medially).

17–13 *Varus* implies medial deviation; *valgus* indicates lateral deviation. For example, at the elbow, a more acute (accentuated) carrying angle is termed *cubitus valgus,* a more obtuse angle is *cubitus varus.* Similarly, at the hip, a more acute angle of inclination of the femur is termed *coxa vara,* a more obtuse angle is *coxa valga.* At the knee, increased angulation between the femur and tibia is *genu valgum* (liable to produce "knock-knee"), decreased angulation is *genu varum* (liable to cause "bowlegs").

18 POSTURE AND LOCOMOTION

18–1 The line of gravity passes behind the hip joints but in front of the knee and ankle joints (see fig. 18–1).

PART 4 THE THORAX

19 THE SKELETON OF THE THORAX

19–1 The manubrium is approximately on the level of T.V.3 and 4. A horizontal section between these vertebrae would pass through the great vessels of the neck, the trachea, esophagus, and upper part of the lungs. A horizontal section through the lower end of the sternum (approximately T.V.10) would be expected to pass through the lower part of the heart and lungs but might easily include a portion of the right lobe of the liver. Thus pericardial, pleural, and peritoneal cavities might be opened in one section.

19–2 The subclavian vein anteriorly and, behind that, the subclavian artery and lower trunk of the brachial plexus lie on the upper surface of the first rib. If a cervical rib is present, angulation of the subclavian artery, brachial plexus, or both over such a rib may cause such features as pain on movement of the arm and numbness of the fingers. This is one example of the "neurovascular compression syndrome" of the upper limb.

19–3 Ribs are counted from above down, beginning with the second, which is generally (but not always) opposite the sternal angle. The first ribs are concealed by the clavicles.

20 THE THORACIC WALL AND MEDIASTINUM

20–1 In inspiration, the diaphragm contracts, descends, increases the volume and decreases the pressure within the thorax, and increases the abdominal pressure. The diaphragm is lower when a subject is in the sitting posture than when in the erect posture, and patients with difficulty in breathing are more comfortable when sitting.

20–2 The vertebrocostal trigone is a variable interval between the costal and lumbar parts of the diaphragm. The pleura may then be in close relationship to the suprarenal gland and kidney.

20–3 Congenital diaphragmatic herniae may occur through the esophageal opening (hiatal hernia), through a gap in the costal part (e.g., from a persistent pleuroperitoneal canal), or through the sternocostal triangle. The development of the diaphragm is complicated (L. J. Wells, Contrib. Embryol. Carnegie Instn., *35*:107, 1954).

20–4 The ventral rami of T.N.1 to 11 are intercostal, that of T.N.12 is subcostal, and those of T.N.7 to 12 are thoraco-abdominal.

20–5 Notching of ribs seen radiographically suggests coarctation of the aorta, i.e., a narrowing associated with an extensive collateral circulation. Enlarged intercostal arteries erode the ribs and show as notches.

20–6 The mediastinum is the interpleural interval. The word *mediastinum* means "middle-standing" (L., *in medio stans*), i.e., a septum or partition, and was formerly used for the structures (e.g., the heart) between the right and left pleurae.

21 THE ESOPHAGUS, TRACHEA, AND MAIN BRONCHI

21–1 The left atrium lies in front of the esophagus in the lower part of the thorax. Dilatation of the left atrium can be detected radiographically because of compression of a barium-coated esophagus.

21–2 The trachea divides at the level of T.V.5 to 7 in a subject in the erect position *in vivo*. This would be approximately behind the middle of the body of the sternum.

21–3 The right main bronchus is shorter, wider, and more nearly vertical than the left, and it is more likely to receive foreign objects.

22 THE PLEURAE AND LUNGS

22–1 The serous membranes line the body cavities and are reflected over protruding organs as their serous coat. The serous membranes are the pleura, pericardium, and peritoneum. The tunica vaginalis testis is a (usually) detached extension of the peritoneum. Serous membranes consist of mesothelium and connective tissue, and they secrete a film of serous exudate.

22–2 Pneumothorax is the presence of air in the pleural cavity. Air may enter from the lung (e.g., from ruptured alveoli) or through the chest wall (e.g., from a perforating injury). The lung then collapses. Artificial (i.e., surgically induced) pneumothorax was formerly used to promote the collapse of tuberculous cavities in the lung.

22–3 The cupola is the cervical pleura over the apex of the lung.

22–4 The right lung is usually heavier, shorter, and wider than the left lung, and it generally has three rather than two lobes.

22–5 The lung and pleura are generally said to cross rib 6 in the midclavicular line and ribs 8 and 10, respectively, in the midaxillary line and then to proceed toward the spinous processes of T.V.10 and T.V.12, respectively.

22–6 A perforation of the lower intercostal spaces should be considered as an abdominal as well as a thoracic wound because the liver, stomach, spleen, colon, kidney, and peritoneal cavity extend to a higher level than the periphery of the diaphragm and the inferior border of the lung.

22–7 A bronchopulmonary segment is the portion of lung supplied by a third-order bronchus.

22–8 The hilar shadows seen on radiography are caused mainly by branches of the pulmonary arteries, although the chief bronchi can frequently be recognized.

22–9 The classic methods of physical examination are inspection, palpation, percussion, and auscultation. The stethoscope was invented for auscultation by Laënnec in 1816.

22–10 The anterior view (postero-anterior projection of the x-rays) is the most frequently employed view in radiography of the chest.

23 THE PERICARDIUM AND HEART

23–1 The transverse sinus of the pericardium is a passage between the venous end (left atrium and superior vena cava) of the heart behind and the arterial end (aorta and pulmonary trunk) in front (see fig. 23–2*A*).

23–2 The apex beat can usually be felt on the front of the left side of the chest, generally in intercostal space 4 or 5, about 6 or 7 cm from the median plane.

23–3 Blood flows from the atria to the ventricles almost horizontally forward (see fig. 23–3) and to the left, especially in the "right heart."

23–4 The sternocostal surface of the heart is formed mainly by the right ventricle. The chamber least visible from the front is the left atrium, of which only the left auricle can be seen.

23–5 In the fetus, the foramen ovale allows relatively oxygenated blood to pass from the right to left atrium (see fig. 23–13). This essential feature of the fetal circulation enables oxygenated blood (from the placenta via the umbilical vein and inferior vena cava) to reach the "left heart" and thence the aorta.

23–6 Hepatocavo-atrial concordance is the agreement as to the side (usually the left) on which are found the suprahepatic part of the inferior vena cava and the morphological right atrium. These relationships are important in the diagnosis of visceral situs and of positional anomalies of the heart. See R. M. Shaher, J. W. Duckworth, G. H. Khoury, and C. A. F. Moes, Am. Heart J., *73*:32, 1967; M. V. de la Cruz and B. Nadal-Girard, Am. Heart J., *84*:19, 1972; and G. Caruso and A. E. Becker, Br. Heart J., *41*:559, 1979.

23–7 The conus arteriosus (or infundibulum) is the outflowing part of the right ventricle. The left ventricle ends in the aortic vestibule.

23–8 The atrioventricular bundle is situated at the lower border of the membranous part of the interventricular septum. The bundle divides, and its limbs straddle the muscular part of the septum.

23–9 In dominance of the left coronary artery, the posterior as well as the anterior interventricular branch arises from the left vessel.

23–10 In a left anterior oblique (L.A.O.) view, the heart appears more symmetrical, the curve of the aorta is "opened," and the x-rays are frequently in the plane of the interventricular septum (see figs. 23–18 and 23–21).

24 BLOOD VESSELS, LYMPHATIC DRAINAGE, AND NERVES OF THE THORAX

24–1 The pulmonary trunk, ascending aorta, and superior vena cava lie in that order from left to right. The right pulmonary artery has to cross behind the aorta and vena cava (see fig. 23–18) to reach the right lung. The right pulmonary artery, therefore, is longer than the left; it is also wider. In contrast, the left bronchus has to cross in front of the esophagus and descending aorta (see fig. 24–1) to reach the left lung, and so it is longer than the right; it is also narrower.

24–2 The ligamentum arteriosum is a fibrous band connecting the left pulmonary artery to the arch of the aorta. It is the remains of a prenatal vessel, the *ductus arteriosus* (see fig. 23–13), which shunts most of the relatively deoxygenated blood from the pulmonary trunk to the aorta (and thence to the placenta for oxygenation).

24–3 The left recurrent laryngeal nerve (from the vagus) winds below the arch of the aorta immediately behind the ligamentum arteriosum (see fig. 24–1). In contrast, the right recurrent laryngeal nerve winds around the subclavian artery. This is attributed to the embryonic position of the nerves, namely caudal to aortic arch 6, the distal portion of which disappears on the right side (allowing the nerve to wind around aortic arch 4, which contributes to the subclavian artery) and forms the ductus arteriosus on the left (allowing the nerve to wind around the ligamentum arteriosum postnatally at its attachment to the arch of the aorta).

24–4 A retro-esophageal right subclavian artery (see fig. 24–1*B*) arises from the descending aorta instead of, as in the vast majority, from the brachiocephalic trunk. It is frequently stated to cause dysphagia. It is attributed to a variation in the development of the aortic complex (persistence of the right dorsal aortic root and disappearance of right aortic arch 4).

24–5 All regions of the body except the right upper limb and right side of the head, neck, and thorax (see fig. 24–5) are drained by the thoracic duct into one of the large veins of the neck, e.g., the left internal jugular vein.

24–6 The phrenic nerves arise from cervical nerves 3 to 5. These segments also supply the skin over the trapezius, and pain may be referred from the diaphragm to the shoulder.

PART 5 THE ABDOMEN

25 ABDOMINAL WALLS

25–1 The transpyloric plane is usually at the level of L.V.1, but it does not necessarily pass through the pylorus.

25–2 The fibers of the external oblique muscle run downward and forward (as in inserting a hand in a pocket), whereas those of the internal oblique muscle go upward and forward. Use is made of this arrangement in the McBurney muscle-splitting incision for appendectomy, in which the three anterolateral muscles are divided successively in the directions of their muscular and tendinous fibers.

25–3 The inguinal (L., *inguen,* groin) ligament, described by Poupart (1695) but previously noted by Falloppio (1584), is merely the lower edge of the external oblique aponeurosis. It extends from the anterior superior iliac spine to the pubic tubercle. Medially, a small part is reflected backward to the pubis as the *lacunar ligament,* and its lateral edge limits a gap (lacuna), which forms the medial boundary of a femoral hernia. A further lateral extension along the pubis and behind the femoral vessels is known as the *pectineal ligament* (see fig. 25–7). It was described by Sir Astley Cooper.

25–4 The superficial inguinal ring is a divergence of fibers (crura) of the external oblique aponeurosis. It lies about 1 cm above and lateral to the pubic tubercle and can be palpated by invaginating the scrotal skin upward along the spermatic cord. An inguinal hernia, especially when indirect (i.e., traversing the inguinal canal), may present through the superficial inguinal ring.

25–5 The deep inguinal ring appears as a slit in the fascia transversalis about 1 cm above and medial to the midpoint of the inguinal ligament. It is the entrance to the inguinal canal, which runs medially and downward for about 4 cm to the superficial inguinal ring. An indirect inguinal hernia enters the canal through the deep ring.

25–6 The chief protection of the inguinal canal is muscular. Contraction of the abdominal muscles narrows the canal and tends to close the rings. The deep ring is strengthened by a loop of fibers in the fascia transversalis (W. J. Lytle, Br. J. Surg., *57*:531, 1970).

25–7 The inguinal triangle is bounded by the inferior epigastric artery, the lateral border of the rectus, and the inguinal ligament. A direct inguinal hernia enters the inguinal canal through the inguinal triangle. Excess fluid within the peritoneal cavity (*ascites*) may be removed (tapped), after the bladder has been emptied, by inserting a cannula immediately lateral to the rectus well above (and hence lateral to the termination of) the inferior epigastric artery (see fig. 25–9).

25–8 The umbilicus is in dermatome T10 and is opposite the lumbar part of the vertebral column, but its level is highly variable. Prof. H. A. Harris, of Cambridge, used to say, "My umbilicus is at the level of the disc between L.V.3 and L.V.4—when I hold it there!"

25–9 *Exomphalos*, or *omphalocele* (Gk, umbilical hernia), is a ventral protrusion of intestine through a large defect at the umbilicus. Many umbilical herniae represent a failure of intestinal loops to return to the abdomen during development. Normally an intestinal loop enters the umbilical cord of the embryo at 6 weeks and is returned to the abdomen of the fetus at about 9 weeks.

26 THE ABDOMINAL VISCERA AND PERITONEUM

26–1 The *coelom*, a cavity bounded by an epithelium of mesodermal origin, usually refers to the body cavity that appears within the lateral-plate mesoderm of the embryo. This cavity becomes divided into the serous cavities, which are lined by the serous membranes: pericardium, pleura, and peritoneum.

26–2 The peritoneal cavity communicates indirectly with the exterior of the body by way of the uterine tubes.

26–3 The transverse colon and mesocolon are fused with the posterior layers of the greater omentum (see fig. 26–5*B*). Because of its protective functions, the greater omentum has been called "the abdominal policeman" (Rutherford Morison).

26–4 The mesentery extends about 15 cm downward and to the right from the duodenojejunal flexure to the ileocolic junction (see fig. 26–6). The intestinal border is many times longer (6 or 7 m.) than the root of the mesentery; hence the mesentery is pleated.

26–5 A subphrenic abscess may occur in any of several subphrenic spaces (some of which are listed in table 26–1), including the lesser sac. An abscess may also be found extraperitoneally (e.g., at the bare area of the liver) or in a subhepatic position (e.g., in the hepatorenal pouch) (see fig. 26–6).

26–6 The lesser omentum comprises the gastrohepatic and hepatoduodenal ligaments and hence extends between the liver (the L-shaped attachment can be seen in figure 28–3) and the stomach and duodenum (see fig. 26–7).

26–7 The lesser sac is the part of the peritoneal cavity that lies (mostly) behind the stomach and lesser omentum. The lesser sac is frequently termed the *omental bursa*, which strictly speaking is a developmental term for the part that lies behind the stomach and in the greater omentum. Figure 26–4 shows the surgical approaches to the lesser sac.

26–8 The epiploic foramen, described by Winslow (1732), is the opening (aditus) from the greater into the lesser sac. It lies immediately behind the free, right edge of the lesser omentum (see fig. 26–3*A*).

26–9 Malrotation of the intestine is a vague term used for anomalies of rotation and fixation. One example is subhepatic cecum. Details of these complicated and disputed matters can be found in R. L. Estrada, *Anomalies of Intestinal Rotation and Fixation,* Thomas, Springfield, Illinois, 1958, and N. Lauge-Hansen, *The Development and the Embryonic Anatomy of the Human Gastro-Intestinal Tract,* Centrex, Eindhoven, The Netherlands, 1960. Even the normal development (see fig. 26–8), however, is not well understood. It is no longer believed that the normal stomach and duodenum undergo rotation (R. Kanagasuntheram, J. Anat., *91*:188, 1957, and *94*:231, 1960).

27 THE ESOPHAGUS, STOMACH, AND INTESTINE

27–1 The foregut-midgut and celiac–superior mesenteric junction is at the middle of the duodenum. The midgut-hindgut and superior-inferior mesenteric junction is near the left end of the transverse colon.

27–2 The gastro-esophageal junction may be closed by contraction of its spiral musculature, even though an anatomical sphincter may not exist (A. J. Jackson, Am. J. Anat., *151*:267, 1978). The anatomy of this region is surprisingly complicated and confusing. See G. W. Friedland, Am. J. Roentgenol., *131*:373, 1978, for a good review.

27–3 A *hiatal hernia* occurs through the esophageal opening (hiatus) of the diaphragm. The hiatus is normally formed variably by the right crus of the diaphragm or by both crura (R. E. M. Bowden and H. A. El-Rambi, Br. J. Surg., *54*:983, 1967). Hiatal herniae are either sliding (in which case the cardiac end of the stomach slides up into the thorax) or para-esophageal (in which case the front wall of the stomach may roll up beside the esophagus).

27–4 *Cardia* is a Greek term meaning either "heart" or the not-so-distant "upper mouth" of the stomach. *Pylorus* comes from the Greek for "gatekeeper." The duodenum (L., twelve) is approximately twelve fingerbreadths long: 12 × 2 cm = 24 cm.

27–5 In *congenital hypertrophic pyloric stenosis* the pyloric sphincter is thickened (hypertrophied). The condition can be treated by incising longitudinally all layers down to but excluding the mucosa.

27–6 The "duodenal cap" is the beginning of the first part of the duodenum as seen on radiography (see fig. 27–2). It is mobile and lacks circular folds.

27–7 The jejunum is shorter and typically emptier, more vascular (redder *in vivo*), and thicker walled than the ileum.

27–8 A *diverticulum ilei* is a remnant of an embryonic structure, the *vitello-intestinal duct,* which connects the yolk sac to the intestine. A useful although not entirely accurate mnemonic is that it occurs in 2 per cent of adults, is 2 inches long, and is found 2 feet from the ileocolic junction. The diverticulum is generally associated with the name of Meckel, but it had been described earlier by John Hunter, another example of the disadvantage of eponyms.

27–9 *Gastrojejunostomy* is a surgical anastomosis between the stomach and the jejunum (e.g., to bypass a duodenal ulcer). The Greek word *stoma* means "mouth" (cf. *colostomy,* a surgical opening of the colon onto the surface of the body). Anatomical nomenclature is an important source of medical terminology in general.

27–10 The most frequent positions of the vermiform appendix are retrocecal and pelvic (K. Buschard and A. Kjaeldgaard, Acta Chir. Scand., *139*:293, 1973). An inflamed appendix may become fixed to the psoas muscle, resulting in pain on extension of the thigh, which stretches the muscle. An inflamed pelvic appendix, because of its close relationship to the uterine tube, is difficult to differentiate from an infected tube (*salpingitis*).

27–11 The small intestine is longer and more mobile. Moreover, it contains circular folds and villi, whereas the colon is characterized by teniae, haustra (sacculations), and appendices epiploicae. Both small and large intestines contain glands and lymphatic follicles.

28 THE LIVER, BILIARY PASSAGES, PANCREAS, AND SPLEEN

28–1 The liver is divided into right and left anatomical lobes along the left limb of the H formed at the porta (see fig. 28–2) and, in front, along the falciform ligament. The much larger right lobe includes the caudate and quadrate lobes. The porta is the "gate" of the liver, and the "vein of the gate" is termed *portal*. For many centuries the liver was represented by a schematic five-lobed organ. It should be appreciated, however, that ancient anatomical illustrators used formula symbols and did not try to represent true-to-life relationships and details. "It is not a matter of all this being false, for the categories of 'correct' and 'false' do not belong to the basic substance of such a formal idiom" (R. Herrlinger, *History of Medical Illustration,* Pitman, London, 1970).

28–2 The liver is divided into right and left functional lobes (i.e., according to ductal and vascular distributions) along an irregular plane close to the right-hand limb of the H (see fig. 28–2). Further subdivisions constitute the hepatic segments (see fig. 28–4).

28–3 The ligamentum teres is the obliterated left umbilical vein. In the fetus, the umbilical vein and ductus venosus convey oxygenated blood from the placenta to the inferior vena cava (see fig. 28–5). In the adult, the ligamentum teres begins at the umbilicus, ascends in the free margin of the falciform ligament (which acted as the mesentery for the umbilical vein), and ends at the porta. By identifying the falciform ligament surgically, it is possible to cannulate the occluded umbilical vein and continue to the portal vein (e.g., for diagnostic purposes).

28–4 The liver is maintained in position by being (1) joined to the posterior abdominal wall by veins (hepatic and caval) and ligaments (coronary and triangular), (2) supported below by viscera (e.g., the right kidney and colic flexure), and (3) suspended by ligaments (falciform and teres) (D. K. Tran et al., Bull. Assoc. Anat., *58*:1121, 1974).

28–5 The "bare area" of the liver is the part of its surface not covered by peritoneum. It is situated posteriorly on the right lobe and is in contact with the diaphragm.

28–6 The lesser omentum may be regarded as the mesentery of the bile duct. Its right free border contains the bile duct, portal vein, and hepatic artery (see figs. 26–3*A* and 28–3). Variations of these structures, both before and after the porta, are frequent, are of surgical importance, and have been well documented. A valuable source of information on surgical anatomy in general is the two-volume work, *Surgical Anatomy,* 5th ed., by B. J. Anson and C. B. McVay (W. B. Saunders Company, Philadelphia, 1971).

28–7 The Greek word for bile is *chole*. Hence the bile duct is the choledochal ("bile-receptacle") duct. *Cholecystectomy* is removal of the gallbladder. "Black bile" was considered one of the humors of the body, and its supposed excess was known as *melancholia*.

28–8 The gallbladder is sought in the angle between the right costal margin and the linea semilunaris (lateral border of the rectus). Pain and tenderness here are found when the gallbladder is inflamed (*cholecystitis*).

28–9 As the bile duct traverses the duodenal wall, it acquires the choledochal sphincter, or sphincter of the bile duct (see fig. 28–6*B*), described by Oddi (1887) but already known to Glisson (1654). Sphincters may also be present around the termination of the pancreatic duct and around the hepatopancreatic ampulla. With regard to the ducts and blood vessels in the biliary region, it should be appreciated that "variation is rampant." Moreover, the arrangement of the muscular apparatus of the choledochoduodenal junction is still disputed. (See J. Delmont, ed., *The Sphincter of Oddi,* Karger, Basel, Switzerland, 1977.)

28–10 The portal vein is formed behind the neck of the pancreas by the union of the splenic and superior mesenteric veins (see fig. 28–8). Portal hypertension (caused by obstruction) results in opening up of portal-systemic anastomoses, e.g., at the lower end of the esophagus. Uncontrolled hemorrhage from esophageal varices can be treated by a portal-systemic shunt, e.g., by anastomosing the portal vein with the inferior vena cava (*portacaval anastomosis*).

28–11 An *annular pancreas* is one that encircles the second part of the duodenum. Its development is obscure. (S. W. Gray and J. E. Skandalakis, *Embryology for Surgeons,* W. B. Saunders Company, Philadelphia, 1972, provides much applied information on congenital anomalies.)

28–12 Long before birth, the dorsal and ventral diverticula, from which the pancreas develops, become fused, as do their ductal systems (see fig. 28–8). The dorsal duct, which drains the body of the organ, unites with the ventral, which ends at the greater duodenal papilla. The main pancreatic duct (figured by Wirsung in 1642) has therefore both dorsal and ventral developmental components. The original termination of the dorsal duct forms the accessory pancreatic duct (Santorini, 1724) and ends at the lesser duodenal papilla. See R. O'Rahilly and F. Müller, Acta Anat., *100*:380–385, 1978.

28–13 The spleen is generally on the left side. The normal asymmetric arrangement of the viscera, whereby the liver and right atrium are on the right side of the body, is known as *situs solitus*. Reversed asymmetry, i.e., a mirror image, is known as *situs inversus*. Although the thoracic and abdominal situs usually match, rarely they may differ (*situs inversus partialis*). Agenesis of the spleen and multiple spleens (*polysplenia*) are usually associated with partial situs inversus.

28–14 Usually the spleen is palpable only when enlarged. Indeed, "the spleen must be at least one-third as big again as normal before it can be detected by clinical methods" (Hamilton Bailey). The spleen lies against ribs 9 to 11, and splenic puncture (e.g., in diagnostic hematology) is performed with due precautions in intercostal space 9 in the midaxillary line during full inspiration.

28–15 A common site for accessory splenic tissue is the tail of the pancreas. Accessory spleens hypertrophy after splenectomy. The development of tissues in situations in which they are not normally found is known as *aberrance* (L., "wandering away"), *ectopia* (Gk, "out of place"), or *heterotopia* (Gk, "other place"). In addition to supernumerary or accessory organs (e.g., spleens), anomalous differentiations are also included (e.g., gastric tissue in esophageal mucosa, pancreatic tissue in the stomach, and endometrial tissue in the ovaries). The origins of these conditions are in dispute.

29 THE KIDNEYS, URETERS, AND SUPRARENAL GLANDS

29–1 A line joining the hili of the kidneys would cross L.V.1 or 2 (see fig. 29–3).

29–2 In approaching the kidney from behind (lumbar extraperitoneal route), one must avoid the pleura (see fig. 26–1*C*). The fibers of the latissimus dorsi are separated or cut, the external and internal oblique muscles are displaced laterally, the transversus aponeurosis is incised, and the quadratus lumborum is drawn medially. The erector spinae, quadratus lumborum, and psoas major are all kept medial (see fig. 29–5). It is also possible to remove a kidney (*nephrectomy*) from in front (transabdominal transperitoneal route).

29–3 A perinephric abscess would be situated in the perinephric, or perirenal, fat between the renal fascia and the renal capsule (see fig. 29–5).

29–4 Examples of congenital anomalies of the kidneys are duplication, ectopia, malrotation, and "accessory arteries." *Renal duplication* arises as a doubling of the ureteric bud. *Renal ectopia* is a failure of the normal relative ascent of the prenatal organ, so that the kidney remains pelvic. An example of malrotation is anterior pelvis, i.e., a ureteric pelvis that failed to shift into its usual medial position. In *horseshoe kidney,* the inferior poles commonly are united, the pelves are anterior, and anomalous blood vessels are present. "Accessory arteries" (e.g., to the lower pole) are normal segmental arteries that have persisted postnatally. They represent extrarenal branching of the main renal artery rather than actual accessory vessels.

29–5 A ureter is likely to be obstructed where it is narrowest, namely (1) at its pelvis, (2) on crossing the pelvic inlet, or (3) in the wall of the bladder.

29–6 So-called *renal colic* is actually ureteric colic, i.e., violent spasmodic pain that passes from the loin to the groin (referred along the course of the genitofemoral nerve). The colic is caused by obstruction, which produces distension and increased muscular contractions.

29–7 The right suprarenal gland lies behind the right lobe (bare area) of the liver and the inferior vena cava; the left suprarenal gland lies behind the lesser sac (separating it from the stomach) and the pancreas (see fig. 26–6). The glands are different and variable in form: the right may be pyramidal, the left like a croissant, although each in turn has been said to resemble a "cocked hat."

29–8 The cortical system comprises the suprarenal cortex and accessory cortical masses that may be found near the kidneys, ovaries, and testes. The chromaffin system (cells that have an affinity for chromates) is usually taken to include the suprarenal medulla, paraganglia (near sympathetic ganglia), para-aortic bodies (near the origin of the inferior mesenteric artery), possibly the carotid body (near the bifurcation of the common carotid artery), and scattered masses in the abdomen and pelvis. Chromaffin cells are generally regarded as modified postganglionic neurons.

30 BLOOD VESSELS, LYMPHATIC DRAINAGE, AND NERVES OF THE ABDOMEN

30–1 The aorta (which comes from the *left* ventricle) ends at L.V.4 to the *left* of the median plane. Hence the right renal artery is longer than the left (see fig. 29–5).

30–2 The caudal part of the foregut (to the middle of the duodenum) is supplied by the celiac trunk. The midgut (to the left part of the transverse colon) is supplied by the superior mesenteric artery. The hindgut is supplied by the inferior mesenteric artery.

30–3 The lower end of the esophagus (*esophageal varices*) and the anal region (*hemorrhoids*) are two important sites of portal-systemic anastomosis.

30–4 The inferior vena cava (which is going to the *right* atrium) begins at L.V.5 to the *right* of the median plane. Hence the left renal vein is longer than the right (see fig. 29–5). The left testicular vein enters the left renal vein at a right angle (see fig. 29–4), an arrangement that may be partly responsible for the much higher incidence of varicocele on the left side.

30–5 Usually the right ureter keeps to the right of the inferior vena cava. Rarely, the ureter may wind around behind the vena cava and may become obstructed. This "retrocaval ureter" is really a vascular anomaly (*pre-ureteric vena cava*) caused by a portion of the vena cava being derived from a channel situated more ventrally than is usual. (For a good bibliography see M. M. Kanawi and D. I. Williams, Br. J. Urol., *48*:183, 1976.)

30–6 Vagal fibers extend indirectly (through the plexuses in the abdomen) as far distally as probably the left colic flexure or the beginning of the descending colon. The sacral parasympathetic distribution (pelvic splanchnic nerves) may extend as far proximally as the termination of the transverse colon.

30–7 The lumbar plexus (which gives rise to the femoral and obturator nerves) is derived from L2 to 4. The sacral plexus (which gives rise to the sciatic nerve) is derived from L4 to S4 (see fig. 30–6).

30–8 The iliohypogastric and ilio-inguinal nerves are best regarded as, respectively, the main trunk and collateral branch of L.N.1. Both nerves supply muscles (internal oblique and transversus) and skin. The iliohypogastric nerve supplies part of the buttock laterally (lateral cutaneous branch) and the abdomen above the pubis (anterior cutaneous branch). The ilio-inguinal nerve, which traverses the inguinal canal, supplies skin on the thigh and scrotum (labium majus). The genitofemoral nerve (L1,2) divides into (1) a genital branch, which traverses the inguinal canal and supplies the cremaster and skin of the scrotum (labium majus), and (2) a femoral branch, which descends behind the inguinal ligament (lateral to the femoral artery) and supplies skin over the femoral triangle.

PART 6 THE PELVIS

31 THE BONES, JOINTS, AND WALLS OF THE PELVIS

31–1 The boundary between the lesser and greater pelves is the pelvic inlet or brim (upper pelvic aperture) (see fig. 31–1*A* and *C*). The greater pelvis is a part of the abdomen proper.

31–2 The visceral (pelvic) surface of the body of the pubis faces almost directly upward, so that the bladder rests on it.

31–3 The subpubic angle is formed by the conjoined rami of the pubes and ischia (see fig. 31–5). It can be estimated *per vaginam*. A narrow angle may cause difficulty during parturition.

31–4 The diagonal conjugate diameter is the distance between the lower border of the pubic symphysis and the sacral promontory. It can be measured *per vaginam* (see fig. 31–3).

31–5 Although the surfaces are not flat, the sacro-iliac joint is considered to be plane (synovial) in type. Interlocking surface irregularities and strong ligaments reduce motion to a minimum. The ligaments become relaxed during pregnancy. As in most joints, degenerative changes (such as marginal lipping) appear with increasing age, but they do not necessarily cause symptoms.

31–6 The pubic symphysis is a cartilaginous joint in which the articular surfaces, covered by a thin layer of hyaline cartilage, are united by the interpubic disc, which consists of fibrocartilage and may contain a cavity. Movement is negligible. The interpubic disc becomes softer during pregnancy.

31–7 The pelvic viscera are supported by the bony pelvis (the bladder rests on the pubis), by the peritoneum and ligaments, and by the pelvic and (recently disputed) urogenital diaphragms.

31–8 The rectovesical pouch is formed by the reflection of the peritoneum from the front of the rectum onto the upper surface of the bladder (see fig. 31–9). In the female, the rectovesical pouch is divided by the uterus and vagina into the shallow uterovesical pouch and the deeper recto-uterine pouch (Douglas, 1730).

32 BLOOD VESSELS, NERVES, AND LYMPHATIC DRAINAGE OF THE PELVIS

32–1 The pelvis is supplied chiefly by the internal iliac arteries, which give numerous branches.

32–2 In the fetus, the internal iliac arteries are continued as the umbilical arteries to the umbilicus and through the umbilical cord to the placenta, to which they convey deoxygenated blood (see fig. 23–13). Postnatally, most of each umbilical artery (see fig. 32–2) becomes the medial (formerly called lateral) umbilical ligament (see fig. 25–9).

32–3 Most of the perineum is supplied by the pudendal nerves (S2 to 4) from the sacral plexus. The pudendal nerve leaves the pelvis through the greater sciatic foramen (see fig. 32–2), appears in the gluteal region, and passes through the lesser sciatic foramen to reach the pudendal canal of the ischiorectal fossa. The pudendal nerve can be "blocked" *per vaginam* in obstetrics, the needle being directed toward the ischial spine.

32–4 The pelvic splanchnic nerves are visceral branches of the ventral rami of sacral nerves 2 to 4. They pass directly to the pelvic viscera, in the walls of which are ganglia for the relay of the sacral parasympathetic fibers. Included are motor fibers that, by way of the inferior hypogastric plexus, supply the bladder, rectum, and colon as far proximally as probably the termination of the transverse colon.

32–5 Lymph from the colon drains into the superior and inferior mesenteric nodes (in accordance with the arterial supply of the colon). The upper part of the rectum drains into the inferior mesenteric nodes; the lower part, together with the upper part of the anal canal, drains into the internal iliac nodes; and the lower part of the anal canal, together with the external genitalia, drains into the inguinal nodes. Details of the lymphatic drainage of the body may be found in H. Rouvière, *Anatomie des lymphatiques de l'homme,* Masson, Paris, 1932.

33 THE URETER, BLADDER, AND URETHRA

33–1 The contracted bladder is thick-walled and lies in the pelvis. The distended bladder projects upward into the peritoneal cavity and is in contact with the anterior abdominal wall. Hence access to a full bladder may be gained extraperitoneally (see fig. 33–1). As a rare congenital anomaly, the bladder may open onto the abdominal wall suprapubically (*ectopia vesicae* or *exstrophy of the bladder*).

33–2 The bladder of a child occupies a higher position because of the smaller relative size of the pelvis and the greater relative size of the bladder.

33–3 The trigone displays a smooth mucous membrane, is more vascular, is prone to disease ("pathological zone"), and may perhaps differ developmentally from the rest of the bladder.

33–4 Lithotomy, or "cutting for stone" (Gk, *lithos,* a stone; cf. lithograph), is the removal of vesical calculi, which were formerly much more common. Perineal lithotomy is a very ancient operation: the suprapubic approach dates from the sixteenth century. An interesting account is provided by O. H. Wangensteen and S. D. Wangensteen, *The Rise of Surgery,* University of Minnesota Press, Minneapolis, 1978, Chapter 4.

33–5 The narrowest portion of the urethra is the external urethral orifice, and the membranous part is also not as readily dilatable as elsewhere. In contrast, dilatations occur near the external orifice (*fossa navicu-*

laris) and in the bulb of the penis (*intrabulbar fossa*). The most frequent site of rupture of the urethra is the posterior portion of the spongy part or the lower portion of the membranous part. The fascial attachments are such that extravasation of urine then proceeds forward into the connective tissue of the scrotum and anterior abdominal wall.

33–6 Urethral glands open into most of the urethra. In addition, the prostatic ducts, prostatic utricle, and ejaculatory ducts open into the prostatic part (see fig. 34–3), and the bulbo-urethral glands open into the spongy part. The female urethra, which receives urethral and para-urethral glands, probably corresponds mainly to the upper part of the prostatic part of the male urethra.

34 MALE GENITALIA

34–1 The testis descends during the last trimester of pregnancy by entering the deep inguinal rings, inguinal canal, and superficial inguinal ring, and then reaching the scrotum. The descent is usually complete at a variable time between the beginning of the last trimester and shortly after birth. The mechanism of descent is obscure, although hormonal regulation is known to be important. The role of a ligament known as the *gubernaculum* has long been disputed. (See, for example, H. Zaw Tun, Anat. Rec., *190*:591, 1978). Failure of descent is known as *cryptorchidism*.

34–2 The tunica vaginalis testis is the lower portion of the processus vaginalis of the peritoneum, which, in the fetus, precedes the descent of the testis from the abdomen into the scrotum. After descent, the upper part of the processus becomes obliterated, whereas the lower portion persists as a closed sac around the testis (see figs. 25–11*A* and 34–2*C*). The processus may remain patent (see fig. 25–11*B*). A collection of fluid in the tunica vaginalis is termed a *hydrocele*.

34–3 Embryonic remnants include the appendix testis, appendix of the epididymis, aberrant ductules, and paradidymis. These are believed to be remnants of the mesonephric and paramesonephric ducts. The appendix of the epididymis may undergo torsion and require excision.

34–4 *Varicocele* is a condition of varicosity of the veins of the pampiniform plexus in the spermatic cord. According to some, however, a varicocele more commonly involves nearby cremasteric veins, which communicate with the testicular veins.

34–5 The spermatic cord, which comprises the ductus deferens and associated structures and coverings, extends from the deep inguinal ring to the posterior border of the testis. The ductus deferens, spinal cord, and thoracic duct are each approximately 45 cm in length.

34–6 The normal prostate, as palpated *per rectum*, is firm and elastic. It is characterized by a median groove ending above in a notch. The median furrow becomes obliterated in carcinoma.

34–7 The surgical approaches to the prostate are transvesical, retropubic, perineal, and urethral (see fig. 34–1). Operations for *prostatectomy* based on each of these four routes have been devised.

34–8 The so-called middle (or median) lobe of the prostate is a term used for the extension that runs forward from the upper part of the posterior surface and lies between the ejaculatory ducts and the urethra (see fig. 34–1). It is generally said to become enlarged in benign prostatic hypertrophy, but it may be that the "central zone" of the prostate is the site of what is really "benign peri-urethral hyperplasia" (J. E. McNeal, J. Urol., *107*:1008, 1972).

34–9 The uvula of the bladder is a mucosal elevation behind the internal urethral orifice (see fig. 34–1). In later life, in the presence of prostatic hypertrophy, the uvula may become accentuated by the middle lobe of the prostate and cause obstruction to the passage of urine.

34–10 The coverings of the prostate, from external to internal, are (1) the prostatic sheath of pelvic fascia, (2) the capsule, and (3), in an enlarged prostate, a "false capsule" of compressed tissue. The prostatic venous plexus lies between the sheath and the (true) capsule.

35 FEMALE GENITALIA

35–1 The ovary is commonly situated on the lateral wall of the pelvis, where it can be palpated bimanually (i.e., with one hand on the abdomen and the other *per vaginam*). The long axis is vertical (see fig. 35–1), and not horizontal as shown in most illustrations, where the broad ligaments and uterine tubes have been spread out (see fig. 35–2).

35–2 The ovary is covered by the so-called germinal epithelium, which is continuous with the mesothelium of the peritoneum. The term *superficial epithelium* is preferable to germinal epithelium because the primordial germ cells are now believed to arise extragonadally. The corresponding covering of the testis is the visceral layer of the tunica vaginalis.

35–3 The uterine tubes, which had been described by many before Falloppio (1561), open into the peritoneal cavity (abdominal opening) and the uterine cavity (uterine opening).

35–4 The broad ligament may be regarded as the mesentery of the uterine tube. It extends from the margin of the uterus to the lateral wall of the pelvis.

35–5 Some Latin terms and Greek roots associated with the genitalia are *testis* and *orchis; ovarium* and *oophoros; tuba uterina* and *salpinx; uterus, hystera,* and *metra;* and *vagina* and *kolpos*. All these are used in various compounds (e.g., salpingitis and hysterectomy) and illustrate the Latin and Greek origins of medical terminology. Many aspects of the uterus are discussed in R. M. Wynn (ed.), *Biology of the Uterus,* 2nd ed., Plenum, New York, 1977.

35–6 The endometrium is the mucosa of the uterus. That of the cervix is sometimes distinguished as *endocervix*. The presence of extra-uterine endometrium (e.g., in the ovary or elsewhere in the pelvis) is known as *endometriosis*. The endometrium of pregnancy is termed *decidua* (L., falling off; cf. deciduous trees, deciduous teeth), because it is shed after parturition.

35–7 The uterus is normally anteverted, i.e., the cervix is directed downward and backward at slightly more than a right angle to the vagina (see fig. 35–4). The uterus is also generally anteflexed, i.e., the body is bent downward and forward at the isthmus. Filling of the bladder tends to push the uterus into a relatively retroverted position.

35–8 In the *Nomina anatomica,* the lower opening of the uterus is termed the *ostium uteri*. This was previously called the "external os" because the cavity at the isthmus (i.e., the junction between body and cervix) was then given the unsatisfactory term "internal os."

35–9 *Hysterosalpingography,* as its name suggests, is the (radiographic) depiction of the uterine and tubal cavities (see fig. 35–3). In addition to demonstrating tubal patency, it allows the detection of various anomalies of the uterus. For examples, the uterus may be partially divided into right and left horns (*uterus bicornis unicollis*). Good accounts of uterine anomalies are given by I. W. Monie and L. A. Sigurdson (Am. J. Obstet. Gynecol., *59*:696, 1950) and by E. Zanetti, L. R. Ferrari, and G. Rossi (Br. J. Radiol., *51*:161, 1978).

35–10 The peritoneum covers the body and supravaginal part of the cervix posteriorly (to form the recto-uterine pouch) but only the body anteriorly (to form the uterovesical pouch) (see fig. 35–4*E*).

35–11 The uterus is supported by the vagina, by muscles (pelvic and, perhaps, urogenital diaphragms), and by ligaments and folds. The uterus is connected to the bladder by the uterovesical fold and to the rectum by the recto-uterine and rectovaginal folds. Fascial thickenings form the lateral cervical, or cardinal, ligament (see fig. 35–5) and the uterosacral ligament. The broad ligaments proceed to the lateral wall of the pelvis, and the round ligaments enter the inguinal canals.

35–12 The epoophoron is a mesonephric remnant situated in the broad ligament (see fig. 35–8). It consists largely of a duct parallel to and below the uterine tube. Embryonic remnants in the broad ligament may undergo cystic dilatation. Good accounts are given by G. M. Duthie (J. Anat., *59*:410, 1925), G. H. Gardner, R. R. Greene, and B. M. Peckham (Am. J. Obstet. Gynecol., *55*:917, 1948), and J. W. Huffman (Am. J. Obstet. Gynecol., *56*:23, 1948).

35–13 The basic surgical approaches to the uterus, as used in *hysterectomy*, are abdominal and vaginal (fig. 35–4*F*).

36 THE RECTUM AND ANAL CANAL

36–1 The rectum begins gradually where the sigmoid colon loses its mesentery, at the level of the middle of the sacrum. The anorectal junction is at the upper limit of the pelvic diaphragm. In surgical usage, however, the rectum is frequently taken to include the upper part of the anal canal, as far down as the pectinate line (see fig. 36–1). It should be noted that the anus (like the pylorus) is merely the orifice at the termination of the anal canal (cf. pyloric canal): "the postern gate of the human citadel."

36–2 *Imperforate anus* is a congenital anomaly in which the anal canal fails (completely or incompletely) to open to the exterior. The rectum may also be abnormal (e.g., it may end blindly at a variable level). Various classifications of anorectal anomalies have been proposed. (See J. C. Goligher, H. L. Duthie, and H. H. Nixon, *Surgery of the Anus, Rectum and Colon,* 3rd ed., Baillière Tindall, London, 1975, Chapter 11.)

36–3 The pectinate line is the line of the anal valves (see fig. 36–1). The anal columns resemble the teeth of an inverted comb (L. *pecten,* a comb). Although its embryological significance is disputed, the pectinate line is generally taken as an important vascular (superior/inferior rectal arteries and veins) and lymphatic (iliac/inguinal nodes) boundary. Moreover, below the pectinate line, the anal canal is very sensitive and the intestinal mucosa soon merges into skin. Below the pectinate line, an anal intersphincteric groove is palpable between the subcutaneous part of the external sphincter and the lower border of the internal sphincter. It may be marked by a "white line," which is actually bluish *in vivo*.

36–4 Hemorrhoids, or "piles," are anal varicose veins. They may be internal (above the pectinate line: superior rectal veins), external (below the pectinate line: inferior rectal veins), or more commonly both. Symptomatic hemorrhoids may be ligated and excised (*hemorrhoidectomy*).

37 THE PELVIC DIAPHRAGM AND FASCIA

37–1 The pelvic diaphragm comprises the levatores ani and coccygei and their covering fasciae. The pelvic diaphragm supports the weight of the pelvic viscera in the erect position. The levatores ani, which have been described as a dilator as well as a sphincter, probably serve for the "fixation and prevention of prolapse of the pelvic viscera passing through the levator hiatus" (S. F. Ayoub, J. Anat., *128*:571, 1979).

37–2 The pelvic diaphragm has a hiatus for the passage of the urethra, vagina, and anal canal (see fig. 37–1).

37–3 Five structures in the body receive the name *diaphragm* (Gk, a partition): (1) the (respiratory) diaphragm, (2) the diaphragma oris, (3) the pelvic diaphragm, (4) the urogenital diaphragm, and (5) the diaphragma sellae. All are muscular or musculotendinous except the last, which is composed of dura mater.

38 THE PERINEAL REGION AND EXTERNAL GENITALIA

38–1 The term *perineum* in obstetrics and gynecology is generally limited to the area between the anal and vaginal orifices. Anatomically, the perineal region is that part of the trunk below the pelvic diaphragm. It underlies the outlet of the pelvis (see fig. 38–1*A*).

38–2 The perineal body (see fig. 38–1*B*), or tendinous center of the perineum, is a median, fibromuscular node situated at the convergence of several muscles, including the levatores ani and the sphincter ani externus. The perineal body may be damaged during parturition.

38–3 The urogenital diaphragm, which is pierced by the urethra and the vagina, includes the deep transverus perinei and the sphincter urethrae bilaterally (see fig. 38–1*F*). However, the sphincter urethrae is not horizontal, and "the concept of a urogenital diaphragm is not borne out" (T. M. Oelrich, Am. J. Anat., *158*:229, 1980).

38–4 The ischiorectal fossa (see fig. 38–2) is the fat-filled space between the obturator internus laterally (overlying the ischium) and the sphincter ani externus and pelvic diaphragm medially (overlying the anal canal and rectum; hence the name *ischiorectal*). The base of the fossa is the skin of the anal region, which allows surgical access to the fossa, e.g., to drain an ischiorectal abscess. *Anorectal abscesses* may be found (1) in the region of the subcutaneous part of the external sphincter (*perianal abscess*), (2) in the ischiorectal fossa (*ischiorectal abscess*), (3) beneath the mucosa of the upper part of the anal canal (*submucous abscess*), and (4) above the pelvic diaphragm (*pelvirectal abscess*). Anorectal abscesses are important because they may give rise to a fistula-in-ano, i.e., a track between the anal canal or rectum and the perianal skin.

38–5 The prepuce is a double layer of skin that covers the glans (see fig. 38–4), and its excision is termed *circumcision* (L., cutting around), an operation that has been used for several millennia in Africa and Egypt.

38–6 The pudendum consists of the external genitalia. The term *female pudendum* is more or less synonymous with *vulva*.

PART 7 THE BACK

39 THE VERTEBRAL COLUMN

39–1 The 24 "presacral" vertebrae are movable. Eighty-nine per cent of people have 24 presacral vertebrae, 6 per cent have 23, and 5 per cent have 25 (P. E. Bornstein and R. R. Peterson, Am. J. Phys. Anthropol., *25*:139–146, 1966).

39–2 The thoracic and sacral curvatures are primary (appearing during embryonic life). The cervical and lumbar curvatures are secondary (appearing during fetal life and accentuated during infancy).

39–3 A typical vertebral arch is characterized by at least seven processes: four articular, two transverse, and one spinous. Lumbar vertebrae have small, additional (mamillary and accessory) processes. The axis has an odontoid process (the dens) but no superior articular processes (merely facets). The atlas has only transverse processes.

39–4 The intervertebral foramina are between adjacent pedicles, and typically each contains a spinal ganglion and a ventral root (or rootlets) of a spinal nerve (see fig. 41–3).

39–5 C.N.8 emerges between C.V.7 and T.V.1. Hence the remaining spinal nerves below leave inferior to the correspondingly numbered vertebrae.

39–6 The key features of cervical, thoracic, lumbar, and sacral vertebrae are, respectively, foramina transversaria, articulation with ribs, absence of both of the above features, and fusion.

39–7 All 33 vertebrae are dorsal. Twelve are thoracic.

39–8 The pars interarticularis, visible on oblique radiographs of the lumbar vertebrae (see fig. 39–9), is the part of the neural arch between the superior and inferior articular processes. The slipping forward of a vertebral body (e.g., L.V.5) is termed *spondylolisthesis* (Gk, vertebral slipping) and usually involves fracture at the pars interarticularis.
Case Report. A 50-year-old nurse complained of numbness of the right big toe, a burning pain down the lateral side of the right leg and foot, bowel and bladder problems, and weakness of the lower limbs. She developed a marked depression over the spinous process of L.V.5. At operation, spondylolisthesis of L.V.5 on S.V.1 was found, with defective partes interarticulares of L.V.5 and protrusion of the L.V.4/5 disc. The leg pains were caused by compression of L.N.5 roots from alterations of the L.V.5/S.V.1 facets. Bowel and bladder difficulties were produced by compression of the cauda equina from gross distortion of the vertebral canal. Decompression by removal of some bone near the pedicles relieved the symptoms (G. Austin, *The Spinal Cord*, Thomas, Springfield, Illinois, 1961).
Numerous case studies on anatomy have been described by E. Lachman (*Case Studies in Anatomy*, 3rd ed., Oxford University Press, New York, 1981) and also by L. K. Schneider (*Anatomical Case Histories*, Year Book, Chicago, 1976).

39–9 The sacral canal contains the dura, cauda equina, and filum terminale. Above S.V.2, a subarachnoid space is also present (see fig. 41–1).

39–10 At birth, a typical vertebra shows three primary ossific areas, one for the centrum and one for each half of the neural arch. (The centrum does not correspond to the whole of the body: see fig. 39–14.) The three primary areas become united by bone in early childhood (3 to 6 years). A number of secondary centers (the ring epiphyses and centers for the tips of the transverse and spinous processes) appear at about puberty.

39–11 Surgical access to the spinal cord and nerve roots is achieved by sectioning the laminae (*laminectomy*) on each side of several vertebrae.

39–12 *Spina bifida* is a developmental anomaly in which the halves of one or more neural (future vertebral) arches have failed to fuse. The spinal cord and meninges (*myelomeningocele*), or the meninges alone (*meningocele*), may protrude through the defect (*spina bifida cystica*). When the defect is skeletal rather than neural, it is termed *spina bifida occulta;* this defect is found in the sacrum in approximately one fifth of the general population (A. C. Berry, J. Anat., *120*:519, 1975), although the incidence is minute in certain peoples (J. I. Levy and C. Freed, J. Anat., *114*:449, 1973).

39–13 The Latin roots *spino-* and *vertebro-* are used with reference to the vertebral column (e.g., in spinotransverse and vertebrocostal). Greek roots with a similar significance are *spondyl-* and *rhachi-* (e.g., in spondylolisthesis and r[h]achischisis, the latter meaning "vertebral cleft" and used for an open spina bifida, especially one that extends along many vertebrae).

40 MUSCLES, VESSELS, NERVES, AND JOINTS OF THE BACK

40–1 The prevertebral muscles lie on the front of the column in the neck (longi capitis et colli, recti capitis anterior et lateralis) and abdomen (psoas major et minor). Although they produce flexion, the main antagonists of the dorsal muscles are the abdominal muscles, such as the recti abdominis.

40–2 The most readily identifiable muscles of the back *in vivo* are the trapezius, latissimus dorsi, erector spinae, and semispinalis capitis.

40–3 The deep muscles of the back produce extension and lateral flexion of the vertebral column, according to circumstances. They are important antigravity muscles in the erect posture and in sitting. They control bending forward and enable the erect position to be regained. At the end of flexion, the muscles are quiet and the column is supported by ligaments and intervertebral discs. Hence the danger to these structures is in lifting "with the back" rather than with the muscles of the lower limbs.

40–4 The "spinotransverse system" consists of the erector spinae and the splenius.

40–5 The most prominent component of the "transversospinal system" is the semispinalis.

40–6 The most important structure in the suboccipital triangle is the vertebral artery.

40–7 The vertebral venous system, which is important physiologically and pathologically, comprises (1) an internal plexus around the dura, (2) basivertebral veins on the back of the vertebral bodies, and (3) an external plexus on the front of the vertebral bodies and on the vertebral arches. For details see H. J. Clemens, *Die Venensysteme der menschlichen Wirbelsäule,* de Gruyter, Berlin, 1961. A classic article is that by O. V. Batson, Am. J. Roentgenol., *78*:195, 1957.

40–8 The 24 presacral vertebrae are separated by 23 intervertebral discs. A variable disc is present between the dens and the body of the axis, and another variable one between the sacrum and coccyx. In addition, depending on age, remains of four discs may be found within the sacrum, and perhaps a further one may be found within the coccyx.

40–9 An intervertebral disc is a fibrocartilaginous joint between two adjacent vertebral bodies. The disc continually changes in structure, "notochordal tissue appears to be absent after 10 years of age," and "in advanced age, macroscopic distinction between nucleus and annulus is lost" (A. Peacock, J. Anat., *86*:162, 1952).

40–10 A "slipped disc" is a herniation of the nucleus pulposus into or through the anulus fibrosus, usually in a posterolateral direction. Pressure on the "lower" nerve roots (the "upper" roots having already left the vertebral canal) may result in sensory and motor deficits, but it is important to appreciate that variations of signs and symptoms are considerable.

40–11 Herniation of the L.V.4/5 disc would be likely to press on the roots of L.N.5 because those of L.N.4 would have already made their exit from the vertebral canal. The pain and sensory loss would be mainly along the lateral side of the leg and on the dorsum of the foot (L.N.5 dermatome), and motor deficiency might include weak dorsiflexion of the great toe (extensor hallucis longus) and atrophy of the tibialis anterior. Similarly, the L.V.5/S.V.1 nucleus would be likely to press on the roots of S.N.1, resulting in pain and sensory loss on the back of the leg, ankle, and sole; weak plantar flexion (gastrocnemius); atrophy of the calf; and diminished or absent ankle jerk.

40–12 Nodding affirmatively involves the atlanto-occipital joints. Shaking the head negatively involves the atlanto-axial joints.

40–13 The cervical and lumbar regions are the most mobile, and they are frequent sites of aches. Fracture-dislocations occur chiefly in these regions. Prognosis is governed by the presence or absence of injury to the spinal cord and nerve roots.

41 THE SPINAL CORD AND MENINGES

41–1 The spinal cord ends at the level of L.V.1 or 2. The range is from one vertebra higher to one vertebra lower: T.V.12 to L.V.3 (A.F. Reimann and B. J. Anson, Anat. Rec., *88*:127, 1944). In the newborn, the spinal cord extends to L.V.3.

41–2 A myelomere is the segment of spinal cord to which a given pair of dorsal and ventral roots is attached.

41–3 Lower myelomeres are at higher levels than their correspondingly numbered vertebrae. Thus, myelomere S1 is opposite T.V.12 (see fig. 41–1) or L.V.1. For details see R. Louis (Anat. Clin., *1*:3, 1978).

41–4 Spinal ganglia are found typically in the intervertebral foramina. Foramina is the plural of foramen: there is no such word as foraminae.

41–5 There are 31 pairs of spinal nerves: 8 cervical, 12 thoracic, 5 lumbar, 5 sacral, and 1 coccygeal.

41–6 The emergence of spinal nerves can be appreciated by remembering that (a) C.N.1 emerges between the skull and atlas and (b) C.N.8 emerges between C.V.7 and T.V.1. Hence all spinal nerves from T.N.1 downward emerge below their correspondingly numbered vertebrae.

41–7 The cauda equina is the bundle of nerve roots in the vertebral canal below those of L1. The term, which means "horse's tail" in Latin, is a translation of a Hebrew term found in the Talmud. Compression of the cauda equina (e.g., from a massive nuclear herniation) may cause pain and numbness (buttocks, back of thighs and legs, and soles), weakness in the lower limbs, and paralysis of the bladder and intestine.

41–8 The spinal cord ends at L.V.1 or 2; the subarachnoid space continues to S.V.2. Between these levels, frequently above or below L.V.5, the subarachnoid space is tapped in *lumbar puncture*. The needle pierces the ligamentum flavum between the laminae and traverses the epidural space before entering the subarachnoid space. Valuable practical information and illustrations can be found in R. Macintosh, *Lumbar Puncture and Spinal Analgesia*, 2nd ed., Livingstone, Edinburgh, 1957. See also figure 41–1.

41–9 The Latin word *spina* is used with reference to the spinal cord (e.g., corticospinal tracts) as well as to the vertebral column. The Latin word *medulla* means "marrow" but is used in English chiefly for the medulla oblongata, the lowest part of the brain stem (and formerly called the bulb). The medulla is continued downward as the spinal cord (*medulla spinalis* in Latin). The corresponding Greek word is *myelos* (marrow), as used in *myelomeningocele*, a type of spina bifida cystica in which the spinal cord and meninges protrude through the vertebral defect. Other examples are *myelography* (contrast radiography of the spinal cord) and *poliomyelitis* (literally gray-matter spinal cord inflammation). The root *myelo-*, however, is used also with reference to the bone marrow (e.g., in the cell termed the *myeloblast*).

PART 8 THE HEAD AND NECK

42 THE SKULL AND HYOID BONE

42–1 The cranium, a Latin form of a Greek term, means the skull. It is sometimes restricted to the skull without the mandible. The skull is complicated: the superb account by Augier in Poirier and Charpy's *Traité d'anatomie humaine* occupies nearly 600 pages! The various measurements used for the skull and for other parts of the skeleton can be found in G. Olivier, *Practical Anthropology*, trans. by M. A. MacConaill, Thomas, Springfield, Illinois, 1969.

42–2 The orbitomeatal plane is a horizontal plane through the upper margins of the external acoustic meatuses and the lower margin of (strictly speaking) the left orbit. It was accepted as a standard at an anthropological congress in Frankfurt in 1884.

42–3 Calcified areas may be found normally in the pineal body and in the choroid plexuses (glomera) of the lateral ventricles.

42–4 The nasal septum is formed by cartilage, ethmoid bone, and the vomer (see fig. 52–2*B*).

42–5 The temporal bone consists of squamous, tympanic, styloid, petrous and mastoid (sometimes considered together as petromastoid) parts.

42–6 The temporal lobes of the cerebral hemispheres extend inferiorly to approximately the level of the zygomatic arches, which are on the orbitomeatal plane (see fig. 43–8).

42–7 The temporal fossa extends from the temporal line downward to the zygomatic arch. It contains the temporal muscle.

42–8 The center of the pterion (parietal-frontal-sphenoid-temporal junction) is about 4 cm above the midpoint of the zygomatic arch and nearly the same distance behind the zygomatic process of the frontal. The pterion overlies the anterior branch of the middle meningeal artery on the internal aspect of the skull, and it corresponds also to the stem of the lateral sulcus of the brain (see fig. 43–8).

42–9 The infratemporal fossa lies behind the maxilla and between the lateral pterygoid plate and the ramus of the mandible. The pterygoid muscles, maxillary artery, and mandibular nerve are among the contents of the infratemporal fossa.

42–10 The occipital bone consists of a squamous part, two lateral parts, and a basilar part.

42–11 The occipital bone resembles a vertebra: basilar part and foramen magnum = vertebral body and foramen; squamous part and external occipital protuberance = laminae and spinous process; occipital condyles = articular processes; jugular processes = transverse processes. This vertebral theory of the skull, first pointed out by Goethe in 1790, was much disputed and exaggerated later. See L. Testut and A. Latarjet, *Traité d'anatomie humaine,* 9th ed., vol. *1*, Doin, Paris, 1948, p. 219; K. K. Singh-Roy, Anat. Anz., *120*:250, 1967.

42–12 The mandibular fossa and articular tubercle belong to the squamous part of the temporal bone.

42–13 The petrous part of the temporal bone is the wedge-shaped prolongation between the sphenoid and occipital bones at the base of the skull (see fig. 42–12). It is highly radio-opaque (see figs. 42–4 and 42–6): L., *petra,* rock, as in petrified. (Cf. "Thou art Peter and upon this rock . . .")

42–14 The main parts of the sphenoid (Gk, wedge-shaped) bone are the body and three pairs of processes or wings: greater wings, lesser wings, and pterygoid (Gk, wing-shaped) processes.

42–15 The cartilaginous part of the auditory tube lies in a groove between the petrous part of the temporal bone and the greater wing of the sphenoid bone.

42–16 The choanae are the posterior nasal apertures. They allow the nasal cavities to become continuous with the nasopharynx. "Posterior nares" would be a contradiction: nostrils are by definition anterior (external) openings.

42–17 The bony palate is formed by the palatine processes of the maxillae and by the horizontal plates of the palatine bones.

42–18 Cleft palate is a failure of union of the halves of the palate. It varies from merely a bifid uvula to an extensive cleft in the hard palate. It is frequently associated with cleft lip. A concise account of these deformities has been given by D. Marshall, Dent. Radiogr. Photogr., *46*:3, 19, 1973. Examples of embryologically oriented articles are those by J. H. Scott, Br. Dent. J., *120*:17, 1966; and M. C. Johnston, J. R. Hassell, and K. S. Brown, Clin. Plast. Surg., *2*:195, 1975.

42–19 The "sphenoidal ridge" is the posterior border of the lesser wing of the sphenoid bone. The "petrous ridge" is the superior border of the petrous part of the temporal bone. These two ridges separate the three cranial fossae.

42–20 The calvaria (L., cranial vault, an area that may become bald, *calvus*) is the skull cap. The plural is calvariae. There is no such word as "calvarium."

42–21 The pituitary region may be compared to a Turkish saddle (sella turcica), which has a high back (dorsum sellae) behind the seat (hypophysial fossa). The front part of the saddle, or pommel, has a median projection (tuberculum sellae). The pituitary region may also be thought of as a four-poster bed, the posts of which are the anterior and posterior clinoid processes. (Clinoid means "bed-shaped" in Greek; cf. clinical, at the bedside.) The hypophysis lies in bed, covered by a canopy (diaphragma sellae).

42–22 The mandible presents a body and a pair of rami. Each ramus ends above in two processes—coronoid and condylar (head and neck)—separated by the mandibular notch.

42–23 Important sites of growth of the skull include the sutures, sphenooccipital junction, nasal septum, and condylar processes of the mandible.

42–24 The main fontanelles are situated at the angles of the parietal bones. The anterior fontanelle is the largest and most important.

42–25 The hyoid bone presents a body and paired greater and lesser horns, or cornua.

42–26 The cranial exits of the cranial nerves are listed in table 43–2.

43 THE BRAIN, CRANIAL NERVES, AND MENINGES

43–1 The major divisions of the brain are listed in table 43–1.

43–2 The pons (L., a bridge), associated with the name of Varolius (1573) but better illustrated previously by Eustachius, appears superficially to bridge the two cerebellar hemispheres. These fibers actually connect one cerebellar hemisphere with the contralateral half of the pons. They are a part of the corticoponticerebellar pathway.

43–3 The roots of the trigeminal nerve are sensory (formerly termed *portio major*) and motor (formerly called *portio minor*). Intermediate fibers are probably accessory to the motor root (R. L. Saunders and E. Sachs, J. Neurosurg., *33*:317, 1970; M. Laude et al., Bull. Assoc. Anat., *60*:151, 1976), although a tactile function has also been proposed (J. Provost and J. Hardy, Neurochirurgie *16*:459, 1970).

43–4 The main parts of the midbrain are the tectum (colliculi) dorsally and the cerebral peduncles ventrally. Each peduncle consists of crus cerebri, substantia nigra, and tegmentum (see fig. 42–16*A*).

43–5 The parts of the diencephalon that can be seen from the surface of an intact brain include the optic chiasma, hypophysis, and mamillary bodies.

43–6 The corpus callosum consists of the rostrum, genu, trunk, and splenium.

43–7 The main lobes of the cerebral hemisphere are the frontal, parietal, occipital, and temporal (see fig. 43–4*A*).

43–8 The central and lateral sulci are important landmarks. The lateral sulcus (Sylvius, seventeenth century) is evident (see fig. 43–8). The central sulcus (Rolando, nineteenth century) is not as easily identified. It begins on the medial surface of the hemisphere, where it can be detected by following the cingulate sulcus backward and upward and then finding the next groove anteriorly (see fig. 43–5*B*).

43–9 The insula is an island (Reil, 1796) of cerebral cortex buried in the depths of the lateral sulcus.

43–10 Some 57 varieties of cerebral cortex, only some of which are known to have discrete functional significance, were numbered by K. Brodmann in 1907 on a somewhat arbitrary histological basis. Areas 1,2,3 (sensory areas), 4,6,8 (motor areas), 17, 18, 19 (visual areas), 22 (speech area of Wernicke), 41, 42 (auditory projection area of Heschl), and 44 (motor speech area of Broca) are functionally the most important (see fig. 43–6*A* and *B*).

43–11 The first pair of cranial nerves, in the accepted but arbitrary enumeration, are the olfactory filaments. It should be noted that the olfactory bulbs and tracts are parts of the cerebral hemispheres and are *not* cranial nerves.

43–12 The two lateral ventricles are followed by the third ventricle, aqueduct (Sylvius), fourth ventricle, and central canal. The lateral ventricles communicate with the third ventricle by the interventricular foramina (Monro, 1783). The fourth ventricle communicates with the subarachnoid space by median (Magendie, 1828) and lateral (Luschka, 1855) apertures.

43–13 Each lateral ventricle consists of an anterior horn, central part, and posterior and inferior horns.

43–14 The chief recesses of the third ventricle are the optic, infundibular, pineal, and suprapineal (see fig. 43–9).

43–15 The lateral recesses of the fourth ventricle curve forward to present near the front of the brain stem, close to cranial nerves 7 and 8 (cerebellopontine angle) and 9 (see fig. 43–1). The end of each recess opens as the lateral aperture.

43–16 The hypophysis comprises the adenohypophysis (infundibular, intermediate, and distal parts) and the neurohypophysis (median eminence, infundibular stem, and infundibular process) (see fig. 43–12*A*).

43–17 The cranial nerves and their components are summarized in table 43–2.

43–18 The chief parasympathetic ganglia associated with cranial nerves are the ciliary, pterygopalatine, otic, and submandibular (see table 43–3), but ganglionic cells are present diffusely in the cranial nerves (J.-A. Baumann and S. Gajisin, Bull. Assoc. Anat., *59*:329, 1975).

43–19 The chief processes of the dura are the falx (L., sickle) cerebri, tentorium (L., tent) cerebelli, falx cerebelli, and diaphragma sellae.

43–20 The tentorial notch is an important gap between the tentorium and the dorsum sellae. It contains the midbrain (see fig. 43–15).

43–21 The most important meningeal vessel clinically is the middle meningeal artery (see fig. 43–8). Middle meningeal hemorrhage is usually caused by an injury (often relatively trivial) associated with fracture of the temporal bone.

43–22 The subarachnoid space is the meshwork in the leptomeninges that contains cerebrospinal fluid (see fig. 43–11). Excessive C.S.F., either within the ventricular system or in the subarachnoid space, is termed *hydrocephaly*. The most frequent cause is a blockage of the normal circulation of C.S.F.

43–23 The most important cisterna is the cerebellomedullary (cisterna magna), which can be "tapped" through the posterior atlanto-occipital membrane (see fig. 41–1).

43–24 The term *carotid siphon* is used for the horizontal U-shaped bend (parts 2, 3, and 4) of the internal carotid artery, as seen on an angiogram (see figs. 43–18 and 43–19*A*).

43–25 The middle cerebral artery may be regarded as the continuation of the internal carotid artery. The term *carotid* (Gk, *karos,* heavy sleep) refers to the stupor that may result from compression of the carotid arteries. Under good conditions, however, one internal carotid artery can be occluded without any residual symptoms.

43–26 The posterior inferior cerebellar artery (often spoken of as "PICA") a branch of the vertebral artery, supplies the medulla behind the olive, an area that includes important nuclei and tracts. Vascular disease of this dorsolateral part of the medulla may result in widespread signs and symptoms (Wallenberg's syndrome, 1895).

43–27 The circulus arteriosus, described by Thomas Willis in 1664, is a polygonal anastomosis between the vertebral and internal carotid arteries (see fig. 43–17). Beautiful illustrations of normal and abnormal cerebral circulation can be found in M. Waddington, *Atlas of Cerebral Angiography with Anatomic Correlation,* Little, Brown, Boston, 1974. In certain mammals (e.g., sheep) part of the course of the internal carotid artery is taken over by a network (*rete mirabile*) that is supplied by an external vessel, e.g., the maxillary artery (P. M. Daniel, J. D. K. Dawes, and M. M. L. Prichard, Philos. Trans. R. Soc. Lond., *237B*:173, 1953; H. Minagi and T. H. Newton, Radiology, *86*:100, 1966).

43–28 The great cerebral vein (Galen, second century) is formed by the union of the left and right internal cerebral veins.

43–29 The confluence of the sinuses is the highly variable junction of the superior sagittal, straight, and left and right transverse sinuses. For variations see H. Browning (Am. J. Anat., *93*:307, 1953); A. Elmohamed and K.-J. Hempel (Frankfurt. Z. Pathol., *75*:321, 1966). The confluence was formerly called the *torcular* (L., wine press) *Herophili* (Herophilus, fourth century B.C.)

43–30 The cavernous sinus consists of one or more venous channels (sometimes a plexus) that extend between the superior orbital fissure and the apex of the petrous part of the temporal bone. Important relations are the internal carotid artery and cranial nerves 3, 4, 5, and 6 (see fig. 43–22).

43–31 The lateral lacunae receive emissary, diploic, meningeal, and sometimes cerebral veins (see fig. 43–11, *inset*). Detailed accounts are available, and these include H. M. Duvernoy, *The Superficial Veins of the Human Brain,* Springer, Berlin, 1975; and J. Browder and H. A. Kaplan, *Cerebral Dural Sinuses and their Tributaries,* Thomas, Springfield, Illinois, 1976.

44 THE EAR

44–1 The tympanic membrane is a sandwich composed of epidermis, a fibrous basis, and mucosa. The membrane is incised (*myringotomy*) in the postero-inferior quadrant of the pars tensa to avoid the ossicles and chorda tympani. The incision would allow the exit of fluid or pus from the middle ear (in *otitis media*).

44–2 The tympanic cavity communicates with (1) the mastoid air cells and mastoid antrum by way of the aditus and (2) the nasopharynx by way of the auditory tube (see fig. 44–3).

44–3 The epitympanic recess, an attic above the level of the typmanic membrane, contains the head of the malleus and the body and short crus of the incus (see fig. 44–4).

44–4 The tegmen (L., covering; cf. integument) tympani is the portion of the petrous part of the temporal bone that covers the middle ear and separates it from the middle cranial fossa.

44–5 The fenestra vestibuli (oval window) is closed by the base of the stapes. The fenestra cochleae (round window) is closed by the secondary tympanic membrane (the lateral layer of which is tympanic mucosa). The wall of the labyrinth is subject to bone disease (*otosclerosis*), and invasion of the fenestra vestibuli causes fixation of the stapes. To establish an alternative sound pathway, an artificial window can be made surgically (*fenestration*) in the lateral semicircular canal.

44–6 The incudomallear and incudostapedial joints (see fig. 44–6) are saddle and ball-and-socket synovial joints, respectively. The stapediovestibular junction at the fenestra vestibuli is either a fibrous or a synovial joint.

44–7 The nervus intermedius is the smaller part of the facial nerve, intermediate in position between cranial nerve 8 and the larger part of 7. The nervus intermedius joins the remainder (motor root) of 7 in the internal acoustic meatus. It contains fibers transmitted by the chorda tympani —taste fibers for the anterior two thirds of the tongue, secretomotor fibers for salivary glands, and some pain fibers—and also fibers for the lacrimal gland (see fig. 44–7).

44–8 The geniculate ganglion contains the cells of origin of the taste fibers in the facial nerve (see fig. 44–7). Similarly, the trigeminal ganglion contains the cells of origin of most of the sensory fibers in the trigeminal nerve, and a comparable afferent significance applies to the ganglia of cranial nerves 8, 9, and 10. Hence these sensory ganglia differ functionally from the parasympathetic ganglia associated with the cranial nerves: the ciliary, pterygopalatine, otic, and submandibular ganglia (see table 43–3).

44–9 The facial nerve (see fig. 44–9) runs (a) laterally to the geniculate ganglion, (b) backward after it leaves the ganglion, and (c) downward to the stylomastoid foramen. It then proceeds forward in the parotid gland (see fig. 47–5).

44–10 The chorda tympani, not recognized as a nerve by Falloppio (1561), is a branch of the facial nerve that crosses the tympanic membrane and joins the lingual nerve to gain the tongue. The chorda contains (1) taste fibers from the anterior two thirds of the tongue and (2) secretory fibers, which synapse in the submandibular ganglion, the postganglionic fibers supplying the submandibular gland (as discovered by Claude Bernard, 1851). The fibers of the chorda tympani run in the nervus intermedius (see fig. 44–7).

44–11 The level of a lesion of the facial nerve is inferred from effects that depend on whether specific branches are intact. In supranuclear lesions (i.e., above the facial nucleus in the pons), the upper third of the face is spared whereas the lower two thirds are paralyzed. In other words, a lesion of the *upper* motor neuron affects the voluntary movements in only the *lower* part of the face. The reason is that the frontalis, orbicularis, and corrugator are innervated bilaterally. In infranuclear lesions, all the facial muscles are involved, and taste and salivary secretion (chorda tympani) may be affected, depending on the level of damage. Interference with the stapedius may cause painful sensitivity to sounds. Below the stylomastoid foramen, complete or partial facial paralysis may result. The common type of unilateral facial paralysis is a lesion of uncertain origin near the stylomastoid foramen: *Bell's palsy* (1821).

44–12 Important signs of facial paralysis are inability to wrinkle the forehead and shut the eyes tightly (in infranuclear lesions) and inability to smile and show the teeth on one side. When an attempt is made to display the teeth, the angle of the mouth is drawn up on only the normal side and the mouth characteristically appears triangular (see fig. 47–4*D*)

44–13 The otic capsule is a layer of dense bone around the perilymphatic space in the petrous part of the temporal bone. In the adult, the capsule is an integral part of the petrous portion of the bone and no longer maintains the anatomical individuality it possessed in the fetus.

44–14 The main parts of the osseous labyrinth are the semicircular canals, vestibule, and cochlea. The membranous labyrinth comprises the semicircular ducts, utricle and saccule, and cochlear duct (see table 44–2). The cochlea is the essential organ of hearing. The vestibular apparatus, concerned with equilibration, includes the utricle and semicircular ducts.

44–15 The lateral semicircular canal becomes horizontal only when the head is flexed about 30 degrees. A paper model of the three canals is simple to construct (J. D. Lithgow, J. Laryngol., *35*:81, 1920).

44–16 The apex of the cochlea points anterolaterally, whereas that of the petrous part of the temporal bone is directed anteromedially (see fig. 44–9).

44–17 In a section through the cochlea, the scala vestibuli is seen to be separated from the scala tympani by the osseous spiral lamina. The cochlear duct is separated from the scala vestibuli and scala tympani by the vestibular (Reissner, 1851) and basilar membranes, respectively (see fig. 44–10*D*).

44–18 The neuro-epithelial areas in the membranous labyrinth are in the ampullae of the semicircular ducts (ampullary crests), in the utricle and saccule (maculae), and in the cochlear duct (spiral organ). This makes a total of six receptive areas in each ear (see fig. 44–11). The spiral organ (Corti, 1851) is the organ of hearing.

45 THE ORBIT

45–1 The lacrimal and ethmoid bones are likely to be splintered in injuries to the rim of the orbit, and the ethmoidal air sinuses are frequently opened. A paper model to show the composition of the bony orbit can be useful (R. O'Rahilly, Anat. Rec., *141*:315, 1961).

45–2 The nasociliary nerve accompanies the ophthalmic artery. Rarely, the ophthalmic artery arises from the middle meningeal artery (J. Brucher, Radiology, *93*:51, 1969), or the middle meningeal artery may come from the ophthalmic artery (O. F. Gabriele and D. Bell, Radiology, *89*:841, 1967). These variations are associated with the anastomosis between the lacrimal and middle meningeal arteries, which is ultimately between the internal and external carotid arteries.

45–3 The central artery of the retina, complete obstruction of which results in blindness, is the most important branch of the ophthalmic artery. Details of the central artery have been published (S. Singh and R. Dass, Br. J. Ophthalmol., *44*:193, 280; 1960), as have also accounts of the ophthalmic artery (S. S. Hayreh and R. Dass, Br. J. Ophthalmol. *46*:65, 165, 212; 1962).

45–4 The abducent nerve, which bends sharply across the petrous part of the temporal bone, is liable to damage from increased intracranial pressure.

45–5 The ciliary ganglion, which lies between the optic nerve and the lateral rectus muscle, is the peripheral relay station of the parasympathetic fibers to the eye (see fig. 45–6).

45–6 Cranial nerve 2, 3, and 6 and the nasociliary nerve enter the orbit within the common tendinous ring (see fig. 45–3).

45–7 The orbicularis oculi (facial nerve) closes the eyelids, whereas the levator palpebrae superioris (oculomotor nerve) ''opens the eye.''

45–8 *Horner's syndrome,* caused by a lesion of the cervical sympathetic nerves, is characterized chiefly by ptosis, seeming enophthalmos, and cutaneous hyperemia. The condition, described by Horner in 1869, had been recognized by Claude Bernard in 1862.

45–9 On looking downward and to the right, if the left pupil fails to descend, the left superior oblique muscle is likely to be paralyzed (see fig. 45–8).

45–10 On looking upward and to the left, if the right pupil fails to ascend, the right inferior oblique muscle is most likely to be involved (see fig. 45–8).

46 THE EYE

46–1 Most of the optical power of the eye is concentrated at the front surface of the cornea.

46–2 The conjunctival epithelium continues as the front layer (anterior epithelium) of the cornea (see fig. 46–3). The combining form *kerato-* refers to the cornea (e.g., in *keratitis,* inflammation of the cornea).

46–3 The aqueous humor in the anterior chamber drains into the sinus venosus sclerae (see fig. 46–3, *inset*), a canal described by Schlemm in 1830.

46–4 When ''drops'' are used in the examination of the eye, the sphincter pupillae and the ciliary muscle are removed from action by parasympathetic paralysis. The dilator pupillae is then unopposed.

46–5 The posterior and anterior chambers communicate at the pupil.

46–6 *Glaucoma,* which may lead to blindness if untreated, is characterized by increased intra-ocular pressure. A narrow iridocorneal angle predisposes a person to ''closed-angle glaucoma.'' However, in another type (''open-angle glaucoma''), symptoms may not develop until late in the disease, when the optic discs and retinae have been seriously affected. The intra-ocular pressure is controlled by medicaments, e.g., miotic drops, which prevent dilatation of the pupil and consequent embarrassment of the iridocorneal angle.

46–7 In obstruction of the aqueous pathway, e.g., in glaucoma, excision of a portion of the iris (*iridectomy*) would keep the iridocorneal angle open and allow adequate communication between the posterior and anterior chambers.

46–8 The ciliary muscle, supplied by parasympathetic fibers from the ciliary ganglion (see fig. 45–6), is involved in accommodation. Contraction causes increased curvature of the lens.

46–9 The ciliary muscle and sphincter pupillae are supplied by parasympathetic fibers (synapses in ciliary ganglion). Sympathetic fibers (synapses in superior cervical ganglion) supply the dilator pupillae and also the orbitalis and superior tarsal muscle. The orbitalis (Müller, 1858) partly closes the inferior orbital fissure (C. Vermeij-Keers, Z. Anat. Entwgesch., *141*:77, 1973). The superior tarsal (or palpebral) muscle arises from the under surface of the levator palpebrae superioris and is inserted into the upper border of the tarsal plate. This unusual association of smooth (tarsal) and skeletal (levator) muscle elevates the upper eyelid. In *Horner's syndrome* (sympathetic interruption), drooping (*ptosis*) of the lid occurs in the absence of the tarsal component.

46–10 So-called *detatchment of the retina* is really a separation between layer 1 and layers 2 to 10. This is the developmental cleavage plane of the embryonic optic cup.

46–11 The macula and fovea (in line with the visual axis) are temporal (lateral) to the optic disc (or nerve head). See figures 45–2 and 46–5.

46–12 A cataract is an opacity of the lens. Such a lens can be removed surgically. Formerly it was merely displaced downward ("couched") out of the line of vision, and this is one of the oldest surgical operations.

47 THE SCALP, AURICLE, AND FACE

47–1 The layers of the scalp are as follows: *s*kin, *c*lose subcutaneous tissue, *a*poneurosis (galea aponeurotica) and occipitofrontalis, *l*oose subaponeurotic tissue, and *p*ericranium. That the initial letters form the word *scalp* is an excellent mnemonic. Deep to the scalp are the skull, meninges and subarachnoid space, and cerebral cortex (see fig. 47–1).

47–2 The antagonists of the orbicularis oculi are the levator palpebrae superioris and the frontalis.

47–3 The buccinator is the main muscle of the cheek, which it keeps taut and free from the teeth. It is said to be concerned with whistling and blowing (L., *buccinator,* trumpeter). The buccinator and the more superficially placed masseter are overlain by the buccal pad of fat, which contributes to the roundness of an infant's cheek and may aid in suckling. The pad and the buccinator are pierced by the parotid duct, which possesses a sphincter at that level (J. Bossy, L. Gaillard, and J.-P. Saby, J. Fr. Otorhinolaryngol., *14*:71, 1965).

47–4 The orbicularis oris is a complicated sphincter that closes the lips and can also protrude them. It forms a ring (L., *orbicularis,* a small ring) around the mouth. Several sets of fibers are included: deeply, fibers from the buccinator; more superficially, fibers from the levator and depressor anguli oris and other small facial muscles; and fibers proper to the lips.

47–5 The facial muscles develop from pharyngeal arch 2 and hence are supplied by the facial nerve.

47–6 Hemorrhage from a cut lip is controlled by compressing both parts between the index fingers and thumbs, because the labial arteries of the two sides anastomose across the median plane.

47–7 The area around the nose and upper lip is termed the "danger area" of the face because it is drained by the facial vein, which is connected with the cavernous sinus. Thrombosis of the cavernous sinus is a dangerous condition.

47–8 The "vertex-ear-chin line" is generally taken to indicate the boundary between the cranial (i.e., trigeminal) and spinal cutaneous territories of the head and neck (see fig. 47–6), but overlapping does occur.

47–9 The two main modes of spinal innervation of the skin are (1) by numbered nerves in dermatomes (e.g., C.N.4) and (2) by named nerves (e.g., great auricular). *Herpes zoster* ("shingles") is believed to be a viral disease of dorsal root ganglia, resulting in a cutaneous eruption that corresponds to the distribution of the dorsal root fibers of the affected ganglion. The term *herpes zoster* in Greek means literally a creeping, girdle-shaped condition, and the word *shingles* comes from the Latin *cingulum,* a girdle. A very instructive series of photographs of successive dermatomes involved in herpes zoster can be found in K. Hansen and H. Schliack, *Segmentale Innervation,* Thieme, Stuttgart, 1962.

47–10 The trigeminal ganglion (Gasser, eighteenth century) is situated in a dural recess, the cavum trigeminale (Meckel, 1748), on the petrous part of the temporal bone. It contains the cells of origin of most of the afferent fibers in the three divisions of cranial nerve 5. Any one of these divisions may be involved in herpes zoster. Specific, temporary pain in the trigeminal area (*trigeminal neuralgia*) may be paroxysmal in type (*tic douloureux*). These conditions are extremely complicated. (See, for example, R. Hassler and E. A. Walker, eds., *Trigeminal Neuralgia,* Thieme, Stuttgart, 1970.) The Latin word *trigeminus* (''triplets'') was given to the nerve because of its three main divisions.

48 THE PAROTID, TEMPORAL, AND INFRATEMPORAL REGIONS

48–1 The facial nerve traverses the parotid gland (see fig. 48–2*B*) and is liable to injury during *parotidectomy*. The parotid gland consists of a superficial and a deep portion (''lobe'') or layer wrapped around the branches of the facial nerve and connected by one or more isthmuses. Examples of the extensive literature are R. A. Davis et al., Surg. Gynecol. Obstet., *102*:385, 1956; J. Winsten and G. E. Ward, Surgery, *40*:585, 1956; and D. H. Patey and I. Ranger, Br. J. Surg., *45*:250, 1957.

48–2 The temporal and infratemporal fossae communicate deep to the zygomatic arch.

48–3 The masseter and the medial and lateral pterygoid muscles have at least two heads each. Actually, the masseter has three (J. D. B. MacDougall, Br. Dent. J., *98*:193, 1955). It has been suggested that the lateral pterygoid muscle comprises two functionally different muscles (P. G. Grant, Am. J. Anat., *138*:1, 1973; J. A. McNamara, Am. J. Anat., *138*:197, 1973), although this concept has also been questioned (R. P. Lehr and S. E. Owens, Anat. Rec., *196*:441, 1980).

48–4 The muscles of mastication are supplied by the mandibular nerve, which contains all the motor fibers of the trigeminal nerve.

48–5 The squamous part of the temporal bone provides the upper surface for the temporomandibular joint (see fig. 42–12).

48–6 The sphenomandibular ligament, of doubtful mechanical significance, develops from the sheath of the (Meckel's) cartilage of pharyngeal arch 1. Even in the adult, it can be traced back to the petrotympanic fissure and the anterior ligament and process of the malleus (J. Bossy and L. Gaillard, Acta Anat., *52*:282, 1963; J. G. Burch, Anat. Rec., *156*:433, 1966), so that it is really a malleo-tympano-sphenomandibular ligament.

48–7 The posterior fibers of both temporal muscles retract the mandible.

48–8 Dislocation of the temporomandibular joint is reduced by strong downward pressure on the back of the mandible (to overcome spasm).

48–9 The middle meningeal artery (see fig. 43–8) is clinically the most important branch of the maxillary artery. Hemorrhage from it may occur in head injuries.

48–10 Pain from disease of a tooth or of the tongue is sometimes referred to the distribution of the auriculotemporal nerve to the ear.

48–11 The afferent fibers that converge on the pterygopalatine ganglion belong mostly to the maxillary nerve and do not synapse in the ganglion (see fig. 48–7).

48–12 The ciliary, pterygopalatine, otic, and submandibular ganglia are basically collections of parasympathetic postganglionic neurons (see figs. 48–7, 48–9, and 48–11). The trigeminal ganglion houses the pseudounipolar neurons that are the cells of origin of most of the afferent fibers in the three divisions of the nerve.

49 THE SUBMANDIBULAR REGION

49–1 The submandibular gland is wrapped around the posterior border of the mylohyoid muscle (see fig. 49–1). The old term *submaxillary gland* dates from the era when the mandible was referred to as the inferior maxilla. The submandibular duct was described by Wharton in 1656. (The parotid duct was portrayed by Stensen, or Steno, in 1662.)

49–2 The anterior and posterior bellies of the digastric muscle are supplied by cranial nerves 5 and 7, respectively. This is associated with their development. The muscles of mastication, the mylohyoid muscle, the anterior belly of the digastric muscle, and the tensor tympani develop from pharyngeal arch 1 (supplied by the mandibular nerve), whereas the facial muscles, the stylohyoid muscle, the posterior belly of the digastric muscle, and the stapedius develop from pharyngeal arch 2 (supplied by the facial nerve).

49–3 The mylohyoid muscles, which form the diaphragma oris, are inserted into a median raphe and the body of the hyoid bone. The disyllabic word *raphe* means a seam or suture in Greek.

49–4 The submandibular ganglion is suspended from the lingual nerve (see fig. 49–3).

49–5 The middle (second) part of the lingual artery is covered directly by the hyoglossus (see fig. 49–4, *interrupted lines*).

50 THE NECK

50–1 The sternomastoid muscle arises from the sternum and clavicle (Gk, *kleis,* key, refers to the clavicle in the form *cleido*) and is inserted into the mastoid process and occipital bone. Actually, the muscle is therefore sternomastoid and cleido-occipital (and there are deeper cleidomastoid fibers). The muscle is the anatomical and clinical key to the neck, dividing it into anterior and posterior triangles. The development of the sternomastoid muscle and trapezius is complicated (J. McKenzie, Contrib. Embryol. Carnegie Instn., *37*:121, 1962), but it is of particular interest because of the double (cranial and spinal) innervation.

50–2 The accessory nerve is tested by asking the subject to shrug the shoulders (trapezius) and then to rotate the head (sternomastoid muscle).

50–3 The most important contents of the posterior triangle are the accessory nerve, brachial plexus, subclavian artery, and lymph nodes. Tuberculous lymph nodes ("glands in the neck") in the posterior triangle were formerly very common.

50–4 The carotid triangle (sternomastoid/posterior belly of the digastric muscle/superior belly of the omohyoid muscle) is named from its relationship to the carotid arteries. The triangle contains the external carotid artery and its three anterior branches (facial, lingual, and superior thyroid), the hypoglossal nerve, and the greater horn of the hyoid (see figs. 50–2 and 50–12). More deeply placed are the superior laryngeal nerve and C.V.3,4. The common and internal carotid arteries, together with the internal jugular vein and vagus, generally lie under cover of the sternomastoid and therefore, strictly speaking, are behind the triangle. A smaller triangle (Farabeuf, 1872) is formed by the hypoglossal nerve and the (common) facial and internal jugular veins.

50–5 The Greek work *eidos* means "like". Thus hyoid means U-shaped; thyroid, shield-shaped; cricoid, ring-shaped; arytenoid, pitcher-shaped; sphenoid, wedge-shaped; pterygoid, wing-like; and clinoid, shaped (fancifully) like a bed.

50–6 The thyroid gland, like the prostate, has a capsule, outside which is a fascial sheath ("false capsule"). In *thyroidectomy,* the gland and its capsule are removed. Preservation of the parathyroid glands can be ensured by leaving the posterior part of each thyroid lobe in place.

50–7 A goiter is a non-neoplastic and non-inflammatory enlargement of the thyroid gland. The condition is endemic in certain regions.

50–8 The recurrent laryngeal nerves are closely related to the branches of the inferior thyroid artery. Because of "the variability rather than the vulnerability" (Berlin) of the nerve, injury may occur during thyroid surgery. Examples of the extensive literature are P. Blondeau (J. Chir., *102*:397, 1971) and J. M. Loré, D. J. Kim, and S. Elias (Ann. Otol. Rhinol. Laryngol., *86*:777, 1977).

50–9 The thyroid gland develops largely as a median diverticulum from the floor of the pharynx. Parts of the embryonic thyroglossal duct may remain as cysts, the pyramidal lobule, and accessory thyroid tissue (e.g., in the thorax). *Thyroglossal cysts* are in, or close to, the median plane, and they may be found at any level between the mouth and the cricoid cartilage. In the presence of a median swelling in the neck, thyroglossal cysts and enlarged lymph nodes should be kept in mind. For a lateral swelling, cysts of pharyngeal pouch origin and tuberculous nodes should be considered. Rarely, the thyroid fails to descend and develops in the tongue (*lingual thyroid*). Examples are given by D. J. Weider and W. Parker (Ann. Otol, Rhinol. Laryngol., *86*:841, 1977) and by M. R. Kamat et al. (Br. J. Surg., *66*:537, 1979).

50–10 *Tracheotomy* is performed usually through rings 2 to 4. Hence the isthmus of the thyroid gland is incised. In an emergency, *cricothyrotomy* is considered preferable for non-surgeons.

50–11 In terms of surface anatomy, the carotid arteries ascend from (1) the sternoclavicular joint, along the anterior border of the sternomastoid muscle, to (2) a point medial to the lobule of the auricle. The common carotid artery divides usually at the level of the upper border of the lamina of the thyroid cartilage. Variations are given by D. K. McAffee, B. J. Anson, and J. J. McDonald (Q. Bull. Northw. Univ. Med. Sch., *27*:226, 1953). The carotid surface line indicates also the approximate position of the internal jugular vein, vagus, and sympathetic trunk.

50–12 The branches of the external carotid artery are listed in table 50–2.

50–13 The glossopharyngeal nerve, which is afferent from the tongue and pharynx (hence its name), is tested by tactile sensation (swabbed applicator) in the vault of the pharynx and sometimes by taste on the back of the tongue (this is difficult to determine, however).

50–14 Below the jugular foramen, the vagus is joined by the internal branch of the accessory nerve, the fibers of which are "accessory" to, and distributed with, branches of the vagus (see fig. 50–19). These accessory fibers are believed to include those (1) in the pharyngeal branches of the vagus that supply most of the muscles of the pharynx and soft palate, (2) in the external branch of the vagus that innervates the cricothyroid muscle, and (3) in the recurrent laryngeal nerves of the vagus that supply the muscles of the larynx (see fig. 50–18). Accessory fibers in the vagus are tested by asking the subject to say "ah": the uvula should proceed backward in the median plane. Laryngoscopy reveals the condition of the vocal folds.

50–15 The hypoglossal nerve curves below ("hypo-") the tongue, the muscles of which it supplies. It is tested by asking the subject to protrude the tongue. A lesion of one hypoglossal nerve would result in deviation of the protruded tongue toward the affected side.

50–16 The middle (second) part of the subclavian artery and the brachial plexus are lodged between the scaleni anterior and medius (see fig. 50–28*B*), where compression (accentuated by a cervical rib) produces one of the neurovascular syndromes of the upper limb.

50–17 The third and most superficial part of the subclavian artery can be compressed or ligated in the angle between the clavicle and the posterior border of the sternomastoid muscle (see fig. 50–3*B*). The collateral circulation to the upper limb (e.g., suprascapular artery with anastomosis around scapula) is generally adequate after ligation of any of the three parts of the subclavian artery.

50–18 The pyramidal interval between the scaleni and the longus colli houses the cupola of the pleura (and therefore also the apex of the lung) and the vertebral artery.

50–19 The sympathetic supply to the head and neck leaves the spinal cord (chiefly segments T1 and 2) in ventral roots and passes through rami communicantes to the thoracic part of the sympathetic trunk. These preganglionic fibers ascend in the trunk to reach and synapse in the three or four cervical ganglia (fig. 50–24).

50–20 The stellate ganglion is a synonym for the cervicothoracic ganglion, i.e., the combined inferior cervical and first thoracic sympathetic ganglia (see fig. 50–24). *Stellate block* (e.g., in the treatment of vascular disease in the upper limb) can be performed through a needle inserted above the middle of the clavicle and directed medially toward the head of rib 1. To digress from the stars to the sun, the term *solar plexus* is a popular name for the celiac plexus.

50–21 The deep cervical lymph nodes are largely concentrated in a chain along the internal jugular vein (see fig. 50–27).

50–22 The phrenic nerve (C.N.3,4,5) descends on the front of the scalenus anterior (see figs. 50–8 and 50–17).

50–23 The main layers of the cervical fascia are (1) investing, (2) pretracheal (more extensive than its name suggests), and (3) prevertebral (see fig. 50–29). Detailed accounts are available: M. Grodinsky and E. A. Holyoke, Am. J. Anat., *63*:367, 1938; and E. S. Meyers, *The Deep Cervical Fascia*, University of Queensland Press, Brisbane, Australia, 1950. Several articles on special regions have appeared, e.g., the parotid fascia (G. R. L. Gaughran, Ann. Otol. Rhinol. Laryngol., *70*:31, 1961).

50–24 The carotid sheath contains the common and internal carotid arteries, internal jugular vein, and vagus.

51 THE MOUTH, TONGUE, AND TEETH

51–1 The parotid duct opens opposite the upper second molar, where it may be injected with iodized oil, thereby outlining the course of the duct radiographically (*sialography*).

51–2 The opening between the oral cavity and the oropharynx is called the faucial isthmus. It is situated between the pillars known as the palatoglossal arches.

51–3 The muscles of the soft palate (except the tensor) are believed to be supplied mainly through the pharyngeal plexus by the internal branch of the accessory nerve. Other cranial nerves (7, 9, 12) may perhaps contribute. The pharyngeal plexus on the middle constrictor of the pharynx is formed by branches of cranial nerves 9 and 10 and by sympathetic fibers. The vagal contribution (pharyngeal branch of vagus), however, consists principally of accessory fibers.

51–4 The sulcus terminalis is the boundary between (a) the oral and (b) the pharyngeal parts of the tongue. The boundary between the innervation by (a) the lingual nerve and chorda tympani and (b) that of the glossopharyngeal nerve is in front of the sulcus (see fig. 51–2). The apex of the sulcus, which may present a minute pit termed the foramen cecum, is the site of origin of the embryonic thyroglossal duct. *Thyroglossal cysts* may be found at any level between the mouth and the cricoid cartilage. Rarely, the thyroid fails to descend and develops in the tongue (*lingual thyroid*).

51–5 The vallate papillae are immediately in front of the sulcus terminalis. *Vallum* in Latin means a wall or rampart. Each papilla, surrounded by a moat, is within the circumference of the wall and not around it: hence the old term *circumvallate* is not suitable.

51–6 The valleculae are depressions, one on each side of the median glosso-epiglottic fold.

51–7 The attachment of the tongue to the mandible is important clinically because, by pulling the mandible forward during anesthesia, one prevents the tongue from falling backward and obstructing respiration.

51–8 The lymphatic drainage of the tongue is mainly to the submental, submandibular, and deep cervical nodes (see fig. 51–4). Carcinoma of the tongue may be spread by lymphatics to the mandible and neck, and local spread may occur to the floor of the mouth, tonsil, epiglottis, and soft palate.

51–9 The anatomical crown of a tooth is the part covered by enamel, whereas the clinical crown is the part that projects into the oral cavity (see fig. 51–5).

51–10 The upper molar roots are closely related to the floor of the maxillary sinus. Hence pulpal infections may cause sinusitis, or sinusitis may cause toothache.

51–11 Most teeth are successional, i.e., they replace (succeed) corresponding milk teeth. The permanent molars, however, are accessional, i.e., they are added behind (are an accession to) the milk molars.

51–12 The milk teeth develop before birth but do not erupt until between ½ and 2½ years.

51–13 The first permanent teeth to erupt are generally the "6-year molars," which are the sixth teeth (i.e., first molars) in each quadrant and which erupt usually at the age of 6 or 7 years.

51–14 In the "symbolic system," the permanent teeth in each quadrant are numbered from 1 mesially to 8 distally. In the "sequential system," the teeth are numbered sequentially from 1 to 16, right teeth to left in the upper jaw, and then from 17 to 32, left teeth to right in the lower jaw. For a description of three systems of such "dental shorthand," see R. Ashley and T. Kirby, *Dental Anatomy and Terminology,* Wiley, New York, 1977.

51–15 Occlusion is any functional relation established when the upper and lower teeth come into contact with each other. Keeping the teeth (Gk, *odous,* tooth) straight (Gk, *orthos,* straight) is the literal meaning of *orthodontics,* just as keeping children. (Gk, *pais,* child) straight is the literal meaning of *orthopaedics*.

52 THE NOSE AND PARANASAL SINUSES

52–1 The anterior and posterior apertures of the nose are the nostrils (or nares) and the choanae, respectively.

52–2 The nasal conchae are three or four shell-shaped, bony projections on the lateral wall of the nasal cavity. *Concha* is the Latin form of a Greek work meaning "shell." The inferior concha is an independent entity, whereas the others are parts of the ethmoid bone (see fig. 52–1). Each concha is covered by a thick, highly vascular mucosa that warms and moistens the incoming air. The inferior concha has been likened to an air-conditioning plant (Negus). Large vein-like spaces ("swell bodies") over the middle and inferior conchae may become congested with blood during a "cold in the nose" (*coryza*) or sometimes during menstruation.

52–3 The main openings into the nasal meatuses are the posterior ethmoidal cells into the superior meatus, the anterior ethmoidal cells and frontal and maxillary sinuses into the middle meatus, and the nasolacrimal duct into the inferior meatus. The sphenoidal sinus and some posterior ethmoidal cells drain into the spheno-ethmoidal recess above the superior concha (see fig. 52–3*B*).

52–4 The olfactory region of the nasal cavity is bounded by the superior nasal concha and the upper third of the nasal septum. It is supplied by the olfactory nerve, which ends in the olfactory bulb: the olfactory bulb and tract are *not* part of the first cranial nerve.

52–5 For general sensation, the nasal cavity is supplied by the ophthalmic and maxillary nerves, the latter largely through branches of the pterygopalatine ganglion (see fig. 52–5).

52–6 The most important arteries to the nasal cavity are the sphenopalatine (from the maxillary artery) and the anterior ethmoidal (from the ophthalmic artery). The most frequent source of nasal bleeding (*epistaxis*) is the junction between septal branches of the superior labial and sphenopalatine arteries (see fig. 52–5).

52–7 The paranasal sinuses develop as outgrowths from the nasal cavity; hence they all drain directly or indirectly into the nose. Thus, *rhinitis* may give rise to *sinusitis*.

52–8 The chief paranasal sinuses are the maxillary, ethmoidal, frontal, and sphenoidal. The maxillary sinus, associated with the name of Highmore (1651), was well known earlier and was masterfully illustrated by Leonardo da Vinci. The functional significance of the paranasal sinuses is obscure (P. L. Blanton and N. L. Briggs, Am. J. Anat., *124*:135, 1969). The maxillary sinus can be irrigated through the nasal cavity (F. N. Ritter, Laryngoscope, *87*:215, 1977). The hypophysial fossa may be approached by way of the nasal cavity and sphenoidal sinus for removal of tumors. Cf. figure 53–4. For details of the sinuses, see O. E. Van Alyea, *Nasal Sinuses. An Anatomic and Clinical Consideration,* 2nd ed., Williams & Wilkins, Baltimore, 1951. The maxillary sinus has been described by B. Vidić in *Orban's Oral Histology and Embryology,* 8th ed., ed. by S. N. Bhasker, Mosby, St. Louis, 1976. This is a useful book for dental students.

52–9 Each ethmoidal labyrinth is the lateral mass of the ethmoid bone, and it comprises 4 to 17 air cells (the ethmoidal sinus). The labyrinth is situated between the medial wall of the orbit laterally and the lateral wall of the nasal cavity medially (see fig. 52–1). A labyrinth (Gk, maze) is an intricate arrangement of communicating passages; hence it is an appropriate name for the ethmoidal labyrinth as well as for that of the internal ear.

52–10 A postnasal mirror reveals the back part of the nasal cavity, as seen through the choanae, and including the posterior edge of the nasal septum and the openings of the auditory tubes (see figs. 52–4 and 53–11*B*). The procedure is termed posterior *rhinoscopy* (Gk, *rhis, rhinos,* nose; as in rhinoceros, from *keras,* horn. Cf. keratin).

53 THE PHARYNX AND LARYNX

53–1 The pharynx is situated in front of C.V.1 to 6 (see fig. 53–1). Occasionally the nasopharynx and laryngopharynx are referred to as the *epipharynx* and *hypopharynx,* respectively. Similarly, the epitympanic recess and the tympanic cavity below the level of the tympanic membrane are sometimes called the *epitympanum* and *hypotympanum,* respectively.

53–2 The nasopharynx is the uppermost part of the pharynx, but (at least in its anterior part) it may also be regarded as the back portion of the nasal cavity (F. W. Jones, J. Anat., *74*:147, 1940; K. Leela, R. Kanagasuntheram, and F. Y. Khoo, J. Anat., *117*:333, 1974).

53–3 The pharyngeal isthmus (between the nasopharynx and oropharynx) is bounded by the soft palate, palatopharyngeal arches, and posterior wall of the pharynx.

53–4 *Adenoids* (Gk, gland-like) are hypertrophied (naso)pharyngeal tonsils on the posterior wall of the nasopharynx. They may cause respiratory obstruction. Their removal (*adenoidectomy*), which was first undertaken in 1868, is generally combined with tonsillectomy.

53–5 The pharyngeal hypophysis, situated on the posterior wall of the pharynx, develops at the pharyngeal end of the stalk of the craniopharyngeal pouch (Rathke, 1838). Like the sellar hypophysis, it is an endocrine gland (P. McGrath, J. Endocrinol., *42*:205, 1968) and contains several types of secretory cells (C. B. González, G. F. Valdés, and D. R. Ciocca, Acta Anat., *97*:224, 1977).

53–6 The so-called auditory tube, described by Eustachi (1563) but known even before the time of Christ, would be better named the pharyngotympanic tube. Its cartilaginous part is a diverticulum of the pharynx that opens behind the inferior nasal concha (see fig. 53–4). The osseous part is a prolongation of the tympanic cavity, into the anterior wall of which it opens. The tube is closed at rest but opens during swallowing and phonation, perhaps by a "milking" action of the levator and tensor (S. Seif and A. L. Dellon, Cleft Palate J., *15*:329, 1978; see also V. K. Misurya, Arch. Otolaryngol., *102*: 265, 1976). A detailed account of the tube is available in J. Terracol, A. Corone, and Y. Guerrier, *La trompe d'Eustache,* Masson, Paris, 1949.

53–7 The faucial (or oropharyngeal) isthmus is bounded by the soft palate, palatoglossal arches, and tongue.

53–8 The pharyngeal lymphatic ring (Waldeyer, 1884) comprises the nasopharyngeal, tubal, palatine, and lingual tonsils. It is presumed to be a protective collar against infections and organisms that might enter through the nose and mouth.

53–9 The (palatine) tonsils are located between diverging pillars on each side of the pharynx, namely the palatoglossal and palatopharyngeal arches. *Tonsillectomy,* an operation described by Celsus in the first century A.D., is now performed either by dissection or by a special instrument known as a guillotine.

53–10 The piriform recess (or sinus or fossa), in which foreign bodies may become lodged, is the part of the cavity of the laryngopharynx situated on each side of the inlet of the larynx (see fig. 53–3).

53–11 A *pharyngeal diverticulum* usually forms posteriorly through the fibers of the inferior constrictor (between the thyropharyngeal and cricopharyngeal fibers). Increased intrapharyngeal pressure is regarded as an important factor in the production of a "pulsion diverticulum" through a weak area ("Killian's dehiscence") between the parts of the inferior constrictor. Moreover, swallowing in the presence of cricopharyngeal incoordination may be important in allowing mucosal herniation through a weak area in the pharyngeal wall (W. S. Payne and A. M. Olsen, *The Esophagus,* Lea & Febiger, Philadelpha, 1974). Regurgitation and difficulty in swallowing (*dysphagia*) may result, so that surgical excision may be indicated. Normally, a sphincteric zone is described immediately below, although also supplemented by, the inferior constrictor (C. Zaino et al., *The Pharyngoesophageal Sphicter,* Thomas, Springfield, Illinois, 1970).

53–12 The motor innervation of the pharynx is chiefly through the pharyngeal plexus, which is formed by the pharyngeal branches of cranial nerves 10 and 9. The motor fibers in these branches are derived from the accessory nerve. The facial nerve may also contribute (J. Nishio et al., Cleft Palate J., *13*:20, 1976).

53–13 The interior of the larynx may be viewed *in vivo* either indirectly through a laryngeal mirror or directly through a laryngoscope (see figs. 53–11*C* and *D* and 53–12). During the nineteenth century, the stethoscope, ophthalmoscope, laryngoscope, gastroscope, cystoscope, rectoscope, and bronchoscope were invented, in that order.

53–14 The larynx ends opposite C.V.6, where the pharynx and larynx become continuous with the esophagus and trachea, respectively.

53–15 The hyoid bone is generally not included as a part of the larynx. The larynx is suspended from the hyoid bone, which is in turn suspended from the base of the skull. The styloid process, usually 30 mm in length, may be as long as 80 mm. The stylohyoid ligament, which connects it to the lesser horn of the hyoid bone, may become partly or even completely calcified, or it may become a chain of ossicles (J. R. Chandler, Laryngoscope, *87*:1692, 1977).

53–16 The arytenoid cartilages are extremely mobile. *Arytenoid* means shaped like a vase: Rebecca at the well with her pitcher balanced uneasily on her shoulder like the arytenoid on the shoulder of the cricoid (T. P. Garry, cited by A. K. Henry).

53–17 The corniculate cartilages (Santorini, 1724) are in the ary-epiglottic folds and on the apices of the arytenoid cartilages, with which they form a backward-projecting horn (or cornu; hence the name). The cuneiform cartilages (Wrisberg, 1786) are also in the ary-epiglottic folds, immediately in front of the corniculate. These cartilages form elevations that may be visible on laryngoscopy (see fig. 53–12). A small, unimportant nodule in the posterior border of the thyrohyoid membrane is known as the *cartilago triticea* (L., grain-like).

53–18 In acute respiratory obstruction, the infraglottic cavity may be entered through the cricothyroid ligament (*cricothyrotomy*).

53–19 The vestibular folds are frequently referred to as "false vocal cords" because they do not produce voice sounds.

53–20 The rima glottidis, i.e., the interval between the vocal folds, is the narrowest part of the laryngeal cavity.

53–21 The presence of a foreign body is the commonest cause of laryngeal spasm.

53–22 Mucosal swelling does not spread below the glottis because the mucosa is closely adherent to the vocal folds.

53–23 Afferent vagal fibers from the larynx (superior laryngeal nerves), trachea, and bronchi reach the medulla. Then a deep inspiration is followed by closure of the vocal folds, forceful expiration, and sudden opening of the vocal folds. Foreign matter is usually removed by the rapidly moving air.

53–24 Unilateral severance of a recurrent laryngeal nerve causes paralysis of the intrinsic muscles, except for the cricothyroid. However, the abductor (posterior crico-arytenoid) is usually affected first (Semon's rule), so that the involved vocal fold remains in the median plane, except when jostled by the normal fold (Chevalier Jackson). The voice is usually hoarse, as was shown experimentally in the dog by Galen in the second century.

GLOSSARY OF EPONYMOUS TERMS

Although eponyms should be avoided, they are in frequent use and a guide to their meaning is useful. This glossary provides a list of anatomical eponyms, many of which are in common usage. The chief source has been J. Dobson, *Anatomical Eponyms,* 2nd ed., Livingstone, Edinburgh, 1962. Biographical notes concerning the workers commemorated in this glossary are given by Dobson, as are citations of the publications in which the structures are described. Separate entries are used here to distinguish two or more workers who have the same surname (e.g., Meckel, Petit).

Adam's apple—laryngeal prominence
Adamkiewicz, artery of—a large anterior radicular artery from an intercostal or lumbar branch of the aorta; it supplies the lumbar enlargement of the spinal cord
Addison's place—transpyloric plane
Alberran's gland—the portion of the median lobe of the prostate immediately underlying the uvula of the urinary bladder
Albini's nodules—tiny nodules on the margins of the mitral and tricuspid valves
Alcock's canal—pudendal canal
Allen, fossa of—a fossa on the neck of the femur
Ammonis (Ammon), cornu—hippocampus
Arantius, bodies of (corpora arantii)—nodules of the aortic and pulmonary valves
venous canal of (canalis arantii)—ductus venosus
Arnold's nerve—auricular branch of the vagus
Auerbach's ganglia—ganglia in the myenteric plexus
plexus—myenteric plexus

Ball's valves—anal valves
Bartholin's ducts—sublingual ducts that open into the submandibular duct
glands—greater vestibular glands
Bauhin's glands—anterior lingual glands
valve—ileocecal valve
Bell's muscle—the muscular strands from the ureteric orifices to the uvula, bounding the trigone of the urinary bladder
Bell's nerve—long thoracic nerve
Bellini's ducts—orifices of collecting tubules of the kidney
tubules—collecting tubules of the kidney
Bertin's columns—renal columns
ligament—iliofemoral ligament
Bichat's ligament—lower part of the dorsal sacroiliac ligament, sometimes known as the transverse iliac ligament
Bigelow's ligament—iliofemoral ligament
Billroth's cords—arrangement of red pulp in the spleen
Blandin, glands of—anterior lingual glands
Botallo's duct—ductus arteriosus
foramen—foramen ovale of the heart
ligament—ligamentum arteriosum
Bourgery's ligament—oblique popliteal ligament
Bowman's capsule—glomerular capsule
glands—serous glands in the olfactory mucous membrane
membrane—anterior limiting lamina of the cornea
Breschet's bones—suprasternal ossicles
Broca's convolution—inferior frontal gyrus of the left cerebral hemisphere
Brödel's bloodless line—the line of division on the kidney, between the areas supplied by the anterior and posterior branches of the renal artery
Brodie's bursa—bursa of the semimembranosus tendon
Bruch's membrane—basal lamina of the choroid
Brücke's muscle—meridional fibers of the ciliary muscle
Brunn's cell nests—epithelial cell masses in the male urethra
Brunner's glands—duodenal glands
Buck's fascia—deep fascia of the penis
Burns' ligament—falciform margin of the fascia lata at the saphenous opening
space—fascial space above the jugular notch of the sternum

Calot, triangle of—cystohepatic triangle
Camper's fascia—superficial layer of the subcutaneous tissue (superficial fascia) of the abdomen
Chassaignac's space or bursa—the retromammary space between the deep layer of subcutaneous tissue and the pectoralis major
tubercle—carotid tubercle on the sixth cervical vertebra

Chopart's joint—transverse tarsal joint
Civinini, foramen of—pterygospinous foramen
Cleland's cutaneous ligaments—cutaneous ligaments of the digits
Cloquet's canal—hyaloid canal
gland—lymph node in the femoral ring
septum—femoral septum
Colles' fascia—membranous layer of the superficial perineal fascia
ligament—reflected inguinal ligament
Cooper's ligament—(1) pectineal ligament; (2) suspensory ligament of the breast
Corti, ganglion of—spiral ganglion
organ of—spiral organ
Cowper's glands—bulbo-urethral glands
Cruveilhier's nerve—vertebral nerve
plexus—plexus formed by the dorsal rami of the first three spinal nerves: "posterior cervical plexus"

Deaver's windows—fat-free portions of the mesentery framed by vascular arcades adjacent to the attached margin of the gut
Denonvillier's fascia—rectovesical septum
Descemet's membrane—posterior limiting lamina of the cornea
Dorello's canal—foramen formed by the attachment of the petroclinoid ligament across the notch at the petrosphenoid junction; it contains the abducent nerve
Douglas, fold of—rectouterine fold
line of—the arcuate line of the posterior layer of the sheath of the rectus abdominis
pouch of—rectouterine pouch
Dupuytren's fascia—palmar fascia

von Ebner's glands—serous glands near the vallate papillae
Edinger-Westphal nucleus—part of the oculomotor nucleus in the midbrain
Ellis' muscle—corrugator cutis ani muscle
Eustachian (Eustachi) tube—auditory tube (strictly, its cartilaginous part)

Fallopian (Falloppio) canal or **aqueduct**—facial canal
ligament or **arch**—inguinal ligament
tube—uterine tube
Farabeuf, triangle of—area of the neck bounded by the internal jugular vein, facial vein, and hypoglossal nerve
Ferrein's pyramids—medullary rays of the kidney
Flack's node—sinuatrial node
Flood's ligament—superior glenohumeral ligament
Folian (Folius) process (processus folii)—anterior process of the malleus
Fontana, spaces of—spaces of the iridocorneal angle
Frankenhauser's ganglion—uterovaginal plexus of nerves
Frankfurt plane—orbitomeatal plane

Galen, vein of—great cerebral vein
Gärtner's duct—longitudinal duct of the epoöphoron
Gasserian (Gasser) ganglion—semilunar ganglion of the trigeminal nerve
Gerdy's ligament—suspensory ligament of the axilla
Gerlach's tonsil—tubal tonsil
Gerota's capsule or **fascia**—renal fascia
Gimbernat's ligament—lacunar ligament
Giraldès, organ of—paradidymis
Glaserian (Glaser) fissure—petrotympanic fissure
Glisson's capsule—fibrous capsule of the liver
Golgi apparatus or **complex**—a system of cytoplasmic membranous organelles or lipochondria
corpuscles—proprioceptive endings in tendons
-Mazzoni corpuscles—corpuscular nerve endings
Graafian (de Graaf) follicle—vesicular ovarian follicle
Gruber, ligament of—petroclinoid ligament
Grynfeltt's triangle—triangle bounded by the posterior border of the internal oblique muscle, the anterior border of the quadratus lumborum, and above by the twelfth rib
Guerin's valve or **valvule**—a fold of mucous membrane in the navicular fossa of the urethra
Guthrie's muscle—sphincter urethrae

Haller's ductulus aberrans—diverticulum of the canal of the epididymis
layer—vascular lamina of the choroid
rete—rete testis
Hannover, spaces of—zonular spaces
Harris' lines—transverse lines in long bones near the epiphysis, sometimes seen radiographically
Hartmann's critical point—site on the large intestine where the lowest sigmoid artery anastomoses with the superior rectal artery
Hasner, valve of—lacrimal fold
Haversian (Havers) canals—spaces in compact bone
glands or **folds**—synovial pads or fringes of synovial membrane consisting largely of intra-articular fat
lamellae—bony layers surrounding a Haversian canal
system—a Haversian canal and its surrounding lamellae, the structural unit of bone (osteon)
Heister's valve—spiral folds of the cystic duct

Henle's ligament—lateral expansion of the lateral edge of the rectus abdominis, together with transversalis fascia and the transversus aponeurosis; it forms the medial boundary of the femoral ring
loop—looped portion of the renal tubule
spine—suprameatal spine
Herophili (Herophilus), torcular—confluence of the sinuses
Hesselbach's fascia—cribriform fascia
ligament—interfoveolar ligament; a thickening of the transversalis fascia (and possibly of the extraperitoneal tissue around the inferior epigastric vessels) extending from the inner edge of the deep inguinal ring upward along the inferior epigastric vessels, toward the arcuate line.
triangle—inguinal triangle
Heubner, artery of—a recurrent branch of the anterior cerebral artery
Hey's ligament—falciform margin of the fascia lata at the saphenous opening
Highmore, antrum of—maxillary sinus
body of—mediastinum testis
Hilton's line—the white line in the anal canal
His, bundle of—atrioventricular bundle
Horner's muscle—lacrimal part of the orbicularis oculi
Houston's fold or **valve**—the middle one of the three transverse rectal folds
Humphrey, ligament of—anterior meniscofemoral ligament
Hunter's canal—adductor canal
Huschke, foramen of—gap in the developing tympanic ring
Hyrtl, porus of—pterygo-alar foramen

Jackson's membrane—a peritoneal fold or adhesion between the cecum or ascending colon and the right abdominal wall
Jacobson's nerve—tympanic nerve from the glossopharyngeal
organ—vomeronasal organ

Keith and Flack, node of—sinuatrial node
Kent's bundle—atrioventricular bundle
Kerckring's valves or **valvules**—circular folds of intestine
Kiesselbach's area—site in the nose of the junction between the septal branches of the superior labial and sphenopalatine arteries
Killian, dehiscence of—presumed weak area between parts of the inferior constrictor of the pharynx posteriorly
Koch's node—sinuatrial node
Krause, glands of—accessory lacrimal glands near the superior fornix of the conjunctiva
Kupffer's cells—phagocytic stellate cells lining the sinusoids of the liver

Labbé, vein of—inferior anastomotic vein of the brain
Laimer, triangle of—posterior divergence between the vertical fibers of the esophagus below the inferior constrictor of the pharnyx
Langer's lines—cleavage lines of the skin
Langerhans, islets of—pancreatic islets
Langley's ganglion—portions of the submandibular ganglion in the submandibular gland
Leydig's cells—interstitial cells of the testis
Lieberkühn's glands, crypts, or **follicles**—intestinal glands
Lieutaud's trigone—trigone of the urinary bladder
Lisfranc's joint—tarsometatarsal joints
ligament—interosseous ligament between the second metatarsal and the medial cuneiform bone
tubercle—scalene tubercle on the first rib
Lister's tubercle—dorsal tubercle of the radius
Listing's plane—equatorial plane of the eye
Littre, glands of—urethral glands
Lockwood's ligament—sling for the eyeball; it is formed by muscular sheaths
Louis, angle of—sternal angle
Lower's tubercle—intervenous tubercle
Ludwig's ganglion—a ganglion associated with the cardiac plexus
Luschka, foramen of—lateral aperture of the fourth ventricle
glomus or **glands of**—glomus coccygeum
joint of—bilateral jointlike spaces between adjacent cervical vertebrae; they are formed by lips on the upper surfaces of the bodies of C.V.3 to T.V.1
nerve of—(1) posterior ethmoidal nerve; (2) sometimes used to refer to the sinuvertebral nerve
tonsil of—nasopharyngeal tonsil

McBurney's point—reputed site of maximal tenderness in appendicitis, "between 1½ and 2 inches from the right anterior superior iliac spine upon a line to the umbilicus"
Macewen's triangle—suprameatal triangle
Mackenrodt's ligament—lateral (transverse) cervical or cardinal ligament of the uterus
Magendie, foramen of—median aperture of the fourth ventricle
Maier, sinus of—common channel into which lacrimal canaliculi open
Maissiat, bandelette of—iliotibial tract
Malpighian (Malpighi) canal—longitudinal duct of the epoöphoron
capsule—splenic capsule
corpuscles or **bodies**—splenic corpuscles
layer—germinative zone of the epidermis
Marcille's triangle—triangle bounded by the medial margin of the psoas major, the lateral margin of the vertebral column, and below it by the iliolumbar ligament; it contains the obturator nerve

Marshall's fold—fold of the left vena cava
vein—oblique vein of the left atrium
Mayo's vein—prepyloric vein
Meckel's cave—cavum trigeminale
ganglion—pterygopalatine ganglion
Meckel's diverticulum—diverticulum ilei
Meibomian (Meibom) glands—tarsal glands
Meissner's corpuscles—specialized sensory nerve endings in the skin
plexus—submucous plexus
Mercier's bar—interureteric ridge
Merkel's corpuscles or **discs**—one form of sensory nerve ending, found chiefly in the skin
Moll, glands of—sudoriferous ciliary glands
Monro, foramen of—interventricular foramen of the brain
Montgomery's tubercles or **glands**—enlarged sebaceous glands projecting from the surface of the areola of the nipple
Morgagni, columns of—anal columns
foramen of—(1) foramen caecum of tongue; (2) sternocostal triangle or foramen
hydatid of—appendix testis
lacunae of—urethral lacunae
sinus of—(1) interval between the superior constrictor and the base of the skull; (2) ventricle of the larynx
Müller's fibers—radial fibers in the retina
muscle—(1) tarsal or palpebral muscle; (2) orbital muscle; (3) circular fibers of the ciliary muscle

Nélaton's line—a projected line extending from the anterior superior iliac spine to the tuber of the ischium
Nissl bodies, granules, or **substance**—cytoplasmic chromidial substance of neurons (rough endoplasmic reticulum)
Nuck, canal of—patent processus vaginalis peritonei in the female
Nuhn, gland of—anterior lingual glands

O'Beirne's sphincter—circular muscle fibers at the junction of the sigmoid colon and rectum
Oddi, sphincter of—spincteric muscle fibers around the termination of the bile duct

Pacchionian (Pacchioni) bodies—arachnoid granulations
Pacinian (Pacini) corpuscles or **bodies**—lamellated corpuscles
Passavant's ridge or **bar**—pharyngeal ridge
Pawlik's triangle—an area on the anterior wall of the vagina in contact with the base of the bladder and distinguished by the absence of vaginal rugae
Pecquet's cisterna—cisterna chyli
Petit's ligaments—uterosacral ligaments
Petit, spaces of—zonular spaces
Petit's triangle—a "triangle of lumbar hernia" between the crest of the ilium and the margins of the external oblique and latissimus dorsi muscles
Peyer's nodules—solitary lymphatic follicles
patches—aggregated lymphatic follicles in the ileum
Poupart's ligament—inguinal ligament
Prussak's space or **pouch**—part of the epitympanic recess between the flaccid part of the tympanic membrane and the neck of the malleus
Purkinje fibers—cardiac muscle fibers of the conduction system, located beneath the endocardium

Ranvier, nodes of—interruptions of the myelin sheaths of nerve fibers
Reil, island of—insula of the cerebral hemisphere
Reisseisen's muscle—smooth muscle fibers of the smallest bronchi
Reissner's membrane—vestibular membrane
Remak's fibers—nonmyelinated nerve fibers
ganglion—autonomic ganglion
Retzius, cave of—retropubic (prevesical) space
veins of—retroperitoneal veins
Riolan's anastomosis—intermesenteric arterial communication between the superior and inferior mesenteric arteries
arc or **arcade**—(1) intermesenteric arterial communication between the superior and inferior mesenteric arteries; (2) the part of the marginal artery connecting the middle and left colic arteries; (3) the mesocolon; (4) the arch of the mesocolon
muscle—(1) ciliary bundle of the palpebral part of the orbicularis oculi; (2) cremaster muscle
Rivinus, ducts of—smaller ducts of the sublingual gland
notch of—gap in the tympanic ring
Robert, ligament of—posterior meniscofemoral ligament
Rolando, fissure of—central sulcus of the cerebral hemisphere
Rosenmüller's fossa—pharyngeal recess
organ—epoöphoron
Rosenthal, vein of—basal vein of the brain
Ruffini's bodies or **corpuscles**—specialized sensory nerve endings, found chiefly in deep tissues

Santorini, cartilage of—corniculate cartilage
caruncula of—orifice of the accessory pancreatic duct into the duodenum
duct of—accessory pancreatic duct
Sappey's plexus—plexus of lymphatics in the areolar area of the breast
veins—venous plexus in the falciform ligament of the liver

Sattler's layer—vascular lamina of the choroid
Scarpa's canals—lesser incisive canals
  fascia—membranous layer of the subcutaneous tissue of the abdomen
  ganglion—vestibular ganglion
  nerve—nasopalatine nerve
  triangle—femoral triangle
Schlemm's canal—sinus venosus sclerae
Schneiderian (Schneider) membrane—nasal mucous membrane
Schwalbe's pocket—depression between the tendinous arch of the levator ani and the lateral wall of the pelvis
  ring—anterior border ring of the cornea
Schwann, sheath of—neurilemma
Sertoli's cells—sustentacular cells of the testis
Sharpey's fibers—connective tissue fibers penetrating bone from periosteum and tendon
Shenton's line—a continuous curved line seen radiographically and formed by the margin of the obturator foramen (superior ramus of the pubis) and the neck of the femur
Shrapnell's membrane—flaccid part of the tympanic membrane
Sibson's fascia—suprapleural membrane
  muscle—scalenus minimus
Skene's tubules or **glands**—the paraurethral glands of the female
Spieghel's line—semilunar line of the muscles of the abdominal wall
Spigelian (Spieghel) lobe—caudate lobe of the liver
Stensen's canals—greater incisive canals
  duct—parotid duct
Stilling's canal—hyaloid canal
Stroud's pecten or **pectinated area**—pecten of the anal canal
Sudeck's critical point—site on the large intestine where the lowest sigmoid artery anastomoses with the superior rectal artery
Sylvius, aqueduct of—aqueduct in the midbrain
  fissure of—lateral sulcus of the cerebral hemisphere
  iter of—aqueduct in the midbrain

Tawara, node of—atrioventricular node
Tenon's capsule—fascia bulbi
Thebesian (Thebesius) foramina (foramina thebesii)—openings of the venae cordis minimae
  valve—valve of the coronary sinus
  veins—venae cordis minimae
Toldt's fascia—fixation of fascial planes behind the body of the pancreas
Traube's space—semilunar area on the chest wall over which the stomach is tympanitic on percussion
Treitz, fascia of—fascia behind the head of the pancreas
  muscle or **ligament of**—suspensory muscle of the duodenum
Treves, bloodless fold of—ileocecal fold
Trolard, vein of—superior anastomotic vein of the brain

Valsalva, sinuses of—sinuses of the aorta
Varolii (Varolius), pons—pons of the brainstem
Vater, ampulla of—hepatopancreatic ampulla
  -Pacinian corpuscles—lamellated corpuscles
  tubercle of—greater duodenal papilla
Verga's ventricle (cavum vergae)—posterior extension of the cavity of the septum pellucidum
Vesalius, foramen of (foramen vesalii)—sphenoidal emissary foramen
  Os Vesalii or vesalianum—a separate tuberosity of the base of the fifth metatarsal bone
Vidian (Guido Guidi or **Vidus Vidius) nerve**—nerve of the pterygoid canal
Vieussens, annulus of—ansa subclavia
Virchow-Robin spaces—perivascular spaces in the brain and spinal cord

Waldeyer's organ—paradidymis
  ring—lymphatic ring of the pharynx
Weber's point—a point near the promontory of the sacrum that is the center of gravity of the body
Weitbrecht's fibers or **retinaculum**—retinacular fibers of the neck of the femur
  foramen ovale—a gap in the capsule of the shoulder joint between the glenohumeral ligaments
  ligament—oblique cord of the proximal radioulnar joint
Wharton's duct—submandibular duct
Wilkie's artery—supraduodenal artery
Willis, circle of—circulus arteriosus
Winslow, foramen of—epiploic foramen
  ligament of—oblique popliteal ligament
  pancreas of—uncinate process of the pancreas
Wirsung, duct of—pancreatic duct
Wolfring, glands of—accessory lacrimal glands
Wood's muscle—abductor ossis metatarsi quinti
Wormian (Worm) bones—sutural bones
Wrisberg, cartilage of—cuneiform cartilage
  ganglion of—cardiac ganglion
  ligament of—posterior meniscofemoral ligament
  nerve of—(1) medial brachial cutaneous nerve; (2) nervus intermedius of the facial nerve

Zeis, glands of—sebaceous ciliary glands
Zinn, annulus of—common tendinous ring of the orbit
  zonule of—ciliary zonule
Zuckerkandl, bodies of—paired para-aortic bodies near the origin of the inferior mesenteric artery

APPENDIX

THE ORIGIN OF ANATOMICAL TERMS

Etymology, the study of the formation of words and the development of their meaning, aids in understanding technical terms, frequently helps their memorization, and generally adds considerable interest.

If it is appreciated, for example, that the *tomé* of *anatomy* means "cutting," the significance of such terms as *tracheotomy* and *lithotomy* becomes clearer. Similarly, to know that *ektomé* means "excision" elucidates such terms as *appendectomy* and *gastrectomy*. A further example is *stoma*, a mouth or opening, as in *colostomy*. (Cf. *stomach*, at first referring to the opening of the organ and later to the entire structure.)

Many anatomical terms have both Latin and Greek equivalents, although some of these are used in English only as roots. Thus the tongue is *lingua* (L.) and *glossa* (Gk), and these are the basis of such terms as *lingual artery* and *glossopharyngeal nerve*. In the case of English, Anglo-Saxon may provide a third form, particularly in everyday words such as hand and foot. A good instance is *book* (Old Saxon), *liber* (L., as in library), *biblos* (Gk, as in bible). The following examples illustrate this trilingual system in anatomy and show some English derivatives from Latin and Greek.

Most terms used in biology and medicine are derived from Latin or Greek (the latter

Anglo-Saxon	Latin	Latin derivatives in English	Greek	Greek derivatives in English
Head	*caput*	capitate	*kephale*	cephalic
Skull	*cranium*	cranial	*kranion*	(see Latin)
Brain	*cerebrum*	cerebral	*enkephalos*	encephalon
Eye	*oculus*	oculomotor	*ophthalmos*	ophthalmic
Organ	*viscus*	viscera	*splanknon*	splanchnic
Heart	*cor*	precordial	*kardia*	cardiac
Lung	*pulmo*	pulmonary	*pneumon*	pneumonia
Liver	*hepar*	hepatic	*hepar*	(see Latin)
Gut	*intestinum*	intestine	*enteron*	mesenteric
Navel	*umbilicus*	umbilicus	*omphalos*	omphalomesenteric
Womb	*uterus*	uterus	*hystera*	hysterectomy
Hand	*manus*	manubrium	*cheir*	surgery (chirurgery)
Foot	*pes, pedis*	peduncle	*pous, podos*	chiropody

usually converted into Latin forms). As a result, particular attention has to be paid to the formation of the genitive (possessive) case and to plural forms. Latin nouns are divided into five classes (termed declensions), and their cases vary, as shown by the following plurals: *papilla, papillae; nucleus, nuclei,* but *ligamentum, ligamenta; os, ossa,* but *dens, dentes; sinus, sinus,* but *cornu, cornua.* Examples of genitive singular are seen in *levator scapulae, arrector pili, semispinalis capitis,* and *articulatio genus.* Instances of genitive plural are *levatores costarum, extensor digitorum, vincula tendinum,* and *confluens sinuum.* Moreover, adjectives must agree with their nouns: thus, *ramus communicans, rami communicantes;* and *foramen ovale, fossa ovalis.*

It will be noted that, although many of these forms are used in English, others are not. For example, *rectus femoris* and *lamina terminalis* are never translated, whereas *plexus brachialis* and *corpus ossis ilii* (body of the ilium) always are. This is a matter of common usage: *canalis caroticus* and *foramen jugulare* are always translated, whereas *foramen lacerum* and *dorsum sellae* are not. In the case of those that are retained in Latin, it is prudent to consult an appropriate dictionary. In this way it will be found that the plural of *diverticulum* is *diverticula* (not *-ae*); that the plural of *meatus* remains *meatus* in Latin or becomes meatuses in English (similarly *sinus* and *fetus*); and that the genitive singular of *epididymis* is (*ductus*) *epididymidis* (extra syllable *-id-*), whereas the (nominative) plural is *epididymides.* Other examples of Greek plurals are *pharynx, pharynges;* and *ganglion, ganglia.*

A particularly valuable and interesting source of information on anatomical terminology is H. A. Skinner, *The Origin of Medical Terms,* 2nd ed., Williams & Wilkins, Baltimore, 1961. Assuredly the most concise account of this subject (only 32 pages!) is to be found in P. H. Yancey, *Introduction to Biological Latin and Greek,* 5th ed., Spring Hill College Press, Mobile, Alabama, 1959. Although these books are no longer available for purchase, they may be found in libraries.

INDEX

Items in this index are listed, for the most part, only under the nouns rather than under the descriptive adjectives; for example, the brachialis muscle will be found not under brachialis but under the heading Muscle. In this edition, also, subentries beginning with prepositions are listed under the prepositions; for example, under the heading Arch, the subheading "of aorta" will be found at "o" rather than at "a."

With few exceptions, eponyms have not been used in the text and are not listed in the index. They are given in the Glossary of Eponymous Terms (p. 535).

TABLE OF MEASUREMENTS*

ORGAN OR STRUCTURE	LINEAR MEASUREMENTS IN CM	WEIGHT IN GRAMS
Appendix, vermiform	9 (3–13)	
Bladder, gall	7–10 × 3	
Body, pineal	0.5–1 × 0.5 × 0.4	0.1–0.18
Brain		*Ca.* 1240 (*Ca.* 1000–1565) F. *Ca.* 1375 (*Ca.* 1100–1685) M.
Bronchus, left main right main	5 2½	
Canal, anal inguinal	3–4 4 (3–5)	
Colon	Ca. 150	
Column, vertebral	⅖ of body height	
Cord, spinal	Av. 45	
Duct, bile cystic nasolacrimal parotid submandibular thoracic	4–7½ 3–4 2 5 5 Av. 45	
Ductus deferens	45	
Duodenum	25–30	
Embryo at 8 weeks	3 C.R.	2–2.7
Epiphysis cerebri	0.5–1 × 0.5 × 0.4	0.1–0.18
Esophagus	25–30	
Eye	2.35 × 2.35 × 2.4	
Filum terminale	15–20	
Gland, parotid pituitary submandibular suprarenal thyroid	 2 × 1.5 × 0.5 6 × 3½ × 2	20–30 0.4–0.8 10–20 3–6 40 (20–70)
Heart		250 (198–279) F. 300 (256–390) M.
Hypophysis	2 × 1.5 × 0.5	0.4–0.8
Intestine, large small	*Ca.* 150 500–800	
Kidney	11–13 × 5–6 × 3–4	145 (120–175) F. 155 (115–220) M.
Liver	Variable	1020–2120 F. 1200–3020 M.
Lung, left right		375 (325–480) 450 (360–570)
Meatus, external acoustic internal acoustic	2½–3 1	
Membrane, tympanic	1	
Newborn	*Ca.* 33.6 C.R.	*Ca.* 3350 (2500–4000)
Orbit	3½ × 4	
Ovary	4 × 2½ × 1	*Ca.* 7
Pancreas	23 × 4½ × 4	110 (60–135)
Pharynx	12	
Placenta	15–20 × 3	*Ca.* 500
Prostate	3 × 3½ × 2	5–20
Rectum	12–15	
Spleen	Variable	55–400
Testis	4–5 × 2½ × 3	25 (20–27)
Thymus	Variable	20–40 (*aet.* 6–35)
Trachea lumen in adult lumen in newborn	9–15 1.2 0.1–0.7	
Tube, auditory uterine	3–4 10	
Ureter	25–30	
Urethra, female male { prostatic { membranous { spongy	4 20 3 1–2 15	
Uterus, multiparous nulliparous	9 × 6 × 3½ 8 × 4 × 2	110 (100–120) 35 (30–40)
Vagina, anterior wall posterior wall	7.5 9	
Vesicle, seminal	4.5 × 1.7 × 0.9	

*Adapted from various sources, including F. W. Sunderman and F. Boerner, *Normal Values in Clinical Medicine*, Saunders, Philadelphia, 1949, and J. Ludwig, *Current Methods of Autopsy Practice*, Saunders, Philadelphia, 1972. F.: female. M.: male. Difficulties in selecting normal organ weights necessitate that considerable caution be exercized before assuming that a figure found is abnormal.

MEDIAN TIMES OF APPEARANCE OF POSTNATAL OSSIFICATION CENTERS IN THE UPPER LIMB*

Bone	Postnatal Centers	50 Per Cent Appear	Fusion Complete Radiographically
Clavicle	Medial end	Adolescence	3rd decade
Scapula	Chief coracoid	*Ca.* birth	
	Base of coracoid	Puberty	Adolescence or later
	Acromion (2), medial border, inferior angle	Puberty or adolescence	
Humerus	Head	*Ca.* birth	
	Greater tubercle	½ F. 1 M.	20
	Lesser tubercle	3	
	Lateral epicondyle	9 F. 11 M.	
	Capitulum, lat. part of trochlea	⅓	13 F. 15 M.
	Medial part of trochlea	9 F. 10 M.	
	Medial epicondyle	3½ F. 6 M.	14 F. 16 M.
Radius	Proximal end	4 F. 5 M.	13 F. 16 M.
	Distal end	1	16 F. 18 M.
Ulna	Proximal end	8 F. 10 M.	13 F. 15 M.
	Distal end	5½ F. 7 M.	16 F. 18 M.
Capitate and hamate		⅓	
Triquetral		2 F. 2½ M.	
Lunate		2½ F. 4 M.	
Trapezium, trapezoid, scaphoid		4 F. 6 M.	
Pisiform		8 F. 10 M.	
Metacarpals	Heads (2–5) or base (1)	1–2½	14 F. 17 M.
Phalanges	Bases	1–3	14 F. 17 M.
Sesamoids		11 F. 13 M.	

* Ages, where different, are specified in years for female (F.) and male (M.).

(See back of this page for Table of Measurements.)